CIVIL ENGINEERING
BUILDING PRACTICE

CIVIL ENGINEERING
BUILDING PRACTICE

Suraj Singh
P. Eng. M.I.E.

CBS PUBLISHERS & DISTRIBUTORS PVT. LTD.
New Delhi • Bengaluru • Chennai • Kochi • Kolkata • Mumbai • Pune

ISBN: 81-239-0450-9

First Edition: 1996
Reprint: 2001, 2003, 2006, 2009, 2010, 2011, 2012, 2014, 2015

Published by:
Satish Kumar Jain for CBS Publishers & Distributors Pvt. Ltd.,
4819/XI Prahlad Street, 24 Ansari Road, Daryaganj, New Delhi - 110002
delhi@cbspd.com, cbspubs@airtelmail.in • www.cbspd.com
Ph.: 23289259, 23266861, 23266867 • Fax: 011-23243014

Corporate Office: 204 FIE, Industrial Area, Patparganj, Delhi - 110 092
Ph: 49344934 • Fax: 011-49344935
E-mail: publishing@cbspd.com • publicity@cbspd.com

Branches:
• *Bengaluru:* 2975, 17th Cross, K.R. Road, Bansankari 2nd Stage, Bengaluru - 70
  Ph: +91-80-26771678/79 • Fax: +91-80-26771680
  E-mail: cbsbng@gmail.com, bangalore@cbspd.com
• *Chennai:* No. 7, Subbaraya Street, Shenoy Nagar, Chennai - 600030
  Ph: +91-44-26681266, 26680620 • Fax: +91-44-42032115
  E-mail: chennai@cbspd.com
• *Kochi:* Ashana House, 39/1904, A.M. Thomas Road, Valanjambalam, Ernakulum, Kochi
  Ph: +91-484-4059061-65 • Fax: +91-484-4059065 • E-mail: cochin@cbspd.com
• *Kolkata:* 6-B, Ground Floor, Rameshwar Shaw Road, Kolkata - 700014
  Ph: +91-33-22891126/7/8 • E-mail: kolkata@cbspd.com
• *Mumbai:* 83-C, Dr. E. Moses Road, Worli, Mumbai - 400018
  Ph: +91-9833017933, 022-24902340/41 • E-mail: mumbai@cbspd.com
• *Pune:* Bhuruk Prestige, Sr. No. 52/12/2+1+3/2,
  Narhe, Haveli (Near Katraj-Dehu Road Bypass), Pune - 411041
  Ph: +91-20-64704058/59, 32342277 • E-mail: pune@cbspd.com

Representatives:

• Hyderabad: 0-9885175004	• Nagpur: 0-9021734563
• Patna: 0-9334159340	• Vijayawada: 0-9000660880

Printed at:
J.S. Offset Printers, Delhi

A BOOK ON CIVIL ENGINEERING BUILDING PROJECT MANAGEMENT GUIDANCE FOR THE USE OF THE EXECUTION ENGINEERS AND CONSULTING ENGINEERS TO BE USED AS A CONTRACT DOCUMENT

IN THE MEMORY OF CIVIL ENGINEER K.L. NAGPAL, LECTURER, ARYA BHATT POLYTECHNIC (FORMER K.G. POLYTECHNIC) DELHI ADMINISTRATION, NCT, INDIA

AUTHOR : ENGINEER SURAJ SINGH
        Chartered Engineer (India) M.I.E.

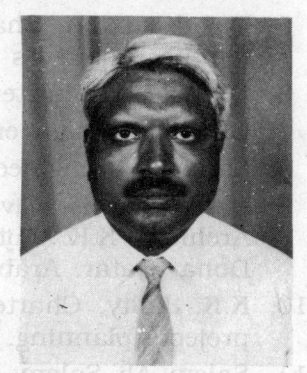

The author
Suraj Singh Chartered Engineer (India)
M.I.E. (Civil Engineering Division) 3373, Delhi Gate,
New Delhi 110 002, Ph. 3261748

Head : Sky Associates Building Engineers
3373, Delhi Gate, New Delhi 110 002
Schematic planning (Town in NCR) 4/1989 onwards
Site Agent : Al Attiyah Contg. & Trading Co., Doha, Qatar, 3/1985 - 2/1989
Project Site Engineer : ECCO (Tripoli) — Joint venture of Govt. of India & Libya, 4/1983 - 10/1984
Senior Site Engineer, Project Site Manager, Site Engineer & Junior Project Engineer : SITAC Ltd. UK, EURODEC & Associates, Ansals — All in Iraq, 7/1981 - 2/1983
Junior Engineer : Delhi Development Authority, Delhi, 12/1974 - 6/1981

IN THE YEAR OF PLATINUM JUBILEE OF IEI. ROYAL CHARTER 1935 AND INCORPORATED 1920. An Institution of India responsible for applications and promotion of general advancement of engineering and engineering sciences called 'Engineering' in India and a professional engineering constitutional body under existing law of Constitution of India

Gratitudes

The author is indebted to the following engineering personalities in the industry for obtaining co-ordinational cooperation during practice in the international engineering field to reach summarisation of system.

1. Nigel Ribinson, Chartered Engineer (UK) MI Struct. E. Project Manager for schools project, Qatar Consultants Group, Doha.

2. Haswell, MRIBA Qatar Consultants Group, Gogarty QS, MICS

3. Mel Lidster, Chartered Engineer (UK), MICE, Resident Engineer & the Project Co-ordinator, Rice, Perry, Ellis & Partners for the project II Royal Guard Army Project 'Barzan camp development for training'

4. Ian Stott, Chartered Engineer, RIBA, Phase II project co-ordinator Rice Perry Ellis & Partners as in 3 & G. Craig RIBA, Co-ordinator Project

5. David Buxton, Chartered Engineer (UK) MICE, Structural Engineer, Rice Perry Ellis & partners for the project as in 3.

6. C. Mclean, Chartered Engineer (UK), Senior Services Engineer, RPE as 3.

7. Billy Civil engineer RPE as in 3.

8. Lorry Civil engineer COW RPE as in 3.

9. Elias Francis, Civil Engineer, Head (Buildings), Er. Khalid, Er. Nicolas & Er. T.V. Thomas, Architect K.K. Jaitly, Zaki. S. Er. Suleman, Er. Faiz — Al-Attiyah Contracting & Trading Co., Doha, Qatar, Arabian Gulf.

10. K.K. Jailty, Chartered Architect (India) as a partner with author in Sky Associates group for projects planning.

11. Salem Ali Salem, Engineering Manager, Electrical Construction Company, Tripoli, Libya — a joint venture of Indian Govt. with SPLAJ Govt. (Libya).

12. Er. K.R. Jain, Chartered Engineer (India), FIE, Chartered Surveyor (India) ECCO, ACE in CEL.

13. Fathe, Civil Engineer, ECCO, Engineer Shafan ECCO, Haz Habib ECCO, Er. Banerjee ECCO, Ahmad Zarook ECCO & many others.

14. Er. A.K. Handa, Proprietor, EURODEC & Associates UK Baghdad.

15. Bhim Sen Chowdhry, Construction Manager, EURODEC.

16. Er. Jainal Abdudin, Govt. of Iraq, Central Establishment for Buildings.

17. Er. D.K. Mehan, Project Engineer & Manager, Jordanian Co. for Dar Al Thawra (UECO).

18. Orkaan Mohammed Ali, Resident Engineer, Dar-al-Thawra Printing Complex, Baghdad, Iraq.

19. Er. K.S. Dashrathi & Er. P.L. Raina, Managers of Sites, Ansals, Shirkat.

20. Chetan Shah, Chemical Engineer, Teyseer Trading & Contracting Co., Doha.

21. Kumar & Mohammad Javed, Electrical Engineers of Al-Kuwair, Doha.

22. Electrical Engineer Prabhakar of Voltage Engineering, Doha, Qatar.

23. Mechanical Engineer Ali Afzal of Doha, Qatar with Alamgeer Associated with Trags and Er. Khalid with Pioneer.

24. The author is particularly thankful to Abdul Salam, a Sales Engineering professional who without fail extended overall financial assistance along with all other arrangements during the manuscript making of this first edition and continuing the same to bring out second major edition in future; and to Engineer M.L. Chauhan who inspired me to write the book. Particular thanks to Dr. Mala Saini MBBS and Ms. Sumitra who extended support.

REFERENCES

(1) Jacuzzi; (2) IEI; (3) Architect Act; (4) IOB; (5) Copyright Act; (6) Constitution of India; (7) IPC; (8) Evidence Act.

Preface

Dear Reader,

Pleasure comes to me bringing out this small book on the subject of the practical project management on the civil and architectural collectively termed as building engineering (specialised discipline) including all the aspects of the building from the conceive stage to the final completion of any project involving the services of all (the design, supervision and the execution) civil, architectural, structural, allied services engineers such as electrical, mechanical, plumbing and drainage & the specialised engineers, etc., to participate in one way or the other under the direction and the guidance of an engineer in terms of contract purpose designated by the owner/user/promoter or the body for which the project has to be carried out. The engineer may be an architect, a civil engineer, a structural engineer depending on the merit of the case and the individual qualifications including the tutorials. 289 sketches have been added for the guidance of the designers and the planners and the students of the engineering and the architecture just to let the reader apprise of what are the overall responsibilities of an engineer and the other engineers by whom the project shall be completed under the leadership of the engineer in one way or the other, directly or indirectly as some body working for the contractor shall also be guided and directed by the engineer as far as the construction contract is concerned.

This book is based on the experience with attainment in the civil engineering discipline on building projects in the various countries particularly on the British and the European typical contracts and the systems which are supposed strictly adhered to the standards. I confide in the readers to find interest in the contents of this book and may make use for practical work. Though this book covers very small part of the discipline but shall comprehensively fit in for the engineer's job in office as well as in field and part dissemination of knowledge and distance continuing education in practice including working legal base.

I shall be glad to get the critics and the suggestions from the well-wishers and the readers and be grateful to all of them.

Note : The public in general may read this issue for their benefit to appraise with the building engineering practical aspect and augment the general knowledge in dealing with the problems of their proposed houses, etc.

Chartered Engineer
SURAJ SINGH
P. Eng. (India)

Contents

DIRECT COURSE
OF
BUILDING PRACTICE
MANAGEMENT CIRCUIT / SYSTEM DIAGRAM

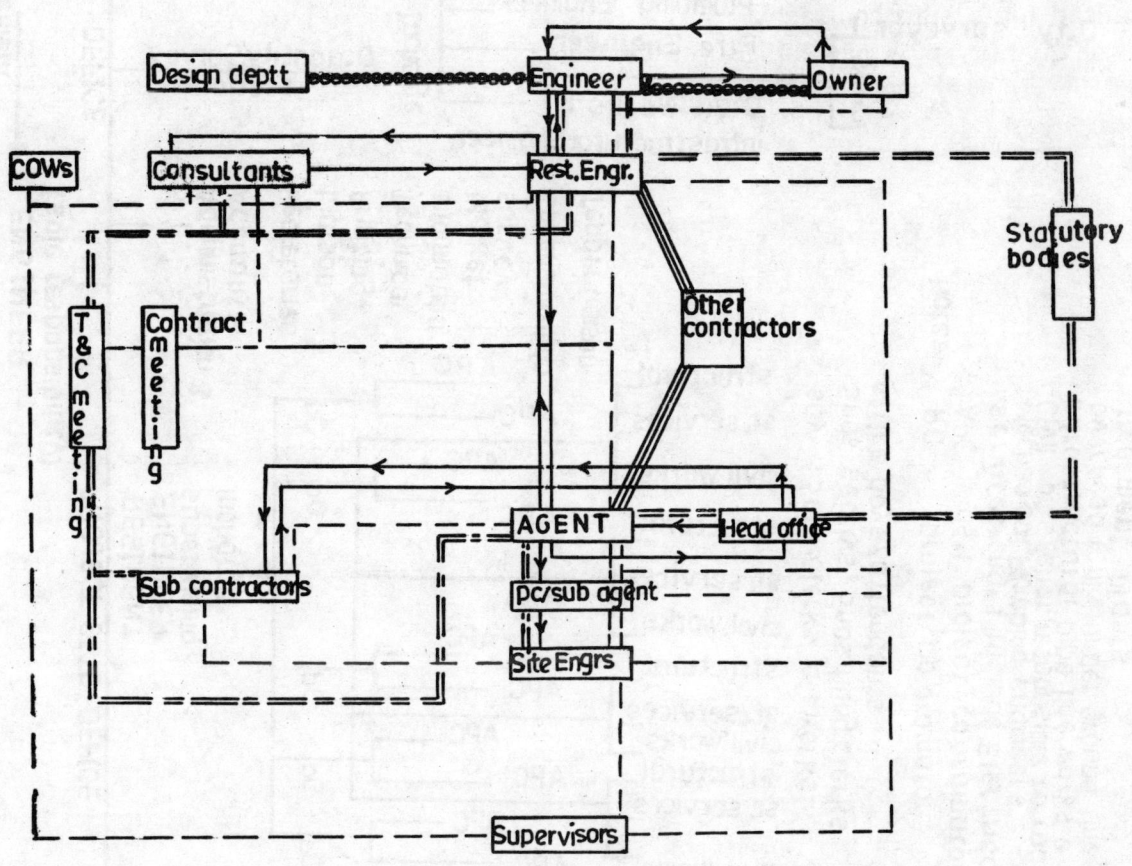

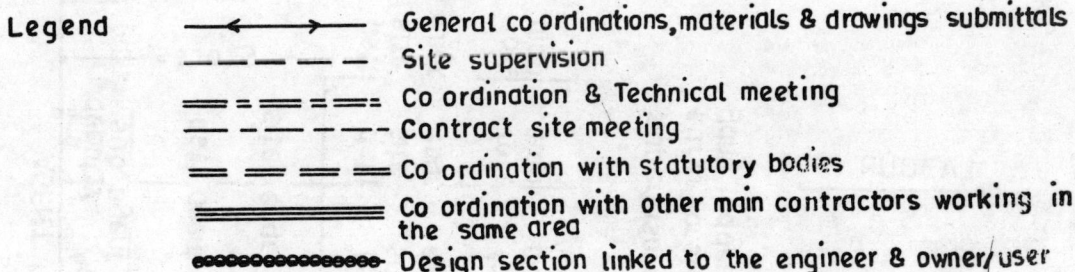

Legend

⟵———⟶ General co ordinations, materials & drawings submittals

– – – – – Site supervision

══════ Co ordination & Technical meeting

— – — – Contract site meeting

═ ═ ═ ═ Co ordination with statutory bodies

══════ Co ordination with other main contractors working in the same area

●●●●●●●● Design section linked to the engineer & owner/user

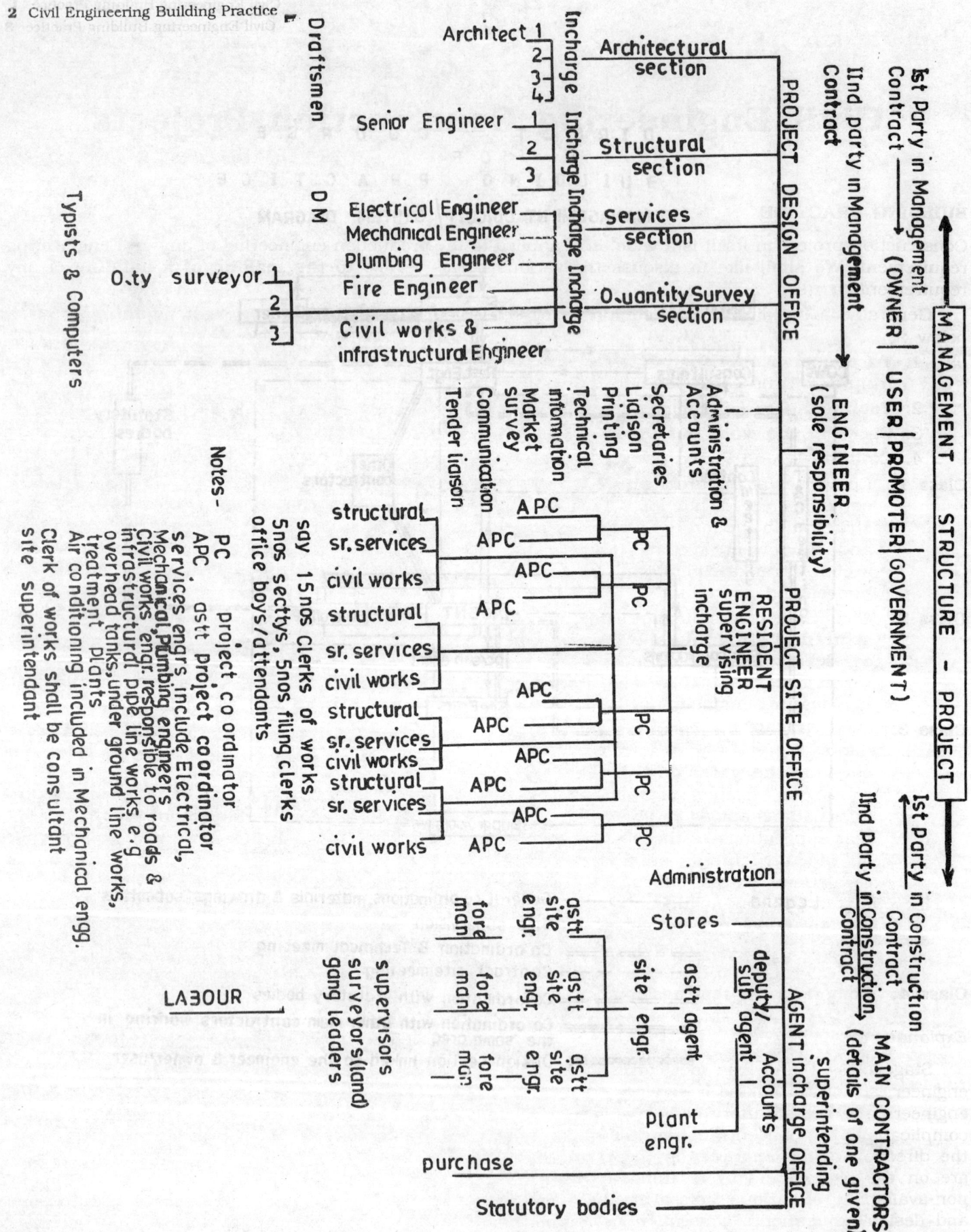

MANAGEMENT STRUCTURE - PROJECT

Ist Party in Management
Contract ———>
IInd party in Management
Contract

(OWNER | USER | PROMOTER | GOVERNMENT)

ENGINEER
(sole responsibility)

Ist Party in Construction
Contract <———
IInd Party in Construction
Contract

MAIN CONTRACTORS
superintending
(details of one given)

PROJECT DESIGN OFFICE

Incharge — Architect 1, 2, 3, 4 — Architectural section

Incharge — Senior Engineer 1, 2, 3 — Structural section

Incharge — Electrical Engineer, Mechanical Engineer, Plumbing Engineer, Fire Engineer — Services section

Incharge — Q.ty surveyor 1, 2, 3 — Quantity Survey section

Civil works & infrastructural Engineer

Draftsmen — DM

Typists & Computers

Administration & Accounts — Secretaries, Liaison, Printing, Technical information, Market survey, Communication, Tender liaison

PROJECT SITE OFFICE

RESIDENT ENGINEER supervising incharge

PC — APC structural, APC sr. services, APC civil works

PC — APC structural, sr. services, civil works

PC — APC structural, APC sr. services, civil works

PC — APC structural, sr. services, APC civil works

AGENT incharge OFFICE

Administration

Stores — astt site engr., astt site engr., astt site engr., site engr.

deputy/ sub agent — astt agent, astt agent

Accounts

Plant engr.

purchase

Statutory bodies

foreman, foreman, foreman — supervisors, surveyors(land), gang leaders

LABOUR

Notes —

PC — project co ordinator
APC — astt project coordinator
services engrs. include Electrical, Mechanical, Plumbing engineers
Civil works engr. responsible to roads & infrastructural pipe line works e.g overhead tanks, under ground line works, treatment plants
Air conditioning included in Mechanical engg.
Clerk of works shall be consultant site superintendant

say 15 nos Clerks of works
5 nos secttys, 5 nos filing clerks
office boys / attendants

Civil Engineering Construction Projects

BUILDING PRACTICE

Construction project in itself is a wide term applied to the production engineering of any civil engineering requirement. We shall like to discuss the various stages related to **the making of a building** of any requirement or use.

Generally, a building project comprises of four **major classes** or **categories of planning** as given below :

1. **To conceive or visualisation of an idea**/a dream to form a structure of optimum use with fruitful returns corresponding the capital investment in a fixed duration.
2. **Technicalities** (all engineering details formations)
3. **Carrying the work out on site** to the final form (execution)
4. **Commissioning** to the use and **operational maintenance**.

Class 1. Comprises of the architectural as well as artistic aspects of vision and various considerations equipping the owner for whom the work/project shall be carried out with an hypothetical structure in the form of **bird's eye view/perspective/model** as felt necessary by the engineer and the owner keeping in view the **co-ordinational opinion** between the **engineer** of architectural discipline and the **owner** finally concluding the proposal matching the owner's budget and use requirements. The indepth discussion for that is not worth here.

Class 2. Comprises of architectural planning and detailings per the **statutory bye-laws** followed by **structural designing** and **detailings** and further enriched by the inclusions of all the **allied services schematic details** regarding **electrical; engineering; plumbing** and **drainage; civil works; mechanical works;** and all the **external** and **internal** works with the proper **landscaping** maintaining the **natural equilibrium** for the users of the proposed structure.

Class 3. Comprises of the operations which appear to the common eyes that something concrete is taking place as this stage includes complete **tendering work, contract award** and the **carrying the works on the site** to the final form and preparation of the BOQs specs, contract conditions, fixation of rates of various items, lot of correspondences and exchanges of informations among various agencies interested to participate in the project for their business, real applications of the visualisation and the dreams on the site to the practical formation of the structures, supervisions, payments, co-ordination, contract disputes, budget escalations, extras, substitutions, additions, deletions, alterations, export and imports of the materials, financial dealings, accidents compensations, exposure of all the skills, investigations, research, experience, confirmation of the theory of engineering and various other factors. In this class it appears that all the actors working on one platform for one goal.

Class 4. Comprises the **testing, commissioning, use** and **maintenance** of the works.

Explanations :

Stage/class 1 and 2 in fact correspond to the mutual interactions among the owner, architectural engineer, structural engineer, services engineers, specialist engineers, civil works engineers, draughting engineers and the statutory bodies which are all indoor activities and of course do not result in much complications but concentration on the paperwork/deskwork and in the drawing and design office under the direction of the **engineer** in the **terms of contract**. Factually, all the approaches and the ideas' are on the papers and every thing appears to be smooth going as a dream. At this stage due to non-availability of various facts a number of assumptions are made subject to the practical availability and design adjustments consequently, you shall find a number of **changes at site** during execution followed by various instructions particularly in some sections where every item is not feasible enough to be carried out per design standards.

Stage/Class 3 : This is the most difficult class and shall be further divided into a number of activities/operations.

a. Preparations of various **bills of quantities** (BOQs).

b. Preparations of all the **building specifications** (Specs.).

c. Converting these into the **tender documents**.

d. **Short listing** of tenderers or **prequalifications**.

e. **Floating the tenders**, their **closing** and **scrutiny**, **negotiations**, **suitability check** and either rejection or **award by the owner**.

f. Site handing over, commencement of construction, supervision and control till practical completion of the contract, the complete site clearance by various agencies working on the contract.

g. Snags, rectifications, practical completions, main work accomplished.

Stage/Class 4 : **Major testing** and **commissionings** of all the systems are carried out to the satisfactions of the specialist engineers of the concerned disciplines for the specified period in the contract. Generally one month commissioning is recommended by the British standards and thereafter the maintenance of the built structures commences that shall be done for a period of 400 days or as specified in different contracts. The final completion shall be issued by the engineer to the contractor on the successful completion of the maintenance period in all respects as specified in the general and the projects specifications of the concerned contracts. The grant of the final completion certificate shall be the day where the contract shall be termed **completed contract**.

The explanations of various activities/operations are given in the followings :

(a) **Preparations of various BOQs :** This is a great deal of technical work potential and performed by the quantity engineers in consultations with the engineers of the different disciplines under the final supervision of the 'engineer' all based on the detailed engineer's designs and planning and the schematic details and project specifications as prepared by the specialist engineers of respective disciplines. The system of forming it depends on what type of construction contract shall be adopted as often in the Middle East and other advanced countries the trend in practice is to adopt the itemised lump sum contract wherein the tenderer offers a fixed sum for accomplishing the whole job according to the scope of the works mentioned in the tender documents based on the rates of all the items priced by the tenderer in the provided BOQs. Practically, this form of contract is of great convenience to all due to in it having potential to reach various settlements of expected disputes of additions, alterations, omissions, etc. It also helps to evaluate the amount of the works done very easily, without measuring the work done on site and recording in the measurement books.

All the disciplines of the building or the structure as the case be are included in one BOQ in the general sequence, e.g., excavations and back fills or in short earth works, concrete works, reinforcements, forms, block works/brick works, metal works, wood works, electrical, engineering, plumbing and drainage, decoration and finishes, miscellaneous and other works, if any, etc., and all summarised to the concluding page sheet of the bill.

The basic advantage in this method is that the costs of individual structures shall be known separately and all the variations could be decided with ease. Valuation too shall not take excessive time as the same is done observing the percentages of the items included in the bill. The hard work done at the time of preparing the bills of quantities is overbenefited by the instant estimation or judgement of valuation and saves a great deal of taking measurements every time for making payments as done in the other types of contracts having only one bill for all the contract jobs. For making the BOQ, the measurements from the detailed drawings should be very carefully taken off by the quantifier in the interest of fair treatment on the job. In precedence to the BOQ, preambles are itemised comprising of all the probable expenses separately to provide an opportunity to the tenderer to price carefully. This may cover all the expected expenditure the contractor shall have to spend on the contract per the clauses in the general as well as specific conditions and any other particular clause.

Clauses may include expenditure on procurement of workers, staff, site mobilisation, any advance money required by the contractor before commencing the job, procurement of machines, third party insurance for the workforce, accident benefits, leave benefits, various other benefits and overhead expenses, transports, messes, drinking water, staff and workers accommodations alongwith the creches, facilities to be provided for the engineer's/resident engineer's office and staff on site, (It may include the engineer's office, overheads, maintenance, car, drinks, etc.) extensions, contractor's staff, tests, expenses on the compensations in case of death of the workperson, timely payments, programmes, working and shop drawings, record/as built drawings, necessary liaison work, co-ordinations, telephones, telexes, meetings, electricity, drainage, fuel, guarding, fencing in the special cases of prohibited sites provision for obtaining the entry passes, air passage in case of foreign employees, bank expenses, weekly and monthly reports, technical experts and all other general and particular expenses per design of the job. It is not essential that the tenderer should price all the preambles as that may be included in the other parts of the bill but marked as included.

(*b*) **Preparation of various building specifications :** This is very much specific part of the execution requiring the use of deep technical knowledge significantly and precisely. Bifurcating to two shall be termed as general specifications and the project specifications. The general specifications relate to the standard specifications published by the public works departments or any other competent organisations, generally known to all the people practicing on contracts of building constructions while the project specifications are the specified items for special uses in the contract designs to serve a special requirement of the user and detailed reference to the engineer's details is a must to be made in the writing of that and preceding the boq nomenclature of the item and after the tender bond as a summary of all the project specifications, the materials required for those and the source of availability for the guidance purposes only but while using the name of the manufacturer or the supplier it must be made clear that the equivalent shall also be accepted to avoid the suspicion of nepotism. The addition of a sentence that the special materials as provided shall be used per direction and instruction of the manufacturer in all the cases as some material shall possess some time warranty for a fixed period and also in case the demonstration is required for the purpose of training the staff of the user, for the purpose of its procedure how to use and maintain for a smooth functioning shall be beneficial to all for the ease of the item's execution. But whatever be included, the length of the writing should be given due concern as very lengthy specifications often invite boredom by the reader who shall close after reading a quarter of that. Better avoid it being a tool of wastage of time and the syllabus of a subject of an engineering degree programme. That is why the engineers engaged in drafting the specifications must be very senior and experienced professionals who can assess the practical use of that, also communicate the maximum in a minimum time to make it interesting for the engineers on site refer that every time for the discussions and executions. He must be very talented, communicative, technically master of the subject and of course patient. Some examples of the specific items are epoxy floorings, X-ray rooms plaster, squash courts plasters, squash courts, kitchen equipments, suspended or false ceilings, carpetting, swimming pools, structural steel, structural steel painting on the surface exposed to weather, armoured plates, sports surfacing coverings comfort systems and high standing systems, engineering systems, electrical systems, mechanical installations, generally the services systems are designed by the construction contractor based on the specifications provided to the contractor, lift installations, pump installations, sports equipments, any other uncovered in the general works or architectural engineering department specifications.

(*c*) **Converting the above documents to the tender documents :** In practice the tender documents comprise of the tender forms, space for the contract forms, various schedules the tenderer shall be supposed to fill for the support of his capacities for the job, form for the qualifications, form for the site superintendence by the tenderer/contractor, provision for the informations to the tenderer, soil investigation report for the guidance of the contractor/tenderer, the brief description of the work required with the purpose and the importance, applicabilities of the general specifications, special specifications or the project specifications, all the bills of the quantities of the complete tenders, the complete design specifications for the systems for which 1st class provision, installations, and functioning is necessitated in the job, the inclusions of the technical graphs and minute details to make the contractor aware to actually what is expected of him, complete technical statistics of electrical and mechanical design and functioning without leaving any doubt to the tenderer, the complete sets of the drawings in all respects as explained before, (architectural engineering; structural engineering; civil work plumbing & drainage; electrical; mechanical; engineering, engineer's details for complete proposals, bar bending schedules according to the adopted codes, the information felt necessary to be told the tenderer for the efficient disposal of the tender being clear all what is supposed to be done, corrigendums, any other instruction to be included by the engineer or the owner, any clarification requested by the tenderers, etc. Practically, it has been observed that the tenderers finalise the tenders at a late time at eleventh hours and a number of clarifications and checks the tenderer should do are missed just believing what is given in the BOQ shall be all ok and lack of technical coordination and ineffective quantity check shall bring the price competitive but may be due to bad luck of the appointed contractor the quantities as found during the construction to be much in excess, the contractor shall definitely be looser on that account and there shall be no way to raise claims. Therefore, the engineer's best effort shall be to prepare the tender with great degree of accuracy and take his own time while the tenderer should be given enough time for the calculation of his price for the best operation or working of the resulting contract which shall be in the interest of all the parties.

(*d*) **Short-listing of the tenderers/pre-qualification :** It is not practicable for a big size contracting work to invite all the interested parties to offer the tender, i.e., an open invitation shall not be preferred. Particularly for a general contracting work, the best approach is to call for the prequalification for the selection in the shortlist for inviting to buy the tender documents at the interest of the party for the general contract which shall mean that the selected contractor after the tender shall carry out the works of all disciplines as a whole either himself or by engaging sub-contractors of the approved category holding the statutory licences but at the risk of the contractor as the contractor shall be answerable and responsible to the engineer and the employer for the proper execution of the contract.

After giving the open invitation for the registration for short listing seeking the information such as general details of the contractor, his organisational appraised details, his financial potential, his technical potential, the experience of the contractor in contracting and of the height of technical degree of the works executed by the contractor, the assets of the company, the running works, the works expected to come, the staffing levels particularly technical, the technical assets of the company such as laboratories, research centres, and whatever the engineer deems to know of the tenderer should be filled or included in the appraisal form, the engineer shall scrutinize the submitted appraisal of the party/candidate and discuss with the employer/owner, later make a list of the selected parties and intimate to them about the shortlisting.

Whenever you have to open/float the tender, send letters to all the shortlisted parties to know about their interest and if interested ask them to collect the tender documents for a fixed amount, the closing date and the amount for the tender bond should be included in the letter.

It is a great advantage for a civil engineer working in the construction of the buildings to work with a general contracting firm/contractor for certain years in his professional life to proliferate the technical knowledge of building construction as on turnkey basis along all the disciplines of the complete building and escalate from the narrow deliberation of supervising only the civil engineering activities. The technical and practical co-ordinations shall provide you/the engineer with great degree of dynamism and an elegant personality with command in the field.

(*e*) **Floating the tenders, closing, scrutiny and negotiations :** Invitations are communicated to the shortlisted parties/companies/agencies in the form of a letter exhibiting the name of the job, the cost of the tender documents, the date of the tender closure, the time of the closure, budgetary reference if belongs to the government bodies, and a tentative date to buy the tender documents with a mandatory condition that once the tender documents are received by the contracting company the submission of the tender of the party shall be necessary or the name of the party shall not be included in the next call of the other works. Sometimes, this clause shall be a burden after looking into the documents and assessing and the party may not be interested to do that job as it would not return them the expected profits, in that case the tendency of the tenderer shall be to offer higher prices so that the tender is not accepted. Some times, it may be fruitful for the engineer to obviate that clause and just write that in case the tenderer does not want to offer the price, the documents be returned but the cost of the documents shall be non-refundable, the tenderer may accept that condition and avoid unnecessary expenditure on closing the tender for nothing. Sometimes, the tenderer may find the job of return but for short terms and he shall have to expand the organisation for a short period and there may be no guarantee for more jobs after that to stabilise the expanded organisation, the tenderer may not like to close the tender. This clause should be strictly adhered to for the projects of national security where importance should be given to instruct the tenderers to close the tender with the return of all the tender documents in the possession as issued may be sealed by the contractor/tenderer as that documents shall be returned to the contractor when the contract is signed.

A specification is also added that in case of national security concern, the tenderer shall not disclose any information about the work or related to that to any/some third person/party directly or indirectly by way of speech or in any/some other form of communication as he shall have the privilege of knowing complete details of the job being in possession of the work tender documents, other than the parties related to for the compilation of the tender closure may be sub-contractors proposed, suppliers, supervisors, quantity engineers, agents or the employees of the contracting company and the tenderer shall make certain that the persons or person or sub-agencies to whom the related information is provided, supplied, communicated shall also not in any form disclose the same to any one not related to the tender whatever the circumstances may be or come what may, failing that the tenderer shall have to face the rejection, cancellation of his tender or even cancellation of the award or the work at any progressive stage on the consequence of the breach of contract and necessary future action of either blacklisting or imposing any/some form of penalty. This is very strict clause and should be taken seriously.

Having received the tender documents, the tenderer's contracts manager/tenderer examines the scope and the contents. Being general contract comprising all the disciplines namely architectural, structural, engineering air conditioning and mechanical, electrical, plumbing and drainage, civil works, external works, etc. It is not necessary that the tendering agency might have all the disciplinary divisions of its own (nice if the agency got that on its own) the contracts manager or the assistant manager shall go through the scope of work of the other agencies to be involved in the job and request them formally to quote their competitive prices. This request should be made to a number of the approved agencies, sub agencies holding statutory licences. The request shall also be made to various agencies in the specialised fields inside or outside the country. For example, the sub-agencies for electrical, mechanical, air conditioning and plumbing and drainage, etc., may generally be available in the country of concern while the other specialist agencies be outside and in many cases they have also their agents in the city. The contracts manager shall make available all the informations to all the said people and receive quotations in due course before the closing of the date of the tender. The contracts manager shall price his own items for example structural, finishes, masonry, excavations, or whatever items he may like to do himself and the portion of the building services, special items on the receipt of various quotations after doing comparison. The comparison may be done by the assistant contracts manager or the quantity engineer or quantity surveyor. Having filled in all the items and the extensions done to the cost column, the sum total of the bill accumulated to the summary incorporating the profit margin, the tender price reached to the final figure, that shall be quoted in the tender form of the documents which is normally included in the beginning pages of the contract forms following the job schedule.

Tender Bond : It is a document carrying a guarantee for a fixed percentage of the tendered amount to be provided at the time of tender submission without which the tender shall be invalidated. In case the tender is accepted and the tenderer does not wish to accept to do the job due to certain reasons, the tender bond money shall be forfeited.

Having reached the final figure to be quoted as tender price, the contracts manager shall fill up the tender sum form, detailing the amount in words and figures both, all the machines to be deployed and the Organisational potential for the job disposal and brief description of planning how or what way the tenderer proposes to carry out the job. It must be remembered that the time of completion is given in the tender documents and the tenderer has to accept the same. The tenderer shall mention in the form of contractor's superintendence the names of all the agents, sub-agents, deputy agents, project co-ordinators, site engineers, assistant site engineers or the deputy engineers, general foremen, junior engineers, land surveyors, quantity engineers, etc., who shall supervise the work headed by the agent and to whom the engineer or the resident engineer shall pass instructions, the educational, technical and practical qualifications, any special skills shall also be given.

There may be many conditions in the contract that the tenderer is supposed to have gone through complete details of the tender before closing and submitting, for example the time of completion, the time of payment, etc. Some contracts conditions may include time for payment as reasonable period after the certification of the valuation of work. The tenderer shall make sure whether he can afford to abide by all the conditions if so he shall not include any condition on his own and the submitted tender shall be treated as unqualified while if there is some condition included by the tenderer of his own before submission, the tender shall be treated as qualified and shall be a part of comparative statement to be prepared by the engineer for the owner's considerations. It is not necessary that the employer is bound to accept the lowest tender or that the tender be accepted, it could be cancelled without assigning any reason whatsoever be the case.

A job may comprise of hundreds of major and minor items; drawings and every drawing carries certain number of reference and a list is provided in the documents. The experienced tenderer shall compare the list of drawings to the drawings delivered to him by the engineer. The tenderer should also include the references of all the addendums and the instructions received before the closing of the tender. Having closed the tender it shall be sealed. All the papers may be sealed by affixing the company's seal, i.e., on the drawings being returned, tender forms, bond, BOQ, specifications and then submitted and a receipt is obtained. This exercise protects the tenderer from the possibility of the employer/engineer to make any concealed changes or the alterations or the additions in the drawings, BOQs and also the specifications. The tenders are opened in the presence of all the tenderers and a statement is prepared called the comparative statement. It is a general practice that the lowest priced tender is likely to be accepted but it is not necessary as a number of aspects from the practical viewpoints are concerned. For example, the lowest pricing tenderer may not be very well reputed, may not be financially sound, may not be having a very good track record, and may be debited with so many plus points. His price may be so low that the practical feasibility of the job is doubtful, the engineer and the owner shall not risk the stoppage of the job at any stage or the delayed progress to jeopardize the owner's business or hamper the financial returns of the capital investments. This depends entirely on the jurisdiction of the engineer and the owner to whom the work be awarded for a smooth and qualitative functioning of the completed works. The process between the opening time of a tender and signing the contract is very flexible and long which may involve a couple of months together.

What actually happens is the engineer first picks up the lowest tender, checks the reputation, past records and if he feels/considers that the price tendered looks reasonable, directs the concerned quantity and the contract engineers to scrutinise the tender contents. All the arithmetical checks are carried out and errors found are to be adjusted. If the checked prices come more than the quoted price, the tenderer is asked whether he is still ready. If the checked price comes higher than the quoted, a reduction is done to the tendered price and the tenderer is informed. Various details are asked for from the tenderer and the sub agencies as mentioned in the tender forms' schedules. A check is carried out whether the tenderer is in a position including all associates to execute the job successfully. Having ascertained all the checks/negotiations, the work is finally awarded to the tenderer by formally communicating to him by posting an appointment letter or award letter and a date is fixed to sign the contract. In case the engineer is not in favour of the lowest priced tenderer, the negotiations are made to the second lowest and the same procedure repeated as explained earlier subject to the employer/owner's/client's prior approval.

Having received the contract award letter, the tenderer becomes appointed contractor w.e.f. the date of issue of the letter for the job in question subject to his formally signing the contract form. The award letter contains the name of the job, budget, references, the date of job commencement, the date of take over of the site, the name of the engineer's representative or the resident engineer, the contract completion date, the contract sum/price agreed, any accepted qualification and the date by which the contractor shall submit his detailed programme to the engineer that is normally 14 days from the date of award. The contractor is also asked to submit the curriculum vitae/bio data of all his site personnel responsible for the site supervision for the engineer's approval. This letter must be signed by the engineer and copied to the owner or the employer, the engineer's representative, consultants, and all the related concerned parties to the job directly.

Having received the award letter, the contractor comes into full dynamism and particularly a period of first two weeks needs every one to be fully alert as the taking over site and mobilisation are the threshold of the job.

Significant to mention that before closing the tender an intelligent contractor or his manager shall visit the site if he is not conversant with the surroundings to make checks of the conditions while sometimes as a matter of process for some urgent and necessary projects the engineer calls on a pre-tender date a meeting to all the tenderers to apprise them of the site, job, various aspects of the job and they are taken around the site in case there are certain restrictions on account of the site belonging to prohibited area zone. This is really a good practice on the engineer's part to save lot of confusions and communication and save time. All the clarifications are given and a final note of the meeting is sent to all for information and confirmation and later forms part of the documents.

(f) **Site handover, commencement of construction, supervision and control till practical completion of the contract, the complete site clearance by various agencies working on the contract :** Whatever explained till now was on the papers or in the form of communication but now the watch makes actions and come into play. It is obvious that the employer/owner must have provided some form of consultancy group under the control of the engineer, that may be his own in case the owner is a govt. department, a statutory body, or a local self government or the employer appoints a group under the control of the engineer/director/the group. Generally, you shall find a consultancy group in operation on every important job. Since engineer is fully responsible for the job control but practically this authority is delegated for actions to a resident engineer or so called engineer's representative or the project manager or the coordinator supported by the assistant resident engineer, structural engineers, civil engineers, services specialists engineers, architects, senior and junior site superintendents of works or the clerks of works and assisted by the administrative assistants.

Immediately after the award letter is in receipt the contracts manager should arrange a meeting with the consultants headed by the resident engineer along with the expected site agent and have a brief about the project, of course the contractor shall introduce himself and the company's achievements and expect the consultancy group to introduce the members. The site layout plan is given and the contractor's agent should table the site establishment layout for an approval. This shall impart an impression on the group that the contractor is not sleeping. This meeting shall result in general discussions and the contractor shall be requested to furnish the details of the curriculum vitae of all the key personnel, material and the shop drawings submittals after the contract documents are issued to the contractor.

The contractor should obtain the contract documents immediately after the award as he is supposed to propose the programmes in a period of 14 days. This meeting is normally held either in the consultant's office or in the engineer's office/employer's office. The meeting is chaired by the resident engineer or the engineer and the minutes are circulated to all the concerned parties. This henceforth, shall be conducted at site and named as site meeting/contract meeting.

It is worth mentioning, the supervision control on the part of the contractor which is of utmost significance to the project should the contractor's superintending staff be the fittest to the job requirements and the right man/person is deployed for the right job, the results will be very favourable, for the project or job shall look flying provided the employer deals fairly in making payments and obliging on the conditions of the contract. Site management particularly on a building construction complex is a tough and hard working continuous operation necessitating intelligence, dynamism, cleverness characteristics on all the members of the executive team/contractor's site team assisted by the resident engineer and his staff. To any contractor or his staff the engineer representative or his staff appears as a boss/bosses, nevertheless, the clever contractor's staff shall make use of their presence for the assistance to help work progress smooth but normally this shall not be found on all the sites due to the lack of experience in the staff of both the parties. It is suggested that both parties should realise that they are on site to achieve a common target to be delivered to the owner in a fixed time. This understanding should be captured by the professional people/engineering staff and the personal unwanted relations developed on site be set aside submitted to the smooth and better working relations. Always there exists a misunderstanding among young engineers but by virtue of obtaining further tutorials in the form of practical experience after being rolled from one job to another and then to another is generated the required understanding naturally making the engineer experienced and matured. You shall always find a matured engineer a patient person and handling the problems without much worries but decisively and expeditiously.

Since the contractor is fully responsible for the complete execution of the contract job to the engineer's specifications, this is a very important status to be disposed off. In case the contract is of general nature wherein all the disciplines operations involved in the building are to be carried out, a great degree of competency is needed for the disposal of the following responsibilities efficiently.

Co-ordinations

1. Architectural/structural
2. Structural/mechanical
3. Electrical/mechanical
4. Structural finishes/electrical
5. Structural/plumbing and drainage
6. Services/finishes
7. External services/external works
8. Services liaison to the statutory bodies
9. Structure/structure

Liaisons

1. Main contractor/sub contractors
2. Sub contractors/sub contractors through main contractor
3. Main contractor/consultants' group and engineer
4. Contractor/other contractors working in the same site
5. Main contractor/engineer
6. Main contractor/local statutory bodies

The above explained descriptions are only a few but there may be many others depending on the nature the size of the job. Now we will discuss the kind of contractor's organisations in normal cases but may vary from place to place. A contractor's organisational layout has been included in the following page/chart for your reference and guidance.

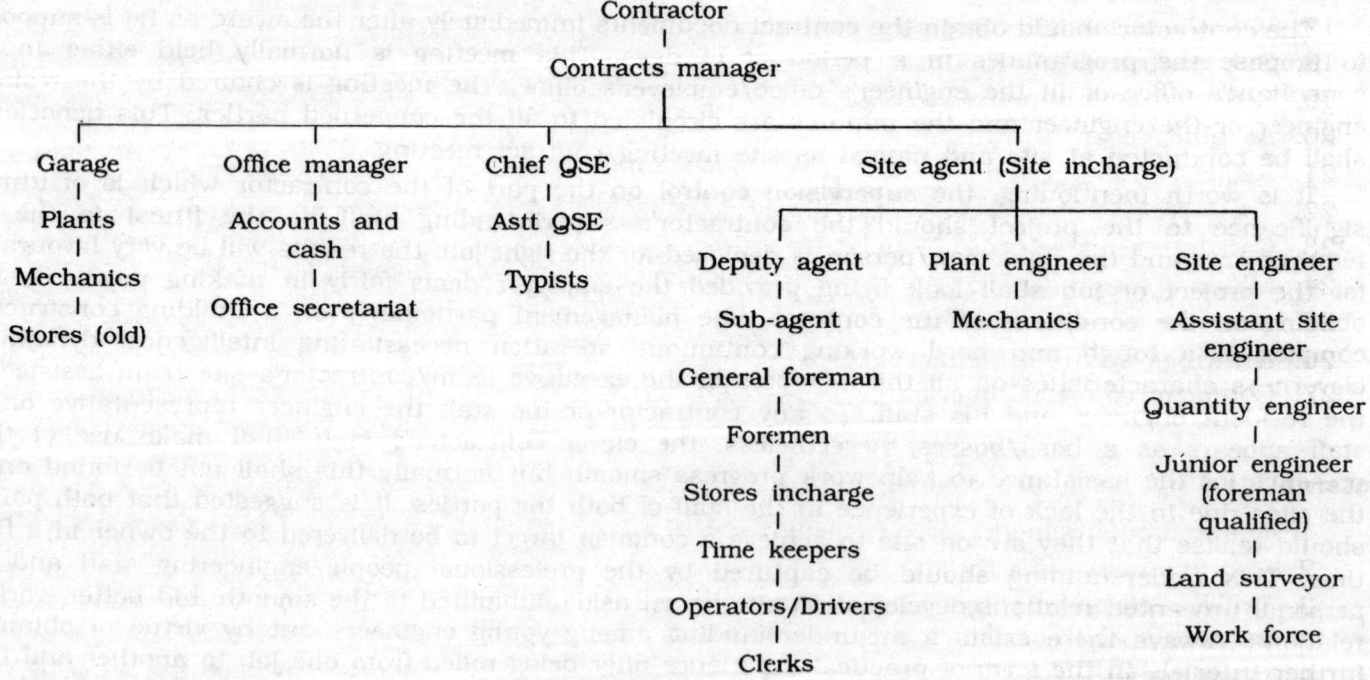

Agent : The contractor's chief executive on the job site is designated as agent or the site agent or the contractor's representative or the site manager for the project. The agent shall be responsible for the job in an overall manner, answerable to the contractor, to the resident engineer and the consultants group and to the engineer also if necessitated. He has to dispose off a group of duties of complex nature and technicalities. The agent on a buildings project must be technically competent, administratively sound and dynamic, energetic, and posses many other secret eyes. He shall be bound to be financially responsible in business as well. Some contractors do not want the contracts managers wherein that case the part of that responsibility shall also be included in the agent's duties. The agent acts as a superintending engineer as far as technicalities and coordinations are concerned which have been explained earlier and other minute duties listed and shall be further elaborated. This means that only holding merely an engineering degree or academic qualification is not sufficient but a specialist tutorial qualification as experience on contracts of buildings to successfully match the job requirements. Since he is the manager of the site, his duties are listed as follows but shall not be restricted to that only.

1. Maintain site diary
2. Daily report submittal
3. Material submittals
4. Shop and working drawing submittals
5. Weekly programmes
6. Quarterly programmes
7. Complete programmes
8. Revised programmes
9. Monthly progress reports
10. Monthly technical and co-ordination meetings
11. Monthly technical and co-ordination meetings internal for the contractors
12. General meetings
13. Special and instant meetings
14. Site memorandums
15. Site instructions
16. Engineer's instructions
17. Valuations for the completed works
18. Sub contractor's valuations
19. Reflected ceiling plans
20. As built/record drawings
21. Requests for inspections
22. Tests reports
23. Concrete pour records
24. Material test records
25. Material status schedules
26. Drawings status schedules
27. Complete co-ordinations and liaison related to allied services/departments/local bodies.

internal dealings with the contractor's organisations

1. Material requirements procurements
2. Coordinations with the garage, stores and workshops, etc.
3. Daily reporting to the head office for obtaining various records pending for submissions, approvals and coordinations
4. Coordination with the main quantity surveyor responsible for the job in case the same is not based on the site for medium size jobs.
5. Coordination with the office manager and the other staff
6. Coordinations with the contracts Manager
7. Solving other miscellaneous labour problems
8. Co-ordination with the other engineering section of the company to take or give any machine or any material involved in the job or workforce adjustments.

Dealings at site

1. Daily programme
2. Daily supervision
3. Daily material order
4. Solving instant problems
5. Solving problems with the consultants and pursuing the matters to solutions
6. Solving staff disputes and making necessary orders in the interest of the job
7. Labour bills for wages and overtimes and leave sanctions
8. Labour working on contract valuations to be kept moving
9. The preparation of the reinforcement bending schedules
10. Proposing alternatives to the resident engineer for discussions and approvals
11. Technical study of the job and instructing the site engineers when required
12. Daily meetings and general liaison as required

Elaboration of the coordinations as pointed earlier

Architectural/structural : Tender/contract drawings comprise of architectural, structural building allied services, civil works and miscellaneous works. Architectural details should be related to the structural details and in case of discrepancies being discovered that should be informed to the resident engineer or the related structural and architectural engineers and the remedial engineer's details obtained. The structural details as included as engineer's details cannot be followed blindly since these drawings are used on architectural requirements and the details as in engineer's design. Having brought that in the notice of the resident engineer, it is upto the resident engineer whether the alteration is to be made in the engineer's details but a site memo should be requested by the agent to confirm the instructions to protect the contractor's skin. In this case there can be no claims raised due to the aberrations as the same should have been examined before closing the tender. The general practice adopted is that alterations/adjustments are done to the structures and recorded as built drawings. Be noted that this can be done in one go on all the parts of the project since it is not practically possible to have to overload the work on site as well as low staff strength on site. The tendency of every contractor remains to maintain the minimum level of staffing on site to make economy but this practice of comparing the drawings should be carried out continuously particularly by the site engineers or the project coordinators of the contractor who are responsible fully technically for the part of project site assigned to him/them. The moment the site engineer discovers any discrepancy, immediately he has to bring that to the notice of the site agent or the coordinator who shall get the discrepancies remedied by the resident engineer/specialists engineers. Site engineer may directly contact the resident engineer/specialist engineer and let the decision taken or finalisation reached be communicated to the agent or the coordinator later so that the site head is kept informed.

To a new structure this coordination may not be that difficult but complexities arise when a structure is existing and the alterations and additions are to be carried out. For example, an existing shed made of structural steel framework and the old walls of cladding and roofed with old sheets is to be dismantled partly, altered and extensions to be carried out in the existing as a new structure in both the directions. The new design incorporates new structural steel framework of different sections in the portion to be extended and alterations in the existing framework. A RCC slab shall be added between the floor level and the eaves level to act as a balcony to the gymnasium below. The existing old walls shall be converted to a sheet cladded wall. This is the case where an existing shed is to be converted to a modern sports club/complex. The main contractor decides to award the structural steel work and the wall sheet cladding work to two parties both in the different countries due to quotations being lowest. The contracts manager presumes that both the contractor/supplier or the sub-contractors are sincere and shall coordinate efficiently. The contract (sub-contract) further provides a clause that the structural steel sub contractor shall supply the material while the erection shall be carried out by the main contractor himself. The sheet cladding subcontractor shall supply as well as erect the finished product supplied himself. On the demand by the structural steel sub contractor the site agent communicates all the technical informations to him. The sheeting cladder sub contractor measures all the details from the site personally. The drawings of the structural steel are submitted for the engineer's approval who in turn comments on the submitted working drawings that the approval is made subject to the coordinations with the cladding work. The two sub contractors are not found very cooperative practically. The cladding man/person submits his drawings and there is a lot of confusion but the consultants and resident engineer approve them commenting that the approval is granted subject to co-ordinations. You can expect what shall be the fate of site during the execution. During the progress many problems arise, alterations and adjustments are done. This kind of job particularly requires very close coordination to avoid any/some loss due to mis-coordinations which shall contribute to time wastage only.

Structural/mechanical/engineering : Engineering works involve the elements of the centralised air-conditioning systems comprising of supply and return air ducts, volume control dampers, attenuators, fire dampers, and various other fittings to pass though either the structural walls or the webs of the structural frame beams coming in the duct routes or the ducts penetrating the beams. It has been made clear that the contract drawings or the tender drawings issued for the services shall be converted into the working drawings from their being the schematic drawings and also in the contracts a clause is included that for the complete airconditioning system the contractor shall remain fully responsible for the design, installation and commissioning and also the maintenance to first class standard of functioning, therefore, the complete responsibility for its functioning goes to the contractor who shall be supposed to design, install, coordinate and commission. Therefore, while designing the size of the ducts the airconditioning contractor must request the agent or the main contractor to provide complete information of all the structural beams, walls and all other details he thinks necessary for the purposes of the coordination, may be the ceiling height, room height, the level of the beam soffits, etc.

He must carefully study the structural details for the fixation of the sizes and the duct routes which shall assist in saving of many reroutings as is often experienced. Simultaneously he shall look into the technical aspects of the discipline for example, the velocity of cool air shall be maintained within limits. Remember no structural engineer shall permit the pass of the ducts through the structural beams of RCC particularly where the large shear forces and bending moments are resisted by the beams. In case the airconditioning contractor inevitably requires the designed size of the duct and the structure is not built up at that time, the structural engineer at his technical will in consultation with the resident engineer and the engineer may alter the size of beam to accommodate the beam in isolation with the duct space provision. But there shall remain no choice once the structure is built. Location of the heavy machines such as chillers shall be done in such a manner that the loads are uniformly distributed on the beams designed for the purpose.

Electrical/mechanical/engineering coordinations : This part of the coordinations is in fact a cumbersome process and particularly when these disciplines are in the hands of different persons/contractors. Since the contract drawings need be converted into the working drawings which shall be drawn on the basis of the details of the equipments to be installed are available. Most of the items may be imported and the economical decision to finalise the purchase order shall take a long time. The conditions in the main contract that the shop drawings be submitted in a period of 60 days or any/some other prescription shall remain of mere description if the parties involved are not ready from the very moment of tendering and the result shall be that these drawings are not completely approved even after couple of months of the job start helping in delaying the job. Always there shall be a pool of submission drawings before every monthly meeting by the contractors to protect them from the minuting of the unsubmitted drawings which is not in a good professional practice.

Reflected ceiling plans : Air-conditioning duct works, diffusers and splitters, electrical light illumaniries details are of much significance, directly related to the coordinations and these plans cannot be finalised in the absence of all the said informations and the technicalities involved. The lame excuses by the different parties to the disciplines shall not solve any purpose if made just to defend the skins and not contribute progressively for the item. The agent shall have to tighten every one involved for that or else he shall be in problems. This item shall involve various other factors such as the sizes of the ceiling tiles which should be in the match with the sizes of the a.c. diffusers, the splitters and the sizes of the illumaniries should match the tile module. It shall also need the provision of equal intensity of illumination every where so the positioning of the ceiling lights be done that way and also the diffusers should not be very near to the illumaniries as a matter of safety of light fittings which shall mean repositioning of the lights and diffusers. Then the sizes of the tiles shall dominate the pattern of fixing the tiles in the ceiling per the architectural requirement. The architect shall not like to see so many cut tiles in the room or in the centre, therefore this is a very complicated exercise and should be done very carefully. There may be the situations where the already laid duct routes clashing the fixing of any of the foregoing items and a blame comes from the related sub-contractors that the reflected ceiling working drawings were not provided in the time and the agent shall be in a problem as the contractor may find claims from his sub-contractor for the compensation for readjustment of the duct routes due to late approval of the ceiling working plans. Agent can realise what shall be his position in that case and may be the same amount debited to him or he may be fired.

Situation is more worse when the false ceiling man/person is not ready in time with the details of his sizes and all other requirements to be passed to the reflected ceiling programmes. Practically, the suspended ceiling plans before making final drawing submittals should be sent to the electrical mechanical engineering contractors for the purpose of obtaining the no objection clearance from them in that case they shall not be in a position of raising any claims against the contractor. After obtaining approval for the positions of the light fittings and the diffusers and the splitters, the drawings are finally sent for submission to the resident engineer and thorough examinations are made by the group consultants. The group consultants shall make various comments on these drawings which should be incorporated at an earliest and the approvals obtained after the making of resubmissions.

This operation requires a great degree of cooperation among three and prompt action on the part of all in the interest of the job but this is seldom found. You shall find that the ducts drawings are provisionally approved with comments and later altered due to false ceiling adjustment and intensity of illumination requirement which is really unpleasing situation and the agent earns a bad impression. This contains a high potential of negligence in case the agent is a qualified engineer and virtually a project engineer and that he is supposed to help self.

The other part of major technical coordination exists between electrical and mechanical items. Engineering contractor installs a machine according to the approved design, the electrical contractor must know the sizes of the adopting cables to the isolators of the machines. Generally, the connections from the machines input to the isolators are done by the mechanical engineering contractor. But the thorough specification must be made available to the electrical contractor to enable him to make the required material for approval and procure in time to avoid any delay. There may be various type of machines in air conditioning and mechanical works for example air handling units AHU, chillers (a combination of compressors evaporators, condensers and fans), pressurisation units pumps, extracts fans, etc. In case the detailed advice is not made in time and both the parties do not agree on exchange of the technical statistical informations to the satisfactions in the system's feasible operationing, a likely delay shall be expected and the main contractor shall be in unpleasing situation.

Structural finishes/electrical : This is a normal coordination and all the informations should be supplied to the electrical contractor or his representative on site whenever asked for, complete set of drawings should be made available to the electrical contractor at the time of making the submittals of shop drawings. Many times the electrical man/person shall have to deviate the routes of cables/wires and change the positions of the switches and sockets depending on the positioning of finishes and furnitures locations. Locations of the switchboard, MCB, transformers, etc., must be explicitly defined. All such changes should be recorded in the **as built drawings**. Light illuminaries should be installed in a geometrical pattern to the architectural items., light should match the ceiling tiles modules. Laying of conduits in the concrete before pouring and inspection.

Structural/plumbing and drainage : This is a common coordination and well known to every civil engineer working on the building projects in the industry. Adjustment of levels to the structure and in foundations, the routing of water pipes through the walls and the floors. Naturally, the drainage pipes at g.f. level are concealed under ground and the invert levels dictate the depths of the footing at certain positions. The pass of a drainage pipe through the footing and the strap beams should be avoided and the same be passed beneath the base level of the footing. Sometimes, it is necessary to take the pipe through the structure, in that case structural engineer should be consulted and extra provision be made for restrengthening the reduced section. Above ground floor level the sanitary pipes generally pass below the slab soffits and sometimes below the beams as well. In every modern building suspended/false ceiling is provided and enough space can be found to accommodate the sanitary pipes. Care should be taken that the proposed space for the air conditioning ducts, false ceiling hanging wires and electrical cables trunkings and trays, etc., are not encroached upon for which a closed coordination of the affected zone is required. The only additional cost to be borne is the cost of providing steel brackets anchored to the slab and the pipes resting on the brackets. In fact internal drainage and water supply is a complicated operation requiring high degree of engineering skill and patience. The approach of the consulting architect is to see the water pipes concealed in the walls/structure or in the finished elements and embedded in the floorings for which due care must be exercised to maintain the technical general specification of the related public works departments which shall be necessary for obtaining the clearance for supplies. In this field sanitary works should be dealt with by an experienced foreman or engineer as the chances of mistakes are high. The material submittals should be done carefully according to programme schedule. Should the item misordered, delivered and even installed, the test results shall reveal the worthiness and the replacement shall jeopardize the contractor and the working of the system. So sanitary and plumbing should be treated a special element in civil engineering constructions for building projects.

Drainage coordination is equally important since the part work is preceded the structural progress. The site foreman and engineer shall take care of the activities and inspections to be done at right time to avoid delay in the construction activities. Providing block outs of the cubicle seats and units at the time of the erection of the form works. Maintaining the top level of the traps flush or 5mm below the finished floor at the located area. Test should be carried out promptly and approval obtained in time.

Services/finishes : In a building every facility is intended to be included, e.g., telephones and intercoms, fire alarm systems, call bell systems, public speaker systems, drinking water fountains (for public buildings) minor irrigation planters water boxes, gas supply, water supply, internal drainage, air conditioning, etc., but not important criterion to accommodate these requirements is to see that the architectural appearance internally as well as externally is not aberrated. Should any deviation is considered essential due to site conditions, prior approval of the resident engineer and the concerned architect must be taken and recorded. It is worth noticing that the idea on the papers many times cannot be practically formed into practicality on site due to various constraints, factors and a compromise is a must. Actually technical coordination drawings are prepared for such situations for all the services and the approval of the working drawings obtained to make a record in as built drawings. There may be occurrence of clashes among services of different disciplines, the only way is to reach an agreement to the satisfaction of all parties concerned as explained. Telephone ducts need per statutory bye laws in some states to be approved and inspected by the authority (even in some countries the telephone duct of standard size e.g. 90 mm dia is supplied by the telephone department to the convenience of connecting to the standard branch). In such cases where the specification stipulates the agent has to procure according to the requirements by contacting the telephone authority officials. The laid duct having duly been fixed, the stack pipe erected, the telephone inspector should be invited for the inspection, approval obtained, the resident engineer and the consultant accordingly informed. Be noted that in such statutory bye laws applicability situation, consultants approval is not final even though the resident engineer grants approval but it has no validity as far as the telephone department is concerned. They shall not accept the connection for the purpose of installation of the telephone line to the user. Once the inspector carries out the inspection, the backfilling may be done and the contractor is relieved of the contract clause operation responsibility due to its being a success. For water supply and drainage the statutory bye laws are different. Similarly, satisfactory coordination should be made for the cubicle pipes, units and kitchen equipments with the finishing items such as tiling, any special flooring so items match the architecture.

Similarly, air-conditioning and electrical II that is finishing items in services should be dealt with the finishes of the building.

External services/external works : External services include sewerage, storm water drainage, water supply to the buildings, irrigation in the landscapings, telephone ducts, electrical ducts, fire fighting pipes and hydrants, connections to the septic tanks, connections from the sub stations to the internal transformers, connection to the substations, external electrification, etc.

External works include the hard and soft landscaping, service roads, sports courts and their surfacings, games fields, security fencings, services yards, boundary walls, gates, steps, swimming pools, tree pits, special works, minor auxiliary buildings, (sub-stations, guard houses, worship building, pump houses, under ground water storage tanks, walk ways, pergolas and sheds, external lighting poles, but not the least) All the under ground pipe works should be carried out prior to any of the above items. All the foundations should be finished before the start of the external piping works. Remember, many services run under ground externally for which proper coordination is a must and many deviations shall occur. For example, in the drawing water line and electrical cable may be shown at the same section but it is impossible and a deviation must be there to shift the electrical cable. Two services pipes may strike, it should be avoided. Proper surrounding to the network should be done to avoid cracking in future when the traffic is put on. The inner levels have to be related to the finished ground level and many times adjusted due to raising or lowering the entire area of the land. Remember the initial survey on the basis of which the drawings are prepared might have been carried out in the past but due to wind movements the surface could be duned or eroded. The new level must have been adopted in the case of structural details as well as mentioned in the structural/structural coordination. You should consider enough time in the programme network/chart for the completion of the external work to maintain a practicable rate of progress. Sometimes, what happens is that the part of the services passing in your area are in the contract of the other contractors in the same site say the infrastructure main contract, in that case you have to wait till that guy carries out his job. But just awaiting his execution merely you cannot solve the problem. There must be a clause in your contract specifications that the coordination with the other contractors working on this project is the contractor's responsibility. In that case you are bound to coordinate with the site agent of that contractor in records, apprise the man/agent responsible, of your programme to let him alter his programme according to your requirement and better if done in the presence of the consulting engineers, nevertheless, the progress of your coordination should be regularly informed to the resident engineer so that if you approach him for help, he may easily persuade the other contractor to extend help to you. But keep the records for your protection. There can be a possibility that the electrical cables are to be provided by the other contractor upto the terminals of your distribution board and therefrom comes in the scope of your work. You make sure that this operation is carried out by that contractor as soon as possible properly coordinated to your electrical contractor's trades programme or else you have to face your sub contractor crying and raising claims for the delay of the activities and you shall find a letter to your office that the electricians are withdrawn temporarily due to nonavailability of clear site and consequently you shall be hammered by your head office. To avoid this, better study the services drawings yourself, keep contact with your sub contractor's foreman, try to persuade the infrastructure contractor, apprise the resident engineer and the consultants. This way you can get the job done keeping infrastructure contractor think of your being vigilant or that contractor may have to face claims from your office. This type of job requires coordinated management, may be your site engineers think that you are always found in different offices taking a cup of tea or coffee all the time and of no use to the company but those innocents are not acquainted of the contract clauses' negligence of which may cost your company a lot and you may be fired in that case. Try to grasp as much as you can in all the disciplines notwithstanding your being a civil engineer or the structural engineer or being the architect even, since the overall responsibility rests on the shoulders of the site agent. You are the superintending head incharge of the project site to do the job and this is part of overall superintendence. This is just an example but it can be in the same sense applied to other infrastructural services too. Sometimes, external infrastructural services do not form part of your contract but the other external works as mentioned do. In that case to commence your operations you have to match your programme either to the infrastructural contractor's programme or in case the infrastructural contractor comes subsequent to your contract then you have to ask that contractor to go the way of your programme for that particular coverage of the job.

All the external levels of man holes, soakways, water chambers, electric man holes or so called draw pits, etc., should be flush with the finished external levels and the invert levels should be calculated accordingly except in the unfinished areas say the grass or lawns where the top level shall be kept raised. Sometimes, there can be infrastructural services below the sports surface or the games fields which is never horizontal but in gradient both ways, then the services infrastructural units should be dealt accordingly.

Liaison to the statutory bodies : At the time of taking over the site you have to request the resident engineer to provide you an authority letter addressed to the statutory bodies for supplying you the record drawings of the site for your use. You shall also write to these departments independentl informing that you have started the work on that site according to contract and the building permit reference. You shall make sure that all the records drawings are in your possession before commencing the activities of the excavations to avoid damaging the existing services and it must be remembered that particularly the excavation shall be the specific responsibility in the duties of the site agent. The agent shall have to take care of all the **way leave forms** as included in the contract forms(way leave is a distance required to be left between the government existing properties and your building exteriors) Should your extreme point comes in between the two limits, it is the responsibility of the agent to intimate the resident engineer & request to decide what to do and at the same time contact the concerned department for remedial action. Even if the consultants and the resident engineer approve the setting out of the building, the contractor shall not be relieved of his responsibility for the same as that shall be considered a civil crime. The contractor shall all the time be aware of the real location of the site. To be called a site agent looks very pleasing but very difficult to succeed. In some countries should be existing services be damaged during excavations, heavy penalties are imposed on the contractors for the rectifications of the services. Similarly, as said earlier the agent shall coordinate to the statutory departments throughout the contract time and the resident engineer kept informed of what the outcome are. On important projects 'as built drawings' are revised every half yearly, you should know the alterations from the site record drawings which are updated. The objectionable materials per bye laws cannot be included in the job should you by chance come to know, immediately inform the resident engineer so that corrective steps be taken in time and alterations or alternatives employed.

Structural/Structural : This looks ridiculous but factually you face in the field. For example, the contract says that you have to take foundations upto a hard stratum of the specified bearing capacity. Excavations reveal that the same be either available at a lesser depth or at the extra depth as shown in the drawings. For the case of being at extra depth than specified, you shall carry the extra excavation till the required bearing capacity is met and adjust the levels in the foundations accordingly. But when the case is reverse that you are meeting the specified bearing capacity at lesser excavation, the intelligent and the practical resident engineer shall ask to raise the formation levels in the foundations to save for the owner. In that case you shall have to adjust all the levels in the foundations and propose for the approval. The adjustment shall be applied to the formation level, the top of the blinding concrete levels, the top of the footings' levels, the bottom of the strip beams levels, the top of the strip beams, the top of the stub columns, and the depth of the block or the brick wall to be constructed between the ground floor slab and the top of the strip beam, the protection of the waterproofing level if provided and also the concerned services pipes as the case be. These proposals shall be approved before the start of the blinding concrete in the foundation. The site agent shall get that done at an earliest to avoid the wastage of time. If that is delayed the agent shall bear the pressure from all directions.

Liaison

1. **Main contractor/Sub-contractor :** It is very essential to keep the process of updating and exchanging the informations on regarding the developments on both sides. The material submittals, shop working drawings submittals made by the sub contractors to the main contractor office must be promptly endorsed for forwarding submittals to the resident engineer duly scrutinised by the contractor regarding the conformation of the contents to the specifications in the contract. Similarly, the submittals shall be received back approved, rejected or approved with comments from the resident engineer's/engineer's office having been scrutinised by the resident engineer and his specialists engineers employed in his office. Sometimes, to save time the submittals which are finally recommended for approval are only forwarded to the engineer for the grant of 'approved' while those rejected are returned back to the contractor directly. The contractor's agent should act promptly on these returned submittals and put them in the process of posting to the related sub-contractors with a covering letter on the contractor's pad commenting about the submittals and instructing the sub contractor to act promptly either for the resubmission, or for incorporating the comments and resubmission or for taking necessary action for the placement of the orders, etc., or issue the required prints of the approved drawing for the distribution purposes, while the contractor should maintain a posting records with him. Should the submittals be suppressed by the agent due to carelessness, the overall effect shall be negative on the

contents items to be placed in procurement process or in carrying the works out from the working drawings.

Per existing practice the main contractor instructs the sub-contractors to issue six prints of each approved shop drawing sealing on that 'Approved' which shall be endorsed by the resident engineer, the specialists engineers and then distributed by the agent to the site engineers, sub-contractor and the resident engineer accordingly.

The agent shall send a copy of the main construction programme to each of the subcontractors to let them fit in the respective trades programmes and be appraised of the time to commence the mobilisation. Having received the programme copies, that should be sent to resident engineer for approval by the main contractor after scrutiny.

The agent shall receive site memorandums during the progress of works regarding the sub-contract items. Should the agent think that the reply of the memo is out of his contractor's scope, he should promptly post the same to the sub contractor's office under cover of his office instructing the sub contractor to act in time for taking the necessary action depending on the urgency of the memorandum contents. The contents can relate to work being carried out, to the deficiency of the submittals, or the miscellaneous nature like the action on the minutes of meeting, etc., not taken in time. Apart from the resident engineer and his staff pressing the sub-contractors for works' progress, it shall be the prime duty of the agent to pursue and instruct all the sub-contractors to meet the requirements of all expectations according contract and press his own engineers as well to clear the site for the continuity of the sub-contractor's operations. Even then if the sub-contractor does not come to right tracks then it is the agent's duty to notify to the head office or the contracts manager recommending the necessary action and if he himself is authorised to do that, he should not hesitate.

The agent should instruct the subcontractor's responsible representatives on site to provide daily returns regarding manpower/work force, material arrived, progress, etc., and request the guy to prepare the inspection request proformas in advance to be sent to the resident engineer's office for approval, for example, a RCC slab is going to be ready for pouring, the agent should take care that the inspections requests sheets regarding electrical, air conditioning, plumbing and drainage block outs, and the miscellaneous services along with building inspection request in all are sent to the resident engineer quiet in advance per the conditions in the contract to facilitate proper inspection of all the disciplines being carried out and the remedial works attended to if any pointed out by the resident engineer and the specialists engineers to avoid unnecessary delay to the slab concrete.

The agent should call frequently the technical and co-ordination meetings with the subcontractors regarding all as explained in the foregoings and record the minutes of the meeting duly circulated to all and the resident engineer. This shall impart an impression of the agent being a good controller of the job and the resident engineer and the consultants shall help him fully as the burden of their own shall reduce to some extent. You will find the representatives taunting on you in amusement that this agent is a problem but don't worry and frankly reply that it is nothing to do personally.

Coordination with the subcontractors summarily described as the travel of mutual informations regarding developments in time and no lapse shall be accepted.

The agent should himself take enough pain to check all the activities of services going on and from time to time, interrogate the foreman on site to know what is going on, shall go on and what are the problems on the site. He shall let you know every thing even if the sub contractor's engineer sometimes shall hide. This information shall assist you to pull the sub contractor but take care that this relation should be secret or the gentle foreman may have to face the loss of trust of his boss, i.e., the electrical or the mechanical engineer of the sub contractors. These foreman guys and even own can help you a lot should you have enough diplomatic tactics in tracing the assistance you are searching for. The team should be cohesive in principle and practical too on site but an agent should be shrewd with eyes in all directions. Apart from approved sub contractors of major disciplines of allied services, you shall have to engage different types of minor sub contractors of course, approved by the engineer. For them the selection should be made in time to facilitate timely approval. Such sub-contractors may be swimming pools sub contractor, (for the design of the system, construction and installations), kitchen equipments sub contractors, sports equipment sub contractor, roofing sub contractor, decorative works of ornamental nature such as GRC panels in the forms of geometrical arches or the design of the motifs, for the floor carpetting works, for the special floorings, rubber granulate floorings, sports squash courts floorings, special wall linings, for the special timber lining works in the walls and the ceilings, metal works, glass wall linings for the courts, sports courts surfacings, squash court plasterings, X-ray rooms plastering, mapple floors, etc., but not the least. This depends on the type of organisation the main contractor possesses. Some have the big organisations and they carry out most of the work's elements of the contract themselves while some do letting many parts of the contract to the subcontractors. Whatever be the case the responsibility of the site agent remains the same.

2. **Sub contractor/sub-contractor through the main contractor** : Sometimes, certainly the situation arises that the site agent shall have to permit the various sub contractors dealing particularly electrical and mechanical equipments to exchange the technical information among themselves when the agent is too busy to spare time subject to the availability of outcome to him in time. Seldom, the agent shall decide the technical and design aspects in the said connections but his remaining fully informed is a must for coordination purposes may be he is asked a question in the site meeting when none of the services engineers shall be present. The agent shall be supposed to reply all the matters put or raised being responsible for the job. Better should all the sub contractors exchange the details themselves to save the time of all the related persons and avoid paper work in correspondences.

3. **Main contractor/resident engineer and engineer** : This is a full time coordination process from the beginning till the completion and this shall start immediately after the receipt of the contract award letter by the main contractor from the engineer. Most of these coordinational duties shall be performed by the company's agent (site diary involving the daily job record, inspections, activities carried out, activities completed, planes, any special remark), (material and drawings submittals and pursuances for approvals, programmes, progress report and review, proposal architectural and structural and miscellaneous, meetings, site memos, site instructions, engineers instructions, valuations, coordinated drawings, as built drawings and records, test results, samples, concrete pouring records, material status, drawings submittals and approval status, etc.).

Apart from above practical coordination e.g., the agent shall finalise the difficulties coming in the way during execution of the job either himself or through the resident engineer and shall propose the alternatives subject to the specified standards and get them approved by the resident engineer and the engineer. Whenever asked by the resident engineer, shall make the method statements available to him for the approval for any/some activity falling in doubt of the resident engineer or the specialist engineer. In case the agent discovers some technical flop or flaw in the drawings which may lead to unsuccessful maintenance or functioning or working of the building shall immediately bring that to the notice of the resident engineer in consequence of that there may be a change in the engineer's details for that part. The coordination of allied services are done through the main contractor and the agent shall coordinate all of them as the same is covered under this title. The explanation of how the documents shall be worked with shall be given under the responsibilities of the company's agent.

4. **Contractor/other main contractor** : As explained earlier, the agent shall keep updated coordination with the other contractors working on the same overall project and particularly those related to his job. Infrastructural contractors shall excavate the roads frequently and bar some portion of the road access at various sections. Similarly, other building contractors may be executing the portion of various buildings and technical advice must be exchanged. Should the other contractor wish to know about the agent's job drawings or any/some technical information, the agent shall be contractually bound to provide him with that to gear up the cooperation. The agent may obtain circulars from the resident engineer from time to time regarding road closures, the joint meetings with the other contractors working on the same site, regarding the programmes on which the agent shall be supposed to act promptly. The agent shall seek approval of the head office should the cost of the project is effected or contract variation is involved. The agent shall not accept the decisions that add to the delay of the contract.

Every contractor wants to help the other to help self and this motto should be maintained. You can borrow some machine or material in urgency and similarly may cooperate in a number of ways to contribute to the contracts positively. The agent should try to promote good working relations on the project site. The agent shall formulate the rules himself how to deal as there is no hard and fast set procedures but naturally gifted common sense and no complex of inferiority or superiority but an attempt to the access of professional cooperation and standards. The agent may visit the office of the other contractors and in turn invite them to his own office as a matter of knowing each other more.

This attitude shall result in avoiding harsh letters and let the agent come to know how the resident engineer and his staff treat that contractor. Should the agent find the different and polite and favourable treatment towards that contractor by the resident engineer and his staff the agent may discuss and request the same treatment towards his company but this should be reflected in a friendly manner. Once every body on site from different organisations knows each other, you will definitely find the job running smooth provided the agent's control on his own organisation is very well gripped and they listen to what you speak. Please refer to the electrical mechanical engg/plumbing and drainage coordinations for the additional explanations.

5. **Main contractor/local statutory bodies :** As explained earlier as a matter of contract clause operation, the agent shall be bound to advise the local water supply, drainage, electricity departments, telephone departments, etc., and any other service necessarily mentioned in the contract, that the contract work of the particular reference awarded wide letter no...has been commenced on site by the appointed contractor...The agent shall also request for the issue of all the records drawings of the contract area at an earliest to avoid any damage to the existing infrastructure. The agent shall move per conditions of the contract regarding obtaining approvals of the working drawings, and the parts of the contract for which the approval of the concerned deptt. is essential, inform the related authorities in time so that the work be done under their approval. In case the agent finds that existing services are being encroached by the contract drgs and necessarily to be re-routed by the statutory departments. That shall also be done by contacting the required official and if required under the law the charges shall be paid to the statutory bodies for the diversion. Remember breaking or causing any damage to existing services or property of the government shall be amounted to an offence and the contractor be liable for imposition of severe penalty for which the agent shall be held responsible and may be fired. The clearance shall also be obtained from the related departments by the agent in the form of no objection certificates for the provision of the supplies to the building services.

6. **Main contractor/engineer :** In case the agent shall have to write regarding a matter involving the contract legal implications, that should be addressed to the engineer only and the copy circulated to the resident engineer. The control of the legal standing matter shall be done by the engineer and the resident engineer shall be responsible for the technical coordinations and implementations of the decisions as taken by the engineer and the practical supervision of the contract works. In fact the resident engineer in the layout levels of the engineer's organisation comes on the same level as the agent in the contractor's organisation. All the claims regarding extras shall be forwarded to the engineer and copies to the resident engineer that shall help the resident engineer to assess in advance and respond to engineer when asked for. Similarly, all other claims such as extensions of times, unforeseen circumstances, non-payments, delayed payments, employer's defaults, major changes required in the specifications, etc., should be addressed to the engineer and copied to the resident engineer.

What a contractor's agent is supposed to do?

As the designation indicates that the person engaged as the agent shall represent the contractor on the project site and he must do what a contractor shall do. In fact contractor is a businessman and shall like to make the maximum profits and delivering the good quality of the job expected of the contract. The agent should not work according to what salary he is drawing but realise the value of the successful performance which shall always be a credit in the profession. Every eye working for the project shall be watching the performance of a new agent so he must be careful.

An agent should be a director, and administrator, a commander on the workforce and the staff, a guide to his subordinates executives e.g., project coordinators, site engineers, deputy and sub agents, general foreman, foreman, assistant engineer, junior engineers, land surveyors, quantity surveyors and general administrative personnel and should have the capacity to use his technical skills if required from time to time. He must have a business mind too as in case the contractor does not employee a 'contracts manager' above the level of 'agent', the head office expects the agent to report directly to the owner, in that case the agent shall need enough market ideas. He has then to select the sub contractors, suppliers himself. It is of course cumbersome job but very essential. Remember agent may be salaried plus a partner in the profit percentage and hence it then is his own interest how much he can earn. Agent must be energetic, dynamic physically and mentally patient, diplomatic, administratively strong, good coordinator, a good corresponder, convincing, a good leader, a sympathetic person to solve the

SITE DIARY

Date ....   Budget ref. ....

Name of the work/Job ....   Contract ref. ....

S. No.	Description	For Contractor/Agent	For Engineer/Resident engineer/Owner
1.	Work planned		
2.	Activities completed		
3.	Activities started		
4.	Man power/work force — main contractor		
5.	Man power/work force — sub-contractors		
6.	Daily temperatures		
7.	Visitors to site		
8.	Test samples		
9.	Activities carried out Main contract Sub-contracts		
10.	Accidents		
11.	Inspections		
12.	Materials arrived		
13.	Machines arrived		
14.	Total machines on site		
15.	Others		

Agent signature   *Resident engineer signature*

worker's problems and see that injustice is not done to anyone, to see the salaries and other payments are made in time, to see that all the safety measures are being adopted to prevent the accidents, to see the transport is available in time to carry the workers to their camp, to see that they are not sick seriously, to see that sufficient messes exist, drinking water available on the site as well as in the camp, to allow them the leave when due or as requested, to send the wage bill in time and all other duties expected of an agent to perform. Remember if the persons working on the site are not happy, the healthy and harmonious working atmosphere shall not be built as labour politics shall be the likely outcome in all the cases. Also to remember that good agent shall be respected by all. You cannot just command respect by face all the time. Apply face in rare cases but the people should fear by looking at you.

1. **Maintaining site diary :** Site diary or daily job record : It is a document submitted to the resident engineer daily detailing the man power, machinery present, material arrived, next day plans, activities started, activities completed, maximum and minimum day temperature, visitors to site, inspections held, materials samples sent to the laboratory for testing, accidents, special remarks, etc. This shall be signed by the contractor's agent and the resident engineer and kept as a record. Some resident engineers like the contractor's agent keep the site diary as tabulated below in **A4 size triplicate** notebook.

The portion on the left side is used by the contractor's agent while the right side is reserved for the resident engineer, his staff, employer and the special visitor on the job from the engineer side. The day sheet is in triplicate, one for the agent, one for the resident engineer and one for the contractor's head office. The benefits in this system is that the agent notifies in the dairy only and the resident engineer or his staff is bound to see daily for the next day's inspections proposed and approval is signed in it. Any instruction or the advise or comment shall be entered in that only.

Some consultants prefer daily job record

It is a standard proforma provided by the engineer and shall carry all the informations as in the site diary except the notices for the inspection purposes and approvals. Separate standard proforma are issued for the purposes of making requests for inspections of various disciplines and the agent is supposed to notify in advance enclosing the related sketch of the proposed inspections. All the resident engineers expect a notification of 24 to 48 hours in advance of the inspection timings so the dealing engineer gets prepared and programmes accordingly not to overlap the inspections on the jobs of the other contractors simultaneously. This condition remains adhered to in the early stages but as the working relations among the contractor's staff and the resident engineer's and his staff get cohesive due to good performance of the job, that may practically/virtually be made flexible depending upon the progress of the works.

2. **Daily report submittal :** This report is submitted to the contractor's head office daily by the agent based on the details given by the site engineers incharges for their respective sites and confirmed by the agent including the next day programmes. Main purpose of this report is to let the head office know what the people have done and what shall they do the next day and also to control the wastage of the time and avoid mismanagement by the site people. This also envisages on the immediate needs on the job and imprest money requirements. This makes the head office conversant/apprise of what is going on the site that the control is good. But this is applicable when the site is near the head office as the agent could visit the head office daily but in the case when the site is far away and the agent cannot go daily to the office in that case the agent himself acts as the head office and does everything on his responsibility to make the site moving.

3. **Material submittals :** Condition in the contract requires that every item to be installed in the job should be approved by the engineer before the same is brought to the site. A proforma is issued to the contractor describing the name of the work, item no. per the BOQ, specifications reference, name of the manufacturer, catalogue or the technical literature, all signed by the contractor or his agent and numbered for the registration reference. This submittal is made in triplicate alongwith sample if required in the resident engineer's office where the concerned specialist examines the validity and the genuinity of the proposal to the prescribed standards per the specifications and in case it conforms, an approval is granted and endorsed by the engineer and the owner otherwise, returned to the contractor rejected or asking to resubmit after more details are available as comments made. The same procedure is carried out in connection with the sub-contractor's material submittals through the main contractor. Once approval is obtained, the approved sample with a reference number on that is kept in the sample room in the office of the resident engineer to facilitate the supervision by the specialists and the supervising engineer and the clerk of works to check the use of the same on the site. The contractor cannot make use of the unapproved materials brought on site in the permanent inclusion for the job. In case some delivery problems exist, the contractor may submit an alternative and obtain the approval prior to bringing in use. The resident engineer is empowered to order the removal of unapproved stock brought to site by the contractor. You will find hundreds of material submittals on building projects comprising of building materials, electrical materials, engineering materials, plumbing and drainage materials, all internal and external infrastructural and materials to be included in the finishes and the landscaping and all the special items involved. It is the duty of the agent to check that unapproved items are not ordered. Sometimes, due to lacking information from the supplier the item cannot be approved in it then the agent should pursue the matter with the supplier to provide the required information to match the required technicality as specified so that the item be finally approved and ordered by the contractor or the sub contractor in time to avoid the delay on the job. Once the material delivery is made to the site the agent is bound to inform the resident engineer through his daily job record about the details of the delivery supported by the authentic documents to enable the resident engineer, specialist engineers and the quantity engineer to have a check on the delivery to impart a pass. Proforma follows.

Material submittal sheet

Name of the client/owner   Engineer

1.	Contractor ref no. . . . .
2.	Name of work . . . .
3.	Ref to BOQ item . . . .
4.	Ref to specs or drg. . . . .
5.	Name of manufacturer . . . .
6.	Technical literature . . . .
7.	Sample . . . .
8.	Remark/delivery . . . .

Signature with seal of contractor   Date

For the resident engineer and engineer

| Architectural | Engineering | |
| Structural | Electrical | Others |

Remarks

| Approved | Rejected | Approved with comments |

| *Coordinator* | *Resident engineer* | *Engineer/Owner* | *Date* |

4. **Shop drawing submittals/working drawings submittals :** As discussed earlier that the services drawings forming part of the contract drawing shall only be for the schematic purposes to guide the contractor what to do per the requirement of the project. Based upon the material submittals, architectural and other structural requirements, the electrical, mechanical and engineering, plumbing and drainage, related builder's works, etc., drawings shall be prepared per grids and section references to a scale of 1 : 20 so that every detail on the drawing shall be exhibited elaborately depicting clearly and properly coordinated to each other and also to the interdisciplinary details. These drawings shall pertain to the finishes and decoration also but comparatively to the allied services drawings the quantity shall be too low. The contract drawing require the main contractor to see that the drawings are submitted within a period of 60 days from the day of contract award but practically this is very difficult on a big size project as the main contractor shall involve many agencies to carry out the allied services works to each discipline each agency and many other contractors for the minor clements consequently, the output shall be sloppy as every agency shall do according to its own interest to save more and the only problem to bring all of them on one table at a time shall be faced by the contractor's agent and really difficult to make them all work in close cooperation to do the smooth work. The architectural and the structural details are made available to these sub contractors dealing with the services and they prepare the drawings. The site agent shall have to prepare the drawings of his own disciplines of the building works such as reflected ceiling plans, coordinated drawings of special structures such as swimming pools on any proposal needed on the site. The drawings regarding fixing details of doors, windows, ladders,or any item specially required to be incorporated due to manufacturer's special instructions are all incorporated in the working drawings.

The dealing engineers submit the drawings in triplicate under standard proforma exhibiting the submittal ref no., name of work, contract specification ref, section in the project, etc., with the agent who further forwards them in duplicate on the contractor's seal duly checked by the main contractor's services engineer to the resident engineer retaining one with him as a record. It is in the good of the main contractor to appoint some experienced services engineer to assist the agent do the job of submissions smoothly and also keep a watch on the drawings whether appropriate or inadequate, to avoid the wastage of time due to unnecessary rejection and resubmissions. Having gone through the drawings, the resident engineer recommends based on the comments of the specialists engineers, may be the architect, the services engineers or the structural engineer, etc., and forwards them to the engineer for the final approval or redirects them to the contractor in case of being rejected for the resubmission as the case be. Once the drawings are approved finally, it is the duty of the main contractor to instruct the sub contractor to produce 6 prints sealed as 'approved' and get them endorsed by the resident engineer and the specialist engineers for the final distribution to the related parties as explained earlier. For his own drawings, he shall produce the prints, get endorsed over the seal and distribute accordingly. The agent shall see that the only approved drawing should be issued to the site engineer for the purpose of job execution and no other.

Shop drawing/working drawing submittal proforma to be placed on the top of the proposed duplicate drawings for the submission to the resident engineer.

Owner	Engineer	
1. Name of the work/contract		
2. Reference to the drawings		
3. Reference to the specifications		
4. Reference to the contract drawing		
Signature and sub-contractor seal	*Signature and main contractor seal*	*Date*
For resident engineer		
Architectural engineer	Electrical engineer	Others
Structural engineer	Engineering	Plumbing engineer
Date		
Remarks		
Approved	Rejected	Approved with comments
Date		*Engineer/Owner*

Note : In case of the **Approved** status please get the required no. of prints (6) endorsed by the resident engineer and the specialist engineer that may be the architectural engineer, senior services engineer or the structural engineer or the electrical engineer or the mechanical engineer or the plumbing and the drainage engineer or the other disciplinary special engineer per the item of submission.

5. **Weekly programme :** Based on the overall programme, quarterly programme, availability of manpower/work force and material, the site agent is supposed to submit a weekly programme in the form of bar chart at the end of the week to the resident engineer to enable him to keep the inspectors and dealing engineers engageable at the time of inspections and prepare the schedule of expectations. It is not very tedious job since site agent can easily judge what operations can be carried out in coming 6 days based on his experience of the same job. Moreover, he can seek the advise of his site engineers dealing the site programme directly on day to day basis. The important thing is to see that the necessary materials and machinery and the manpower are made available to assist the engineers to give the results. Proforma follows.

		From				To	

Name of work ....

S. No.	Description Day	I	II	III	IV	V	VI	Remark
1.	Form work	Bldg X _____						In case something in particular to be depicted for the comments of the consultants or special concrete programme is required, use this space/column.
2.	Rft. placing		Bldg X _____					
3.	Pouring concrete				Bldg X _____			
4.	Excavation	Bldg Z		_____				
5.	Blinding		_____ Bldg Z					
6.	Carpentry	In shop _____						
7.	Metal work	In shop _____						
8.	Electrical I fix	In slab _____						
9.	Mech I fix		In slab _____					
10.	Plumbing I fix			_____ In slab				
11.	Block/brick walls		In bldg Y				_____	
12.	Finishes				In bldg ZZ _____			Overtime work on weekly off days
13.	Decoration				In XZ _____			
14.	Testing	Bldg A pipeline					_____	
15.	Fabrication				For shed _____			
16.	Additionals	All related works and curing						

Signature of Site agent

CC — All site personnel to plan and programme accordingly
Head office for information

6. **Quarterly programme :** Based on the overall programme a quarterly detailed programme is prepared by the agent for engineer's approval. In fact it is not very essential to have it but should the resident engineer requires that due to the complexity of the project, the contractor is contractually bound to prepare that. It is not a complicated task but the activities are further detailed for a period of 13 weeks and only those items as are covered in the main programme should be included.

The rate column should be filled in based on the period allowed for completing a particular activity. Advisable here is to allow a little more time for a particular activity to compensate for the time lose due to contingent problems of non availability of the material or workforce. The bottom portion exhibiting the details of the proposed availability should be clearly exhibited/depicted. Average of proposed availability of the tradesmen drivers, operators, labourers, etc., be shown. Machines of various types must be shown in the case of mechanical constructions method techniques are involved e.g., excavator, shovel, fork lift, stationary or mobile concrete pumps, tower cranes, mobile telescopic cranes, tipper trucks, dumpers, rock breakers, riggers, road rollers, pavers, compactors, etc. Proforma follows.

Break up of the detailed nos. of artisans such as steel fixers, masons, tilers, block builders, carpenters welders, plumbing fitters, air conditioning duct workers, foreman, electricians, helpers, etc., should all be written that shall assist you as an instrument to suggest or inform needs in the following period or in near future and form on the basis the weekly programme and placement of local material orders and arrange the machine and different men/work force and materials.

7. **Complete programme :** This envisages on the formation of the whole project work on one programme in the frame of either bar chart or in the critical path method network or the programme evaluation and review technique. CPM and PERT are good techniques for the framing of the programmes

Quarterly programme

Contract No.			DOS								DOC		
Name of work			Date of start							Date of completion			

S No.	Quarterly programme Description of Item	Rate of prog. per week	Week 14 to week 26 — Weeks													Remark and progress monitoring
			14	15	16	17	18	19	20	21	22	23	24	25	26	
1.	(Building A)															
	Foundations	50%	▬	▬												
	G.F. Works	50%			▬	▬										
	Cols GF-FF	50%			▬	▬	▬									
	Slabs	30%						▬	▬	▬						
	Plastering	50%									▬	▬	▬			
2.	(Building L)															
	Excavation	50%		▬	▬											
	Blinding	80%			▬	▬										
	Foundations	45%				▬	▬	▬								
3.	(Pump room)															
	Foundations	50%						▬	▬	▬						
4.	(Football pitch)															
	Compound wall	33%									▬	▬	▬			
	Machines															
	Manpower															

on the big and complex projects where most of the durations the interaction of the operations and the activities control shall be of much significance. But these programmes shall not be of some use to the common site supervisors and foremen who are in the direct control of the job and sometimes even to the engineers on sites as technician engineers who must understand all this. It is therefore, advisable to frame the bar charts for the individual buildings and the structures and the infrastructural works as these shall be easily understood by all on site in addition to the network systems so that a combination of the two shall be used at site for the purpose of the progress monitoring. Interpretation shall be made easy this way and a combined programme of all the structures in the form of the bar schedule or chart be constructed for the whole contract. This shall make the programme review and progress convenient and easy for the contract site meetings. Proforma follows.

In addition to the said chart the proposed manpower/work force and mechanical power shall also be included for the guidance. The items as included in the exhibited bar schedule are only for the purpose of the guidance and the agent shall have to decide the items himself depending on the nature of the job, the contents of the items in the BOQ the details of the job in the plans and the related engineer's details. The main point is the experience of the dealing agent.

Similarly, the bar schedules for all the other structures should be prepared and got approved by the resident engineer and the engineer along with the combined unified overall job bar schedule representing all the details covering the date of start to the date of completion period. There shall be a number of charts depending upon the number of the buildings and their sizes and also the nature of the job. The unified chart shall exhibit the details of manpower/work force, machineries and the equipments to be engaged on the job for the whole period of the contract, the quantity may vary per requirement and per the contractor's experience. In the early stages this activity shall appear to be very

Bar Chart programme for one building																				
Bar chart programme			Building A							Date of start Date of contract completion									
S. No.	Description of activity	Rate %	Weeks																	
			1	2	3	4	5	6	7	8	9	10	11	12	13	14	15	16
1.	Take over and mobilisation	50			Progress indicator Programme indicator														
2.	Survey and excavation	50																
3.	Blinding	100																
4.	Foundations																	
5.	Back fill + Ground floor																	
6.	GF-FF cols																	
7.	FF slab																	
8.	FF-SF cols																	
9.	SF slab																	
10.	Electrical 1st fix																	
11.	Plumbing 1st fix																	
12.	Engg. 1st fix																	
13.	Block work																	
14.	Plastering																	
15.	Rendering																	
16.	Roofing																	
17.	Doors, windows wood works																	
18.	Metal works																	
19.	Staircase																	
20.	Flooring																	
21.	Electric 2nd fix																	
22.	Plumbing 2nd fix																	
23.	Mech. 2nd fix																	
24.	Testing & commissioning																	
25.	Snags and handover																	

hard but the benefits are assured at the time of calculating the monthly progress review for the meeting. Once the agent is accustomed to the use of this instrument, the project progress of various activities shall be on the tips to monitor the progress with the programme.

In the commencement stage of the job hardly a few building say 2 or more be started simultaneously, means at that time the agent shall be dealing with only those related programme charts and most commonly in the mid way of the completion time all the programmes shall come into work operation. It shall be beneficial for the agent to compute the weekly progress percentage on every chart at the end of the activity and transfer the information to **time period** versus **programme** percentage graph.

Based on the total cost of the project or the building in question, the agent shall allot a percentage of the cost to all such individual programmes and the activities to make a total 100 percentage. It shall enable the agent to reach the total summary progress percentage while reviewing the progress for the site meeting.

Allocation of percentage to a particular activity

Mathematically, it is easy to say to divide by the no. of activities in the chart a figure of 100 but an intelligent agent should consider a number of factors. Say some activities which are directly done on site shall consume more time while the others that may not be directly in his supervision or supplied in completed form to be fixed only, shall definitely consume much less time for the agent's time consideration purpose. For example, RCC cast in situ, foundation excavatrons and their preparations in the rock formations, finishes on site, decoration and roofing, etc., should be allotted more weightage of time factor. The items such as doors, windows, metal works and all ready made items be less weightage in the programme time. Activities of the allied services are generally distributed as Plumbing, electrical, an engineering I & II and testing and finals. Here 'I' refers to the first stage of the work and the 'II' related to the installations of the related items and the 'final' refers to the complete testing and commissioning operations, etc. For example, electrification 'I' means laying conduits in the concrete before the concrete pour and drawing wires after the concrete. 'II' indicates fixing of points (switches and sockets) and illuminaries.

Similarly, fabrication and fixing of the aluminium air conditioning ducts insulated with the glass fibre and painted with the vapour barrier seal with all fittings such as bends, attenuators, volume control dampers, fire dampers for the supply and the return of the cooled air for the system, laying of the chilled water pipes of copper fixed on the steel brackets, installation of the equipments, etc., come under 'Fix I' of engineering while the fixing of the diffusers and the splitters, grilles, return air space, connecting the machines AHUS and chillers to their isolators, supply and the operational functioning systems come in the 'fix II.' Well it is all upto the contractor's/consultant's team how to manage the distribution of the activities.

Actually sometimes, the contractor should further prepare based on the time periods for these activities shown in the main programme, their own trades programme (technical details) exhibiting the details of all the operations minutely and review the progress on that basis. The agents shall find the three trades programmes, i.e., electrical, engineering, plumbing and drainage all to be submitted for the approval by the engineer. Every activity as included indicates two bars, one shaded and one firm. The shaded bar indicates the programmed progress while the other the practical progress. The later shall be shaded from time to time depending upon the quantity of the work done. This shall make you aware how far or behind you are of the programme.

8. **Revised programme :** The programme of the individual building or more buildings shall fluctuate either ahead or behind schedule. No body bothers in case the progress comes ahead but in the other case where the building is behind schedules due to many constraints say the employer does not release payments in time, the contractor does not want to invest more funds, the importation of the material is not made in time due to the lack in the issue of the letters of credits, financing banks do not cooperate, major changes are induced in the design due to unforeseen or any/some other reason, the progress lags behind programme considerably and the resident engineer and the contractor conclude that the programme schedules in no way can be met, hence the revision of the programme to a new completion date is insisted wherein the agent reframes the whole thing as before and it itself shall make a big exercise. Many resident engineers do not insist on the revised programmes but go on monitoring on the original programme for the consideration of the progress review and to include the delay in the ratification of the extension of the time which shall form a complex calculations for the quantity and contract engineers.

9. **Monthly progress report**

10. **Monthly technical and coordination meeting with the resident engineer**

11. **Monthly technical and coordination meeting with the sub-contractors**

12. **Contract meeting/General meeting/Site meeting**

The said titles are interrelated and finally related to the contract or the site meeting or the progress review meeting.

According to the contract the engineer/resident engineer holds the meeting to review the monthly progress and discusses all the problems related to the contract and termed as the **contract meeting** or the **site meeting** or the **progress review meeting**.

Contract meeting involves the technical and the coordination work done for the project activities, the progress achieved, materials arrived at site, materials ordered, expected time of the arrivals, any possible delay, valuations, payments, quality controls or any other kind of assistance required by the contractor, alterations and proposals, etc.

In fact generally, one week ahead of the contract meeting, the resident engineer likes to hold that meetings as a part of the contract meeting possibly to avoid the presence of every party and also to reduce the longitivity of the contract meetings. So practically, the technical and coordination meeting needs the presence of the electrical subcontractor, engineering sub contractor, plumbing and drainage sub contractor, or any other services sub contractor or the related concerned approved engineers of the disciplines, in case the main contractor carries all the disciplines himself, and the contractor's agent and from the resident engineer's side the representation is done by the electrical, mechanical, plumbing engineer as well as project coordinator headed by the resident engineer.

The details of the materials submittals, shop drawings submittals, their approval or the status schedules, pending submissions and the possible dates for the same, materials ordered, materials delivered, ETA, exchange of the technical informations, (EME) general problems related to the main contractors/building works, presence of the manpower/work force, works of the fabrications carried out, remedial works, trades programmes, and the relation of those programmes to the main approved programmes, progress achieved and comparison scaled on the approved programmes, actions required shall be all discussed in that meeting in total and recorded, minuted and circulated to all presents, engineer, the owner, the contractor and the sub contractors.

In the first coordination meeting the resident engineer should ask the agents to schedule the materials and the working and the shop drawings to be proposed for the engineer's approval on which shall depend the contents of the future meetings. (The agenda minuted shall be given in the pages following how shall that be recorded by the resident engineer).

Leaving the technical and coordination meeting to be conducted by the resident engineer shall not prove to be in the good of the agent and the contractor as well. Every resident and his engineers expect the agent of the contractor or the site coordinator to deal with the coordination themselves as that goes to the responsibility of the contractor. An intelligent and experienced agent shall start carrying out this work himself. He should request all the parties in relation to the meeting to attend the internal technical and coordination meeting based on the same agenda as with the resident engineer and this should be held regularly to cover up all the issues to be taken up by the resident engineer and the consultants and all the correspondences and instructions should be included in the agenda. The agent should conduct this meeting exactly to the same way as the resident engineer does but leave the technical decisions on the resident engineer and the specialist services engineers. All the actions to be taken on the site memos, site instructions, received from the resident engineer's office should be recorded and the parties concerned pressed to follow up the matter to comply with per contract obligations at an earliest. The status of the materials orders and the delivery schedules, status of the working drawings submission and approval schedule, technical problems in the coordinations, financial aspects of the items and variations, additions and alterations, etc., and any other business or the miscellaneous work and the date of the next meeting should all be recorded in the minutes of the meetings and copies circulated to all present in the meeting, the resident engineer and the consultants, the head office of the sub contractors, and the head office of the contractor. This method shall impart agent a situation where he shall be informed for all the allied services happening and make himself answerable of any quarry by anyone and command the sub contractors and also subcontractors shall not be in a position to befool him, of course, it shall provide the agent an opportunity to acknowledge/acquire some technical and engineering aspects of the job and a need to go through the drawings of the allied services to find more and more from both the schematic as well as the working drawings. The agent may be in a position sometimes even to cross examine the subcontractors on his own and whether the subcontractors and the resident engineers are fair to the contract or not and enjoy the respect of all concerned. Minutes proforma follows.

MINUTES
Technical and Coordination Meeting
(Contractor's chair)

Contractor's name

Name of the work   Contract No.

Sub : **Technical and the coordination meeting No.**

Venue   Site   Date   Time

S. No.	Description						Action
	Present : **Name** **Level** **Representing**						
	A Contracts manager Contractor						
	B Contractor's agent Contractor						
	C Electrical engineer Sub-contractor						
	D Mechanical engineer Sub-contractor						
	E Plumbing engineer Sub-contractor						
	F Services engineer Special sub-contractor						
1.	**Apologies :** None						
2.	**Minutes of the last meeting**						
	Agreed by all/Amendment made to the entry No.............as follows						
3.	**Material**	**Total**	**Submitted**	**Approved**	**Balance**	**Remark**	
	Electrical						
	Mechanical						
	Plumbing						
	Others						

Details of the main materials

Electrical engineering

a. Conduits — Approved and delivered to the site in parts

b. Drawing wires — Approved and placed in order

c. Light fittings — A type, B type, K type approved Type X Y Z to be submitted

Electrical sub-contractor

d. Fire alarm system — Approved with comments, the sub-contractor to provide the clarifications sought immediately

Electrical sub-contractor

e. Call bell system — Approved, material en route from France

— do —

f. Smoke detector — Rejected, the contractor to resubmit
Date of submission to be two weeks

Main contr. Sub-contractor

g. Chillers isolator — Submitted to the main contractor to be forwarded to the resident engineer

Main contractor

h. Light poles 12 m and 18 m high — Approved. Locally available/the order to be placed in one week time

Sub-contractor

i. Switch and sockets — Approved, locally available

—

j. Main circuit-breaker — Approval, under scrutiny. To be pursued.

M. contr/sub contr

k. Distribution boards — Approved en route to stores | Sub-contr
l. Water proof lights for the swimming pools — Pending submissions within 4 days

Mechanical engineering
a. Aluminium ducts — Submitted, approved, under fabrication | sub-cont.
b. Glass fibre insulation — Approved, locally available | — do —
c. Volume control dampers — Approved, delivered
d. Attenuators — Approved, delivered on site
e. Fire dampers — Submitted, more details sought, to be submitted in 10 days | Sub contr.
f. Filters — Submitted for approvals, to be pursued | M. contr/sub
g. Chillers — Under submission to the main contractor in 3 days | sub contr
h. Air handling units — do — | — do —
i. Chilled water copper pipes — Approved, to be ordered | sub contr.
j. Extract fans — Partly submitted and approved, order placed | sub contr.
Remaining to be submitted in 4 weeks
k. Pressurisation units — To be submitted in one month | sub contr.
l. Condensers and pumps and all the other items — To be submitted in 12 days | sub contr.

Plumbing and drainage engineering
a. S.W. pipes and fittings — Approved and on site
b. C.I. pipes — Approved and on site
c. SCI pipes — Approved and on site
d. WC Asian — rejected, to be resubmitted in one week | sub contr.
e. WC European — rejected — do — | sub contr.
f. Wash hand basin — Approved, to be ordered | sub contr.
g. Sinks — Approved — do — | — do —
h. Urinals — Approved with comments, clarification sought, awaited from the manufacturer | — do —
i. Water pipes and fittings — Cold water pipes approved and on site
Hot water pipes to be submitted in 3 days | — do —
j. Water tanks — To be submitted next week | — do —
k. All other pending items

Shop and working drawings					
Discipline	Total	Submitted	Approved	Approved with comments	Balance
Electrical					
Engineering					
Plumbing + drainage					
Special services					

	Details	
	Electrical : Generally, all the drawings are submitted and approved and only 35% of the major drawings of the following descriptions are pending due to the non-availability of the mechanical engg, technical information related to the machines, pumps, chillers and AHUs, etc.	Engg. sub contr.
	Mechanical : Most of the drawings are pending for submission, only 15% so far have been submitted. Remaining to be submitted in one month time. This is deviating the programme. Recovery programme is requested, the sub contractor promises to submit in 2 days.	sub-contr.
	Plumbing and drainage : 50% drawings submitted, none approved due to technical reasons. Main contractor instructs the plumbing engineer to pursue the matter personally with the resident engineer and sort out the rejected drawings at an earliest to avoid delay. Contractor's agent to assist in this.	sub- contr. main contr.
	Swimming pool : System submitted for approval, under scrutiny in resident engineer office. Main contractor to pursue the matter with the resident engineer	— do —
	Reflected ceiling plans : Under preparation in the agent's office and expected to be submitted in 15 days.	main contr.
5	**Trades programmes**	
	Electrical : To be submitted in 3 days	sub contr.
	Mechanical : Under preparation, to be submitted in 10 days	sub contr.
	Plumbing : Under preparation, to be submitted in 10 days	— do —
	Pool system : Under preparation, to be prepared in 10 days	Main contr.
	All the sub contractors admitted that they are in the receipt of the main contractor's approved programmes of all the buildings and main combined and coordinated programme for the guidance of the sub contractors	All sub contrs.

6. **Progress**

Discipline	Programmed	Progress	Remark
Electrical	15%	10%	lagging behind
Mechanical	10%	5%	do —
Plumbing, drainage	20%	20%	per schedule

7.	**Co-ordination**	
	a. Electrical engineer sought information about the chillers and AHUs and co-ordinate technical statistics. Mechanical engineer to supply in 3 days	Elect. + Mech. sub contr.
	b. Electrical engineer demanded the availability of certain air conditioning equipment layout drawings to relate his electrical system. Delay due to mech. sub contr. to take 20 days. Agent insisted upon the mech engr. to resort effective measures.	Mech. sub contr.
	c. Swimming pool engr. wants the clearance of related mech., equipment, drainage and plumbing details. He is held due to that. Agent requested the mech, and plumbing engineers to conclude the matter expeditiously in the interest of the job.	Mech. + Plumbing sub contr.
	d. Actions on memo no. .... dt.	
	e. Action on memo no. .... dt.	
	f. Action on the instruction no. .... dt.	

8. **Financial :** No extras till now
9. **Next meeting :** – .... Day ...... time

Signature Agent
Distribution — All present
                 Resident engineer for information
                 All sub contractors for action
                 Head office for information

Monthly Progress Report — for site meeting

Owner ....
Contractor ....
Budget ref. ....

Date of commencement ....
Date of contract completion ....

S. No.	Description of building/ Structures/BOQ Ref./ programme bar	% this month	% last month	% combined programme this month	% combined programme last month	Programmed	Remark
1.	Mobilisation + establishment	100	50	0.05 × 100	0.05 × 100		+ weeks
2.	Bldg. A	10	5	0.2 × 10	0.2 × 5		– weeks
3.	Bldg. B	5	–	0.3 × 5	0.3 × nil		+ weeks
4.	Bldg. C	–	–	0.25 ×	0.25 ×		
5.	Bldg. D	–	–	0.08	0.08		
6.	Bldg. E	–	–	0.08	0.08		
7.	Bldg. F External services + other work	–	–	0.04	0.04		
	Total						

Overall programme is + weeks. The critical state to lag in buildings..................
Measures to be taken
To increase workforce; to shift machine from bldg. ahead on programme to the bldg. lagging behind; to bring in more sub-contractors, etc.
Special remark : Due to delayed payments by the employer, the cash flow situation of the contractor has worsened to the non-fulfillment of the commitments. The banks stopped issuing the overdrafts. The engineer is requested to pursue the owner to release the valued payments in time or the work is afraid to be further delayed and may be out of control later due to time overrun.
Overall statement : Having overall picture of the job the contracts manager/site agent should mention the status of the job according to his observation say the progress % against the programme schedule indicating the lag or advancement ....
'Any other comment can be added depending upon the contract and the exact case'
Enclosed details

Note : The said tabule is an indicator of the progress summary as an overall picture of the job that shall be abstracted only after the preparation of the individual progress charts as have been elaborated in the following page which shall indicate the detailed percentage of each activity on every building stating the progress is ahead or the behind of the scheduled programme. There shall be as many progress charts or tabule as many programmes. Now it shall depend on the experience and the understanding of the agent preparing the statements how far he may reach the reality and the approximations.

S. No.	Description	Programmed %	Progress this month %	Progress last month %	Ahead/Behind weeks	Remark
	Progress Tabule — Building A					
1.	Excavation					
2.	Blinding					
3.	Bitumat					
4.	Footings					
5.	Foundations					
6.	Backfill & GF					
7.	Cols. GF-FF					
8.	Slabs FF					
9.	Cols. FF-SF					
10.	Slab S floor					
11.	Block work GF					
12.	Cols SF-TF					
13.	Slab TF					
14.	Block work SF					
15.	Plumbing 1st fix					
16.	Plumbing II fix					
17.	Electrical 1st fix					
18.	Electrical II fix					
19.	Mechanical 1st fix					
20.	Mechanical II fix					
21.	Testing and final					
22.	Wood work					
23.	Metal work					
24.	Plastering					
25.	Rendering					
26.	Roofing					
27.	Painting					
28.	Doors and windows					
29.	Flooring + carpetting					
30.	Snagging + handover					
	Total					

Statements : Programmed percentage — %
Progress percentage — %
Progress lags behind by   weeks/ahead by   weeks
Carried over to the abstract of progress

Note : This type of the progress tabule shall be prepared for the complete bar charts or the individual programmes to be incorporated with the summary of the complete progress.

STATEMENTS

For the building and services on site
Work force — Building

S. No.	Designation	Weeks				
		I	II	III	IV	V
1.	Agent					
2.	Engineers					
3.	Staff					
4.	Masons					
5.	Carpenters					
6.	Steel fixers					
7.	Riggers					
8.	Operators					
9.	Technicians					
10.	Helpers					
11.	Drivers					
12.	Watchman					
	Total					

Work force — Allied services

S. No.	Designation	Weeks				
		I	II	III	IV	V
1.	Plumbing engineer					
2.	Plumbers					
3.	Helpers					
4.	Electrical engineer					
5.	Electricians I					
6.	Electricians II					
7.	Helpers					
8.	Mechanical engineer					
9.	Duct workers					
10.	Installers					
11.	Helpers					
12.	Other services people					
	Total					

For the machines and equipments

S. No.	Name	Weeks				
		I	II	III	IV	V
1.	Excavator					
2.	Rock breaker					
3.	Generator					
4.	Mobile crane (telescopic)					
5.	Tower crane					
6.	Shovel					
7.	Concrete pump					
8.	Bulldozer					
9.	Dumper					
10.	Tractor					
11.	Sand blaster					
12.	Welding m/c					
13.	Vibrators					
14.	Trucks					
15.	Others					
	Total					

Machines of the sub-contractors for allied services

S. No.	Name	Weeks				
		I	II	III	IV	V

Material submittal and delivery schedule status updated

S. No.	Description of the material	Submittal reference	Date of submittal	Rejected/ Approved	Date of the approval	Details of the material placed in order	Expected time of arrival (ETA)	Date of actual arrival	Material sample inspection Approved/ rejected or any other order given to the contractor	Damages, if any, in transportations	Delivery problems	Remarks

Note : Material should be entered in the above tabule per the approved list of the material submittals and no change shall be made in that without the approval of the engineer. Similarly, the services materials status shall be prepared for the technical and the coordination meetings as explained earlier in the related discussions on the matter.

List of the materials (date-wise) brought to the site as a consequence of the local or regular supply

S. No.	Name	Unit	Date	Source
1.	Cement	Mt		
2.	Blocks 200 mm	nos.		
3.	Blocks 100 mm	nos.		
4.	PVC 1000 g sheet	no.		
5.	Agg. 20 mm	cum		
6.	Agg 10 mm	cum		
7.	Sand c./dune	cum		
8.	Plywood	no.		
9.	Polystyrene			
10.	Scantlings			
11.	Structural steel	Mt		
12.	Reinforcing steel	Mt		

It is also good if you provide a list of details to indicate the conditions of the weather during the reported period. Temperatures should be included.

It is also good should you include in the report the details of the accidents occurred during the reported period.

Include in the report or a few days earlier the detailed valuations and claims through your quantity engineer which shall be taken up during the meeting in the financial part of the agenda of the meeting.

So all these documents shall generally be prepared for every site/contract meeting and be posted in single package at least 48 hours before the time of the meeting to the resident engineer to let him go through the documents also.

Package of the documents for the site meeting

1. Summary sheet of the overall progress on the top
2. Progress reports of all the individual programmes/or the buildings as the case be in the individual sheets and placed second from top.
3. List of workforce and the machinery deployments during the progress period and be placed third from the top.
4. Material submittal and delivery schedule according to the approved list and be placed fourth from the top.
5. List of the materials during the regular supplies in the progress period and be placed fifth from the top.
6. List of the weather conditions during the reported period following.
7. List of any accidents following.
8. Valuation to be carried personally and generally one day before the meeting discussed with the resident engineer's quantity engineer and finalised.
9. Minutes of the previous meetings should be taken with you while joining the meeting as the contents of this meeting shall be related to the contents of the previous minutes of the meeting.

Should you be ready with all the above documents and satisfactory answers for the raised problems, shall successfully attend the meeting without hesitation or being let down and enjoy the meeting.

Since the scope of the discussions for which the meeting is going to be held more or less shall confine to the above documents until any special discussion or problems about delay or mismanagement of performance. In the start of the job it shall appear a difficult task to prepare all such documents but later after attending a few meetings things shall be repetitive and easy and the remaining time on the project shall be easy going as this part of the site management is concerned. In some case the easy going resident engineer and his project coordinator may depend upon the informations provided by the agents for the preparation of his own report to the engineer through the resident engineer.

General meeting/site meeting : This is actually held once a month to monitor the progress on the contract and chaired by the engineer/resident engineer, and those present in the meeting are the specialists engineers of the resident engineer, the project coordinator, the quantity engineer, and from the contractor, the agent, the project coordinator and the site engineers, and the quantity engineer and if wished the employer's representative or the employer himself and the discussions are based on the progress report documents submitted by the contractor agent.

Agenda of the contract meeting/site meeting/general meeting

1. Agreement on the recorded minutes of the last meeting by the resident engineer and before this the apology if any received shall be recorded.
2. Progress in details of the whole project scope in the contract under meeting involved, including the progress of the allied services referring to the minutes of the co-ordination and the technical meeting minutes already issued by the resident engineer and on the queries of the allied services co-ordinations issues, the contractor's agent and the dealing project co- ordinator for the resident engineer shall be supposed to respond due to the sub contractor's engineer would not normally be present once they have attended the technical and coordination meeting.
3. Detailed discussions on the material submittals, approvals and deliveries and any other related problems on the same and shall base upon the material submittal, and delivery schedule as submitted by the contractor duly updated.
4. Detailed discussions on the shop/working drawings in connection with the building works if any.
5. Quality control on site and the remedial measures taken in cases of any complaint by the resident engineer and the quality engineer on site or the supervising engineer and the related staff of the resident engineer.
6. Any alteration, addition or deletion required or requested by the contractor or by the employer.
7. Valuation of the works done and the extras if claimed by the contractor and payments.
8. Remedial measures to control the work progress.
9. Any other discussions or business.
10. Next date of the meeting.

The contents of the meeting may vary every month and shall depend on the type and nature of the project. The minutes of this meeting shall be drafted and recorded by the resident engineer or his project co- ordinator and circulated to all those present, to the engineer, owner, contractor and all related people on site.

Special and instant meetings : These are the meetings of minor nature and can be held between the contractor's supplier and the sub contractor, between the sub contractor and the consultant, between the agent and the sub contractor, between the agent and the resident engineer dealing with the finalisation of the minor matters or obtaining related information for the suitability of the items/systems proposed for the submittals. The meeting may be held on a particular issue separately between the agent and the resident engineer for example, some dispute arises on site regarding the method of working and the site engineers/foremen and their not being in a position to explain or the resident engineer calls on the agent and the concerned staff to discuss the matter and agreement, or a compromise is made how to carry out the item on site. The resident engineer can request or ask the agent to furnish the method statement to be approved by the resident engineer and then to be adopted for the execution of the activity on site. Another example, that a lapse occurs on the site on quality and the resident engineer holds the work as a particular activity, the matter is then taken among the site engineers and the assistants and the way is sorted out.

Confidential meetings may be held in case the resident engineer/his staff's complaint about one engineer/foreman spoiling the job according to the consultant and ways are sorted out by transferring the engineer or the foreman to less responsible job. There may be a meeting when some subcontractor does not turn to the job expectation and the performance be poor, the agent is called upon to discuss and take measures for the improvement.

There is some emergency on the job and the contractor is called upon to discuss the matter or all the contractors are called upon to sort the matter out. Sometimes, the aberrations are applied to the programme due to unforeseen conditions. Sometimes, the owner wishes to delete or alter the part of the contract job and the reaction of the contractor is necessary on the possible situation of already procured material or the expenses the contractor incurred on the same.

Any confirmation made during these meetings must be recorded and the following **site memo** be issued by the resident engineer to confirm the discussions and the solutions decided.

The contractor
....
Sub : Building X, Y
Sir,
Agent/resident engineer discussion held dt. .... confirms the followings.
The alteration to the programmes shall be done as follows.
a. X building to be postponed till further instruction
b. Y building be augmented with the workforce
Signature of Resident engineer
CC — Engineer, owner, site engineer, file, contractor

Site memorandum : A letter written to the contractor on site by the resident engineer on any matter regarding the contract job execution for actions/records with a reference No. and the date of issue.

Project	Resident engineer/consultant
	Site memo
Job No.	Date ....
M/s Contractor	
....	
Sub : **Programme building X/A**	
Dear sir,	
The programme in the form of bar chart submitted dt...is approved in principle with minor comments made on the copy of the programme enclosed. Please ensure the incorporation of those comments and issue 3 copies for our use.	
Signature of Resident engineer	
CC — Engineer, owner, file	

On a big project the agent may receive hundreds of the site memos in a year or half and he shall have to make a hundred reply. The agent shall receive in site memos on various issues like the material submittals, shop drawing submittals, execution problems on site, warning instructions on the contract and many other controlling and miscellaneous factors. The agent shall have to go through all these memos very carefully and reply in accordance to the requirements and the facts on the contract and delay on the reply may cost a lot to the contractor.

Site instructions

These are the instructions serially numbered in the form of letter shown below intimating the intention of the engineer/owner to delete or deviate or add or alter or any other change to be instructed in the execution of the job involving variation or no variation on the contract. The main point is that it is issued by the resident engineer in consultation or the verbal approval of the engineer since the real instruction that is the engineer's instruction shall take time to be formally issued. In the site instruction a note shall be added by the resident engineer regarding the constitution of financial variation on account of the instruction as shown in the instruction proforma below.

Resident Engineer/Consultants

Site Instruction No. ....   Date ....

M/s Contractor
....

Sub : Building No. ....
Contract reference ....

Dear sir,

Carry out the following items as given below.

Signature of Resident Engineer

Should the contractor consider a cost variation as extras, a variation order be submitted and the approval or the affirmative response be obtained from the engineer before executing or carrying out the work or/otherwise it shall be assumed that there is no variation and the contract price shall be paid only.

CC — Engineer, Owner, File

It is in the interest of the contractor to advise the engineer/resident engineer that the variations are involved in the Site Instruction number...dated... The details of the variation involved shall be submitted as and when ready and seek further confirmation whether the instruction be carried out or negated. The contractor should commence in executing only on obtaining the confirmatory letters in the form of letter or telex or any other form of communication.

Engineer's Instruction : In fact, the process provisionally operates by the issue of the site instruction but contractually, only the Engineer's Instruction is treated valid for the purpose of claims as variations. But the problem is that practically, the Engineer's Instruction takes months to be issued due to the engineer's office being always busy therefore, for the efficient execution of the works the site instructions precede the Engineer's instruction. No variation can be approved based on the site instruction but engineer's instruction. Many site instructions may be combined in one engineer's instruction. These instructions are issued in the same proforma language as the site instruction except the note of information about the constitution of the variation before the execution of the instructions. These Instructions are posted to the contractor directly from the engineer's office.

Valuations

Valuations are prepared for the claims of the monthly works done by the contractor. Valuations are prepared for passing the works done by the sub contractors monthly. This is a major type of valuations and usually prepared/worked out by the quantity engineer/quantity surveyor who visits the site monthly or sometimes on a big site remains based on the site and notes the percentage of the works carried out in references to the contract bill of quantities in case of lump sum itemised contracts in consultation with the site agent/site engineers or collect the bills of measurements taken in the contracts of schedule of rates type. Generally, itemised lump sum system of contracts is adopted in many countries and it is easy to operate. The easiest way is to summarise all the bills of quantities separately into a few heads and the quantities separately into a few heads and the quantities and the cost are entered. For example, a sample has been given on the next page for the reference.

All the BOQs are added and the cost computed. Then a list of materials on site account is prepared and the amount added to the cost. The enhancement on any account per contract condition or documents if any, shall be added to the computed sum. The payment and the retention moneys in respect of the last valuation and this valuation are adjusted respectively as deductions. The net valuation shall then be agreed by both the quantity engineers of the resident engineer and the contractor and signed by the resident engineer as a recommendation for the valuation payment to the engineer who shall dispose the same to the employer/owner having been satisfied himself for the contents signed by the resident engineer in the form of certificate as request for payment. Generally, in this type the quantities are not considered upto two places of decimals but a little approximation shall not matter much.

| Work .... | | Contract Ref. .... | | | Budget Ref. .... | |
| Date .... | | Valuation No. .... | | | Month .... | |
S. No.	Bldg. A BOQs	Unit	Qty. in BOQ	% done	Rate/unit	Cost/Amount
1.	Excavation	cum	500	70		
2.	Blinding	sqm	200			
3.	Bitumat	sqm	100			
4.	Founds.	cum	80			
5.	GF Slab and Backfill					
6.	GF Col-FF Slab					
7.	F.F. Slab					
8.	Block work					
9.	Plaster work					
10.	Decoration					
11.	Furniture					
12.	Flooring					
13.	Roofing					
14.	Plumbing					
15.	Electrical					
16.	Mechanical					

And the other BOQs followed in the cummulations and the cost of the works carried out reached. Include the materials on site and make deductions, etc.

Valuation to the sub-contractor : Generally, the sub contractors submit their claims of the works executed monthly to the main contractor and on the basis of these claims, the main contractor may include his percentages of the disciplines carried out and reach the amount of the valuation. The payments to the sub contractors shall be made on the basis of the sub contract and not on the main contract as the amount of the worked items may differ in the two. Some times what happens that the contractor shall have to pay more to the sub contractor than what is received from the engineer/owner for those items due to some adjustments made at the time of tender while in some cases this may be reversed. In some cases the contractor shall make the same amount pay to the sub contractor. A general condition may be included in the sub contract that the payment to the sub contractor shall be made only after the receipt of the concerned valuation money from the owner and in a period of a minimum of 7 days or any other specified period of the receipt at the convenience of the two. In case the main contractor is delayed for being paid for long, all the related parties shall have to suffer due to that.

Valuation to the minor sub-contractors : Generally, the agent/main/sub contractor entrusts or sublets the items/itemised jobs to different trades persons working on the same site or outsiders as a minor sub contract for example, plastering, tiling, form shutters steel fabrication and placement in the slabs or columns, roofing, etc., all depending on the convenience of the main contractor or the site agent or the site engineers.

The measurements of all such minor contracts are certified by the site engineers and passed by the site agent for the purpose of the valuation to be prepared by the quantity engineer who shall finalise every thing in that connection and the contracts manager shall make payment order on that.

In some cases when the trades persons are already on the contractor's roll and draw the wages, the amount drawn shall be deducted from the valuations.

The same principle applies to the supplies made to the site by the supplier for whom the site agent shall certify the receipt of the quantities in good conditions per the supply contract and the payment be released. In case, there is some damage or short supply in the consignment, the site agent shall immediately inform the office manager as well as the contracts manager so that the necessary action for the claims on the agency in insurance and the party in supply be raised in time.

Reflected Ceiling Plans : These plans are a necessity in the building where the services running in the ceiling are to be concealed. The services may be air-conditioning, electrical supplies, plumbing and drainage or any other specified.

The co-ordinated drawings of all the services to be accommodated in the space between the suspended or the false ceiling and the soffit of the suspended slab properly and the ceiling tiles of the modular sizes be geometrically patternised. Hence, to prepare these drawings the contractor/agent has to see that the light illuminaries, the conditioned air supply grills and diffusers, splitters and return grilles, etc., are all fixed in the tiles superficially in a symmetrical pattern to the satisfaction of the resident engineer the electrical engineer the air-conditioning engineer, the architectural engineer. Preferably, the modular size of the tiles should match the sizes of the illuminaries and the other elements to fit in the ceilings.

These drawings shall be prepared to a scale of 1 : 20 for the clear depiction of all the items and every point should be intelligently decided. The quantity of the conditioned air and the intensity of spread and the intensity of the illumination, etc., should conform to the specified designed data as included in the detailed specification of the contract and the system designs by the specialist. Should any cut tiles are necessitated, that may be provided in the periphery of the room. Symmetry should be provided all round the ceiling wherever possible but in major areas of different geometrical pattern say a foyer of a building of irregular shape where many ceiling bulkheads shall be installed for the sky lights, symmetry may be broken. These drawings are very important as shall effect the route of the air conditioning duct supplies and make the mechanical engineer difficult going and he shall cry in case the rerouting or the reelevationing is required for the adjustments in case the finalisation of these drawings are delayed. An experienced man only should be entrusted with this work under the supervision of the agent. After the drawings are approved by the resident engineer, the copies must be provided to all the concerned people, electrical, engineering, plumbing, false ceiling site engineers, resident engineers and his staff, etc., for the site use.

There may be many types of the ceiling tiles for example, the metal ceiling, the mineral fibre ceiling needled structure or the plan structure, perforated tiles with cellular structure. Perforated tiles need a black paper to be placed in the ceiling concealed to hide the surface of the slab above from the eyes.

Record drawings/As built drawings : Execution based on the working and shop drawings shall many times be deviated due to the practical or real co-ordination on site. In that case the rerouting of the approved infrastructural pipes in the network, or the a.c. supply and the return air ducts or the deviation in the conduits shall all be accepted by the resident engineer and the services engineers if technically found correct the reroute/aberration.

Similarly, in the building works due to a number of the constraints on site the changes shall be made architecturally or structurally depending upon the merits of on the spot problems. All those real changes done in the respective disciplines on the job must be recorded in the drawings for the permanent use of the user. The drawings incorporating such records as signed by the contractor's agent and approved by the resident engineer and the engineer shall be called the record drawings or the as built drawings in the language of the contract and forms the part of the contract and the contractor shall be bound to produce these drawings without which the practical completion shall not be granted by the engineer to the contractor. These drawings inform the future user about the maintenance of the works to be done for the user, about the way the building has been done and where shall the man responsible to maintain find the required information from. It is better that the alterations applied on the site are side by side recorded to avoid the unpleasant burden of the record pools to be incorporated in one go at the end by the time the engineers or the agent or the consultants may have forgotten the changes and refer back to the approval records to make the entries transferred to drawings one by one. For the recording of the changes the agent shall go through to the site instructions, site memos, engineers instructions, any other records if required and this exercise if done by the new agent or the site engineer in case of the old being transferred or left the organisation shall prove to be a monotonous and unproductive mountainous job and exposed to many misrecordings. It is also advisable should the contractor manage to get the negatives of drawings from the engineer by request and take out his own prints to expedite the work but that shall depend upon the trust the contractor enjoys from the engineer. The contractor should not feel that the staff engaged on the production of these drawings is a wastage in the duration of the job and later regret when due to the absence of these drawings the handing over is made a problem by the engineer as the record drawings alone forms a very important activity in the contract.

Requests for inspections : Contractually, the contractor is bound for the approval of every item included in the permanent works on the job to be obtained by the contractor from the resident engineer or the engineer for each operation carried out on the job.

Various proformas are available in very simple forms which are used for the specific and general inspections and signed by the agent/site engineer incharge/other sub contractor's engineers through the record of the contractor's despatch to the resident engineer who shall depute the required engineer or the specialist for the inspection within the prescribed time per the contract conditions or the established practice on site. Having inspected the requested portion of the job the concerned inspecting representative shall comment on the request format and grant the approval or the rejection as the case be or approve subject to the carrying out of the comment on the site. It is the agent's duty to get every item executed approved on the job.

Test report : On site tests are carried out for various items, e.g., concrete cubes, concrete slump, water pipes works, sewerage pipe works, back fill earth work compaction test, water tests of water proofing torches on the roofs, water test of water retaining systems in the structures for example, swimming pools, water tanks for storage, septic tanks, operational tests of the complete air-conditioning systems for the specified period, all the installed equipments' test per specifications.

All the test records should be kept approved by the resident engineers. It is the duty of the contractor to arrange for the tests whenever necessary and the request be made to the resident engineer and the services engineers or the specialist engineers as the case be for the observation of the test being conducted. Once the test fails, it should be rearranged in the similar manner. It also forms part of the contract conditions.

Material test records : Before the material can be put to use it should be approved by the resident engineer based on the test reports and when the bulk supplies of the materials is made the tests shall be conducted per the requirement of the specifications and approved for the use for the permanent inclusions. Certain materials need testing are : 1. Sand, 2. Aggregate, 3. Cement (opc/src/or as specified), 4. Water, 5. Bearing capacity of the soil, 6. Backfill materials, 7. Bricks or blocks, 8. Reinforcement bars, 9. Road materials, 10. Bitumen concrete, 11. Concrete cubes, 12. Mortar samples and any other material specifically required by the resident engineer/engineer.

Concrete pour control : Now-a-days most of the buildings are constructed of the structural frames of RCC particularly M20 and M25 mix concretes and also M30 mix. Before the commencement of this activity a trial mix to design the proportion of the ingredients of the concrete is conducted by the contractor/supplier in the presence of the resident engineer/structural engineer depending on the availability and the source of the constituent materials and the admixtures used. Three or more samples are taken and cubes are filled and slumps and the temperatures are recorded. As specified cubes are tested at the 7th day and the 28th day and if the strength matches to the recommended working strength per codes, the resident engineer on the recommendation of the structural engineer approves the trial mix design. The material submittal is made formally supported by the test documents and complete details of the supplier or the source and the approval is granted for the material submittal. The ingredients for the mix cannot be altered throughout the job.

Ready mixed concrete : For major pours the ready mix as approved by the resident engineer is brought from the central batching and mixing plant by transit mixers/trucks and be placed with the help of a telescopic mobile crane, tower crane or the concrete pumps stationed at one place or mobile on the pouring spot or the place of pouring. Batching plant may be situated many kms away from the site and therefore, the transit time must be controlled carefully to avoid the start of the setting time before the compaction of the concrete. The agent should permit a fixed time for transit in consent with the resident engineer and the structural engineer and once agreed, the time should strictly be adhered to. The time of transit mixer being loaded and the time it reaches on site must be strictly controlled. The time of the loading is usually written by the plant operator or the foreman on the voucher/invoice along with the quantity of the concrete in the transit truck. Say the time interval has been fixed as 35 minutes for some site. As soon as the truck arrives on the site the agent or his man deputed for the purpose of concrete supply control and if the transit time exceeds the prescribed approved time the load be rejected and sent back at the cost of the concrete mix supplier otherwise the slump be taken and the consistency checked. If the consistency comes within the specified range and the mix be allowed to be poured and the temperatures as well as the concrete cubes be taken per requirements. The slump shall generally be a maximum of 75mm in case of pour by the mobile crane or the tower crane and the extended upto 100 mm in case of the pouring be done by the pumps in ordinary concretes but the same may be increased further if the plasticisers or the superplasticisers are added to the ingredients with approval of the engineer/resident engineer. The interval of the transit truck reaching the site be decided in advance at the time of placing the supply order by the site agent or the site engineer depending on the means of raising the concrete from the mixer to the place of placing and the batching plant manager be instructed accordingly. There may be the case that the truck may keep on waiting due to the other truck being unloaded and the mix in the awaited truck heated up loosing the slump in that case the pumping and taking out becomes too difficult and there may be the need add a little water in the mix for which the record be made in the pouring record and the engineer's representative or the structural engineer present on site be informed. The quantity of extra water entered in the concrete shall be compared with the special cube results of the concrete of the increased water cement ratio but generally, it has been experienced to keep the concrete working the increased water cement ratio may not bring the reduction in the strength. Having the concrete being poured, it should be vibrated for the thorough compaction and spread to the required level. In hot weather conditions, after 20-30 minutes of the levelling the saturated hessian bags or in rolls should be spread on the levelled concrete to prevent the cracks formation due to the high velocity winds full of dust and due to the evaporation of contents of water. For further precaution to retain the saturated hessian from getting dried out, a pvc membrane should be spread over the saturated hessian. Meantime, the slumps and the cubes be kept on taking and the temperatures recorded and the date of the cubes with the reference of pour element be recorded. All the informations shall be then transferred to the concrete pour record register or the proforma in a regular manner recording the serial no. of the pour, the date of the pour, time of pour, the quantity of pour, the place of pouring, the temperatures both the weather and the concrete, the means of pouring, source of pouring (by the site mixing or by the batching plant supplying source), and the space be kept in the columns of the test results on the 7th and the 28th day. This recorded information must regularly be sent to the resident engineer alongwith the copies of the supplier's records the same day or the next day for the verification and a record.

Before starting a major pour say in a big slab either the pour in one go to a construction joint shall be required. In the later case the place or the position of the construction joint should be agreed in advance with the resident engineer or the structural engineer and the preparatory works be carried out accordingly. At the place of the joint the stop end shutter should be fixed with the continuing reinforcing bars piercing through. The next pour away from the stop end shall be done after a lapse of minimum of 24 hours with the prior approval of the resident or the structural engineer and the surface of the stopped end be clearly scabbled to let the concrete aggregates be exposed for the proper bondings of the new concrete with the old one and done per specifications.

Sometimes, additional stop ends may be required due to the contingent conditions of the pour such as break down of the mixing plant, break down of the concrete site mixer or the concrete pump or the crane installed for the raising of the concrete to the placement site. In that case the stop ends are urgently required and the preparation immediately done after having a discussion with the engineer's/resident's representative on site for which the information shall be given to the resident engineer later or the next day if the pouring is carried out at late hours or in overtimes. In case the pour programme is postponed or advanced due to nonavailability of the concrete or due to any other reason, the engineer's representative or his man on site shall be re-requested for the instant inspection before the commencement of the pour. Raising of concrete by pump requires a high value of the slump and if some admixtures be recommended for the high rise tower concrete to reach a value of slump as high as 200 mm may be acceptable provided the strength of the concrete in the short span of time as well as long span of time is not reduced than the specified according to the codes as in practice in the area. The main problem on sites in the pump concretes of the low slumps is that of the chokage or the blockage of the concrete pump's boom due to the concrete being getting stiff and the receiver of the pump shall not work efficiently. May be, the pump is cleaned thoroughly in that case before the start of concrete again/next batch. Therefore, the value of high slump is recommended for the concretes being poured by the pumps while in the case of manual placing or by the mobile or the tower crane or any other method the low value of 75mm maximum is permitted. Mobile pump has been considered good for the concreting purposes on the buildings of medium height generally. In the case of engaging the mobile concrete pumps, the repositioning of the pump is easy and does not consume much time to recommence the pour from a second or the repositioned spot and the control too is good. The stationary pumps if installed shall require rerouting of the complete pipe system for the purpose of shift of the spot which consumes comparatively more time than in the case of the mobile pump. In the case of the mobile pump the end of the disposal side remains free and the workpersons can hold that by the anchorages applied at the end in the form of ropes to prevent the pipe from dancing due to the hydrostatic and hydrodynamic pressure applied at the boom. The concrete should not be poured directly from the boom to the inside of any beam but first be taken down on the slab forms just adjacent the beam forms and then with the help of the wooden or the steel spades be poured into the beams so the buldging of the beams or so called opening of the concrete form is avoided. The gang to be regularly engaged for the purpose be trained and should not be altered frequently in the interest of the activity go smooth. During the pour the care must be exercised that the electric conduits are not misplaced or broken or the blockouts as provided for a.c. or the plumbing and drainage be intact or the concrete spacers are not out of their positions and persons to look after all these operations including the form carpenters be kept on vigil all the time duration of the pouring.

You must have arranged a group of the concrete vibrators alongwith the vibrator operators usually trained people and two air compressors at least to use one for the operation and the other be standby. In place of the pneumatically operated vibratos the petrol or the diesel operated vibrators can also be brought in use for the convenience of the contractor. Over vibration of the concrete must be avoided. Concreting of columns may not be done fast therefore, it is not preferable to use concrete pumps for the purpose of pouring but there is no limitation to the use of the pump, it depends upon the discretion of the agent and the engineers. In case the sizes of the columns are quiet convenient for the concrete pumps' chute pipe to go in easily and vibration may not cause any problem, keeping a drop of maximum of 2 m the columns can be poured· by the pump also. Site mixing can also be done for the minor pours subject to the approval of the resident engineer which must be in hand in the form of mix material submittal. Crane shall be the best means to raise and place the concrete using a skip and a circular rubber chute attached to this so the pipe passes into the column easily keeping a drop of 2 m at the most to prevent the formation of segregation of the concrete. Concrete in the columns shall be done gradually avoiding the instant hydrostatic pressure to act on the forms. Generally, hydrostatic pressure acts as a resultant at a height of one third the concreting height in one go, therefore, that point must be braced strong enough to resist the expected resultant pressure to prevent the forms from opening to let the column buldge form the verticality.

The bottom portion of the column is liable to honeycombing due to the improper compaction by the vibrator hanged from a height of 3-4 m from the bottom due to the negligence of the foreman and the vibrator operator, therefore, due care must be exercised on this issue and the concerned foreman be given the personal responsibility as far as this honeycombing is the case of concern.

For the foundation concreting the process is comparatively easier and be resorted to the pumping in a case a number of footings are to be poured in one go. Whatever be the source and the transportation and the raising of the mix, proper records of all the pours be maintained to ensure the prescribed quality control as explained earlier. The concrete cubes should be immersed in the water tanks placed under a shed particularly constructed for the purpose immediately after being taken out of the moulds the next day. In case the cracks appear on the surface of the concrete, with the approval of the resident engineer the resin's grout be applied or if the cracks development appear to be of very minute size as usual the case of the hair cracking, the cement mortar grouting may also do in certain cases but to the satisfaction of the resident engineer or the structural engineer and the agent.

All the sections prior to the next pour should be scabbled next day or the complete laitance removed by the use of the wire brushes immediately after the concreting is over. For laying the flooring or the roofing the surface of the slab should be roughened by scabbling before the concrete gains the full strength. Once the concrete gang has been trained to the satisfactory performance, no disturbance/shunting be allowed till it is out of control due to many reason depending on the site control.

Definitely, certain soils contain the sulphates and then the **sulphate resisting cement** (SRC) is brought in the use in such soils. In that case all the foundations and the structures in contact with the soil shall be concreted with the SRC concrete and those not in the contact with the soil be concreted with OPC mix. The two different concrete shall never be mixed due to the chemical detrimental and deteriorational effect in the structures. The stacks of the aggregates and the sand should be properly piled on a strong and a good platform protected from the direct effects of the rain and the sun. The cement may either be stacked in the silos with the help of the cement transporting tankers or on small plants the storage be done according to the specified arrangements where the cement be placed in the mixer's hopper by the men engaged for the purpose and similarly, the aggregates and the sand be loaded into the hopper and the specified quantity of water be poured into the mixer from a top mounted calibrated tank while the weights of the aggregates of 20 mm, 10 mm and the c. sand shall be controlled by a dial gauge fixed to the mixer. The bottom underneath of the hopper is generally filled up with the surging aggregates and the sand passing out from the sides and the edges of the hopper during the loading operations carried out by the men and these extra ingredients shall show the erroneous reading of the dial gauge therefore, a care shall be exercised all the time to clear the annular space between the ground and the hopper bottom. The calibration of the dial and the water tanks be carried out to be fit all time.

Material Status Schedule : As previously detailed in the progress report for the monthly site meetings, an abstract statement containing summarised details of submissions, rejections, approvals, ordering, delivery problems and the deliveries made for all the scheduled materials to be included for the permanent works shall be called material status and delivery schedule or the material status statement.

Working drawings and the shop drawings : As previously detailed in the progress report documents, it is an abstract of all the drawings submitted for the approval, rejected or approved or under the process of approval. The works cannot be carried out without the approval of these drawings. This schedule provides a comprehensive information for drawing nos., contents, and the date of submission, approval and the issue for all the shop drawings for a particular discipline on the contract.

Complete co-ordination and liaison : It should be clear that the details of the technical relations, the suitability, and the accommodation in a given space for all the allied building services and the structural details adaptability in relation to the architectural requirements, all grouped together correspond to the co-ordination. The disciplines of various types of co-ordination has already been explained earlier.

Liaison : All the links to be made in connection with the execution of the job and as required under the operation of control conditions to reach a successful completion. Communication to the statutory bodies, other contractors and the correspondence made to various categories of people linked to the contract job are referred to liaisoning.

Internal dealings with the contractor's organisation

1. **Material requirement :** The agent has to order on the requisition forms to buy the locally available materials apart from the materials to be imported in bulk quantities and to be specially procured from a foreign country or the source. There might have been an earlier agreement with the material supplier at a fixed rate per unit of the item but actual supply is ordered from time to time for example, cement, reinforcement, concrete blocks, bricks, aggregates, or other standard materials such as roofing materials and the polystyrenes to be put in the cavity walls and the roofs. On the requisition are also ordered by the agent for the purchase of consumable items included in the itemised rates e.g. scantlings, plywoods and the timber planks, form work, binding wires, hessian rolls, pvc membranes, nails, tools for the work, equipments, water, diesel, spare parts of the machines to be used on site, safety measures for the persons working on site, miscellaneous expenses for the site staff and on the workforce camps, maintenance of the vehicles, etc.

The agent passes the requisition and sends to the office manager who provides the requisition to the officer in purchase section to supply the required materials to the site or an imprest money is given to the agent or the purchase man to directly purchase the ordered material from the market and then submit the vouchers along with the material requisition order for the reimbursement. All the invoices shall be signed by the site agent as well as the receipt of the material.

The materials which are required on the job in the bulk quantities, having been approved by the engineer on the material submittal forms are placed in order by issuing letter of credits by the head office on the material purchase order received from the agent/contracts manager. The materials are supplied on site and verification of the delivery received in good condition is given to the head office by the agent after comparing that with the list of the materials ordered and the materials delivered. All the expenses made are debited to the accounts of the job to know the real and actual expenses on the job against the valuations.

2. **Co-ordination with the garage :** All the time it is not feasible or practical to keep garages and the workshops on the project site. Therefore, a centralised garage and a workshop/service centre and the stocks are kept at one place and all the transactions are made through the office by all the agents in a contracting company or the organisation. A plant manager is made incharge of the garage and the workshops to whom the agents place their orders about the posting of the machines and other major equipments and plants and the maintenance of the vehicles, trucks and other transports whenever required to site and charge accordingly. In such situations the plant manager may be answerable to the head of the company or the managing director and there shall be witnessed many logger heads

between the agent and the plants manager. The plant manager may be overloaded because he is supposed to supply the equipments and the machines whenever required and his technicians are often found working overtimes. Generally, the company's light and heavy transports remain under his control for the maintenance posting and other controls. Agent has to order the issue of the aggregate trucks, sand, or any other materials by verbal orders and the same is supplied to the site. This is a usual practice when the garages and the site are not far distant, say the site is away from the city while the offices and the garage are the in city or in the vicinity of the city and close contact is not a difficult job and the trucks may cover the distance in fraction of a hour or a hour. Good relations between the agent and the plants manager can help the job to a great extent. Sometimes big contractors own the stone crushers themselves and the aggregates are directly supplied from the crushers to the site on the order of the site agent to the plants manager. Similarly, sometimes the big companies engage their own batching plants for concrete mixing and for asphalt concrete for which plants manager can help the agent a great deal.

3. **Daily reporting to the head office :** In case the site is near by the head office a daily visit of the agent is advantageous to making different correspondences in connection with the various letters to which the reply has to be sent, to deal with the matter of the material and drawing submittals, to see the other people if required for something. In case the sites are far away from the office, the agent may visit the office once a week or the other frequency. The agent has to do the co-ordination with the heads, quantity engineers and quantity surveyors to let them apprise of the progress on the site. Passing of the sub contractors' valuations can also be undertaken during these visits.

4. **Co-ordination with the main quantity surveyor :** Generally, a main quantity engineer or the surveyor for the section deals with the final confirmation of the amount to be passed for the subcontractors' valuations. Site agent and the quantity engineer or the surveyor are jointly responsible for any payment to be made to the sub contractor or the supplier since the rates' secrets are known to the agent and the quantity engineer or the surveyor only. It is therefore necessary, that the two should have the close co-ordination. The agent has to direct the quantity engineer the preferences of the sub contractors' valuations to be done due to the shortage or the limited budgets available sometimes and that is known to the agent or the contracts manager only and also the priority of the job. In fact, the sub contracts of the trades who are mostly from the rolls of the contractor are paid monthly salaries can be kept pending for the valuations being paid while the outsiders who are regular sub contractors of the entire discipline or the trades people are kept paid as and when required. The deductions of the wage bills already paid to the trade men working on the sub contracts shall be deducted from the valuations based on the special details received from the site timekeeper and signed by the site engineers and the agent. The quantity engineer shall seek the complete informations of the works carried out by the sub contractors from the site agent in time to keep himself fully informed of the developments on the site. Where the quantity engineer and the agent act in unison and harmony with great degree of co-ordination, the contract paper work for the head office shall never be a head ache.

5. **Co-ordination with the office manager, staff and other agents :** The agent's part responsibility is the co-ordination and dealing with the office manager to get the clearance of his demands for the site, labour bills, miscellaneous accounts, settlements, office secretary to assist in correspondence typing work and despatch, sending telexes and a making of international communication as and when required by the agent. The agent shall have also to interact with the other agents of the company for the purposes of the movement of the machines from one section to the other and also the use of the specialist groups of certain trades from one section of the company to the other section of the company, the cost of that shall be charged to the job according to the fixed wages or the fixed rent. The agent should be friendly to the counterparts of the other sections to get the maximum technical and miscellaneous benefits.

6. **Co-ordination with the contracts manager :** In the companies that employ the contracts manager above the agents to reduce the work load of business from the shoulders of the agents and to let him involve fully for the technical and site controls of the job, the agent is supposed to report to the contracts manager frequently regarding all the site matters. The authority of the financial sanctions in that case rests with the contracts manager and the agent has to depend on the contracts manager to solve the problems regarding the machineries, materials and the workforce arrangements.

Some organisations want/expect the agents/contracts manager to be responsible for the calculation and compiling the tenders also. This makes the contracts manager or the agent as the case be fully competent and responsible for making required profits. In that case the agent's status reduces to a project engineer(site) if he is technically qualified engineer or a construction manager and to assist him in the designing work on site a consulting project engineer is attached or appointed for him to councel the technical work for him as far as the designs are concerned. So the nature of his co-ordination varies from company to the company. Otherwise, the agent is supposed to be directly responsible to the owner of the company or the head. In fact, the contracts manager acts as a director.

7. **Miscellaneous labour problems :** The agent has to solve all the related problems of his workforce on site for example, transportation on overtime, messing facilities, drinking and non-drinking water supply, safety during operations, payments of the wages in time, leave applications' sanctions and the clearance of the dues, look after the welfare of all the persons working, that a proper treatment is given to all the people. In case of accidents immediate medical aid is imparted with and to see that the victim is taken to the applicable hospital and necessary information be given to the insurance agency through the office manager and also to the resident engineer. Also, it should be watched that the labourers do not fight on site or play. For their amusement a separate place should be assigned by the company and necessary entertaining facilities be provided.

Should the workpersons be provided with the due, the harmony shall exist on site and the good labour management relation shall help run the site or the project efficiently. There is a possibility that the personal frictions generate among the workers, foreman, and site engineers and the work supervisors. In such circumstances agent has to act too carefully to let the misunderstanding not generate among his staff and the workpeople and he should act impartial in all the dealings and should not allow any injustice to any worker. The agent must all the time enjoy the regards and the respects of the people with him so that every one working on site submits to his commands. If the right man is not engaged for this job he may have to face a number of problems. A man if not right for the right job shall not find pleasure in the job and the agent due to his being very experienced on the constructions using his virtues shall fit the unfit person somewhere else and get the job done. The unfit person for one job cannot be straightway rejected in that case the agent shall see that the recruitment of the person unsuitable for the particular job might be by mistake and must try the same person from one job to the other till he is not made fit all the way and then recommendations of his removal made. The art how to extract maximum from the unskilled; semi skilled is really difficult and generally not available in the engineering books but by the tutorial experience of the engineer this art naturally comes in. In fact the field work is a study like tutorials and a training for an engineer and that is the reason that every engineer must have the experience of the site for converting or making a consultant, the practical designer, a person of visions and so many more which depend on the personality and the performance of the engineer in practice.

8. **Co-ordination with the other sections of the company :** Big companies are divided in sections say the building section; the road section; the infrastructure or the civil works section or any other section depending upon the status of the company. The company may have its own sections of the services when they carry out the entire works themselves. The companies may be having the structural and the architectural design sections when the package deal contracts are entered in for a scheme. The contracts manager may like that the part of the road or the infrastructural work being a part of the site to be carried out by the other section of the company from the point of view of the economy or the policy of the company as a whole. In that case the agent shall have to co-ordinate with all the sections of the company for the required discipline of his site for obtaining the technical assistance and all other assistance for the operation of the contractual obligation for the project.

Hence depending upon the nature of the company he is working for, the agent shall act accordingly to dispose the responsibility of coordinating among all the sections from time to time. With the different contracts manager and the different site agents and the dealing staffs inclusive of the other site engineers and the foreman. The agent shall have to deal with the other specialist engineers also if the company employs for the purpose to get things cleared from the engineer when there is some requirement as such. Better a personal dealings and working relations shall help a lot in the successful co-ordination work and to move the site with the maximum men and machines and materials.

Agent's dealings at site

1. **Daily programme :** Though mostly detailed programmes depend upon the way the site engineers, deputy and assistant agents and the coordinators as well the general foremen sort out depending upon the availability of the man power/workforce or the machineries and the material to be used and the job priority but all need the directives and the guidance of the agent as he is the head man on site and knows every thing more than what the others know. That is the reason that the agent plans all the activities for the site in a week advance and informs all the site personnel so that the possibilities of following the programmes are explored by the site engineers and the sub agents in time. Even then the changes shall be made depending on the require-ments on site and a number of factors involved. In all the daily duties at the end of the day the agent shall with the consultations of the site engineer circulate the programme for the next day and include all the urgent and ordinary activities to let the site staff plan accordingly the deployment of the workforce and the machines as best as possible. Some times, one engineer may not release the machine to the another engineer due to the hindrance in the job but the agent has to advise the concerned engineer holding the machine or the people or the material to release and send to the other engineer's site per requirement of the emergency and in case the engineer does not listen then the agent is bound to instruct and order the denying reluctant engineer to do the same or the matter shall be reported to the head office. The agent shall arrange for the material and machine and any other item necessary for the following day from the plant manager. The agent may order the ready mix concrete if the request for inspection for the related site engineer has been approved and the site engineer reports of his being ready and wants the concrete at a fixed time in a planned way. The agent shall instruct his sub agent or the site engineers and then the further passage and the implementation of the decision comes under the duties of the site subordinates or the engineers responsible for the execution and the assistants deployed to assist the site engineers.

2. **Daily supervision :** Agent is not supposed to be all the time on the site supervision but once or at least twice a day he should be at all the corners of the job to let him apprise about what has been going on as he has to write the daily job report and instead of depending on the communicated informations better for him to visit the works himself so that there should be no misunderstanding in the technical data to be referred to in the report which may be in case of collecting the data from a delegated man. Therefore, the agent should himself collect the data and make the precise explanations. Site engineers often are overloaded and you should not depend upon them to inform you as they are always busy in running and managing the complete technical works on site and the additional work may detract them from doing the very important part of the job. They are in fact doing the more practical job such as levels check, layouts setting, site supervisions, etc. For the general reports of non-technical nature say the details of the workforce available/present tradewise and the temperatures; the materials delivered, a clerk can be deputed who shall collect all these informations from the site and also from the sub-contractors and the details of the activities going on, all the dealings being made with the resident engin-eer or his engineers should be entered by the agent on the basis of days happenings. The machines leaving or reaching the site are already known to the agent as without his permission no machine can be released or brought to the site. Whether the machines remain idle or busy working, the agent shall know during his inspection or so called the visits on the site twice or thrice a day and the activities as well. This way the daily job record shall not make a problem for the agent to compile and once the routine is made then it shall be very easy since the activities of the previous days are known and in the civil engineering projects the activities continue to run for a number of days once started hence the previous report shall partly guide the agent for the days' activities and the same thing shall also apply to the works of subcontractors.

Apart from the above agent is supposed to examine whether the activities as on going at site resemble to the specifications and should he find any deviation, agent should ask the engineer or the man responsible to rec-tify the same to avoid the rejection by the resident engineer. The agent is a helping hand to the site engineers from the higher level as to help the site engineer means he helps himself as he is responsible to the contract as a whole as far as the site is concerned. He is supposed to be the most experienced person on site and hence the top responsible for all the operations going on. The agent's duty is not limited to the site control only but also according to the professional conducts he should also train the junior staff by providing them with the opportunities and guide them wherever necessary and help the juniors with the spirit of technical personality development so the juniors may be eligible for the higher responsibilities in future carrier. The agent should provide the opportunity to the good site engineers to let them attend the site meetings and take part in other management jobs to apprise them of the site management functions which shall indirectly be a help to himself as sometimes the agent is sick then he may trust his trained subordinate to act in his place to keep the job moving without much worries. The agent may have to go on routine leave after the accomplishment of the contract period, there must be someone to take care of the job temporarily as no employer shall like to bring a senior engineer for so short a period, therefore, the trained engineer may be of utmost use at that time. A good and intelligent agent shall be liked by all present on site and in the office. Well, there is no limit of supervision. He shall have to sometimes supervise the job directly when the site engineer or the assistant engineer are busy somewhere else and the foreman on site or some skilled sub contractor or the trades men seek some information from you so that the work is not slowed down due to the awaited technical information and by providing the information the agent shall assist the site engineers as well the job but later the information be conveyed to the dealing engineers as to what was told the related people.

The agent shall take care of what people say and the other times not. The agent should visit the resident engineer and staff frequently to know what are the problems on the site to find a way to solve the same as soon as possible. Never criticise your staff all time but absorb and let them realise at the right time of the mistakes. Once they know that you were right and wanted to help them as well as the job and there is nothing personal, they shall listen to you and you will be successful in commanding over them diplomatically. Other way an agent must know about the site all the time and every happenings either directly or through some confidential person and must reply promptly and reasonably whenever asked by the consultants or the resident engineer or the men in the head office so that every one should realise that you do not sleep but are clear of all the problems on site and possess the capacity to clear all those hurdles coming in way.

3. **Daily material order :** Day to day use items are purchased from the cash imprest provided at the site under the control of the agent and it is a usual practice to buy limited items required on site on the daily consumption basis or for a particular activity. Petty items say the material for the office use say the stationary, tea, coffee, or the drinks meant for the consumptions of the resident engineer or the specialist engineers or the visitors or all the miscellaneous items needed shall all be covered in this activity of the agent. The dealing purchase man or the men from the site engineer if he needs something to be procured for site shall all be submitted to the site agent for a purchase order and after obtaining the sanction the purchase man shall proceed for the buying.

4. **Solving the instant site problems :** There is no particular specification of a problem on site and of course, there is hardly a day when there is no problem either small or big on site. These problems are so diverse in nature that the agent shall have to apply his common sense to solve all these problems and also use the technical knowledge in case the problem is of high magnitude as far as the job is concerned, and also there may be problems of the administrative nature created by the site staff of your own or the head office or the sub-contractors or by any body else which the agent on site shall be supposed to handle. Generally, the technical and non technical problems are sorted out at site but the contractual matters shall be referred to the contracts manager for further consideration with the recommendations and actions after personal discussions. An intelligent agent shall not like his site problems to be discussed in the main office as this is a question of his reputation and the self efficiency.

5. **Solving the problems with the resident engineer and the consultants :** Generally, the problems that occur on sites conform to the noncompliance of the specified quality controls, method controls and keeping the site tidy. Methods should be agreed in advance and the quality control actually depends on the performance of the foreman, the site engineer's supervision and overall control of the agent. Generally, the problems exist where the clerk of works and the site engineers are young. The site agent should explicitly instruct all his staff the ways of quality control without adding to the cost of construction. Very much strictness/strictures shall also not bear the results but a constructive approach to the problem. The agent must have the capacity to instruct his own people and at the same

time convince the resident engineer and his men.

6. **Solving staff disputes and ordering :** Since there is a team work on site and no one is entirely in a position to control the activities alone for the complete execution of the job on the contract. This is also true that on a big contract where many site engineers are in the postings and large number of activities go on, always results in the formation of groups and to keep the spirit, the site agent has to fix the targets for every group to achieve in a certain time, consequently, a progress race is generated among the different groups. The site engineers and the foremen responsible for the job that lags behind the target rate do not feel happy and the question of the personal achievements arises. Every engineer and the foreman on site want to see him ahead of others or his target and this may be a reason that the frictions start developing among the different group. Due to the over racing the equipments and the machines fall short of requirements and sometimes the shortage of workpersons mar the progress as the company cannot provide immediately every thing to all the groups economically but expect the groups to use the items in the sequence to the planned and programmed lines. The results may be like that sometimes all the groups want the supply of the concrete from either source simultaneously requiring the pumps or the mobile cranes and the double number of labour and practically the suppliers cannot provide double numbers of the items giving rise to the heated site engineers frictions with the benefitted other one if the other gets the supply. Similarly, sometimes the major delivery arrives from the customs for which prior information delayed in reaching the site and in odd hours resulting the non-availability of the unloading workers and no engineer accepts to spare a few workers from the running activities and posing a practically annoying problem to the agent but agent cannot keep the trailer waiting due to the limited visa of the trailer drive. In that case the agent shall have to arrange somehow the manpower either by stopping the operations and if possible taking the people from outside if feasible on the payment basis from the imprest.

Every officer on site wants to find himself ahead of others and perhaps this may be the reason leading to the development of the misunderstandings and disputes on site. One foreman wants to see the machine working on his site to continue irrespective the urgent requirement on the other site. Similarly, site engineer worries for his site only. It generally happens that people are short on priority building due to the excessive absentees or some other reason on a particular day and the priorityless building on the other site got more people that day due to no absentees, neither the site engineer nor the foreman of that site shall be ready to release the people for the urgent site if requested by the other site engineer. Similarly, the disputes arise when the concrete is to be poured at several places in one day and the trailer full of load arrives. In all such contingent problems, the agent has to carefully interfere and find the solutions and order the related or the reluctant site engineers and the foreman. For avoiding such daily disputes it is better for the agent to hold daily meetings at the end of the day calling the site engineers and the middle level supervising staff on the site to his office on site and chart out the distribution of the workpersons, the machinery, the material and the priorities. In some cases the site engineers and the foreman be busy on sites for the whole day and there is no chance for the agent to call them in that cast the agent should make a note of the priority of the job and go to the site engineer at the end of the day and explain to the engineers what to be done next day and how to adjust the available sources. Agent shall act as a judge on the job and think of the interest of the job as he is the man who programmed the job schedule approved by the resident engineer and the engineer and it is the duty of the agent to keep the programme on line. The agent's excuses shall bear no value to the company or the resident engineer or the consultants except they may sympathise with him only. The administrative part of the agent is that he should make the site people to listen to his command, therefore, the agent must be a qualified and experienced engineer in the specialist field. All the disputes and the complaints should be properly investigated by the agent and try to bring all on site to the integrated spirit of the team work to successfully execute the contract.

7. **Labour bills for the wages, overtime and the leave salary :** As explained earlier agent has to certify the wages to be paid to the workers deployed on his site and their overtime bills. The agent has to sign the wage bills for all the workers along with overtimes. These should reach the accounts in time to enable them, i.e., the office manager and the accountants and the accounts clerk, the check and release the wage payments in time. Similarly, the applications of workers who want to go on leave be recommended and sanctioned per the authority of the agent and the contracts manager according

to the company organisations and the rules. The agent shall sign the leave applications for the forwardance based on the recommendation of the site engineer and the foreman. No worker shall leave the site without the permission of the site agent as the site agent must sign the card of the worker on the day of his leaving. The site agent can disregard the instructions from the head office to relieve some workers temporarily and take the matter to the contracts manager or the head office incharge and ask for the reasons for the removal of his workers from the site without his consent and if there is enough ground for that and the agent is convinced, he may agree for the release of the workers or may refuse otherwise.

8. **Labour working on contract :** It is a common practice to give incentives to some good workers to let them work on fixed rates for an item so that work progresses fast and question of overtime as necessary in the usual case is not required. Such minor contracts need be valued in time so that the payments are made in a reasonable time indeed they are paid wages which are deducted from the valuations. These valuations are prepared by the agent or the site engineers, signed by the site engineer and the trades sub contractor and certified by the agent for a disposal to the quantity engineer through the contracts manager.

9. **Preparing the reinforcing schedules :** Though this does not come in the duties of the site agent to provide the rebar schedule but he must know that thoroughly. Whenever site engineers are busy or the agent feels that the site people are overloaded due to the various activities going on simultaneously and the foreman of the rebar or the sub contractor for the rebar bending and placing comes to the agent that he is not getting the required schedules from the site engineer even by many requests and his work is held up, the agent should take initiative himself and give him the work for some days so that the spirit of the site engineers are not broken.

10. **Proposals for the alternatives to the resident engineer or the consultants :** To keep the work moving smooth and check various constraints, the agent must make proposals to the resident engineer so that the time saving can be achieved and simultaneously, the contractor is benefitted. There may be a problem to procure a particular material the agent should propose an alternative. Similarly, method statement should be proposed for the betterment of the job. If the proposals appeal, the experienced resident engineer in consultation with his specialists engineers & recommendations does not mind granting approval provided it conforms the standards and does not involve any additional cost but welcome if saving shall be made to the contract.

11. **Technical study of the job and instructions :** Though the minor technicalities of the site are carried out by the site engineers even then the agent must know about the details more than the site engineers know which shall help the agent to check the mistakes done by the site people. The site engineers and the site staff must feel that the agent shall point out their mistakes due to keeping eagle eyes and hence every one on the execution shall be careful to exercise his duties and add to the quality control. The requirements make the site agent to go deep into the details of the drawings and the specifications at site or at home. The agent shall pass instructions in case site people are not able to produce the required results due to the lack of the knowledge.

12. **Daily meetings and the general liaison :** As explained earlier the agent should meet the concerned people daily, may be the site staff, the resident engineer or the consultants or the sub contractor's representatives, or the other contractor in relation to the liaison or communicate on telephones. The agent should not fail in any kind of liaison works verbally or the documentary, formal or informal. Every activity must run passing the agent's attention if the agent wants to make a reputed figure at site and the company and in the considerations books of the resident engineer and his staff and the sub contractors as well.

Apart from the responsibilities mentioned above, an agent or the contracts manager's most important responsibility is the cost planning/circulation to an optimum level. No contractor shall like to capitalise or invest the maximum to a minimum profit return but minimum to a maximum profit returns. For example, the following details exemplify the needs of the cost planning.

1.	Duration of work	20 month
2.	Tender price	35 million
3.	Building works	15 million
4.	Engineering and plumbing	10 million (7 million + 3 million)
5.	Electrical	5 million
6.	External and civil works	5 million

ANALYSIS OF THE PROJECT (COSTWISE) BASED ON THE PROGRAMME

Cumulative cost (monthwise)	0.5	1.0	2.0	3.5	5.0	6.5	8.0	10	12	15	20	25	30	35	millions
Month	0					5			10		15	16	17	18 19 20	months
External											1.0	1	2	3 4 5	millions
Plumbing			0.2	0.75	1.0	1.0	1.3	1.3	1.4	1.7	2.0	2	2.5	3	millions
Engg.				0.5	1.0	1.5	1.5	2.1	3.0	4.5	4.5		6	7	millions
Electrical		0.02	0.5	0.3	0.75	1.0	1.0	1.2	1.2	1.5	1.8	3.5	3.5	4 5	millions
Building	0.5	0.98	1.85	2.5	3.5	4.0	5.0	6.0	8.0	10	13.5	14	14	14.5 15	millions

Programmed break up of the cost to the bar chart network

A practical breakup of the monthwise expected cost according to the approved programme has been exhibited above for the various disciplines in connection with the project.

Building See that the first 5 months of the commencement investment do not seem to be high. If you spend one million you may circulate the amount upto a minimum of 6 month getting returns from the valuations. Be clear that most of this amount shall go to the mobilisation, excavations, blinding concrete and the foundations. Further, if you see upto 10th month, you have to procure the building materials such as concrete, steel rfts., shuttering, etc., which are often available on the credited ledger in the market and you avoid 100% investment on that hence recirculation is not difficult if the supplier trusts you. This means if you invest just half a million more you may successfully reach upto 10th month.

Between 10 and 15 month where the money is required to be paid to the sub contractor of your own, finishing items and the imported item/items to be delivered only after issuing the letter of the credits, you shall need enough money to be invested e.g., say 2.5 million as advance payments to be made to the materials and the workpersons and local available materials can be brought on credit. Similarly, circulation can be maintained upto the end. You should have noted that an investment of about 3.5 million has completed the work of 15 million.

Electricals Solid investment required for the electrical items is from 5-10 months in the range of 3/4 of a million. To purchase illuminaries, transformer, etc., a further amount of L.C., etc., about 0.75 million is further required and the remaining met out of re-circulation of the money.

Engineering It needs an investment in the period of 10-15 months and may be of the order of 2.5 million on L.C. to procure the material and the rest to be met out of recirculation of money.

Plumbing It needs an investment of money upto 0.75 million from 10-15 month on LC and the remaining to be met out of the re- circulation.

External Needs an investment at the last when the money from other items of building comes back.

This means an optimum planning depends upon how the overall working plan has been framed. In most contracts the main contractor does themselves the works of the building's items and the external landscape while the services are sublet to the sub contractors so the main contractor's investment remains low but if the sub contractor requests the main contractor has to invest for the sub contractor.

The elaborate breakup of the cost to be invested and the recurring valuations as depicted should be made before the commencement of the project. In fact, this shall act as an instrument to the contracts manager/agent to 'fund the project' guidance and every one in the company knows at what time the investment from the current balance or from the other sources at the calculated or the fixed interests is necessary. The agent has to do hard work in the beginning but shall remain peaceful and without panic till the end provided the running payments are obtained in the time. Monthly expenses from the job ledger shall indicate whether the job is running in profit or running in loss by the comparison of the valuation done to the cost programme curve. Suitable recovery measures could be taken then. Agent must take total help of a quantity surveyor or engineer to make such records. In case the payments are delayed or the job is delayed due to some other reason raising the claims shall be assisted by this instrument. Interested parties for the sub-contract can be asked to do the related discipline at a given cost instead of inviting the quotations after keeping a little margin.

Sometimes, in some countries the authority shall like to award a job directly to a company of repute instead of inviting tender and seek the approval of the company at a fixed cost by the authority. In that case this procedure shall help to know whether the offered price is reasonable or loosing. The company may not accept should it be loosing by requesting to enhance the price to a calculated figure. While comparing the quantities, cost and the expenses, the profit percentage shall narrow down or may go negative.

	Bldg		Elect		Engg		Plum		Extrl.		
S. No.	I	V	I	V	I	V	I	V	I	V	In millions
1.	0.5										(I) indicates
2.	0.5	0.5	0.2	0.02							investment
3.		0.98		0.15							(V) indicates
4.		1.85	0.75	0.3			0.25	0.2			the valuation
5.		2.5		0.75				0.75			of works
6.	0.5	3.5		1	0.3	0.5		1			Serial
7.		4		1		1		1			numbers
8.		5		1.2		1.5		1.3			indicate the
9.		6		1.2		1.5		1.3			month of the
10.		7	0.75	1.2		1.5		1.3			contract
11.	2.5	8		1.4	2.5	1.8		1.35			
12.		10		1.5		2.1		1.4			
13.		11		1.6		2.5	0.75	1.5			
14.		13.5		1.8		3		1.7			
15.		13.8		2.5		4		1.8			
16.		14		3.5		4.5		2	1.2	1	
17.		14.2		3.8		5		2.2		2	
18.		14.5		4		6		2.5		3	
19.		14.8		4.5		6.5		2.8		4	
20.		15		5		7		3		5	
Tot. Invt.	4.0		1.7		2.8		1		1.2		Total investment
		2.5		0.75		1		0.45		0.75	Profit element 15% $\left[\dfrac{5.45}{35.0} \times 100\right]$

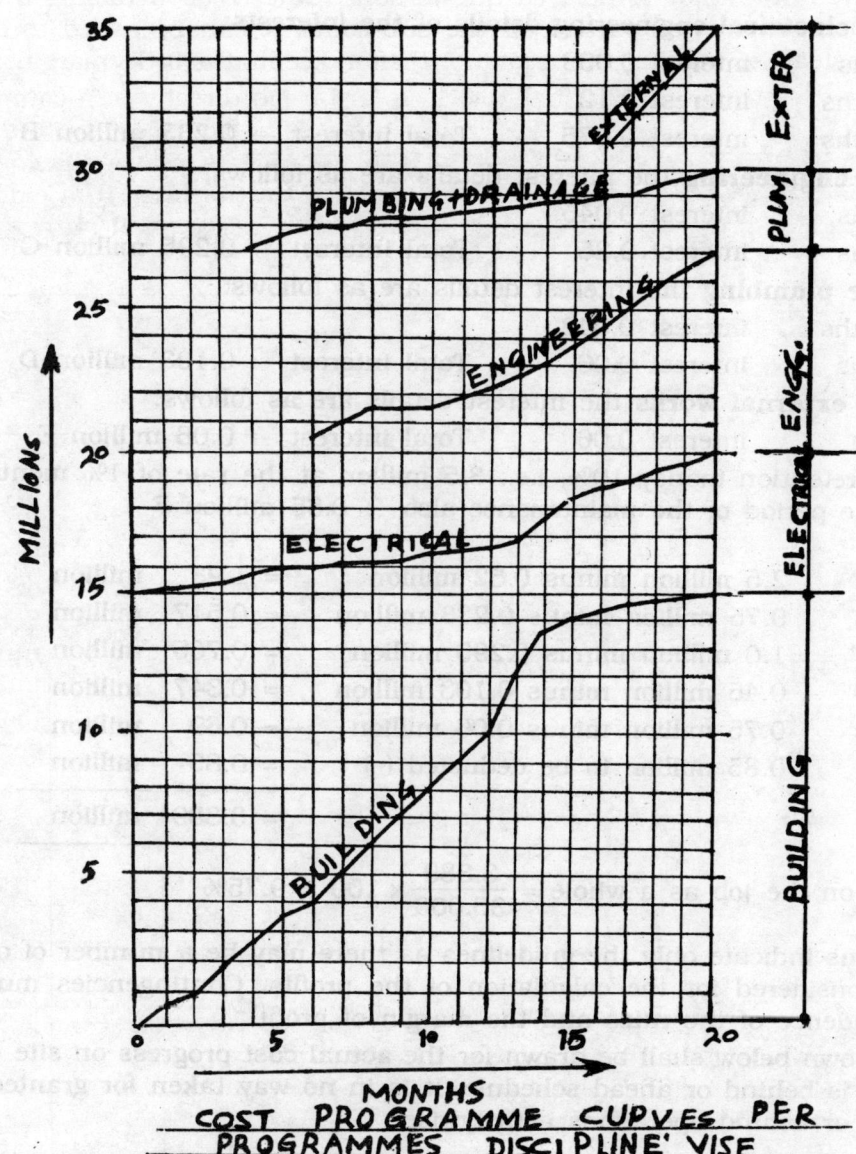

COST PROGRAMME CURVES PER
PROGRAMMES DISCIPLINE' VISE

Total Investments has been 10.7 million for a profit of 5.45 million including the amount that shall be paid as the bank interests which shall be deducted from the gross profit to reach an applicable percentage of net profit for the project as a whole. The details of the interests have been given on the following page for your reference.

For the discipline of the **building** the interest details are as follows:

0.5 million for 20 months	interest 0.1	
0.5 million for 19 months	interest 0.095	
0.5 million for 15 months	interest 0.075	
2.5 million for 10 months	interest 0.25	Total interest → 0.52 million A

For the discipline of the **electrical** engineering details of the interests:

0.2 million for 19 months	interest 0.038	
0.75 million for 16 months	interest 0.12	
0.75 million for 10 months	interest 0.075	Total interest → 0.233 million B

For the discipline of the **engineering** the interest details are as follows:

0.3 million for 15 months	interest 0.045	
2.5 million for 10 months	interest 0.25	Total interest → 0.295 million C

For the disciplines of the **plumbing** the interest details are as follows:

0.25 million for 17 months	interest 0.043	
0.75 million for 8 months	interest 0.06	Total interest → 0.103 million D

For the discipline of the **external works** the interest details are as follows:

1.2 million for 5 months	interest 0.06	Total interest → 0.06 million E

Interest amount on the retention money 10%, i.e., 3.5 million at the rate of 1% monthly for 24 months say including the period of the maintenance also → 0.85 million F

Net profit in the case 'A'	2.5 million minus 0.52 million	= 1.98	million
Net profit in the case 'B'	0.75 million minus 0.233 million	= 0.517	million
Net profit in the case 'C'	1.0 million minus 0.295 million	= 0.705	million
Net profit in the case 'D'	0.45 million minus 0.103 million	= 0.347	million
Net profit in the case 'E'	0.75 million minus 0.06 million	= 0.69	million
	0.85 million to be deducted (–)	= 0.85	million
Net profit on the job		= 3.389	million

Percentage of the profit on the job as a whole = $\dfrac{3.389}{35.000} \times 100 = 9.75\%$

The above calculations indicate only the guidelines as there may be a number of other factors and the constraints to be considered for the calculation of the profits. Contingencies must be there and main thing is the dependence of the rates and the margin of profit.

Similar curves as shown below shall be drawn for the actual cost progress on site to know whether the costwise the project is behind or ahead schedule. It is in no way taken for granted to consider for the project to be ahead or behind operations activitywise.

What a deputy agent or sub agent should do

When the size of the project is big or considerable and the agent cannot find time daily to be aware of every activity personally due to the situation of some portion of the site to be far from the agent's office, it shall be better to deploy some senior site engineer to act as a deputy agent to look after that portion independently and report to the agent. In that case the site execution control of the section under consideration in consultation with the agent goes to the sub agent/deputy agent as the case be. Sometimes, the co-ordination work on the site is excessive and alone the agent may not care of that, so for his assistance a coordinator in the capacity or the level of a sub agent shall be deployed and let the agent take overall responsibility. In that case often the desk work is done by the sub agent per directives of the agent and the agent only signs finally.

A sub agent can also be appointed to take independent care of machines on a big site and take decisions of replacements per agent's directives. For a particular special activity for the whole project a sub agent can be employed to help the project. There is no strict hard and fast rule to the appointment of a sub agent. In fact they are vary rare. In the absence of the agent, the deputy agent shall officiate. Sometimes, a deputy agent or sub agent is responsible to take measurements of all the sections of the buildings either himself or from the site quantity surveyor or engineer and prepare the claims and valuations. The nature of duty for a deputy or a sub agent vary from time to time being administrative one time and the technical the other and overall sometimes for the remotest part of a project under construction.

Sometimes, a qualified engineer may be deputed as a deputy agent or sub agent to prepare the rebar schedule to the steel fixer man per drawings and details to assist the overloaded site engineers. Sometimes, he may be deployed for the preparation of the working drawings, record drawings, etc., to assist the agent in his works.

What a site engineer should do

Site engineer is in fact the real executive person on site on whom rests the main responsibility of implementations of the decisions either taken by the resident engineer, his staff or by the contractor or the agent. He is the technical executive person or the executive engineer and responsible for full responsibility as far as the drawings, details, and the specifications are concerned. He has to see that the works carried out on site from bench marks taken over to the finished building are done correctly and under his supervision and with the supports of the deputy engineer or the assistant engineers, general foreman, junior engineers and the foremen and the other site staff and the workpersons. The site engineer is fully responsible to arrange the workforce, the materials and machines for his site though the agent and see that no barrier comes in his way. He has to arrange access to the site, water supply for construction, concrete manufacturing arrangement, day to day distribution of job through the junior engineers and the assistants, prepare the inspection requests, calling the inspections' people from the resident engineer's office when required, checking the quality of his works produced, maintenance of all records and sending to the agent for further reporting to the resident engineer, getting the quantities of the works measured through the site quantity surveyors, making daily job records of his section or site, checking the workpersons' out put or the progress daily, engagement of the people on overtime if needed, meeting the agent regularly to discuss the programme and the priorities, to advise the agent on the purchase of the materials, to advise the agent of the different methods of a work, to advise the agent to make alterations through the resident engineer and the specialists engineers.

Site engineer has to see that the electrical, air conditioning and the plumbing engineer and foremen turn on site to do the job alloted to them so as not to obstruct building works in case some obstructions remain or the works of those disciplines are not taken up simultaneously, in which case the agent should be fully informed through the site engineer incharge. Site engineer should know complete specifications meant for the job and do accordingly within the tolerances as specified in the specifications. Whenever some discrepancy is discovered by the site engineer, that should immediately be brought to the notice of the agent and the decision awaited. In case he finds agent overbusy, he may discuss that with the resident engineer or the consultants and seek the decision and inform the agent accordingly. Site engineer should find out good people who can be entrusted with the petty contracts and discuss with the agent for finalisation of the sub-contract. What site engineer expects is that he should get enough material, machines and the workpersons in time to run a given work to the programmed lines. He has to check that people are not absent assisted by the time keeper. In case the separate time-keeper is not attached with the site engineer then the foreman shall assist to check the absentees on the site. On small sites, the senior engineer experienced on construction sites can be entrusted as site agent cum site engineer to report to the contracts manager. Site engineer should attend the contract site meetings and if possible also the technical and the co-ordination meetings. The minutes of the meetings should be circulated to him to apprise of the situation. Site engineer must command over the workpersons and all the sub-ordinates attached to him. Since the site engineers are generally young blood and tempered quickly due to the nature of the job but it is better to have a little absorption. Just getting tempered at a worker or labour and crossing the attendance is no intelligent solution but shall generate a feeling of insecurity among the workpersons. Too often shouting also does not help. Better let the shouting business be taken care of by the site foreman or the assistants. Deep study of the drawings and specifications and thought to find better and easy ways to execute the job may help to be successful engineer. Site engineer must promote team spirit for the site and coordinate with the other site engineers for the exchange of men, material, machines and technical and engineering informations. Most of the site engineer's time should be devoted to the field work and come to office for discussions, drawings, etc. He must check all the surveys carried out by the land surveyor or the foreman or the technical assistants provided with the site engineer for his supports.

What a general foreman should do

As the term explains itself a foreman who knows all the works of building and has enough knowledge who understand drawings, must have long site experience can be a right hand to the site engineer and the site agent.

General foreman must be a dynamic person to control all the foreman (carpenter foreman, reinforcement foreman, masonry foreman, concreting foreman, etc.) and command all the workers. If a good general foreman is available, the job of site engineer and the agent gets easy. All the details are provided to the general foreman by the site engineer and he arranges for the run of the details and the distribution of the daily jobs to the workpersons according to the job requirements on the instructions of the site engineers based on site agent's directives. A good general foreman may be rough but if worth is considered an asset at site. Sometimes, his advices are beneficial to the job. General foreman's job is never a sit in office's chair but run here and there to handle the job correctly and see that no workperson undesirably sits idle. He has to act as a teacher to the workpersons and the other junior foremen by demonstrating a particular activity. General foreman and junior engineers on site are often found at loggerheads but since general foreman is senior, the junior engineer has to realise. Instead, the junior engineer has the opportunity to learn the practical aspects of what he studied or never studied.

General foreman, must possess the knowledge of mechanical operations and advise from time to time the agent and the site engineer his ideas. His idea should not be straightway rejected or turned down. He has the authority over the time keepers, the storekeepers, and the junior staff on site to question and they are supposed to answer. Agent can make the general foreman his eyes. A general foreman must seek the help of site engineers all the time as engineer most of the time remains busy in drawings. Then the general foreman should ask the site engineer to expedite the levels, lines so that the idle people are engaged. He has to see that the job is moving faster. He has to make the engineer more and more busy by his own efforts to find the jobs. If the site engineer is absent, it is the general foreman who should take care of the site temporarily.

What an Assistant engineer should do

In fact an assistant engineer on a building project assists the site engineer technically e.g., regarding drawings and details, details of the forms and the reinforcement layouts, levels controls, quantity surveying and valuation measurements, fixing lines to the building, supervising the concrete mixings, laying and assisting the site engineer in his job. He has got complete technical responsibility. Generally, for big buildings assistant engineers are deployed to assist the site engineer and he must follow the instructions coming from the site engineer.

What a foreman should do

Many categories of foremen are necessary on a job.

1. *Carpentry foreman*: to engage the carpenters and helpers to erect the forms and set the lines and check the plumb, supports, etc., all the related jobs.

2. *Bar bending foreman :* to instruct the bar benders how to cut and bend the steel and raise place in the position on site to the satisfaction of the engineer and the resident engineer, general foreman and the structural engineer.

3. *Concrete foreman :* to supervise the concreting work carried out by the concrete gang and arrange every requirement for the same in advance. To see that the levels are fixed in the required positions, vibration is properly done, slumps and cubes are taken. Curing is carried out in time and concrete remedial works if required are done to the instructions.

4. *Mason foreman :* to supervise and arrange the materials and men for the masonry job and their finished items such as tiling, floorings, etc. He has to check the alignments of walls, verticality and inquire the assistant engineer or the general foreman about the openings, lintels, etc. He has to check roofing and plasters and the renderings.

5. *Joinery foreman :* Joinery and the furnishers are very delicate jobs and experienced and understandable man is necessary to fabricate doors' frames, window frames and produce other items such as cupboards, shelves, almirahs, ceiling borders, wardrobes, etc., and make them fit at site. There are other foreman too who shall be required from time to time e.g. road foreman may be one for the filling and the formation levels while the other for the carpetting the surface in asphalt concrete.

Some times specialists foreman are required for the special jobs in different disciplines for a short period who are generally very trained in the respective activity they perform.

What a plants engineer or a foreman should do

The position of a plants engineer is always separately designated and like general foreman on small jobs. He may be designated a plants engineer or a plants foreman and supposed to maintain and services the plants such as batching plants, earth moving machines such as excavators, rock breakers, bulldozers, grader, payloaders, shovels and all other machines given to his responsibility such as mobile cranes, tower cranes, concrete mixers, stationary or mobile concrete mixers, trucks and equipments as compressors, vibrators, and all the transports meant for the site. He has to make them available as required by the agent, general foreman, site engineer on the job.

What is an office manager

At the site office, the agent's head administrator is the office manager vested with the responsibility to carry out the paperwork, correspondences, ordering materials processing, receiving and checking the accounts and preparing pay sheets. He should have under him a pay clerk, order clerk, correspondence secretary and accounts clerk, who in turn control other staff such as invoice checkers, stores keepers, messengers, tea boys, car drivers, watch man, etc. The office manager must be smart and efficient as he has to see the smooth flow of material to site and ascertain that, and also the accounts to be maintained properly to be audited by a chartered accountant. He has to take care of thousands of invoices and the vouchers and cash through the cashier. Office manager must be a man of integrity sharply sensitive to any mishandling of cash.

Time-clerk, time-keeper and store-keeper

Time-keeper has to check the attendance and time of arrival and departure of workpersons and fill up wagecards and maintain timesheets of machines also.

Store-keeper has to maintain the materials records in the store, receive them when arrived and control the issue to site and inform the agent and the office manager when the stores exhaust and keep proper records of all tools and plants and equipments at site.

What a mechanical engineer should do

Mechanical engineer shall be employed by the contractor when the engineering discipline is carried out by the general contractor himself or be with the sub contractor when the engineering works are sublet. In either case the mechanical engineer has to dispose the duties as mentioned below.

1. He has to go through the bills of quantities of the mechanical/engineering and all the schematic drawings and propose the schedules of items for making submittals to the resident engineer/engineer through the main contractor/agent for approval.

2. He has to go through the contract drawings and the contract specifications for the engineering disciplines and check the design and form his calculations to match the project requirements and reach his system which shall be proposed for the engineer's approval based on the proposed material submittals and shall submit the complete schedules of working drawings and the calculation of designs to the resident engineer through the contractor's agent.

3. He has to start making the material submittals according to material schedule submitted for the engineer's approval one by one and obtain the approval of the engineer through the contractor/ agent and take necessary action in the placing of the orders in respect of the approved materials which must fit in the trades programmes to be prepared later based on the contractor's main approved programme supplied to the sub-contractor.

4. He has to do the design of the a.c. system per the contract requirements detailed in the project specifications while the contract drawings shall act as a guide or scheme only and the condition in the contract requires the contractor to design, provide, install, commission, maintain all as a complete system to the first class standards as specified in the project specifications. For this purpose the mechanical engineer has to co-ordinate with the site agent to acknowledge the informations of the structural designs as the beam sizes, slab thickness, position of the walls and the levels of the reflected ceilings from the schematic plans of the ceilings, grilles, air space to be provided for the return air, positions of the light illuminaries, etc., so the space available for the accommodation of the a.c. ducts and the probable routes for the supply and the return air are located easily without clashing with other disciplines.

5. The knowledge of beam depths, position of walls and the openings for the windows and the ventilators shall equip the mechanical engineer with the knowledge of the obstructions liable to come in his way and let him alter the section of the ducts or the routes and at some spots one duct comes crossed over the other and to be adjusted in the available accommodation for which the ceiling height shall dictate the size of the section. The sizes shall include the thickness of the glass fibre insulation and the vapour barrier over that. If coordinated well in the beginning a lot of many contingent clashes at site during carrying out the job shall definitely be avoided. This requires a great knowledge on the part of the mechanical engineer to avoid the remaking of his working drawings after being rejected. He has to work patiently and with concentration. In practice, there shall be a number of shop drawings submitted a few days or on the same day of the meeting of contract and no drawings for a couple of months so it be recorded in the meeting that the drawings are in progress but really this is not a professional way. The result of that rash submission is very clear that most of that lot shall be rejected and remain pending by the sub contractor for a couple of weeks and really no gains on the job. The material submittal delays shall add fuel to the fire really as the drawings cannot be finally approved without the availability of the complete approved details of the material to be included in the system. The ultimate looser on account of the non-approval of the drawings shall be the electrical men whose some of the drawings shall not be submitted without knowing the details of the mechanical related details drawings and shall make the mess of the whole show. Because of the non-availability of the mechanical statistics the electrical engineer shall go on crying and the main contractor regret to appoint the guy as the sub-contractor for the discipline and the agent shall be facing a challenge to bring all the disciplines together. If the delays are increased steadily in finishing the submissions the recovery programmes shall be required from the mechanical engineer to cope up with the lost time and the sub contractor shall have to burn the mid night electricity to complete the lagging behind work.

6. Having finished the submittals, the mechanical engineer has to finish the trades programmes based on the main contractor's approved construction programme which shall be supplied by the contractor's agent to the sub contractors. The details in the trades programmes must include all the activities and fit in to the line of the main programme and the details of the workpersons,

materials and the equipments and machines to be used shall be given.

7. He shall attend all the technical and the co-ordination meetings all the time even though the work of his discipline has not started yet. If required he has to come to advise the agent to provide blockouts for the a.c. ducts wherever required in the floors, roofs, or the walls.

8. It is also true that a.c. duct and insulation work shall start only when sufficient work area is made available by the contractor for the a.c. man to start the activity and covered area is needed to start the work. This means that the walls and the plaster be finished so that the insulation and the vapour barrier is not spoiled by the mortar. In actual practice, a.c. 1st fix starts after a couple of months of the start of the contract.

9. After the mobilisation the coordination should be in such a way that a.c. men are not withdrawn from the site intermittently due to the unclear site.

10. The mechanical engineer has to deploy a mechanical/a.c. foreman full time on site to supervise all the activities and coordinate with others on site.

11. Mechanical engineer should issue the approved prints of the working drawings to all concerned and advise the agent from time to time clear the work area.

12. Most important part of his coordination is with the electrical engineer to whom he has to exchange the complete technical statistical data of the equipments and the layouts of the plant. Both the engineers in fact shall agree over the adaptability of both the systems to each other. Once both engineers agree and the resident engineer and the services specialists agree then major head ache of the coordinations is over. The ways for both to go ahead with the execution of the job shall be cleared for the supply and the installation of the disciplinaries.

13. Mechanical engineer is responsible to get the system approved by the resident engineer and the specialists engineer and the engineer and also for the post installation successful testing and commissioning operations for minimum of a month or the other specified period as the contract clause states. All the equipments shall be tested and records made in the presence of the resident engineer and the services engineers.

14. The agent and the mechanical engineer should act in unison to see that the ceiling system is not closed before the a.c. system and any item related to that is inspected.

15. Mechanical engineer has also to advise the agent to make the various plinths for the machines and the equipments before these arrive on site and also the pedestals for the chilled water pipe supports. This shall assist both in conveniently carrying out the elements of the equipments and machines installations as well as builders works involved in the same.

16. Clearance of all the mechanical engineering items from the resident engineer shall be the responsibility of the mechanical engineer.

17. The mechanical engineer shall also make sure that the equipments arrive in time to the site so that the other disciplines are not delayed.

18. The mechanical engineer shall provide all the information for the contract meetings to the contractor's agent.

19. Similarly, a few months before the completion of the project practically or as specified in the specifications of the project, the mechanical engineer shall provide all the record drawings or as built drawings to the agent for the onward movement to the resident engineer/engineer for the approval of the records. This shall be treated as an important work and should be done simultaneously as the progress on site.

20. Lift installations shall be an important part of the mechanical discipline because the lift manufacturer may not start the manufacturing of the lifts without taking the accurate measurements from the site. So the lift installation time shall be properly coordinated without presumption of being optimistic. Sometimes, due to negligence in finalisation and approval of lifts the complete buildings handing over and completion shall be jeopardised and held up.

What an electrical engineer should do

Electrical engineer shall be employed by the contractor when the electrical discipline works are carried out by the general contractor or be with the sub-contractor as the case be in either case the electrical engineer shall dispose the following responsibilities.

1. He shall go through the electrical bill of quantities and the schematic drawings as a basic requirement of any engineer and prepare the schedules of the items for making of submittals to the engineer through the resident engineer to whom the site agent shall submit on the receipt from the sub contractor.

2. He shall go through the contract drawings thoroughly as well as the contract specifications for the electrical and the related mechanical disciplines and ascertain the adaptability of the systems for the design and the loads requirements for the use.

3. He shall schedule out the details of the shop drawings and enlist the same in an orderly manner for making the submittals and send the schedule to the agent for forward disposal to the resident engineer for approval.

4. He shall start making the material submittals according to the approved list and obtain the approvals in time from the resident engineer and the consultants.

5. He shall start making the shop drawings for submittals according to the approved list of contract documents and shop drawings to the contractor's agent for onward disposal to the resident engineer.

6. He shall propose the trades programmes for electrical disciplines on the job and fit in to come in line to the approved contractor's overall programme supplied to him by the agent. He shall clearly exhibit in that the details of the trades persons, equipments and other details as professionally required to be checked.

7. He shall ensure that the electrical job on site starts in time and in the co-ordination to the building activities being in progress and the items such as the conduits, ducts, junction boxes, and all the items to be concealed in the ground or in the floor slabs or in columns or in the walls, etc., be procured immediately and the execution done in sequence to the building operations. The deployment shall take place immediately after the foundations are over to provide the concealed ducts in the ground floor.

8. Few of the electrical drawings unrelated to the a.c. works should be ready on site duly approved for the purpose of the site inspections.

9. He shall see to engage one of the trained foreman on the site all the times to supervise the activities going on and obtain the written approvals of the electrical consulting engineer and the resident engineer.

10. He shall co-ordinate with the agent on site all the time and make himself aware of the proposed operations and the programmes and also of the internal architecture on the job to position the points of lights and the switches and sockets accordingly. His job of coordination with the architectural/structural is not that much complicated as that of engineering.

11. What he needs to coordinate are finished floor levels, finished height of the false or suspended ceilings to check that the trunkings and cable trays etc., do remain in the ceiling and the space for shot firing the ceiling system wires is available. He has to know the possible elevations of the fittings which are given somewhere in the project specifications or the drawings if not let the same be received from the resident engineer.

12. The main coordination shall exist as discussed before in the responsibilities of the mechanical engineer on job, the adaptability of the machines or the equipments to the cables as provided or to be proposed by the electrical engineer that these should accept the load the machine shall deliver as output or input. He has to satisfy the mechanical engineer for the adaptability and both the engineers shall agree and councel for the agreement by the resident engineer and his specialists. Electrical engineer knows about the technical data of the machines and the mechanical engineer the size of the cables and the quality as well as the systems so the two systems match each other. Generally, the electrical drawings are left pending particularly of the equipments and the machines due to the non approvals of the engineering drawings.

13. Electrical engineer has to finalise the ordering of the materials per approved programmes and

include his material approval and delivery schedule during the technical and the co-ordination meetings the contents of which have already been explained earlier.

14. He has to attend all the co-ordination and the technical meetings and include progress report for the submittals and the works and complete details of the job i/c man, material and the percentage of works and drgs, and the material submittals.

15. He shall issue all the approved prints of the drawings to the concerned people as explained before.

16. He shall be responsible for the inspections of the job carried out by his people. All the equipments, fittings are supposed to be test approved by the resident engineer and the consultants on the approval record sheet.

17. He shall act in unison with the contractor's agent to get everything inspected before the ceiling is closed, to the satisfaction of the resident engineer and the specialist resident engineers and the architectural engineer.

18. He shall co-ordinate with the agent to make available the foundations of electrical poles for the external electrifications as a part of the contract.

19. He shall co-ordinate with the agent to make the transformer plinths and minor m/c plinth ready and the bases of the switch gears, cable trenches as specified in the contract details.

20. He shall provide all the informations for the contract site meetings to the agent.

21. He shall co-ordinate with the agent to make available rooms for the stores, etc.

22. He shall prepare all the records drawings and as built and submit for approval in time as specified in the contract documents to the site agent for the further disposal to the resident engineer.

What a plumbing and the drainage engineer should do

The plumbing engineer may be employed by the main contractor should the work of this discipline be carried out by the contractor/or the sub contractor if the work is sub let. In either case the plumbing and the drainage engineer shall dispose the following responsibilities for the contract contents, or the scope of his job.

1. He shall go through the contract documents thoroughly, i.e., read the BOQs, the contract drawings, the project as well as the other specifications and schedule out a list of the proposed materials for the submission as well as the shop drawings for approval.

2. He shall promptly start making submittals based on the gathered informations according to the approved schedule of the materials and the drawings approvals. He shall keep in mind that his requirement shall be the first immediately after the excavations of the foundations are over as the works of the underground drainage shall be a parallel operation to the foundation operation. And all the related drawings and the approved materials must be available on site to proceed for that activity.

3. For the approval of the ground floor drainage systems the plumbing engineer shall have to co-ordinate with the site agent for the final levels taken in the building and the invert levels as may be changed due to the adjustment in the finished external levels. Routes may not be followed as given in the working drawing or in the schematic drawings but depending upon the positions of the foundations a movement be permitted that may be recorded in the as built drawings.

4. He shall mobilise the people, men, material and the required equipments as soon as the foundations start so that no delays are caused to the sub structural activities due to the drainage and make all arrangement for the testing of the laid pipes for obtaining the approval per specifications. The approval shall be the responsibility of the plumbing and the drainage engineer.

5. He/his foreman deployed fulltime has to make arrangements for the tests of the pipes lines for all the sections as required by the resident engineer and the specialist engineer.

6. He shall submit the trades programmes of his disciplines to depict clearly the detailed activities based on the main approved programme in the possession of the sub contractor as supplied by the agent of the contractor fit in with the time of the activities according to that exhibiting complete details of the manpower equipments and all other details.

7. Having received the material approved through the contractor's agent, he shall make the arrangements for the procurement of the materials per programmes.

8. Having submitted all the drawings relating to the sanitary, drainage, storm waters, hot and cold water supply, irrigation system for the plants, he shall distribute the copies to all the related people on site as explained earlier.

9. All the materials ordered should be delivered to the site per programme and be explicitly detailed in the material status submitted with the documents of the co-ordination and technical meetings to be held monthly or fortnightly if required.

10. He shall attend all the technical and the co-ordination meetings.

11. He shall submit the details of the shop drawings in the format of the shop drawings submittal and the approval status.

12. He shall see that the items delivered for plumbing and sanitary fix II which shall be fixed at the time of finishing items in progress before the final testing and commissioning are properly stored in consultation with the agent who shall arrange for the same.

13. He shall make necessary dealings to coordinate the works with the other disciplines on site for example main infrastructural contractor working in the same project and also the other sub contractors.

14. There could be many changes in the routes, levels, of the laid pipes systems which he shall ascertain to have been recorded in parallel to the execution so as not to miss any deviated record.

15. He shall act in co-ordination where the builder work is necessary and to be done by the agent and also advise the agent for the clearance of the site to get more scope for the operations of his disciplines of the drainage and the plumbing.

What a road engineer should do

Road engineer though required in the last stage of the contract or the project and particularly in the organisations having a separate road section, the work is carried out by that section being specialist in that discipline. The road engineer on the call by the agent shall act in unison and provide all the informations in connection with the material submissions such as aggregates, asphalts concretes, etc., to the agent for the formal submission to the resident engineer and the civil works engineer.

The materials should be approved before being brought to the site. After the setting out of the road lines and the levels to reach the formation levels, the road engineer shall be requested to inspect and give approval for the road activities to commence through the site agent. Back fill and the compaction shall be carried out and the samples of the compacted ground or the fill shall be taken per the direction of the resident engineer or his road specialist and the density results as shall be communicated to the resident engineer in original by the site agent. After that all the activities of the road structure shall commence. All the activities of the construction of the road structures, the road surfacing, the lining, etc., shall be carried out under the supervision of the road engineer by the road foreman and all the approvals from the resident engineer and the civil works or the road specialists shall be taken by the road engineer through the agent.

Road engineer should see before mobilisation of his gangs and the machines and equipments that the necessary works on kerbs fixings, fencings, infrastructural activities involved in the contract and all the minor structures have been completed from the agent and when he gets the clearance from the agent to carry on the job, the road engineer shall start the mobilisation. He shall also advise of his requirements before the commencements of the road activities. Sometimes, there may be no need for a special road engineer and the site agent and the site engineer may deal with the execution of that part of the contract themselves.

Specialists works

In every project you shall find the services contractors for the execution of the special disciplines or the activities or the systems on your job.

1. For the constructions of a swimming pool/pools complete system as a whole including the design part of that.
2. For the treatment plants.
3. For the overhead water tanks and the other water tanks.
4. For the sports equipments supplying and fixing of these items per specification.
5. For the sports surfacings of the high standing and the comfort systems.
6. For the chain link fencings or the other fencings for the security.
7. For the ornamental works, decorative works or the works related to the cultural heritage.
8. For the work of the arboriculture and the green house.
9. For the supplying and fixings the special kitchen equipments in all respects i/c cold stores.
10. For the special rooms such as X-ray room plaster, squash court room construction, special floorings such as rubber special floorings, wall linings of the halls, works of the glass reinforced concrete, for the works in the electronic equipment rooms.
11. For the structural steel works and the works of a particular trade as specified.
12. For the wall and the roof cladding elements.
13. There is no limit for the special elements in any contract but where the coordination is necessary shall be seen by the agent and the related engineers themselves.

For example, the co-ordination among the swimming pool contractor, plumbing and the drainage contractor and the electrical contractor alongwith the architectural disciplines are necessary. Coordination among the tiling work and the under water light and the distribution of the supply system and the disposal of the used water all need the close co-ordination for which a co-ordinated drawing shall be provided by the site agent for the approval and the assistance of the site engineers.

Similarly, the co-ordination for the construction of the water supply and the sewage treatment plants are required.

Agent shall act promptly with full action programme of his own to deal with these people for a successful completion of all the operations to the programmed bar. He must ascertain that all the drawings needed for the complete execution of these operation must be approved with the interdisciplinary co-ordination of services to avoid the clashes and the wastage of the valuable time.

Engineer and his supervision

It is desired to explain in brief the meaning of the term engineer as the people generally misunderstand that. Actually, in the terms of the contracts this designation applies to the person/official/government department head/ professional organisation of the building engineering/architectural engineering firm to whom the owner or the user entrusts the responsibility of the project to his budget availability from the conceiving stage to the final completion which shall involve all the operations of the inception, architectural plannings and the detailings, structural planning, designing and the detailings, schematic services design and the detailings, the design of all the related civil works and external works and the preparation of the bills of quantities and the contract contents and the general and the project specifications, tendering and the award and the control and the administrative management and the site supervisions of the contractor's executions, finalisation of all the payments, budgetary controls, acceptance or the rejection of the job, testing commissioning of the completed jobs per the conditions and the clauses in the specifications, maintenance supervision, and grant of the final completion certificate for the purpose of the release of all the retention moneys to the contractor, and the activity not covered in the said explanations that the preparations of all the detailed drawings for the purpose of obtaining the statutory building permits.

For example, a promoter appoints an architectural or the building engineering firm or the group to plan design and supervise the execution of the job, the appointed firm or the group shall be nominated as 'engineer' in the legal terms and the head man responsible for the group shall be liable to the legal implyation as the man responsible for the design and supervision and the stability of the building as a whole. Similarly, any government or the funding body wants some projects to be executed for the welfare of the masses or for any other purpose, the government department of the civil engineering discipline shall be entrusted to carry out the job of the project and the head engineer of the department of the rank or the designation of the engineer in chief/the engineer member/chief engineer as the case be, shall be designated as 'engineer' in the legal terms. That guy needs not necessarily be the civil engineer or the structural engineer. The architect or the civil engineer or the structural engineer are all eligible for being designated as the engineer in the legal terms. Should the entrusted department does not possess the required calibre or the staff to look after the above explained duties, a group shall be invited to look after the same responsibility and the appointed firm/group shall be termed 'engineer' if the work is completed to be on the firm's responsibility and in some cases the department may wish to appoint the resident engineers only for that purpose.

So after the owner appoints the firm or the group as the engineer and the contract is entered in between the engineer and the promoter or the owner, the head of the firm shall be the party in the contract personally while in the government department the engineer in chief or the engineer member or the chief engineer shall sign on the behalf of the government. When the department appoints an engineering firm as the resident engineer, the nominated engineering firm shall be the project manager supervision.

The big projects cannot be designed and supervised by one man alone as engineer. Therefore, in practice what happens that the engineer responsible for the project appoints under his control a resident engineer for the control of the site execution with a group of the specialist engineers and the other anchoring or the supporting engineering and the administrative and the ministerial staff under the direct control of the resident engineer. All the works as carried out on site shall be of the technical nature and to check the works being done according to the engineer's details or not. Whatever shall be done by the resident engineer be considered by the engineer. In legal terms all the responsibilities of the project shall be shared by the engineer exclusively.

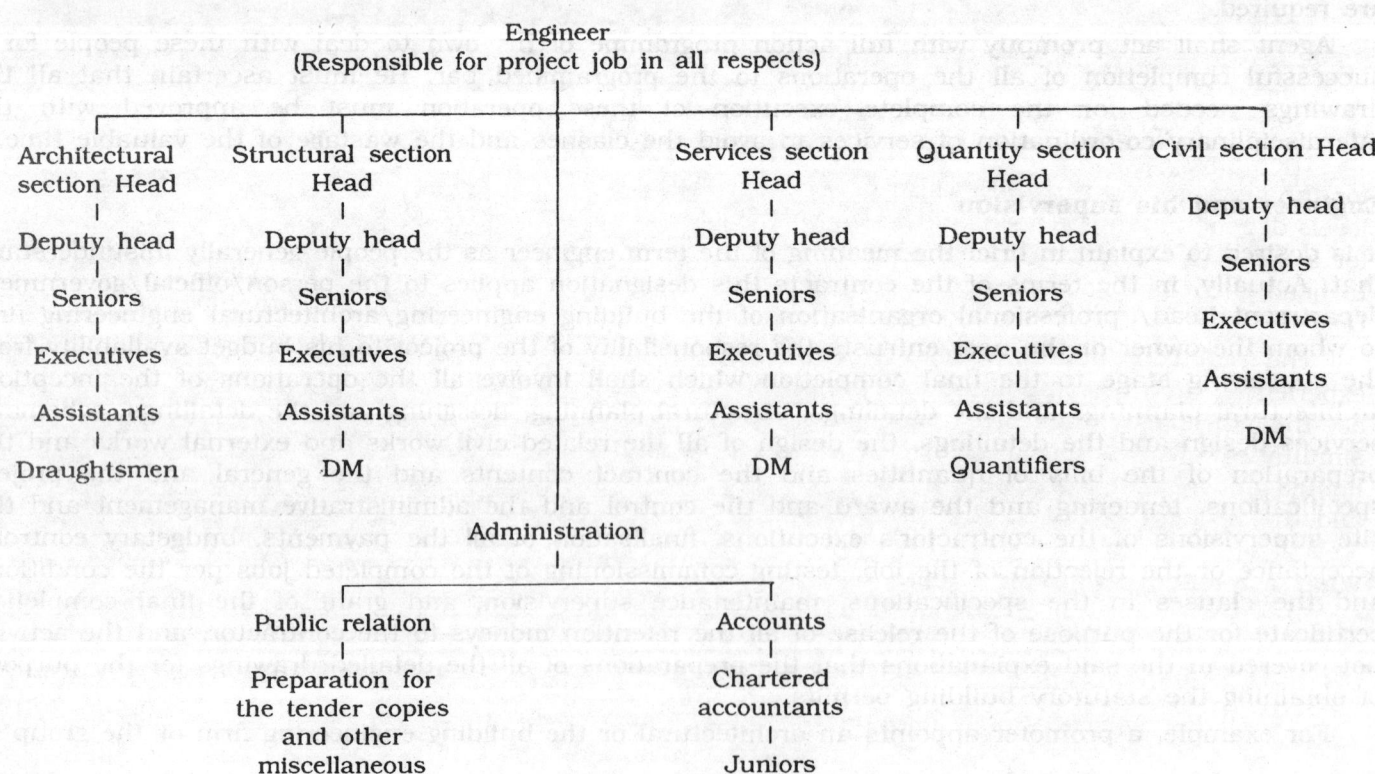

The above organisational structure of the engineer's office shows in different levels of the working for the project under the control of the engineer for the purpose of the guide line only but may vary from the project to project and requirement to the requirement. The above shall be applicable to a very big size of the project. The number of the personnel shall depend on the quantum and the duration of the design period of the job. The engineer shall also employ the resident engineer and the staff which shall be the structure in the organisation of the resident engineer office at site.

Resident engineer and supervision

A resident engineer shall supervise the contract operations on behalf of the engineer alone in case the job is very small but for the big contracts of the constructions, the engineer or the owner as the case be according to the contract between the engineer and the owner or the promoter, shall provide the supporting staff on site who shall act as consultants in the supervision of the construction activities of the operations being carried out by the construction contractor according to the contract documents. These supporting supervising engineers shall include the assistant resident engineer or the project coordinator, the architects, the structural engineer, the electrical engineer, the mechanical engineer, the plumbing engineer, the civil engineer, the quantity engineer, the accounts and the secretarial sections and the site assistants as the inspectors or the clerks of the works in the different ranks all as forming one team of the supervisors.

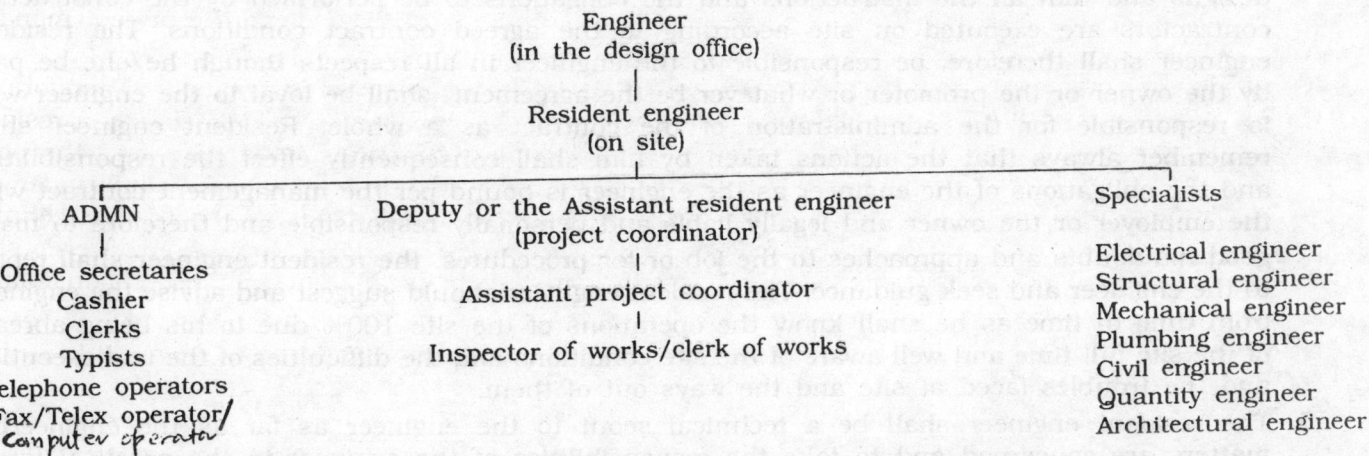

The above site organisational structure for the office of the resident engineer is only for the purpose for the guidance on a medium size project and shall vary in the strength of the placements in different levels depending on the needs of the supervising works. There may be a number of the resident engineers for a vary big size job and a head or the chief resident engineer for a very big size job may be placed to head them and to report to the engineer for the job. In that case the chief resident engineer shall take the instructions from the engineer and supervise the job per contract documents.

The real coordination works of different phases may be carried out in cooperational way. The experienced architect appointed as specialists engineer, the structural and the civil engineers appointed as the specialist engineers may act as project coordinators for the independent phases and simultaneously the specialist for the other phase on the medium size projects where the funds are limited and also the quantity of the operations is of the controllable size. It depends on the engineer and the resident engineer how to organise the management structure.

What a resident engineer should do

1. A resident engineer shall act on behalf of the engineer to ascertain that the works as carried out by the different construction contractors on site or their sub contractors are all per the contract documents (bills of quantities, all the drawings, working drawings, materials approved, all the specifications, manufacturer's instructions and all the related details provided by the

engineer) and according to the contract approved construction programme in a fixed time or duration as laid down in the contract schedule. A resident engineer must be a qualified and a competent engineer of the architectural, structural or the civil engineering discipline on the civil engineering projects as he shall have to apply all the skills he is supposed to acquire during the professional study and the training at the supervision of the construction contractor's operation under his charge.

2. The job of the resident engineer starts even before the award of the construction contracts and continues till the final completion certificate is granted by the engineer on the recommendations of the resident engineer's inspection.

3. The resident engineer shall finalise all the survey details of the plot or the land and fix all the bench marks ready, all the topographic coordinates ready before the joining of the construction contractor on site or at least be having the complete details with him to pass to the construction contractors immediately after handing over the site to let the related people commence the construction operations without wasting a single day. The resident engineer shall be the chief executive of the engineer on the site and a counter part of the agent of the construction contractor.

4. The primarily, the main job of the resident engineer shall be to see that the construction contractors do the job according the details in the documents given by the engineer and his designs and that all the instructions and the obligations to be performed by the construction contractors are executed on site according to the agreed contract conditions. The resident engineer shall therefore, be responsible to the engineer in all respects though he/she be paid by the owner or the promoter or whatever be the agreement, shall be loyal to the engineer who is responsible for the administration of the contract as a whole. Resident engineer shall remember always that the actions taken by him shall consequently effect the responsibilities and the obligations of the engineer as the engineer is bound per the management contract with the employer or the owner and legally liable and personally responsible and therefore to make good his doubts and approaches to the job or for procedures, the resident engineer shall report to the engineer and seek guidance. The resident engineer should suggest and advise the engineer from time to time as he shall know the operations of the site 100% due to his being abreast of the site full time and well aware of the site conditions and the difficulties of the real execution and the troubles faced at site and the ways out of them.

5. The resident engineer shall be a technical scout to the engineer as far as the engineering matters are concerned and to take the responsibilities of the engineer in the practical terms but not legally. The engineer cannot pass his responsibility to the resident engineer or his staff for the actions taken by them as all the actions shall be treated as if taken by the engineer in the legal terms of the contract with the owner. As the contract between the owner and the engineer states that the engineer shall use all his skills and judgements to the design and the construction of the project or the job as a whole. The engineer shall be considered negligent failing to check what the resident engineer and the staff do during the supervision of the works executed by the contractor and any negative consequential effect shall be borne by the engineer in contract terms. The resident engineer cannot be held responsible for the failure of a structure in legal considerations due to the engineer's being completely responsible and liable for the design. The resident engineer shall all the time bear in mind that he shall be responsible to the engineer and not to the owner/user and therefore, should not act on the owner's opinions in the technical matters and seek instructions from the engineer only even after being instructed by the owner.

6. The resident engineer shall see that all the documents have been issued to the construction contractor in time so as to avoid any claims due to the delayed issue.

7. The resident engineer shall see that the site has been cleared and handed over to the contractor per date of the commencement.

8. The resident engineer shall see that all the site staff have been introduced to the contractor/agent/engineers for which a preliminary site meeting before the hand over of the

site shall be conducted by the resident engineer and recorded with due circulation of the minutes of the meeting and the decisions taken if any.

9. The resident engineer shall see that the contractor has proposed the curriculum vitae of the site supervising personnel and superintending incharge such as the site agent, the deputy agent or the project coordinators, the site engineers, deputy engineers, the site assistants and the foremen and surveyors, etc., and that the contractor has proposed the site establishments for which the layout of the site offices of the contractor's offices as well as the resident engineer's offices shall be included for the purpose of approval by the resident engineer/engineer.

10. The resident engineer shall see that the site mobilisation is done as soon as possible.

11. The resident engineer shall see that the contractor has proposed programmes for the approval by the engineer in the prescribed limit per condition or specification.

12. The resident engineer shall see that site bench marks, points of reference or the lines of the coordinates (general) have all been explained or shown to the contractor's agent so the agent may proceed for the setting out of the buildings and layouts.

13. The resident engineer shall see that the surveys of the existing site levels are jointly done by the agent and the man from the resident engineer's office as deputed for the purpose failing that the resident engineer shall be bound to do the survey works before the commencement of the excavation and that must be brought to the notice of the agent. If the resident engineer does not do that he shall accept the levels given by the agent.

14. The resident engineer shall make sure that the agent is aware of the job coordination working systems with the resident's office. The resident engineer shall explicitly explain to the agent how to fit in with the systems of handling the coordination on the project for the materials and the drawings submittals, site inspections, reports, records' maintenance, etc., and the required standards proformas be given to the agent for use.

15. The resident engineer shall see that the contractor has finalised the appointments of the probable sub-contractors to be deployed on the job for the allied services and the details in the forms of the sub-contractors curriculum vitae shall be asked for in complete package for the engineer's approval as soon as possible.

16. The resident engineer shall act patiently to see that the actions on the part of the agent are to the acceptable standards or the guy needs the guidance. The resident engineer shall give the agent time to acquaint himself with the systems and the site and to under stand each other and both may know each other thoroughly. In the beginning the agent does not know how strict is the resident engineer and the resident resident does not know about the calibre of the agent and the competency on the job.

17. The resident engineer shall guide the agent on the matters of the contract operations and about the area and the standards to be maintained in advance.

18. The resident engineer shall see that the contractor starts making material submittals, shop drawing submittals, programme submittals in the time if not, shall instruct and advise the agent to be careful in the operation of a particular contract clause requiring the submissions in a prescribed time limit to avoid delay and record the bad performance on the part of the contractor.

19. The resident engineer shall also ascertain the workperson strength of the contractor, the mechanical potential on the site, the material arrivals and the financial state, whether enough to run the site or fall short of requirement per the contract's promises/obligations.

20. The resident engineer shall check with the specialised engineers attached, all the details of the submittals done by the contractors in different phases.

21. The resident engineer shall inspect the plants and machines with his assistant and ascertain the working conditions and the working efficiency on site.

22. The resident engineer shall inspect the central mixing plant for the concrete or the batching plant or the plants on or off the site with the specialist structural engineer and the coordinator and assess the functioning of that to grant approval to be used for the job which shall include

all the accessories, calibration of various gauges, the supporting structure of the batching plant whether manually controlled or by the use of the computer and the number of the transit mixers that may be used at the time of major pours.

23. The resident engineer shall see that the concrete ingredients are all stored properly per specified standards on site or the batching plant and properly protected against the inclement weather conditions.

24. The resident engineer shall see that the machines do not pose the danger to the human life during the operations. That there shall be no chance of accidents.

25. The resident engineer shall see that the safety measures are taken by the contractor per the construction site safety standards as a part of the contract documents.

26. The resident engineer shall ask the contractor to send the samples selected by him or the assistants to the laboratory regularly for the technical examinations and submit the test reports from the laboratories (already approved by the engineer).

27. The resident engineer shall discuss the programme of the contractor jointly with the agent and his assistants and comment on that for the further incorporation with or the revision of the same as the case be till the same is approved.

Having scrutinised the submissions of the materials, working drawings, programmes the resident engineer shall recommend the approval to the engineer formally.

28 The resident engineer shall remember that the final approval of all the submittals shall be entrusted to the engineer only per contract.

29. The resident engineer shall prefer to advise in case some discrepancy is discovered in the documents immediately to the engineer for the remedial measures.

30. The resident engineer shall refer to the engineer any irregularity if found in the contract documents or the contract itself or the ambiguity of language or the clause or any contractual or the technical mistake as pointed out by the quantity or the contract engineer for clarification immediately.

31. The resident engineer shall coordinate the works of various contractors, to agree detailed programmes of the works, to see that all the necessary instructions are passed to them in time, and should expeditiously furnish the inquired details or the site clarifications by the contractors.

32. The resident engineer shall see that the approved materials are ordered by the contractor in time and the time of arrival as has been shown as ETA in the status tallies to the programme requirements and all the permits have been obtained by the contractor or should help the agent obtain the same.

33. The resident engineer shall see that the third party insurance has been done for all the work persons on site as required in the labour law of the area.

34. The resident engineer shall see that the specifications in regard to the materials are complied with and the workmanship standards are maintained.

In the case of non-compliance of the specification and the poor or the unacceptable workmanship standards the resident engineer shall issue the site memorandums to the contractor through the agent and copied to the engineer and the record. He shall issue the remedial notes in case the major rectification is required on site and the agent has been found to be ineffective on the actions to be taken on the site memos.

35. The resident engineer shall see that building layouts, alignments, levels, shall be surveyed to the design requirements.

36. The resident engineer shall see that additional technical instructions as found necessary to be passed to the site are issued in time promptly.

37. The resident engineer shall see that any clarification requested by the site agent are issued promptly.

38. The resident engineer shall see that all of the staff engaged for the purpose of the inspections

are efficiently acting to the contractor's requests and the supervision is carried out to the standards within the prescribed limits.

39. The resident engineer shall ascertain that the proper coordination is done among the electrical and mechanical sub contractors under the control of the agent and shall help the agent by the engagement of the specialists engineer.

40. The resident engineer shall help the agent on his request to adjust the timings of concreting and the overtime works.

41. The resident engineer shall hold the monthly technical and the coordination meetings and also the monthly progress review meetings jointly called the contract site meeting wherein the progress of the works as compared to the approved programme, status of the material submittal and the approval status, working drawing submittals and approval status, the distribution of the working drawings, technical decisions pending, the contractual matters, materials delivered to the site, financial matters such as payments made to the contractors, the claims, and all other matters left to be covered in the contract meetings. All the matters shall be referred to the engineer in case relates to be decided by him on the contractual or the design matters.

    The minutes of the meetings should be prepared and circulated by the resident engineer to all those present in the meeting, the contractor, the engineer, the owner and related.

42. The resident engineer shall issue the site memos to the agent/contractor, site instruction and prepare the engineer's instruction for engineer's signature.

43. The resident engineer shall issue all the remedial notes to the contractor for doing the remedial works on the rejected items of the contract.

44. The resident engineer shall recommend to the engineer after scrutiny done by the quantity engineer of the valuations for the release of the payments.

45. The resident engineer shall see that the measurements are taken of all the completed works depending upon the type of the contract.

46. The resident engineer shall insist on the contractor to update all of his claims for the extras or the deletions based on the engineer's instruction so the same be included in the valuation.

47. The resident engineer shall act as a channel to all the claims and disputes and should provide the facts to the engineer.

48. The resident engineer shall keep all the records of the measurements and tests and bring the plans in conformity with the work being done on site.

49. The resident engineer shall see that the finished works are free from defects, tested and set in proper functions of the operation.

50. The resident engineer shall ascertain the final cost/value of the works under various contracts and advise the engineer on the variations on the cost. (expected cost)

51. The resident engineer shall keep the engineer informed of all the developments taking place on site as explained and not limited to that only.

52. The resident engineer shall witness all the operations, tests in the a.c. system's testing and commissioning and also as in electrical and plumbing and drainage systems with the help of the specialist engineers.

53. The resident engineer shall check room by room the satisfactory completion of all the elements included in the construction per contract descriptions and keep records of the acceptances of all the rooms one by one based on the comments of the specialists as the same records shall form part of the taking over from the contractor for the practical completion for the purpose of the maintenance.

54. The resident engineer shall see that the agent, the deputy agent, the site engineers, and all other responsible people are on site during the execution of the operations to dispose the duties satisfactorily in the interest of the job otherwise shall give the opportunity to them for the improvements and make the personal discussions with the agent and should pursue the matter with the engineer to ask the contractor to take off any of the personnel from the site due to being inefficient and replace.

55. The resident engineer shall listen to the contractor/agent calmly and look into the proposals and suggestions and if found conforming the standards recommend to the engineer for the approval subject to no additional cost but if possible make a saving.

56. The resident engineer shall request the contractor's agent to provide the method statement for a particular disputed operation or the activity on site and on being satisfied, permit the same to be applied on the risk of the contractor for making good in case of unsatisfactory performance done on site.

57. The resident engineer shall issue the authorisation letter to the contractor for the purpose of coordinations with the statutory bodies/organisation on the request of the contractor. The same way he shall assist the agent in all such matters that shall be helpful in the job though not a part of his responsibility but to develop the best working relations on site in the interest of the contract.

58. The resident engineer and his staff should not accept the gifts and the bribes of any kind or any other gratification which may equally amount to the acceptance as a bribe except the seasonal greetings and the best wishes as on the birth days or any other successful day or on the festivals from the social and the professional viewpoint.

59. The resident engineer shall ask the inspector or the clerk of works to meet daily and report the progress on the site diary of the resident engineer and the activities in progress and discuss the problems and if possible in the presence of the agent.

60. The resident engineer shall request the site agent to keep him informed of all the developments of progress and the daily job record, weekly programme, etc.

61. The resident engineer shall see that the specialists engineers take the timely actions to scrutinize and solve the technical disputes occurred on site due to any reason.

62. The resident engineer shall not consider the specification volumes as the religious guide as not to permit any variation or the aberration. Actually, practical point should inevitably be considered and should the specifications are not deviated, the contractor's suggestions may be considered depending on the merits of the case.

63. The resident engineer shall ascertain that the contractor is not being harassed and no injustice done to him and the contractor should give the same in return.

64. The resident engineer shall maintain a personal diary and make the entries worth remembrance on that and the meeting with the important persons in connections to the job and the notes in brief of the discussions held and what was kept pending, or any significant point which shall help in future for the reference and recollection in case some point comes up later for the discussions.

65. The resident engineer shall see that proper filing system is maintained by the secretary. It is worth mentioning here that resident engineer, contractor, and all concerned must be explicitly instructed and informed by the engineer regarding the powers and the authorities delegated to the resident engineer for the contract operations so the correct approaches be made by the contractor and all the linked/concerned persons on the contract.

66. The resident engineer's main duty is to watch and check the workmanship and the material used. The resident engineer shall act in the manner as an agent of the engineer and therefore, in all major matters he must refer to his superior particularly in the disputed case. The resident engineer shall always bear in mind that the engineer is responsible for the actions taken by him and his staff and in addition the engineer's responsibility becomes unlimited when he is specialist of the same profession as the job per law. This means should the resident engineer acts negligently in founding a building which results in a collapse, the engineer who himself is a consulting engineer will be held responsible and the owner/user may sue the engineer for the loss of the building and all the incidental damages, any passers-by injured or dead to be compensated and may be that all the wealth/properties/assets of the engineer be attached if found guilty/negligent to duty.

Where the engineer is in the employment of the employer the employer shall be responsible for all the actions of the agent and the servants.

In case the resident engineer issues any instruction to the contractor and the contractor considers that the instruction constitutes a variation on the higher side as extras and the instruction is silent for that extra, it is upto-the contractor to carry out or not to carry out the instruction or inform the engineer and the resident engineer about the constitution of variation as extra in the referred instruction and should await for the response from the resident engineer or the engineer as it may be in writing.

67. The resident engineer has no powers to sanction extras other than those which are unavoidable due to site conditions and for which also resident engineer shall seek permission of the engineer and adjustable against the prime sum provision. It is also said that the engineer and thus his resident engineer acts as an arbitrator between the employer and the construction contractor and applicable in so far as the contract conditions permit other wise the unsolved disputes shall be referred to the arbitration by the appointment of an arbitrator on the consent of both the parties to the contract per the Indian law/National law of the arbitration or the related country where the work is carried out.

68. The resident engineer should make impartial decisions and act accordingly and shall not accept the opinions of the employer or the contractor but listen to them and form his own options based on the suggestions. Exactly like the engineer, the resident engineer is appointed for his special skills in engineering and the knowledge which he shall be bound to apply to the matter out before him and all the decisions must be based on that proficiency and consequently, the decisions taken by the engineer in the absence for the reference to a clause to appoint an arbitrator clause shall be binding on the employer and the construction contractor and be considered final.

The employer hires an engineer for the services of making his building and in so far as the engineer acts honestly and with ethics of the profession and apply his least expected skill with the normal competence, the employer must accept the decisions made by him.

69. A resident engineer has to keep in mind or in his view that no contract exists between the construction contractor and the engineer but the employer and the construction contractor and the works have to be carried out to the entire satisfaction of the engineer and the construction contractor has to the execution of the contract of construction so far engineer does not act or has not acted fraudulently/deceitfully/in collusion with the employer/beyond the warranted authority or under suspicion.

70. A resident engineer should not misrepresent knowingly, without belief in truth or carelessly and he should exercise the same virtues as engineer in supervision or supervising, measuring works, extra works, preparation of post contract drawings, estimates, variation orders and within the terms of agreement between the employer and the engineer. A resident engineer must note the requirement of the employer so that the engineer be guided to conform with the regulations in case the employer being a statutory body and take care not to accept favours from the contractors and let the matter slip through his hands.

Failing to remind the engineer for the operation of the penalty clause in case of the delay or any other reason and/or accepted qualifications conforming to the conditions where the clause shall be and is operative to that imposition, final certificate is issued or recommended, shall count to a negligence on the part of the engineer for not deducting the penalty or any other money from the valuation payments.

71. The resident engineer should not accept sub-standard materials/workmanship unless engineer signals the clearance for that. The engineer has to present to the employer or the user or the owner a completed contract in all respects and manners.

72. Resident engineer must question variation at the time of proposal of alternatives by the construction contractor and shall avoid the misunderstanding on the part of the construction contractor that he expects extra payments. In case the construction contractor is not notified about the defects it shall be understood that the work in general is satisfactory as far as the workmanship standards are concerned.

73. In case the employer takes over a part of work before the complete practical completion of the

contract as a whole, the resident engineer has to check the amount of the deterioration due to the wear and tear of the taken over works. The construction contractor shall not be supposed to maintain the part of the handed over works for the additional period as the final completion shall be given in one go.

74. Resident engineer must be careful in forwarding the estimates of the additional works to the employer directly or through the engineer as the responsibility for taking care shall go to the engineer ultimately.

75. Resident engineer should refer matters on which he does not possess the special skills to the engineer or his specialists and await the instructions and the comments. He shall see that his assistants are fully instructed all the times and every junior engineer draws as much experience from the job as possible to enhance his learning withstanding errors by human but should assist him, but should the assistant reiterate the mistakes then the resident engineer shall take the necessary actions.

Contractual/Legal aspect

1. Resident engineer is not a competent engineer as far as the management contract is concerned but an agent of the engineer for the engineer and cannot exercise his judgements but to exercise the supervision control for the job designed by the engineer.

2. The resident engineer cannot deviate the drawings, act as an arbitrator, sanction extras, and act as final judge in disputes regarding the workmanship and his signatures are significant to the engineer only.

Practically, on site and per general conditions of the contracts, for the site operations all the responsibilities of the control are fully delegated to him. The resident engineer is treated as the chief person on site and without him nothing can move and multitude of the problems are solved by him but all on the responsibilities of the engineer. Does it not look ridiculous as the post carries so much attraction for the engineers in the field but the values in real terms is nil.

(a) *Architectural engineering consultants* : Resident engineer on big jobs must have architectural engineers with him to check the works related to that discipline on site whether conforming to the design standards and no aberrations take place. The architect has to make certain the workmanship of various finishing items per specifications being carried out and to suggest or advise the resident engineer on various aspects of alterations in the finishes in case the engineer opts for making savings. Architect has to issue the post architectural instructions through the project coordinator and the resident engineer. The architect may be delegated to work as project coordinator for a particular section contractor on the job in which case the architect shall dispose all the duties as resident engineer for that particular phase of the job and report to the resident engineer, The architect shall supervise the works frequently and collect all the informations during the inspections and from the clerks of works and for the disputes discuss with the resident engineer for settlement. In case of the absence of the structural details and required urgently, the project coordinator shall request the structural engineer/consultant for the same and dispose the details to the contractor through the resident engineer with a covering memo. Similarly, all related details as required during the execution shall be provided to the construction contractor in the post contract times for the purpose of the operations as missed during the design stage by the engineer but are not susceptible to claims.

(b) *Structural engineer/consultant* : Resident engineer on big jobs must have structural consultants with him to assist him in supervising the operations of that discipline and make available any structural clarification/details for the additional works/other details required on site/to scrutinise any proposal set forthwith by the contractor/agent. The structural engineer shall check all the materials from the structural viewpoint. In case of being delegated the responsibility as project coordinator also for a particular contract on the job, shall dispose all the responsibilities of the resident engineer and report to the resident engineer for the advise and taking the instructions.

(c) *Civil works engineer/consultant* : Resident engineer on a big project of complexes shall have with him a civil works engineer looking after the infrastructural discipline as well as the road and surfacing operations on the job and also be the project coordinator for a particular contract on the same job. He shall see into all the related items and act on behalf of the resident engineer for the contract looked after by him and report to the resident engineer for the advise and taking the instructions whenever required.

(d) *Mechanical engineer/consultant*: Mechanical engineer is required to look into the disciplines related to the mechanical engineering for the assistance of the resident engineer and shall check all the machines', and equipments' submissions alongwith the system and all the drawing submittals for work. The mechanical engineer shall recommend all the submittals for the approvals and supervise all the specialised operations on the site in person as the clerk of works of civil engg. may not be having the expertise for the same and shall issue the details as and when requested by the contractor or the resident engineer or the project coordinator and test all the systems of his discipline per the contract clause.

(e) *Electrical engineering consultant*: A resident engineer shall have electrical consultants with him to look after the electrical disciplines, submittals as applicable on contract, supervise, issue the instruction to the contractor through the resident engineer, test, inspect and all other duties.

(f) *Plumbing engineering consultant*: A resident engineer shall have a plumbing engineer to assist the resident engineer depending on the size of the job and check all the submittals and test and instruct as applicable.

(g) *Quantity engineering consultant*: Every resident engineer shall have quantity engineering consultant with him to investigate the valuations proposed by the construction contractors for the works carried out on the contracts every month, to check the extras, to advise for the variation on costs to the resident engineer, to make estimates for the additional works required, to check all the analysis of the rates proposed by the construction contractors or the agents for the items claimed as extras on contract, to operate the contract clause dealing with the extension of times, imposition of the penalties, to advise the resident engineer on all the contract conditions, to measure the quantities on site, to check the materials delivered by the contractor on site, to examine the invoices and the dayworks presented to him by the contractor's agent/quantity engineer, etc., but not limited to that.

Experienced quantity engineers or surveyors are found to be very competent and usually designated as the contracts engineer or the surveyor/claims surveyor and the complete knowledge of the contracts shall be a great help to quantity engineer, shall also assist the resident engineer and the engineer.

Quantity engineer has to look after the costing and examination of the financial aspects related to all the disciplines.

(h) *Office staff*: A resident engineer needs sufficient staff to assist him in filing records, typing dictation accounts maintenance, and for the drinks to be served to the staff and the visitors, for refreshments.

(i) *Inspectors/clerks of works*: The task of continuously supervising the works is done by the clerks of works or the inspectors for the resident engineer. In fact generally, the inspectors are well experienced persons in the trade and stand at the same level as the full fledged foreman of the construction contractor's organisation. The clerk of works must possess the skill of the building works and advise the resident engineer or the coordinator whether something is done wrong, why it is wrong and what way that is to be corrected. The COW must be able to judge the quality, workmanship and the finishes of the works and advise the resident engineer/consultants whether the same conforms with the specifications and the usual trade practice/conventions and to the contract requirements. COWs are generally aged persons and it is really tough for the contractor to aberrate his instructions and should the best inspectors the resident engineer have, a lot of site headaches shall be taken over by him for supervision control. Young COWs or the inspectors on the other hand are found problematic to match the standards as the degree of flexibility is low and practical experience lacks. The guy may not be enough tactful as the job needs. He may quick think of a better way but not be able to get it done his way and the conflicts generate. Good inspectors are rare and resident engineer should backup good inspectors as they have a very difficult task to perform or undertake and should their instructions be countermanded by the seniors or the resident engineer, the authority of the inspector/COW shall be destroyed over the contractor.

A bad inspector may upset the good working relation between the agent and the resident engineer and consequently, the resident engineer shall have a difficult problem in hand. That inspector must be tried to move from one section to another till he finally settles or adjusts or otherwise the services be withdrawn from the site.

An inspector is supposed to go around the site frequently and every part of the site must be inspected by him under his charge daily and the activities supervised at close by him and the comments be made in the daily dairy.

Inspector shall report of all the machines working/idle, workpersons present on site, materials delivered, activities going on and completed and all the report in general. This report should be given to the resident engineer or the coordinator every day to apprise him of the works conditions.

Whenever required, the resident engineer shall arrange for the overtime supervision of the operations by the deployment of the inspector.

Frequently, in the hot seasons of high day temperatures the concrete pouring activities are undertaken either in the early morning hours or at the night hours and the resident engineer must deploy the required supervising staff for that and the COW is a choice. Practically, all the operations for the preparatory works for the concrete pours are checked by the COWs on behalf of the resident engineer and for the assistance of the specialists engineers. Preparatory works shall include the level controls, form erection, placement of steel reinforcements, electrical conduits and the routes, blockouts for services and all miscellaneous. The final approval of the pour cannot be given by the inspector as it is a structural matter and the authenticity lies with the structural consultant only. The indoor duty of the inspector is very limited and most of the times he remains on the sites. The inspector shall not be included in the site meetings. Inspector is supposed to follow the instructions of the resident engineer, the coordinator and the consulting engineers from time to time. An inspector deals on site directly with the foreman, gangers, assistant engineers, site engineer but he should not officially pass any instruction to any of them but advice as far as the deviation for the job is concerned but for the supervision of the works per drawings the inspector has full authority to act. He should not interfere the way the site engineer or the agent manage the programme but ask for the programmes to ascertain his own timings. A good inspector may help run the job smoothly and a relief to the resident engineer and others.

Working Relations

(a) The resident engineer and the contractor's agent must have good and trustful working relations. In the beginning both are unknown to each other and want to know each other. A contractor's agent shall like to know the general characteristics and behaviour of the resident engineer towards the job and the same way the resident engineer shall like to know about the agent whether the guy is technically intelligent and practically dynamic. It should be kept in the view that both persons are professional members and have to protect their clients and the interest of the clients within the professional limits/codes of conduct and the guidance. Once the resident engineer realises that the trust can be placed on the agent and the agent realises that the resident engineer is reasonable enough and does not harass unnecessarily and is a practical engineer then the job has a good luck and the contractor shall be lucky for the execution of the job.

Agent will like to receive all the instructions before time of the work is carried out and not after the start of the work or in the mid or when the activity is completed. Agent shall never expect the resident engineer to interfere in the contractor's internal administration and control measures and in the control of the sub contractor or staff without the permission of the agent. Similarly, the resident engineer or his staff shall not discuss about the job with the foreman or the ganger.

(b) Agent expects the resident engineer to explain to him about any complaint before channelising so that the agent could take care of corrections if the aberrations had taken place. No agent prefers a resident engineer who is always rigid, always limiting to the specifications and the procedures on every small matter and does not bother about the practical aspects of the constructions. Both the top engineers of the respective parties to the contract should make one thing very clear that they are working for a common goal and the resident engineer should adjudicate in the way right and fair to both on the technical skill and knowledge acquired from the profession.

(c) A resident engineer shall observe the performance of the contract in the early stages of the contract during the submissions of the materials and the shop and the working drawings which shall naturally be carried out by the agent. The agent shall also be judged by the resident engineer the way the mobilisation of the site is carried out and the site excavation is carried out, liaison is done with the statutory bodies and the contract construction programme is submitted in due time as specified in the contract conditions. This is very important period of at least two months on site from the contract commencement for an agent to demonstrate his capabilities of administration and consolidating the position which shall impart positively on the contractor's status.

Once the resident engineer realises that the agent is the right appointment done by the contractor, he may relax a little and the contractor too as the head man available on site is reliable and the right man for the right job.

(d) In case the agent does not come upto the standards or the expectations, it shall be a matter to worry the resident engineer and the contractor. Every resident engineer wants the agent to keep him informed of developments through the daily job records, weekly programmes, replies of the site memos, and progress reports and similarly, the agent expects the resident engineer to be prompt to provide the informations and arrange for the inspections in the requested times. Both should be professionally sound to push the work. Should an agent be an experienced and very well technically qualified, he can train a new resident engineer or the coordinator how the job be done.

Resident engineer should avoid commenting on the agent as incompetent or fraudulent on contractor or the inefficient as this shall help none on the job.

(e) In case the disputes arise regarding the works rejected and the contractor does not want to take measures for the rectification, the resident engineer should ask the agent to go to site with him and on the site the resident engineer should not speak/point out but let the agent inspect himself and comment whether the rejection is OK or not OK. If the agent himself says that workmanship is not to the mark and promises to take the remedial measures or if asks the resident engineer what is not to the mark or what is done wrong, then the resident engineer shall explain to the agent about the mistakes and asks whether the agent is ready for the remedial works. If the agent does not agree or agitate the resident engineer shall not be excited but just conclude that please think over and we shall take up the matter some other time. May be the agent propose some other measure please discuss that and try to convince the agent or be convinced what the agent proposes and let him carry out to the contractor's risk. Failing all the above, the resident engineer shall have no choice left but to issue a remedial note for taking up the correction of the wrong items. The matter shall also be taken to the engineer in case of the disputes but the resident engineer should remember that every day the problems being tabled on the engineer's desk shall annoy the engineer a great deal and make him consider the immaturity of the both engineers of the contractor and the resident engineer, therefore, the best thing is that the minor disputes be solved on site as far as possible.

(f) Resident engineer should not immediately react in case he finds the agent making frequent mistakes or be incompetent. He should let him make the mess of the site and take note of the losses the agent made to the contractor and then ask the contractor confidentially that this is the efficiency of your head on site but this he should intimate much before to the agent that mistakes be avoided. No contractor shall like the earning of so many losses frequently due to his staff and replace by more competent person. The other side of the picture should also be realised by the resident engineer that the agent be facing a lot of problems and not getting the support and the backings from the contractor and the site staff and also nobody appreciates the efforts made by the agent, and also the resident engineer should not all the time open the specification books for showing to the agent for the reference as the agent also knows the same but it may amount to the insulting of the agent and depreciates his honour.

(g) In fact resident engineer should help the agent as much as he can do because he is the person to pull the job ahead and on schedule. He is the person to anticipate the problems to be faced and the way to find solve the problems. The agent shall not resent to meet resident engineer as and when required to discuss and solve the problems and to adhere to the specifications for certain items of utmost importance but shall expect to be relaxed on some items of low importance. So the resident engineer should interact with the agent in a way neither too lenient nor too harsh.

(h) Other than the resident engineer's and the agent's interactions and cooperations the agent has to coordinate and interact with the other specialists and the consulting engineers, architect, structural engineer, civil works engineer, the services engineers and the COW. For the pursuance of the approval of the submittals, proposals and various other types of coordinations as desired according to the duties of the agent and to be accountable to the decision-making and implementations on the execution on site.

(i) Sometimes, the agent is so much personalised that the resident engineer and the consultants have so much trust in him that even the contractor is of no reliability and the presence of the agent and taking the responsibility shall help solve the problem. His informations make the resident engineer dependent in long run and his absence for a long time during vacation or his having gone for good is felt by the resident engineer and staff. In fact good agents having experience and dynamism keep resident engineer and the consultants busy by making proposals and finding out the discrepancies in the drawings and requesting for the clarifications and instructions and the post contract details.

(j) On various occasions the resident engineer and the consultants shall find the visitors mostly from the contractor's side with cheerful mood and with gifts to greet on the day but it is against the professional conduct to accept the gifts as it shall amount to acceptance of bribe indirectly for any future favour and unfair treatment towards the owner as having accepted any/some gift or gratification, resident engineer may not be impartial in his judgements on the job and may loose the trust. Once the trust is lost, the resident engineer shall be suspected all the times. It is also not civil to refuse the courtesy calls all the time but on special occasions and should be celebrated within limits.

Say a successful triumph or victory on the job, simple function of some kind such as opening of a site by the contractor, celebrations of some personnel's personal occasion such as child's birth day, marriage, resident's staff birth day or other ceremony, or the farewell party when someone departs from the job for good or the promotion of some one, etc., and shall depend on the common understanding of the resident engineer and the associated staff. If some of the contractor's site senior personal seeks accompany someone for a personal drink, or invites the guy to his home on the personal level shall also not amount to the bribery. Actually, these actions do promote cooperation on site in the interest of the job.

The records to be kept in resident engineer's office

1. The site memos. issued to the contractor/agent.
2. The site instructions issued to the contractor/agent.
3. The daily job record from the contractor shall be recorded.
4. The weekly programme as received from the contractor's agent.
5. The material submittals, approvals and recommendation and disposal to engineer.
6. The shop drawings submittals, approvals, recommendations and disposal to engineer.
7. The engineer's instructions prepared and disposed to engineer for approval.
8. The minutes of the coordination and the technical meetings.
9. The minutes of the progress review monthly meetings/contract meetings.
10. Abstracts of the progress from all the contractors and disposal to the engineer.
11. Intimation to the contractors for various aspects on the contract clause operations.
12. Returned submittals to contractors after approvals.
13. Records of various tests and inspections made by the consultants.

14. Internal communications with the consultants.
15. Communication with the engineer.
16. Communications with the project design consultants.
17. Special records of the detailed designs.
18. Record of the materials delivered to site and samplings.
19. Measurement records and valuations.
20. Variation orders prepared, under scrutiny and passed.
21. Daywork records of the contractors.
22. General coordinations among the contractors' records.
23. Correspondence with the nominated sub contractors.
24. Records of all the consultants personal files and staff records.
25. Communication with the head office.
26. Communication with the planning authority.
27. Communication with the statutory bodies.
28. Communication with the employer/owner/user.
29. Communication with the contractor's head office.
30. Communication with the contractor's agent.
31. Communications with the special advisor.
32. Communication with the special supplier's and contractors.
33. Miscellaneous communications.
34. Records as confidential files.
35. Drawings registers.

Financial files

1. Current claims for the main contractor.
2. Day works current claims.
3. Claims passed.
4. Day works and extras passed.
5. Engineer's certificates and the correspondences.
6. Variation orders passed.
7. Variation orders pending or in the process of scrutiny.
8. Contractor's invoices and claims.
9. Additional works expenditures.
10. Misc. files.

Supervising records files

1. Inspector's daily returns.
2. Site diary and the weather records.
3. Resident's engineer diary.
4. Weekly and monthly reports.
5. Instructions to the contractors.
6. Sketches issued to contractors.
7. Remedial notes and memos for actions.
8. Complaints.

1. *Site memo* : It is a regular correspondence made by the resident engineer regarding all the matters occurring on site under contract, may it be site work, may it be submittals may it be

mobilisation and inspections or quality, may it be material delivery, etc.

2. *Site instructions :* Instructions particularly given for an addition, alteration, amendment, etc., within contract condition and no cost variation as an extra. It can be treated invalid if not explicit and the agent may ask for a cost variation order before proceeding for the execution of the instructions.

3. *Daily job record :* Agent submits every day the complete activities and the progress achieved on site, workforce available, machines available, material delivered, and specials if any, all to the information to the resident engineer and gets remitted back as approved.

4. *Weekly programme :* Agent sends a weekly programme on the last day of the week to the resident engineer for his informations regarding the intentions of the agent for the proposed activities.

5. *Submittals :* Agent makes the material, shop drawings submittals for the formal approval of the engineer and after the approval that is sent back to the agent.

6. *Included in the 5th*

7. *Engineer's instructions :* The real instructions come from the engineer's office regarding any additional work, alteration, omission, etc., constituting the variations and the resident engineer has to prepare the same for the disposal to the engineer's office for approval.

8. *The coordination and the technical meeting :* This meting is held monthly and forms part of the contract site meeting and complete details of this meeting are the technical and the coordinations of all the building disciplines including the allied services. The minutes are prepared by the resident engineer and all the matters regarding the submittals, deliveries, progress, technical and non-technical disputes, financial matters and correspondence, etc., are covered under the chairmanship of the resident engineer with the assistance of the project coordinators, and circulation of the minutes is done to all related.

9. *Contract site meetings :* This meeting is also held monthly and should be chaired by the engineer but generally, conducted by the resident engineer and the presents are coordinator, consulting engineers, contractor's agent, coordinator or the site engineers , quantity engineers, the owner's agent given privilege for the attendance. Agenda includes the complete progress review, different statuses, approvals, site problems, quality positions, and all other aspects related to the contract execution on site and recorded by the resident/coordinator and circulation of the minutes to all the related/relevant parties.

10. *Abstract progress review :* The resident engineer reviews all the progress reports as presented by the agent or the contractors and summarises into an overall progress on the job as a whole and sends to the engineer with his comments alongwith the separate progress review of the different phases of the job to apprise the engineer and the owner for the practical progress status of the job and to take measures, if found necessary.

11. *Intimation to contractors :* The resident engineer has to intimate to the agent from time to time should the agent fail to operate the different clauses of the contract and prevent adverse effect on the job.

12. *Returning submittals :* Any submittals either rejected or approved has to go back to the contractor promptly.

13. *Tests and inspection records :* All the tests and inspections carried out by the inspectors, agents, consultants, etc., are to be recorded in the formats of the tests reports from labs or from site.

14. *Internal communications :* Resident engineer has to communicate to all the consultants regarding all the important matters and hold meetings from time to time. Similarly, the consultants communicate with the resident engineer to clarify, advise, and seek instructions.

15. *Communication with engineer :* Resident engineer has to write a lot to the engineer regarding developments on site and what actions and instructions are necessary. This shall involve both technical and non technical matters as without getting clearances from the engineer, the resident engineer cannot issue the instruction to a contractor.

16. *Communication with the project designer :* Resident engineer may write to the designer for a particular problem directly or through the engineer depending upon the problem severity. Generally, clarification and design informations are required while to incorporate changes with.

17. *Special records of the detailed design :* If feasible, the resident engineer should keep the special

design records of the project to enable him to refer while the necessity arises to save time.

18. *Material delivered to site and sampling :* The resident engineer has to keep all the records of the materials delivered to site and get all the samples from the agent to match the material to the approved sample.

19. *Measurement records :* All the measurements taken on site should be recorded separately for reference.

20. *Variation order :* All the copies of the variation orders passed, under preparation, etc., should be in a comprehensive file.

21. *Day work records :* All the records of the dayworks as presented by the contractors and verified by the COW should be filed for future references on claims.

22. *General coordinations among contractors :* All the main contractors coordinate frequently through the resident engineer or directly whatever be the case, the records submitted by them should be recorded/filed properly.

23. *Correspondences with the nominated sub- contractors :* This is a special case where selected contractor works under main contractor. This record should be filed properly.

24. *Records of all consultants and staff :* The approved CVs, and track records of all the consultants and staff should be filed separately and complete details enclosed.

25. *Communication with the head office :* Resident engineer has to correspond with the head office from time to time for intimating the contract status and progress, etc., to apprise the head.

26. *Communication with the planning authorities :* This is not done very frequently but whatever be the contents be recorded properly.

27. *Communication with the statutory bodies :* All the correspondences made to the government bodies or organisations regarding the commencement of the contract, record drawings or the execution of the services, handing over, diversions, etc., should be recorded properly.

28. *Communication with the employer :* This may be sometimes done and proper records made subject to the engineer being advised.

29. *Contractor's head office :* Generally, matter related to the contract shall be communicated to the head office of the contractor from engineer's office to be recorded properly.

30. *Contractor's agent :* All communications from and to him be recorded properly.

31. *Communication with the specialist advisor :* Resident engineer may directly receive the letters from the specialist advisors or send to them through the engineer or else, all should be recorded properly.

32. *Communication with the specialist supplier :* Sometimes, a consultant may like to verify some technicalities of the material to be supplied by him. All this types of correspondences should be recorded.

33. *Miscellaneous :* The communications not included in the said records should go into this record.

34. *Confidential file :* Shall depend on the resident engineer how to utilise this record.

35. *Drawing register :* Very important documents for the registration of all the drawings issued to the contractors from the beginning to the completion of the contract.

Financial files as mentioned are all maintained by the quantity engineer/surveyor section and called for whenever required by the resident engineer or engineer.

Supervising record — The description of all the mentioned records given in the previous explanations and not worth mentioning again and, of course, self explanatory.

Formats for the following standard records have been given in the proceeding pages.

1. Engineer's Instruction. 2. Minutes of the technical and coordination meeting. 3. Minutes of the contract site meeting. 4. Abstract of project progress report. 5. Remedial note. 6. Confidential letter from engineer to contractor. 7. Variation Order. 8. Snagging list. 9. Day works. 10. Progress charts/graphs. 11. Various correspondences for the completions' issue between engineer and contractor. 12. Site Clarification. 13 General Inspection request. 14. Concrete pour inspection request. 15. Daily Job Record. 16. Accident Report.

EMPLOYER/OWNER/USER

ENGINEER'S INSTRUCTION NO. ....

Ref ....   Date ....

M/s Contractor

....

Job reference/contract ref. ....

Sir,

You are instructed to carry out the following works as included in the site instructions no.........
dt......dt......dt......dt......dt......

1. Building A   Omit item ....   Add item ....
2. Building B   Add RCC   Delete PCC
3. Building C   Alter the item and area in drg. to the following specifications.

Please submit your variations immediately.

Signature   Signature
Engineer   Resident engineer

CC — Owner, resident engineer, contractor, quantity engineer for adjustments, file

The above format clearly exhibits that engineer's instruction can be issued jointly for the contents of a number of site instructions but the site instruction shall be issued separately for every instruction and with the reservation of the contractor's will/wish to do the job in case that instruction constitutes a cost variation but in the engineer's instruction the contractor has been left with no choice but to carry out per the contract conditions. The contractor shall propose the claims on the basis of the engineer's instruction only.

The site instruction shall be issued immediately for the immediate execution of the job and the contents shall be similar but the engineer's instruction shall take a considerable time for being released by the engineer but generally the clearance for the variations are given on the basis of the site instruction to avoid the wastage of time.

RESIDENT ENGINEER/CONSULTANTS/DEPARTMENT

Reference............   Date........

Minutes of the meeting

Name of the work   Budget........
Contractor   Contract reference.........
                                                         Venue.........

Sub : **Coordination and the technical meeting number**
Apologies —

Present	Name	Level	Representing
1.		Resident engineer	Engineer
2.		Coordinator	Res./Engr/consultants
3.		Mechanical engineer	do
4.		Electrical engineer	do
5.		Plumbing engineer	do
6.		Contractor's agent	Contractor
7.		Coordinator	do
8.		Sr. site engineer	do
9.		Services engineer	do.
10.		Electrical engineer	Sub contractor
11.		Mechanical engineer	do
12.		Plumbing engineer	do
13.		Special services engr.	Special sub contr.

Apologies as shown above may be taken in the description of the minutes or as above

Distribution
1. All present
2. All the sub contractors through the main contractor
3. Main contractor
4. Engineer
5. Owner for informations
6. Quantity engineer of resident engineer
7. Inspector
8. File

No.	Description	Action
1.	Apology — Last minutes to be agreed **Materials** — Details of the material submittals	

Discipline	Total No.	Submitted	Approved	Balance	Remark
Plumbing					
Electrical					
Mechanical					

Details of the major materials to be given as mentioned in the internal coordination minutes of the technical and coordination meeting conducted by the agent.

2. **Drawings** — Details of the working drawings submittals

Discipline	Total No.	Submitted	Approved	Balance	Remark
Plumbing					
Electrical					
Mechanical					

Details of the major drawings to be given as mentioned in the internal coordination minutes of the technical and the coordination meeting conducted by the agent.

3. **Progress** —

It includes the programme also. Discipline-wise details to be given.

Discipline	Programmed progress	Actual progress	Ahead/behind	Remark
Electrical				
Mechanical				
Plumbing				

Reasons for the delay

Electrical —

Mechanical —

Plumbing —

Remedial measures —

Electrical —

Mechanical —

Plumbing —

4. **Actions** on the **site memos** and the **instructions**

Give complete references to the memos and the instructions and the actions taken on them by the related people to be in action/included in action

5. **Technical** and **coordination**

In this take all the points related to the electrical, mechanical, plumbing and any other coordination necessary one by one for the actions and the decisions. This portion of the meeting shall form major part of the technical and coordination matters being discussed and shall generally involve all the major parties to the contract.

6. **Financial** — Discussions for extras, deletion, additions, etc.

7. **Any other business** — Discussions other than above

8. **Next meetings**

Fix the date and time for the next meeting

Resident engineer/project coordinator's signature

ENGINEER/OWNER/DEPARTMENT

Minutes of the Meeting

CONTRACT SITE MEETING NUMBER ....

Ref.		Contract ref.		Date

Name of work

Present	Name		Level	Representing
1.			Engineer/rest.	Owner if owner not present/engineer
2.			Project coordinator	Engineer's rep
3.			Architectural engineer	Engineer's rep
4.			Structural engineer (full or partly)	Engineer's rep
5.			Other consultants	Engineer's rep
6.			Contractor's agent	Contractor
7.			Contractor's coordinator	do
8.			Sr. site engineer	do
9.			Quantity engineer	do

Distribution —

Engineer, owner, contractor, those present, staff for contract, file

No.	Description	Action
	Apology, if any —	
	Agreement on last meeting's minutes —	
1.	**Progress**	
	Record in this portion the percentages of the various activities as agreed with the contractor's agent during the meeting or a little before the start. This shall be a long sheet as to cover all the items according programmes. All the statuses of the activities shall also be recorded as **ahead/delays**.	

Programme	Activity	Programmed %	Progress % this month	Progress % last month	Ahead/ behind in weeks
Building A					
Building B					
Building C					
Building D					

and continued till all the programmes are covered up in the format.

Abstract of the **progress** versus **programme**

Programme	Programmed progress %	Actual progress %	Ahead or behind weeks
A	20%	18%	–2 weeks
B	5%	10%	+2 weeks
C	–	5%	+2 weeks
D	–	–	nil
E	5%	10%	+1 week

and continued till all the buildings or programmes completed

Overall — The contract is... % complete against ...% programmed revealing a lag of... weeks ahead/behind

Remedial measures — Record the remedial measure in case of the progress being lagging behind to the recovery per programme as suggested or agreed by the contractor or the agent and accepted by the resident engineer.

Services brief

Just incorporate the major abstract of the technical and coordination meeting held last as the same shall form a part of this meeting for the submittals and progress.

2. **Materials**

   In this portion all the details of the various building materials approved, pending for approval, rejected, delivered, ETA, and the delivery problems, alternatives proposed, all per the material status submitted by the contractor shall be included.

3. **Workpersons/Work force** — From the report of the agent, record the number of work people on site which shall tally with the resident engineer's own record.

4. **Machines** — Record all as given in the agent's report duly ratified.

5. **Contract matters** — All the matters covered in the contract for the purpose of the operation on the contractor's part.

6. **Finance**

   In this part are covered the valuation details of the jobs, the extras, additions, non-payments problems dayworks, and all remaining issues shall be recorded.

7. **Quality** — Record the consultants' and the specialists' reports on quality and the contractor's responses on that and the actions to be taken.

8. **Structural** — All problems related to the structure.

9. **Architectural** — All related matters

10. **Allied services** — All related matters

11. **Information required**

12. **Clarifications** and the **last meeting**

13. **Any other business** not covered

Next meeting —   Date   Time

Signature resident engineer/Engineer   Co-ordinator

It is not necessary that engineer shall sign this minutes for the purpose of circulation, resident engineer and the project coordinator would do.

The abstract of the **progress for the complete job** to be compiled by the resident engineer for the attention of the engineer and the owner.

Name of the work —
Budget references ....   Date of start ....

No.	Contractor	Contract	Contract duration Days	Time elapsed Days	Programmed %	Progress %	Ahead/ Behind	Remark
1.	M/s V	Phase I	500	250	35	29	–6 weeks	
2.	W	II	400	180	25	18	–8	
3.	X	III	400	200	40	35	–4	
4.	Y	IV	300	100	20	16	–4	
5.	Z	V	450	200	40	35	–3	

and include all the phases and the related details.

Overall statement — Give the statement contract vise to let the owner and the engineer be apprised fully of the job.

Comments — Comment on the progress contract-wise.

Problems, if any, for the **attention of the engineer** and the **owner**.

Signature of resident engineer
CC — File

ENGINEER/RESIDENT ENGINEER
Remedial Note No...

Date ....   Ref. ....

Contract : ....
Please rectify the followings mistakes as noted on site

1. Building A   Column is out of plumb. Give the grid reference.
2.   B   Floor is not in correct slope. Refer grid.
3.   C   Slab developing cracks. Refer grid.
4.   D   Wall is not correctly concreted. Demolish that. Refer grid.

The said works shall be reconstructed/demolished/dismantled/be remedied as follows.

Resident engineer/engineer   Project co-ordinator

CC — Engineer/owner

Quantity engineer of resident engineer/engineer
This remedial note shall remain in force till the remedial actions are not carried out and the
payment shall not be released for the items included.

Remedial works carried out   Inspected and approved.

Contractor's agent   The remedial note no. ...... is released

Inspection sheet duly approved enclosed for reference   Resident engineer/engineer

                                                    Payments be released

CC — Owner
    Engineer
    Contractor
    Quantity engineer of the resident engineer/engineer for necessary action.

Engineer
CONFIDENTIAL LETTER

Contract : ....
M/s Contractor
....

Job : ....
Sir,
My resident engineer .... for the said job has reported that your site engineer/site
agent/any other responsible personnel .... has created a lot of nuisance on site and
deemed inexperienced to the present job requirements/responsibility and we do not want to delay
the contract as considerable time is being wasted under the supervision of the incumbent
personnel in carrying out the remedial works from time to time. It is also in your interest to save
the losses done or definitely would be done. We shall no more treat him as a competent
respnsible personnel in the level appointed by you. Please take necessary action to deploy
additional personnel in the void position as the approval of the incumbent stands invalid with
immediate effect.
Thanks
Sincerely your's
Engineer
CC — Resident engineer for information and n.a.
    Owner
    Site confidential file
    Contract personnel file

Engineer

Variation Order No...

Date ......

Job : ....

Budget ref. ....

Contract Reference ....

Amount involved in this variation order ....

Total contract cost ....

Reference to the Instruction — EI No. ....

A variation of additional/deductable cost of ..... In words........................ for the instructed works considerations is approved per condition of contract clause No. .....

The final cost of the contract after the inclusion of the cost involved in this VO comes to a value of......In words.....

Engineer

Resident engineer

CC — Owner

        Contractor

        Resident engineer

        Quantity engineer of RE

        Files.

The said cost as finalised in the variation order shall depend upon the details given by the contractor's agent/QS and verified by the resident engineer/QS and the figures arrived at an agreement by both and certified by the resident engineer shall be considered for the inclusion in the variation order.

Snagging formats

Building ....   Drawing ....

S. No.	Room	Builder work	Plumber work	Electrical work	Mechanical work	Air-condng. work	Finishes	Consultant's remark
1.	15	Completed	done	fittings incomplete	—	grilles not fixed	touch up not done	no
2.	21	OK	OK	OK	OK	OK	OK	OK
3.								
4.								
5.								
and carry on till all completed.								
Resident engineer/Project co-ordinator								

Day Works

Name of work ....

Contract and contractor ....,......

Ref. to instruction ....   Section ....

No.	Description of item	Unit/Qty	Hours	Rate	Cost
1.	Removal of the existing girder of RS joist				
	Mobile crane 30 MT with operator	No./1	6 hrs		
	Foreman assistant	No./1	6 hrs		
	Semi-skilled workmen	No./4	6 hrs		
2.	Welding the steel rods to the existing braces				
	Welder	No./1	2 hrs		
	Helper	No./1	2 hrs		
3.	Similarly enter all other details as be				

Site agent/Site engineer   COW/Co-ordinator/resident engineer

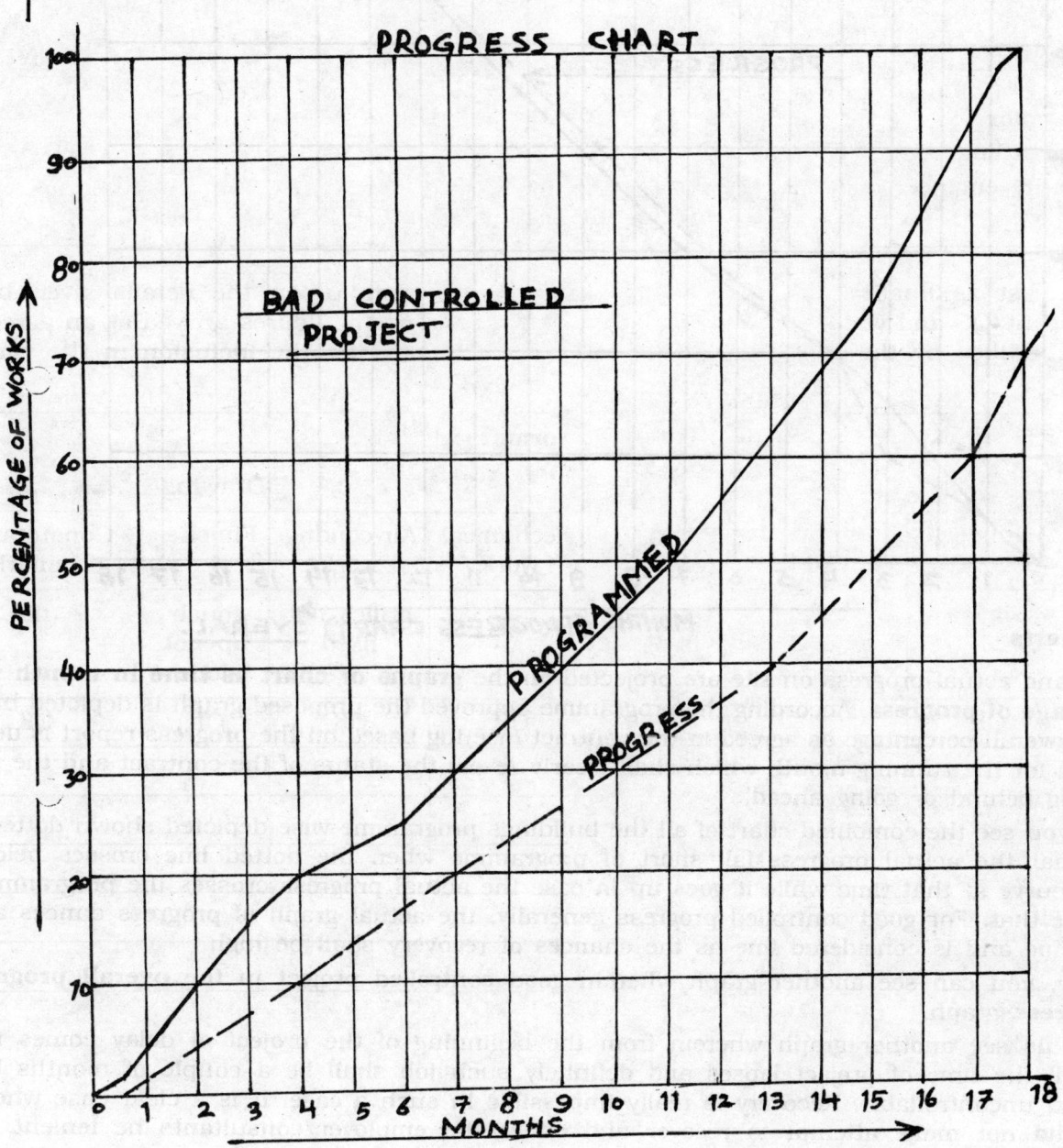

PROGRESS CHART

BAD CONTROLLED PROJECT

PERCENTAGE OF WORKS

PROGRAMMED

PROGRESS

MONTHS

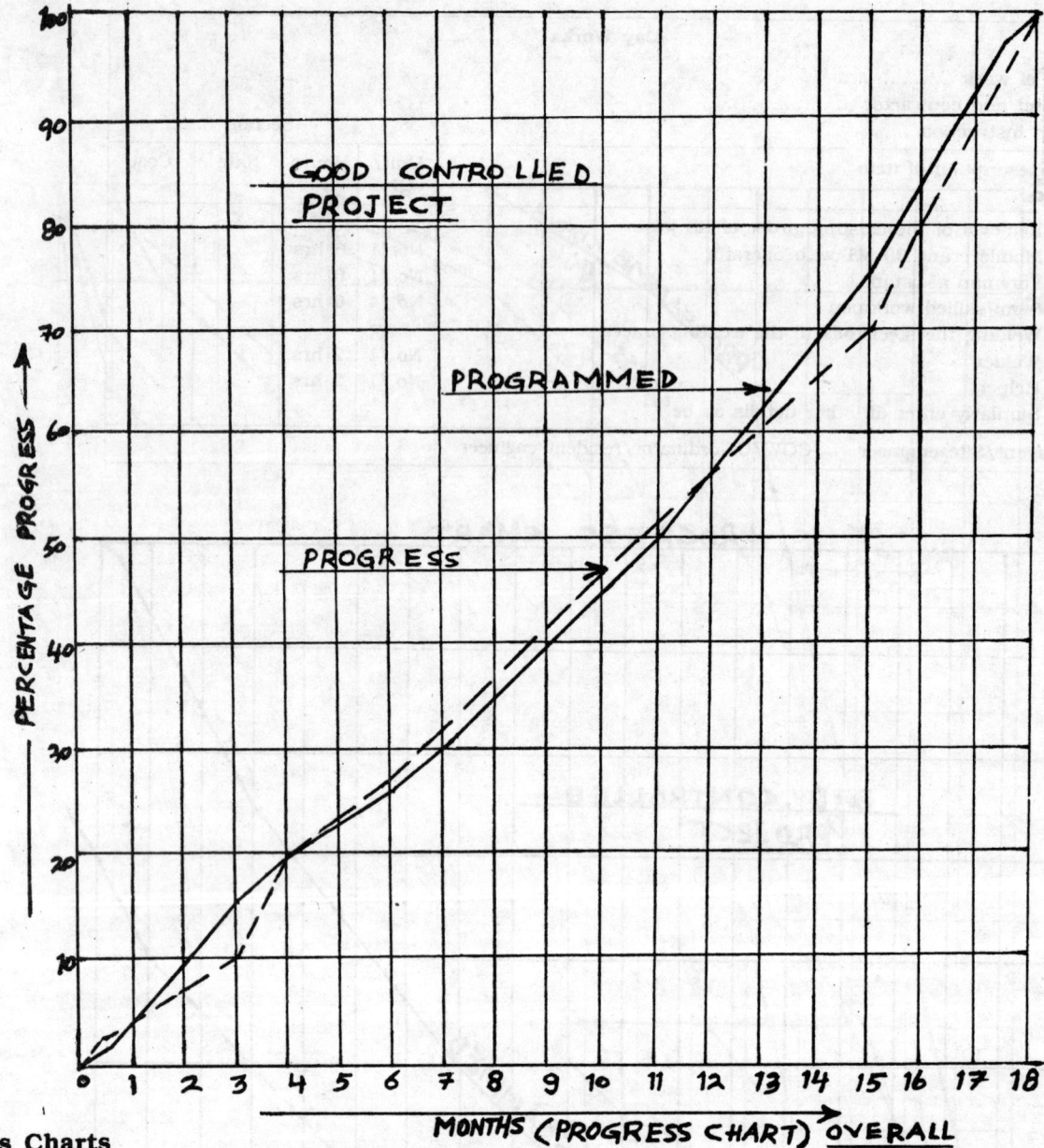

Progress Charts

Programme and actual progress on site are projected on the **graphs or chart as time in month versus the percentage of progress.** According the programme approved the proposed graph is depicted building vise and an overall percentage as agreed in the contract meeting based on the progress report is depicted on the graph for the running month which shall clearly reveal the status of the contract and the project either lagging behind or going ahead.

Should you see the combined chart of all the buildings programme-wise depicted shown dotted lines indicating that the actual progress fall short of programme when the dotted line crosses below the programme curve at that time while it goes up in case the actual progress crosses the programme and travels above that. For good controlled progress generally, the actual graph of progress dances around the programme and is considered fine as the chances of recovery shall be high.

Similarly, you can see another graph wherein good controlled project in the **overall programme** versus **progress** graph.

Now let us see another graph wherein from the beginning of the project of delay comes in and continues till the time of project lapses and definitely such job shall be a couple of months behind schedule and uncontrollable. Recovery is really impossible in such a case. It is a clear case where the contractor did not make attempt to recover and either the employer/consultants be lenient. If the employer be strict, the contractor should be squeezed by the engineer, and the employer.

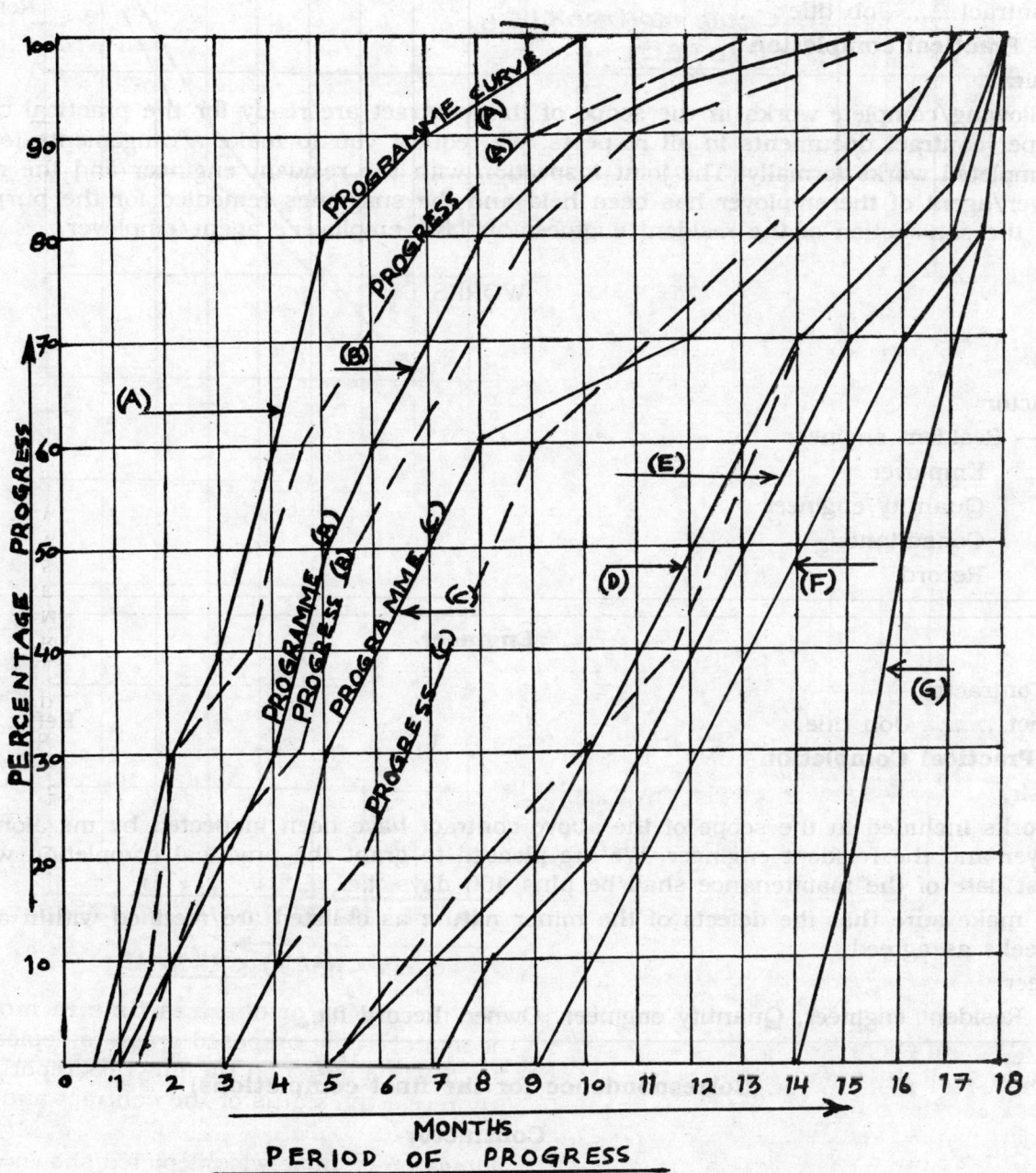

PERIOD OF PROGRESS
PROGRAMMES AND PROGRESS BUILDING VISE

(Contractor to correspond to engineer for getting practical completion)

Contractor

The engineer ....
Sub-contract........ Job title   Ref.........
Sub — **Practical completion**
Dear sir,
The following/complete works in the scope of this contract are ready for the practical completion
w.e.f...per contract documents in all respects. We request you to make arrangements to take over
the completed works formally. The joint inspection with the resident engineer and the related
employer/agent of the employer has been held and the snaggings remedied for the purpose of the
use to the satisfaction of the resident engineer and the employer's agent/employer.

WORKS

Contractor
Copy — Resident engineer
       Employer
       Quantity engineer
       Consultants
       Record.

Engineer

M/s Contractor
Contract .... Job title   Ref. ....
Sub : **Practical Completion**
Dear sir,
The works included in the scope of the above contract have been inspected by me alongwith the
employer and the resident engineer. We are pleased to grant the practical completion w.e.f.... and
the last date of the maintenance shall be plus 400 days, i.e., ...
Please make sure that the defects of the minor nature as enlisted are rectified within a period of
two weeks as agreed.
Engineer
CC — Resident engineer, Quantity engineer, Owner, Record file.

(Correspondence for the final completions)

Contractor

The engineer
Contract .... Job title   Ref. ....
Sub : **Final completion**
Dear sir,
The maintenance period of the above contract shall expire on ...Please issue the final completion
certificate as there are no construction defects as experienced during the maintenance period.
Contractor
CC — Employer, Resident engineer

Engineer

M/s Contractor
Contract .... Job title   Ref. ....
Sub : **Final completion**
Dear sir,
Your ref.
As requested by you the final inspection has been carried out and the defects in the enclosures have been given in details on the works constructed by you. Please carry out the remedial works to enable the reinspection at an earliest and let me know about the progress of the remedial works/rectification of defects to the satisfaction of the resident engineer.

Engineer
CC — Employer, Resident engineer, Quantity engineer

Contractor

The engineer
Contract .... Job title   Ref. ....
Sub : **Final completion**
Dear sir,
With your reference as above, the major defects as included in the enclosed list...are the developments due to the wear and tear but the other defects which are clear due to us have been remedied/rectified and the resident engineer has inspected and signed the site diary as enclosed. Please be kind enough to issue the final completion certificate at an earliest.

Contractor
CC — Owner, Resident engineer, Quantity engineer

Engineer

M/s Contractor
Contract .... Job title   Ref. ....
Sub : **Completion of the maintenance and the contract completion**
Dear sir,
Your reference ....
This is to certify that the said contract no......has been completed fully and in all respects per the contract documents as well as conforming the drawings and specifications and successful performance of the period of maintenance for 400 days. No penalties of any kind have been imposed on the contractor/contract as the extension of ...days had already been granted to the contractor for the delayed period. This certificate shall be treated as the completion certificate of the contract except for the items which provide guarantees for a fixed period of five years and the bank guarantee for securing the same clause has already been given by the contractor.
Wish you all the best.
Thanks.

Engineer
CC — Owner, Quantity engineer.

Engineer

Site Clarification

Contract — Job title

Contractor ....   Section ....

Date ....   Drawing/Specs ....

Here are written the contents for which the clarification is sought

Agent

Resident Engineer

Details of the clarifications are given in this space

Project co-ordinator/Resident engineer   Date

The above proforma shall be utilised purely for the site clarifications' purposes in the cases of the misunderstanding in the drawings or the BOQs and not for any claims.

General Inspection Request

Engineer

Contract ....   Discipline ....

Contractor ....   Element ....

Date ....   Section ....

Please inspect the followings :

Site engineer/Site agent   Time of Inspection ......   Date of inspection ......

Resident engineer

Inspection comments shall be entered in this space

Approved   Rejected   Approved with Comments

COW/SPECIALIST/PROJECT   CO-ORDINATOR/RESIDENT ENGINEER

Note : 24 hours shall be required in advance for this general inspection request to be received in RE office.

Engineer
(Concrete Pour Inspection Request)

Contract ....
Contractor ....
Element ....
Section ....
Position of the allied services details ....
Details of the inspections to be made ....

Date ....
Conc. Grade ....
OPC/SRC ....
Quantity ....
Means ....

Details of the check list to be inspected by the consultant

Machines for the concrete raising —
Equipments for the concrete handling —
Details of the concrete vibrators —
Details of the compressors or the generators —
Details of the work persons to be engaged —
Details of the technical staff to be deployed on the concreting supervision —
Details of the trades persons to support the pour —
Details of the curing arrangements —
Details of the curing compound/Water for the concrete curing —
Details of the tools specifically needed for concrete finishes —
Arrangement for the concrete site sample tests —
Order of the pour to be depicted in the enclosed drawing —
Details of the construction joints as agreed/to be agreed —
Timings adjustments of pour —
Concrete order placed/To be placed —
Other details —

Site Engineer	Site agent	Approved/Rejected/Comments
COW	Structural consultant	Project co-ordinator/Resident engineer

Note : 48 hours notice shall be required for the inspection of structural concrete pours.

Daily Job Record
(Main Contractor)

Contract .... Job title

Contract ref. ....

Date ....

Work persons available on site —

Technical —

Non-Tech. —

Trades persons —

General —

Total —

Materials arrived —

Materials removed —

Machines arrived —

Machines removed —

Weather : Temperatures in the morning, noon, end of the day — humidity — general

Details of the activities in run, completed, started with approximations in the quantities to be given

Details of the activities for the next day plan

Details of all the machines available on site in working order and idlings if any

Visitors to site —

Inspections held and tests, if any —

Accidents, if any —

General —

Site Agent

Recorded

Verification COW/Inspector i/c

Project co- ordinator   Resident engineer

Similarly, the agent shall receive the daily job record from the sub-contractors in the standard proforma of the sub-contractor's daily job record duly signed by the representative and enclose them with the main contractor's report proforma duly verified for the contents. This report shall be kept as a permanent record for the job in the resident engineer's office as well as contractor's office for the reference.

Accident report proforma

Contract .... Job title

Contract reference ....

Date ....   Date of accident ....

Reference to the section in contract site —

Name of the victim —

Details of the victim's employment —

Status of the injury —

Name of the site supervisor incharge —

Name of the site engineer —

Actions taken by the contractor immediately —

Details of the informations communicated to the insurance agencies —

Benefits as required per labour law by the victim/survivors in case of death —

Police report, if any, with complete details of the FIR and actions —

Any of the employee of the contractor under arrest/on bail —

Assistance sought to be rendered by the resident engineer/ Engineer for the mediation —

Future safety measures augmentation required —

Give the recommendations given by the site safety specialists —

Time by which the contractor shall arrange the measures —

Duration by which the operations shall be effected —

Other general safety measures in action on site —

Site agent

Proforma for the records of the **infrastructural chambers** which shall be installed in the **sewerage** system, **water supply** system or the **miscellaneous**.

Engineer

Contract .... Job title
Contract reference ....   Date of inspection and record ....
Reference to the discipline ....
Section ....
Standard reference, if any ....

Technical checks during inspections

Description of the chamber —
Number of connections coming in —
References to the inlets —
Reference to the chamber this connection leads to —
Reference to the chamber this connection comes from —
Sizes of the chamber in details of the shape —

External top finished level of the cover —
Invert level of the chamber —
The finished ground level of the surroundings —
The details of the benchings/other finishes —
The details of the floorings and sides finishes —
The details of the access to the base of the chamber —
Reference and the details included in the contract documents —

Contract details/variations, if any, and the reason for the same —

Comments at the time of the testing —

Comments after the remedial of the observations after the test and final clearance —

Agent   Senior specialist   Resident engineer/   Taken over/Handed over
                                   Rep. from statutory deptt./Municipality

Testing and commissioning of the **building services systems** per **contract conditions**

Engineer

Contract .... Job title
Reference ....
Sub-contractor ....
Discipline ....
Section ....
Design requirements per specifications —

Date of the commissioning for the test —
Name of the engineer's rep. present at the time of start —
Name of the contractor's agent/rep at the time of start —
Date of the last day of the test —
Details of the test specific observations during the month/test period of commissioning.

Day	Description of machine with the daily average performance and comments by the specialists	Acceptable standard Yes/No	Remark
1			
2			
3			

and carry over the table till the complete testing and commissioning is accomplished per the detailed specifications and the contract conditions.

Agent/rep. Senior services engineer/ Discipline service engineer Resident engineer/ Rep. from statutory deptt./Municipality

OPERATIONS/ACTIVITIES FOR THE CONSTRUCTION OF A SWIMMING POOL BELOW GROUND

Before proceeding for the construction make sure that all the working and the co-ordinated drawings have been approved and ready for use, Ref. General informations of Jacuzzi. 289

1. Marking/Setting out of the central line and establishing a bench mark and approval of the setting out works/details on ground by the resident engineer.
2. Excavation of the required area i/c working area approximately 1.5 m - 2.0 m around the pool area depending on the nature of the soil to be excavated, i.e., rock, soft soil, hard soil, etc.
3. Keep a control on the levels of the formations constantly and reach the nearer levels. Probably this massive earth work in excavation should have been carried out by the use of earth moving machines except the last or the termination part of the earthwork which shall usually be dressed manually according the dictation of the formation level profile.
4. Prepare the profiled surface for laying blinding concrete on compacted earth.
5. Let the resident engineer inspect the compacted base and obtain the approval.

6. Fix the elevational marks for reaching the top of the blinding concrete.

7. Blind the surface to obtain a smooth surface/topping.

8. Mark the profile of the raft and all central/grid lines so that the marks are visible as projected having the raft been poured/laid. References transferred for work feasibility.

9. Check the details as supplied in the shop/working drawings by the swimming pool supplier/installer whether the intake water pipes' system shall be laid in the portion below the raft or to be laid in the raft structure as generally, are laid in raft.

10. Should the system be laid in the raft structure, proceed for the water proofing system as specified that is "spreading the bitumat membrane layers duly overlapped and make the surface durable and let that be inspected and approved by the resident engineer and then proceed for applying the water proofing protection layer/membrane/coat, i.e., a coat of 25 mm thick cement mortar in the ratio of 1 : 3 so that the bitumat treatment system be protected during the preparatory and pouring operations of the RCC raft"

11. Let the plumbing guy start laying the water supply pipes and the fixing of the water disposal traps per the details provided in the approved working drawings or the shop drawings and positioned on the site to the approved co-ordinated drawings and the test conducted according to the specifications and the approval obtained.

12. Reinforcements to be laid for the raft must be cut and bent according the structural drawings without damaging the pipes of water supply during the placement of the rft. Care must be exercised for preventing the depositioning of the pipes either by fixing the pipes post installation of the reinforcements or before the commencement.

13. Simultaneously, the forms of the raft periphery and the wall starter be carried out and the water stops/water bars as specified in the engineer's details in the structural drawings be installed and the welding of the connections in the form of T, butting or the overlapping as feasible shall be done per the specifications and the instructions and the approval of the resident and the structural engineer obtained. The top marks of the concrete levels shall be fixed to the details as well as the fixing of the starter bars of the walls be done.

14. Let the preparatory works for the raft concrete be inspected and approved by the resident engineer for the permission of the pouring.

15. Let the concreting operations preferably from the batching plant be carried out using the grade of concrete mix as specified or generally M30 SRC/OPC as the case be and be completed in one go and the surface finished rough and the joints for taking the wall concrete be wire brushed or scabbled later and curing compound if available and permitted, spread or the slab be cured by spreading the saturated hessian rolls per the general specification of the concrete/s.

16. The operation of the reinforcing steel placement in the walls shall be started per engineer's details in the design and the completed works shall be inspected and approved stage by stage of the progress as the position of the construction joints dictate the quantity of the work in one inspection. Construction joints shall be provided per details or as directed by the resident engineer in consultation with the structural engineer. The forms of the walls shall be fixed per the positions of the joints. Make use of the common sense to repeat one shutter form plate to the complete construction of the pool shelf and balance tank. During construction you have to place in position the water bar/stop, the sump pipe in the balance tank, take down pipes to the ring main from the overflow drains, the block outs for the underwater flood lights, etc., all as detailed in the engineer's details and the working drawings as approved by practically co-ordinating in a coordinated drawing form. The concreting preparatory works are ready for the engineer's/resident's inspection and approval. The concreting shall be carried out according to the general specifications & approved. The main precaution to be taken is to avoid the displacement of the pipes during the pour for which one plumber shall have to be engaged/deployed on duty during concreting.

17 For the roof forms of the balance tank the care to be exercised is that the form be conveniently **struck after the** curing period is over.

18. Having completed the structural works duly cured, a water test is to be carried out according to the project specifications as prescribed in the specifications for the water retaining structures which may vary from 72 hours to 168 hours retain water in full capacity condition and let the leakage be noted/observed by the agent and the resident engineer alongwith the structural engineer. Should some leakage be observed, mark the spots inside and outside by the colour marker. Outside of the tank shall pose no problem for marking but for the inside, the location measurements be taken which shall be transferred after the tank is dewatered. Now the leaking spots are ready for the application of the remedial works using the epoxy mortar and the injections as the case be. The epoxy compounds must be approved prior to the application by the resident and the structural engineer. Having remedied the defects let the inspection be carried out by the resident and the structural engineer. If they are happy let the further operations be carried out or start for the second test by filling the tank shelf till no leakage state/condition is obtained.

    It is very essential to be remembered by the site engineer and the agent should the water test is not cleared in time, that may jeopardise all the following operations of the swimming pool construction, therefore, a great degree of care must be exercised by the site engineer and the agent during the concreting operations.

19. Ask the plumber/plumbing foreman to finish the ring main system and get the approval.

20. Commence the water proofing bitumat application on the walls on the external side/surfaces according to the specifications and obtain the approval.

21. Simultaneously, carry out the operation of the wall tiling underlayer mortar spreading which shall generally be thick layer. You have to fix on site the actual level of the drain top, the top of the walls, etc., to enable the tiler to understand where the first tile from the top is to be fixed. The top should never be a cut tile line as architecturally looks awkward. This operation is very slow, precise and time consuming but very important and requires constant supervision by competent person/supervisor.

22. Having the external waterproofing application been approved, proceed for the protective system on the bitumat in the form of the block/brick wall or some other treatment as specified in the engineer's details leaving the holes for the pipes installed and then after a few days backfilling can be started per specifications after clearing the space and having been inspected by the resident engineer.

23. Remember, that the finishing operations are being carried out inside the pool while the pool services operations outside the pool shelf. The installation of the sand filter on the already constructed plinth, construction of the plinths for the pumps, construction of the acid chambers, construction of the over head water tank to be installed in the filter room (generally, the OHT shall be provided on the steel frame), trenches construction for the pipes, electric cupboards, etc., should all be in progress. The sand filter plinth shall take the load as designed and the design of plinth shall be requested in case not available. The concrete shall generally be filled in the sand filter per the specification of the supplier of the filter.

24. Construction of the pump house and the plant room are of common nature as in usual buildings so the details are not included in this list.

25. The piping operation shall take considerable time so arrange accordingly.

26. Fix the prefabricated drains as supplied with holes, without holes and corners halves, made of clay finished properly. The thickness of the mortar joint should not be less than 12 mm or as specified.

27. Having completed the finishing operations, i.e., floor and wall tiling, fixing of the surface drains, flooring in the balance tank and usual application of a coat or two of acrylic bituminous emulsion paint inside the walls of the tank, piping work, installation of the equipments and plants, fixing of flood lights, wiring and connections, grouting of the space between the surface of light and the blockouts in concrete in the tank walls to be done with epoxy concrete, fixing external isolators and transformers, all the accessories such as steps ladder in the tank wall, jumping stand/diving board the placement of the PVC grilles resting on the aluminium supports

bore on the drains projections, as a whole the system shall be tested for a successful run of the pool.

28. Surroundings shall be compacted to receive the synthetic drainable green carpet spread to specifications of the manufacturer.

29. The external works shall be done according to the layout of the swimming pool which may include the construction of a compound wall with gates and certain pergola as well as some chain link fencing all per the contract documents. The swimming pool is ready for use.

30. **Specific points about tiling** Sometimes, the engineer designs the tiling work in different colours and according to the lanes system in the tanks to be used for the sports purposes. In that case the tiling work should be carried out according to those architectural details and the tiles ordering should have been done accordingly. The pattern of the tiling should not be disturbed in any case. Before taking off the finishing operations a co-ordinated drawing of the tank should be prepared in which the positions of the gratings, spacings of the water sprinklers, positions of the underwater flood lights, and all other details felt necessary be included based on the module of the available tile sizes and the symmetry be maintained in all the cases. Any cut tile required should be at the bottom of the pool. May be the drain as fixed takes two tiles on the edge and the same followed vertically and reflected in the floor as well.

31. **Electric lights** The grouting of the space behind the light shall be done with epoxy concrete only.

32. It is in the interest of the agent to make ready the coordinated drawings comprising of all the details of the tile work, sump pipes, underwater lights, positions of the water inlet and exit system, the positions of the surface drains, and all other accessories in hand duly depicting the foregoings dimensionally. This needs the agent/site engineer to make available all the informations in advance, i.e., the size of tiles, size of drains, approved working drawings of the related plumbing and drainage, and the electrical engineering disciplines, etc. This detailed item needs the concentration on the part of the site engineer/agent/resident engineer/consultants and the co-ordinated drawings finalised in time so that every one understands clearly what shall be done to achieve the goal. Once the things are explicit the operations as explained shall run very smoothly.

33. Actually, this part of the job needs the agent's personal attention and contribution as the site engineer may not find this much time or be entrusted to a sub agent if the agent is also busy but come what may all be done to facilitate the smooth functioning of the swimming pool system as a whole to the architectural and the leisurely satisfactions.

General Activities In a Building

Available

Co-ordinates and the grid plan for the layout/arrangements/general survey details
Cellar floor layout
Ground floor layout
First floor layout
Second floor layout
Additional floor layouts
Roof floor layouts
Roof layout
Details of the architectural engineering constructions
Details of the structural engineering and bar bending/reinforcement schedule/structural steel sections
Details of the allied services engineering and the working drawings

1. Set out the building per the co-ordinates and the grid plan and obtain the approval. Go ahead with the excavations with due respect using the records drawings if any till the final levels are reached either for the footings or for the pile caps. Do the complete piling works according to

the pile layouts drawing and let the piles be approved after the test and finish the piling operations per detailed codes and project specifications and found all the pile caps. In case piling is not needed, prepare the excavated surface of the footings for the formation levels and do the compaction in the complete area of the footings and the strip/plinth beams and get the resident engineer's approval and blind the compacted surface to the levels as defined in the drawings 'the top of the blinding surface'. Refer SK1 - SK11.

2. Transfer of the axii to the foundations/footings and beams/raft/pile caps as the case be and finish the foundations' operations such as spreading the pvc visqueen 1000 gauge, the reinforcement placements for the footings/raft/piles/beams, etc., and erect the usual forms and fix the marks for the concrete pour and pour the concrete per specifications and the starter bars for the columns let go in the pour.

3. Apply water proofing treatment as specified and get approved.

4. As mentioned in '2' in case pvc is not to be applied, apply the bitumat and a CM 1 : 3 protection on that before fixing the forms and the reinforcement.

5. The surface of the concrete where the next concrete has to be placed shall be wire brushed so that the complete laitance be removed/scabbling done after the concrete has set in, which is not preferred by the engineers.

6. Strike the forms after the time specified/next day/48 hours as the case be and cure.

7. Carry out the operations of the form erection in the starter and pour which shall act as a base for the super structure and the columns. (stub columns).

8. After the curing of the concrete has been over, apply the dry surface of the concrete to remain in the contact of the soil of the backfill with the waterproofing acrylic paint and bituminous treatment on that as specified in the engineer's details. All the footings, stubs, beam sides, etc., shall be treated and the side of the stub walls in the foundations over the beams.

9. Apply the block/bricks wall protection for the treatment applied or with some other alternative protection material as agreed with the resident engineer.

10. For the forms of the footings, the wall which is supposed to protect the treatment could be used as a form for the footings subject to the approval of the resident engineer and time saving be made.

11. After the protection is done, backfill shall be carried out for which the approval of the resident engineer shall be taken for the space being cleared and to make sure simultaneously, that the underground works of the plumbing and drainage as well as other parts of the allied services completed duly tested and approved.

12. Should the space exists between the finished slab level and the top of the plinth beam, to be filled with the block/brick wall duly waterproofed and protected.

13. Backfilling be carried out for the ground slab/1st slab, and compacted if required.

14. Preparatory works for the columns between the GF and FF should simultaneously be in progress when the g.f. slab operation is going on (g.f. indicates ground floor).

15. Finish the ground floor compaction and all the related concealed pipes and the conduits for all the services as found necessary per details duly approved and get ready for the blinding operation. Blind the surface to the required levels to the specified thickness and smoothen the surface to the standard.

16. Apply an isolation membrane in the form of pvc 1000 g visqueen on the blinded surface and arrange for the reinforcements in the form of prefab mesh BRC of approved make.

17. Erect the edge forms on the perimeter, and install the steel mesh approximately in the middle of the slab.

18. Install the compressed filler/flexcell all around in the profile of plan to form a movement joint.

19. Call the electrical man and let him finish the conduits laying and get approved.

20. Call the plumbing man and let him finish the sanitary blockouts and get approved.

21. Make the blockouts for the telephone PABX if any.

22. Make complete arrangements in final for the works prepared. Check all the movement joints and see that the compressed filler is in the position and the pvc visqueen 1000g goes through in between the filler and the blocks or the concrete surface and the good grip exists.

23. Let the inspection be carried out for which the request must have already been sent to the resident engineer in advance time as specified, and approval for the concrete pour obtained.

24. Pour the concrete as specified and arrange for curing by either using the curing compound spray or flooding the slab with water the next day on the already spread saturated hessian the same day after the concrete starts setting for the protection of the cracks particularly in the dry hot weather conditions.

25. Complete all the columns upto the beam soffit level but better during concrete pouring if a few cms be left as the masons shall definitely fill the concrete above the mark and forget to remove that and consequently, at the time of fixing the beam soffit the same shall be chipped out consuming manpower to be wasted/applied for no gains. (Ref. SK12 - SK42.)

26. Make sure that all the dowels for the lintels or the beams for the transfer of the bending moments from the beam to the columns at top are all in position before concreting the column till top level. Arrange for the curing either by the curing compound or by wrapping the saturated hessian protected by the PVC membrane to retain the water as much as possible from being evaporated and keeping continuous cool and top also wrapped or by pouring the water all the time on that. *When the curing compound is used, first saturate the surface of the column well and then apply the compound as instructed by the manufacturer for the retention of the water on column surface. Curing compound has been found effective where the scarcity of the construction water exists but the application should be done carefully to make sure that the perfect compound film forms on the wet surface of the column concrete and the top be saturated with hessian. When the compound is used, a second coat of degradation compound shall be applied if the plaster or the other finishes are to be bonded to the surface during the period in which the film shall remain in existence depending on the manufacturer's catalogue or generally, it may be 6 months of the day of the application of the curing compound.*

27. Go ahead with the operation of the form erection of the beam soffit, sides and the suspended slabs soffit or any other element. (A part of the preparatory works.)

28. Let the fabrication of the beams stirrups, other detailed bars, etc., go on in yard of steel foreman.

29. Fix the level marks on the columns to enable the foreman fix the beam soffit levels and the whole slab levels using the inverted level method or the other method of reduced levels if possible. (Refer to SK45, 50 — Typical forms of suspended slabs.)

30. Make sure that the proper releasing agent has been applied on the shutters before going ahead with the raising of the fabricated reinforcement to the slab shutters. Releasing agent should be of the quality as approved and be dried before fixing or placing the reinforcements.

31. When the placement of the steel bars is over the electrical man, plumbing man, mechanical man for the installation of the allied services as needed at this stage, i.e., the fixing of the conduits and the boxes, the blockouts for the sanitary and the engineering to be in position as detailed in the relevant approved services drawings).

32. Remember, before fixing the sides of the beams and the soffits, the axis lines of the slab should be located up and the positions of all the beams to be checked accordingly. Close the sides of the beams and to the lines and fix the concrete spacers to the correct positions and of specified thickness and bring the reinforcement work to pourable standard. Inspect yourself thoroughly and see if something is missing or undone. Keep a hole in the beam for the final cleaning and to be closed before commencement of the operation concreting. (Refer SK1 - SK9)

33. Having satisfied yourself of all the elements and disciplines normally, call the resident engineer on the requests of already despatched sheets of the concrete pour, electrical, plumbing and drainage, and engineering for the inspection to obtain the pour permission. (Refer proformats)

34. Engage a steel fixer on the slab forms and carpenters to look after the dispositioning of the spacers and the binding wire and the loosening props respectively during concreting on the prepared and the approved works.

35. Arrange for the items as included in the check list and be available at the time of inspection as those included in the request form. Check list shall include ready mix or the mixer on site, pump or the tower or the mobile crane, water and hose pipes/pumps, concrete vibrators 1½" and 1" and also standby, compressors, wheel barrows according to the requirements, hand shovels, concrete dumpers/trolleys when the concrete is to be transported from the site batcher, concrete skips, teeth hand shovels, hessians impregnated in the water drums full of water, polythene rolls, handy plastic buckets, wire brushes, spirit levels, artisan's tools for masons and for carpenters, steel fixer's pliers, slump cone arrangement, cubes 6 no. or more concrete thermometer, weather, thermometer, measuring tape, etc. (Refer list of artisans' requirements)

36. Now make a start and supervise every operation to the specifications of the concrete laying and maintain all the records taking slump, temperature, cubes, transportation time of the transit mixers, etc., and all shall be entered in the concrete record register tabulating as Pour No., section, grid, floor, element, quantity and concrete grade, temperatures, slums, time of concrete arrival, time of departure of transit mixer, pour placement temperatures, 7th day and 28th day cube compressive strengths, remarks of the laboratory technician, remark of the consultants/structural engineer, details of the supply invoices, source of the supply, date of form striking specifically, date on which form is struck, date on which curing stopped, details of the disputes in concrete elements, results of the hammer tests or the static load tests or any other tests as demanded by the engineer/resident engineer and all other informations the site agent and the engineer/resident engineer deem fit to be included for the specific references to be used by the future inquiry commissions during miscarriage of structural elements or the structure as a whole.

37. Clear all the laitance and spread saturated hessian and the PVC or the curing compound. Clear the site for 12 hours.

38. Inspect the surface next day and visually examine the formation of cracks if any, if considerable cracks are found/discovered contact the resident engineer for further inspection and advise. Minor hair cracks do not create major problem. Grout the cracking surface with the recommended resin by standard pumping of chemicals or any other remedial measure as specified to resortion.

39. Cure well and construct the upstands if any.

40. Carry out the works of columns from FF-SF and before this the axis should have been transferred from the ground to the FF and the setting of all the columns done accordingly. Construction of the 100mm kickers are advised for better work. (Refer standard sketches)

41. In case any facie works are detailed, that shall be carried out prior to the commencement of the beams and the suspended slab works after the finish of the columns' structures. (Refer sketches)

42. Similarly, the structural concrete frame can be raised to any levels subject to the erection of a strong and well gripped scaffoldings all around the building profile provided with the working platforms railed and safety nets provided.

43. Some specifications suggest the stripping of the slab soffit shutters after 3 days of casting provided the props remain intact but practically, it is very difficult operation to take out the shutters. After the specified time for the striking forms lapses, the deshuttering shall be carried out carefully and the site cleared of all debris. *The green surface should be closely inspected to find out the honeycombing, any void in concrete, opening of shutters of the beams or the slab, or any other defect, and the remedial works carried out as instructed by the resident engineer/structural engineer after being informed and holding the inspections.*

It is too common in the field that the foreman or the other irresponsible personnel or even less calibred site engineer to protect the skin shall hide the defect by plastering in rich cement mortar and roughening with the hessian cloth but it is really unsuccessful cheating as the same shall be visible as a separate surface and the site engineer shall be caught by the resident engineer or the structural engineer during visit or round and shall add to the loss of trust on the contractor's part which should really be avoided. It is therefore, better to resort to the approved concrete remedial measures to maintain the professional trust.

Measures

I **Honey combed surface** Remove the weak concrete till the hard concrete base is reached and let that be inspected by the resident engineer. Shutter the surface in the original profile, grout in rich mortar/cement slurry, pour the concrete of the richer mix manually, compact well and close the opening till hardening. This type of defect is generally observed in the columns near the bases.

II **Void in the surface** Due to inadequate compaction in the last batch before long interval of time and the delivery of the new/next batch due to any probable reason and also the foreman's negligence in supervising the operation to the specifications, sometimes, some voids are left. Chip the weak mass of the concrete out & expose the reinforcements and prepare the surface and let it be inspected by the resident/structural engineer and pour the new rich concrete to the specifications. In case the chipping and the concreting be not feasible due to the member being thin, apply the epoxy mortar approved by the engineer according to the manufacturer's instructions in layers not exceeding 15-20 mm thickness all specifically meant for. Excessive use of the epoxy is also not good due to the extra cost/expenses to be borne by the contractor as well as the different modulus of elasticity value in case voluminous epoxy being permitted.

III **Bulging/Opening of the forms** This is also a common defect as a result of providing inadequate supports to the vertical sides of forms and mispositioning the braces/scantlings. The bulged surface shall be chipped out to the plumb. In real bulged surface the chances of the reinforcement to buckle are rare if the later placed well.

44. As the skeleton of the building is cleared from the buildings to buildings in any complex or floor to floor in any building, the following activities in the sequence such as walls construction (ordinary and cavity) either in the bricks or in concrete blocks, be carried out simultaneously fabricating opening frames in the shops or the factory.

45. The wall cladded between the columns shall be connected to the columns by inserting wall ties shot fired in the concrete in the fourth course repeatedly. In the cavities walls externally, the cavity ties shall be installed per specifications. The cavity may be insulated by providing in the intruded polystyrene of the specified thickness and the density or any other material approved by the engineer.

46. Carry out the plastering as usual and to the detailed specifications.

47. Make sure that the fabrication of the a.c. duct works are in progress in the shop. (Engineering/air-conditioning shop)

48. As explained in the activity no. 46, take care that the electrical conduits and the switches and the sockets all run in the wall during construction or the chasing by the electric grinder shall be carried out before the plastering is done. After enough area of the plastered surface is available to permit the engineering sub- contractor to commence the mobilisation, contact the foreman (engg) and let the engineering 1st fix activity be commenced. Ducts shall be raised and installed duly insulated with glass fibres and painted with the vapour seal per the engineering specifications.

49. Very close watch on the co-ordinations shall be needed at this stage.

50. Let the electrical guy finish the operations of fixing the trunkings and the cable trays fixing all per electrical specifications.

51. Simultaneously, the operations in the stairs must have been finished or progressing. (Refer sketches)

52. Major part of the building construction in civil engineering discipline is over as RCC frame skeleton, plastering, block walls in cladding and partitions, etc., have been completed. The first fix of the allied services shall go as routine.

53. Mobilise yourself for the commencement of the architectural finishes operations such as fixing the frames in the openings, marking of the levels for the flooring, fixing in positions all the

door frames, and all the activities by that the building shall be granted the **status of weather proof**.

54. Carry out the flooring operations in the dry areas.

55. Let the plumbing guy fix his installations' traps in the wet areas.

56. Carry out the operations of skirtings, wall tilings and flooring works according to the details as approved in the coordinated/working drawings per architectural requirements.

57. Generally, the ceiling (suspended) man shall shot fire the wire hangers at this stage to avoid inconvenience later when the ducts are in position.

58. Drawing wires in the conduits cast in the suspended slabs and in the walls shall be commenced. All the switches and the points boxes shall be installed.

59. Hanger wires for the light illuminaries shall be shot fired to the soffit according to the **co-ordinated drawings of the false ceilings/Reflected ceiling plans**.

60. Roofing works shall be carried out simultaneously.

  I Marking **plinths for the machines'** bases and the **scabbling of the slab surface**.

  II Construction of the **plinths wherever required**.

  III **Scabbling** the **completed surface of the roof slab** if not done before.

  IV Cleaning and laying mortar layer 1 : 6 or any other mix as specified in the **compartments distribution** of which already agreed with the resident engineer in an overall gradient of 1.5% or 50 mm minimum to 200 mm maximum as the case be.

  V **Triangular sectioned** mortar shall be spread along the edges of the roof to facilitate the turning of the **roof torch water proofing membrane** to be sealed in the chase.

  VI Application of the waterproofing roof torch with the specified overlaps and before that all the supporting chiller pipes if needed shall be provided with the supporting steel brackets embedded in the mortar and shot fired to slab, as also all the electrical isolators as well as all other requirements be completed before the torch is started, completed, inspected and approved.

  VII Let the waterproofing roof torch as laid over the entire roof including the machines' plinths and chased in the parapets, be put to 72 hours water test by flooding the entire torch with 25 mm minimum standing water all over or as agreed with the engineer or specified particularly. If no leakage is visible from any part of the slab, the test shall be approved. In case considerable leakage is noticed, the water from the roof shall be drained out (all the water outlets must have been water sealed before allowing the water being flooded) and the remedial works to the roof torch be carried out till the satisfactory test is carried out and approved.

  VIII Once the roof torch is approved, it is liable to be damaged therefore, it shall be covered immediately. All the machines shall be installed before the commencement of the roof torch application work. This shall mean that the plinth portion of the machine be waterproofed too and in that case the remaining portion shall be treated later.

  IX Polystyrene (extruded) heavy duty or as specified or any other material approved by the engineer shall be laid on to and be covered by the aggregates/paving slabs/flagstones as smooth finish tiles of concrete to act as walk way on the roof and the sealant provided over the chased roof torch in the form of mastic or the polysulphide or as specified. The aluminium flashing/paint as specified be applied for the solar reflection.

61. To remember here is that the false ceiling tiles cannot be installed before the roof water test is cleared, therefore, so far only suspension system could be done/installed.

62. All the duct works, insulations, fittings such as bends, volume control dampers, fire dampers, attenuators, filters, etc., should be in position and approved.

63. Similarly, all the cable trays, trunkings, wiring for the fire alarm, for the smoke detectors, and all the systems needed per specifications, above ceiling plumbing pipes and joints, chilled water pipes running above the false ceiling, water pipes, gas pipes, and all other services as detailed to be accommodated in the above ceiling spaces should be brought to the completion and the clearance obtained by all the discipline incharges from the resident engineer so that the final works in the system of suspended ceiling be commenced.

64. Once the works above the ceiling level are inspected and approved, let the false ceiling man go ahead with the fixing in positions the approved tiles as detailed in the approved **reflected ceiling plans**.

65. The supply and the return air grilles, diffusers, splitters, and light fittings/illuminaries all be installed in parallel of the fixing ceiling tiles subject to the availability of the materials.

66. Fix the door panels, windows generally supplied in the finished form, and close all other openings simultaneously, as in principle the false ceilings and other finishing items cannot be left exposed to weather hence it is better to render the building weather proof before starting the weather susceptible operations such as ceiling tiles, special paints, special floorings (rubber, epoxy, carpets, etc.) and make the building weatherproof to start these activities.

67. Fix skylights and the ceiling bulkheads as detailed in the engineer's details and the bulkheads' frame be screwed to the slab. The frame shall be made of the MS section or the fabricated rolled sections welded together to form hanger and the asbestos screwable sheets be provided in the internal linings of the bulk heads. All the skylights must be closed from the top. The ends of the ceiling tiles meeting at the bulkhead shall be trimmed well and a separated timber piece well decorated shall be provided per architectural engineering consultant's approval.

68. All the floorings such as marble chips tiles, cement concrete floors, mosaic floors, timber floors for squash courts if any, epoxy floors, etc., to be completed.

69. All wall linings to be finished as specified and equipments installed for the sports, and any special purposes. Remember at this stage the structural work load is finished on the building, hence the workforce to be diverted to the external works for the drainage, water mains, fire mains and hydrants, irrigation's water supply, etc., and most of the workforce should have been diverted much before.

70. There are elements of internal finishes and decorations and external infrastructure and landscapings left to be completed.

71. Painting on the internal surfaces as specified shall be carried out and the architectural (architrave) beads of the openings finished.

72. The cubical units to be completed in all respects and light fixtures to be provided in position per details shown in the reflected ceiling plans.

73. The last item such as final touch to wood works, paintings, etc., and laying of fixing carpets, fixing the matwells, as detailed shall be carried out.

74. **The testing and commissioning** of the air conditioning system, electrical engineering system, plumbing and drainage systems shall be carried out as specified in the project specifications and all the observations recorded in the presence of engineer's/resident engineer's/specialists consulting engineer's all as formulated.

75. Almost all the items of the building have been executed except the snags and the final touches. Divert most of the workforce to the other areas.

76. The kerb stones work, manholes, compaction, fillings, subgrade preparations, etc., be completed and the road structures be constructed.

77. Snagging as advised by the resident engineer/engineer/employer to be remedied and the building shall be almost completed practically.

In the said/above/foregoing explanations, an attempt has been made to bring to the notice of the engineers some of the activities or the operations as experienced in the field and it is not feasible to write each and every operation/activity/nomenclatures but it shall certainly act as a guide to provide wide range of activities in general for the engineers on the threshold of the engineering carrier/fresh or semi experienced engineers and the practicing architects and engineers and the engineers who have to co-ordinate the allied services in the field and fit in the management team.

Sports courts surfacing

1. Courts line marking

2. Courts painting in 2 coats orthogonally to each other
3. Resispeed 2 coats applied orthogonally
4. Resin 1 coat applied orthogonally to the surface below and above
5. Surfacer 1 coat of cement paint mix and painted perpendicular to the primer.
6. Primer 1 coat applied uniformly on the prepared base of the concrete or the asphalt concrete base
7. Prepared base with the defects rectified using patch binder.
8. Concrete or the bitmac wearing surface as the case be.

Details of the propriety works

1. Mark the court's surface on the proposed **base of concrete or the bitmac**.
2. Make good all the uneven spots and unacceptable surface by spreading **the patch binder** supplied by the manufacturer per his instructions.
3. **The workforce** for this particular item of court's surfacing should **wear special shoes as not to damage the painted surface, uniform for accepting the paint splashes. Safety glasses, hand gloves, etc.**
4. **The equipments required are an air blower, paint squeezes, hard hand scrapers, hard sand paper, paper tape/masking tape, games line instrument, small drum, light paint mixer, etc., and the timings of the operations to be decided by the temperatures of the area and the range permitted as given in the manufacturer's instructions for the temperature controls at the application time.**
5. Having prepared the surface by **the patch binder** the last day, apply **one coat of primer** from the ready mixed supplied paint and spread by the arm squeeze in the direction parallel to the small side of the court and go on repeating till the application covers the whole surface of the court. The applier has to keep on moving with squeeze held by one hand at its top and the other near the mid.
6. Next day stick the paper or adhere the paper tape along the periphery on the line marked so that the applied paint material does not come outside the periphery but goes on the paper tape after crossing the line.
7. Clean the surface by the air blower and take out all the loose materials if any and clean all the dust accumulated. **Should the weather be inclement, do not proceed.**
8. Mix the surfacer paint with the cement and water in the required proportion and immediately start the application on the primed surface by the arm squeezes but in a direction perpendicular to the direction of the application of primer and continue till the whole surface is completed. Only mix as much surfacer as is needed, finish the court because the extra quantity if left shall be wasted.
9. Next day clean the surface and make good any of the defects observed during application of the surfacer and proceed for the application of the resin coat in the direction perpendicular to the previous coat. Movement of the squeeze shall remain as before.
10. Next day, clean again and apply **one coat of resispeed** in a direction perpendicular to the direction of the top coat of **the resin** and then apply the second coat of the **resispeed** perpendicular to the first coat.
11. Next day, clean the surface of the resispeed again and check, should some repairs be needed and make good with resispeed if so.
12. **The paint coats** shall be applied very carefully. Apply the 1st coat perpendicular to the previous coat of resispeed and finish the complete surface very smoothly making a note that no overlapping signs during the movement of the squeezes are visible. After a few hours or when the 1st coat dries out completely, apply the second or the top coat to a pleasing and very very smooth outlook.

13. Make sure that all the **posts' holes/sockets**, etc., for holding the net frame are embedded and grouted in the positions **before the surfacer application**.

14. Next day take out the paper tape and a straight line periphery of the sports court surface is available to render a pleasant look.

15. **Court's lining** Start marking the layout carefully and mark the centre line of the courts linings. Install two rolls of 50 mm wide paper tape on the game line instrument and move the instrument, following the lining's central line marks till the paper tape is spread on the complete linings. Now take a small brush and apply the white/yellow or any other colour specified and supplied with the consignment in the space between the two paper tapes adhered to the painted surface in the complete length of the court's linings.

Apply the paint in two coats as provided in the specification or as instructionally specified. Next day, remove the paper tapes and the court lining marks are ready.

16. Generally, **green** or **red colours** are used in **the courts** painting while **the white** or **the yellow** shall be used for the **linings**.

17. The sizes of the courts and the lining design shall be governed per provisions made in **the codes of the International Sports Association**, the details for that shall be supplied by the engineer or **the standard approved society of sports**.

18. The court shall not be open to use the same day.

19. Install the equipments as specified.

20. Erect the chainlink fencing or any other fence approved, the details to be given in the engineer's design.

21. There are generally two systems in sports court surfacings

I **High standing system**

II **Comfort system**

The thickness of No. 1 shall be much less than as in No. II.

Epoxy flooring 1.5 mm and 3.0 mm thickness

Shall be provided in the situations where the resistance to the wear and tear, to the acids, to the abrasions, to the oil and where huge strong stores are required.

1.5 mm thick

1. **Surface etching** The concrete surface to be given an epoxy covering shall be etched either by the scabbling hammer or by spreading a solution of 10% concentration of Hydrochloric acid mixed in water three times the quantity and tested before the application whether the solution etches the surface. After spreading the solution, immediately apply the clear water to dispose off the acid solution after etching to prevent any further damage to the concrete.

2. Get a heavy duty mixing apparatus electrically operated and the epoxy paint be stirred in the container of the paint. The person handling should use the eye protection, and wear gloves. Adhere paper tapes to the confining limits of the floor area and apply the paint on the etched surface and spread the silicon powder covering all the surface and also in excess before the paint dries out.

3. After specified hours per manufacturer's instructions remove the extra powder from the surface and apply 2 coats of stirred paint as specified and no one shall be allowed to enter the area that day.

4. Next day take out the paper tape.

3.0 mm thick

1. Procedure as mentioned in 1 & 2 above holds good for this item also but instead of spreading silicon powder (silicon oxide) spread Aluminium oxide.

2. After a few hours as specified per manufacturer's instructions remove the extra aluminium

oxide and apply one more coat of epoxy paint and spread the silicon powder before the paint dries out.

3. Remove the powder in excess and apply the first coat and then final coat. Remove the paper tape from the floor area and the floorings are ready.

The above flooring is very expensive and should be used to limited extents only.

Concrete Auxiliaries	Resin Compounds/systems
Plasticisers	Repair compound
Super plasticiser	Grouts
Releasing agent	Sealers
Curing compound	Primers, Bonding
Air entraining and foaming agents	Protective coatings
Water proofers	Flooring systems
Surface retarders	
Set retarders	
Accelerating agents	
Grouts and Mortar admixtures	
Bituminous coatings	
Speciality products	

The above concrete auxiliaries & the resin systems shall be used in different quantities & ratios & form the characteristic of the structural concrete meeting concrete states namely improvement in the workability, enhancing the initial setting time of the concrete, quick setting, improvement in the striking the concrete forms/deshuttering operation, improved curing methods, attaining the concrete leak proof, concrete repair systems, etc.

The above systems and auxiliaries shall be strictly used per instructions of the manufacturer after the approval given by the project engineer/owner's appointed engineering technocrat.

Knowledge of different types of contracts

1. Package deal contract
2. Bills of quantity contract
3. Schedule of rates contract
4. Lump sum contract
5. Itemised lump sum contract
6. Bills of quantity and percentage rate contract
7. Cost plus percentage contract
8. Cost plus fee contract
9. Labour rate contract
10. Target contract
11. Construction by direct labour
12. Sub contract

1. **Package deal contract** This contract between the employer and the contractor covers the job of designing and construction of works and during tendering the contractor has to take into consideration all the preliminary designs required as that shall form part of tender and should meet the requirement of the employer. Only reputed contractors should be entrusted to work on package deal since the big and reputed contractors engage their well qualified and design experienced/tutorialed organised staff. Should the contractor be paid properly, a qualitative services in designing and execution is obtained.

The two specialists, i.e., design specialist and the construction specialist work on one platform leading to a fruitful combination of skills but a degree of mistrust can arise as the employer may doubt the genuinity of the contractor in carrying out the works to the standards to make savings. To avoid that, the employer should engage the supervising consultants for additional expenses vested with full authorities as the engineer possesses/enjoys.

The tendering for a job in this type of system is a tedious job while the invitation of tenders by the employer is easy. The only difficulty in this type appears when all the tenders are received with different project designs and not necessarily the lowest quoted tenderer offers the best design. The design may appear best to the employer but exceeds his budget. So to meet all these possibilities, the employer shall need the services of an independent advising engineer all the time in case the employer does not possess this skill. The tenderer too shall hesitate to venture.

2. **Bills of quantity contract** In this type of contract the tenderer offers the job based on the sum reached by pricing individual item mentioned in the bill of quantity/BOQ calculated from the drawings. The measurements shall be accurate as far as possible. The contractor is not bound to execute the work exactly to the details in drawings in case some item is not included in the bill of quantity which if necessary, shall be paid extra. This category permits the variation in the quantity as the payment to the contractor is made based on the works measured item-wise on site. In case the work done on site amounts the same as given in the bill quantities, the employer shall pay the same price as the tendered sum. Alterations are permitted as the same rates of payments and chances of disputes between the contractor and the employer are rare. Tender can be closely compared item-wise. Properly drafted, a contract based on bills of quantities and fixed item price can prove the best as it provides the opportunity for alterations, quantity variations to some extent and leaves way open to engineer to exercise his responsibility properly.

3. **Schedule of rate contract** For the contracts where item cannot be measured from drawings and actual quantity not anticipated, a schedule exhibiting the item but not the quantity and more items may be included for the temporary works due to contingent conditions and uncertainties prevailing. Quantities may be given for guidance only. There should be no guarantee that all the items provided in the schedule will be carried out. Contractor should consider all the items provided in the schedule as provisional only. The rate should be quoted in such a manner that a great degree of variation of quantities is acceptable. The contractor if sensible, shall quote the rate of all the items in schedule considering the overheads and other expenses included in every item instead of spreading all over as in the case of fixed rate BOQ case.

4. **Lump sum contract** This is a contract wherein a fixed price is tendered for completely doing a work based on the drawings, specifications and site conditions. A quantitative work is to be done before inviting tenders. Engineer may provide a bill of quantity for the purpose of assisting the tenderer to apprise of items to be carried out but not as a guarantee to job quantity. Should there be any number of items not included in the BOQ, the contractor is supposed to carry out all. This category is useful where no alteration in the works are expected and no great risk attached and job is small. The only advantage is no measurements are required for making payments to the contractor.

5. **Itemised lumpsum contract** This is a contract wherein a fixed price is tendered for complete construction of a job based upon the drawings, specifications, design, any particular items as well. All the bills of quantities are enclosed in the documents including the preambles and the tenderer is supposed to fill the prices in all the items to reach the tender sum. The tenderer has to check and ascertain that the bill of quantities cover all the items according to drawings and specifications and if found something missing, should inform the engineer or include the price of those items somewhere in the bill. The tenderer has to make sure for all aspects of the architectural, structural and allied building and other services and the conditions specified in the scope of work. The advantage in this category is that for every item a price is quoted by the tenderer and if not quoted, it is understood that the same has been included in some

other item. All the calculations are checked during the scrutiny and in case of arithmetic error discovered, adjustment is done and the tenderer is informed accordingly.

In case the amount of the total sum comes more than the tendered amount, the tenderer is asked whether he is still ready to do the job and on the other hand the tender sum is corrected to the calculated lower value. Similarly, all the items are checked and the corrections are made accordingly. The engineer can·ask for the break up of any items. Engineer or owner has the freedom to make any alteration, deletion, additions or changes in the designs and accordingly, variations are considered. In case the engineer feels that an extra item is to be ordered and the contractor's price for that item is more, he can request the contractor to provide analysis/break up of the items and on receipt of the same, a reasonable rate can be fixed for the extra item but the engineer should be reasonable. The contractor cannot refuse the execution of the extra item as he is bound by the prevailing conditions in the contract.

In case order is given for an item not included in any of the bills then engineer can ask for an analysis of that item based on the current market rates. This type of contract is very practicable in carrying out the works of very big sizes as accommodates alterations and the changes in the designs. Payments on the material procured on site is also made easy as the individual item rates are present. In case of rejections or condemnation, the stoppage of reasonable payment is ordered. It is also easy for the contractor to raise claims due to the availability of all the item rates. The works are not measured in details but the office work for the preparation of the contract documents is really voluminous and extensive.

6. **Bills of quantity and percentage rate contract** In this contract the tenderer has to quote a percentage below or above the rates given in the bill of quantities and the cost expected. Actually, these rates of items are based on the market rates of yester/past years. The contractor shall be paid on the actual measurements of the items executed/carried out. There is a scope to alter the design and the variation of the quantities to a limited extent but the chances of claims are rare other than increase in the cost of labour and material more than the specified limit. There is no guarantee that the quantities included in the bills shall be definitely carried out but since the quantities are based on the drawings in most cases there is not much difference in the carried out quantity and the bill of quantity. In fact, public works departments and the other similar bodies adopt a schedule of rates of all the items based on the rates of a particular year from time to time and the same remains in practice for many years. The tenderer has to calculate the current rates of all the items and find an average percentage to be applied to all rates and a percentage below/above of the schedule of rates is calculated and quoted in tender. The lowest values in the rates higher than schedule and highest percentage in the rates below the schedule of rates are generally accepted. The engineer and the contractor should see that all the items have been carried out to avoid any disputes and the rate quoted is tolerably justified.

7. **Cost plus percentage contracts** This contract provides payments of all the costs of labour, materials, machineries, overheads and a percentage of profit above that. This type is very cumbersome process as maintenance of accounts for all the items is difficult and the contractor shall not be efficient due to getting enhanced profit as the actual cost goes up. (Not recommended)

8. **Cost plus fee contract** The contractor is paid the actual cost of labour, materials and machines but his overheads and profits are fixed. This fee is actually tendered in the competition and may be negotiated between the employer and the contractor. This type is really useful to increase the efficiency and to tackle difficult projects. Engineer and the contractor can act jointly to produce the best quality job and provide ways to adopt different methods to carry out the works. The contractor shall expect the attractive fee in all cases.

9. **Labour rate contract** In this case all the materials, machines and plants, etc., are supplied to the contractor and the later is supposed to mobilise labour, supervise, finish the job in time, to the drawings and the specifications of contract. The tenderer has to quote the labour rates including profits for all the items in the schedule of items. The contractor is paid based on

actual measurements on site. There may/may not be a binding for the quantities to be carried out but this type is good as far as the materials and machines and the payments are made in time to the labour contractor. The contractor has to fix his own rates for the items as for further payments to the people engaged on the overtime shall have to be made since 8 hours working may not be enough for finishing the quantities. Generally, this type forms part of the sub contract/s with the main contract trusted to the contractor's artisans for efficiency

10. **Target contract** In this type of contract cost is borne by the employer and a fee is paid to the contractor for a target sum and if the target is increased, the fee is reduced and vice versa. All the documents, methods of constructions and the specifications must be clearly drawn up and no contingencies are acceptable. In case, due to negligence of the engineer the targets cannot be met, it shall be unfair to deal with the contractor. (Not recommended for big jobs)

11. **Construction by direct labour** Promoter/employer may carry out the works by using his direct labour, own machines, material, engineering and other staff who are in permanent employment, particularly for the moderate size of jobs. It is possible to augment the manpower and machines at any time temporarily to meet the demand to keep up the programme by hiring from outside sources. The method of employing a regular workforce helps the employer retain skilled groups all times. For example, water works, sewage work, other public works, always retain expert gangs with them. But for the big jobs where competition always positions itself, it is the contractor's staff highly efficient and qualitative to control properly and finish the job in time. The employer's staff may not be that fast/proficient and qualitative and not reliable for very big jobs. A contractor can fire his staff or workperson in a day if found inefficient while the men in regular employment in government/public sector/local self government bodies may not be. So this class of contract system is good for maintenance works/jobs.

12. **Sub contract** Undertaking by a contractor as a subsidiary to a main contractor to carry out part of the main work of the main contract with prior approval of the engineer or a party appointed by the terms in the specifications as a specialist firm or a given firm nominated sub contractor to be paid by the main contractor.

Some miscellaneous contract terms

(A) Prime cost sum

Provision of a sum in the bills of quantities to cover any item for the nominated sub contractor or supplier together with the opportunity for the main contractor to add a percentage or a lump sum to cover the related works and profits.

(B) Provisional sum

A reserve sum/fund for doing a particular item quoted by the engineer in the bill and not by the contractor which may or may not be spent wholly or in part. The amount to be paid to the contractor shall be agreed later with respect to the bill rates for the works actually carried out. The main contractor is entitled to carry out the work included in the provisional sum if the work is of general nature otherwise, the same shall be carried out by the sub contractor to be paid by the main contractor under his supervision. The main contractor has no ground to complain in case the whole of the work under a provisional sum is taken from him and given to others or not at all carried out.

(C) Price enhancement on material

Some contracts include the provision of extra payments or reduction on account of rising or falling state of material prices. To operate this condition, the tenderer should include a list of current material prices as on date of tender in the absence of that the operation of the condition shall be difficult. The variation shall apply to all materials enlisted subject to an excess over the specified tolerable variations may be 5-10% to be covered under the contractor's risk. Now to check/investigate the claim under this clause, the engineer shall be provided the exact date of the enhancement and the details of the materials consumed after that alongwith the records of the payments made to the approved material supplier and the authenticated receipt of the supplier and the quotation of the supplier on the date of the receipt of the tender by the engineer or the tender closure day. The difference between the new rates and the tendered rates for the quantities purchased and consumed by the contractor per the statements on the site verified by the contract quantity engineer/surveyor shall be payable to the contractor with due discounts and further ratification by the engineer's quantity consultant for the real consumption of the quantity after the date of the enhancement. The extra shall be paid only subject to the adherence to the quality control. Suppose the tendered item does not match the requirements of the specifications but the material brought is costlier meets the specification, extra shall not be paid.

Also when the supplier whose price the tenderer quoted is out of the business and the new source is higher in rate, a marginal extra may be paid based on the quotation by the new supplier of the date of the tender or nothing paid otherwise. Extra may be paid in case the government increases the prices of the material due to budgetary requirements.

(D) Enhancement on wages

Should the rise and fall clause of variation applies and in that case nationwide increase in the wages shall be paid to the contractor after the date of enhancement subject to contractor's claims supported by the paysheets/rolls duly signed by the workpersons. This cause is valid as a policy change of the government ordering increase and not when the contractor allows the increase on his own.

(E) Extras

(i) **Lump sump itemised contract** This carries very few chances of extras as the contractor is bound to complete the job in the contract scope in all respects as required in the drawings and the specifications and other contract documents i/c the boq. Variation arises when the engineer/owner wants to omit/delete/add/alter the design in which case a variation order is to be passed by the engineer. Extra shall be applicable on additional work other than specified in the contract documents as a whole.

(ii) **Bill of quantity contract** Generally, you will find extra items and the substitute items in this because all the items shall be measured as actual done on site and any item not covered in the BOQ shall be paid separately on the agreed rate. Sometimes, the engineer/employer wants to substitute one item by another so that the cost to be paid is also substituted. It may be a saving on the job or the additional expenses. Any item ordered by the engineer other than the BOQ shall also be considered an extra item. Sometimes, unforeseen conditions not generally anticipated compel the contractor to extra expenses which may or may not be compensated depending on the contract conditions plus profit and delay if any.

All the alterations, additions or the omissions from the original BOQ should be covered under variation orders for the purpose of valuations and accounts. Variation order is only issued by the engineer on behalf of the owner and the resident engineer has no power for that. In case the contractor has altered the specifications to the super standard on his own, no variation shall be allowed. All extras should be shown separately normally as follows :

1. Extra bill items and substituted items
2. Daywork records
3. Unforeseen conditions claims
4. Enhancement in the wages and material prices

Any other extra work not included and essential for the job should be brought to the employer's notice and permission obtained by the engineer.

(iii) **Fixation of the extra price** The bill of quantity prices shall be used in case where the work is of similar nature and carried out under similar conditions to the work in BOQ. In case the engineer feels that the price in the boq is unreasonable under the ordinary circumstances as in the boq item, the engineer shall fix such price for those circumstances. The engineer may order the additional/substituted work to be paid based on the daywork basis, in which case all the details of the actual execution shall be taken from the site through the agent. The unit rates for machine's engagement., workforce engagement and all miscellaneous shall be included in the calculations based on the approved schedule of rates being used in the area/or authentic.

(iv) **Unforeseen circumstances** The payment towards this claim shall be based on the nature of claim from the contractor and the applicability in the contractor's agreement with the employer and also the details of accounts and all calculations exhibiting extras as expenses presented alongwith the claims.

(F) Loaded price

Some times items in the BOQ are so calculated by the contractor that the profit element stands very excessive than usual and that price in no way can be used for the extra payments.

(G) Claims

Claims for the extra payments presented by the contractor

1. Adverse physical conditions or hindrances or obstructions which could not be reasonably foreseen.
2. Not being in agreement with the rates fixed by the engineer for the additional/altered work.
3. Progress of work effected due to the delays caused to the contractor.

The above claims may be presented in the engineer's office any time after the occurrence of the claim/event and before the contract completion/termination of the contract provided the notice for the intention of the claims has been served on the engineer by the contractor within a period of one month w.e.f. the date of claim occurrence due to the unforeseen condition, in the absence of this notice/claim, the admission of the claim after the specified period of notice shall not be allowed and the claim discarded. All the provisions of the general and the specific conditions shall be kept operative along with this claim.

The engineer should be provided with all the facts and the estimates presented to him in support of any claim. The contractor should present all his records and accounts for the inspection by the engineer/resident engineer/owner which shall definitely reveal the evidence should the contractor has lost the amount due to unfair payment or the delay caused to him. The diaries of the resident engineer and the inspector's daily returns shall provide enough assistance to exhibit that the work was slowed down due to the difficulties and how many workpersons were employed and how the additional expenses were made on plants. The contractor's agent's daily job record shall itself reveal the part of the facts.

(H) Delay claims

Delay must be real and the cause be outside the contractor's liability and risk and caused loss or extra expenditure to the contractor which could be made no good elsewhere. A period during which a contractor cannot employ the workpersons and the machines at their normal intended output according to the nature and the quantity of work per agreed programme or under any rearrangement of programme is termed as a delay.

Delay can effect a part or whole of work. It can be partial or full. Say consequent upon the engineer's instruction a machine is stopped and has to wait for the alternative instructions and till the instructions come, no alternative operation is available for the machine, the stoppage shall amount to a delay and idling of the machine.

Similarly, when the work to be done by the other contractor under another contract does not clear the work area of your contract in time and the work of the contract is hindered or not started, a delay shall be counted for that part of the contract area.

An extension of time is not necessarily a delay claim as it could be caused due to the additional works and the rates permitted for those items in the contract variations include all the overheads.

Similarly, the cost of unforeseen conditions or enhancements are paid under separate heads. Claims for the delays can be raised when —

1. One contractor delays another
2. The work is delayed due to the alterations of the design and the cost if any involved.
3. Claims made due to misrepresentations are admissible should the contractor raise the matter immediately it is discovered and not after the work is carried out.
4. Claims raised for the plants remaining idle due to the order of the stop work by the engineer for the alterations in the designs cannot be admitted should the cost of the same has been included in the variation order passed.
5. Claims for the loss of profit due to the extended time barring the contractor to deploy the staff for the organisation of a new work supposed to commence after the completion date of this contract, shall not be admitted as far as this contract is concerned but the contractor may claim damages on this account.
6. Claims for any material lost by the engineer's instruction due to the change in the design is admissible if not included in the variation order for that instruction.
7. Claims for the inclement weather conditions are not admissible since the contractor is supposed to be acquainted with the meteorological records of the area per construction contract clause.
8. Claims for the natural causes generally never occurred but during contract unfortunately occurred causing the losses to the contractor shall be claimed on the insurance agency.
9. Claims for any loss caused to the contractor and the circumstance is not under the employer's liability shall go to the insurance agency or govt. agency.

The above detailed points shall be read in conjunction to the delay and the claims clauses forming part in the specific conditions and the general conditions of the contract.

Practical completion When all the disciplines' operations of a building or the buildings and other works included in the contract scope are completed and approved, the contractor has to notify the engineer/resident engineer/Owner for the inspection and takeover. The resident engineer shall inspect all the disciplines and the elements room by room and mark clearance or no clearance in the snagging list alongwith the comments made therein for the attendance of the contractor. All the inspections are carried out and the snagging list is given to the contractor to carry out the remedial works and let the resident engineer know for the inspection. Similarly, all the inspection chambers, man holes, etc., are inspected and the report is filled in the standard format. Having done the remedial works, a final inspection shall be carried out for the practical completion. It is upto the employer to take over all the buildings at one time or one by one subject to obtaining the occupation certificate from the statutory authorities. In the latter case, practical completion shall come into effect/force when the building is taken over even the date of completion may be far away for the contract as a whole for the considerations of the maintenance period. Every contract includes a maintenance period in the specifications for which the contract amounts include the cost to be paid to the contractor for the protection and maintenance. **The final completion** shall be granted only by the engineer when the period of maintenance has been over & no defects appear in the works maintained after the completion of the works practically. The retention of the contractor's security money/amount shall be released only after that subject to the submission of long term guarantee of the warranted items carrying terms more than the maintenance period. These items may include sports surfacing for the courts, water proofing torches for the roofing, the glazed curtain walling and all other specific items of the different disciplines as the case be.

CIVIL ENGINEERING PROJECT MANAGEMENT
DEFINITIONS OF THE PARTIES INVOLVED ON BUILDING PROJECTS

A. PROMOTER/USER/BUYER/GOVERNMENT BODIES/OWNER

The individual person, proprietor, group, financier, etc., who desires to invest the funds/money for certain civil engineering project for a definite purpose, for a definite profitable business or for personal use or for the use for a certain part of the community or the society or for a commerce, etc., as the case be and undertakes to take the complete responsibility of the finances and the consequences on the same. The promoter appoints the engineer for the project to design and supervise the execution and the design to be in accordance with the prescribed requirements and needs specific to meet the target within a specified budget and defined time. Promoter also appoints the contractor for the purpose of the construction executions of the scheme based on the recommendations of the engineer for the project. In both the contracts, i.e., I. **The Management Contract** (*between the promoter and the engineer*) and II. **The Construction Contract** (*between the promoter and the execution contractor*), the promoter stands as the party I while the engineer and the contractor for execution as party II in the respective Contracts. The promoter owns the responsibility and the liability of all the money payments for the services rendered by the two parties on the job. In the contract with the construction contractor the promoter shall be statutory termed as 'EMPLOYER' of the execution contractor and in the contract with the project engineer termed as the 'CLIENT' of the 'ENGINEER'.

B. ENGINEER FOR THE PROJECT

Immediately after being appointed *engineer in all professional respects* for the scheme of the project, *the individual professional in engineering/group/agency* as the case be, comes into effect to legally enjoy the status of being termed/designated 'ENGINEER' for the project, i.e., to own the complete technical & engineering obligations and the *professional responsibilities for the complete designs and the supervision* over the construction execution which resembles the status of the *engineer-in-chief* or the *chief engineer* in lieu of that in *the government/equivalent sector* and *the promoter* and *the engineer* are *bound on the management contract* and the payments are made accordingly to the engineer by the promoter for the engineering services rendered/to be rendered by the engineer to the promoter for the project. *Since the payment on the contract is made in a fixed accepted % of the cost, the engineer owns the complete liability of the adverse consequences of the engineering design failure of his works/services and shall have to face the prosecution and bear all the damages caused to the promoter. The engineer shall enjoy complete administrative control of job engg. (Engineering from the inception to completion - Management)*

C. EXECUTION CONTRACTOR/CONSTRUCTION CONTRACTOR

Immediately after being issued with the award letter by the promoter, the tenderer enjoys the designation of the *construction contractor* and shall be bound to carry out all the project works for the promoter based and *specified per engineer's details and instructions* and put forward *all the claims of works to the promoter.*

For Information only   TYPICAL   **(Management Contract)**
Contents of our offer as engineer for the proposed township promotion project

Civil engineering project management contract agreement

1st party   Promoter.....................

2nd party   M/s Sky Associates (Building Engineers) of 3373, Delhi Gate, New Delhi 110 002 headed by the Chartered Engineer (India) Suraj Singh M 46231

TITLE/NATURE of the professional services to be rendered by the engineer to the promoter

To conceive, incept develop, design (complete engineering details, i.e., architectural, planning and detailing, structural designing and detailing, all building allied internal and external services schematic structural networks and quantity engineering), prepare submission drawings to obtain the building permits from the concerned relevant statutory authorities/departments, drawing/drafting tenders for the construction contracts, administer, manage and supervise as consulting group/consultants all the construction operations at the project site in the national capital region of the state of .... and operate the terms and conditions of the construction contract's execution till the completion with the commissioning of all the jobs as included in the project details the estimated cost of which at the current market rates of 91/92 fiscal year amounts to 60-70 billions of Indian rupees and subject to the inflationary growth in the near future.

The guide line break up of the engineer's responsibilities included or intended to be included in the scope of this contract

1. Schematic layout plans of different residential, commercial and all other components of the proposed satellite town project.
2. Schematic architectural plans' layouts of all building and relevant structures
3. Architectural and structural developments and the projections of the complete formation mechanisms of all the buildings related to the project.
4. Preparation of the contract forms and the contract legal documents
5. Preparation of the guide to the management for the civil engineering project administration and control based on the UK institution of civil engineers' guide lines.
6. Preparation of the application petition to the obtaining of a no objection certificate to be moved by the promoter on the competent authorities of the lands control administration of the state government comprising all the technical layout sizes of the lands requirement schedule and the other features necessary for that (Technical)
7. Site survey supervision and soil investigation interactions post land acquisition.
8. Schematic plans of layouts review and adjustments settings, modelling making supervision and the submission of the layouts for the approval (Technical services only).
9. Schematic structural plans of all buildings/structural layouts/arrangements
10. Architectural developed plans, elevations, sections and other details
11. Structural developed plans and other structural details
12. Architectural and structural working details
13. Complete allied services schematic details
14. Foundation designs and details
15. Submission of the drawings for the building permit (Technical services only)
16. Quantity engineering details for the tendering purposes
17. Project site works supervision and control of administration over construction execution.
18. Post practical completion maintenance supervision for 400 days only.
19. Preparation of the preliminary schematic reports, chapters on the project reports, representation in the related meetings for the financial technical support to the promoter in the National Capital territory of Delhi/any other area or State/country etc.

Engineer's exclusive right

Sky Associates as a group consultants reserve the copy right of all the drawings and the related documents prepared by the group and the same cannot be used in any manner by any person/firm under his/their seal for this or any other project without obtaining the written permission from the engineer/technocrat/group.

The duration of the said title/professional responsibilities shall be termed for eight calendar years (English) from the day of the formal appointment of the firm/group as Engineer. The probability of any practical extension shall be dealt separately.

The term 'ENGINEER' as applied per British/international codes of professional practice and competency for the civil engineering project management contracts shall apply to this contract for the foregoing professional title/services/responsibilities in all the related communications for and on the job contract and for all legal implyations and implications.

Terms and conditions to govern this contract

1. The engineer shall councel all the professional services as defined under the title on this contract on behalf of the promoter/1st party to this contract and also as covered and ethicated under the civil and architectural codes of practice and the institutions' codes of conducts and also the statutory regulations and requirements.

2. The follow up for obtaining the land use approval in the name of the promoter and all the monies to be paid towards this operation/proceedings, the processing of the lay outs approvals and all the monies to be paid towards this operation/proceedings and the fee and other expenses to be paid for the building permit shall be borne by the promoter the 1st party but the technical services shall be done by the engineer as included in the scope of the duties of the engineer.

3. The cost for the land surveys, the soil investigations and the explorations and all the other tests to be carried out before the designs of the project, shall be borne by the promoter, the first party.

4. This contract shall be operated on the guide lines laid down by the Institution of the civil engineers London UK specifications for the construction execution contracts in **the lump sum contract forms based on the itemised bills of quantities for the individual buildings and the related structures and all the other allied infrastructure operations and the utilities** as included in the planning of the project.

5. The tenders shall be floated by the engineer in different phases as agreed by the 1st party and scrutinised after being closed. The 1st party shall award the construction contracts only after having been recommended by the engineer.

6. The monthly running payments to the construction contractors for the valuations of the works executed/done by them shall be released only after the issue of the payment recommendation certificate by the engineer to the 1st party.

7. The facilities for the engineer and his site staff offices shall be provided by the construction contractors as included in the related bills of contract's quantity.

8. The complete tender and the contract documents shall be prepared by the engineer on behalf of the 1st party, the promoter for the job.

9. The 2nd party 'Engineer' shall charge a professional service fee of ...... % on the total cost of the project from the conceive stage to the completion stage and the release structure for that shall be as given below :

   a. 25% at the time of appointing the group as engineer based on the estimated project cost.

   b. 2nd 25% at the time of scheme layout submission for approval from the competent authorities also based on the project estimated cost.

   c. 3rd 25% at the time of issuing tenders to the interested parties also based on the project estimated cost.

   d. Remaining amount of the fee shall be distributed proportionately to the progress of the execution of the project on the site and the fixed sum be released at fixed interval of three months and shall also base on the real expense on the project.

10. a. The 1st party, the promoter shall release the structured payments immediately after the notice being served by the engineer on the 2nd party and the term as included 'immediately' shall mean a maximum time of seven calendar days of the issue of the notice by the engineer on the promoter's desk

    b. In case of the delayed payments by the 1st party to the construction contractors or the engineer or either of them, the construction contractors to be appointed later post completion of the tendering process, all the consequential damages shall be borne by the first party.

   c. Should the project/job duration be elongated by the 1st party due to any of the 1st party's constraints, the proportional extra cost as compensatory fee for the extended length of time shall be paid by the 1st party to the engineer.

11. The engineer shall act as an independent judge for this title job to dispose the technocratic responsibilities to the best and the optimum benefits of the job keeping in view the prospective interests of the 1st party the promoter simultaneously, treating the construction execution contractors fairly and professionally.

12. The legal disputes as often arise on the construction execution contracts shall be reserved to be covered under a clause in the construction contracts referring these disputes to an independent arbitrator to be mutually appointed by the 1st party and the relevant construction contractor/s according to the provision made in the Indian Arbitration Act in force for the construction contracts in the field of the civil engineering projects in case the parties in disputes do not accept the engineer's judgement.

13. The 1st party the promoter may deal in person with the engineer or may delegate any of the 1st party's authorised attorney/representative/agent/director, etc., and to this effect an authorisation shall be issued by the 1st party the promoter to the engineer clearly defining the powers invested with the agent/representative and the absent definition of the invested powers shall be construed as the authorised agent enjoys complete powers to deal with the engineer in all project and the professional respects but the promoter shall reserve exclusive right to replace, add, change, terminate, promote, etc., for the dealings with the engineer but prompt advise shall be made available to the engineer in the form of a confirmatory communication so that the unnecessary delays are not caused to the project progress decisions. In case of the engineer's assessment of the promoter's agent to be unmatching to the required professional standards to interact with the engineer in the relevant discussions and the decision making process, the promoter shall accept the engineer's recommendations for the replenishment of the deficit characteristics by either adding a subsidiary agent and/also by a substitute depending exclusively on the wish of the promoter's management. Nevertheless, the engineer shall take the decisions in all matters subject to the professional permissibility degrees and the standards. It would be preferable should a competent technocrat/engineer be deployed for that purpose.

14. The engineer undertakes to assist in providing all the services to the 1st party, the promoter utilising all the skills acquired by the members/partners of the engineer during their working/experiencing/being tutorialed/trained overseas for a couple of years on the building architectural engineering disciplines among various nationalities on European's contracts and technologies to properly utilise and apply to Indian state conditions.

15. No material shall be permitted to be installed for permanent inclusions without the prior material approval to be obtained by the contractors/suppliers from the engineer and the 1st party the promoter and the same conditional clause shall also apply to the detailing works of the technicalities to be run in all disciplines. All the shop working drawings shall be produced by the construction contractors and submitted to the engineer for the approval.

16. Meetings

   a. Monthly co-ordination and the technical meetings as well as the monthly progress monitoring and review meetings shall be conducted and chaired by the engineer or the resident engineer contract-wise during the construction contracts durations of the erection works and the 1st party be advised accordingly for the schedules.

   b. The 1st party shall enjoy the privilege to attend all such meetings for which the minutes of the meetings shall be circulated regularly

   c. The engineer shall attend all the promoter's project board meetings in the level. ENGINEER MEMBER.

17. The engineer shall make use of the Indian standards codes of practice where the engineer deem

fit in the interest of the job and to make savings to the 1st party with strict compliance not to obviate the standards.

18. The engineer shall keep the 1st party the promoter regularly informed and advised of budgetary controls and the alterations as required from time to time. The additions or alterations outside the title scope of this job shall be considered only after the prior written approval of the 1st party, the promoter.

19. The engineer shall reserve the right to advise the recommendations or rejections of any contractor or supplier on the ground of poor performance or the track record or professional gross misconduct, or meeting or not meeting the standards as required of the construction contractor or the supplier as the case be. In case the promoter does not honour the engineer's recommendations then the engineer shall be relieved of all the adverse results and the consequences by the applications of the aberrations of the engineer's recommendations on the job and a permanent record shall be referred on the project diary for the future investigative references for the inquiry commissions.

20. The 1st party the promoter may advise/suggest to the engineer of the bona fide intentions but the engineer shall not be bound to apply the same on the obligations of the duties as engineer. On the vice versa the engineer enjoys the reserve right and privileges to advise the promoter, recommend, suggest to the promoter on any professional matter according to the professional requirements and standards of practice in the international sphere of the civil engineering discipline's projects executions and the designs. In case the 1st party the promoter does not accept the recommendations, etc., of the engineer as a part of the engineer's obligations per contract and the promoter goes the unrecommended ways, the engineer shall not be bound to any liability as a consequence of any of the inclement results and the adverse gains of that on the contract job.

21. The project shall be administered based on the engineer's written guide to the contract management forming the copy right of the engineer and the legal competency of the project shall fall under the jurisdiction of the National Capital territory of Delhi courts of law/area of the job contract.

Promoter/1st Party   Engineer/2nd Party

CIVIL ENGINEERING CONTRACT

GENERAL CONDITIONS

General Conditions of Contract For Construction

1. **Scope** These conditions are incorporated for the applicability and the legal implyations to the civil and the architectural engineering construction contracts existing between the employing party (owner, promoter, user, govt. deptt.) and the constructing party the contractor appointed for the job of construction according to the scope of contract, by the employer based on the written offer (tender) in the category of any of the construction contracts types as defined in the project management guide for the execution and the consulting engineers forming the part of contract documents.

2 **Project management guide** A book/document forming the part of the contract documents detailing the complete guidance of the various parties involved to execute the job from the engineer's desk to the issue of the final completion of the contract. The project shall be operated in the framework of this guide.

3. **Definitions**

   a. **Owner/employer/1st party** The party appointing the construction party based on the acceptance of his tender in any category of contracts and promising making payments in lieu of the works carried out by the constructing party shall be called the first party or the employer or the owner or the user.

   b. **Contractor/2nd party** The party whose written offer has been accepted by the employer/1st party, for the construction of a job defined in the tender documents for the tendered amount/rates/cost/total sum or any other negotiated amount as agreed by the first party, to the contract conditions and documents in a fixed period of completion time shall be called the 2nd party/contractor for the job.

   c. **Engineer** The person/group/firm as appointed by the 1st party to this contract for undertaking the complete scheme/project from the conceive stage to the final completion shall be called engineer for this project. The engineer shall be an independent person/party to this contract.

4. **Liabilities**

   a. **Contractor**

      i. Contractor shall take over the site from the engineer/resident engineer on the date of site hand over to the contractor/the date of the contract commencement and shall commence the mobilisation works followed by the contract scope per contract documents as agreed between the owner and the contractor.

      ii. Contractor shall complete all the elements of various disciplines as defined and detailed according to all the tender, contract documents and per the specific contract conditions or the letter of explanations.

      iii. Contractor shall handover to the employer/1st party the completed works in all respects at the time of practical completion for the employer's use.

      iv. Contractor shall protect and maintain all the works during the maintenance period as defined in the specific conditions/letter of explanations till the final completion of all the works is certified by the engineer.

      v. Contractor shall be liable for the expenses to be incurred on the works, workpersons, materials, machines and all other contingencies and overheads as specifically detailed in the specific conditions/letter of explanations.

      vi. Contractor shall be liable for all the execution responsibilities/liabilities and safety of all during construction.

      vii. Contractor shall be liable to the execution of the works to the satisfaction of the engineer and the owner in an overall consideration and adhere to the contract documents as agreed by him before signing this contract.

   b. **Owner**

      i. The owner/1st party/user shall be liable for making all the payments for the approved valuations of the contractor's works in a reasonable time of 30-90 days of the date of valuation certificate to the contractor's account.

      ii. The owner shall be liable for making all the payments for the extras as approved and the variations orders issued by the engineer and forms the part of the valuations.

      iii. The owner shall be liable to grant the extension of time to the contract completion duration as recommended by the engineer based on the contractor's claims.

      iv. The owner shall be liable to face arbitration in case of disputes remaining unresolved between the owner and the contractor and be liable to make payments as awarded by the arbitrator and decreed by the high court per Arbitration Act provision in force.

c. Engineer

i. Engineer shall be liable to arrange for the complete supervision of the job, administer the contract, provide all the informations requested and required by the contractor till the works are completed.

ii. Engineer shall be liable for presenting the completed contract to the employer/1st party according to the specific conditions of contract or the letter of explanations and all the liabilities per management contract existing between the owner and engineer. Engineer shall be sole responsible for design part of the job.

5. Responsibilities

a. Contractor

Responsible for the complete execution of the job according to the contract conditions and the documents in a fixed time based on the approved programme, on the direction and the administrative control of the contract of construction and supervision by the engineer/owner.

b. Owner

To pay the contractor in time for the works executed by him per contract documents till the contract is completed. There should be no default in the payments.

c. Engineer

To direct, administer the contract and supervising all the operations on the contract site per contract documents and approve or reject the works. It is added that the design has been done by the engineer and therefore the 100% responsibility for that part goes to the engineer till the engineer on supervision and on the design is the same person or the firm.

Sometimes, the engineer responsible for the supervision is different who shall be having the same responsibility as the engineer vested with the complete responsibility of the contract administration, contract operation control and the site operations supervision, etc.

6. Powers

Contractor

a. Contractor shall have the power to take over the site of the contract scope from the engineer having entered into the contract with the employer or the owner.

b. Contractor shall have power to request the engineer to issue all the contract documents and all the related details of the site.

c. Contractor shall have the power to obtain the record drawings from the statutory bodies on the authorisation given by the engineer.

d. Contractor shall have the power to commence and complete the works.

e. Contractor shall have the power to obtain the clearance of the contract site for the commencement of the related operations.

f. Contractor shall have the power to serve the notices on the engineer and the owner in case the control of the contract is not feasible and justifiable.

g. Contractor shall have the power to carry out the job and let the same be approved by the engineer or the representative failing which the contractor shall continue the job.

h. Contractor shall have the powers to get the payments of the works done by him by proposing the monthly valuations or as agreed.

i. Contractor shall have powers to use the allocated areas of the site allocated to him for the purposes of contract operations and shall reserve the rights to secure the site with the prior approval of the engineer for the system.

j. Contractor shall have the power to refer the unresolved disputes to the arbitration, the arbitrator appointed by the mutual consent of both contract parties in case the decision given by the engineer is not acceptable to them. For that a notice shall be served by the contractor on the engineer and the owner not later than one month of the day of the dispute arises.

k. Contractor shall have all the powers over his organisational staff levels and of the sub-contractor/s and appointing any employee for the job.

l. Contractor shall have all the powers as vested on him by virtue of specific conditions of contract/letter of explanations and included in management guide wholly.

m. Contractor shall have all the powers to raise claims according to contract conditions

n. Contractor shall have all the powers to hand over the completed site/works to the owner/employer and obtain completion finally for the retention money release.

Owner

a. Owner shall have the power to get the works awarded to the contractor, done according to the contract documents and supervised by the engineer of the owner's choice.

b. Owner shall have the power to enter the site any time during the site being on.

c. Owner shall have the power to add, delete or omit any item to the contract.

d. Owner shall have the power to change the design of works in parts or full to the new needs.

e. Owner shall have the power to hold/stop the payments on the engineer's recommendations.

f. Owner shall have the power to impose any penalty on the contractor according to the provisions made in the contract specific conditions/letter of the explanations should the works are not completed and handed over in time without the grant of the extension of the delayed period and may claim any other damage on that account if provided specifically.

g. Owner shall have the power to terminate the contract or rescind the contract and employ another agency to carry out the remaining portion of the contract on the risk and cost of the contractor in case of major defaults on the part of the contractor in carrying out the contract works quantitatively or qualitatively as also provided in the specific conditions of the contract or the letter of explanations.

Engineer

Since engineer shall be fully responsible to the design, administer and supervise the scheme/project as a whole, has the powers virtually conferred on him regarding issuing the directive instructions, post contract technical working details, supervision on the operational fronts, direction as a whole, control of the costs, recommend the release of the payments of the works done by the contractor, etc., all as included in the management contract existing between the engineer and the owner/user/contractor's employer till the construction contract is given a completion certificate finally by the engineer and all as covered under the specific conditions of the contract/ letter of explanations.

7. **Insurance**

a. The third party insurance of all the employees working on the contract shall be made or done or provided by the contractor as also included in the specific conditions of the contract or the letter of explanations.

b. The engineer shall insure all his employees for the same coverage and protections.

c. The contractor shall also get insured all the operations done or to be done on site if permissible under the regulations and the existing insurance rules.

8. **Legal binding**

The construction contract as exists between the employer/owner/user as a party to the contract in the capacity of 1st party and the construction contractor in the capacity of the 2nd party to the contract shall be legalised/bounded according to the procedures and the laws in force in the area of the works.

The contract shall be based on a written offer by the 2nd party/contractor and not without that and any amendments made by both parties during negotiations or as a qualifying condition shall be agreed before the contract agreement being signed or entered in legally as without the written acceptance of the amendments by the owner, the contract shall not be treated valid and not form part of the contract and contract documents as aforesaid as far as legalities and implyations of the contract laws are concerned.

THE SPECIFIC CONDITIONS OF THE CONSTRUCTION CONTRACTS
OR
EXTENSIONS/EXPLANATIONS OF THE GENERAL CONDITIONS
COMMUNICATING INTENTIONS OF THE OWNER/EMPLOYER
AND THE ENGINEER AT THE TIME OF SIGNING
THE CONSTRUCTION CONTRACT
AND FOR THE CONSIDERATIONS OF THE LEGAL IMPLYATIONS
AND IMPLICATIONS AND THE LEGAL MEANING
TO AND FOR THE APPLICABILITY OF THE LAW
IN CONTINUATION OF THE GENERAL CONDITIONS
OF THE CIVIL ENGINEERING CONTRACT

INDEX PAGE

CONTENTS OF THE Letter of explanations/specific conditions of the agreement between the Promoter and the Construction execution contractor forming contract forms

1. **SCOPE OF WORK** Construction of Township phase ....
....
....

2. **FIRST PARTY** First party means the promoter/employer/owner/builder, for whom this work is to be carried out/erected, this part is to be treated to be in the status of the governing body as in the government departments/construction contracts. In this case M/s ....
....
....

shall be termed First party in this contract.

3. **SECOND PARTY** Second party in this contract means the construction agency/contractor whose offer/tender/direct award/by any other means to do the job as specified in the scope of work in clause I above per complete contract documents, i.e., the architectural; structural; civil; allied services schematic details; engineer's details; general details; standard drawings; specific drawings; all schedules; bills of quantities; project and general specifications; all communications and instructions from the issue of the tender documents to the time of signing the contract/this contract; all the working drawings proposed by the contractor for the engineer's approval (duly approved by the engineer) and all other details & related documents supposed or implied to be a part of this contract documents as an inclusion for the execution of the technicalities or the engineering part, etc., for a fixed sum (Lump Sum) as given in the contract form in the appendix...has been accepted.

   In this case M/s or Mr/Mrs/Agency/Firm..
....
....

shall be called/termed the second party to this contract.

4. **ENGINEER** Engineer means the person/firm/group of building engineering consultants architectural engineering consultants/technocrative consultants as appointed by the First party as in clause 2 'To conceive; design; direct; administer; and supervise the complete construction execution contract operations (operations as included in this contract) to be carried out by the second party per contract documents as defined in the clause 22 of this contract till the final completion of this contract. Engineer vests/owns with/the responsibility/obligation to present the finally completed contract to the first party in all respects of the scope of contract works. In this case M/s SKY Associates building engineers headed by the Chartered engineer (India) Suraj Singh MIE shall be the ENGINEER for contract.

5. **SITE CONDITIONS** The second party shall be deemed to have satisfied itself as regards existing roads; railways; or any other means of transport in operation to and from the proposed contract site/area; communications and access to the job site/area; the nature of the soil and the excavations (soil investigation report as included for the guidance only); all the conditions in which the construction execution agency has to execute/carry out the contract job; all the facilities to procure or obtain the required materials as a part of specifications; general as well as particular materials; all the concrete ingredients and the availability of water; availability of the workperson power and arrangements of the equipments and machines to be engaged on the job; and must ascertain itself of feasibility, ability and capacity before signing this offer due to the work being very prestigious, time bound and no claims of any type corresponding the foregoings in this para shall be admitted once this contract comes into the operational effect. The 2nd party should include/make provisions of all the related/relevant expenses in the bill of preambles or some where else. Should the 2nd party does/do not provide the same in the required provisional space/columns, it shall be assumed/construed that the same has been included somewhere else in the other bills of quantities and no claims shall be admissible later for that omission/condonation, irrespective of the additional expenses to be incurred by the

2nd party, the contractor.

6. HANDING OVER THE SITE

i. The contractor shall be handed over the site in his/its contract immediately after the finalisation of the contract/acceptance of the tenderer's offer as the case be and the actual date of commencement shall be counted from the same day as of handover practically subject to the contractor been efficient to take over from the engineer and in case the contractor intentionally delays the take over due to the contractor's internal problems protecting the agency to mobilise then the date of commencement shall be counted w.e.f. the last day of the time given in the work award/acceptance of the contractor's offer/tender irrespective of the actual day of the take over by the contractor and the contractor shall not raise any claim for the extension of the time due to the failure in taking over during the defined/notified period.

ii. There may be an instance where the contractor shall be handed over a part of this contract area and the balance later due to some unavoidable reason, an attempt shall be made to adjust the contractor's construction programme accordingly or in case of this adjustment not working, a proportional to the contract capacity based on the quantities, date ahead of the actual hand over date in the same percentage of the contract time as the percentage of the un-handed over area shall be fixed as the monitoring date for the contract completion as far as this part of the contract is concerned subject to the parties to this contract being in agreement on the recommendations of the engineer.

7. MOBILISATION

i. The 2nd party shall mobilise the site within a period of 30 days but this period may be reduced to 14 days or two weeks in case of small size jobs/medium size jobs depending on the area as well as nature of job, from the day of handover and shall include all the site establishments of the 2nd party and the standard offices with the specified facilities as specified in the general specifications, alongwith the provision of the complete engineer's and the resident engineer's offices, establishments of the complete workperson accommodations to the entire satisfaction of the labour laws per area codes, procurement of all the necessary machines and the equipments to commence the job in full swing all subject to the engineer's inspection or his resident's inspection. The contractor shall erect a security fence in consultation with the resident engineer/engineer but the same shall be accepted subject to the alterations as and when required to let the other agencies working in the same area of this contract and the surroundings areas so that the advancement of the other contractors shall not hinder due to this contract.

ii. The engineer shall hold a pre site handover meeting to introduce each other and in that the contractor must table the complete site mobilisation sketches and obtain the approvals. This shall save the unusual wastage of time of a couple of days. How to proceed on the job shall also be outlined.

8. PROGRAMME

i. The 2nd party shall propose the detailed programme of construction in the networks of CPM (Critical Path Method) and the bar charts exhibiting weekly break up, the rate of progress as proposed, the man power/work person power and the machines to be deployed, the levels and strength of the supervisory staff and all other informations as considered useful for the monitoring purposes during the progress review. This programme should be proposed in a period of 14 days w.e.f. the day of commencement of contract building vise and also an overall comprehensive programme of the job contract works.

ii. The individual programmes shall be made a tool to calculate the quantity of the job done and any lag ahead or behind or per schedule while the comprehensive programme shall indicate the overall status of the contract. All the recovery programmes if necessitated shall be founded on these programmes. Having received the contractor's construction programmes, the engineer shall go through, discuss with his group and the contractor's agent and after amendments shall grant approvals. Once the approval is granted, the contractor shall be bound to move accordingly or the engineer/his representative/resident shall have every right to question and interfere till the satisfying answers or assurances are given by the 2nd party to replenish the losses if any aberrations take place on the job due to that.

iii. Should there be major deviations from the approved programme for a certain period in the contract due to any reason, the contractor shall bring that to the engineer's notice and produce the evidence by producing another programme for that period convincing the engineer that no overall loss in the programme shall be inculcated and there should be no concern of recovery, otherwise the recovery programme shall be given by the contractor.

iv. In addition to the bar charts as explained above the contractor shall be asked to produce the regular quarterly programmes completely detailed for the purpose of site inspectors' adjustments or any other reason.

v. The contractor shall also regularly provide the engineer representative the weekly programmes at the last day of the week's morning to enable the resident engineer deploy the supervising staff accordingly and also provide an idea to the contractor's agent for his guidance throughout the week for the deployment.

9. **CONTRACTOR'S SUPERINTENDENCE** The contractor shall propose the detailed CVs, i.e., curriculum vitae of all the staffing levels at control, execution, supervision on site which shall include the proposed site agent, deputy agents, sub agents, co-ordinators, site engineers, assistant site engineers, junior engineers, general foreman, land surveyors, foremen incharges, and the quantity surveyors, etc., in a complete package alongwith the testimonials and 6 photographs duly signed by the contractor.

The CVs shall be scrutinised by the engineer and if found to the acceptable standards, an approval shall be granted subject to the observance of job performance. The CVs may be invalidated at any time if the marked performance is not met and in that case the contractor shall be asked to deploy another person/staff and submit the CVs for approval. No staff of the cadres from control to the on site supervision shall be allowed to work without the approval of the engineer.

10. **SUB CONTRACTOR** The contractor has to produce the CVs of all the sub-contractors proposed on the job which may include engineering, electrical, plumbing and drainage, any specific item such as swimming pool systems, sports courts surfacing, etc., and may depend upon the main contractor how he/agency wishes to move the job. Whatever be the contractor's channels the complete packages of all the CVs shall be submitted by the contractor for the approval and this package shall include descriptive appraisal of the agency proposed, its capacity, past works successfully carried out supported by the evidences, licences granted by the statutory organisations/authorities confirming the proposed agency's discipline's specialisation, financial status of the agency the details of the running works in hand, the detailed CVs of all the staff levels as on the same pattern per clause 9 above. Unapproved sub contractors shall not be permitted to work on the job. The approval shall be granted only after the scrutiny of the submissions and communicated accordingly. In case of unacceptance, the main contractor shall immediately propose another agency for further scrutiny and approval.

11. **SUB CONTRACTORS**

i. This clause shall include all the proposed regular suppliers of the important materials to be used for the permanent inclusions for which an approval/must be obtained.

ii. In this category may be included the suppliers of the materials such as cements, reinforcing steels, concrete blocks or the bricks, concrete ingredients as aggregates, source of water, claddings, doors and windows suppliers, lift manufacturers and any other agency/category as required by the engineer/resident engineer from time to time and should be submitted on the standard forms as described some where else in the management guide (part/document of contract).

12. **MATERIAL SUBMITTALS**

i. The contractor as well as the sub-contractors through the main contractor shall make submittals of all the materials as listed in the material list/schedule of materials for the permanent inclusions, within a period of 14 days of the day of contract commencement. Every material shall be submitted separately in triplicate on the standard form/proforma detailing the names of the material, b.o.q. reference, specification, name of the manufacturers/suppliers, IS/relevant codes references, literature and samples, etc. The submission shall be sealed and signed by the contractor/agent and numbered for the reference back. The material shall be submitted in the office of the resident engineer on site and a receipt obtained.

ii. The approval/rejection/resubmission, i.e., approval with comments shall be communicated to the contractor's box provided in the resident engineer's office. It is in the interest of the contractor to check from time to time about the status of the approval. The material should not be placed in the purchase order before the approval is obtained. The unapproved material or the materials not conforming to the sample as approved shall be ordered to be removed from the site immediately.

iii. All the samples of the approved materials shall be kept in the sample room under the key of the resident engineer to facilitate the easy inspection of any material being used or being brought to site any time.

13. **SHOP/WORKING DRAWINGS**

i. The contractor and sub-contractors through the main contractor shall submit in duplicate all the shop working drawings in a period of 60 days at the most from the day of the contract commencement. The drawings shall include all the disciplines such as engineering, electrical, plumbing, drainage, pool services, lifts, reflected ceilings, fixing details of various buildings materials as asked for by the engineer or the representative, and all the other relevant drawings not covered in this para.

ii. In the absence of the shop/working drawings duly approved and sealed by the resident engineer, the work shall not be allowed to be carried out and if carried out shall not be inspected and no payment shall be made for that item done. The details of the shop and working drawings procedures have been elaborated in the contents of the management guide as a part of this contract documents yet a brief procedure is outlined hereunder :

a. Produce the technical details of all the routes and their sizes and complete technological details in co-ordination to the services schematic engineer's details issued as a part of tender/contract documents and also co-ordinated to the other various relevant disciplines and the specific design requirements as specified in the general and the project specifications.

b. The details of your design and co-ordinations must conform to fit in according to the material submittals as approved or likely to be approved otherwise the system you are proposing shall not work successfully and you shall have to redo the whole exercise at your cost.

c. Make submittals of your proposal in triplicate to the main contractor on the standard form for the shop/working drawings submittals duly sealed and signed and the main contractor in turn/counter forwardance of the submittal to the resident engineer shall scrutinise at that/his level seal and sign in duplicate and retain one copy with him in the custody of the site agent.

d. The contractor's services engineer/engineer of the relevant/related field as approved by the engineer shall pursue the case in person with the resident engineer's specialist engineer and get the approval of the drawing.

e. In case of the rejection of the drawing due to any technical reason, comments shall be made on the returned drawing for the resubmissions and after comments being incorporated with the scrutinized requirements, approval shall be given.

f. The drawings shall be marked by the resident engineer/engineer as follows.

I. APPROVED II. APPROVED WITH COMMENTS III. REJECTED

Approved shall mean that there is no need for the resubmission and the contractor may go ahead with the execution of the related portion of the contract. The sub-contractor shall submit the six prints of the approved drawings to the main contractor duly sealing 'APPROVED' on every copy/print and the main contractor shall get it done sealed and signed by the resident engineer/specialist engineer of the relevant/concerned discipline. Having the prints been signed by the resident engineer, the main contractor shall issue two copies to the resident engineer, retain two with him/agent and issue the remaining two to the sub-contractor.

Approved with comments shall mean that the drawings must be resubmitted duly just incorporating the comments as made. Once the comments are incorporated, resubmittal shall be done similarly as the original submittal using the same ref. no. just advancing the suffixes. The drawing shall be considered finally approved only when it bears on it the 'APPROVED' mark. The 'REJECTED' shall mean there are major faults, snags, discrepancies and should also be resubmitted with the original ref. no. with advanced suffixes. The inspection of the works carried out in the sub title/heads of the items of services shall only base on the approved working/shop drawings and in no case on the basis of the schematic tender/contract drawings. So the urgency of these approvals of the shop/working drawings must be given weight by the contractor/sub-contractors in the interest of the job and also to keep the approved construction programme in line with the actual execution.

14. MATERIAL

i. No material of any kind on this very important job which shall be engineered on the international standards covering the advanced building technology being applied to this country's/state's conditions, shall be permitted to be brought on site or included in the permanent structures. Alternative use of the materials shall be permitted in the exceptional cases where the contractor produces the evidence that the procurement of the material in question is outside the control other than the fiscal reasons which is the contractor's risk and the contractor shall make the fresh submittals of the alternative materials mentioning on the alternative submittals, enclosing the reasons and evidences. Once the alternative material is approved, the previous approval shall be superseded and the contractor shall not be permitted to use for the same item both the materials. So in doing this the contractor must exercise due care of his own risks before pricing this offer.

ii. Similar conditions shall apply to the approved working drawings. The contractor has to carry the work out accordingly till the site conditions restrict the approved system or there is some clash among various disciplines for which separate solutions shall be found out in each case on site and required amendment done in the system by re-routing or re-levelling or adjusting in the elevation but that shall be recorded in the 'as built drawings/record drawings.

15. SAMPLES

All the samples of the approved materials shall be placed in the sample room of the resident engineer's office under his lock and key on site. The cost of all such samples shall be borne by the contractor as well as of the room for which the contractor must make provision in the offer to meet all these expenses and any other later on related expense.

16. MATERIAL DELIVERY

i. All material deliveries procured on site shall be regularly informed to the resident engineer by the contractor alongwith the supporting source documents enclosed in original alongwith the daily report/job record of the contract. The engineer/resident engineer shall have every right to inspect the delivery and ask for the delivered material samples. The contractor must assist the engineer or his resident engineer in performing the inspection without any obstruction so that the material acceptance be given efficiently subject to the proper storage and protection according to the manufacturer's instructions or the standard methods of protection and storage or as directed by the resident engineer/engineer. In case that is not done, the material arrived shall not be counted/included in the material on site accounts for the purpose of valuations.

ii. In case the material arrived on site after inspection is rejected by the engineer or the resident engineer, prompt action shall be taken by the contractor or the agent to remove the delivery off from the site as soon as possible after the issue of the site memorandum/inspection, etc.

iii. In case the engineer/resident engineer desires the samples to be taken to the approved laboratory for further verification or the regular/routine examination, the samples shall be sent/dispatched to the nominated and approved laboratory duly collected by the resident engineer from the contractor/agent and a covering letter addressed to the laboratory by the resident engineer/engineer and a reference quoted in the daily job record in respect to that. Any/some action as necessarily required in the test reports of all the samples and as directed by the resident engineer/engineer, shall be promptly and actively taken by the contractor/agent to satisfy the engineer. The contractor shall make provisions for all these expenses in the offer.

iv. Even after doing all this, the contractor shall not be relieved of the responsibilities, obligations and risks of that on the job. There may be some materials that necessitate the regular tests at every delivery or at a fixed interval, the contractor should promptly perform the tests through the approved nominated laboratory and send on the report to the resident engineer for information and necessary actions, if any.

17. **ALTERNATIVE PROPOSALS** The contractor if so wishes may propose any alternative details for the substitution of the engineer's details for some particular section of the job which may render saving of time, saving of money to the contract with no drawdown in the maintenance of the accepted or implied standards, but no additional/extra cost. The same cost as in contract may be accepted in case the proposal appears to be better than the engineer's details and making saving of time but this shall not be admissible every now and then but subject to the concrete proposals in the beginning with competent and due practical demonstrations and no alterations accepted in this clause shall be admitted to any extra variation or make any legal ground for arbitration/suit.

18. **WATCH AND WARD OF SITE** The contractor shall make all necessary arrangements for the safety on site for all men and women working in the contract area, all machines used on site, all equipments on site perishable and imperishable, all safety measures/precautions/cautions in all the construction operations/activities such as the shoring of the excavated trenches, to provide necessary scaffolding or platform for the persons to stand and work, to provide a safety net on high rise structures/buildings, during all the working hours and in all the preparatory works carried on off/on hours. The contractor shall engage proper guards on the site, and erect/deploy the necessary hoardings to indicate/illustrate the contract area. The contractor shall take all necessary means not limited to the foregoing/said in this para/clause to the satisfaction of the engineer/resident engineer and the safety standards statutory acceptable.

19. **SITE ARRANGEMENTS** The contractor shall make all necessary arrangements for providing ample drinking water for the work force on site, toilets with the septic tanks to the specified standards/or any other alternative arrangements to prevent the health hazard to break out, electric supply from the main feeder of the local/state electricity board/body/department and also the required number of standby generators for the work as well as for the use of the work force. In addition to that the tea, coffee or the accepted refreshments as agreed per labour contracts and the provision of the necessary messes for the work force food, provision of creches for the children of the work force if employed that way, the provisions of the medial facility, medical first aid on every part of the contract site, a trained compounder all the time deployed on the site alongwith the regular visit by a competent doctor, and a transport stand by all the time awaiting any emergency call in case of major or minor accidents occurring or any mishappening taking place, regular transport to bring and take back all the workforce from and to the camps, and all these facilities must remain in working order all the time efficiently. These arrangements as explained in the para are not the end/or limited to, but shall also match the standards and the real requirements of site as statutory defined by the labour laws

as well to the satisfaction of the labour departments and also the engineering profession acceptable standards. The contractor should make provision for all these expenses in the offer/tender.

20. **SITE DIARY** The contractor shall maintain a site diary in triplicate in the A-4 size all the time to be kept on site in the custody of the site agent and shall be asked any time by the engineer or any of his resident/representative. The diary shall be filled up by the site agent/contractor's agent daily recording the planning, the activities commenced, the activities completed, the work force on site, the sub-contractors and their similar details, the material arrived on site, the machines arrived on site, the machines removed from the site, the tests conducted, the inspections held, the accidents occurred, the approximate quantity of the works carried out, (by the contractor and the sub contractors both), the details of the visitors to site, the approvals required the next day or the days to pour in, the weather conditions, the maximum and minimum atmospheric temperatures, and any other happening left uncovered in the foregoing description, e.g., meeting held. The site diary is very important documents/records, helps both the contractor and the engineer in the future when there is any argument about the conditions on site, or the work carried out on certain date/day, the doubts about the time of arrival of the materials or the visitors inspection times, the engineer or the resident engineer may make it useful for any clarification as the general comments in the site diary of the resident engineer and that of the site agent shall be similar. Even some notes could be written in that to make use for the future benefits to recollect/recall the happenings. This document shall not be made a tool of evidence in the arbitration but should be granted a degree of standard practice professionally.

21. **RIGHT TO SITE** The employer, i.e., the first party, the engineer and all his staff shall have the exclusive right to enter the site and inspect the operations any time. The contractor shall provide all necessary means to assist these people to accomplish the inspections.

22. **CONTRACT DOCUMENTS** All the documents forming part of this contract as a whole. These shall include the contract bond, (performance bond), job schedule, scope of work, the amount priced for completing the job, all the qualifications as agreed before the finalisation of this contract, all the general conditions, the specific conditions, any assurance accepted by the contractor in writing before signing the contract, all the general specifications, the project specifications, the manufacturer's specifications for the approved materials, all the communications from the tender being open to the closing day/date in the form of telex/facs/or in general meeting records of all the tenderers in the engineer's office or on site, any communication issued from the engineer's office before the closure of the tender, all the changes made before the closure in the bills of quantities, all the architectural, structural, services which shall include engineering, air-conditioning, plumbing and drainage, civil works, bar bending schedules, the general and specific details in the form of A-4 size books (volumes), all the working drawings to be approved later, and all the other matters as defined or described anywhere else in these documents and not restricted/limited to the as said/foregoing but referred to the implied meaning/s and the professional intention/s of the engineer at the time of forming the contract, etc. These documents shall remain in force till the final completion is granted to the contractor by the engineer at the time of successful completion of the maintenance period of the contract which is 400 days in this contract after the grant of the practical completion of the contract but excluding the items forming/carrying the long term warranty/guarantees upto 5 years or more as the case be which shall be treated separately for the purpose of the contract as those shall remain in force till the successful completion of the warranty/guarantee. All these documents except the working drawings (a part of the contractor's work) shall be issued to the contractor immediately after the award of the contract. The contractor should ascertain that all these documents in duplicate are handed over to contractor/him in time. In case of additional copies of the documents, the contractor shall have to pay the additional cost of producing the copies, to the engineer and shall make a request for that.

23. **CONTRACT DOCUMENTS ON SITE**
    i. The contractor shall keep one full/complete set of the contract documents on site in the custody of the agent who shall make the same or any part of that available to the engineer or any of engineer's representatives whenever, that is asked for/requested for the purpose of consultancy or for reference.
    ii. The negative copies of the documents shall not be issued to the contractor as those form the copy right over the documents of the engineer as also all the project details shall be covered in the copy right hence only to be used for the execution of this job/contract. It is advisable to the contractor to produce more prints himself of these documents for the use of the middle level officers on site through the agent as these officers may spoil the original issue/contract issue, but this reproduction of the documents shall be carried out under permission of the engineer only and not without that in any case.

24. **COPYRIGHT AND SECRECY** The contractor or any of the contractor's staff shall not reveal the details of this contract to any person not related to this contract directly or indirectly. The contractor shall also not let anybody else copy out the concepts and technical details for the purpose of another professional use which may mar the prospects of the employer's market and engineer's practice. In such a violation of the clause, severe concern may rise to the level of rescinding the contract at any stage of execution and get the job done by another agency at the risk and cost of the appointed execution contractor/II party/original contractor.

25. **STATUTORY REGULATIONS** The contractor shall keep the local bodies/organisations informed regarding the commencement of various works and arrange from time to time through the assistance of the engineer or the representatives to obtain all the services record/as built records, which are very necessary for the contractor to excavate the foundations or the new services routes safe and protect the existing routes of any services from being damaged by the excavation. Remember, this comes under the duties of the contractor's agent or the contractor personally to safely carry out the excavations. The contractor shall liase with the local departments throughout the contract durations to obtain the necessary approvals for various executed/carried out/as built items from time to time to expeditiously accomplish the contract without delay. The contractor shall also liase with the other main contractors and sub main contractors working on the other phases/main contracts, to co-ordinate with them regularly to avoid any work clash or programme clash. The expenses for all these activities/items as explained in this clause para in the foregoing should be included in the tender/offer.

26. **WAY LEAVE FORMS** This means the set backs to be left unused/uncovered/unoccupied and exclusively for the access of the concerned government departments and is very common and special term in the construction contracts. At the conceive/design stage it goes to the responsibility of the engineer to check that the forms are not violated or do not contravene the statutory bye-laws as in force at the time of the project design. These may include the width of the access road to the project site from the main highway, space to be provided on both sides of an electric high voltage overhead pilon, or any other specified requirement and existing property on the land at the time of the design which the land surveyor shall mark in the plane tabling detail of the plot survey/topographical details. But there may be various instances at the time of the contract execution that some property comes in or near the way of the contract area in which case the contractor shall have to contact the concerned local department for the bye-laws or the removal of the same or whatever it may be, that shall be brought to the notice of the engineer. In case the way leave item is brought to the notice of the engineer in time, the engineer may take necessary steps to clear the way for the contract project to run/move unhindered in the interest of the job and contractor too.

27 **MAN POWER/WORK FORCE** The contractor shall have to arrange the workforce of all the trades including that of the sub-contractors as agreed in the construction programmes schedules from time to time and the engineer or the resident engineer shall have the authority to question about the deficit workforce/power. Complete habitation for all the work persons shall be arranged by the contractor in all respects as also included in respective clause 19 of these conditions. The contractor shall take the complete workforce off the site to their homes

immediately after the practical completion is granted to the contractor by the engineer/party I except the force required to maintain the works. The formal practical completion shall be granted only after the site is cleared in all respects to the satisfaction of the engineer and the employer. The contractor should make provision of all these expenses in his offer.

28. **THIRD PARTY INSURANCE** The contractor shall arrange for the third party insurance of all the workforce on site and these documents be given to the engineer as a matter of contract record. The contractor shall include all the probable expenses on the operation of this clause in his offer/tender.

29. **ACCIDENT BENEFITS/COMPENSATION** The expenses to be incurred on all the accidents on site according to the provisions in the accident prevention acts/codes and the disability compensation benefits according to the relevant labour acts in force during the construction duration, as a liability of the contractor/party II, should be included in the offer at the tendering stage as a part of tender sum.

30. **SAFETY ON SITE** All the measures and cautions as required by the operation of law shall be observed on site and come under the direct responsibility of the site agent and the agent's senior staff on contractor's behalf till the payment problems are not posed/created by the contractor to let the agent and the senior staff dispose the obligations and their responsibilities successfully towards the operation of this clause/condition. All persons on site shall wear the hard hats and standard work uniform on which the names of the work persons/employees alongwith the name of the company/contractor/party II must exhibit. Gum shoes shall be worn by all persons working on any item where the damage to the legs/feet is susceptible. All the nails after use as remain embedded in the scantlings, battens, parts of the forms, or in the soil of the ground or in the working and the surrounding areas shall be frequently removed by the contractor as it is very dangerous and always given little care. If the nail punctures through the foot or pierces through, it shall take a number of days of the person to recover fit to work/job and also may create danger of the victim to be septic. The engineer shall not tolerate at any cost this carelessness or neglect or negligence. Refer to the site safety measures please. (Law and codes)

31. **ACCIDENTS** In case of occurrence of the accident the contractor's agent shall at once take the action to transport the victim patient to the hospital and inform in details in the standard proforma about the accident to the engineer and the insurance agency for the prompt and efficient follow up actions.

32. **NON-PAYMENT COMPLAINTS** In case of any complaint of non payment by the contractor to any person working on this contract directly or indirectly is received in engineer's/resident engineer's office, an immediate explanation of the contractor shall be called on and in case of unsatisfactory reply, the complaint claimed amount shall be recovered from the contractor's valuations and paid to the incumbent/complainant.

33. **WORKING HOURS** The net working hours on site shall be eight hours a day and a convenient lunch break in addition. The working days in a week shall be six. Should the contractor wishes to prolong the working hours on site due to contractor's tight programme or to recover the lost time, the contractor shall inform at least 48 hours in advance of the intentions so that the arrangements of the site supervising engineers/clerk of works are properly arranged. In case the contractor does not inform in time, the activities/items carried out in the absence of the engineer's representative shall be liable to rejection depending upon the nature of the items involved. The preparatory works for the concrete pour shall however, not cause any problem as far as the forms of the suspended slabs are concerned. Preparatory works shall include the erection of forms of suspended slabs, beams, fabrication of steel reinforcement/rebar, levelling, alignments of forms, reinforcements placements and miscellaneous services elements, etc.

34. **REFRESHMENTS** The contractor shall include in the price of contract all the minor expenses for the tea/coffee/cold water/soft drinks, etc., during inspection time when the supervising engineers/engineers representatives/COWs are on site for this purpose for hours together off the office or the market. This gratitude courtesy shall not be considered as bribe.

35. **BRIBE** The contractor shall not bribe the engineer, his resident engineer, or any of the engineer's staff directly or indirectly in any manner, by way of gifts or personal favours or promises which may smell the smoke of the bribe. On festivals or special ceremonies of the engineer's staff, the greetings shall be accepted only. Parties can be attended as far as the triumph over the programme or over certain complicated portion of the job or some complicated disputes are solved or sometimes, in tension/strain/depression/monotony the engineer's staff needs the company of the professional counter part/colleague of the contractor say the agent, or the brilliant site engineer, etc., but that should limit/confine to the personal drinks. In case it is proved that the contractor tried to bribe any of the employer's staff or the engineer's staff or the engineer or threaten to give a favour or may cause damage to the engineer or any of the engineer's staff or the staff of the employer/the party I, the contract may be liable to be rescinded and the remaining work of contract shall be carried out by the appointment of another agency or by the employer's own staff at the risk and the cost of the contractor in this contract. The contractor is advised to include all the probable expenses in case the contractor feels to spend or loose on account of the operation of this clause of no bribery in the tender sum or/otherwise, the contractor shall repent on this contract.

36. **PROGRESS**

    i. Once the contractor's construction programme is approved by the engineer the contractor shall be bound to carry out the works accordingly and the engineer or the resident engineer shall enjoy every right to question the contractor of the activities/operations being carried out in contravention to the approved construction programme. The contractor shall monitor the progress according to this/the approved programme as the engineer shall too do the same way. The contractor shall have to give the reasons in case the contractor aberrates the programme and make sure that the overall programme is not adversely effected or goes out of line leaving the progress far behind/uncontrollable/unrecoverable at the end of the programmed completion period. The contractor shall be responsible for weekly monitoring at contractor's level.

    ii. In case when there is a major aberration/deviation on the approved programme lines, the contractor shall have to reproduce the whole programme or if that is not necessary per engineer's opinion, the recovery programme shall be asked for for a particular section of the construction programme or even the revised programme to substitute the distorted period and the both programmes shall jointly be employed/used for the monitoring and the progress review.

    iii. The contractor should bear in mind that this contract project is very much qualitative, time bound and fabricated/framed/manipulated excuses shall not work to convince the engineer or the employer/party I. The contractor must be clear and firm on what the contractor states in writing in this connection, if found false, it shall amount to the breach of trust due to the trust being very important factor professionally in these types of project contracts and in fact the win of trust of the engineer and the employer is an asset for the contractor in this discipline/field/branch of the civil engineering projects, which may/shall be lost before being gained and then the contractor shall be given a third degree treatment/supervised suspiciously all the time. Therefore, the contractor shall have to abide by the approved programme till/unless the restraints/troubling factors go out of practical control because of the unforeseen circumstances exclusively e.g., despite concentrated efforts and the placement of the material purchase orders in time per programme, the deliveries of the reinforcing steel, cement, etc., do not come in time from the approved source or the payments of the valuations are not made to the contractor in the reasonable time as agreed at the time of signing this contract.

    iv. Monthly Technical and Co-ordination meetings and also the monthly Progress review and monitoring meetings shall be conducted by the engineer/resident engineer and both the meetings jointly shall be termed in this contract of execution the contract meeting/site meeting wherein, all the aspects of the contract technical as well as non technical (general business and administration) shall be discussed among all the parties (employer, contractor, engineer) and recorded as discussions and decisions. Sometimes, major decisions shall also be taken in such site meetings/contract meetings if necessary according to the situation.

v. From the contractor's side/behalf, the representatives shall not be below the level of the site engineer incharges/executive engineers headed by the site agents and from the engineer's side/behalf the specialists engineers and the project co-ordinator headed by the resident engineer/engineer and also from the employer's side/behalf employer in person or the authorised agent/director or above. This meeting shall be chaired by the resident engineer/engineer. The complete agenda of this meeting/s has/have been included in the contents of guide to the management forming a part of the contract documents.

37. **INSPECTIONS** The contractor shall notify in writing to the resident engineer on the standard proforma alongwith the photocopy of the drawings of the related/relevant portion of the contract requested for inspection and the discipline 24 hours in advance of the scheduled inspection time for the general items and 48 hours in advance for the inspections for the concrete pours in the structures. The details of the standard proforma has been included with the management guide as a part of this contract. The approval, rejection or rectification as the case be, shall be granted on the proforma itself. In case of the rejection of any inspection the site agent/the site engineer shall resubmit the request for inspection. The approval shall be invalid in case the contractor does not carry out the work within 24 hours of the approval and in that case the reinspection shall be requested by the site agent/site engineer on the same proforma containing the previous approval.

38. **NOT OBTAIN APPROVAL**

i. In case the contractor does not obtain approval of any item formally and carry out the works on the contractor's risk, the payment/valuation shall not be made for that part of the job and this may also effect adversely the following items in the sequence of the unapproved items executed by the contractor at own risk. That is why, it is very important duty/responsibility of the agent and the site engineers to make sure/ascertain that the works as carried out are inspected and approved accordingly by the resident engineer to avoid unnecessary disputes. The post execution inspection could be held if the resident engineer is convinced of the definite problems with the contractor's staff due to one reason or the other reason in which case the resident engineer may ask the supervising engineer to inspect on the formal humble request of relevant contractor's agent confined to certain technically/professionally implied limitations/merits/drawfore depending on the individual case but that shall not be adopted as a convention/regular practice but seldom (0.1%).

ii. In case the disputes arise due to the unapproved works/items and the resident engineer is not convinced, the resident engineer shall ask for the specified test check and ascertain/make sure the standards of the works carried out, all at the risk and the cost of the contractor and the resident engineer having satisfied, the approval shall formally be given otherwise, the order shall be passed by the resident engineer to enforce the remedial action on the standard proforma 'REMEDIAL NOTE' as included with the guide to management, which shall remain in force untill the remedial actions are taken and a request be made formally by the contractor's agent to the resident engineer and approval is granted by the resident engineer after the successful inspection of the remedied works on the standard inspection request proforma. After the approval of the remedial works the enforcement of the remedial note shall be released/invalidated, and the contractor may claim the valuation for the items covered under the enforcement of the remedial actions legally.

39. **DISPUTES**

i. In case disputes arise out of the unapproved works for which the payment cannot be made, the decision of the engineer shall be binding on the contractor and the engineer's staff.

ii. In case of the disputes between the employer and the contractor due to any reason, the engineer shall act as an independent judge to settle/resolve the disputes and the engineer's decision shall be final and binding on both the contractor and the employer until the matter is referred to the arbitration according to the provisions of the Indian Arbitration Act/country's law.

iii. In case of disputes arising out of site supervision, the resident engineer shall be empowered/authorised to resolve the disputes.

iv. In case the disputes involve the extra finances, the resident engineer shall take the matter to the engineer for further instructions.

40. **METHOD STATEMENT** The resident engineer can ask for the method statements for the execution/carrying out of any item or a part of the contract and the contractor shall provide the complete details of the same. The contractor shall work according to the method statement after being approved by the resident engineer but this approval shall not relieve the contractor of contractor's responsibilities for the correct execution of the item. The elaboration of the method statement has been given in the management guide (contract part). The investment of this authority with the engineer/resident engineer is mandatory.

41. **INSTRUCTIONS**

i. On this job the following types of instructions shall be brought in use and shall be issued from the engineer's or the resident engineer's office.

SITE INSTRUCTIONS; REMEDIAL NOTES; ENGINEER'S INSTRUCTIONS; AND MISCALLANEOUS INSTRUCTIONS; SITE MEMORANDUM, ETC.

ii. Site memos, site instructions and the remedial notes shall be issued from the resident engineer's office/site office as well as the miscallaneous instructions such as inspections requests or daily movements on the job.

iii. Engineer's instructions shall be signed by the engineer only and shall form the basis of all the additions or deletions or extras payments on the contract job. The payments of these items shall not be considered finally if the engineer's instruction is not issued in time. Refer to the management guide for the detailed meanings of these items/terms.

iv. The engineer or the resident engineer shall instruct the contractor through the agent and no other official of the contractor be instructed directly by the resident engineer except in a situation when the contractor acts/works as the agent self duly approved.

v. The co-ordinators, the sub agents, the site engineers, the assistant engineers etc., may be orally/verbally instructed but must be confirmed by the resident engineer in writing to the contractor/contractor's agent. All the instructions to the contractor shall be given in writing under the titles as described above in this clause/condition (i) and should the contractor feels that any of the instructions given by the engineer/resident engineer constitutes any variation (extra or saving to the contract) in the cost/price on the job, the contractor should defer the work included in that instruction and request the resident engineer/engineer in writing informing/advising about the contract variation and the cost involved. If the engineer issues the instructions to carry out the instructed item, the contractor shall go ahead with the execution of that item/s as the contractor shall definitely be paid for the variation of any extras involved. In case of deductions, the contractor may give valid reasons why the instruction should not be carried out to the design requirement as the employer enjoys every right to change, add, delete, adopt new design etc.,

vi. Any claim for the extras done/executed should base on the engineer's instrution only followed by an amount in formal variaiton order.

42. **MATERIAL LIST** At the time of the tender submission the tenderer/contractor be advised to include the current price list of all the materials to be used/included on the contract job to enable the engineer to know/apprise the market rates at the time of tendering/calculating price and shall be made use of resolving the contractor's claims based on the material cost enhancements greater/more than a tolerable percentage on the basic market rate at the time of tendering and on such a long duration contract job, there shall/may be reasonable appreciation in the prices of materials due to inflation and many other macro economical reasons. In case the tenderer/contractor failes to include the priced list of materials of the time while tendering, no claim on that account shall be admitted towards the enhancements/extra expenditure probably incurred by the contractor except for the state/government controlled items for which contractor shall produce all the evidences in support, and the relevant

documents and the different consumption statements time versus progress graphically to grossly prop and support the claims. The claims shall be admitted/considered only after the documents and the details provided being verified/ratified by the resident engineer in complete co-ordination with the engineer's/resident engineer's/contractor's quantity engineer's and the contractor shall include the rates of various items according to the contractor's intention, to obtain the material approvals of from engineer.

43. **VALUATIONS** The value of the works carried out by the contractor on this contract shall be evaluated monthly prior to the day of the contract site meeting for which the contractor shall submit the monthly works done valuation in the form as agreed between the resident engineer/quantity engineer/surveyor and contractor's agent/quantity engineer/surveyor. After the valuation is recommended by the resident engineer based on the check and certification by the resident engineer's/engineer's quantity engineer/surveyor, the engineer shall issue the 'request for payment' certificates to the employer/party I in this contract who in turn shall release the certified value of the monthly works carried out by the contractor/party II in this contract, to the contractor within a reasonable time as defined somewhere else in the contract documents as implied/agreed by both the contract parties.

44. **VARIATIONS** All the variation orders issued by the engineer for the deletion ommissions, additions or for any/some other reason shall be included by the contractor in the monthly valuations regularly for the convenience of the keep up of the cost record and cost planning control.

45. **CLAIMS FOR EXTRAS** In case of extras the contractor shall provide all assistance in measuring the items of works carried out, in respect of that in the presence of the agent or the delegated contractor's official for the purpose and enable the preparation and give a pass to the variation order, Generally, the variation order details are prepared by the contractor and verified/ratified by the resident engineer after verification/ratification/inspection and work measurements by the quantity engineer/surveyor.

46. **CONTRACTOR'S RECORDS** For any claim/s raised by the contractor regarding the cost increase/enhancement, due to labour/work force wage rise/increase, material rate rises/enhancement, the engineer/resident engineer shall have the authority to demand/ask for the complete documentary details of the records in relevance to the claims raised. These documentary details may/shall include the actual payments made to the work persons/workers for the effected periods, the consumption statements of the materials included with the claim and the real proof of the delivery and the source which shall be compared with the recorded daily job records of the contract. Should the contractor hesitate for the production of those documents, the claim shall not be processed and returned to the contractor observing the deficiencies of the presentation of the claim and shall not at all be admitted unless all the requirements of the clause are met by the contractor and also it shall put a barrier to reach the arbitration for the same claim by the contractor.

47. **DELAYED PAYMENTS** In case of the delayed payment beyond the agreed reasonable time by the 1st party (employer) to the II party (contractor), the engineer shall use the virtually vested authority granted by the professional priveleges due to being a legal engineer for the contract job, to request and advise the employer of the contractor/engineer's client to make the payments as certified to the contractor. Even then, if the payment is not made by the employer to the contractor, the contractor may resort to the contractor's rightful legal actions under this contract and according the sanctity of the legal provisions/remedies prevailing in such and similar cases and duly covering under any/some formal qualification/condition included/added/agreed/approved at the time of tendering and the same being accepted by the employer forming a part of this contract/part of this contract documents.

48. **RATES AND DEDUCTIONS**
    i. Since this is an itemised lump sum price contract in which all the items as included in the bills of quantities shall be priced in the usual way and extended to the bill summation

further carried to the summary of all the bill of quantities of the building and other structures, the check up of all the entries of the prices/rates shall be made by the engineer as also the sums.

ii. **Adjustment to the rates** In case the sum in the bill of quantities goes on the higher side, the proportional reduction shall be applied to the rates of the indiviual items to maintaining the priced sum as that was and is.

iii. In case the total sum of the bill of quantity comes on the lower side than calculated, the reduction shall be applied to the priced/quoted sum and the contractor shall be informed of the reduction and then this depends on the tenderer to still carry out the job. The tenderer may refuse in that case to carry out the work wherein the tender bond provided by the tenderer shall be forefeited/lost/confiscated by the employer and a record shall be entered.

iv. **Deductions** 10% security money shall be deducted from all the valuations and kept with the employer/1st party until the final completion is granted by the engineer to the contractor/II party or relatively linked to a contract qualification or some warranty of items.

v. All the advances made to the contractor shall also be deducted from the following valuations.

vi. **Performance bond** The contractor shall submit a performance bond of a fixed amount for the whole period of the contract to secure the 1st party of the contractor's default on the contract. The amount to be mentioned in the tender documents.

vii. **Tender bond** The tenderer shall submit a tender bond of a fixed amount as mentioned in the tender invitation letter.

49. EXTRAS

i. In case of the additional requirement of the job in any portion under the scope of this contract and not included in the contract documents, the engineer shall instruct the contractor to carry out the job as needed due to technical or some other reason based on the same rates of the related/relevant/corresponded items as in the bills of quantities. In case the contractor does not accept that rate for the additional quantity/items justifying the contractor's denial due to the low rates quoted by the contractor for those items and the replenishment due to the low quoted rates covered somewhere else in the bills of quantities, the engineer shall ask the contractor to produce the detailed analysis of rates and reach a reasonable figure after the thorough scrutiny and the contractor shall/should accept that with a degree of flexibility as the rigid attitude in such a case shall help none. After the agreement of the rate, the quantity shall be measured and the valuation for the variation computed for an order.

ii. **Day work** There shall be some extras/additions/substitutions for which the actual measurements may not be possible/susceptible due to many constraints being unclear because of imperfect parameters of the item involved, consequenty, causing problem in fixing a mutually acceptable rate of item by both the parties and the resident engineer/engineer not being in capacity or potential of perfect calculations and measurements. In such a situation, the clause/condition of observing the daywork shall apply/work successfully wherein, the engineer shall ask/demand the contractor to observe strictly all the operations of the relevant item and present/produce the daily report in respect to that particular activity/or the item, including the number of work persons deployed and the time for which they worked daily, the quantities of all the materials consumed, the materials brought in use for the preparatory works, the durations/time for which the machines and the equipments brought in use and the hours for which the supervisory staff worked, all signed by the agent, site engineer in charge and the engineer's/resident engineer's clerk of works, verifying/ratifying the contents of the day work. On the basis of this report/record, the valuation of the particular variation order shall be computed as accurately as possible to the satisfaction of all the parties involved.

iii. Since this is a lump sum itemised bills of quantities contract, there may be the instances that the items shown in the contract drawings are/may not be included in the bill of quantities or the other way or in the project specifications some activity or technicalities are mentioned but not included/coverd in the either drawings or the specifications both general and project or the bill of quantities, shall not give way legally/professionaly to the admittance of claims due to extras/extra. The whole set/pool of the contract documents as defined in clause 22 and also due to the implied conditions of the nature of the contract, the conventions of the profession in the civil engineering discipline corresponding to this contract, shall be treated as a unit of one document read combined/monolithically for the purpose of the contract administration and execution for the successful completion of the contract job according to the scope of the contract works. The tenderer shall go through the complete contents of the contract very carefully in a professional way and confirm itself of all the documents using/utilising the services of a chartered quantity engineer. Particularly, this contract shall reveal various implied meanings for the contents which should be interrelated regarding all the documents. For claims also, the services of the chartered quantity engineer should be arranged. Unnecessary writing letters/correspondences for raising claims by the contractor/s observing/discovering/pointing out any/some of the discrepancies of the bill of quantities or in the other documents shall result in no gains to the engineer or to the contractor or the employer and contribute only to the production of the unproductive paper work and severe wastage of time of both side professionals.

50. **OTHER EXTRAS OVER CONTRACT** All other extras not covered in the scope of this contract shall not be ordered by the engineer to the contractor/s without obtaining the written documents/authorities from the employer/engineer's client which shall also be acceptable to the contractor. In such a case of additional extras over the contract scope, on the acceptance by both the parties to the contract, a confirmation shall be issued by the employer to the contractor/s and agreement by the contractor in turn shall make that a part of this contract and the variation be processed.

51. **DELETIONS**

    i. The first party shall have the reservations to omit in full or in parts any work or part of a work at any time/stage after the contract comes into effect and this reservation shall remain in force till the completion date of the contract practically. The first party shall not assign any reason/s for this act/thing as above and the second party/the contractor shall not raise any claims on this clause/condition/s except for the damages caused to the contractor because of the execution of this clause and by the operation of the law of the country.

    ii. The material brought to site for that job as detailed shall be paid to the contractor/second party on the claims raised by the contractor including all the consequential charges and profit margins of the contractor and due settlement/s reached.

    iii. Any other money/amounts paid on the ordering of some materials for the deleted work and the same being non refundable according to the business transaction clauses of marketing practice or the contractor shall loose some of the cost/money in case of the cancellation/s of the purchase order/s or any/some other damage/s caused to the contractor/s or the contractor raises some valid point that may construe to be reasonable to the engineer, the payment of costs on those amounts shall be made to the contractor/s by the employer, failing which the contractor/s the second party to this contract shall have the lawful right to move the arbitraiton per provisions made in this contract.

    iv. No claim for the contractor under this clause because of the purchase of any machine or equipment making an asset of the contractor shall be admissible under the consideration/s as a genuine claim but if utilised for the deleted works, hourly claims for the works done be admissible as a part of the item.

    v. No claim shall be admitted for the deployment of a particular group of work person for that due to the workpersons being adjustable on site and reallocation of workforce be done.

    vi. These claims as considered above shall also be guaged/related/corresponed to the other conditions/clauses of this contract provisions comprehensively.

    vii. In some cases the quantity of the omissions may be considered and the advocated plea of the contractor consists of justifying the quoted rates being on much higher side than usual and the higher amount will be utilised towards the expenses to be incurred on the other parts of the contract job and if it appears genuine to the engineer, thorough scrutiny of the items and cost shall be carried out by the engineer and then if found suitable per the contract conditions, compensatory extra variation/s shall be ordered/instructed to make a justification on the contract.

52. QUALITY AND QUANTITY

i. The contractor shall be completely responsible for the contract's job execution quantitatively and qualitatively to the entire satisfaction of the engineer according to the engineer's details, plans and the specifications and including all the contract documents etc.,. For the purpose of the quality control the contractor shall employ its technical personnel/staff/officials under the superintendence of a competent engineer/chartered engineer (civil and architectural) levelled as contractor's site agent completely responsible for the execution of this contract job and all the methods as described in the guide to management shall be conducted on site regularly, open to checks by the engineer or the resident engineer or any delegated representative all the contract job duration. The contractor shall/should employ/deploy very experienced/qualified/chartered technocrats and the technical assistants of all the required/included contract disciplines to make sure/ensure the delivery of the goods in the stipulated/required time/contract completion period/duration and to the prescribed and the defined/specified standards otherwise, the contractor shall suffer per contract obligations in the case of a lenient/sloppy attitude in any of the quantity and the quality.

ii. The contractor shall make use of all the feasible modern construction methods and make use of the modern construction machineries and equipments as far as possible to accelerate the rate of work progress on this contract. The contractor shall handover to the owner/party I all the buildings as a whole in the scope of this contract including all the ancilliary and auxilliary structures in the completed form in all respects of the contract and the civil engineering profession to the confinement of this contract and ready for commissioning and permanent maintenance.

53. TESTS

i. The contractor shall arrange from time to time all the tests technically and necessary per standards/specification or directed by the engineer/resident engineer for the works performed and also as advises/recommendations according the relevant codes of the discipline concerned. The contractor shall form/establish a fully modern and updated engineering test laboratory/laboratories for conducting the required tests on site which shall be witnessed by the resident engineer or the representative specialist/s engineer/consultant/s. All these tests observations shall be recorded by the contractor in a proper and specified manner/style and the engineer/resident engineer/specialist shall examine the test results to the prescribed/accepted standards.

ii. In case the arrangements for certain tests cannot/could not be made feasible in on the site laboratory/laboratories, the contractor shall propose in writing some independent laboratory for those test unable to be conducted on site, the approval to that request shall be given by the engineer after scrutiny of the proposed laboratory standards and details as provided elaboratively on the standard format/application by the contractor for the grant of the approval of an independent laboratory.

iii. Tests for all the duct works/pipe lines shall be carried out on site per general/project spcifications/provisions in the relevant codes/existing standard practice/technical requirements/any other convention. The records of all these/those tests conducted shall be forwarded to the engineer's resident for necessary actions on the contract perview.

iv. The resident engineer/engineer shall reserve the right to ask for/direct random testing for any item/s executed/carried out at any time during the execution/carrying out of the entire contract work.

54. CONTRACOTR'S DUES The contractor shall make regular payments to all the work persons

employed/deployed/engaged, employees sub- contractor/s, suppliers, and all the parties in deal with the contractor's business for this contract job in time so that the delays due to this cause do not come in the way of advancing progress on the programmes lines simultaneously linking this clause/conditions to the contractor's own valuation payments/dues to be paid by the employer in time as agreed after the certification of the valuation of all the approved works in the reasonable time accoring to the contract otherwise, the clauses of damages shall apply.

55. **HOLIDAYS** The contractor shall notify to the resident engineer/engineer in advance regarding the observance of the holidays other than the weekly off days. In case of programmes of partly working on overtimes in those holidays, the contractor shall in advance 48 hours intimate the engineer/resident engineer of the programme so that the necessary arrangements could be made for the supervisions. Night works could be permitted in shifts if the contractor so wishes to expedite the retarded progress of the contract/advance the contract.

56. **WEATHER/CLIMATE**

   i. The contractor shall ascertain/make sure about the weather conditions of the region of the contract work from the meteorological departments and also the past history/records of at least five decades and price the tender accordingly as no kind of claims due to that reason shall be admissible on the contract for the engineer's consideration/s. The claims due to the inclement/adverse weather conditions of the devastating nature/s never appeared/occured in the prescribed considerable period of five decades for which the claims for the extension of contract time claims may be accepted, but all the damages of the material shall be claimed from the insurance agencies.

   ii. At the tender stage the contractor shall assess all such risks and price the offer/tender accordingly. Every rain day shall bring a letter from the contractor's agent's office informing stoppage of works due to rain/thunderstorm/dust storm/sand storm, which are entirely unpleasing correspondences/letters and wastage of time for both the resident engineer and the contractor's agent. The contractor's agent shall be supposed to be a person/official/engineer of a high dignity and a chartered engineer possessing certain competency who must value own time and the time of the other professional/s.

57. **UNFORESEEN**

   i. All other circumstances not covered above and also out of the control of the contractor, shall be termed/defined/called as unforeseen circumstances. If the disturbances on the job is caused due to the reason of the unforeseen circumstances, the contractor shall get the extension of time on contract time. The contractor shall insure itself against all such contingencies.

   ii. Unforeseen circumstances may include call of BUNDS/GENERAL STRIKES by the political parties; illegal labour unrest on the job not due to the contractor's default; natural disasters; striking public transport system; break up of war or equivalent; or unrest directly or indirectly effecting negatively the works production not under the control of the contractor entirely, social unrest in the community effecting the work evidently etc., and all other circumstances other than included in the unforeseen circumstances, the contractor shall anticipate those itself being experienced contractor/business person in the civil engineering construction discipline/field which shall not at all be considered for the purpose of the extension of time claims, admittance and consideration.

58. **CLAIMS OF EXTENSIONS** The contractors have a tendency to raise the claims of extensions of the contract time based on unrasonable ground/s every now and then. The contractor/s be clear that the following guide lines shall be controlling the clause for the admissibility for the claims for obtaining the extension.

   a. If the site is not cleared by the employer or any other contractor/s working in the same area/vicinity of this contract area, encroaching the area under this contract either in full or in part.

   b. If the contractor is not supplied/issued/provided with the details of the contract documents in time immediately after the contractor's application, and any of the decisions to be taken by the employer or the engineer is not communicated to the contractor in time as a result or consequence of which the programmed progress is jeopardized. This shall include the information required and the confirmation/s on the tacit approvals of any of the items.

c. The contractor is not paid the certified valuations' money by the employer in a defined reasonable time as agreed depriving the contractor's circulation of the capital investments and the sources of the finances.

d. Major additions are ordered by the engineer/employer on the job exceeding the defined or accepted cost variation limits per contract specifications or provision for that or any other specific qualification/s or generally, accepted or implied term.

e. Stoppage of the work by order of the engineer in full or in parts due to the design changes or any other reasons but should be in documentary form.

f. Unforeseen circumstances out of the control of the contractor or employer.

g. Coordinations' problems caused to the contractor by the other contractors on the same major project site area on the other phases provided this contractor was efficient in managing complete coordination as and when required and the other parties/contractors expected to respond for coordination did not seriously take the coordinating problems consequently, contributing to the time loss of this contractor.

h. Any other valid reason/logic forming the ground for the extension of contract time the engineer deems fit or acceptable for the admittance/consideration based on the merits of the case of the consideration.

i. All other claims not included or covered under the said/foregoing guide lines shall be inadmissible for the acceptance for the contract time extension consideration and if the contractor even then wishes that justice has not been given to the contractor, the contractor may refer the case to the arbitration per clause provided with this contract.

59. **PENALTY** A fixed penalty shall be imposed on the contractor for not practically, completing the works in all respects according to the contract documents per day of the delayed period and be deducted from the contract valuation/s if the extension of the contract time is not recommended by the engineer and approved by the employer. The penalty amount per day is given in the contract schedule.

60. **CONTRACT RESCINDING/TERMINATION** If the contractor does not give the satisfactory performance on the contract and site as defined by the engineer's and the contract standards, construed to be legally and contractually constituted as the contractor's responsibilities, liabilities and the professional obligations, and even after the contractor being warned by the engineer a number of times/at the most thrice to let the ignorant/inefficient contractor avail the opportunity to improve the contract performance to avoid the unpleasant complications and legal disputes keeping in view the interest of the job and also of the parties involved in the contract, the engineer shall be forced to recommend the employer of the contractor to terminate or rescind the construction contract and the balance/remaining/undone scope/quantity of the contract to be completed by appointing some other competent agency/or the employer itself, at the risk and the costs of the original contractor/agency. The advise/recommendation of the engineer shall be accepted by the employer of the contractor in all cases per management contract.

61. **CONTRACT COMPETENCY AND ARBITRATION** Engineer shall be an independent party to this contract to carry out and decide the deals, financial and non financial claims and counter claims and the decisions/recommendations taken by the engineer shall be binding on the contractor and the employer of the contractor/client of the engineer. In case of the occurrence of the major disputes wherein, the contract parties do not come to an agreement and the acceptance of the engineer's recommendations/judgements/decisions, the contract parties shall be at full liberty to refer the disputes to a mutually appointed arbitrator/group of arbitrators according the provisions made in the Indian Arbitration Act/country's relevant laws or Acts.

62. **LIAISON OF CONTRACT** The contractor shall keep the engineer/the resident engineer all the

time throughout the duration of the contract working informed of all the activities, operations, progress, happenings or the mishappenings, handover and completion accordingly and any loss or damage caused to the contractor due to the misinformation/s shall be the contractor's risk. The contractor shall similarly, liase with the 'all statutory bodies, authorities, departments, organisations, corporations, committees, institutions, etc., and all the main contractor's and the sub contractors on the same project job under the control of the engineer/chief resident engineer/resident engineer. For more details of the liaison responsibilities/obligations please refer to the guide to the civil engineering project management, a part of contract document on this job.

63. **STOP WORK** The order to stop work shall be given by the engineer through the resident engineer/chief resident engineer whenever there is any design change or alterations or the additions or omissions being considered in some parts of the contract or also if the adverse effects by the wrong doings of the execution performances or unacceptable performances of the contract works are revealed or unearthed with a/an probable/imminent expectation to degrade or deteriorate the employer's reputation in the public e.g., concrete cubes fail and the concrete in the structure or else to be dismantled for recasting. Immediately upon the receipt of such instructions from the engineer or the resident engineer, the contractor shall stop the works in the subject sections of the contract. The contractor may raise the variation claims for the design changes but for the wrong performance, the contractor shall be liable for the recovery payments to be made to the employer on account of the loss of the professional reputation and the contract time if any. There is no definition of some specific nature for making such counter claims but shall depend on the merits and demerits from case to case and the extent of the damage as worked out by the specialists.

64. **DISCIPLINARY ACTION** In case the contractor or any of the contractor's approved staff/personnel does not care for the contract's agreed procedures for carrying out the contract execution and the works/operations are operated without the approval of the engineer/resident engineer, the reservations of the right to exercise/impose/take/resort to the required necessary action/s shall rest with the engineer who shall decide the disciplinary action/s against the contractor under intimation to the contractor/s and the employer of the contractor and this action in this clause/condition may/shall include removing/withdrawing of any of the contractor's staff/personnel including the site agent or in the extreme case when the water crosses over lips situation is created, the contractor shall be given a fixed time/served with a notice to improve the contract's performance grossly/perspectively/in an overall consideration, or/otherwise, the termination or the rescinding of the contract may be considered to be on the cards which shall depend on the degree of the poor/unacceptable performance and its frequency as the engineer knows/is well aware that no contractor shall want/like to execute the poor works but there shall always be certain bad elements in the whole lot of the contractor's personnel/contract team polluting/contaminating the site environment/atmosphere of the contract works performances. It depends upon the contractor how far the contractor shall be able to win the professional working trust of the engineer and the employer.

65. **EMPLOYER CONTRACTOR RELATIONS** The contractor shall not discuss any technical and the engineering matter/problem with the employer of the contractor including the financial aspects as that may create mixing/confusion and proliferate the misunderstandings between the engineer and the contractor's employer as well as between the engineer and the contractor. In case the contractor wishes to suggest to the employer something beneficial to the job, the contractor shall propose the idea to the engineer or the resident engineer as the engineer is the lawful adviser and the specialist of the job in contract who shall consider the suggestion/proposal/advices of the contractor from the professional, engineering, commercial, financial and feasibility basis. Contractor should realise and keep in mind that the goal/target of the engineer and the contractor is to accomplish the job correctly in all respects, that is why, a generation of good working relations on the job shall be very important factor for all the parties involved and interested for the successful completion of the job with the minimum disputes.

66. **AS BUILT/RECORD DRAWINGS** The contractor shall record all the aberrations/deviations of the working/shop and the contract drawings to the works carried out and this shall be done in parallel to the works execution to save piling up of drawings in hundreds at the end of the contract execution engineering as without the approval of these drawings in respect of all the disciplines in this contract, the execution/construction engineering shall not be considered completed practically. This is an advice to the contractor to save the inconvenience for all the parties concerned/relevant and help make the work easy for all as without the record/as built drawings, the completed works cannot be maintained and the completion shall not be granted finally and practically. It is in the interest of the contractor to perform this part of the job side by side. If the agent is transferred, the next man shall find it extremely difficult and sometimes even improbable task to perform.

67. **TESTING AND COMMISSIONING**

i. The completed services of all the disciplines shall be tested at the end during the period of the practical completion for the successful performance/s for at least one month continuously and if found yielding/failing at any stage during this period of one month, the counting of the one month period shall begin/commence again from the day of retest commencement. Particularly, the engineering and the electrical disciplines shall be of the top significance to be protected in this clause/condition and shall be tested in the same style and the manner until the successful testing and the commissioning shall reach/accomplish and shall be recorded by the specialists engineers of the engineer and the relevant responsible engineers of the statutory bodies/authorities per the requirements under the state laws. The contractor shall arrange for all means necessarily required for accomplishing the purpose of the testing and the commissioning including the water and the electricity from any source and shall include all the expenses in the offer at tender time.

ii. All the services disciplines shall be designed on the first class installations and functioning which requires an effective checking of design requirements of the concerned system and accordingly production of the drawings on the contractor's part as the tender documents and contract documents comprise of schematic details for guidance and the specifications provide the real requirements of all the systems and that is why, the contractor shall include this criteria also at the time of calculating the price of tender.

68. **CONTRACTOR'S INTERNAL COORDINATION**

i. The engineer advises the contractor to hold the contractor's own coordination and the technical meetings with the contractor's sub contractors regularly to solve various problems on site under the contractor's chair and circulate the minutes of the meeting to the resident engineer for information only. This shall make the contractor's staff more efficient about the job and also the gross time of the meetings with the resident engineer and the consultants be curtailed considerably as all the disputes and the problems shall already be cleared with the solutions ready in hand.

ii. Similarly, the contractor may hold these meetings with the other counterparts working under the engineer/resident engineer on the same major job and inform about the problems and any of the confirmations as made after these meetings to the resident engineer in time so that the necessary actions shall be promptly and timely taken by the resident engineer/engineer/contractor/s.

69. **GUIDE TO SITE MANAGEMENT** The contract documents include a practice guide to this contract project management particularly for those professional engineers and the professional contractors who are not well conversant with the British and international practice on the itemised lump sum contracts and the procedures which are considerably easy as far as the contractor's part is considered whilst the cumbersome works have to be done by the engineer in the pre tender works to prepare various sizes' immense volumes to produces of the contract details. It shall be advised that the contractor's personnel on site in any level of supervision and control should go through this book/document before commencing work on site. This guide book/document comprises with the details and descriptions as titled under.

1. Categories/Classes of planning
2. Explanation of classes
3. Preparation of various bills of quantities
4. Preparation of various building specifications
5. Converting the above documents to the tender documents
6. Shortlisting/Pre qualification of tenderers.
7. Floating tenders, closing, scrutiny and negotiations
8. Tender bond
9. Site handover, commencement of construction, supervision and control till the practical completion of the contract, complete site clearance by the various agencies working in the contract area
10. Coordination, liaison
11. Contractor's organisational structure levels
12. Contractor's agent
13. Internal dealings with contractor's organisation and dealings at site
14. Elaborations of coordinations
15. Liaison
16. What a contractor's agent is supposed to do
17. Liaison
18. Agent's dealings at site
19. What the other site personnel of the contractor should do
20. What the engineer and the engineer's personnel should do
21. Correspondences proformas
22. Swimming pool operations
23. General activities in building constructions
24. Sports courts surfacing and epoxy flooring
25. Knowledge of different types of contracts
26. Some miscellaneous contract items/terms
27. Claims and delay claims
28. Practical completion

70. MAINTENANCE

i. This contract includes the post practical completion maintenance of the completed works by the contractor for a period of 400 days under the engineer's or the resident engineer's supervision. There may be certain items of a fixed warranty for a number of years which shall be dealt with separately according to the project specifications and the contractor remains bound for those items till the warranty remains in force. For all other works having the maintenance being successfully completed, the contractor shall be paid the dues by the employer on obtaining the completion certification from the engineer's office except for the warranted items for which the security or any other guarantee shall be needed by the engineer for the employer.

ii. Final completion shall be certified by the engineer only on the successful completion of the maintenance as specified and the contract shall be termed completed at that time. Engineer shall present the completed contract to the employer of the contractor/client of the engineer.

The final completion shall be granted only by the engineer when the period of maintenance has been over and no defects appear in the works maintained after the completion of the works practically. The retention of the contractor's security money/amount shall only be released after that day subject to the submission of long term guarantee of the warranted items carrying terms more than the maintenance period. These items may include sports surfacing for the courts, water proofing torches for the roofing, the glazed curtain walling and all other specific items of the different disciplines as the case be.

NEW SITE

General Information about commencing a new building project site

1. Take over the site from the Engineer/Resident engineer
2. Take over all the contract documents at an earliest
3. Inform the relevant statutory bodies about commencement of the contract
4. Request the Engineer to get you the authorisation to obtain the record/As built services routes of the contract area from the relevant departments
5. Attend the 1st contract and introduction meeting
6. Obtain the approval of the site mobilisation and the establishment plan proposed
7. Start mobilisation which includes construction of the Agent's office with all modern amenities, Deputy agent's office, site coordinator's office, and the site Engineer's offices and the site stores, site cabins for the Assistant engineers/site foreman/General foreman, supervisors, all the amenities for the work persons accommodation according to the site conditions and the contract requirements. Entrust all these operations to the responsibility of the general foreman or a site engineer. Prefabricated portacabins may be employed for site offices. Also construct immediately the facilities for the Resident engineer and staff if covered in the scope of your contract. Try to finish this part of the contract job as soon as possible as the real work commencement depends on this base.
8. Simultaneously formulate the job programmes/bar charts/CPM/PERT as the case be for which the general quantitative informations be collected from the contract documents as defined earlier in some other parts. Site layout plans, schematic plans, number of structures involved in contract, number of BOQs, quantities of different materials, all the elements and the sub heads, priority criteria in contract, duration of contract, contractor's potential of resources, proposed potential in connection with workpower, machineries, investment cost on job, shall all assist you to formulate the gross idea about the dream programme. Use your tutorialed experience under your hand in personal studies and make up the dream programme and then step by step formulate the real bar chart duly co-ordinating all the activities together avoiding the clashes among them. When all the activities are clash free and all the charts are coordinated well, fix in the details of the resources supporting the programmes, discuss with the company/management if required and make submission thereafter. Discuss with the resident engineer and the specialist engineers dealing with the project to help clear the approval.
9. Meanwhile apply for the curriculum vitae of all the supervising personnel including your proposed sub contractors to make submittal for the approvals.
10. Make submittals for the proposed materials and the working and the shop drawings in all the relevant engineering disciplines and obtain the approval.
11. Let the site survey team go fast to establish layouts for the 1st structure/s included in the programmes and try to get them approved in respect to the consonance of the way leave forms.
12. Keep site on technically and try to mobilise more and more workpower to make the site run fast or at least in line with the programme bar for the first few months at least in all respects to know the general nature of the resident engineer, the staff and the support of your contracting organisation.
13. Keep full coordination in the schematic and working technological engineer's details to avoid the most probable mistakes on sites.
14. Perform and let the other officers perform all of their duties and responsibilities as described in the different designations of the site personnel.

15. Setting out of a building and excavations though look easy but are the bases of the structure correctness hence be regarded with importance and the agent is obliged to complete these activities directly under his/her supervision according to practice and professional responsibility

16. Once one of the site is mobilised, explore the second and thereafter third and so on.

17. The time you think that the site is in the grip, please smoothly bring the subordinates to the direct supervision and install yourself on the superintending position by making rounds and keep the complete informations from the site from some reliable source with you.

18. You keep in mind that you are not dealing with the children but with budding engineers so act accordingly and let them also develop their skills as you shall have mostly to deal with the higher levels with the engineer's staff and the sub-contractor's staff. So administrate, be diplomatic and hidden politician so as to utilise this art to extract the most possible from all the sources to the benefit of your project.

GENERAL REQUIREMENTS IN THE STATE OF THE ART

IN GENERAL

Masonry

Trowels all types and sizes
Straight edge Aluminium float
Straight edge timber float
Mixing troughs
Mortar mixer
Concrete mixer
Water level with pipe
Spirit levels of all sizes
Mallet
Hammers of different weights
Plumb bob
Measuring steel tape
3 m, 5 m, 15 m, 30 m
Plaster floats
Power floating equipment/s
Floor grinding machine
Carborundum stones of coarse to finest category
Grinder electrically operated
Shot firing gun and bullets
Gum shoes
Hand gloves (rubber and cloth)
Safety glasses
Mason's scaffolding/pipe scaffolding
Empty drums
Timber planks and battens
Nylon ropes/Jute ropes
Drinking water container/Thermos flask
Vibrators 1″, 1½″ required dia nozzles
Portable generators petrol operated
Compressors with pipes and accessories
Hand shovels and hand spades

Scrapers
Teeth shovels
Wire brushes and hand brushes
PVC membrane 500 gauge, 1000 gauge
Material hoisting arrangements
Hand chisels
Scabbling hammer
Guniting equipment and accessories
Mortar spreader on wall for rough plaster
Concrete nails made of steel
Hard hats
Tile cutter manually operated and mechanical
Block cutter mechanically operated
Electric generator
T-square big and small
General scaffolding materials
pipes, jack, bracings, safety net
Gunny bags, curing compound, degrading compound/s
Electrical bulbs and wires, switches, boards
Thread line Nylon
Plastic buckets
Pans/Taslas
Bamboo buckets
Aggregates measuring timber box
36.5 × 30 × 30 cm with 2 handles
Hand wheel borrows both sizes with tyres
Ladders (timber, aluminium)
Tools kit, torch

Carpentry

Hammers

Craw hammers
Plumb bobs
T-square
Saw both types
Sawing m/c electrical
Spirit levels long
Water levels and pipes
Hand planners
Power planner
Carpenter table
Measuring tape 3 m, 5 m, 15 m, 30 m
Safety glasses
Nails wrought iron 4″, 3″, 1″
Nails steel 4″, 3″
Timber battens 3″ × 3″, 4″ × 4″
Timber planks 20 mm thick
Steel plates and backings (perforated)
Plywood marine/commercial
   18 mm, 12 mm thick 8′ × 4′
Wall ties/bolts
Plastic sleeves pipes
Pencils for marking
Work making shop in shed
Electric generator/supply
Material hoisting arrangements
Steel files to sharpen saw teeth
Crow bars
Thread line Nylon
Chalk marker
Hard hats, Hand gloves (rubber, cloth)
Hand augers and electric drills
Form oil and brushes
Hand brushes
Drinking water jug/Thermos flasks
Clamps/Shikanjas both lengths
Scaffoldings and safety nets
Tools kit, Ladder (timber, aluminium), torch

Steel fixing

Cutting machine electrical
Hand cutter equipment
Bending dyes
Working table in shed
Gas cutter
Transportation arrangements
Unloading arrangements/off loading
Timber or a good impervious platform for the
   storage of the material

Binding wire rolls
Pliers and scissors
Raising equipments
Precast concrete covers/spacers
Plastic covers/spacers
Measurement tapes, 3 m, 5 m, 15 m, 30 m
Chalk sticks
Hand gloves (rubber, cloth)
Hard hats
Welding equipment and accessories
Electric generator
Drinking water jug/thermos flask

Miscellaneous

Wrench
Spanner set
Grease pump
Oil pump
Oil drums
Oil tank
Generator shed
Fuel tank
Hydraulic oil
Water tank and pump
Electric testers
Electric instruments
Voltmeter, Ammeter
Fuses, DBs, Switch board
Control panel
Cables and wires
Fitter's vice and table
Ducting shop with all relevant tools and
   arrangements
Hand drill
Electric drill
Grinder
Squeezes
Brushes
Hand shovels
Hand axes
Hand pans
Hand spades
Plastic buckets

Machines in construction

Tower cranes with slings
Mobile cranes with slings (tyres or crawler
   mounted) (telescopic)
Site generators upto 20 kva

Excavator with tyres
Rock breaker with adjustable
bucket replacement system
Shovel
Payloader
Transport trailor trucks
Site concrete batcher
Batching plant
Central mixing plant with cement silos
Truck mixers various capacities
Stationary concrete pump
Mobile concrete pump with boom
Hydraulic trucks/Dippers
Buses
Cars/Jeeps
Pick ups
Concrete trolleys/dumpers
Fork lifts
Concrete skips/buckets with gate
Winch
Compactor, compacting roller
Pneumatic roller
Road roller
Bulldozer/Angle dozer
Bitumen spray tanker
Grading machine/grader
Sand blasting equipment
Paint Sprayer
Drag line
Dewatering pumps/Miller pumps
Centrifugal or diaphram pump
Submersible pumps
Jack hammers drills with all the compressed
    air tubes adaptable to the compressors with
    all types of drill bits for rock manual
    excavations and final surface dressing of the
    rock breaker excavated work
Curing compound spray gun/pump
Mastic gun for gun grade compound
Rigging machine
Pile driving machine, accessories and equipments
Pile load testing arrangements
Mechanical shop fully equipped
Pre cast concrete hydraulic pressure yard
Company yard
Crushing plant
Pre-mix plant

Site establishment

Slump test cone and rod for tamping

Steel cubes 150 mm × 150 mm (6 no. each set)
Weather thermometer
Concrete thermometer (maximum/minimum)
Spirit level
Measurement tape small/scale
Concrete cube storage in water tank
Cube tank shed
Traxlometer (soil density measuring more than
    200 mm fill after compaction)
Proctor test equipment
Micrometer to measure the paint thickness on
    steel structure
Vernier calliper
Screw gauge, Moisture Penetration
    meter/instrument for plaster
Fully equipped civil engg test laboratory for
    concrete, aggregate, soil, sand, steel, and all
    the miscellaneous test arrangements
Sampling arrangements of the approved
    materials
Big site stores/stocks
Kitchen, lavatory,
Conference room
Telephones, Telex, Fascimile
Photocopiers
Typewriters
Central control room
Clerical room
Filing system, diary, despatch
Drawing room fully equipped
Parking under shed
Guard room
Reporting room
Offices for all site senior personnel
Accounts office
Coordination office
Labour control office
Workers accommodations with full facilities
Stationary according to requirements
Theodolite with tripod
Level with tripod
Surveying measurements tapes
Levelling staff
Markers and stakes
Level books and reference books
Two way radio system to control
Computer room for a big site, if necessary
Medical attendance for first aid room

Establishment of the sub contractors' offices and all facilities as for you

All similar facilities for the engineer and staff if in contract subject to Engineer's/RE approval

Site sign board for site entry and offices

Site diary

Workers site uniforms/dresses with names of individuals and company, workers identity cards/gate passes

Name of company on every machine, vehicle, office, transportation, etc., for identification

Materials

Piles with shoes

Pile caps of concrete R.C.

Ply wood 18 mm

Planks 20 mm

Nails 100 mm, 75 mm, 40 mm

Blocks of concrete 400 × 200 × 200 mm, solid/hollow

Battens 75 mm × 75 mm, 100 mm × 100 mm, 100 mm × 75 mm

Aggregate 20 mm, 12 mm, 40 mm

Sand C, Dune Sand/Fine sand

Cement SRC/OPC in bags/Cement in Silos

Water

Bitumat in rolls

Bitumat protection/Bitumat

Reinforcing steel 32 mm, 25, 22, 16, 12, 10, 8 mm and tor/Mild as specified

Binding wire

Supporting jacks and pipes

PVC membrane 1000 g, 500 g

Electrical conduits PVC/metallic with junction, etc.

Sanitary pipes and jointing material

Compressed filler/flaxcell

Acrylic bitumen paint

Lime hydrated

Door frames Timber/Aluminium/Steel

Window frames Do

Hold fasts

Connecting screws and plugs

Expanded metal laths

Expamet mesh

Curing compound

Degrading compound

Concrete repair compounds resin systems for injections

Epoxy material for concrete repairs

Floor tiles as specified

Wall tiles as specified

Skirtings as specified

Doors shutters/Leaf as specified

Windows and ventilators Aluminium/Steel Glass blocks as specified

Stair rails steel/timbers/ornamental as specified

Parapet rails ornamental/ordinary as specified

Ceiling tiles as specified

Special flooring materials as specified

Structural steel sections to engineer's details

Rolled steel joists, nuts bolts

Rivetting arrangements, base plates

Structural steel paint as specified

Acrylic primer

Hoisting/lifting points

Structural steel rafters

Prefab steel mesh BRC

8 mm, 10 mm, 150 × 150 mm c/c

Concrete nails made of steel

Cavity wall ties

Wall ties

Intruded polystyrene

Light cavity filling material, if specified

Wall bolts for concrete wall and PVC sleeve, epoxy mortar

Shot firing bullets

AC ducts material as specified

Aluminium

Glass fibre

Duct paint

Duct hangers with metal angles and connecting system (fixing system)

Duct fittings as VCD, bends, FD, Attenuators, etc.

Air Handling units (chiller units, extract fans)

Copper piping system

Plant rooms for machines

Pressurising unit

Isolators and control panels

Supply duct line

Return duct line

Water supply lines

Machine bases for installation

Screen Walls around the machines as specified

Protection meshes

Compressors, evaporators, etc.

Condensers

Base frames on the ground or the plinths for the pipes

Fresh Air chambers prefab or to be constructed on site

All relevant electric connections and accessories

Grills and diffusers, splitters

Filters for carbon/a.c. filters

Operation manuals and controls

Control rooms

Lift rooms

Lift beams

Lift car with guides and connections

Hydraulic arrangements

Lift doors

Lift controls, escalators, travalators, conveyer belt

Floor diaphrams

GI pipes and fittings

Insulation for the pipes

Copper pipes

Pipe paint

Cold water supply and hot water supply system with fittings

Drinking water system, non-drinking water system, drinking water fountain

Wet area tiles

WC, bath, bidets, sinks, wash hand basins, steam baths, hydromassage baths, Tapes, cisterns, waste pipes

Sanitary duct, water tanks,

Man holes, inspection chambers

Septic tanks, municipal connections,

Soak aways/pits,

Rain water stouts, Rain water pipes,

Rain water outlets,

roof aluminium flashings, gargogyles, roof skirtings, sealant, water tank with fittings and beams, suspended ceiling system, angles, flats, wires, air gap material, ceiling tiles, plain finish/needle finish, ceiling bulkhead of rolled steel frames finished with the asbestos tiles or sheets connected by screws and washers well trimmed, sealant for the sky light/roof light, ceiling edge piece, acrylic primer on walls, wall paint, screed concrete for floors, flooring tiles, wood skirtings, plastic plug and screw, trimmer piece with nails for carpet gripping, underlay carpet, carpet, epoxy floor material, paper tape, matwell materials, recessed connections in the floor screed, entrance steps risers/treads, patio doors, glass doors, curtain walls, external doors on rollers, duct

door, reception counters, external texture paint to be applied by special brushes in roller shape, wall lining material, special flooring material, ceiling lining material, GRC panels in the ornamental motifs of different shapes connections of these designs to the structure, other external finish materials, marble stones, sand stones, mineralite, double glazed solar reflector windows, louvers metal, claustra blocks for screen walls, fly mesh, master antenna, kerb stones, precast, plinth protection material, flagstones, materials for the external services, pipes, thrust blocks, connections, man holes, draw pits, inspection chambers, covers light duty/heavy duty, perogolas, sandtex paint, sports surface materials, road fills, road structural materials, seating and benches, sweat soil, tree pits, tree guards, for grass fill sweat soil, sand fill area, benches, irrigation system for the internal as well as external of the building, other landscaping materials, metal gates, boundary wall material, decorative joinery material, plantings beds, raised planters, fountains, water supply and circulation to the fountains, pumping systems, electrical supply materials, all services of the swimming pool, plant room, filters, acid chambers, pumps, bases, water tank, surface drains, surface grills, changing rooms, toilets, pergolas, pool carpet, underwater electric flood lights, flood lights, ladders, diving boards, vacuum cleaning equipments, safety equipment, water test kit, chemical feed pump, showers, thermometers, unions, commercial pool fittings, compound wall, chain link fencing, gates, discharge connection manhole to the municipal sewerage system, granolithic floor material, Neoprene, MS lamina, RS angles and flats, materials for the production of the grooves pockets for the machines

Hand drier, toilet paper holder

Ablution system, gully traps, P-S traps, floor traps, discharge pipes, water filters, Lead, ropes for joints, jointing gaskets, CI, SCI, Asbestos, SW pipes, RCC pipes, thrust blocks

Urinals, Drinking water fountains

Showers, Dish washer, Slope sink,

Double bowl sink, Cleaner's sink,

WC for females, WC for handicapped,

Kitchen equipments, Cold storage (all specified units)

Gas supply system with accessories, Sprinklers,

wet risers, high pressure water spray or foam sprinkler system, fire hose reel, nozzles, static tanks, automatic sprinkler, carbon dioxide system, fire alarm, fire detectors, fire fighting external system, fire hydrants, water pipe line in the ground, supply from the main tanks, fire doors, exit doors, fire lift, separate connection to fire lift for electrical operation, sub station, transformer, switchgear, cable trenches, steel covers, transformer base/plinth, high voltage switch room, low voltage switch room, stand by generator, distribution panels, control panels, PBX/PABX, cable trunking, cable trays, cables, wires. conduits, connections of fixing to walls, junction boxes, switches, Isolators, control panels, electric lights and illuminaries, light poles, light hangers, tubes, bulbs, speaker or audio system, fume cup board, roof torch, roof mortar, roof mat/extruded polystyrene, roof flagstones/tiles, roof aggregate, sky lights, foundations, levelling mortar, grouting material, machine connection bolts, Ironmongery/fixtures and fastenings, Estuncheons, butt hinges, back flap hinge, strap hinge, garnet hinge, parliamentry hinge, rising butt hinge, narmadi, aldrop bolt, barrel bolt, flush bolt, latches, hasp staple, tower bolt, cup board lock, mortice lock, rim lock, lever handle, pad lock, door handle, lever handle, kick plate, door closer, water supply and sewage disposal systems and their treatments.

Categories of productive workforce

Mason — Block builder, tiler, plasterer, flooring, special jobs, concrete repair, grouting, roofing

Carpenter — Form work, joinery, decorative works, Assistants for above artisans, Helpers to above

Steel fixer

Astt fixer

Helper

Welder

Fitter

Astt fitter

Plumber

Astt plumber

Cable jointer

Astt jointer

Electrician

Assistant

Duct moulder

Assistant

Duct painter

Duct erector

Assistants

Mechanic general

Mechanic diesel

Mechanic auto

Fabricator

All assistants to above mechanics and fabricator

Machine operators

Machine attendants/astts

Drivers,

Puncture repairer

Watch man

Rigger

Road gang

Concrete gang

Transportation of material gang

Advance supply material gang

General support gang

Time keeper

Medical attendant

Site tiding gang

Store keeper

Painters

Roofing special gangs

Special floor gangs

Sports surface floor gang

Heavy earth work gang

Riggers, scaffolders

Charge hand

Tandels/mates

General category workers

For general guidance

Plates 1-11 Nos.

SK - SKETCH

Description of the enclosed plates

Plate 1   Preparatory form work for suspended floor slab in progress

Columns poured upto beam soffit level approximately (a few cm below)

Centring and form material lying around

One slab ready for pouring and being given final touch

Concrete mobile pump taking position by adjusting the pump boom and its own body to cover the maximum area of the proposed pour without shifting the station if possible but not necessary

Rock breaker machine is excavating the rocky area

Claustra blocks screen wall is visible in the sub station building in view

Form of capping beam of a boundary wall is ready for concreting

Water filled drums can be seen on the floor near the proposed pour

Steel reinforcement for the connection to the sill beams have been left in the form of dowels

Bitumen painted surface is clearly visible in view

Surface of the ground floor already poured is visible in view

Scabbled joints for the beams have been graphed in view

Plate 2 Concrete pump (mobile) receiving ready mix concrete from the transit concrete truck of 6 cum capacity and pump by reciprocating action pushes the mix to the pouring place through the boom as graphed in view. Levelled surface of poured and compacted concrete is visible in view

Planter beds to contain the sweat soil have been graphed in view

Form material after being struck lying astray in view

The boom of the rock breaker machine is being seen

Plate 3   Concrete pouring operation in progress by pump
Poured concrete levelled, people controlling the rubber pipe and compaction
Electrical conduits of PVC visible and connected between junctions and bends and finally
approved, concrete covers have been positioned so as to render the reinforcement placement
good, Construction joint made before the commencement of the operation, small steel columns
originating from this level for the facie/parapet support in view

Plate 4   On one part of site steel reinforcement fabrication in progress in columns, curing in progress on the previous poured slab where hessian is shown spread, Final cleaning operation of the completed preparatory works of the proposed slab are going on, concrete pump has arrived on site, concrete bucket/skip is kept on site only for emergency, concrete did not arrive yet, gang people are getting ready, small roller compactor is also lying there but not needed at the moment, it appears that site is not working to the full potential due to the off timings and concrete is being poured on over time, rock hammer is working behind the building, blocks stacked

Plate 5   You can see the poured column upto the beam soffit level, some columns exhibiting surface defects, dowels of the beams are in position, some form works in completed form are visible, some preparatory works for the mid level beams are going on, generator shed and generator are visible, one man hole walls are also in the view, site offices in the view, tower crane is also in the view, you can see the completed boundary wall with the movement joints, a compressor near the crane is in the view

Plate 6   View of concrete pour operation from near the pouring spot, people of concrete gang holding the rubber/boom pipe and placing the concrete uniformly, the pipe of the vibrator through which compressed air is supplied is controlled by the gang person, the compressor installed on ground, see the concrete covers, well placed steel reinforcements, properly fixed electrical conduits of PVC duly connected, some block outs in the right position for the sanitary pipes/otherwise for engineering items, rock breaker also in the view from behind, plastic bucket is there

Plate 7   In the view you can see the completed laying operations of the sewerage line of the external works under the backfilling stopped operation, the construction of the manholes walls have been over, view also shows in the block outs fixed during the ground floor operation of pouring for the sanitary or the other services connections, view also includes some portion of the pre fab BRC mesh on the left, works on the columns in progress, beams preparatory works in progress, made up form plates graphed, form and centring material in the view, tower crane in the view, crane rope in view,

Plate 8   Preparation for final pouring of slab over, pump arrived, steel column fabrication going on, crane and rock breaker in view, sub station in view

Plate 9   Some foundations, straps and reinforcement, boundary wall with joints

Plate 10   Some construction machines

Sketches 1 to 50

Layouts and Preparatory Form Works
Typical for General Knowledge

First Hand Written

Not to Scale

Sketch No.

1. Setting out plan of a building with reflected slab grid plan.
2. Setting out plan with footings and plinth beams and form plan.
3. Form plans while the engineer is on the higher levels for inspections.
4. Setting out plan for land surveyor and reflected form plan.
5. Typical plan of setting of a complex architectural building along with reflection of forms.
6. Setting out plan on commencement while on 0-0 level.
7. Reflected form plan of Sketch 6.
8. Setting out plan on ground for engineer's inspection.
9. Reflected form plan of Sketch 8.
10. Typical form work plan of a footing in one stage and a stub column in another stage.
11. Section ABB CCA for Sketch 10.
12. Setting out plan of columns size 600×600 mm spacing 5000 mm c/c both directions and typical preparatory, checking and inspection of a simple column.
13. Typical details for the checking and inspection of a column form.
14. Typical detailed elevation of 100 mm column side.
15. Setting out of columns from the general arrangement and layout plans of any building in contract.
16. Setting out plans for the columns at any level from general arrangement details using string/thread.
17. Typical form work of a rectangular column.
18. Elevation of 1000 mm size plate of a column form.
19. Typical details in plan of L-shaped column form.
20. Elevation of Sketch 19.
21. Elevation of 900 mm side plate.
22. Typical form work plan of T-shaped column.
23. Elevation of Sketch 21.
24. Elevation of 1200 mm side plate.
25. Elevation 300-600-300 mm side.
26. Typical plan details of + shaped column.
27. Elevation of Sketch 26, 300-600-300 mm.
28. Typical plan of a wall form work.
29. Typical sectional elevation of 150 mm thick wall (RCC).
30. Typical elevation 150 mm thick R.C. wall form work.

31. Typical elevator/lift well 2000 × 2000 mm assumed plan details.
32. Lift form work/elevation (A_1).
33. Lift form work/elevation (B_1).
34. Lift form work/elevation (C_1).
35. Lift form work/elevation A.
36. Lift form work/elevation B.
37. Lift form work/elevation C.
38. Elevator/lift door (void) form typical.
39. Typical form plan of a regular pentagonal column 500 mm side.
40. Typical form plan of a regular hexagonal column 500 mm side.
41. Typical form plan of a regular octagonal column 500 mm side.
42. Typical form work plan of a column turned obtuse.
43. Typical details of the form work of different stages of concrete casting in situ for a complex facii virtually operated in a modern building.
44. Typical details of the form work to be erected for same facia above a window opening for air-conditioned accommodation in a modern building.
45. Typical details of form work of facia in facet.
46-48. Typical form details of a staircase.
49-50. Typical slab form details.

Lay out or Setting out plan of a building structure for the purpose of guidance to the newcomers

Reflected slab grid plans while the engineer is on the forms of floor to check

Section X-X

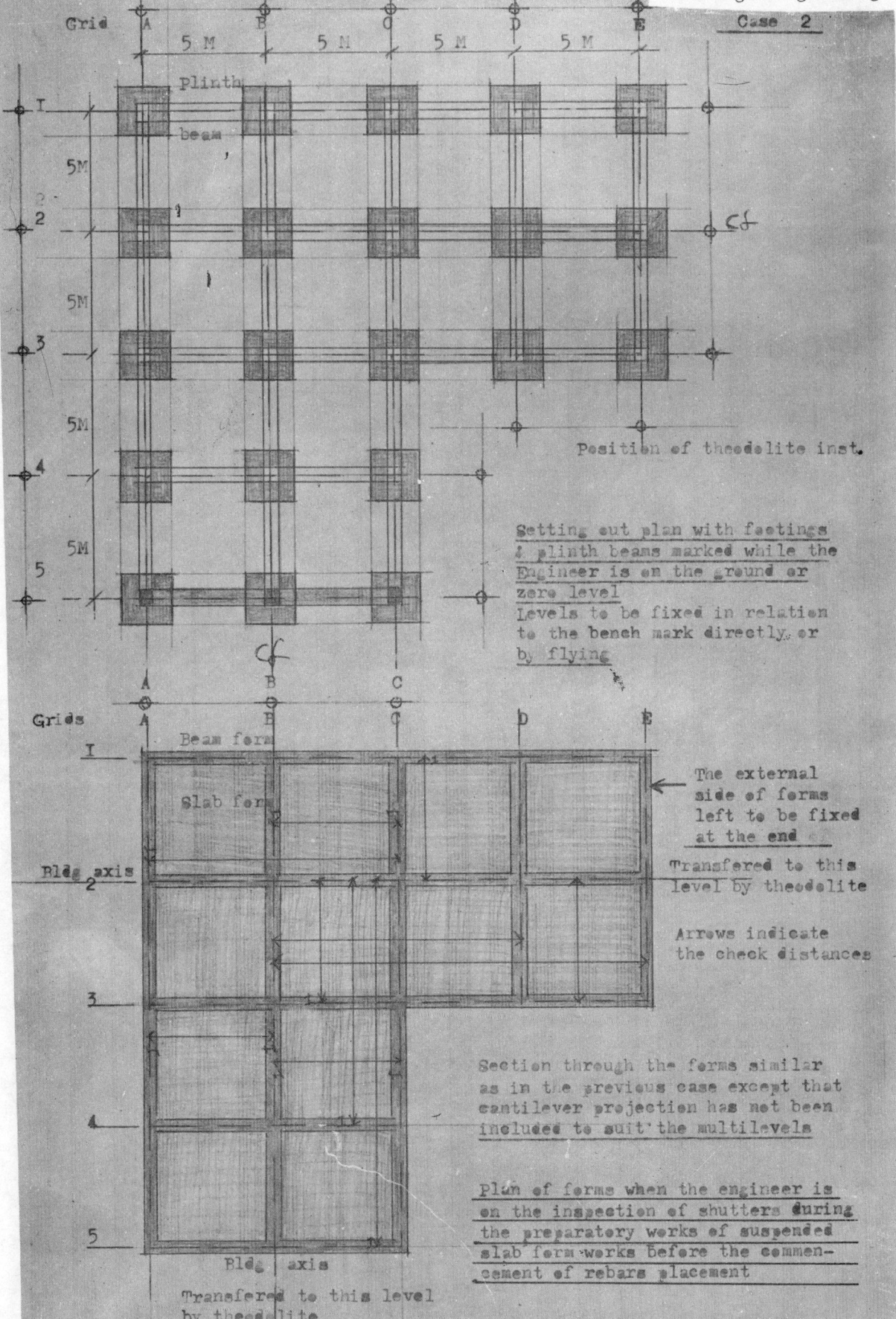

Grid   A   B   C   D   E   Case 2

5 M   5 M   5 M   5 M

Plinth

beam

Position of theodolite inst.

Setting out plan with footings
& plinth beams marked while the
Engineer is on the ground or
zero level
Levels to be fixed in relation
to the bench mark directly, or
by flying

Grids   A   B   C   D   E

Beam form

Slab form

The external
side of forms
left to be fixed
at the end

Transfered to this
level by theodolite

Arrows indicate
the check distances

Section through the forms similar
as in the previous case except that
cantilever projection has not been
included to suit the multilevels

Plan of forms when the engineer is
on the inspection of shutters during
the preparatory works of suspended
slab form works before the commencement of rebars placement

Bldg axis

Transfered to this level
by theodolite

Note- The external forms of the vertical side of beams shall be placed
after placing & fixing the external beam reinforcement for the
convenience of ease of work & maintenance of cleaning. This is in
the discretion of the site engineer/agent

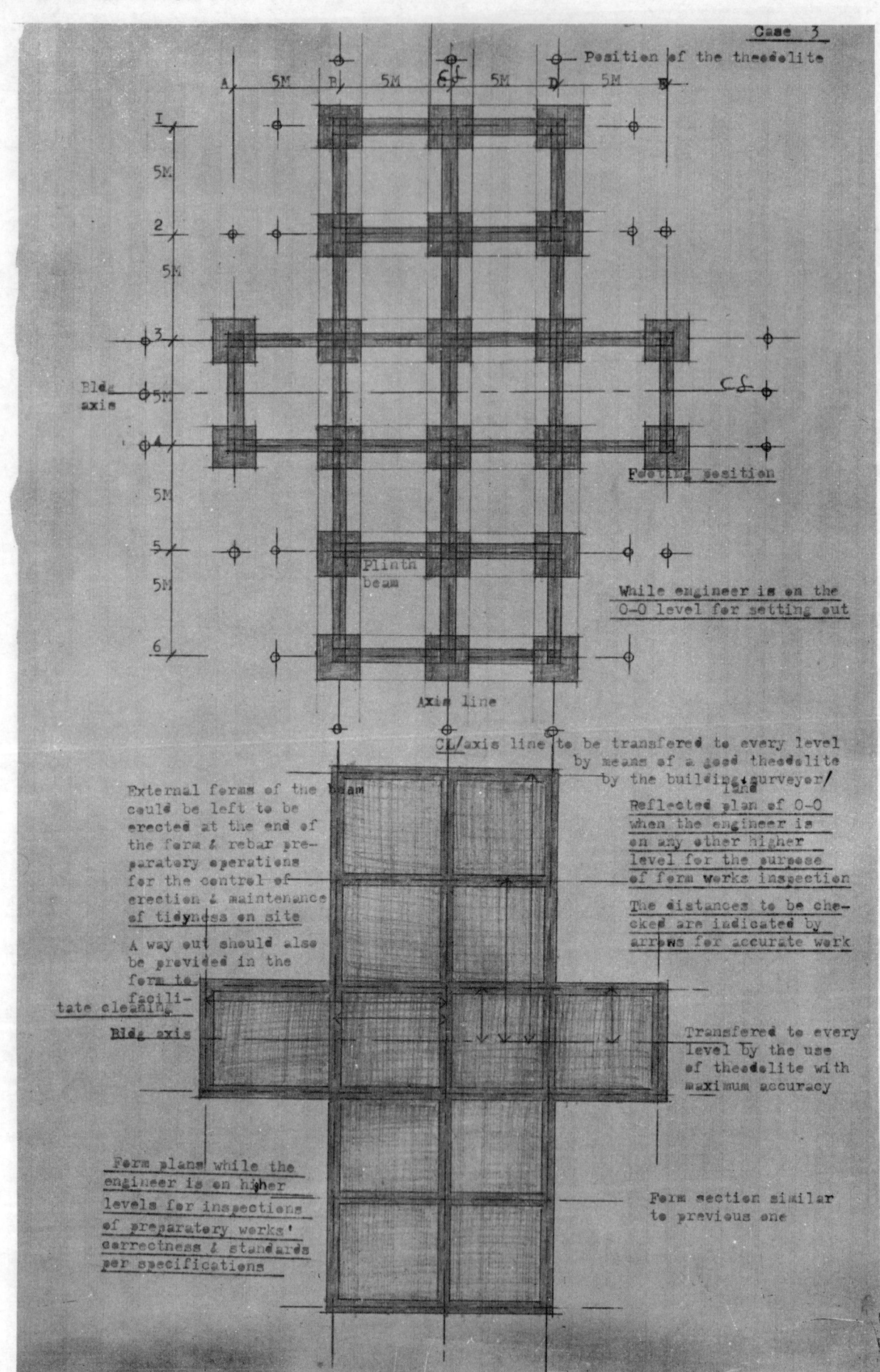

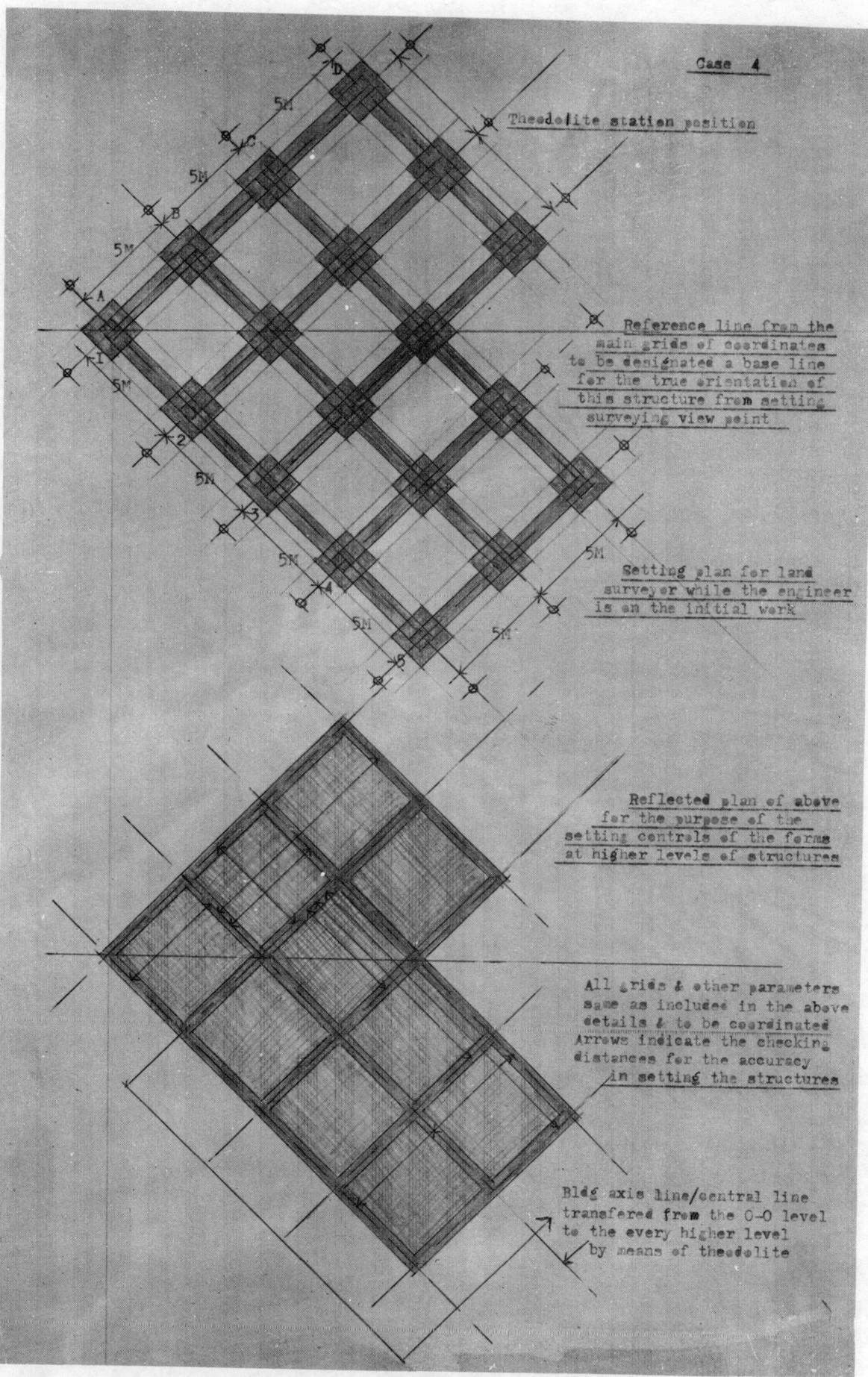

Case 4

Theodolite station position

Reference line from the main grids of coordinates to be designated a base line for the true orientation of this structure from setting surveying view point

Setting plan for land surveyor while the engineer is on the initial work

Reflected plan of above for the purpose of the setting controls of the forms at higher levels of structures

All grids & other parameters same as included in the above details & to be coordinated. Arrows indicate the checking distances for the accuracy in setting the structures

Bldg axis line/central line transfered from the 0-0 level to the every higher level by means of theodolite

5M 5M 5M 5M 5M 5M 5M 5M

A B C D

1 2 3 4 5

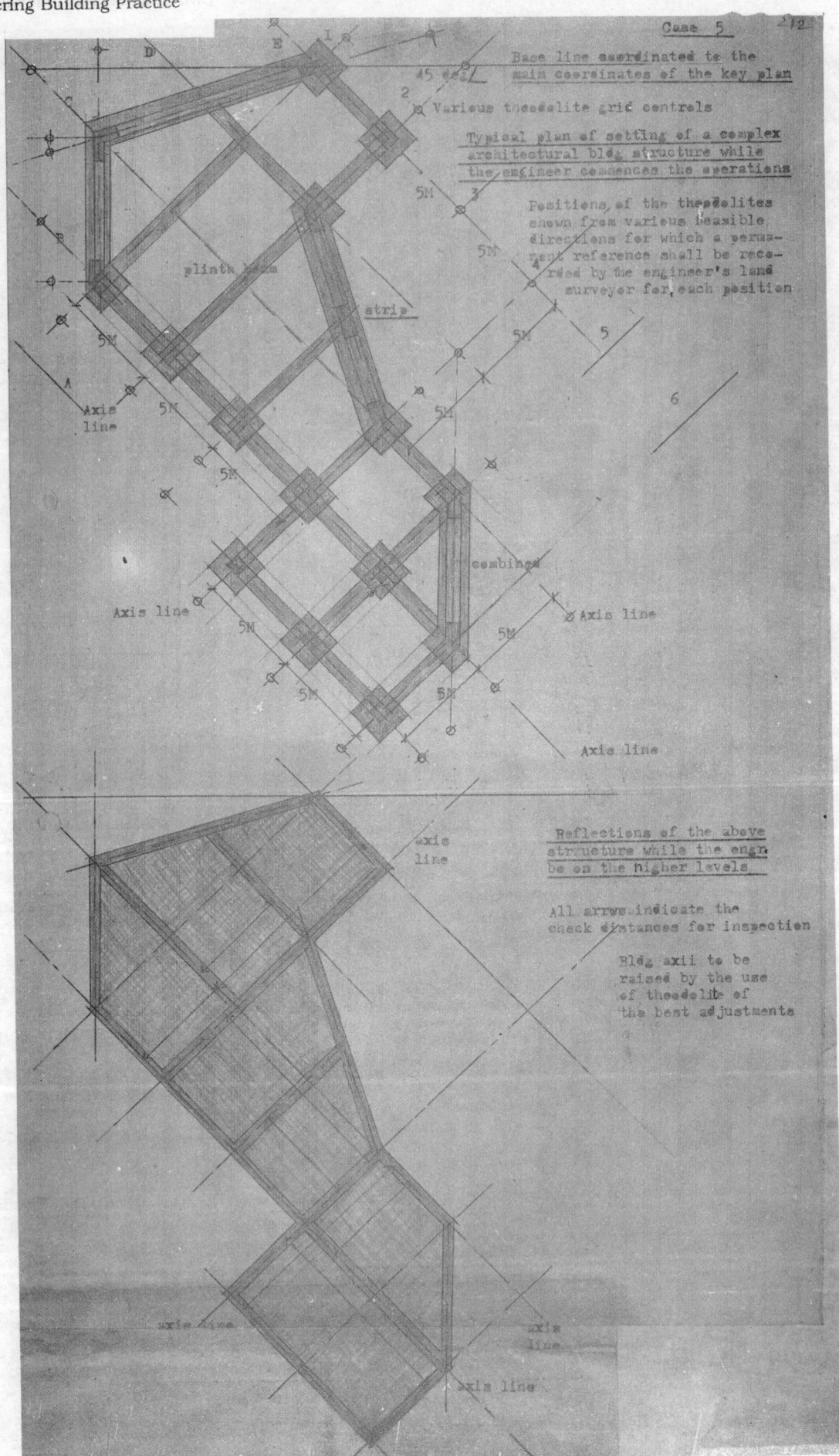

Case 5

Base line coordinated to the main coordinates of the key plan

Various theodolite grid centrals

Typical plan of setting of a complex architectural bldg structure while the engineer commences the operations

Positions of the theodolites shown from various feasible directions for which a permanent reference shall be recorded by the engineer's land surveyor for each position

plinth beam

strip

combined

Axis line

Reflections of the above structure while the engn be on the higher levels

All arrows indicate the check distances for inspection

Bldg axii to be raised by the use of theodolite of the best adjustments

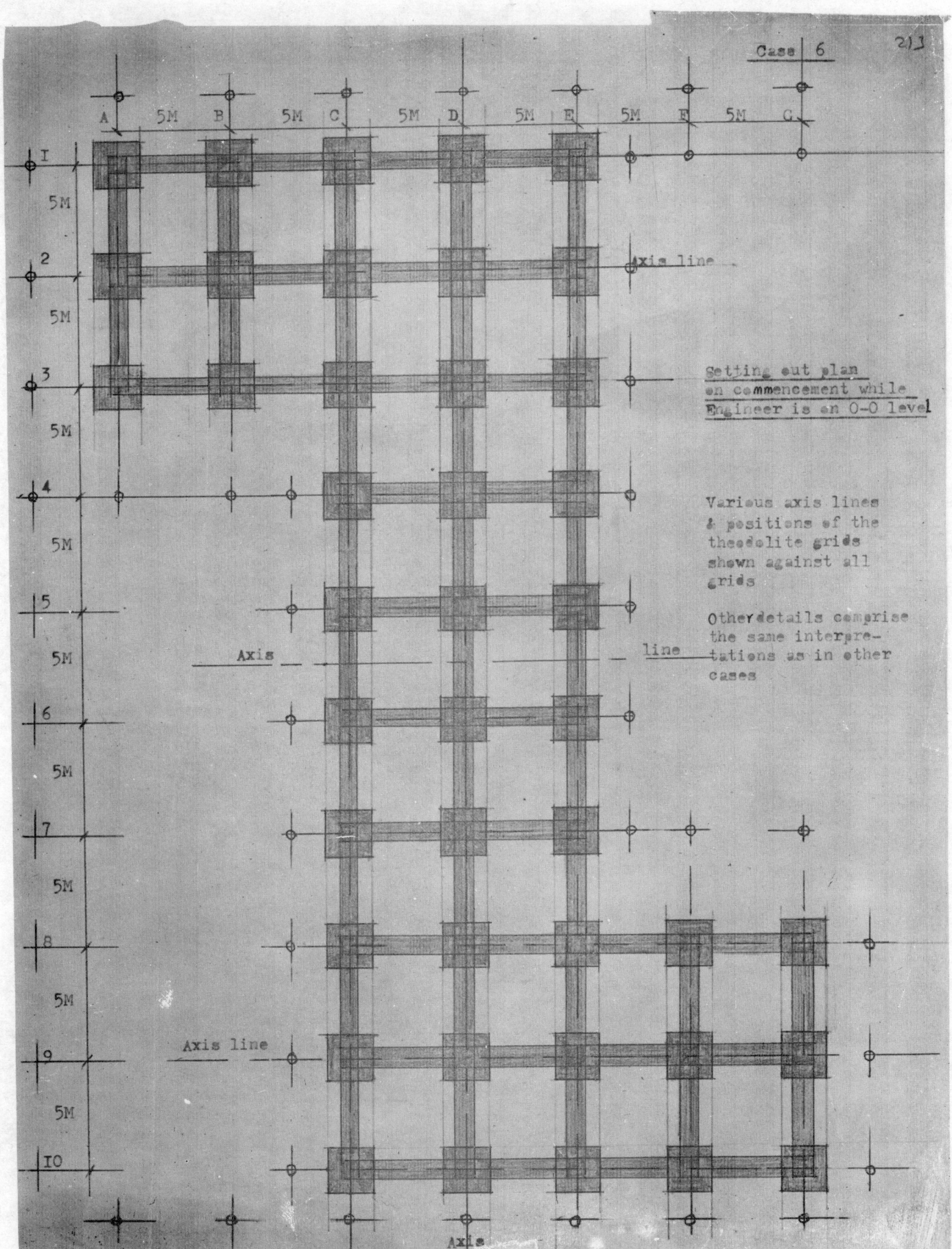

Case 6

213

Setting out plan
on commencement while
Engineer is on 0-0 level

Various axis lines
& positions of the
theodolite grids
shown against all
grids

Other details comprise
the same interpre-
tations as in other
cases

Axis line

5M 5M 5M 5M 5M 5M 5M

A B C D E F G

Axis line

Axis line

Axis line

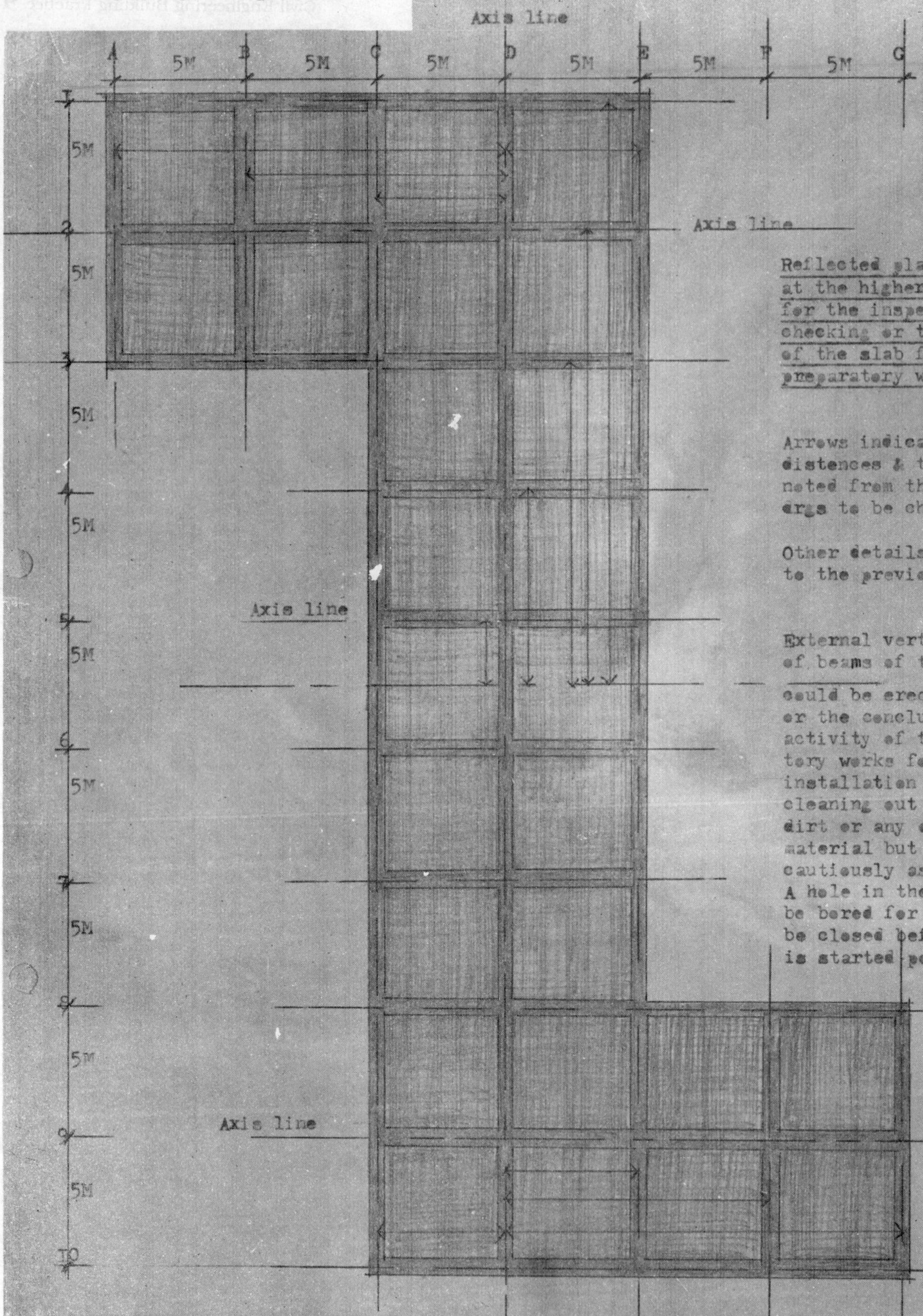

Reflected plan of the case
at the higher levels
for the inspections &
checking or the controls
of the slab forms & the
preparatory works

Arrows indicate the
distances & the measurements
noted from the contract
area to be checked

Other details are similar
to the previous cases

External vertical forms
of beams of the periphery

could be erected as a last
or the concluding operational
activity of this prepara-
tory works for the ease of
installation or erection &
cleaning out the saw dust,
dirt or any other unwanted
material but shall be closed
cautiously as specified.
A hole in the beam side could
be bored for cleaning but
be closed before the concrete
is started poring.

All the axii of the building to be raised or transfered by a theodolite
to every floor; the references of all the theodolite grid points or
the stations to be recorded by the concerned surveyor for the ease of
efficient relocation in case destroyed during excavations by the machines.
The site engineer or the site agent should be very serious about this
control as this shall assist the work till the structure is finished
& also contribute during the execution of the finishing operations.

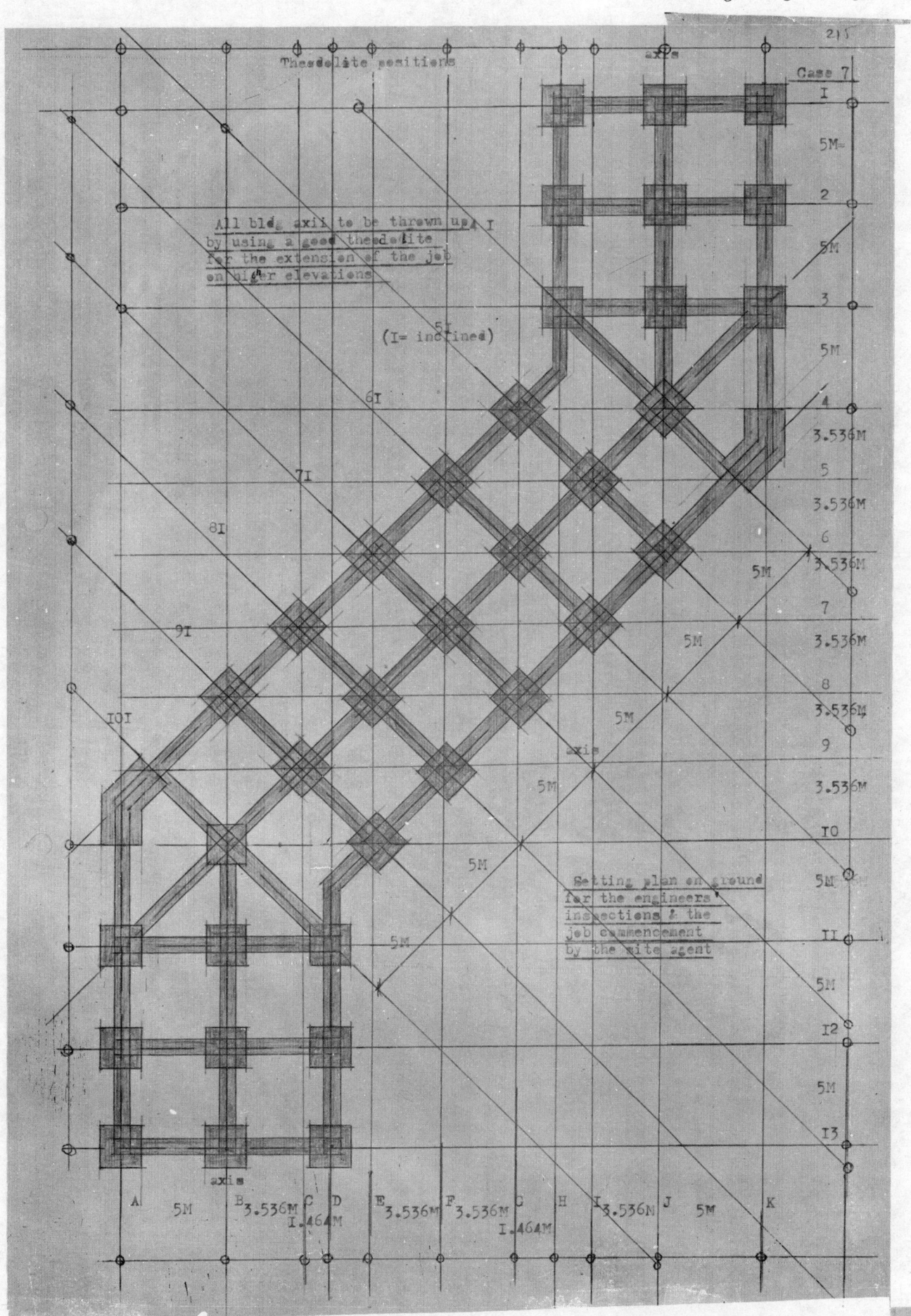

All bldg axis to be thrown up
by using a good theodolite
for the extension of the job
on higher elevations

(I= inclined)

Setting plan on ground
for the engineers'
inspections & the
job commencement
by the site agent

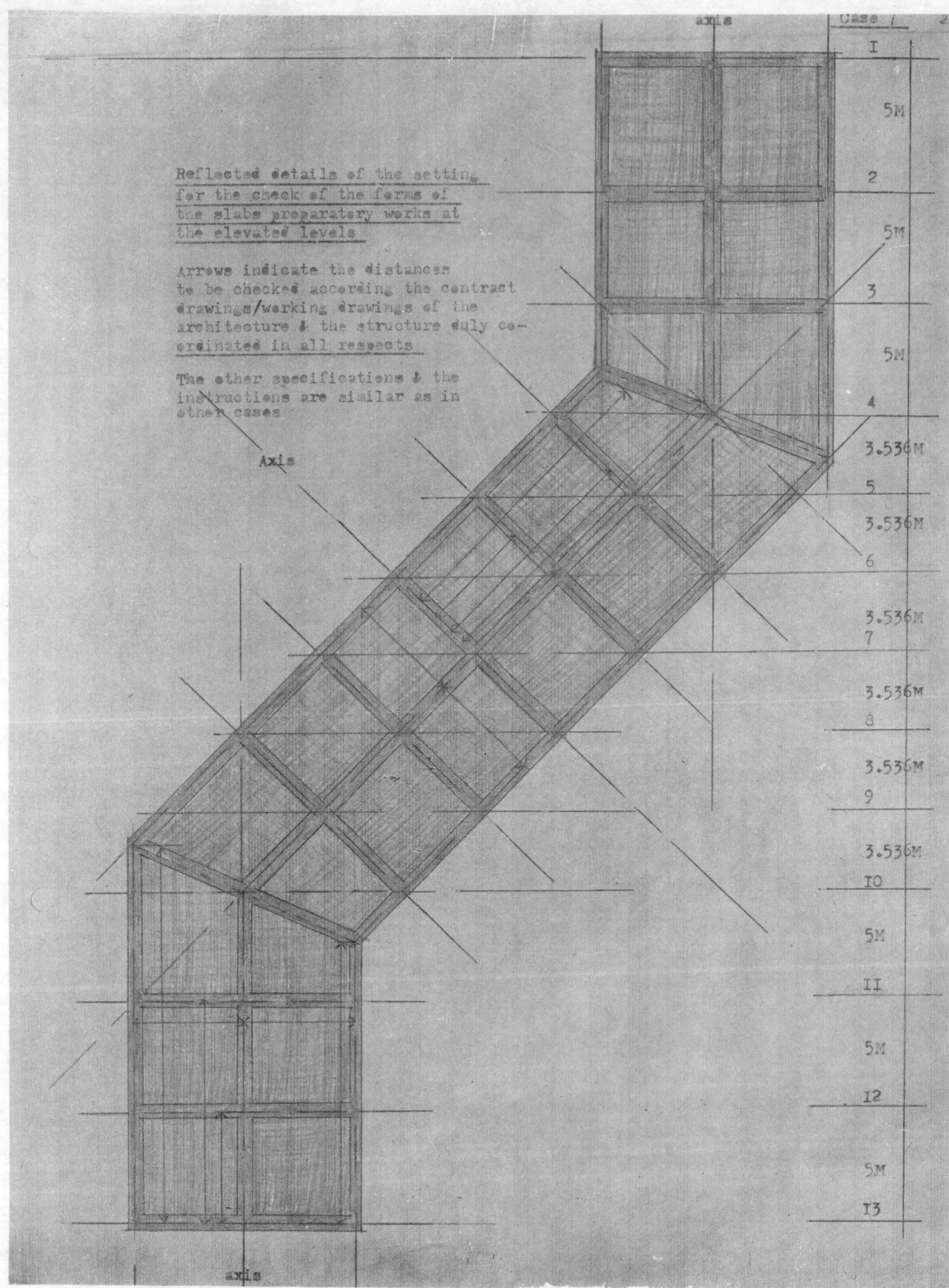

Reflected details of the setting
for the check of the forms of
the slabs preparatory works at
the elevated levels

Arrows indicate the distances
to be checked according the contract
drawings/working drawings of the
architecture & the structure duly co-
ordinated in all respects

The other specifications & the
instructions are similar as in
other cases

Axis

axis   Case   2

I

5M

2

5M

3

5M

4

3.536M

5

3.536M

6

3.536M

7

3.536M

8

3.536M

9

3.536M

IO

5M

II

5M

I2

5M

I3

axis

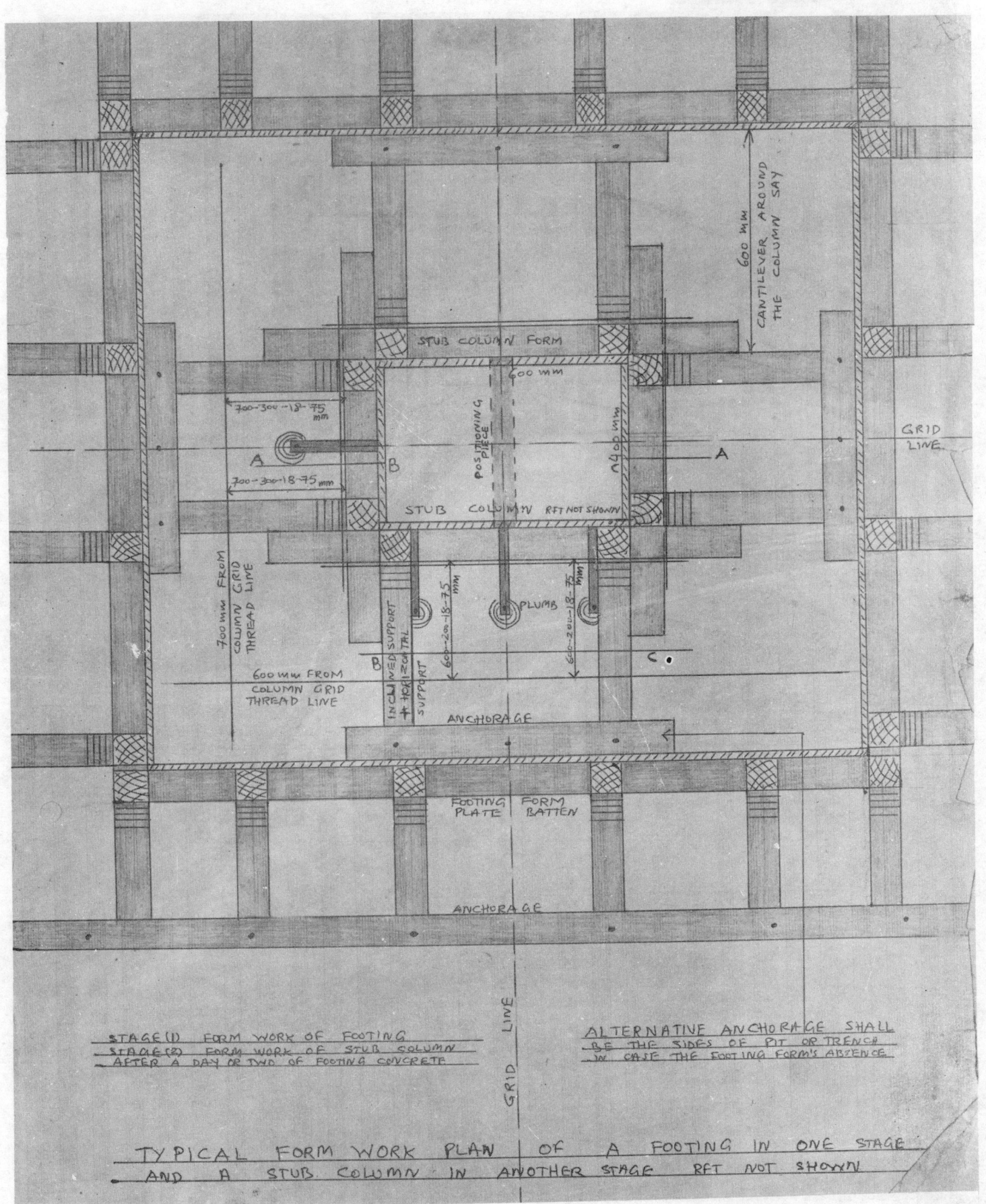

STUB COLUMN FORM

600 MM

POSITIONING PIECE

700-300-18-75 MM

A

B

A

GRID LINE

700-300-18-75 MM

STUB COLUMN RFT NOT SHOWN

600 MM

CANTILEVER AROUND THE COLUMN SAY

700 MM FROM COLUMN GRID THREAD LINE

INCLINED SUPPORT + HORIZONTAL SUPPORT

B

600-200-18-75 MM

PLUMB

600-200-18-75 MM

C

600 MM FROM COLUMN GRID THREAD LINE

ANCHORAGE

FOOTING FORM PLATE | BATTEN

ANCHORAGE

GRID LINE

STAGE(1) FORM WORK OF FOOTING
STAGE(2) FORM WORK OF STUB COLUMN
AFTER A DAY OR TWO OF FOOTING CONCRETE

ALTERNATIVE ANCHORAGE SHALL
BE THE SIDES OF PIT OR TRENCH
IN CASE THE FOOTING FORM'S ABSENCE

TYPICAL FORM WORK PLAN OF A FOOTING IN ONE STAGE
AND A STUB COLUMN IN ANOTHER STAGE RFT NOT SHOWN

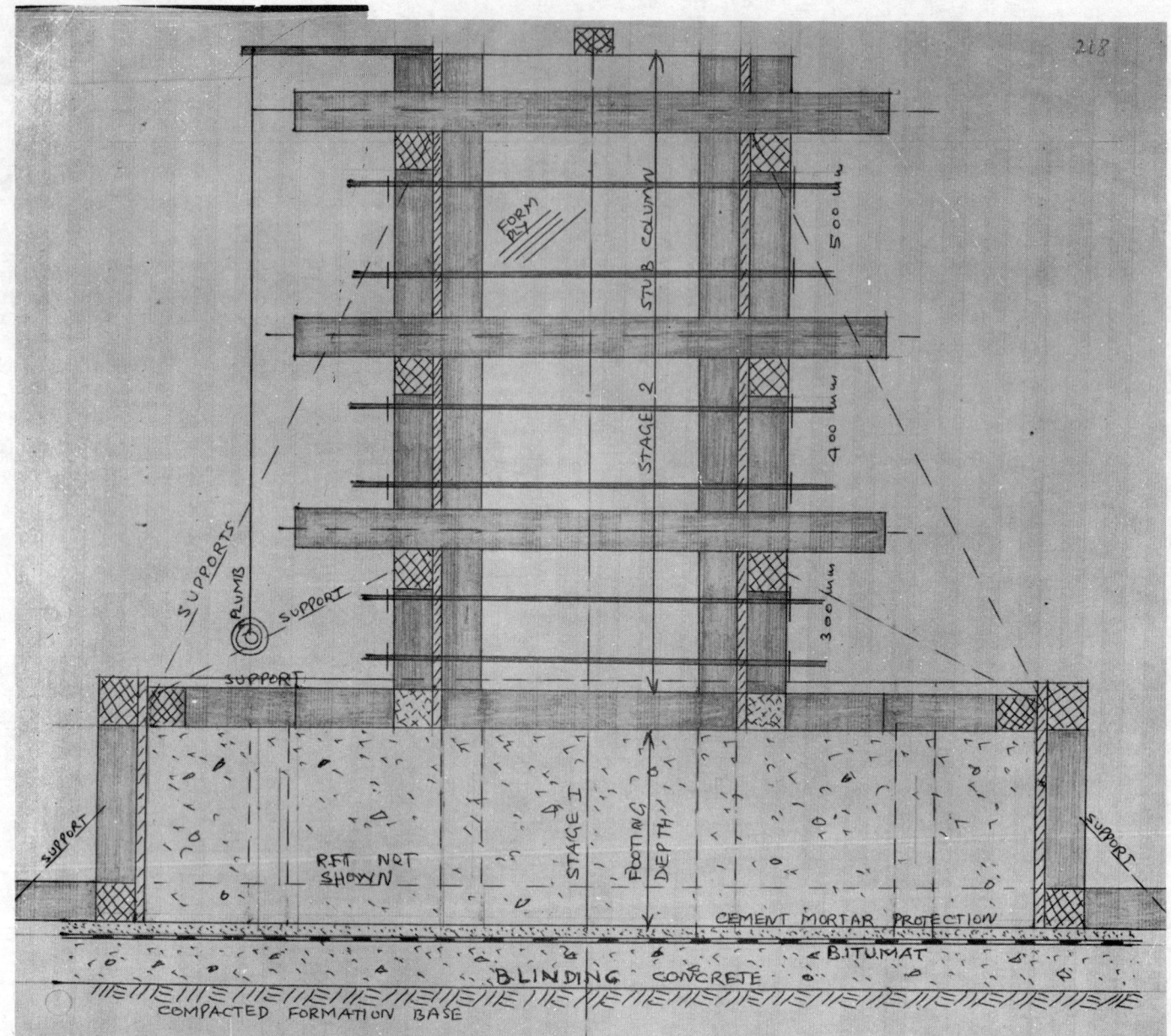

SECTION A B B C C A

<u>Details</u> above indicate the form works of the stub column i.e column portion from the footing top to the plinth level or a little above in one sketch but shall be done in two stages on site. First stage shall work to the pouring the footing & the 2nd shall work to the pouring of the part column.

The details of the foundation treatment in general has also been shown from the practical point of view. The supporting jacks or the scantlings have not been exhibited in full.

The work on every foundation or twin column foundation shall dicate its existence according the Engineer's details from site to site. Site engineer/Agent in consent with the Resident engineer enjoys the competence to decide about the method adoption.

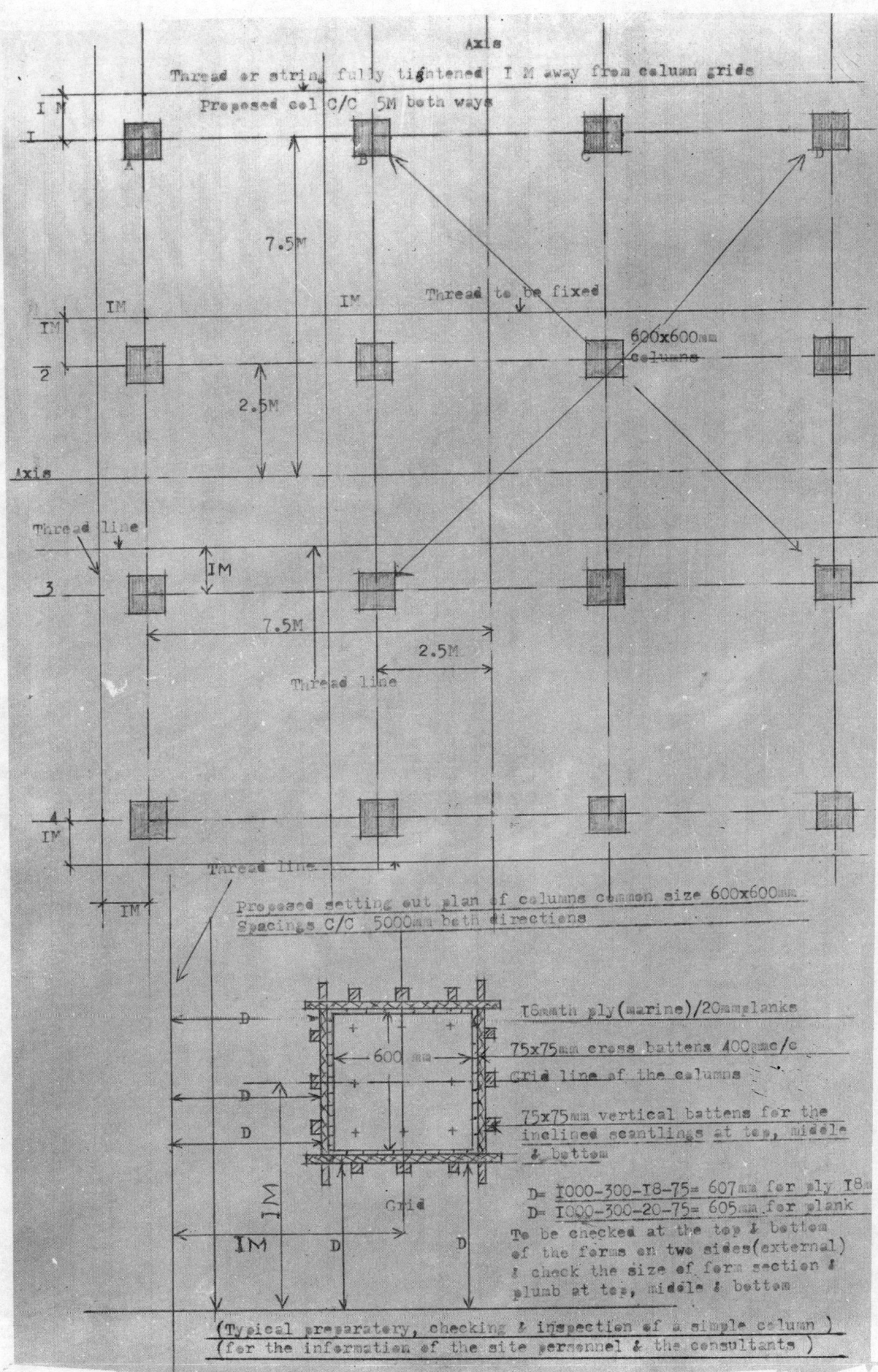

Axis

Thread or string fully tightened 1 M away from column grids

Proposed col C/C 5M both ways

I

A   B   C   D

7.5M

1M   1M   Thread to be fixed

1M

2   600x600mm
columns

2.5M

Axis

Thread line

3   1M

7.5M

2.5M

Thread line

4   1M

1M

Thread line

Proposed setting out plan of columns common size 600x600mm
Spacings C/C 5000mm both directions

18mm th ply(marine)/20mm planks

75x75mm cross battens 400mmc/c

Grid line of the columns

75x75mm vertical battens for the
inclined scantlings at top, middle
& bottom

600 mm

D

D

D

1M

Grid

1M

D   D

D= 1000-300-18-75= 607mm for ply 18mm
D= 1000-300-20-75= 605mm for plank

To be checked at the top & bottom
of the forms on two sides(external)
& check the size of form section &
plumb at top, middle & bottom

(Typical preparatory, checking & inspection of a simple column)
(for the information of the site personnel & the consultants)

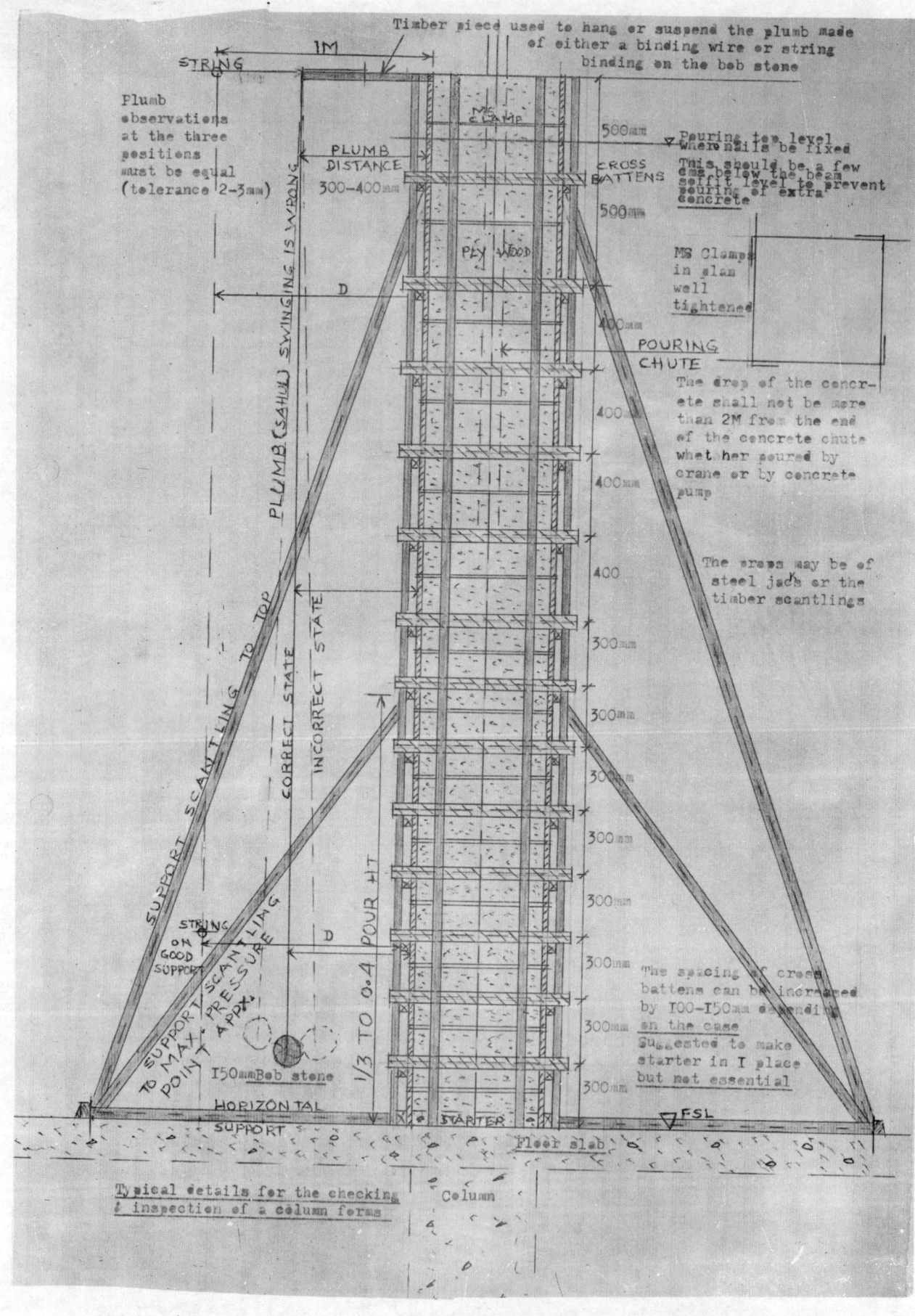

Timber piece used to hang or suspend the plumb made of either a binding wire or string binding on the bob stone

1M

STRING

Plumb observations at the three positions must be equal (tolerance 2-3mm)

PLUMB DISTANCE 300-400mm

MS CLAMP

500mm

Pouring top level where nails be fixed. This should be a few cms below the beam soffit level to prevent pouring of extra concrete

CROSS BATTENS

500mm

PLY WOOD

MS Clamps in slab well tightened

400mm

POURING CHUTE

The drop of the concrete shall not be more than 2M from the end of the concrete chute whether poured by crane or by concrete pump

400mm

400mm

The cross may be of steel jack or the timber scantlings

400

300mm

300mm

PLUMB (SAHUL) SWINGING IS WRONG

SUPPORT SCANTLING TO TOP

CORRECT STATE

INCORRECT STATE

300mm

300mm

300mm

300mm

The spacing of cross battens can be increased by 100-150mm depending on the case suggested to make starter in I place but not essential

D

STRING ON GOOD SUPPORT

SUPPORT SCANTLING TO MAX PRESSURE POINT APPX.

1/3 TO 0.4 POUR HT

D

150mm Bob stone

300mm

300mm

HORIZONTAL SUPPORT

STARTER

FSL

Floor slab

Typical details for the checking & inspection of a column forms

Column

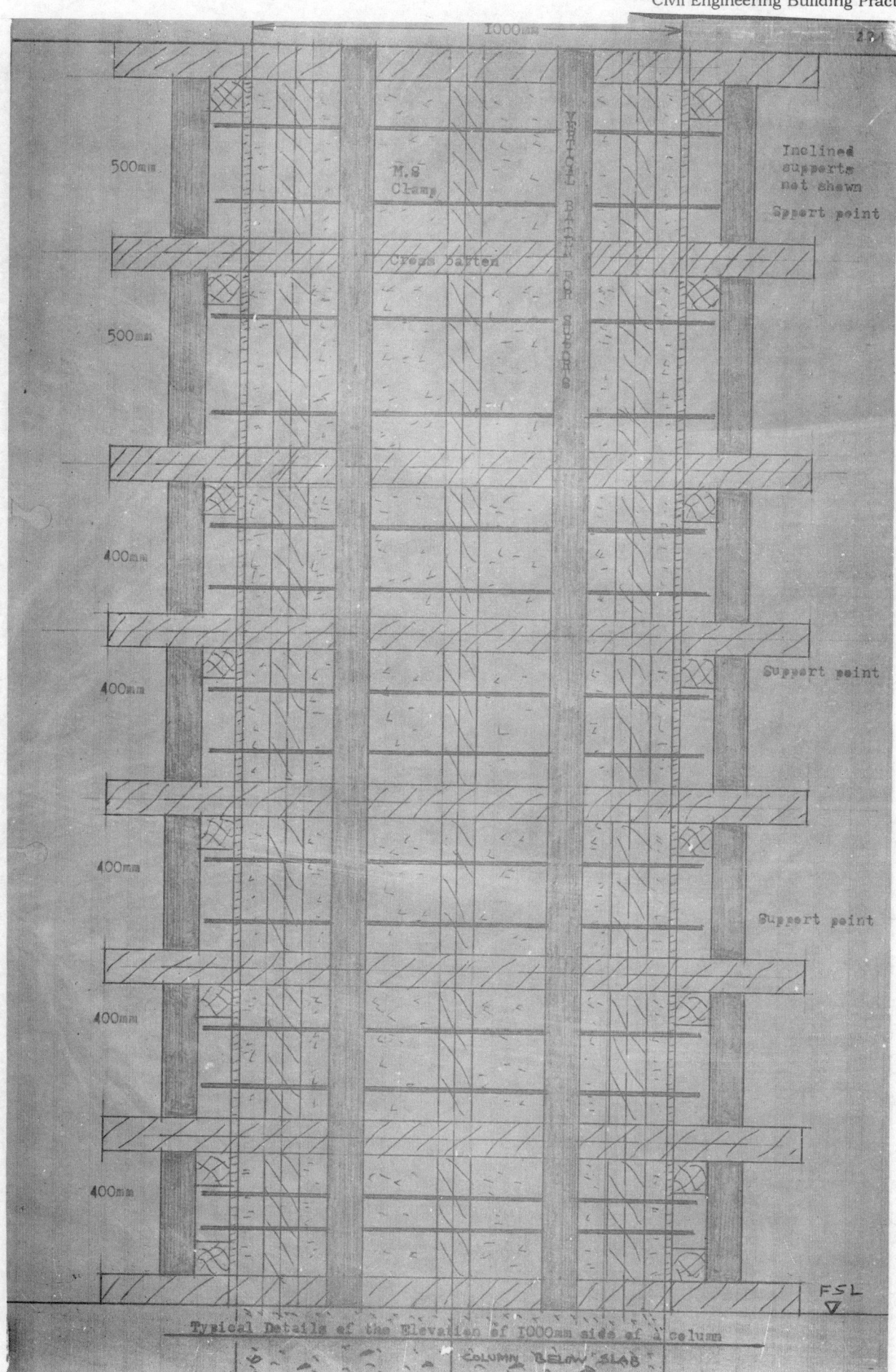

Typical Details of the Elevation of 1000mm side of a column

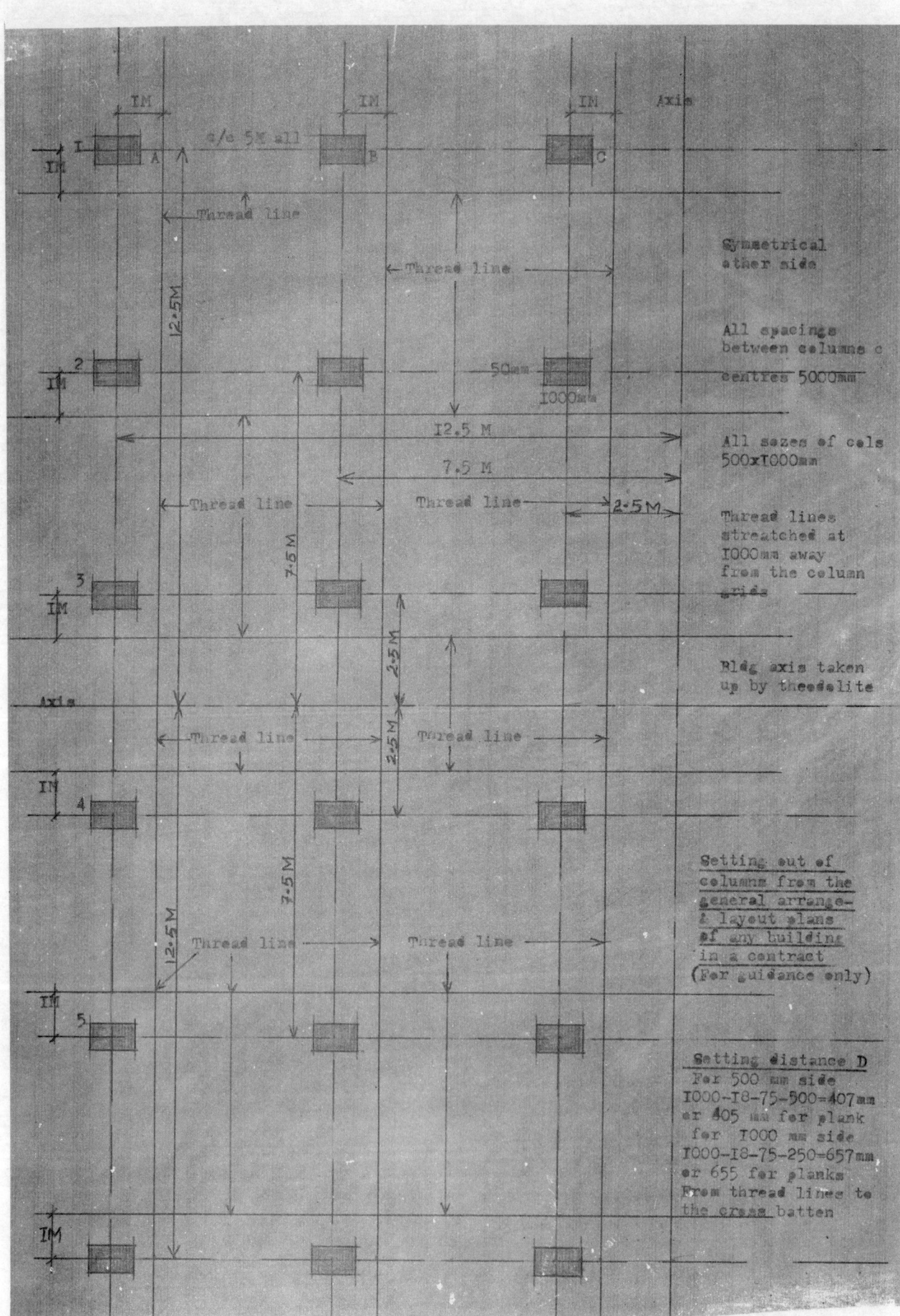

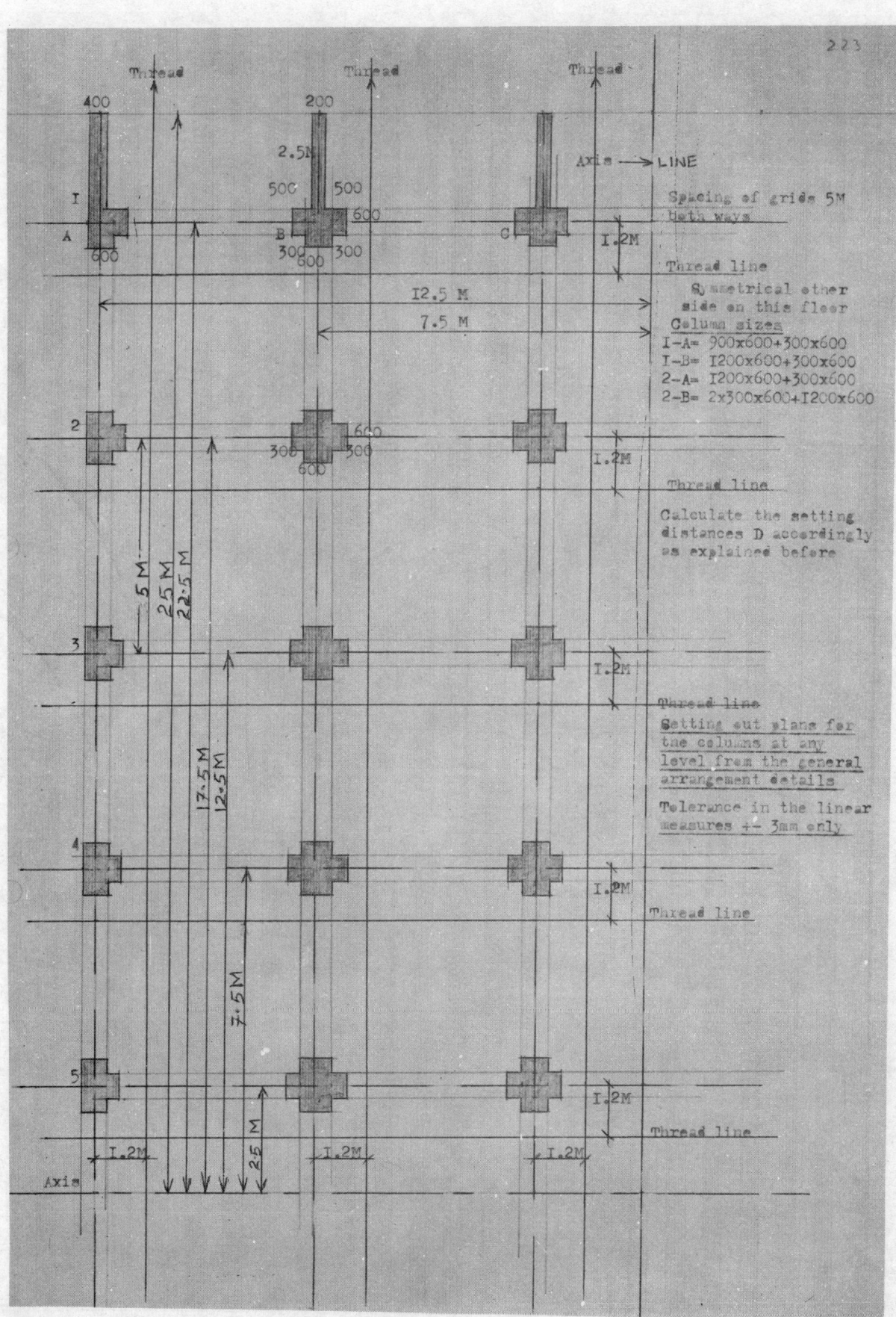

223

Thread   Thread   Thread

400   200

                      2.5M   Axis ⟶ LINE

                      500   500

I   600   Spacing of grids 5M
A   B   C   both ways
600   300   300
                      600   Thread line
                                                          Symmetrical other
                                                          side on this floor
                          12.5 M   Column sizes
                          7.5 M   I-A= 900x600+300x600
                                                      I-B= 1200x600+300x600
                                                      2-A= 1200x600+300x600
                                                      2-B= 2x300x600+1200x600

2   600
                        300   300   1.2M
                        600
                                                      Thread line

                                                      Calculate the setting
                                                      distances D accordingly
5 M   as explained before
25 M
22.5 M

3   1.2M

                                                      Thread line
                                                      Setting out plans for
17.5 M   the columns at any
12.5 M   level from the general
                                                      arrangement details

                                                      Tolerance in the linear
                                                      measures +- 3mm only

4   1.2M

                                                      Thread line
7.5M

5   1.2M

2.5 M   Thread line

1.2M   1.2M   1.2M

Axis

Typical form work plan for a rectangular column

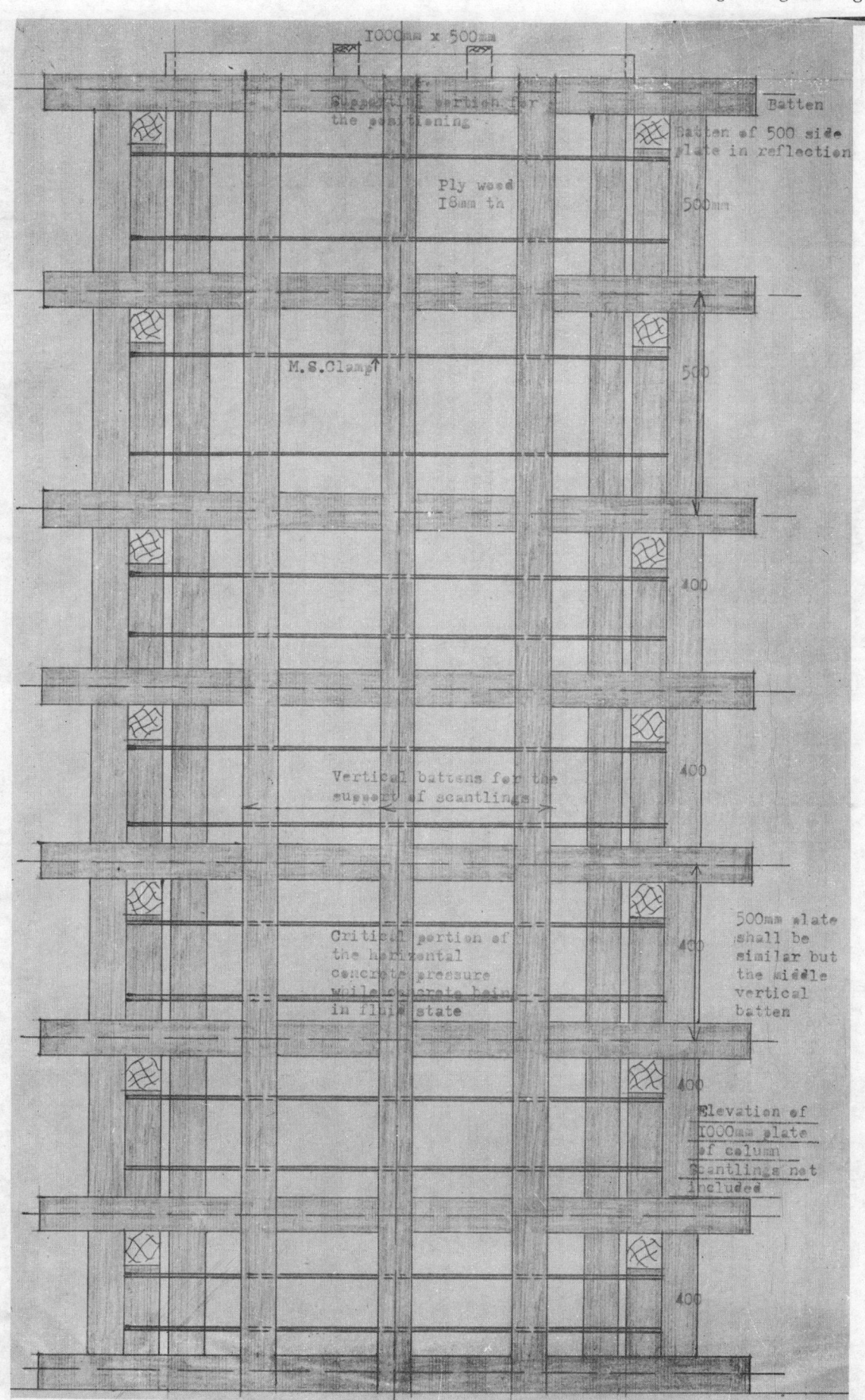

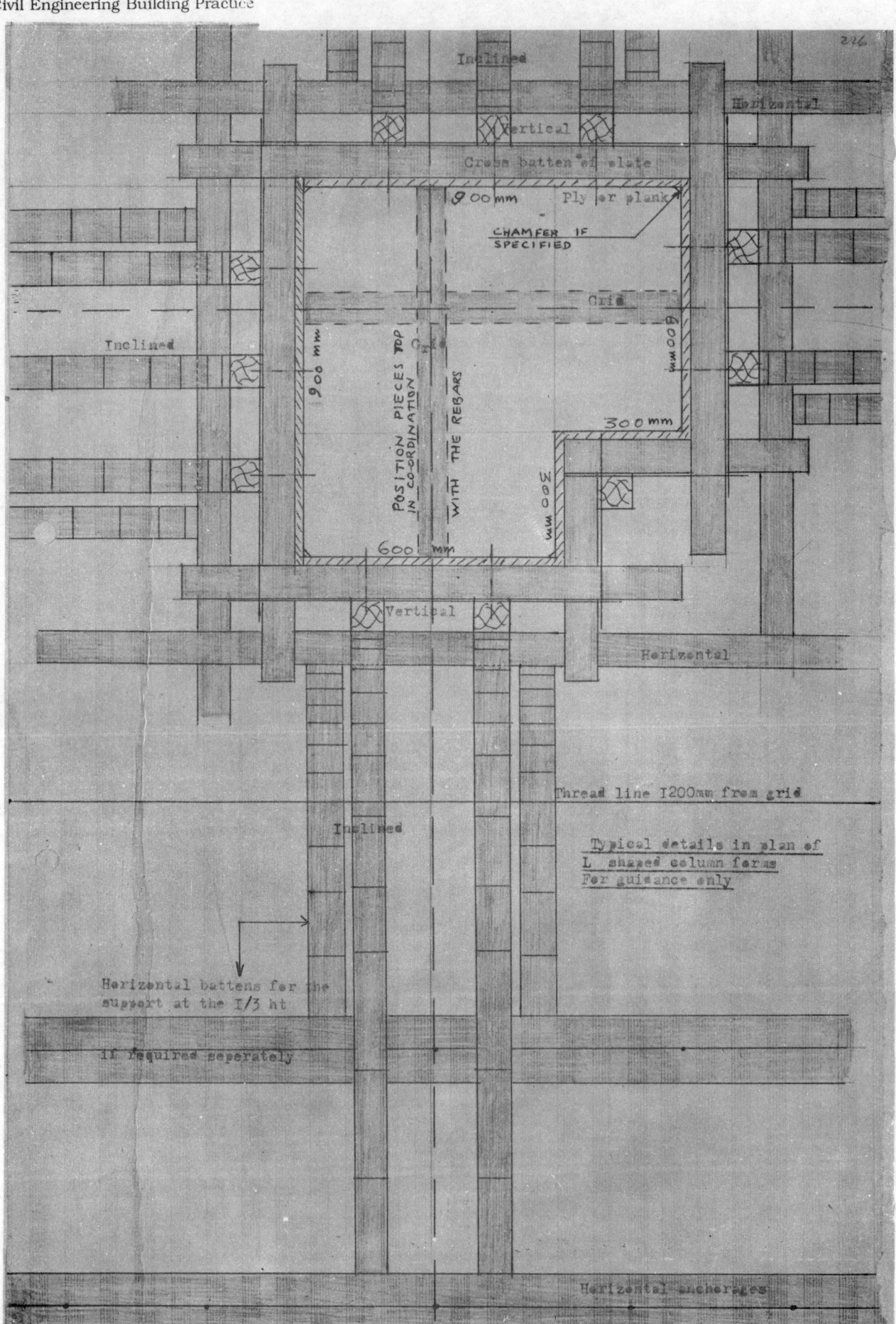

Typical details in plan of
L shaped column forms
For guidance only

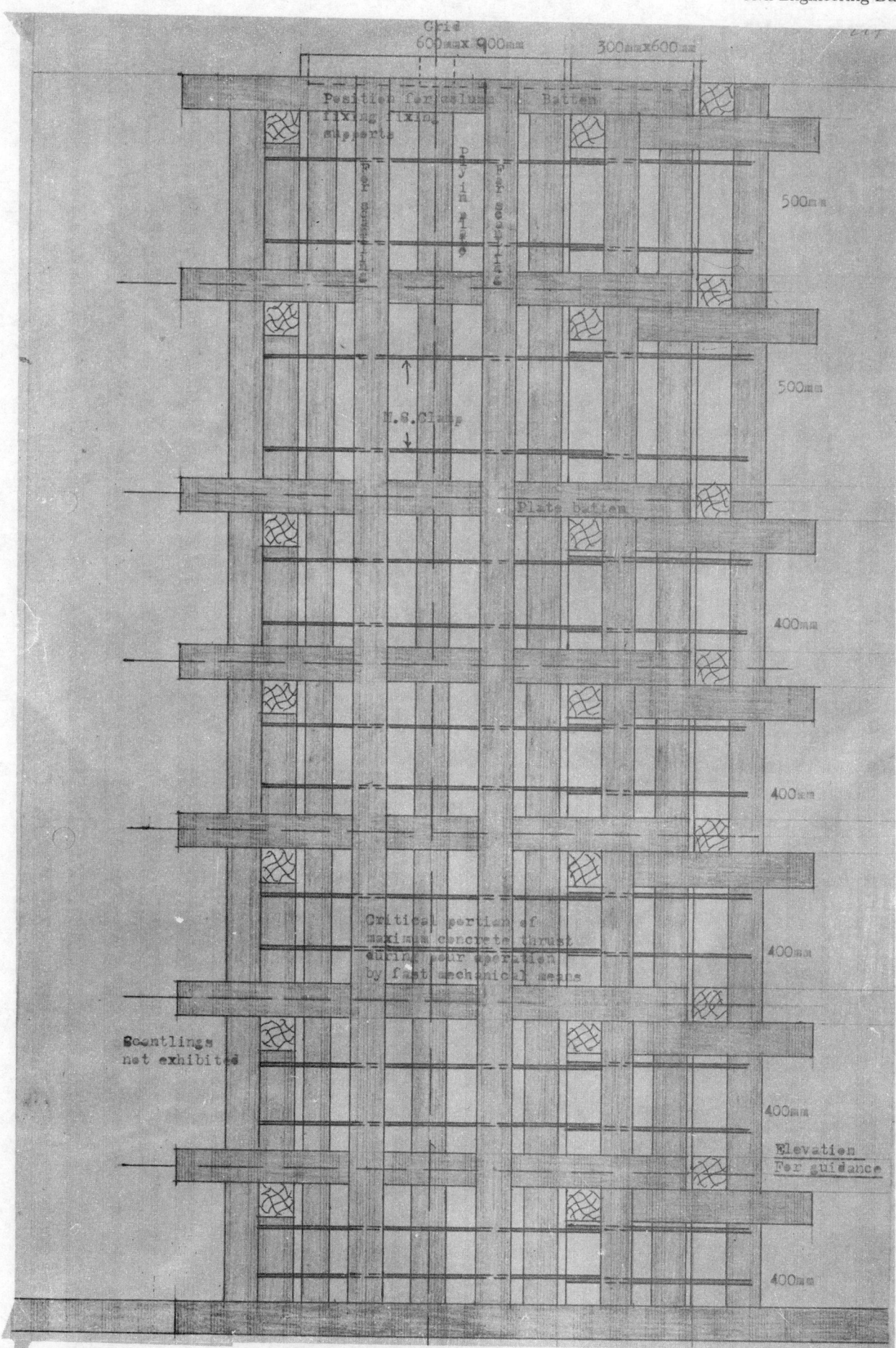

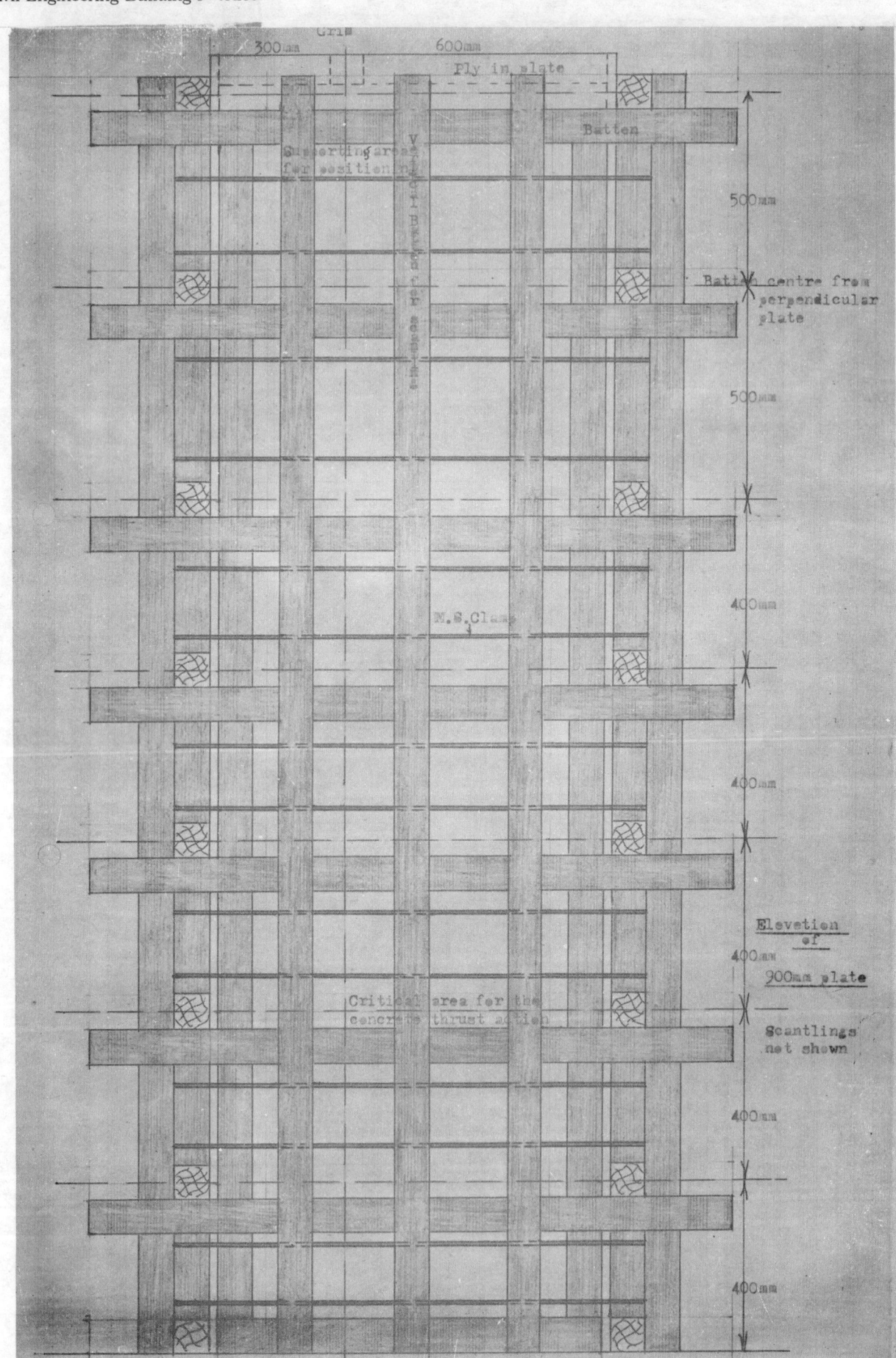

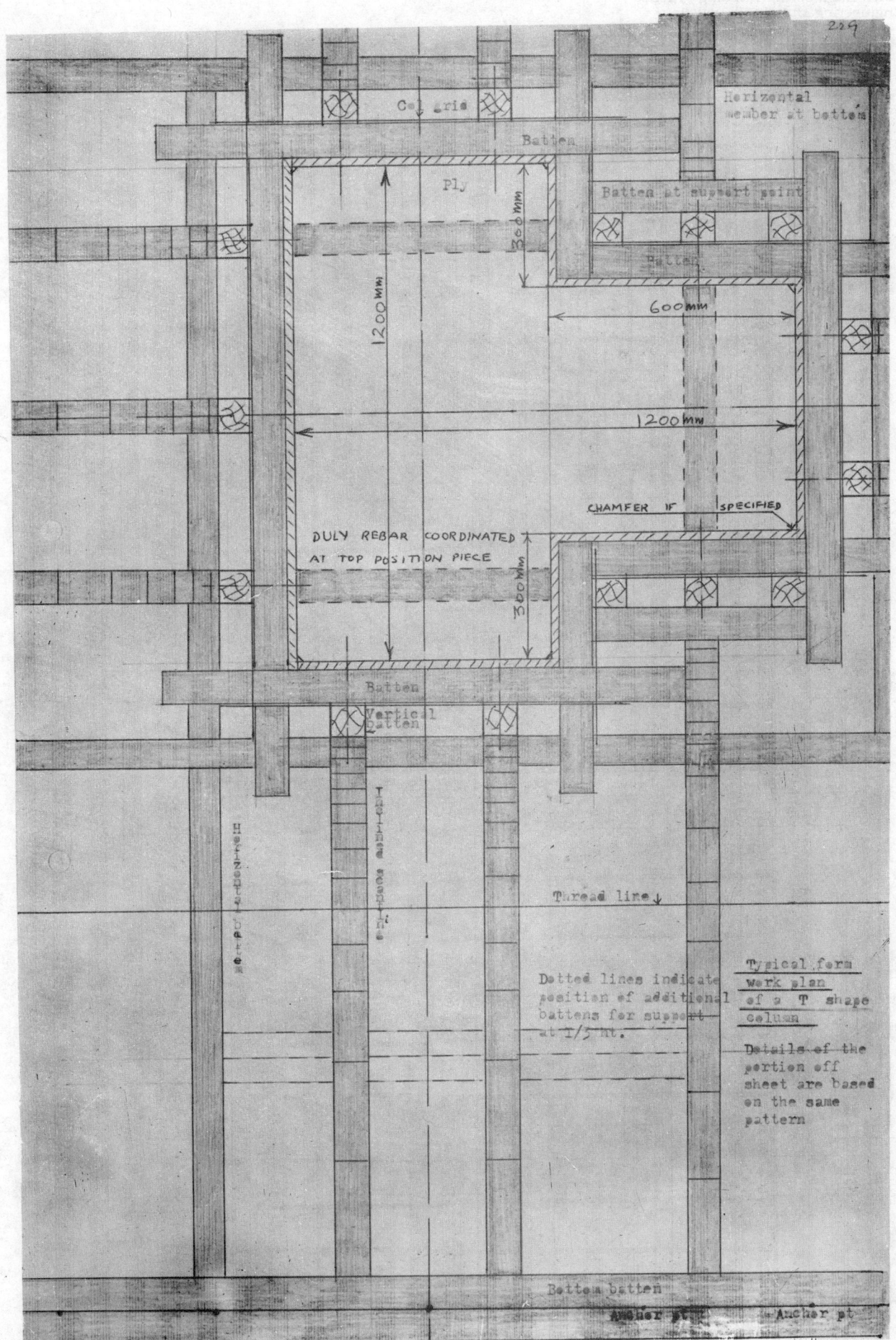

Typical form work plan of a T shape column

Details of the portion off sheet are based on the same pattern

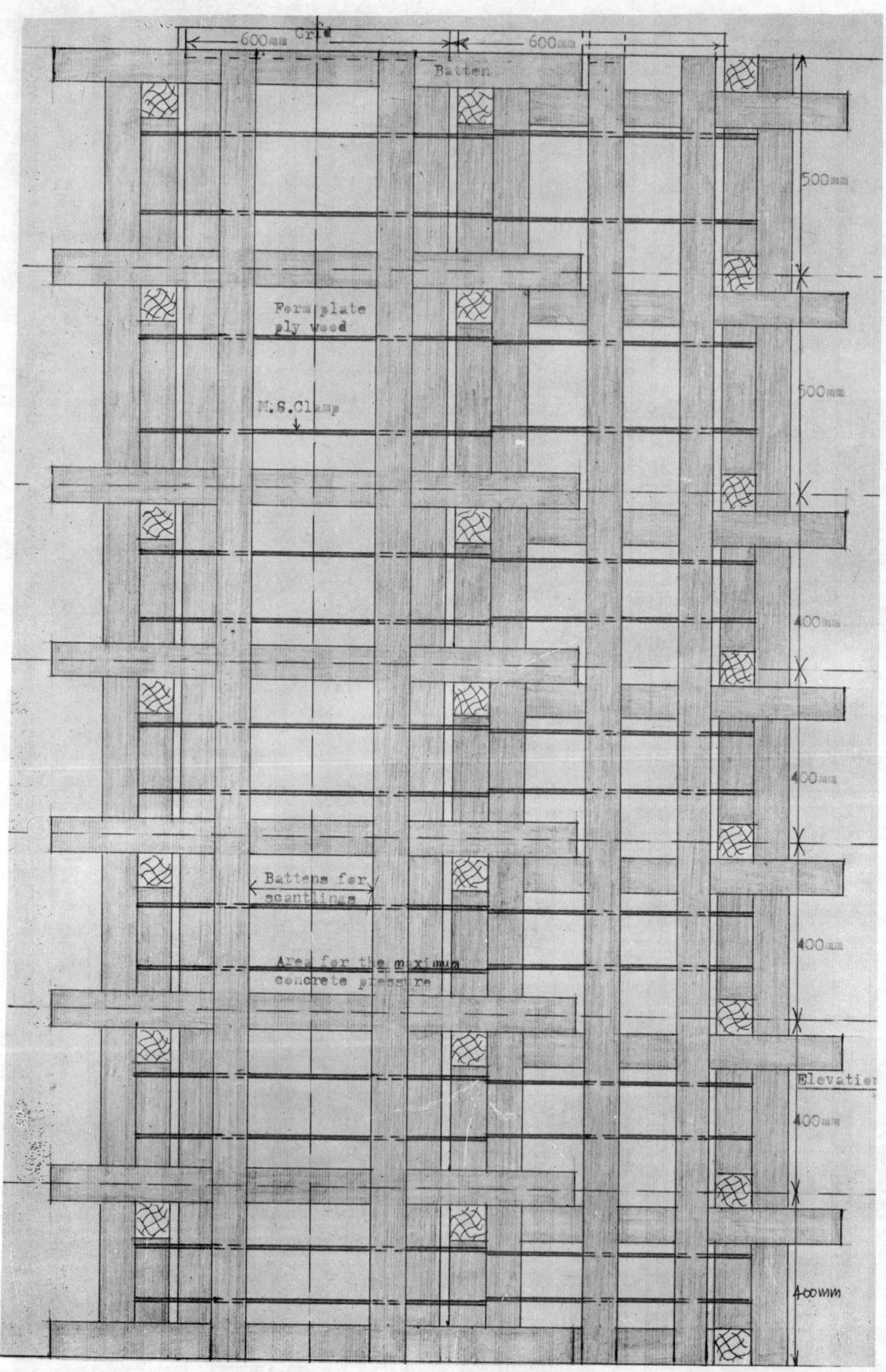

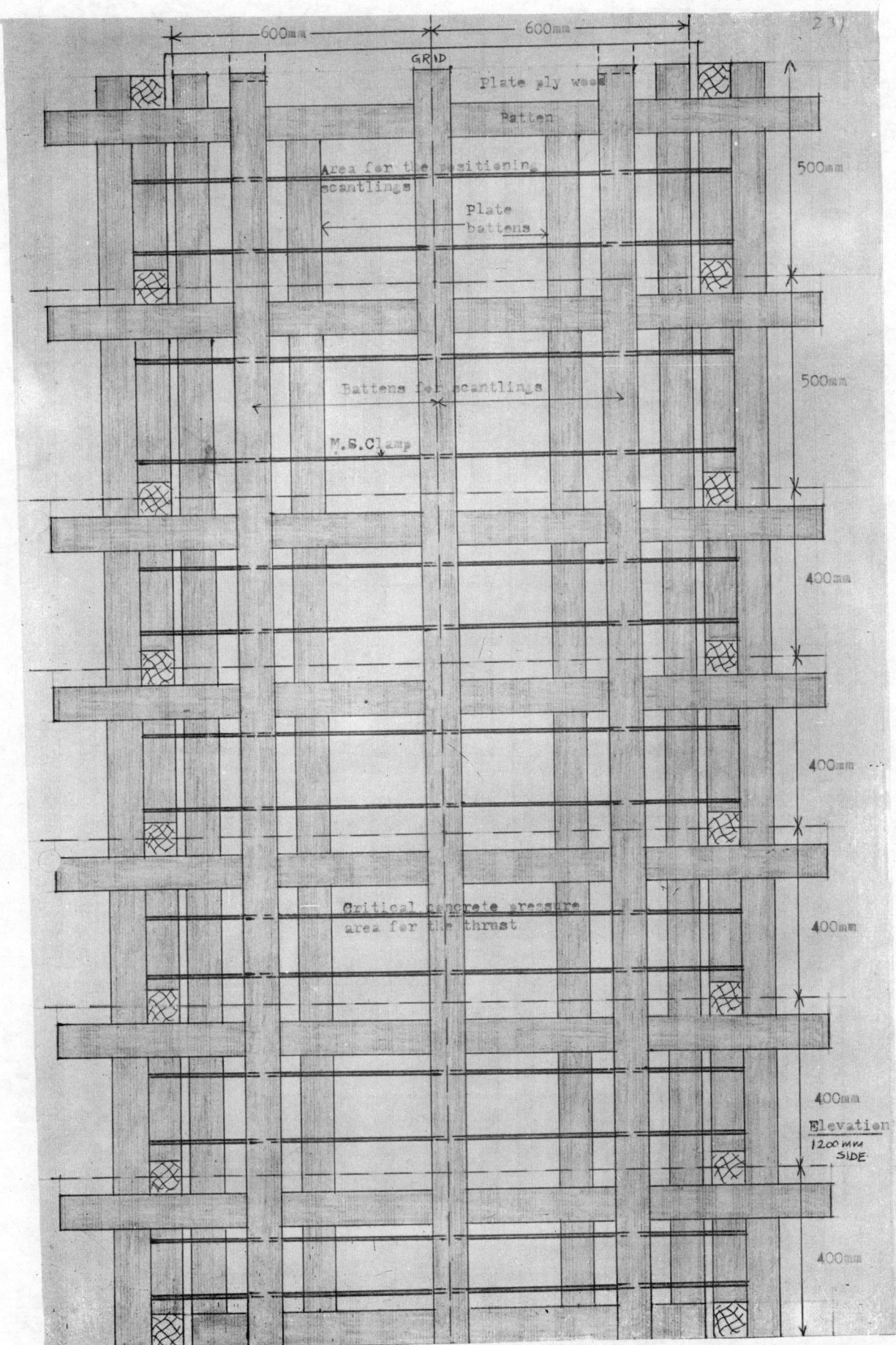

Elevation   300-600-300mm side

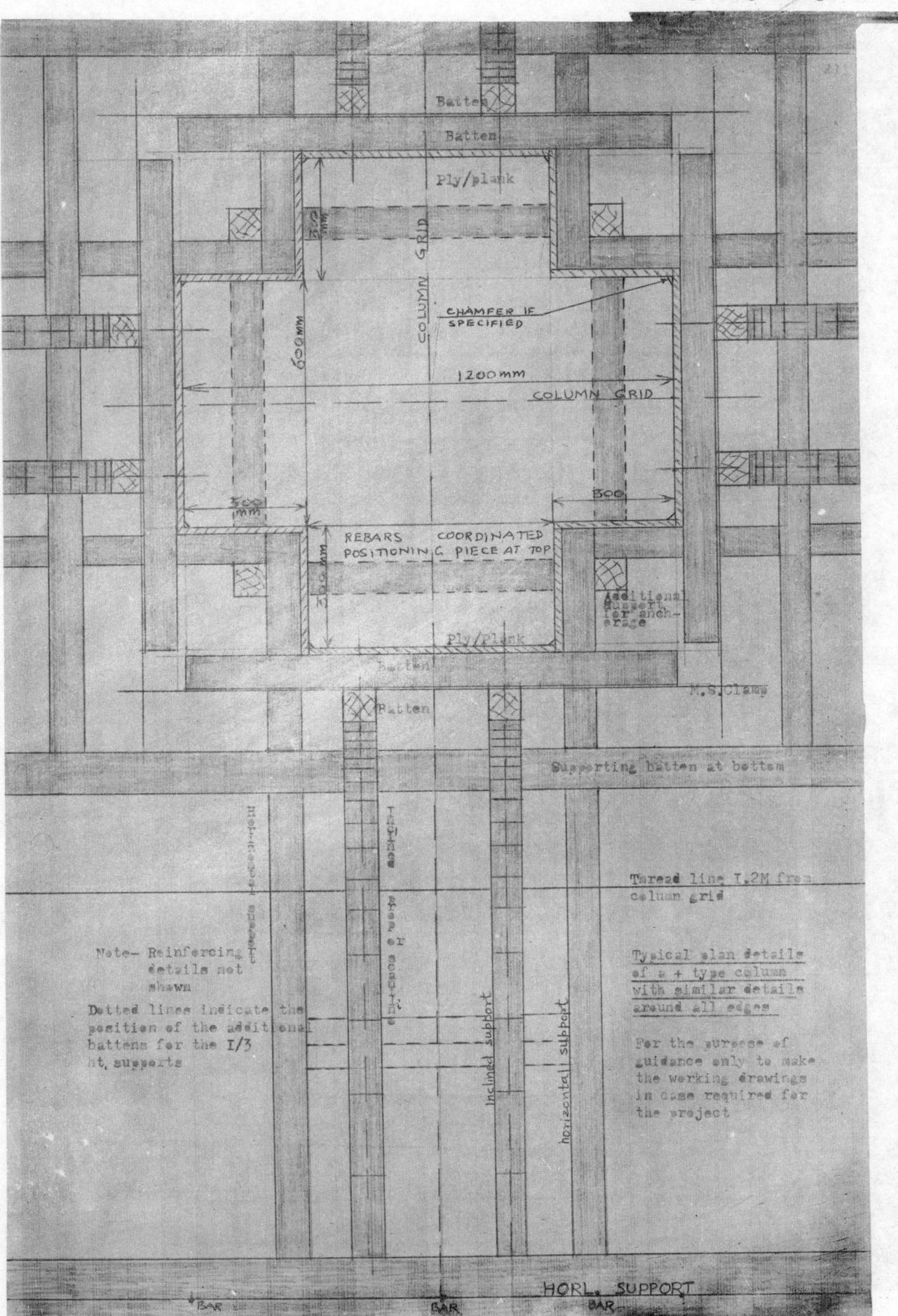

Batten

Batten

Ply/plank

COLUMN GRID

CHAMFER IF
SPECIFIED

600 mm

1200 mm

COLUMN GRID

300
mm

300

REBARS   COORDINATED
POSITIONING PIECE AT TOP

Additional
bars and
anchor-
age

Ply/Plank

Batten

M.S.Clamp

Batten

Supporting batten at bottom

Thread line 1.2M from
column grid

Typical plan details
of a + type column
with similar details
around all edges

For the purpose of
guidance only to make
the working drawings
in case required for
the project

Note— Reinforcing
details not
shown

Dotted lines indicate the
position of the additional
battens for the 1/3
ht. supports

Inclined support

horizontall support

HORL. SUPPORT

BAR

BAR

BAR

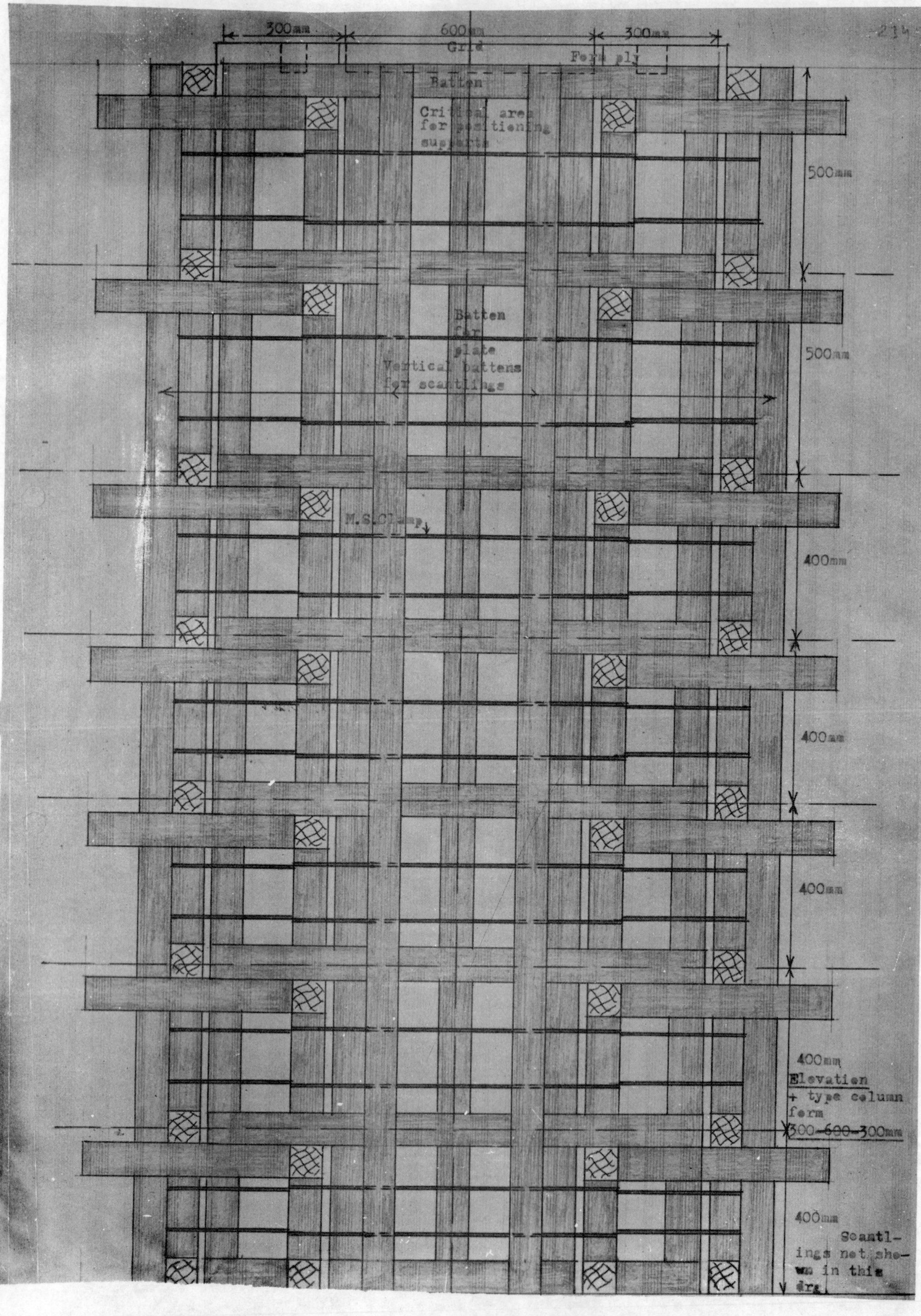

300mm
600mm
300mm
Grid
Form ply
Batten
Critical area for positioning supports
500mm
Batten for plate
Vertical battens for scantlings
500mm
M.S.Clamp
400mm
400mm
400mm
400mm
Elevation
+ type column form
300-600-300mm
400mm
Scantlings not shown in this drg.

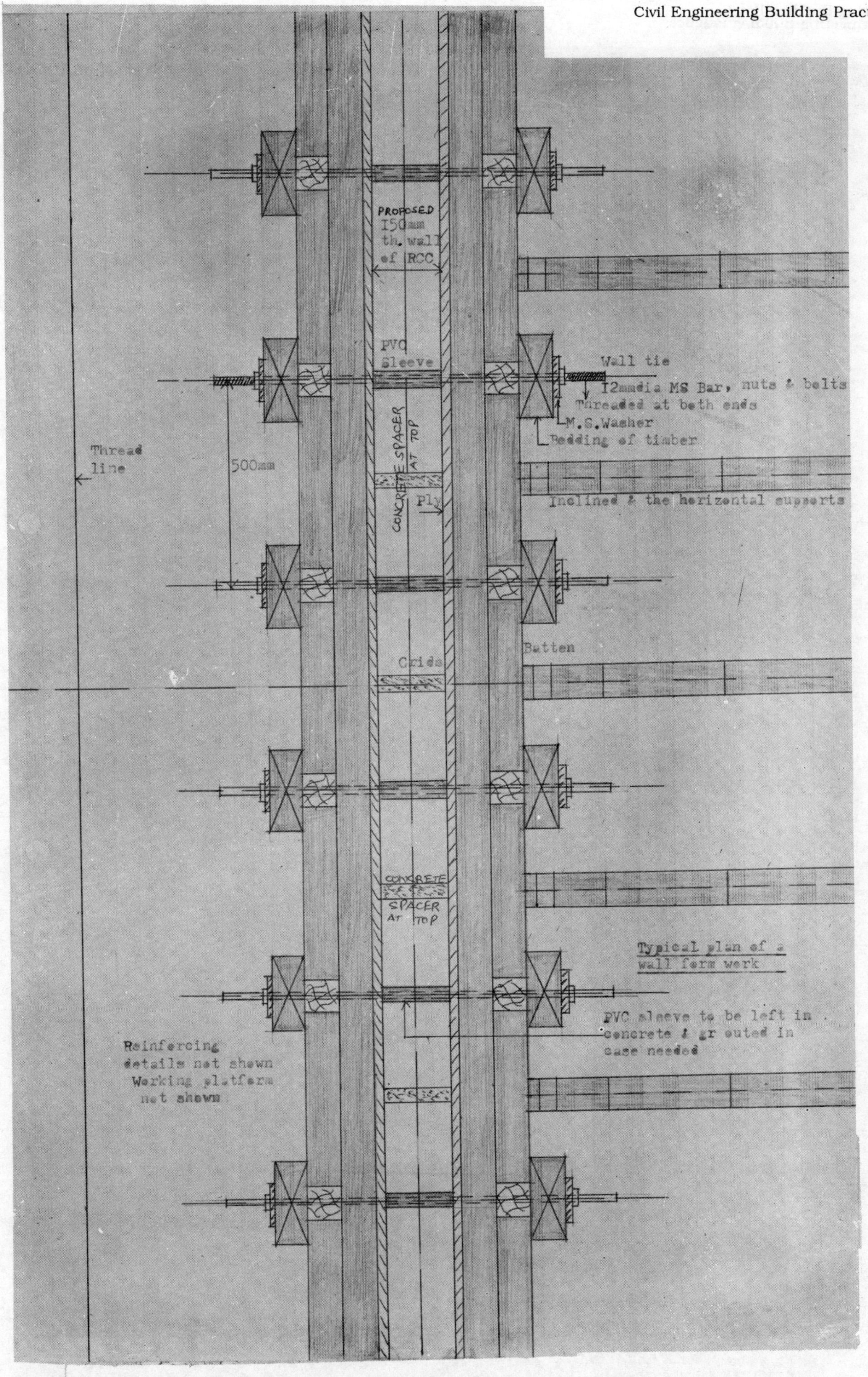

PROPOSED
150mm
th. wall
of RCC

PVC
Sleeve

Wall tie

12mm dia MS Bar, nuts & bolts
Threaded at both ends

M.S.Washer

Bedding of timber

CONCRETE SPACER AT TOP

500mm

Ply

Inclined & the horizontal supports

Thread
line

Grids

Batten

CONCRETE SPACER AT TOP

Typical plan of a
wall form work

PVC sleeve to be left in
concrete & grouted in
case needed

Reinforcing
details not shown
Working platform
not shown

236

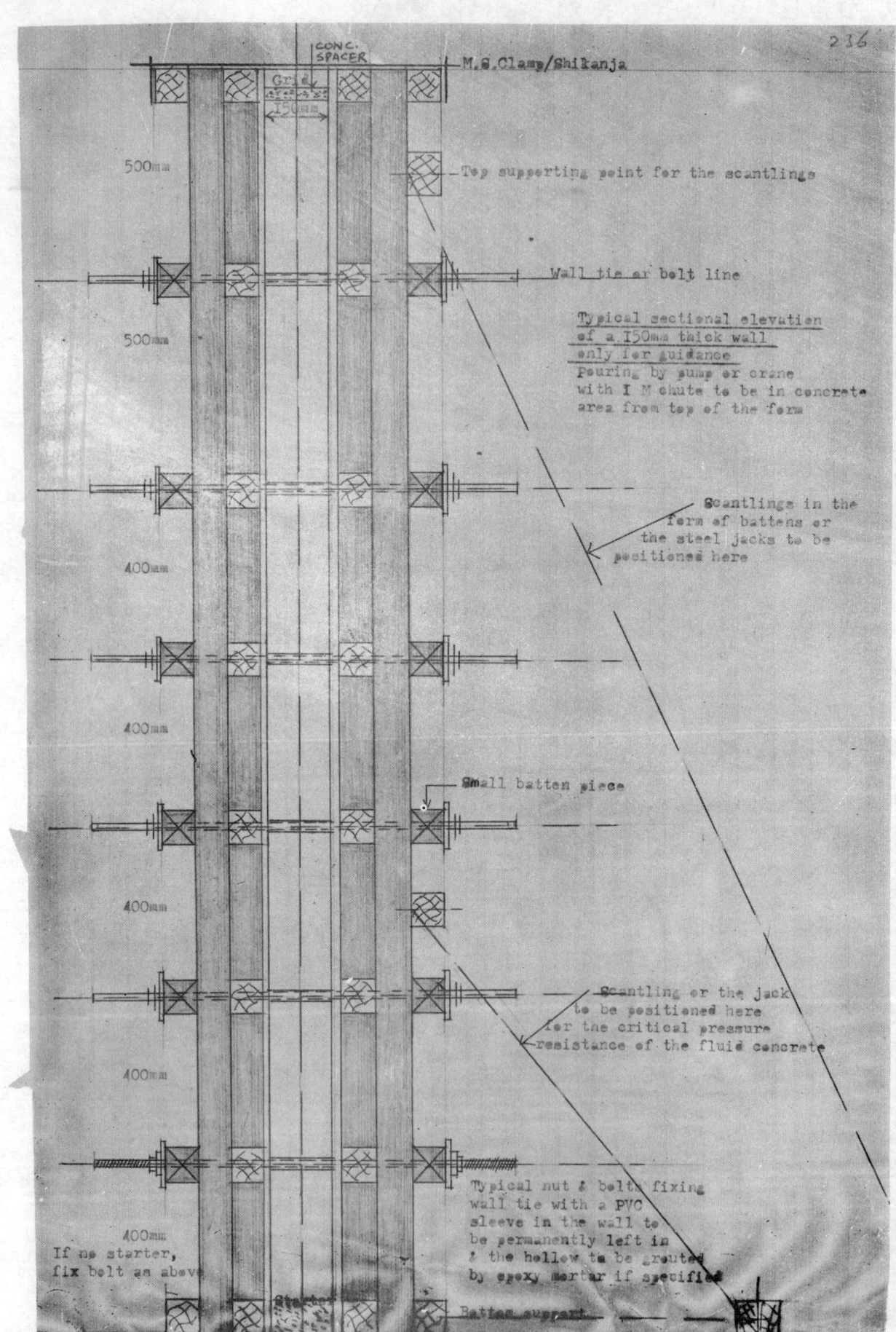

CONC.
SPACER
Grid
150mm

M.S.Clamp/Shikanja

500mm

Top supporting point for the scantlings

Wall tie or bolt line

500mm

Typical sectional elevation
of a 150mm thick wall
only for guidance
Pouring by pump or crane
with I M chute to be in concrete
area from top of the form

400mm

Scantlings in the
form of battens or
the steel jacks to be
positioned here

400mm

400mm

Small batten piece

400mm

Scantling or the jack
to be positioned here
for the critical pressure
resistance of the fluid concrete

400mm

400mm

Typical nut & bolt fixing
wall tie with a PVC
sleeve in the wall to
be permanently left in
& the hollow to be grouted
by epoxy mortar if specified

400mm
If no starter,
fix bolt as above

Batten support

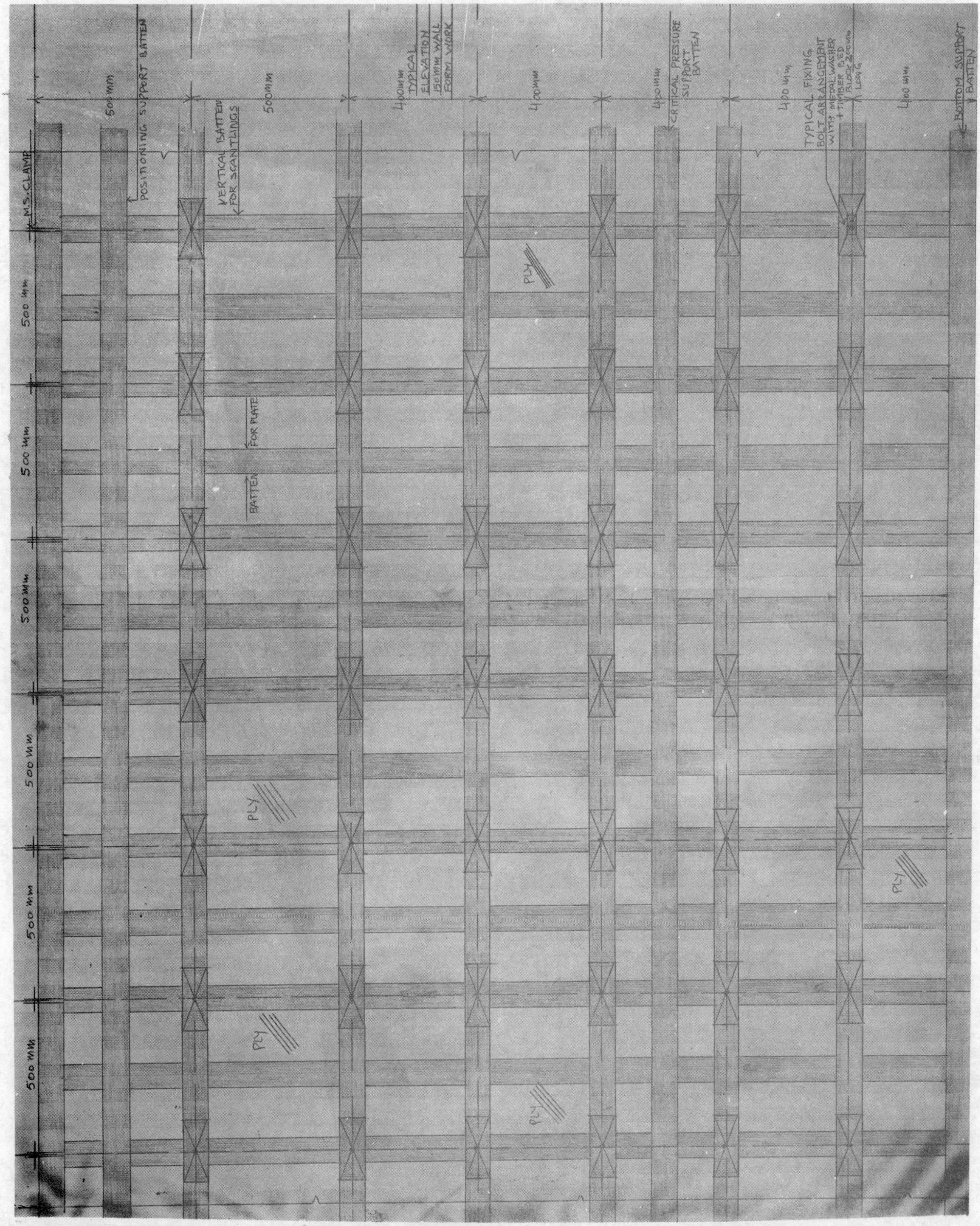

M.S. CLAMP

500 mm

500 mm

500 mm

500 mm

500 mm

500 mm

500 mm

500 mm

POSITIONING SUPPORT BATTEN

500 mm

VERTICAL BATTEN FOR SCANTLINGS

BATTEN FOR PLATE

500 mm

400 mm

TYPICAL ELEVATION 150mm WALL FORM WORK

400 mm

PLY

PLY

400 mm

CRITICAL PRESSURE SUPPORT BATTEN

PLY

400 mm

TYPICAL FIXING BOLT ARRANGEMENT WITH METAL WASHER + TIMBER BED BLOCK 200mm LONG

400 mm

PLY

400 mm

BOTTOM SUPPORT BATTEN

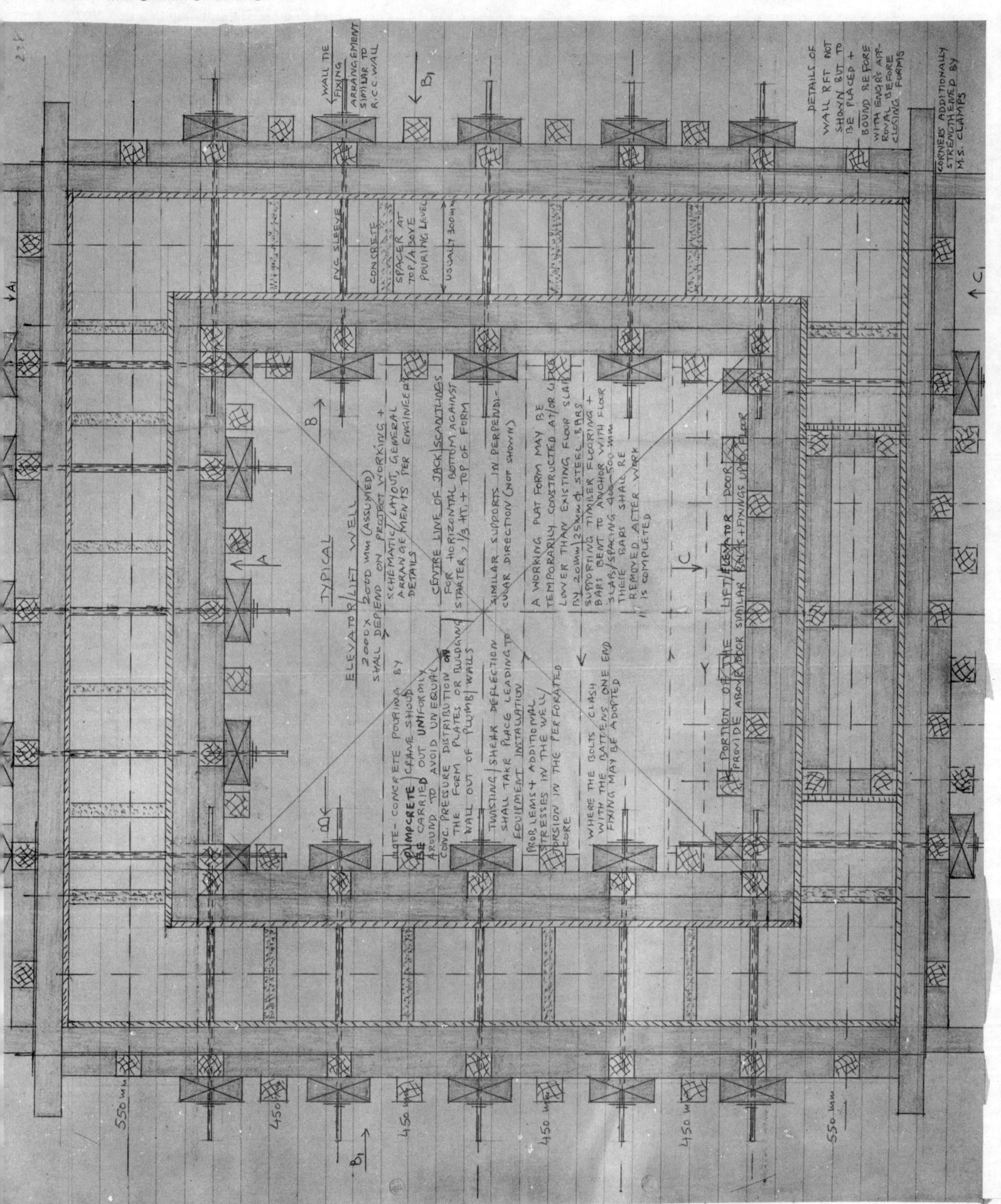

WALL TIE
FIXING
ARRANGEMENT
SIMILAR TO
R.C.C. WALL

DETAILS OF
WALL RFT NOT
SHOWN BUT TO
BE PLACED +
BOUND BEFORE
WITH ENGRY APP-
ROVAL BEFORE
CLOSING FORMS

CORNERS ADDITIONALLY
STRENGTHENED BY
M.S. CLAMPS

PVC SLEEVE

CONCRETE

SPACER AT
TOP /ABOVE
POURING LEVEL

USUALLY 300 mm

CENTRE LINE OF JACK / SCANTLINGS
FOR HORIZONTAL BOTTOM AGAINST
STARTER > 1/3 HT + TOP OF FORM

SIMILAR SUPPORTS IN PERPENDI-
CULAR DIRECTION (NOT SHOWN)

A WORKING PLAT FORM MAY BE
TEMPORARILY CONSTRUCTED AT/OR USING
LOWER THAN EXISTING FLOOR SLAB
BY 20 mm/25 mm ф STEEL BARS
SUPPORTING TIMBER FLOORING +
BARS BENT TO ANCHOR WITH FLOOR
SLAB/SPACING 400-500 mm
THEIR BARS SHALL RE
REMOVED AFTER WORK
IS COMPLETED

TYPICAL
ELEVATOR/LIFT WELL
2000 × 2000 mm (ASSUMED)
SHALL DEPEND ON PROJECT WORKING +

SCHEMATIC/LAYOUT. GENERAL
ARRANGEMENTS PER ENGINEER
DETAILS

NOTE:- CONCRETE POURING BY
PUMPCRETE CRANE SHOULD
BE CARRIED OUT UNIFORMLY
AROUND TO AVOID UNEQUAL
CONC. PRESSURE DISTRIBUTION ON
THE FORM PLATES OR BUCLING
WALL OUT OF PLUMB/ WALLS

TWISTING/SHEAR DEFLECTION
SHALL TAKE PLACE LEADING TO
EQUIPMENT INSTALLATION

PROBLEMS + ADDITIONAL
STRESSES IN THE WELL
TORSION IN THE PERFORATED
CORE

WHERE THE BOLTS CLASH
WITH THE BATTENS ONE END
FIXING MAY BE ADOPTED

A PORTION OF THE LIFT/ELEVATOR DOOR
PROVIDE ABOVE/BELOW SIMILAR BOLTS + FIXINGS UPTO FLOOR

A1

B1

C1

550 mm

450 mm

450 mm

450 mm

450 mm

550 mm

B1

A

B

C

C

Y

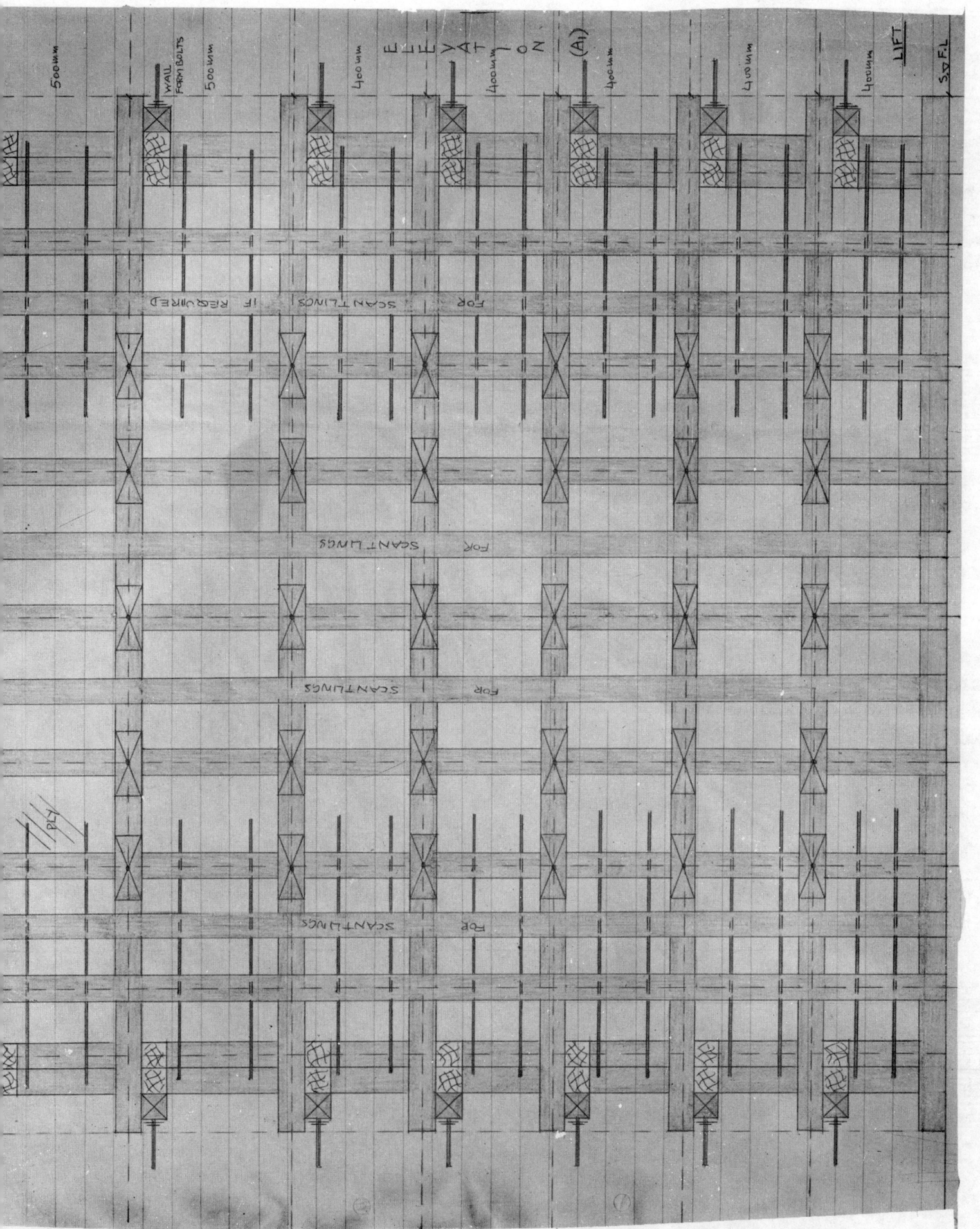

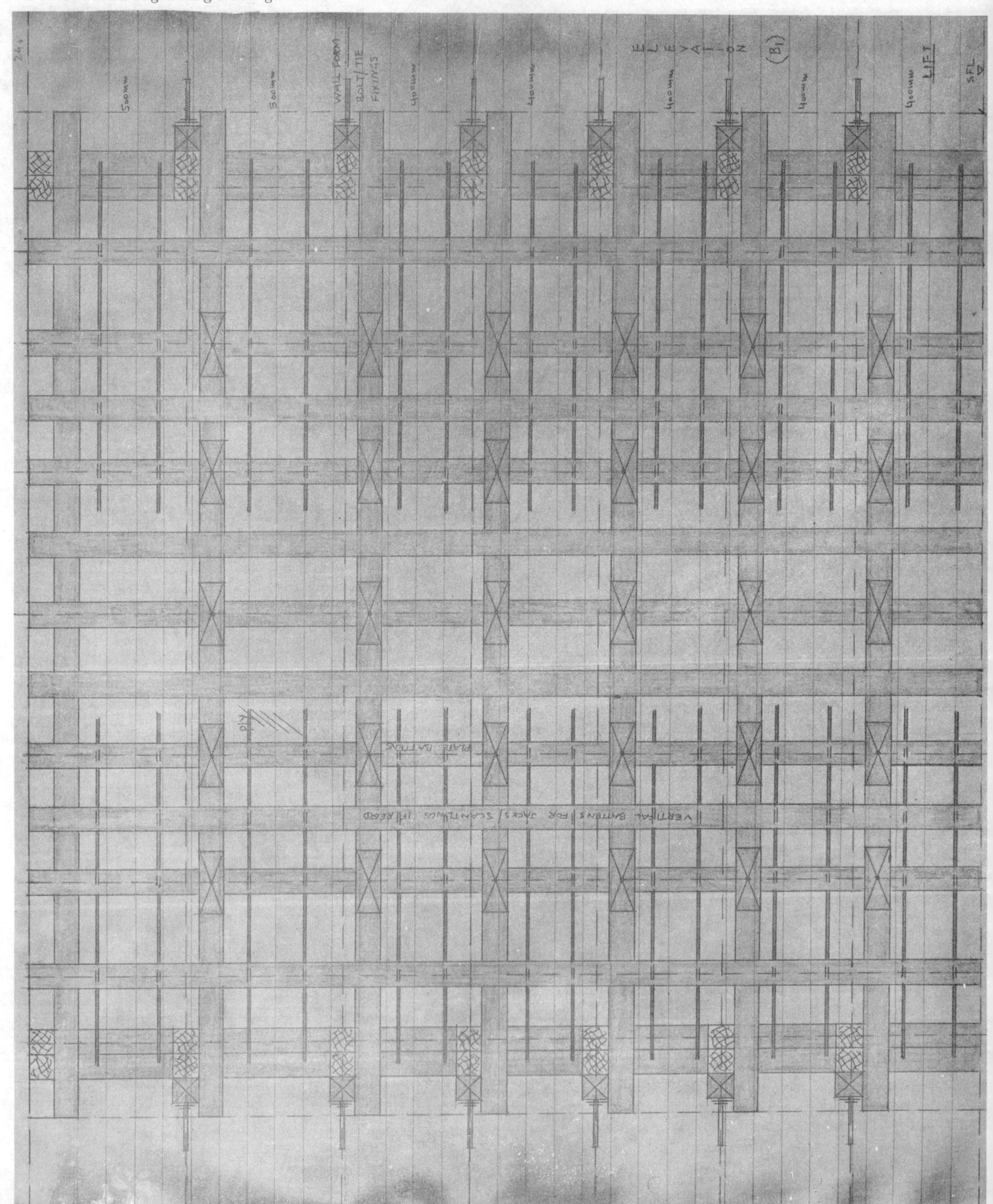

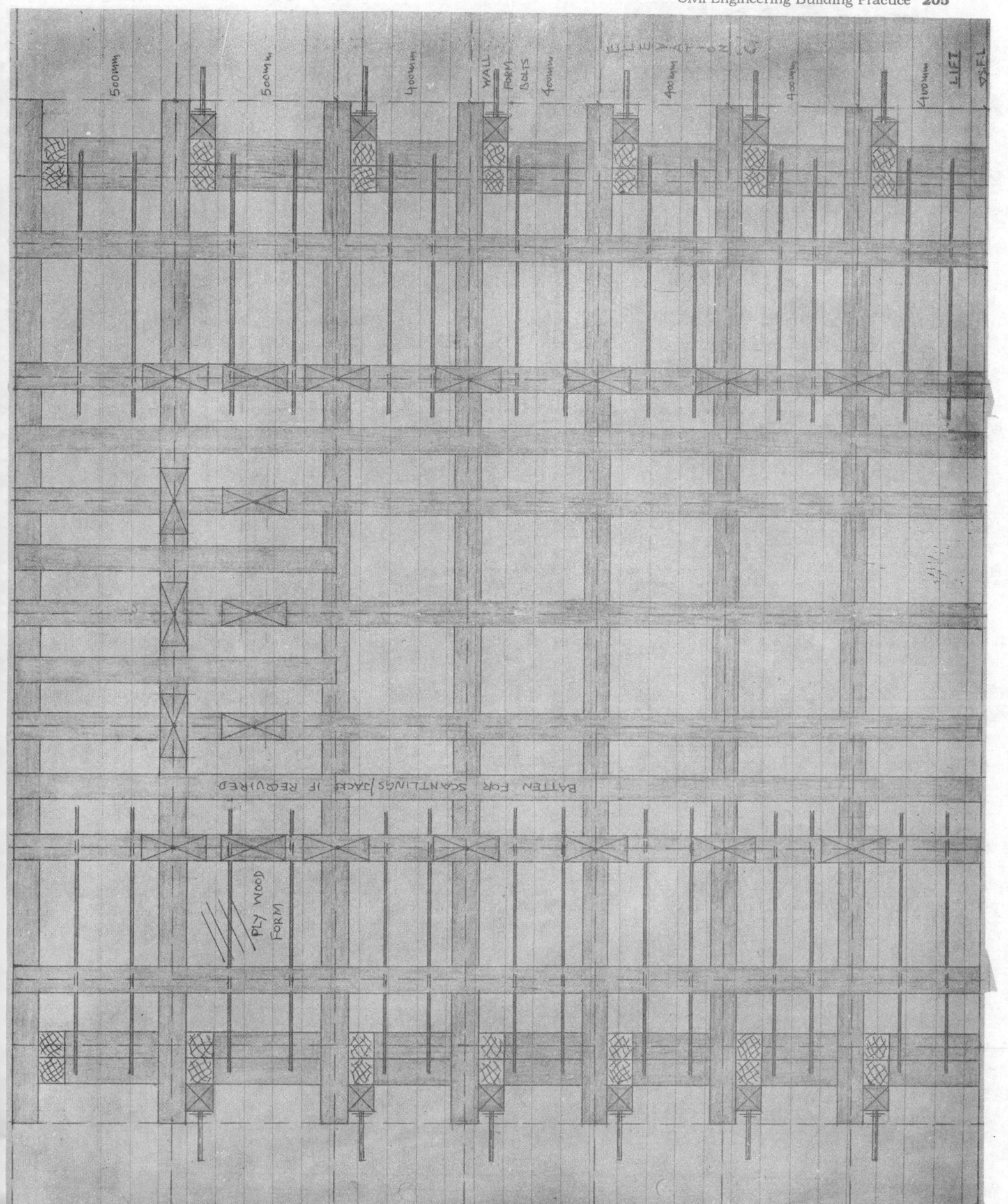

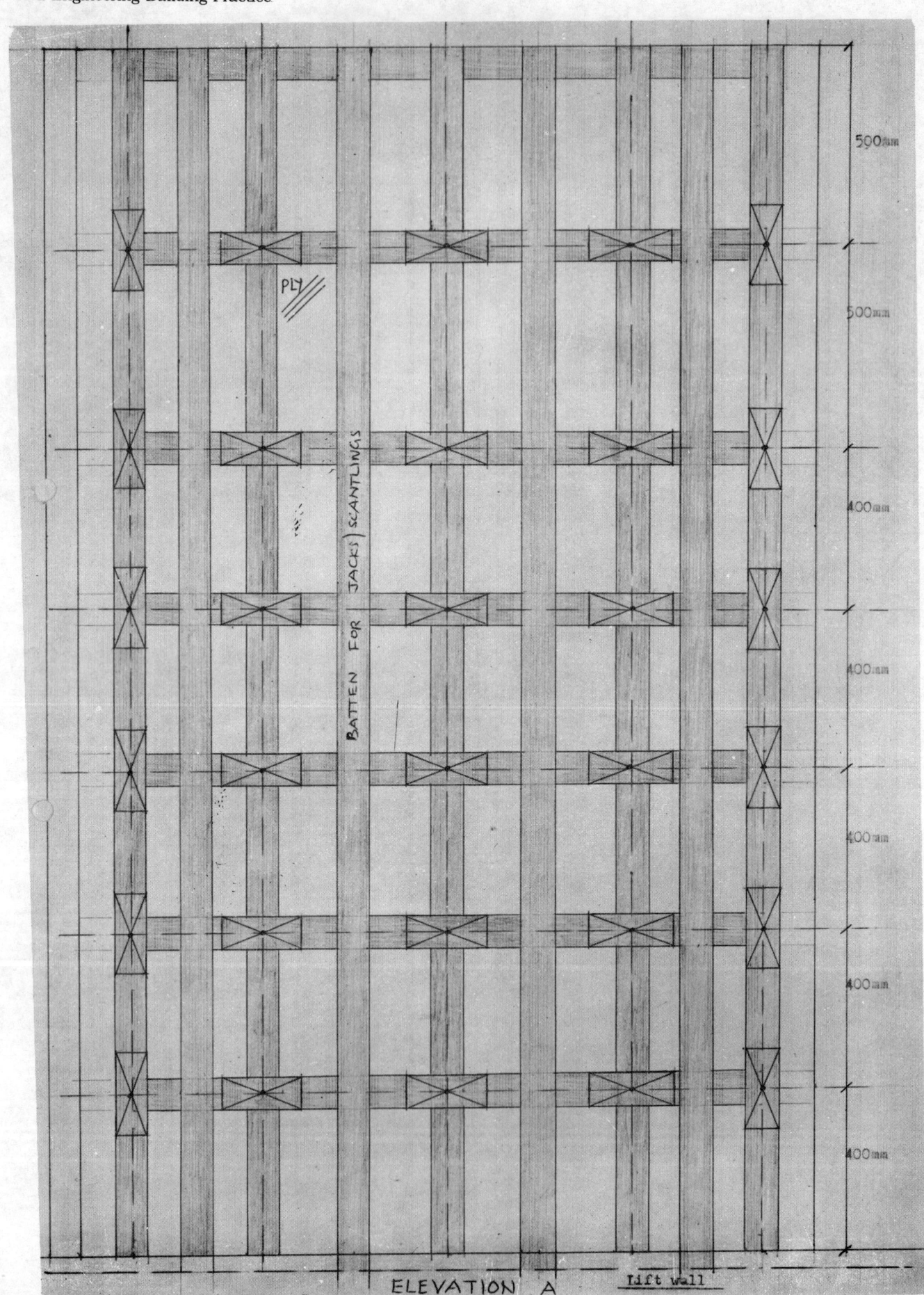

PLY

BATTEN FOR JACKS) SCANTLINGS

500mm

500mm

400mm

400mm

400mm

400mm

400mm

ELEVATION A   Lift wall

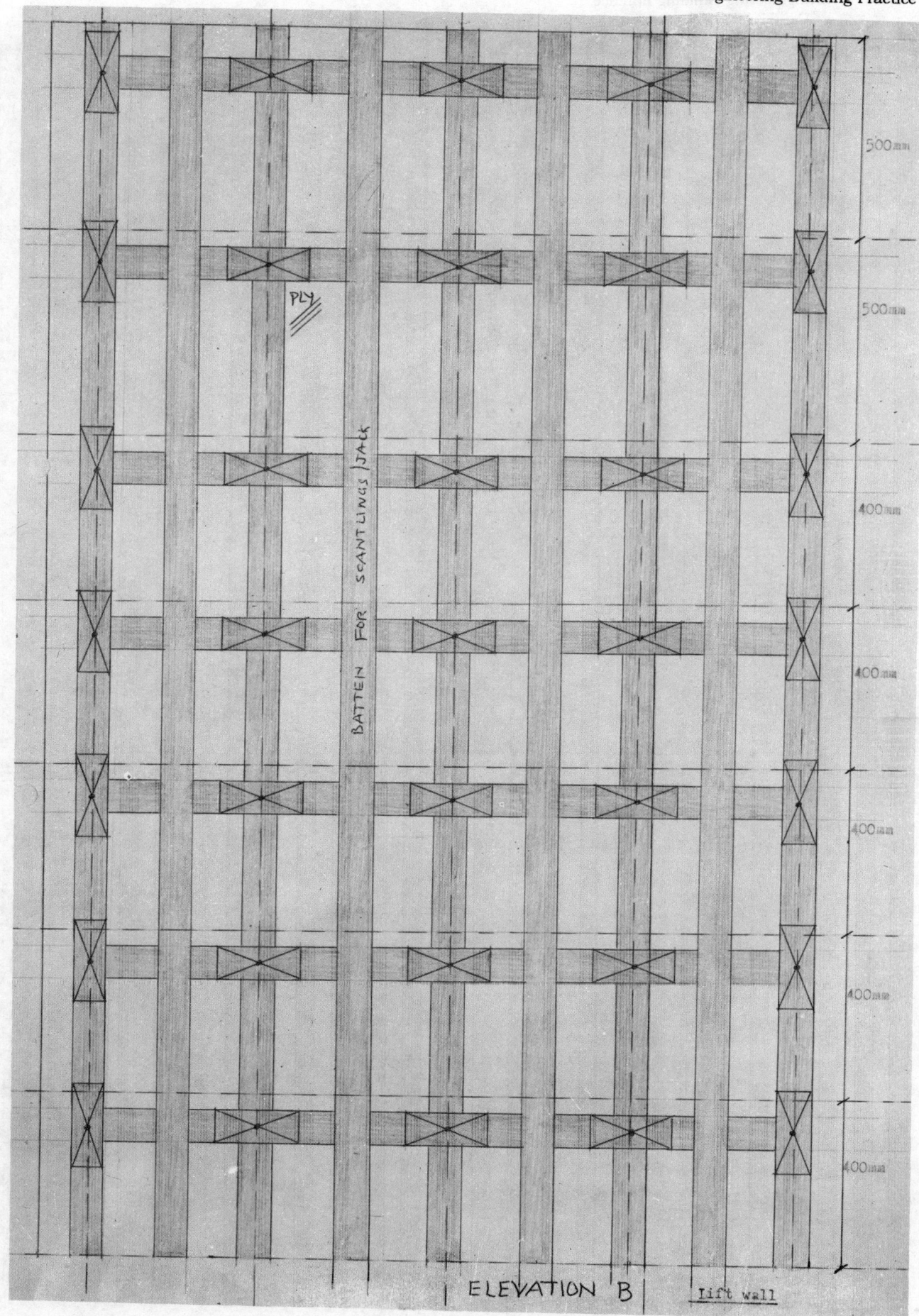

PLY

BATTEN FOR SCANTLINGS/JACK

500mm

500mm

400mm

400mm

400mm

400mm

400mm

ELEVATION 'B'   lift wall

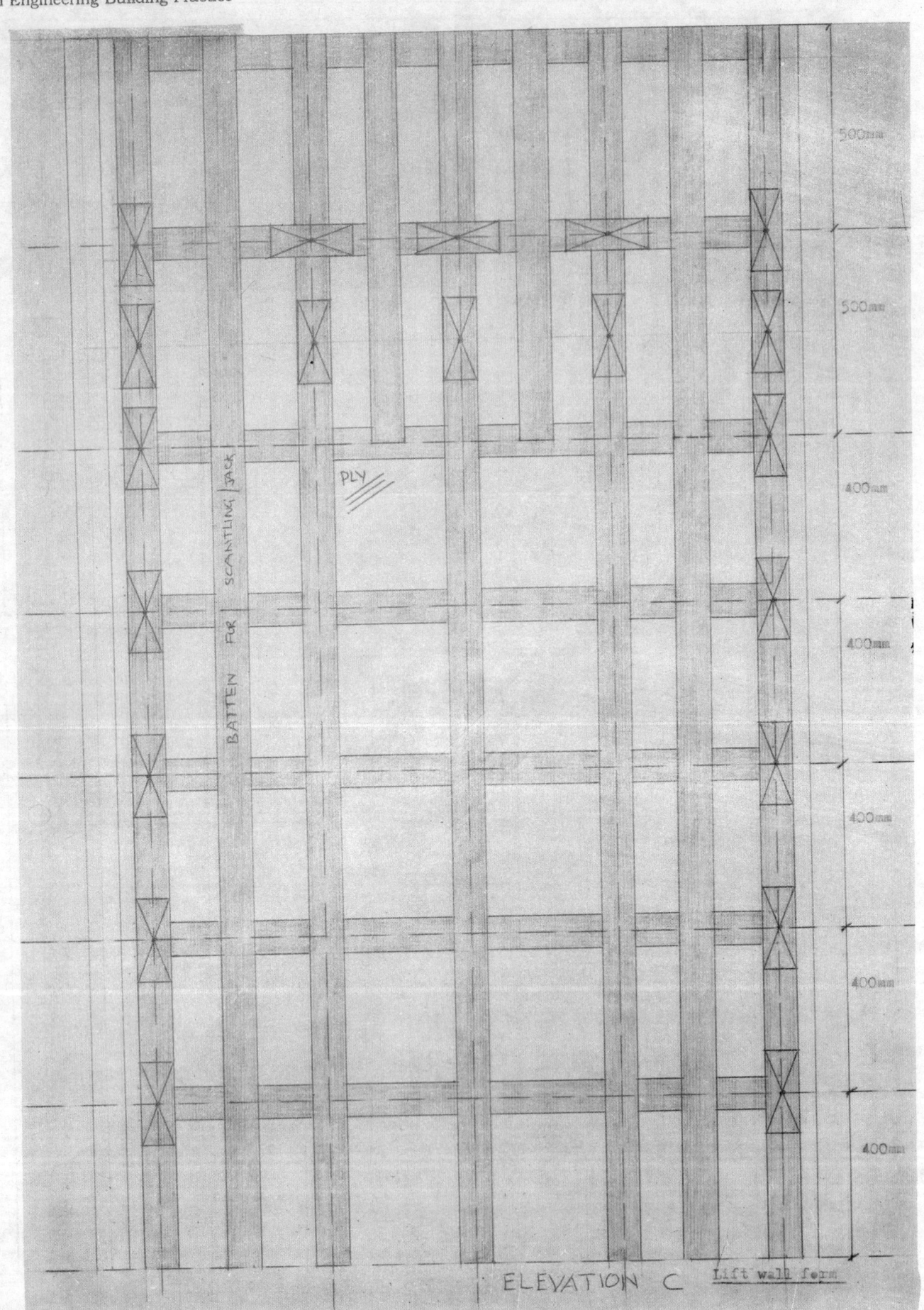

ELEVATION C   Lift wall form

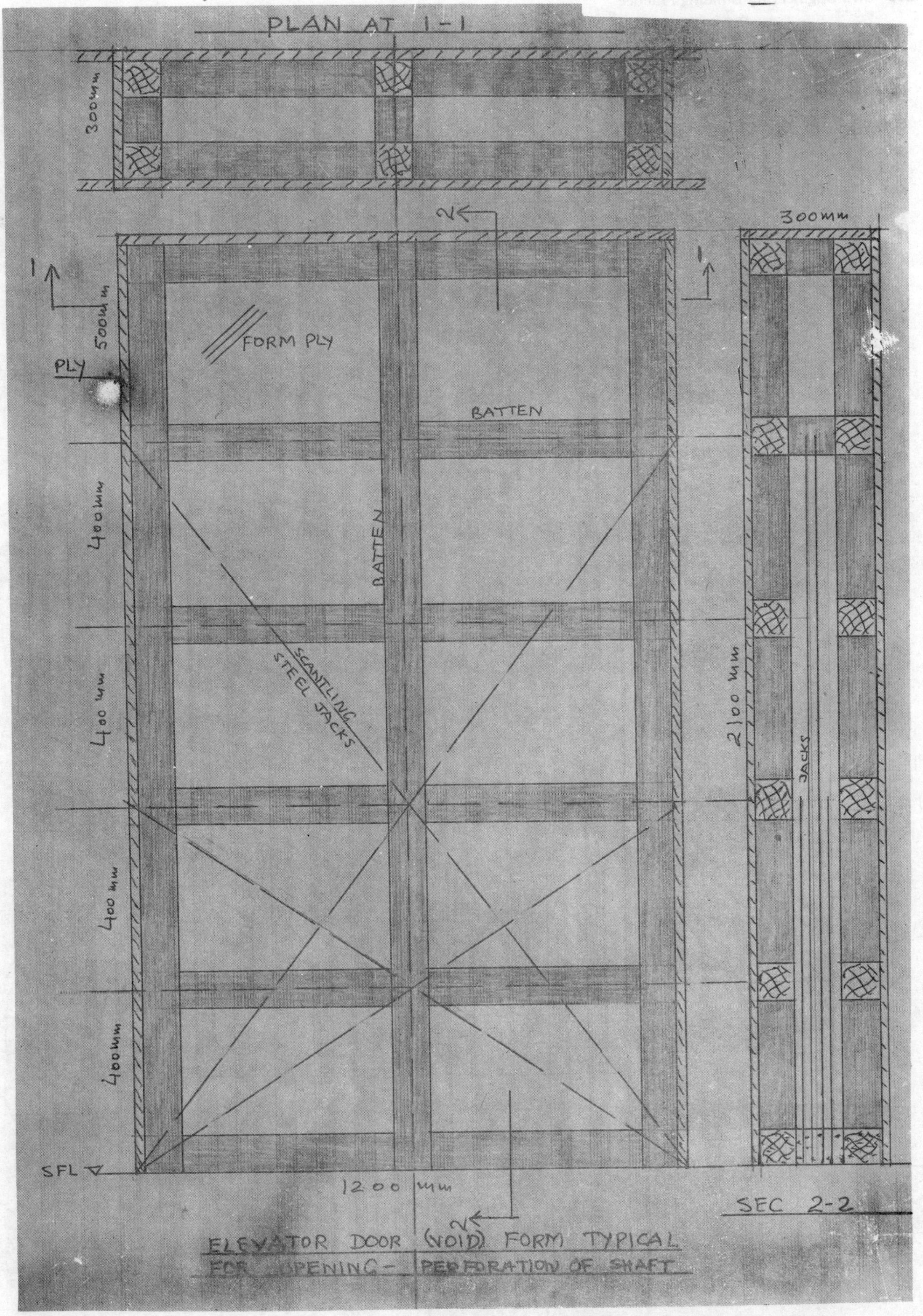

PLAN AT 1-1

300mm

FORM PLY

PLY

500mm

400mm

400mm

400mm

400mm

BATTEN

BATTEN

SCANTLING
STEEL JACKS

SFL ▽

1200 mm

300mm

2100 mm

JACKS

SEC 2-2

ELEVATOR DOOR (VOID) FORM TYPICAL
FOR OPENING - PERFORATION OF SHAFT

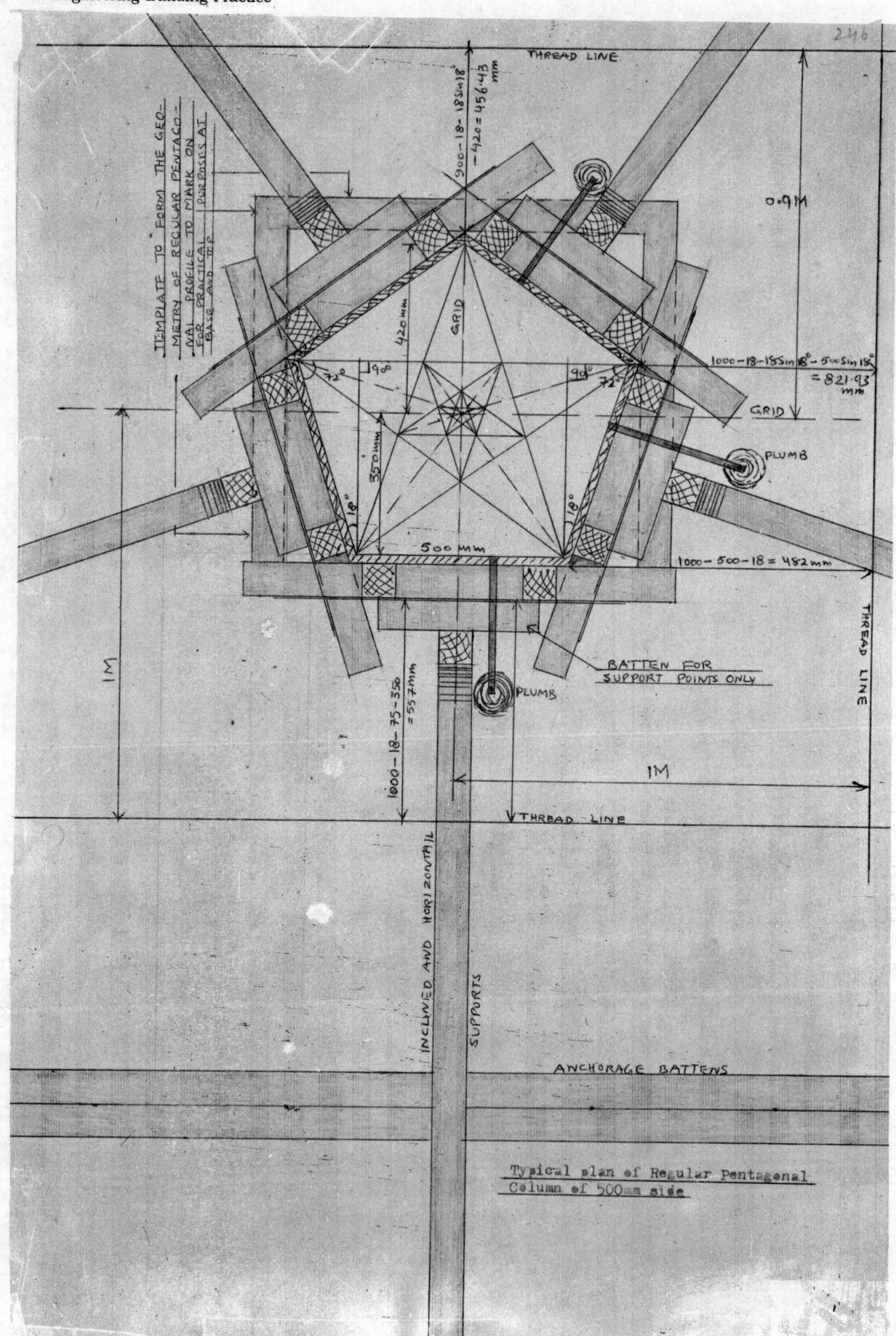

TEMPLATE TO FORM THE GEO-
METRY OF REGULAR PENTAGO-
NAL PROFILE TO MARK ON
FOR PRACTICAL PURPOSES AT
BASE AND TOP

THREAD LINE

900-18-18 Sin 18° = 456·43 mm

0·9 M

GRID

420 mm

72°   900   90°   72°

1000-18-18 Sin 18° - 5x Sin 18°
= 821·93 mm

GRID

350 mm

PLUMB

18°   18°

500 MM

1000 - 500 - 18 = 482 mm

THREAD LINE

BATTEN FOR
SUPPORT POINTS ONLY

1000-18-75-356 = 557 mm

1 M

1 M

PLUMB

THREAD LINE

INCLINED AND HORIZONTAL

SUPPORTS

ANCHORAGE BATTENS

Typical plan of Regular Pentagonal
Column of 500mm side

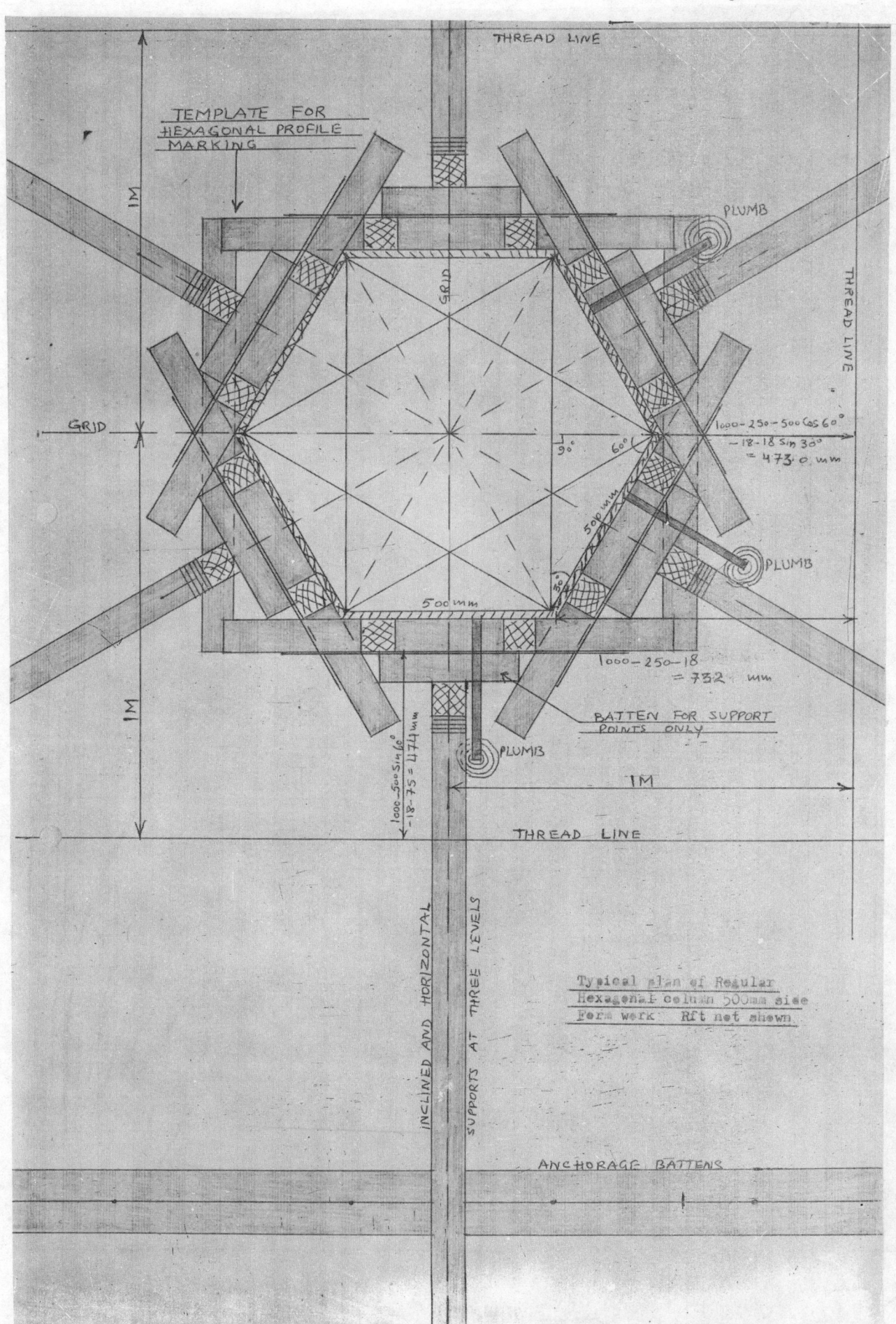

TEMPLATE FOR HEXAGONAL PROFILE MARKING

THREAD LINE

PLUMB

GRID

$1000 - 250 - 500\cos 60°$
$- 18 - 18\sin 30°$
$= 473.0$ mm

PLUMB

$1000 - 250 - 18$
$= 732$ mm

BATTEN FOR SUPPORT POINTS ONLY

500 mm

$1000 - 500\sin 60°$
$- 18 - 75 = 474$ mm

PLUMB

THREAD LINE

INCLINED AND HORIZONTAL

SUPPORTS AT THREE LEVELS

Typical plan of Regular Hexagonal column 500mm size Form work Rft not shown

ANCHORAGE BATTENS

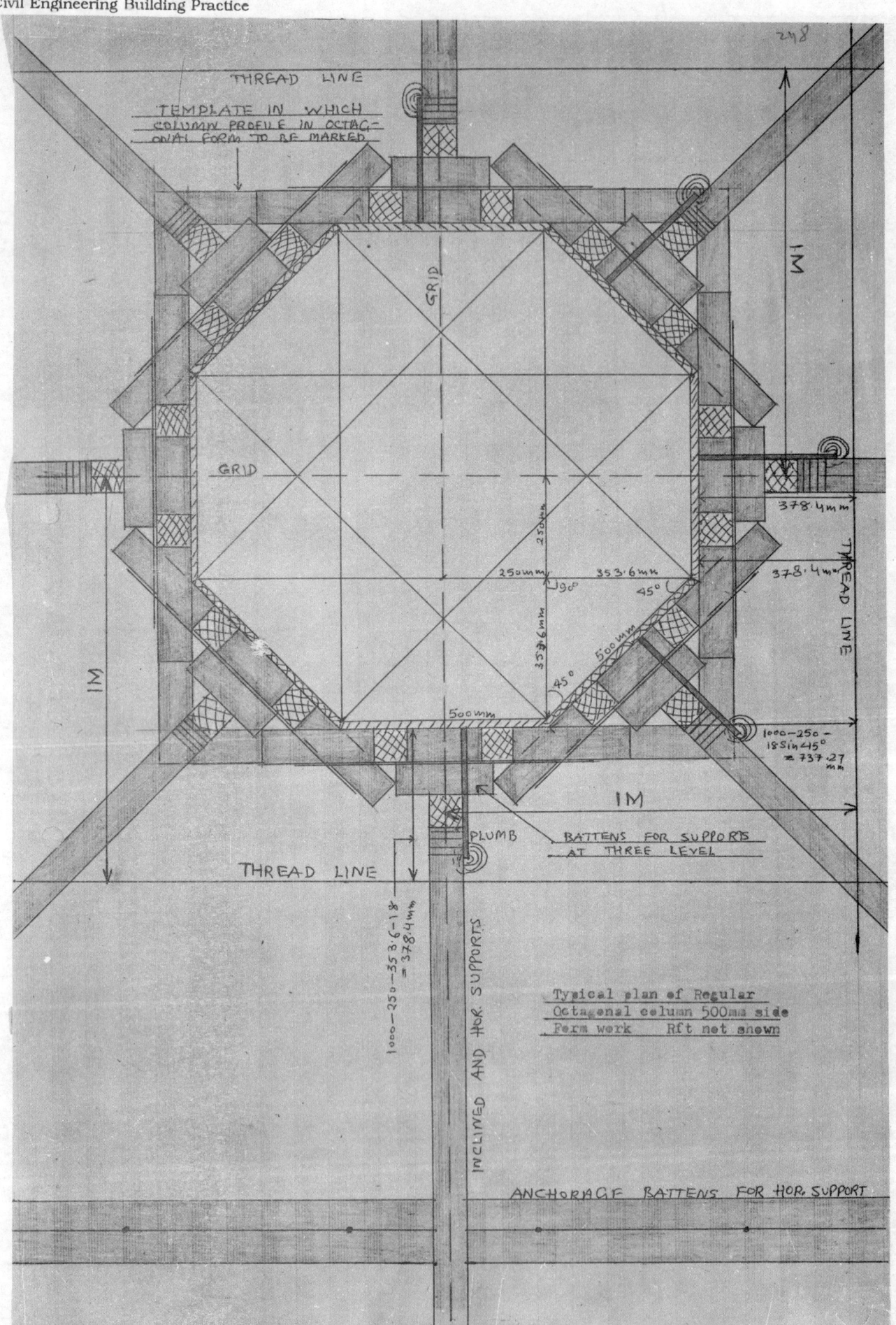

THREAD LINE

TEMPLATE IN WHICH
COLUMN PROFILE IN OCTAG-
ONAL FORM TO BE MARKED

GRID

GRID

1M

378.4mm

378.4mm

THREAD LINE

250mm

250mm

353.6mm

250mm

90°

45°

353.6mm

45°

500mm

500mm

1000-250-
18 sin 45°
= 737.27 mm

1M

PLUMB

BATTENS FOR SUPPORTS
AT THREE LEVEL

THREAD LINE

1000-250-353.6-18 = 378.4mm

INCLINED AND HOR. SUPPORTS

Typical plan of Regular
Octagonal column 500mm side
Form work   Rft not shown

ANCHORAGE BATTENS FOR HOR. SUPPORT

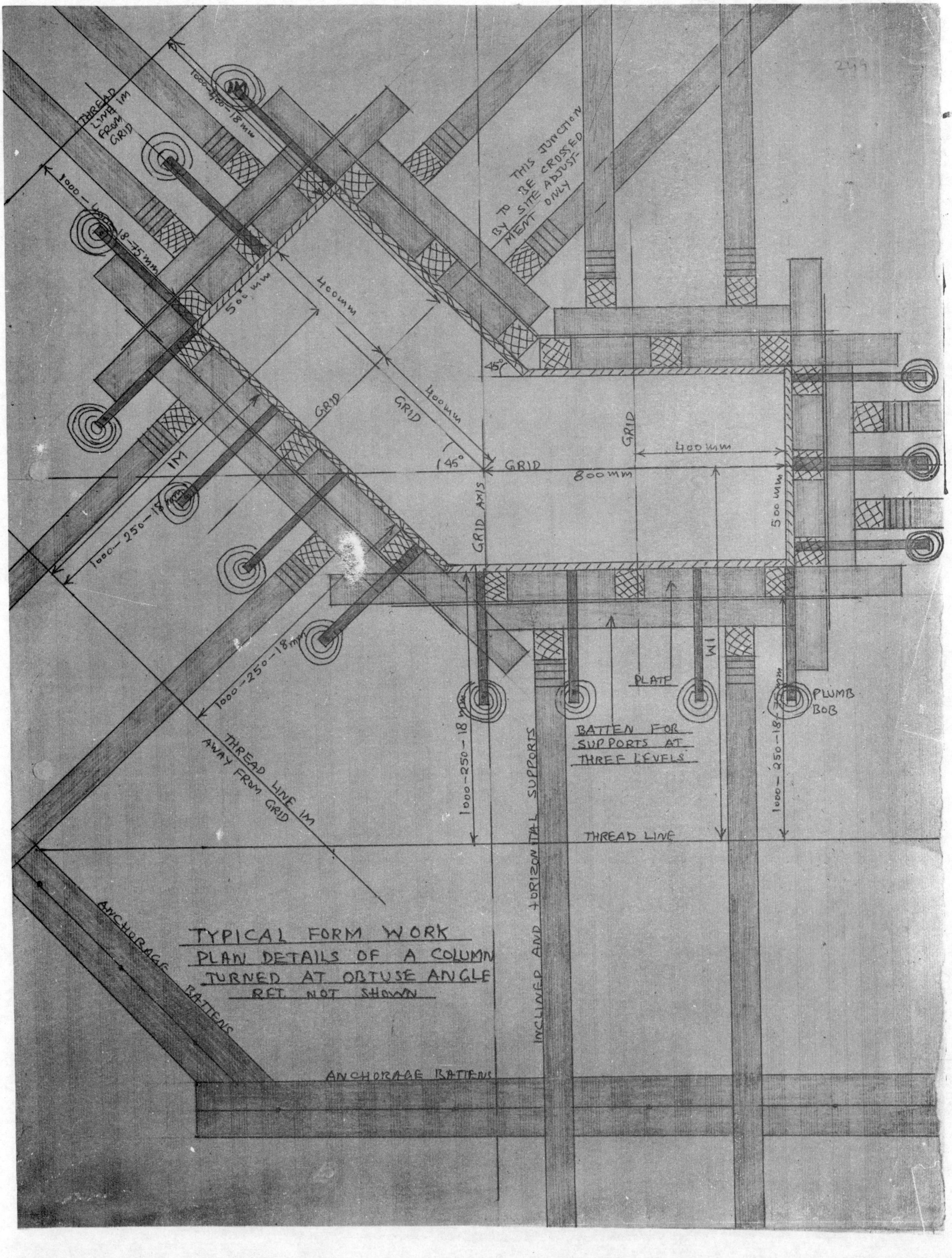

TYPICAL FORM WORK
PLAN DETAILS OF A COLUMN
TURNED AT OBTUSE ANGLE
RFT NOT SHOWN

THREAD LINE 1M FROM GRID

1000 – 250 – 18 MM

500 MM

400 MM

400 MM

GRID

GRID

1M

1000 – 250 – 18 mm

1000 – 250 – 18 mm

THIS JUNCTION TO BE CROSSED BY SITE ADJUSTMENT ONLY

45°

GRID

GRID

45°

GRID AXIS

GRID

800 MM

400 MM

500 MM

PLATE

PLUMB BOB

1M

1000 – 250 – 18 MM

1000 – 250 – 18 MM

BATTEN FOR SUPPORTS AT THREE LEVELS

INCLINED AND HORIZONTAL SUPPORTS

THREAD LINE

THREAD LINE 1M AWAY FROM GRID

ANCHORAGE BATTENS

ANCHORAGE BATTENS

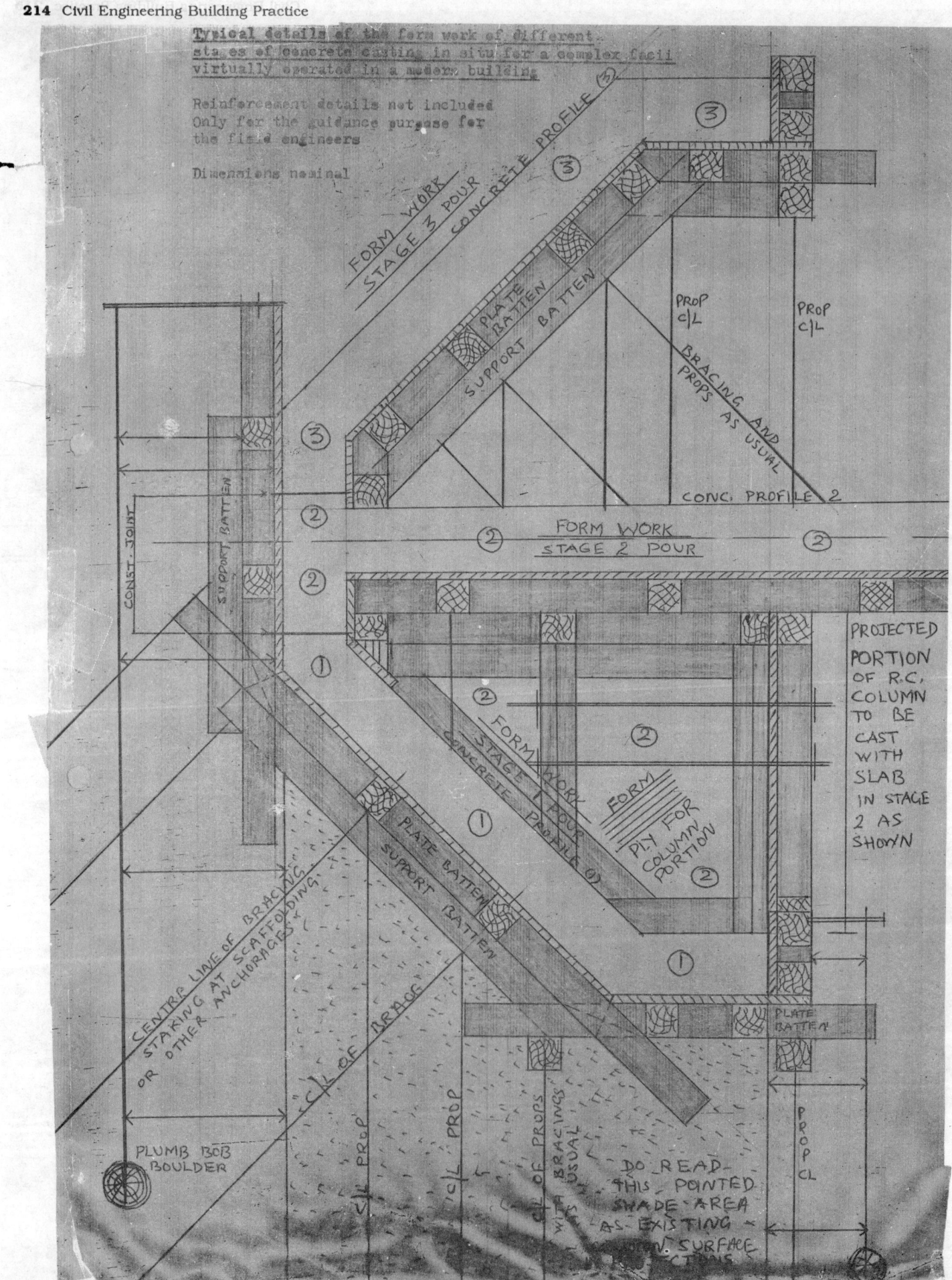

Typical details of the form work of different
stages of concrete casting in situ for a complex facil
virtually operated in a modern building

Reinforcement details not included
Only for the guidance purpose for
the field engineers

Dimensions nominal

FORM WORK
STAGE 3 POUR
CONCRETE PROFILE ③

PLATE BATTEN
SUPPORT BATTEN

PROP c/L
PROP c/L

BRACING AND PROPS AS USUAL

CONC. PROFILE 2

FORM WORK
STAGE 2 POUR

CONST. JOINT
SUPPORT BATTEN

FORM WORK
STAGE 1 POUR
CONCRETE PROFILE ①

FORM
PLY FOR COLUMN PORTION

PROJECTED PORTION OF R.C. COLUMN TO BE CAST WITH SLAB IN STAGE 2 AS SHOWN

PLATE BATTEN
SUPPORT BATTEN

CENTRE LINE OF BRACING STAKING AT SCAFFOLDING OR OTHER ANCHORAGES

CL OF BRACE

PLATE BATTEN

PLUMB BOB BOULDER

PROP
c/L PROP
CL OF PROPS
BRACINGS AS USUAL

PROP CL

DO READ THIS POINTED SHADE AREA AS EXISTING SURFACE

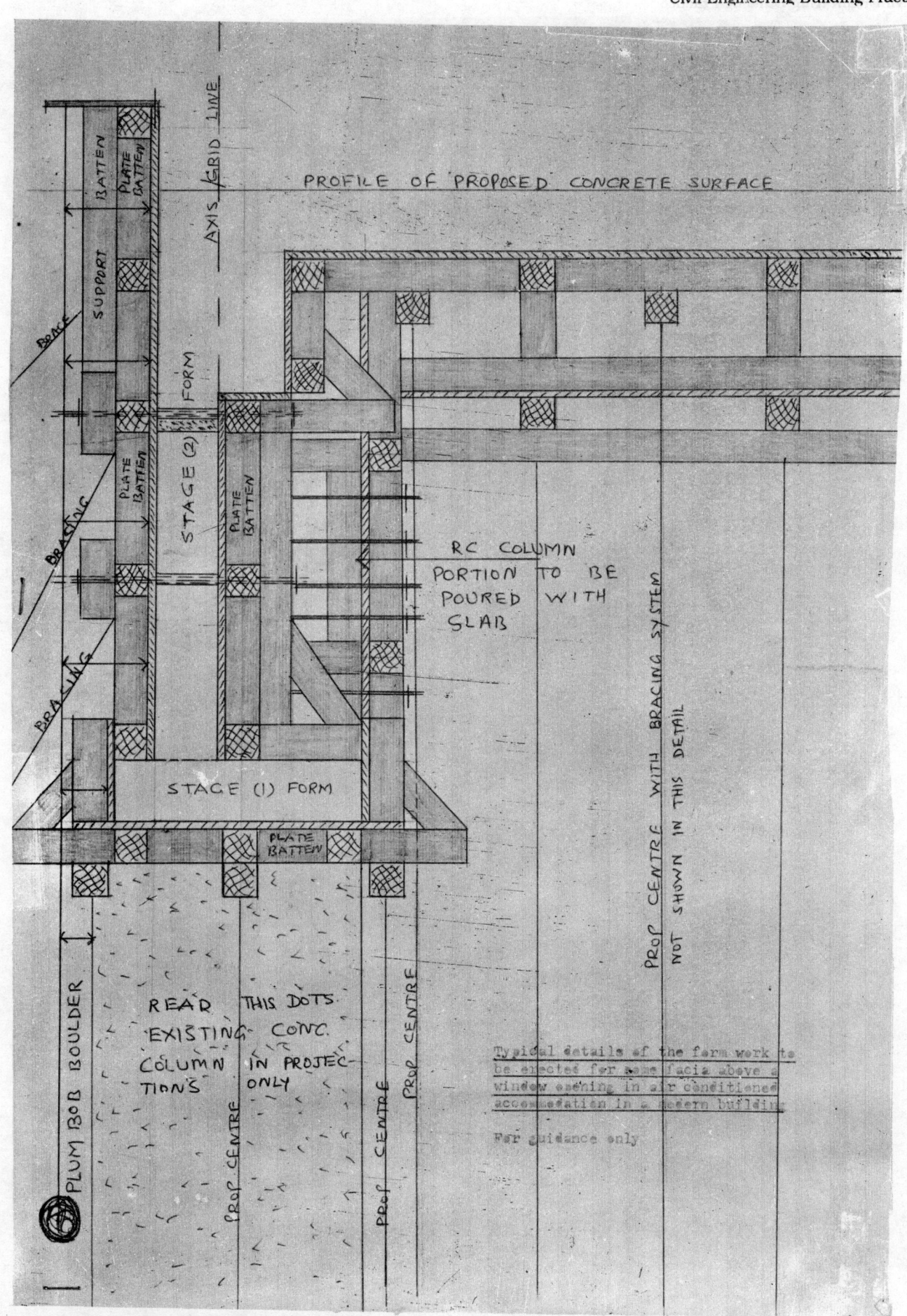

PROFILE OF PROPOSED CONCRETE SURFACE

BATTEN

PLATE BATTEN

SUPPORT

BRACE

AXIS GRID LINE

STAGE (2) FORM

PLATE BATTEN

BRACING

BRACING

PLATE BATTEN

STAGE (1) FORM

PLATE BATTEN

RC COLUMN PORTION TO BE POURED WITH SLAB

PROP CENTRE WITH BRACING SYSTEM NOT SHOWN IN THIS DETAIL

PLUM BOB BOULDER

READ THIS DOTS EXISTING CONC. COLUMN IN PROJEC-TIONS ONLY

PROP CENTRE

PROP CENTRE

PROP CENTRE

Typical details of the form work to be erected for some facia above a window opening in air conditioned accommodation in a modern building

For guidance only

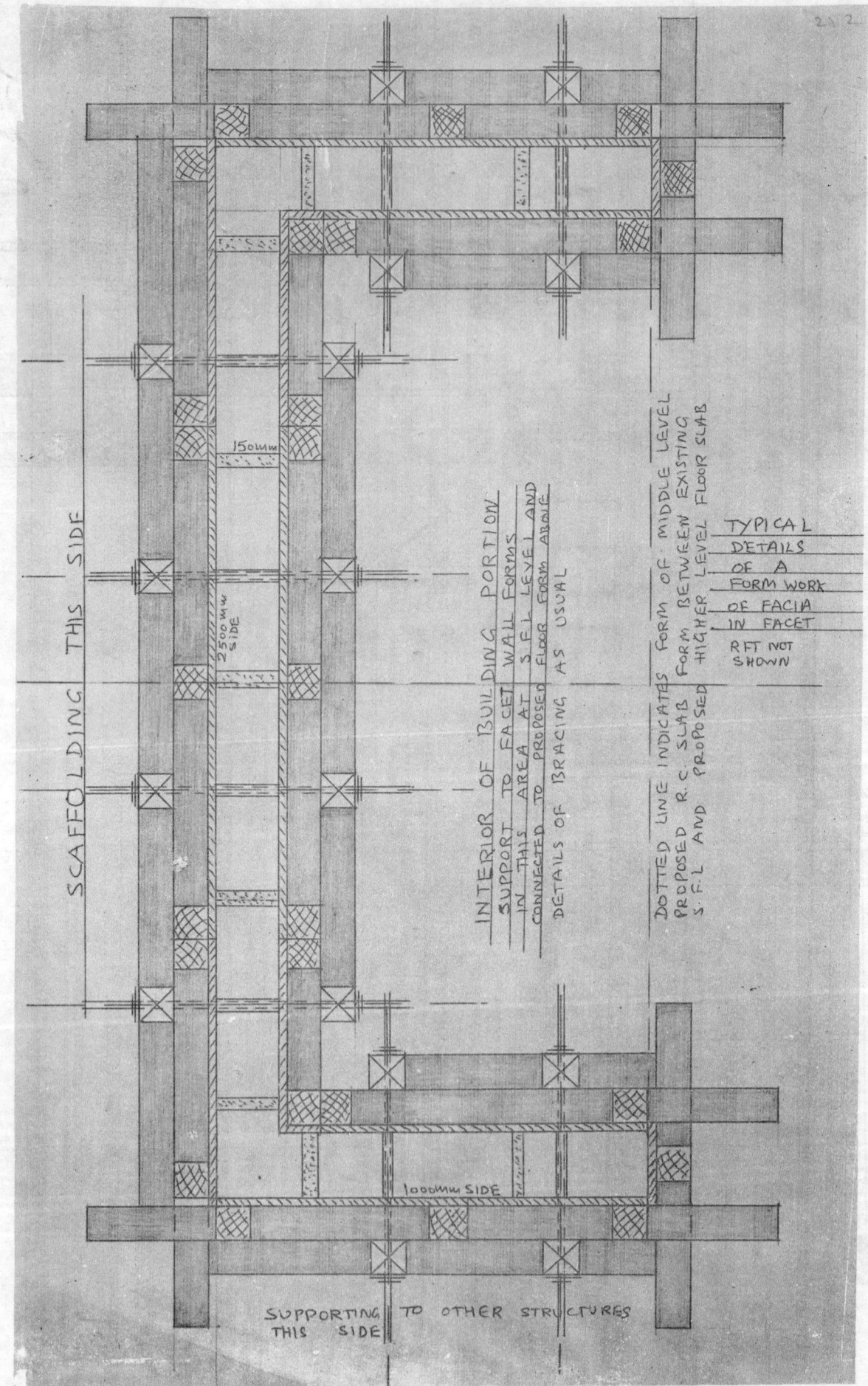

SCAFFOLDING THIS SIDE

150mm

2500mm SIDE

1000mm SIDE

SUPPORTING TO OTHER STRUCTURES THIS SIDE

INTERIOR OF BUILDING PORTION

SUPPORT TO FACET WALL FORMS IN THIS AREA AT S.F.L LEVEL AND CONNECTED TO PROPOSED FLOOR FORM ABOVE

DETAILS OF BRACING AS USUAL

DOTTED LINE INDICATES FORM OF MIDDLE LEVEL FORM, BETWEEN EXISTING PROPOSED R.C. SLAB HIGHER LEVEL FLOOR SLAB S.F.L AND PROPOSED

TYPICAL
DETAILS
OF A
FORM WORK
OF FACIA
IN FACET

R.F.T NOT
SHOWN

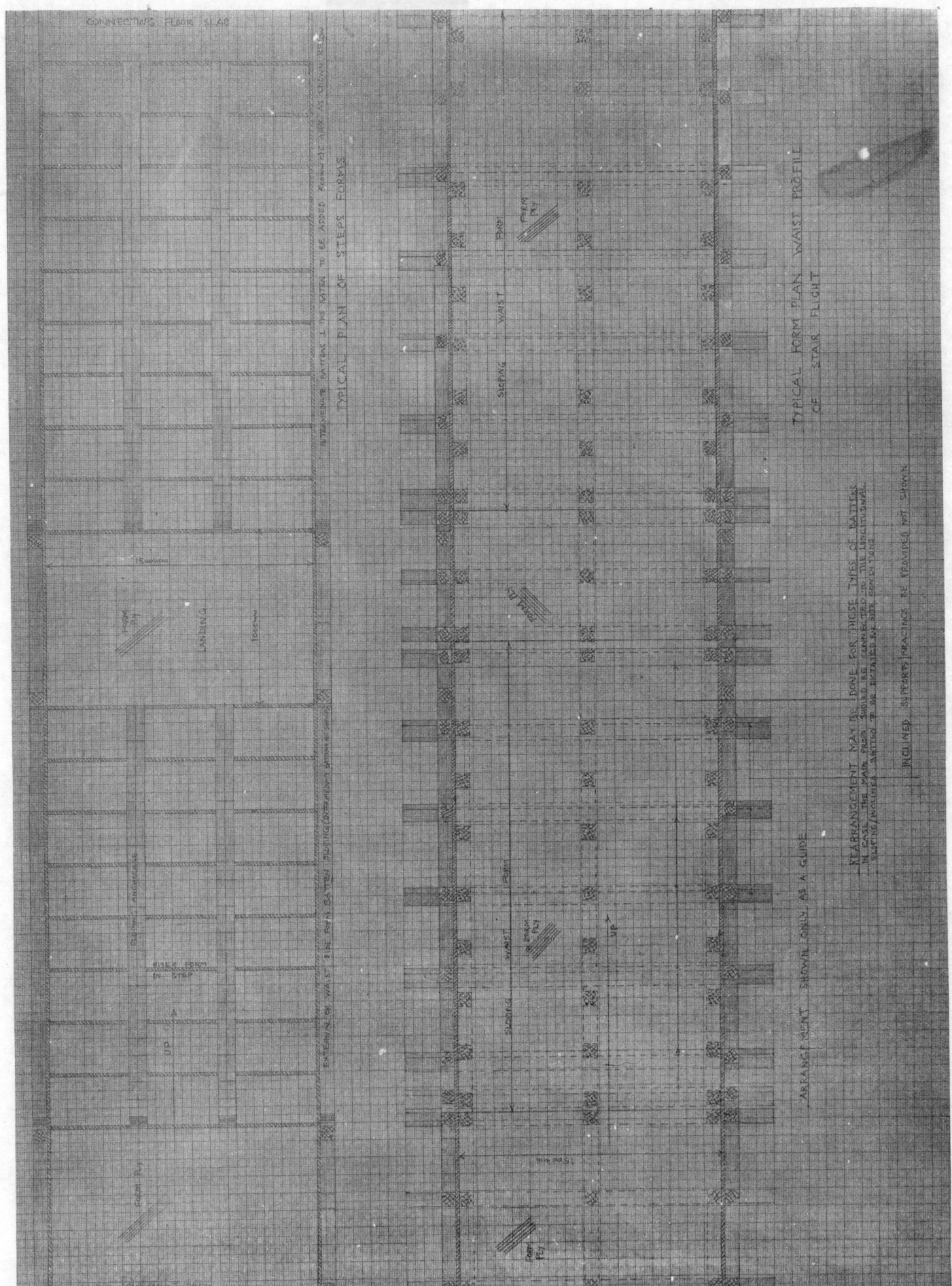

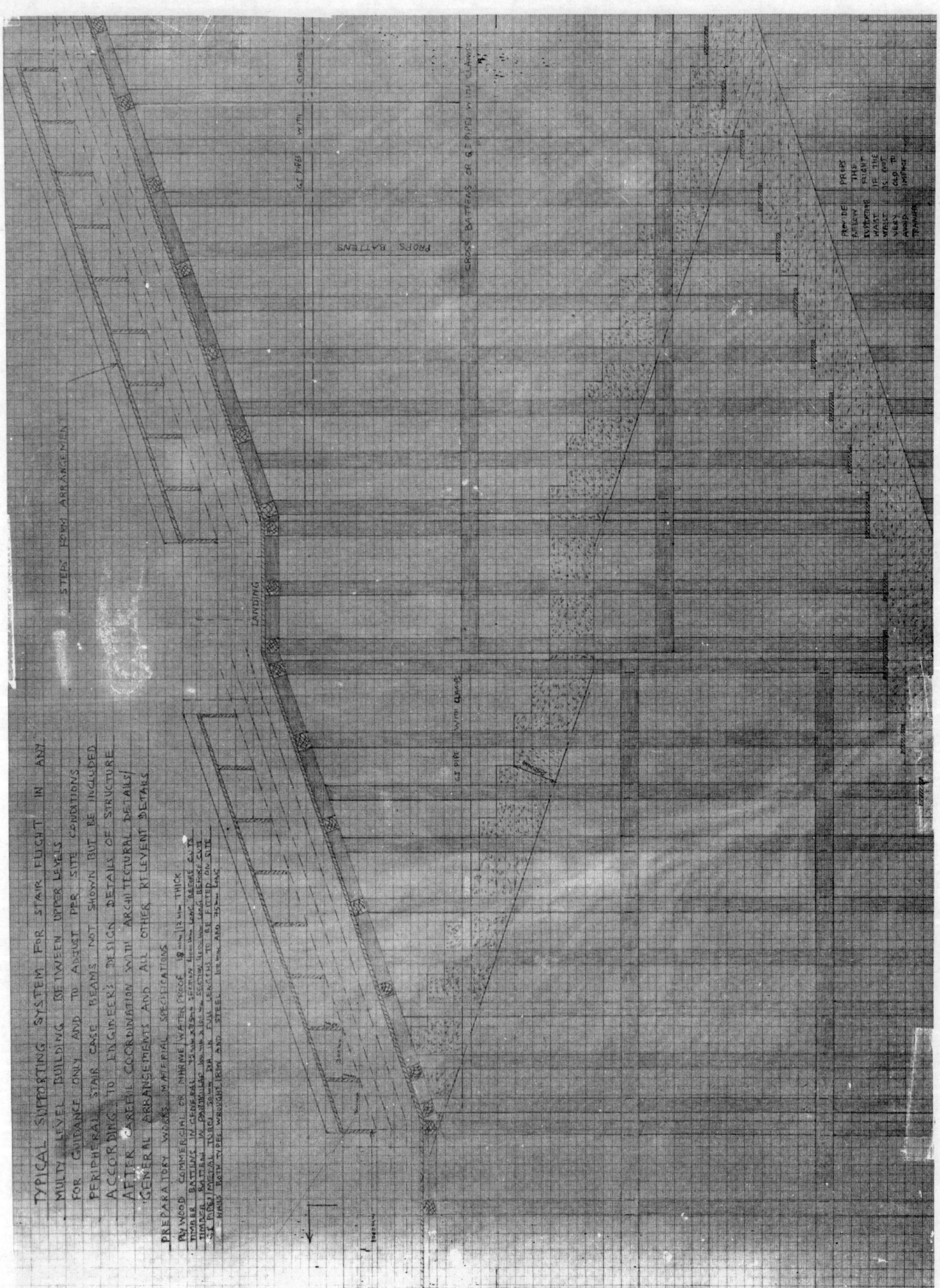

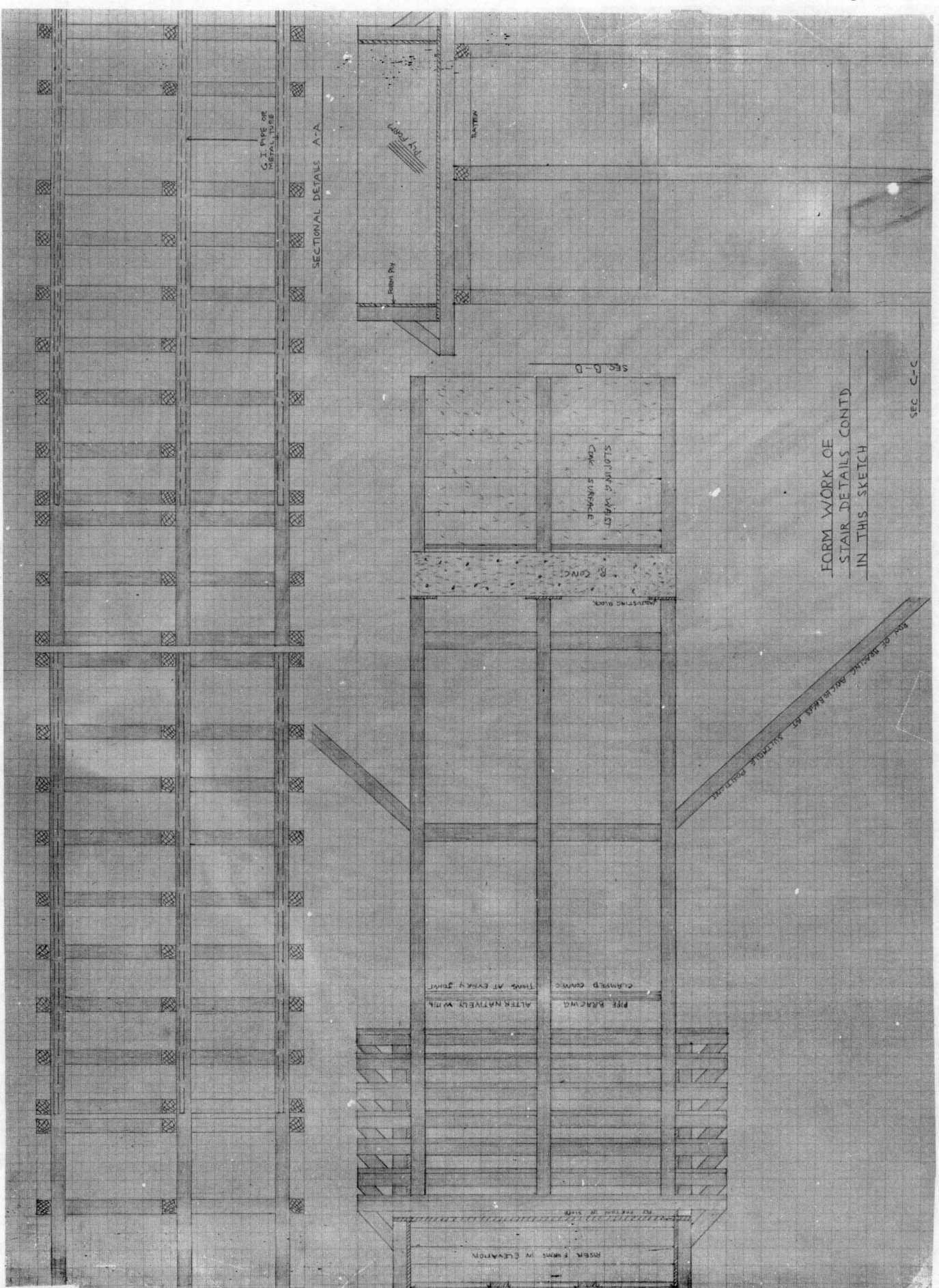

FORM WORK OF
STAIR DETAILS CONT'D
IN THIS SKETCH

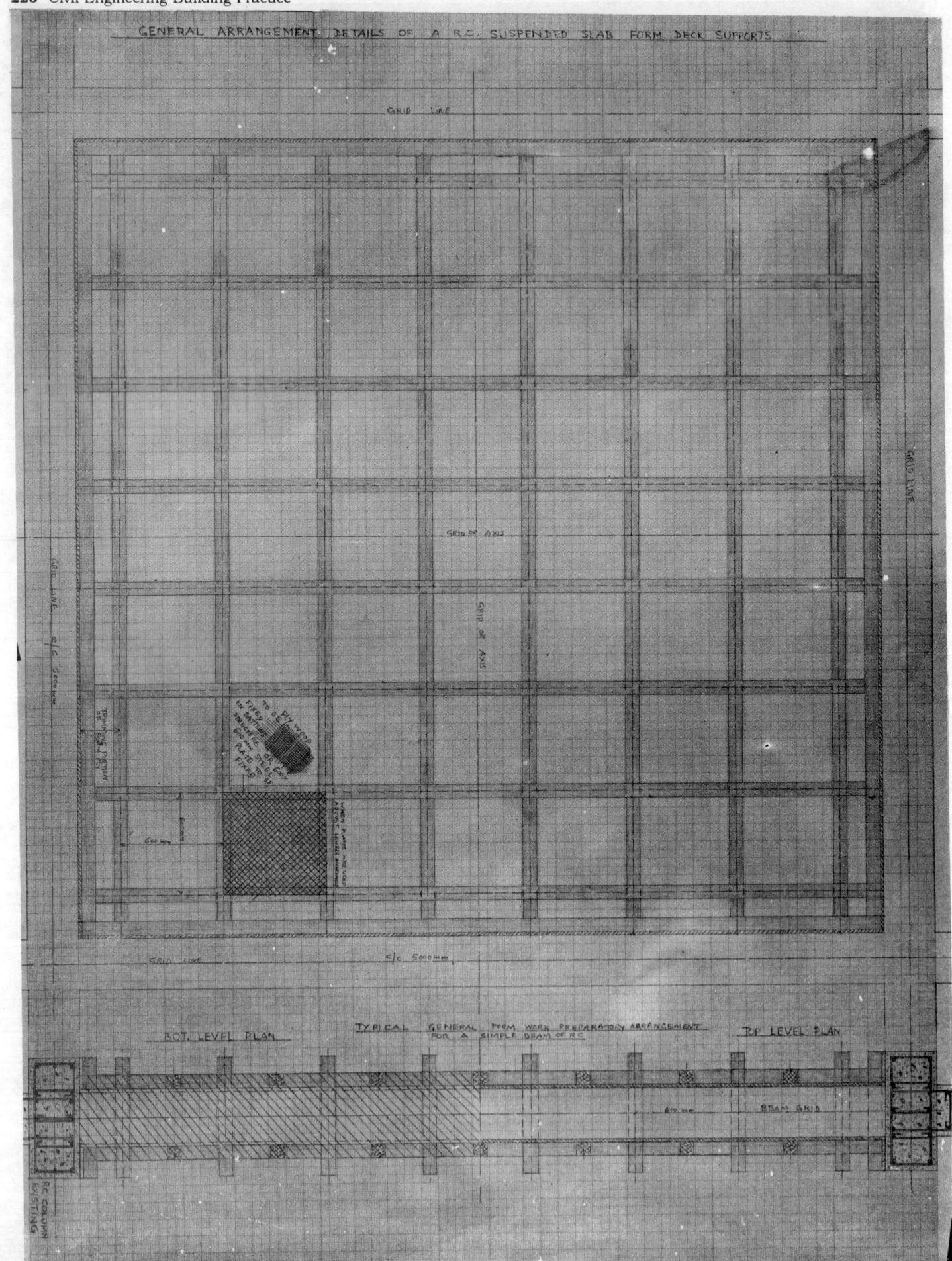

GENERAL ARRANGEMENT DETAILS OF A R.C. SUSPENDED SLAB FORM DECK SUPPORTS

GRID LINE

GRID LINE

GRID OF AXIS

GRID OF AXIS

GRID LINE

C/C 5000 mm

600 mm

BOT. LEVEL PLAN

TYPICAL GENERAL FORM WORK PREPARATORY ARRANGEMENT FOR A SIMPLE BEAM OF R.C.

TOP LEVEL PLAN

BEAM GRID

R.C. COLUMN EXISTING

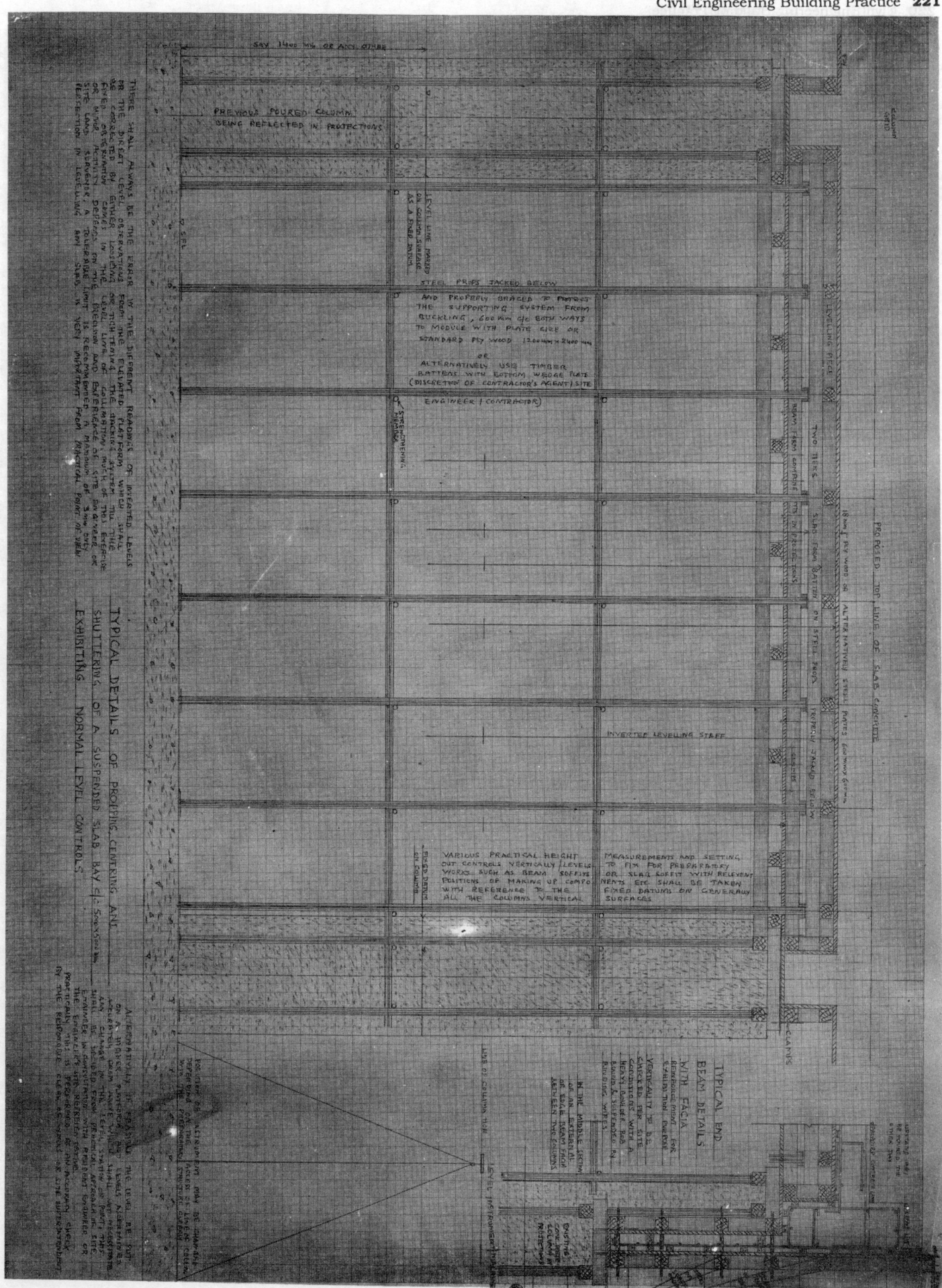

Sketches 51 to 134

Informative Sketches for General Knowledge
First Hand Written
Not to Scale

51. Aerial plan of an imaginary aeroplane-shaped hotel.
52. Cellar plan.
53. Ground floor.
54. First floor.
55. Aerial plan of an institution hostel accommodation.
56. Line plan of Sketch 55.
57. Line plan of a single-storeyed set of offices.
58. Typical details — pergola, kerbs, guard house.
59. Typical details — compound/boundary wall, planter beds and tree pit.
60. Typical details of floor junctions.
61. Typical details of 5 kg light, bulk head frame, aluminium patio door, wardrobe.
62. Typical details of aluminium framed, aluminium paneled glazed window.
63. Typical details of timber door frames.
64. An ordinary mosque.
65. Ordinary electric sub-station.
66. Hotel building aerial plan.
67. Cellar parking for Sketch 66 (line plan).
68. Ground parking for Sketch 66 (line plan).
69. First floor line plan for Sketch 66.
70. Second floor to forth floor line plan for Sketch 66.
71. Fifth to fourteenth floor line plan for Sketch 66.
72. Fifteenth floor line plan for Sketch 66.
73. Sixteenth floor line plan for Sketch 66.
74. Seventeenth floor line plan for Sketch 66.
75. Function/ceremony hall line plan for Sketch 66.
76. Commercial complex aerial view.
77. Cellar line plan for tower building in Sketch 76.
78. Ground floor line plan for tower building in Sketch 76.
79. First floor line plan for tower building in Sketch 76.
80. Sixth floor to thirteenth floor line plan for tower building in Sketch 76.
81. Section 'AA' for tower building in Sketch 76.
82. Elevation 'A' for tower building in Sketch 76.
83. Elevation 'B' for tower building in Sketch 76.
84. Elevation 'C' for tower building in Sketch 76.
85. Elevation 'D' for tower building in Sketch 76.

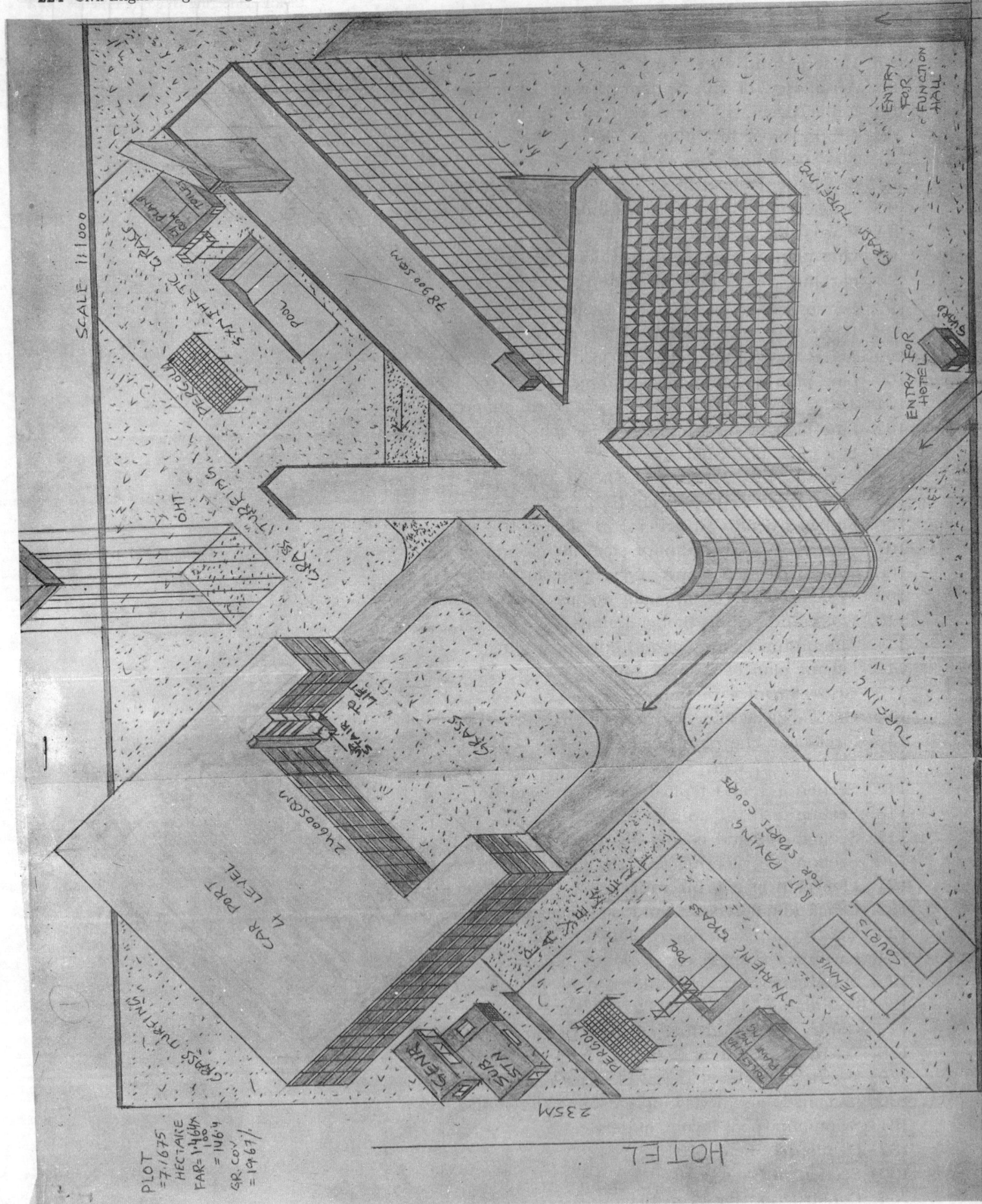

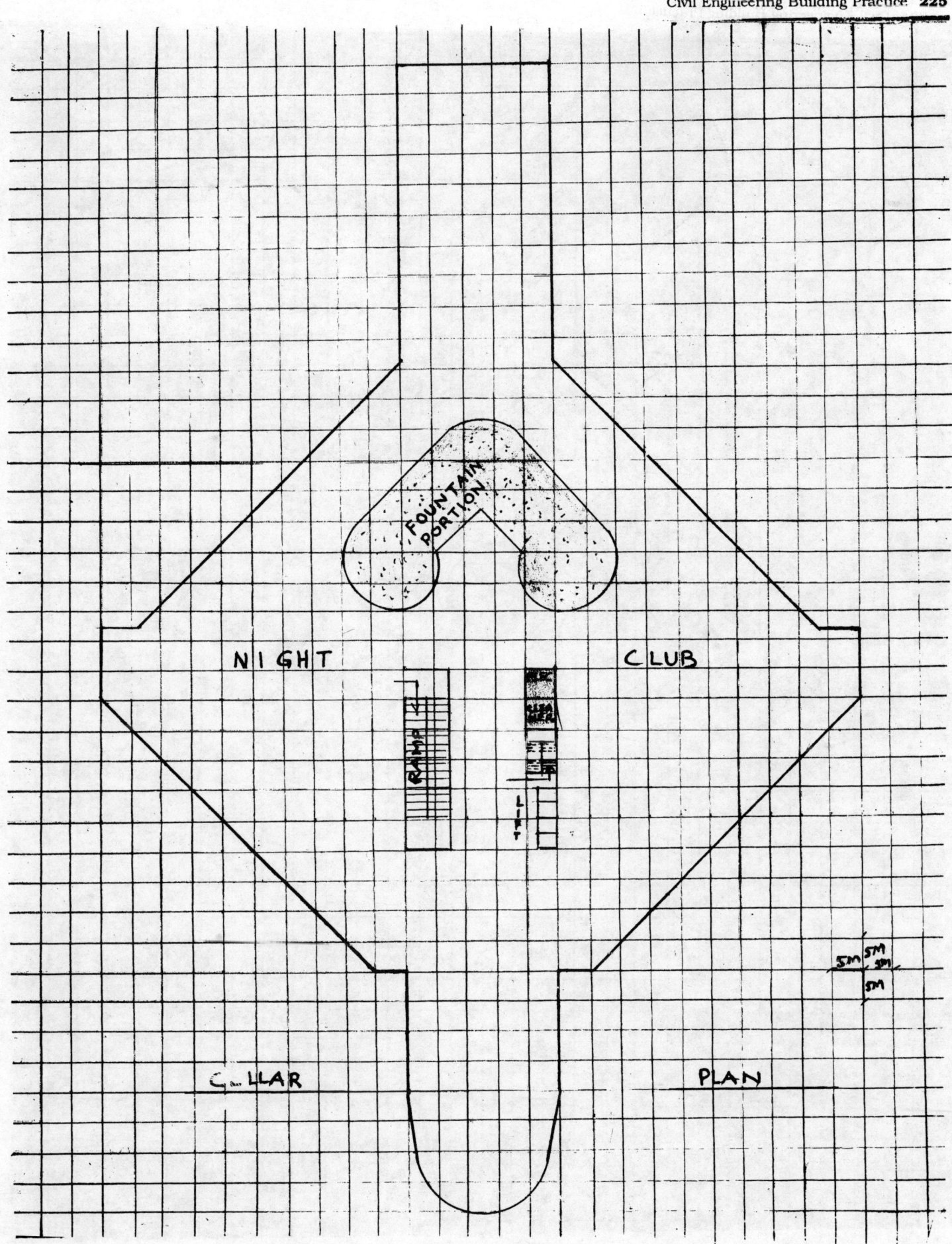

WELCOME

FOUNTAINS IN

DANCE DISC

STAGE LIGHTS

KIT

REST

WET

ELECT CLEANER

LIFTS

UP

RECEP.

WET

GROUND   FLOOR

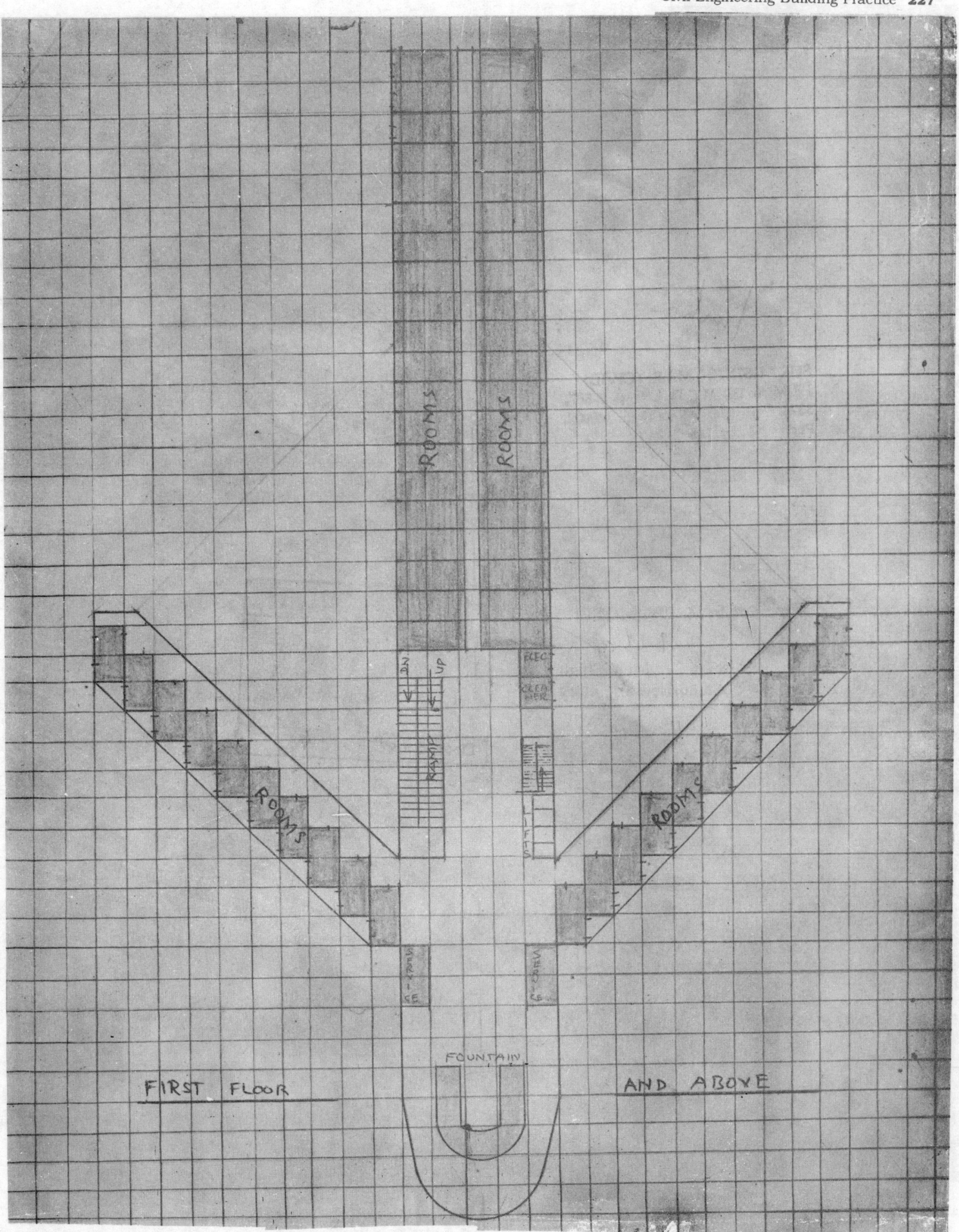

FIRST FLOOR   AND ABOVE

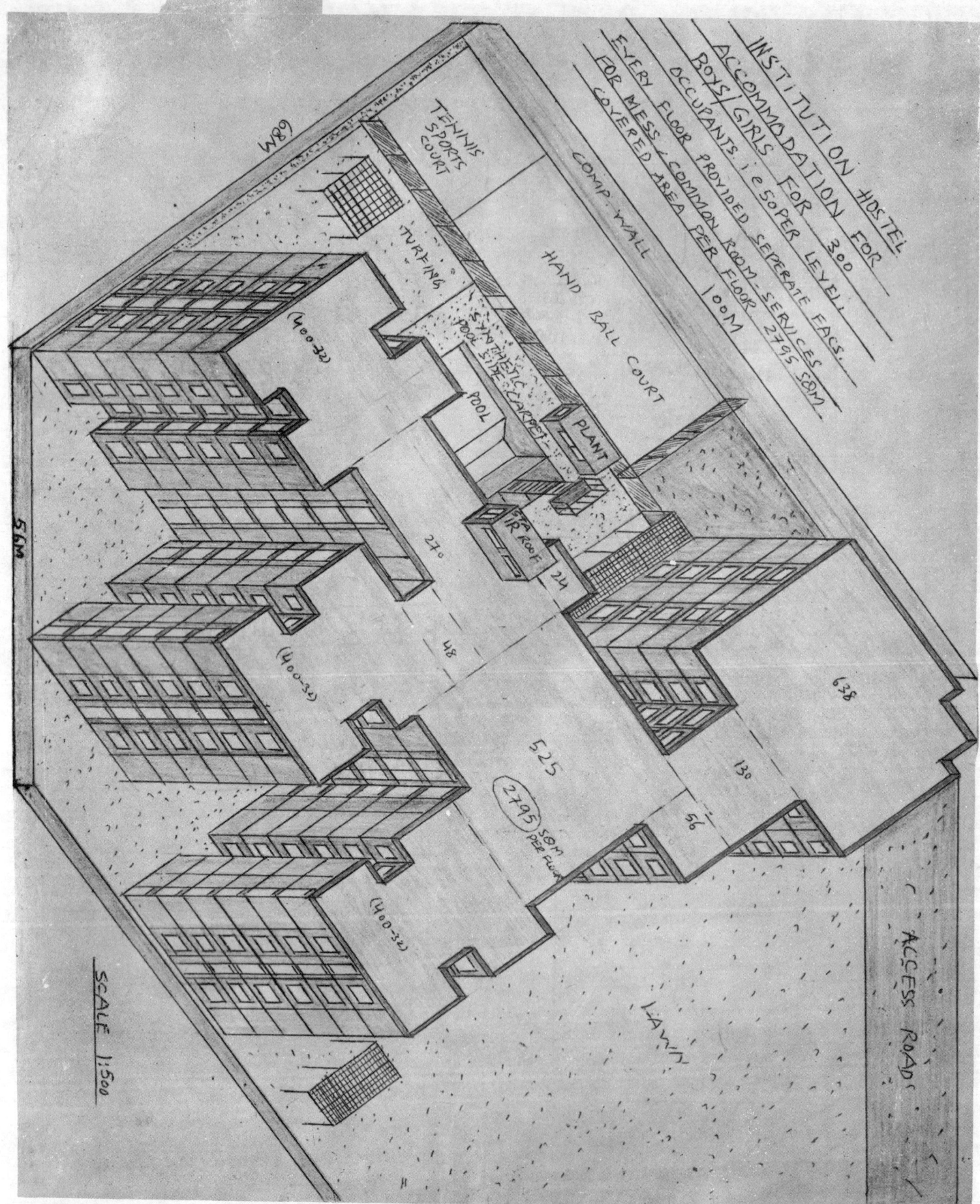

INSTITUTION HOSTEL

ACCOMMODATION FOR

BOYS/GIRLS FOR 300

OCCUPANTS i.e 50 PER LEVEL

EVERY FLOOR PROVIDED SEPERATE FACS.

FOR MESS & COMMON ROOM-SERVICES

OCCUPANTS i.e 50 PER LEVEL

COVERED AREA PER FLOOR 2795 S8M

100 M

COMP. WALL

HAND BALL COURT

TENNIS SPORTS COURT

TURFING

SYNTHETIC POOL SIDE CARPET

POOL

PLANT

R.C ROOF 24

27°

48

(400-32)

(400-32)

(400-32)

68M

56M

638

525

2795 S8M PER FLOOR

13°

56

LAWN

ACCESS ROAD

SCALE 1:500

LINE PLAN FOR BACHELOR ACCOMMODATIONS FOR OFFICERS. FOR HOSTELS

SKY LIGHT INCLD. ONLY IN ONE STOREY BLDG - NOT IN MULTI STOREYED CASE

A SET OF OFFICES ONE STOREY

DETAILS OF
PERGOLA TYPICAL '3M SPAN'

PERGOLA
KERB
GUARD
HOUSE
Typical

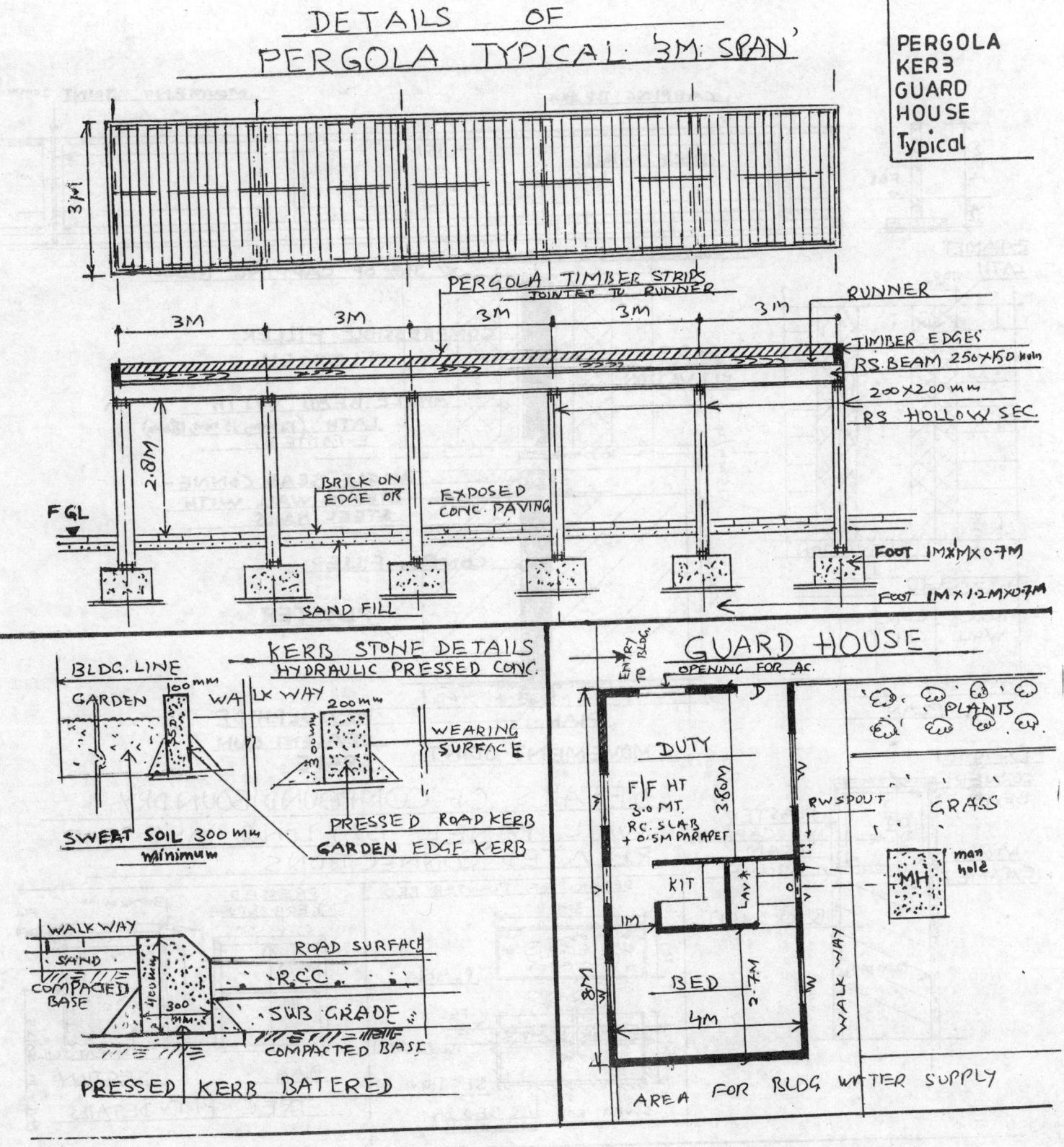

3M

PERGOLA TIMBER STRIPS
JOINTED TO RUNNER RUNNER

3M   3M   3M   3M   3M

TIMBER EDGES
RS. BEAM 250×150 mm

200×200 mm
RS HOLLOW SEC.

2.8M

BRICK ON EDGE OR   EXPOSED CONC. PAVING

FGL

FOOT 1M×1M×0·7M

SAND FILL

FOOT 1M×1·2M×0·7M

KERB STONE DETAILS
HYDRAULIC PRESSED CONC.

BLDG. LINE

GARDEN   100mm   WALK WAY   200mm

WEARING SURFACE

SWEET SOIL 300 mm
minimum

PRESSED ROAD KERB
GARDEN EDGE KERB

WALK WAY   ROAD SURFACE

SAND

COMPACTED BASE   R.C.C.

300 mm   SUB GRADE

COMPACTED BASE

PRESSED KERB BATERED

GUARD HOUSE

ENTRY TO BLDG.   OPENING FOR A.C.

DUTY

F/F HT
3.0 MT.
RC. SLAB
+ 0.5M PARAPET

3.6M

KIT

LAV

1M

BED

2.7M

4M

3.8M

PLANTS

RWSPOUT   GRASS

MH   man hole

WALK WAY

AREA FOR BLDG WATER SUPPLY

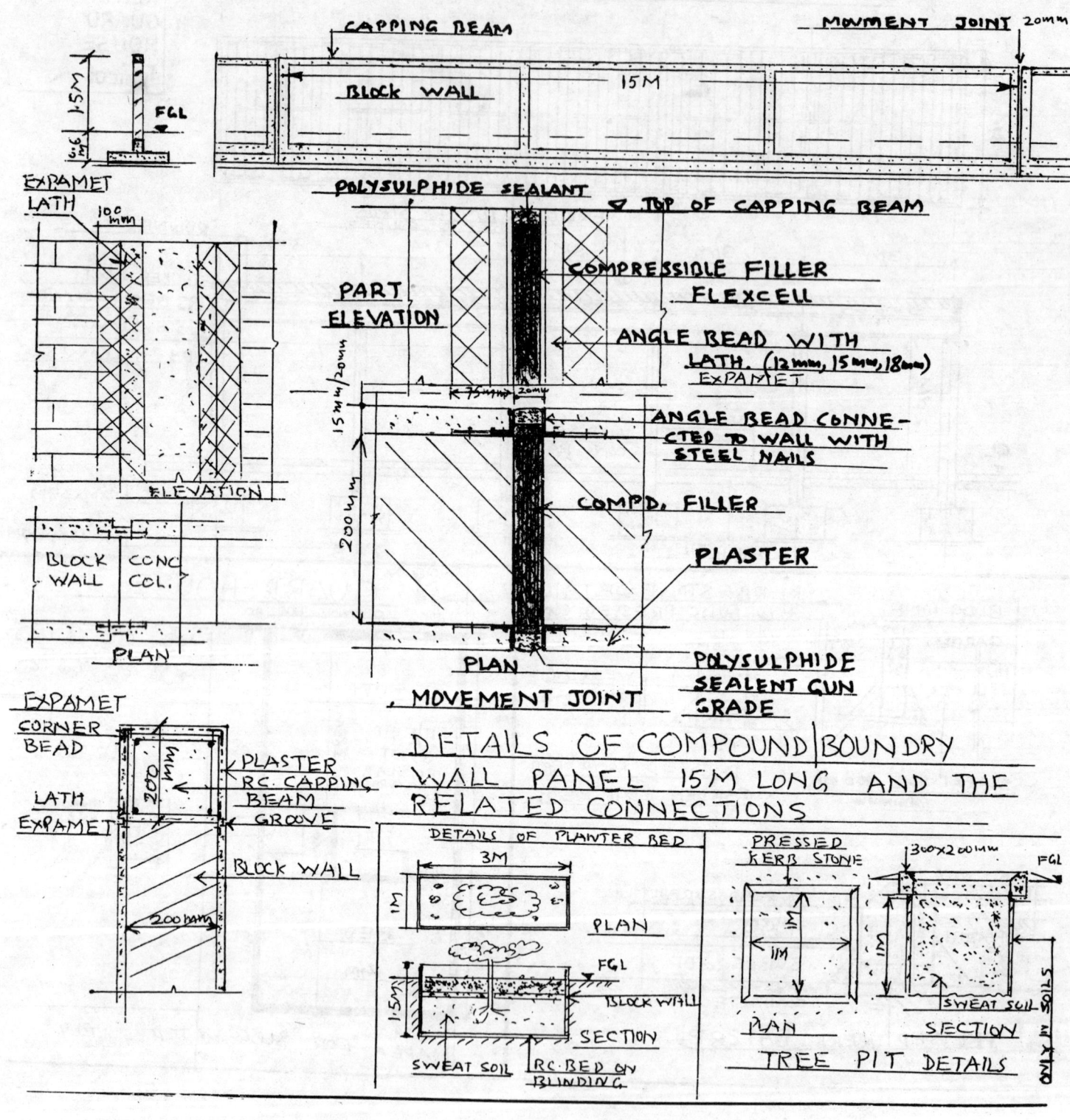

CAPPING BEAM   MOVMENT JOINT 20MM

BLOCK WALL   15M

1·5M   FGL

EXPAMET LATH

100 mm

ELEVATION

BLOCK WALL | CONC COL.

PLAN

POLYSULPHIDE SEALANT   ▽ TOP OF CAPPING BEAM

PART ELEVATION

COMPRESSIBLE FILLER FLEXCELL

ANGLE BEAD WITH LATH. (12mm, 15mm, 18mm) EXPAMET

15MM/20MM

75MM   20MM

200 MM

ANGLE BEAD CONNECTED TO WALL WITH STEEL NAILS

COMPD. FILLER

PLASTER

PLAN   MOVEMENT JOINT

POLYSULPHIDE SEALENT GUN GRADE

EXPAMET CORNER BEAD

200 MM

LATH EXPAMET

PLASTER RC. CAPPING BEAM GROOVE

BLOCK WALL

200 MM

DETAILS OF COMPOUND/BOUNDRY WALL PANEL 15M LONG AND THE RELATED CONNECTIONS

DETAILS OF PLANTER BED

3M

1M

PLAN

1·5M   FGL   BLOCK WALL

SECTION

SWEAT SOIL   RC BED ON BLINDING

PRESSED KERB STONE

300x200MM   FGL

1M

1M

1M

PLAN   SECTION

SWEAT SOIL

ONLY IN SOILS

TREE PIT DETAILS

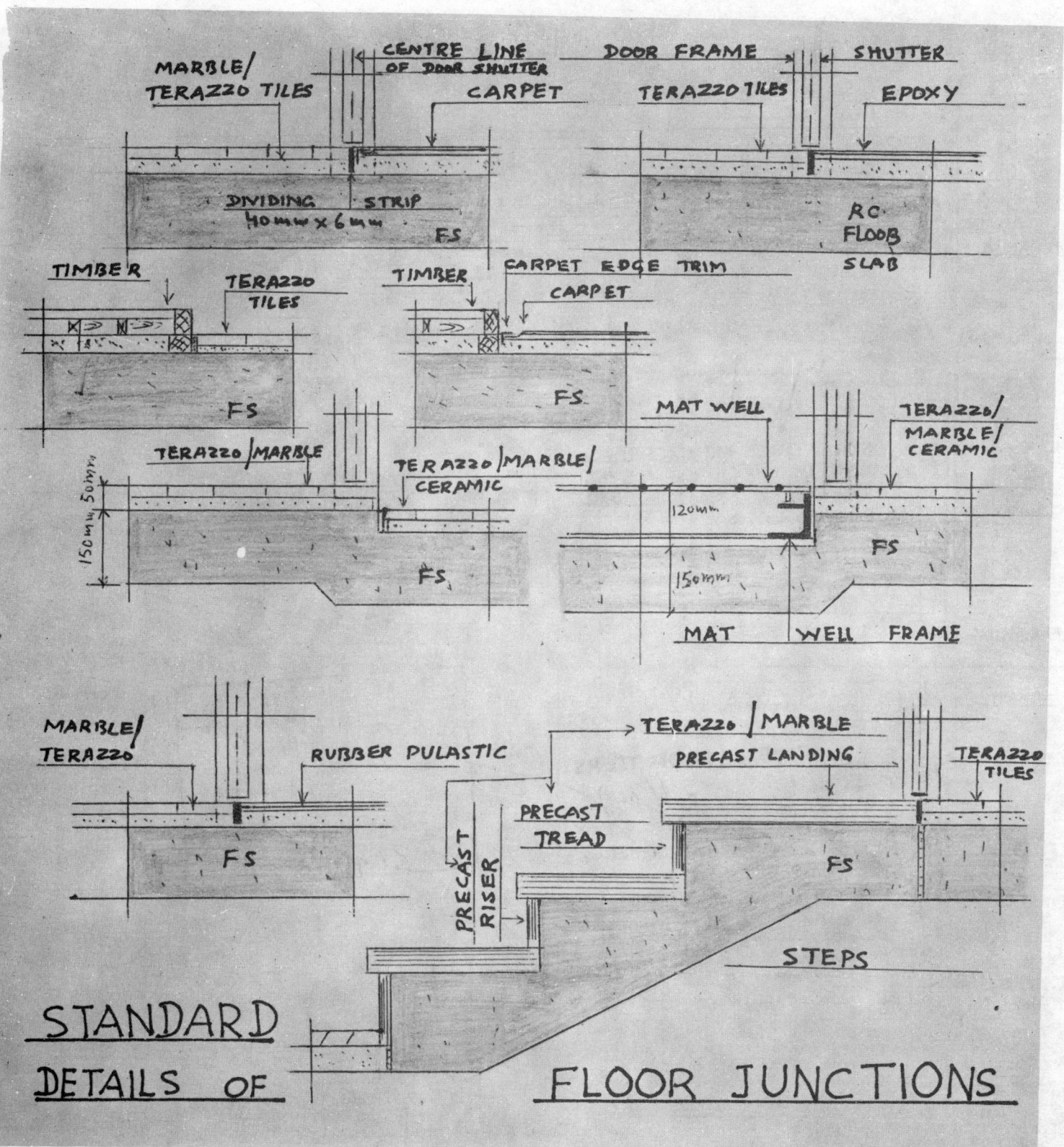

MARBLE/ TERAZZO TILES

CENTRE LINE OF DOOR SHUTTER

CARPET

DIVIDING STRIP 40mm X 6mm

FS

DOOR FRAME

TERAZZO TILES

SHUTTER

EPOXY

RC FLOOR SLAB

TIMBER

TERAZZO TILES

FS

TIMBER

CARPET EDGE TRIM

CARPET

FS

TERAZZO/MARBLE

150mm, 50mm

TERAZZO/MARBLE/ CERAMIC

FS

MAT WELL

120mm

150mm

MAT WELL FRAME

TERAZZO/ MARBLE/ CERAMIC

FS

MARBLE/ TERAZZO

RUBBER PULASTIC

FS

TERAZZO/MARBLE

PRECAST LANDING

PRECAST RISER

PRECAST TREAD

TERAZZO TILES

FS

STEPS

STANDARD DETAILS OF FLOOR JUNCTIONS

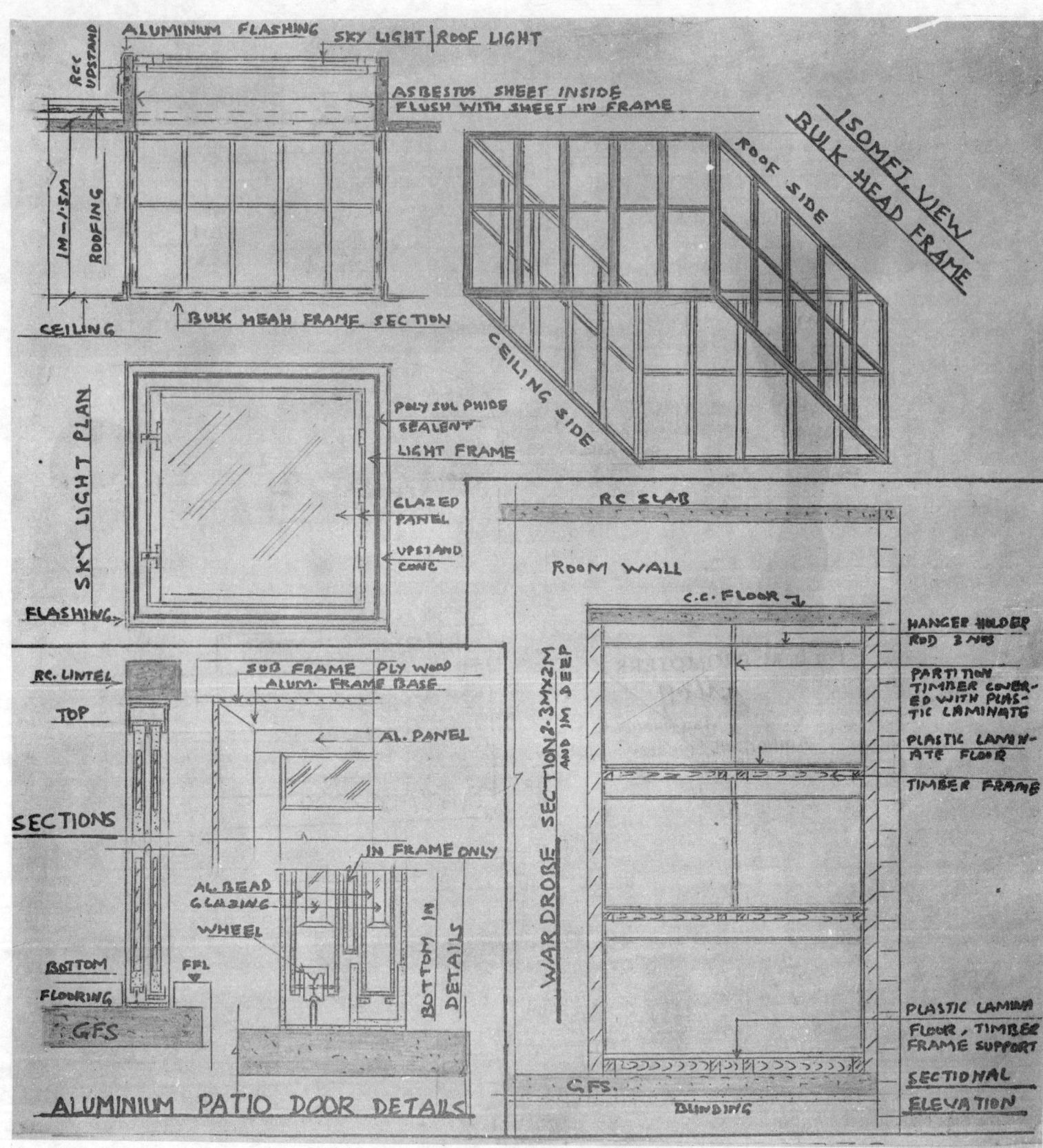

ALUMINIUM FLASHING   SKY LIGHT | ROOF LIGHT

RCC UPSTAND

ASBESTOS SHEET INSIDE
FLUSH WITH SHEET IN FRAME

1M — 1·5M

ROOFING
ROOFING

CEILING

BULK HEAH FRAME SECTION

ISOMET. VIEW
BULK HEAD FRAME

ROOF SIDE

CEILING SIDE

SKY LIGHT PLAN

POLY SULPHIDE
SEALENT

LIGHT FRAME

GLAZED
PANEL

UPSTAND
CONC

FLASHING

RC SLAB

ROOM WALL

C.C. FLOOR

HANGER HOLDER
ROD 3 NOS

PARTITION
TIMBER COVER-
ED WITH PLAS-
TIC LAMINATE

PLASTIC LAMIN-
ATE FLOOR

TIMBER FRAME

RC. LINTEL

TOP

SUB FRAME PLY WOOD
ALUM. FRAME BASE

AL. PANEL

SECTIONS

IN FRAME ONLY

AL. BEAD
GLAZING

WHEEL

BOTTOM

FLOORING

GFS

FFL

BOTTOM IN
DETAILS

WARDROBE SECTION 2·3M X 2M AND 1M DEEP

PLASTIC LAMINA
FLOOR , TIMBER
FRAME SUPPORT

SECTIONAL

ELEVATION

GFS.

BLINDING

ALUMINIUM PATIO DOOR DETAILS

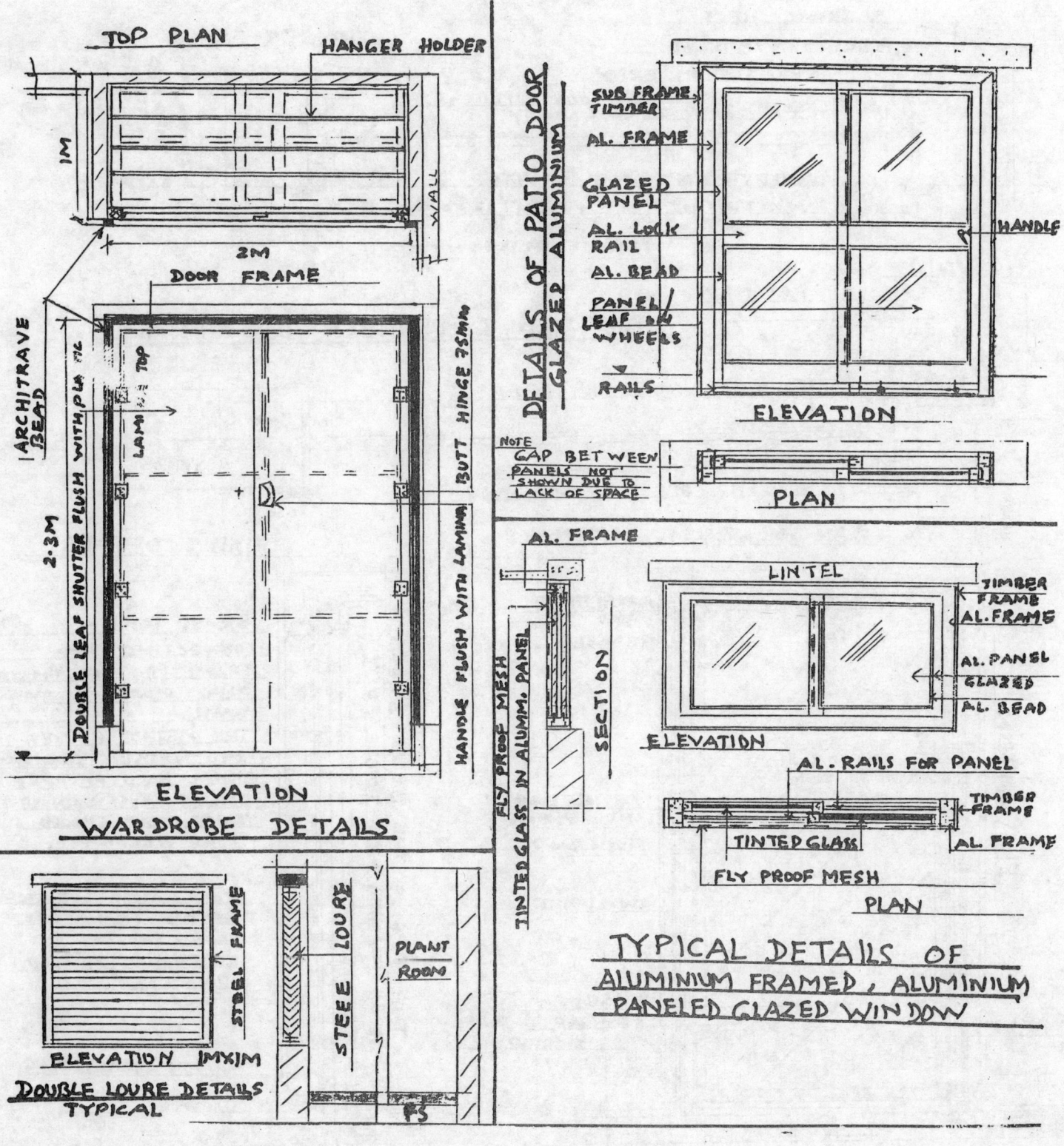

TOP PLAN   HANGER HOLDER

1M

2M

DOOR FRAME

ARCHITRAVE BEAD

DOUBLE LEAF SHUTTER FLUSH WITH PLY 8'C

LAMM TOP

2·3M

HANDLE FLUSH WITH LAMMM

BUTT HINGE 75mm

WALL

ELEVATION

WARDROBE DETAILS

ELEVATION 1MX1M

DOUBLE LOURE DETAILS TYPICAL

STEEL FRAME

STEEL LOURE

PLANT ROOM

FS

DETAILS OF PATIO DOOR
Glazed Aluminium

SUB FRAME TIMBER

AL. FRAME

GLAZED PANEL

AL. LOCK RAIL

AL. BEAD

PANEL LEAF ON WHEELS

RAILS

HANDLE

ELEVATION

NOTE
GAP BETWEEN PANELS NOT SHOWN DUE TO LACK OF SPACE

PLAN

AL. FRAME

FLY PROOF MESH

TINTED GLASS IN ALUMM. PANEL

SECTION

LINTEL

TIMBER FRAME
AL. FRAME

AL. PANEL

GLAZED

AL. BEAD

ELEVATION

AL. RAILS FOR PANEL

TIMBER FRAME

TINTED GLASS

AL. FRAME

FLY PROOF MESH

PLAN

TYPICAL DETAILS OF
ALUMINIUM FRAMED, ALUMINIUM
PANELED GLAZED WINDOW

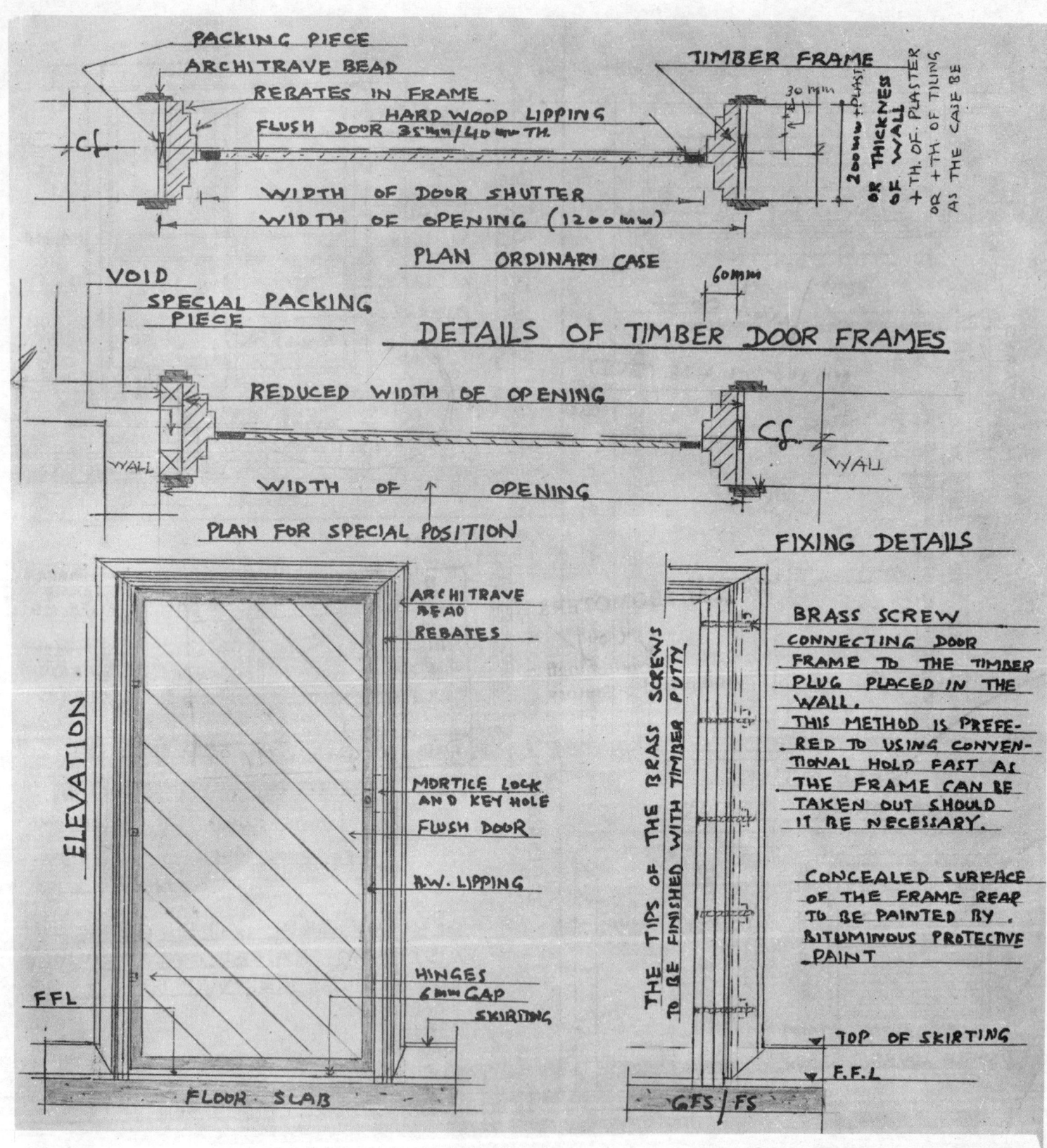

PACKING PIECE
ARCHITRAVE BEAD
REBATES IN FRAME
HARD WOOD LIPPING
FLUSH DOOR 35mm/40mm TH.
WIDTH OF DOOR SHUTTER
WIDTH OF OPENING (1200mm)

TIMBER FRAME

30 mm

200 mm PLAST + TH. OF WALL + TH. OF PLASTER OR + TH OF TILING AS THE CASE BE

C.L

PLAN ORDINARY CASE

DETAILS OF TIMBER DOOR FRAMES

VOID
SPECIAL PACKING PIECE
60mm
REDUCED WIDTH OF OPENING
WALL
C.L
WALL
WIDTH OF ↑ OPENING

PLAN FOR SPECIAL POSITION

FIXING DETAILS

ELEVATION

ARCHITRAVE BEAD
REBATES

MORTICE LOCK AND KEY HOLE
FLUSH DOOR

R.W. LIPPING

HINGES
6mm GAP
SKIRTING

FFL

FLOOR SLAB

THE TIPS OF THE BRASS SCREWS TO BE FINISHED WITH TIMBER PUTTY

BRASS SCREW
CONNECTING DOOR FRAME TO THE TIMBER PLUG PLACED IN THE WALL.
THIS METHOD IS PREFE- RED TO USING CONVEN- TIONAL HOLD FAST AS THE FRAME CAN BE TAKEN OUT SHOULD IT BE NECESSARY.

CONCEALED SURFACE OF THE FRAME REAR TO BE PAINTED BY BITUMINOUS PROTECTIVE PAINT

TOP OF SKIRTING
F.F.L

GFS/FS

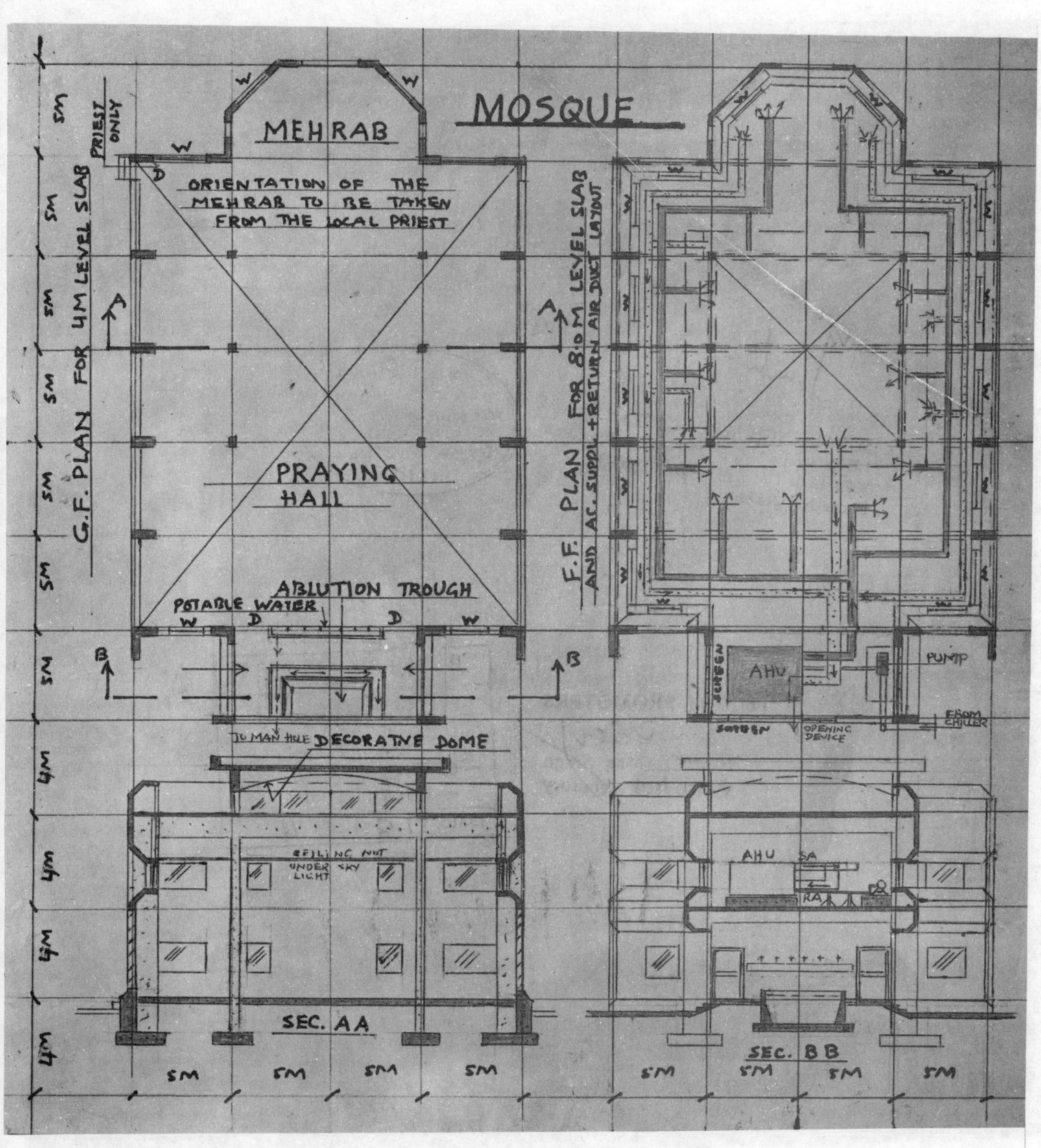

MOSQUE

G.F. PLAN FOR 4M LEVEL SLAB

PRIEST ONLY

MEHRAB

ORIENTATION OF THE MEHRAB TO BE TAKEN FROM THE LOCAL PRIEST

PRAYING HALL

ABLUTION TROUGH

POTABLE WATER

F.F. PLAN FOR 8.0M LEVEL SLAB AND A.C. SUPPL + RETURN AIR DUCT LAYOUT

SCREEN

AHU

PUMP

SCREEN

OPENING DEVICE

FROM CHILLER

JU MAN HOLE **DECORATIVE DOME**

CEILING NOT UNDER SKY LIGHT

SEC. AA

5M   5M   5M   5M

AHU SA

RA

SEC. BB

5M   5M   5M   5M

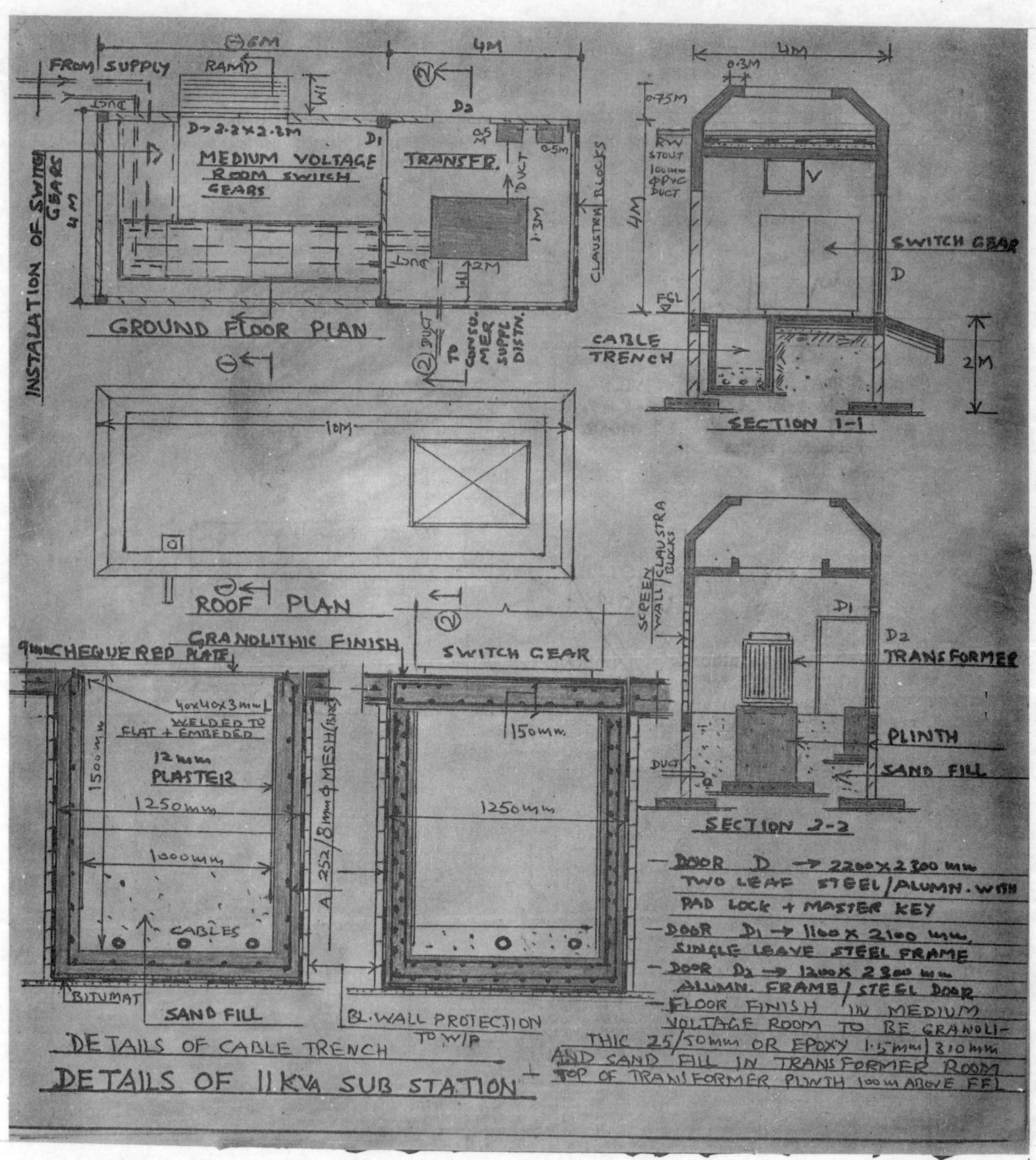

INSTALLATION OF SWITCH GEARS

FROM SUPPLY · RAMP · 0.6M · 4M

D → 2.2 × 2.2M
MEDIUM VOLTAGE ROOM SWITCH GEARS
TRANSFR.

GROUND FLOOR PLAN

ROOF PLAN — 10M

SECTION 1-1
SWITCH GEAR · CABLE TRENCH · 2M · 4M · 0.75M · 0.3M
RW STOUT 100mm Φ PVC DUCT · FCL

SECTION 2-2
SCREEN WALL CLAUSTRA BLOCKS · D1 · D2 TRANSFORMER · PLINTH · SAND FILL · DUCT

DETAILS OF CABLE TRENCH
9mm CHEQUERED PLATE · GRANOLITHIC FINISH
40×40×3mm FLAT + EMBEDED · WELDED TO
12mm PLASTER
1500mm · 1250mm · 1000mm · A 252/8mmΦ MESH (BRC)
CABLES
BITUMAT · SAND FILL

SWITCH GEAR
150mm · 1250mm
BL. WALL PROTECTION TO W/P

DETAILS OF 11 KVA SUB STATION

- DOOR D → 2200 × 2300 mm
  TWO LEAF STEEL/ALUMN. WITH
  PAD LOCK + MASTER KEY
- DOOR D1 → 1100 × 2100 mm,
  SINGLE LEAVE STEEL FRAME
- DOOR D2 → 1200 × 2300 mm
  ALUMN. FRAME / STEEL DOOR
- FLOOR FINISH IN MEDIUM
  VOLTAGE ROOM TO BE GRANOLI-
  THIC 25/50mm OR EPOXY 1.5mm / 310mm
  AND SAND FILL IN TRANSFORMER ROOM
  TOP OF TRANSFORMER PLINTH 100m ABOVE FFL

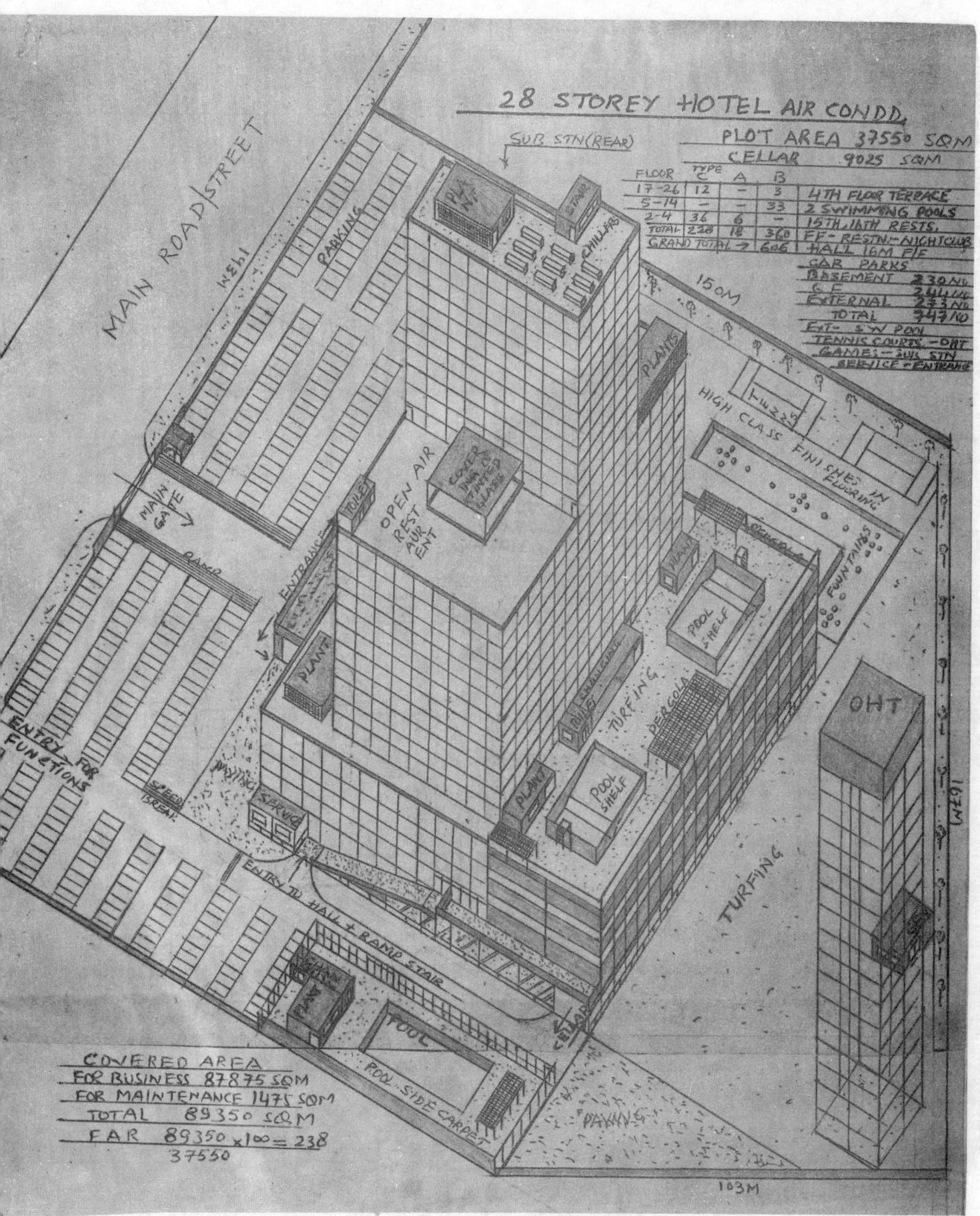

28 STOREY HOTEL AIR CONDD.

SUB STN (REAR)

PLOT AREA 37550 SQM
CELLAR 9025 SQM

FLOOR	TYPE C	A	B	
17-26	12	—	3	4TH FLOOR TERRACE
5-14	—	—	33	2 SWIMMING POOLS
2-4	36	6	—	15TH,16TH RESTS.
TOTAL	228	18	360	FF - RESTN:- NIGHTCLUB
GRAND TOTAL → 606				HALL 16M F/F

CAR PARKS
BASEMENT 230 NO
G.F. 244 NO
EXTERNAL 273 NO
TOTAL 747 NO
EXT:- S W POOL
TENNIS COURTS - OAT
GAMES - SUB STN
SERVICE - ENTRANCE

HIGH CLASS FINISHES IN FLOORING

150M

164M

103M

MAIN ROAD/STREET

PARKING

MAIN GATE

ENTRY FOR FUNCTIONS

SPEED BREAK

PLANT

TOILET

OPEN AIR RESTAURENT

COVERED REST.

CHILLERS

STAIR

PLANTS

PLANT

POOL SHELF

FOUNTAINS

PERGOLA

TURFING

POOL SHELF

OHT

ENTRY TO HALL + RAMP STAIR

SERVICE

CELLAR

PLANT

POOL

POOL SIDE CARPET

TURFING

PAVING

COVERED AREA
FOR BUSINESS 87875 SQM
FOR MAINTENANCE 1475 SQM
TOTAL 89350 SQM

$$FAR \frac{89350 \times 100}{37550} = 238$$

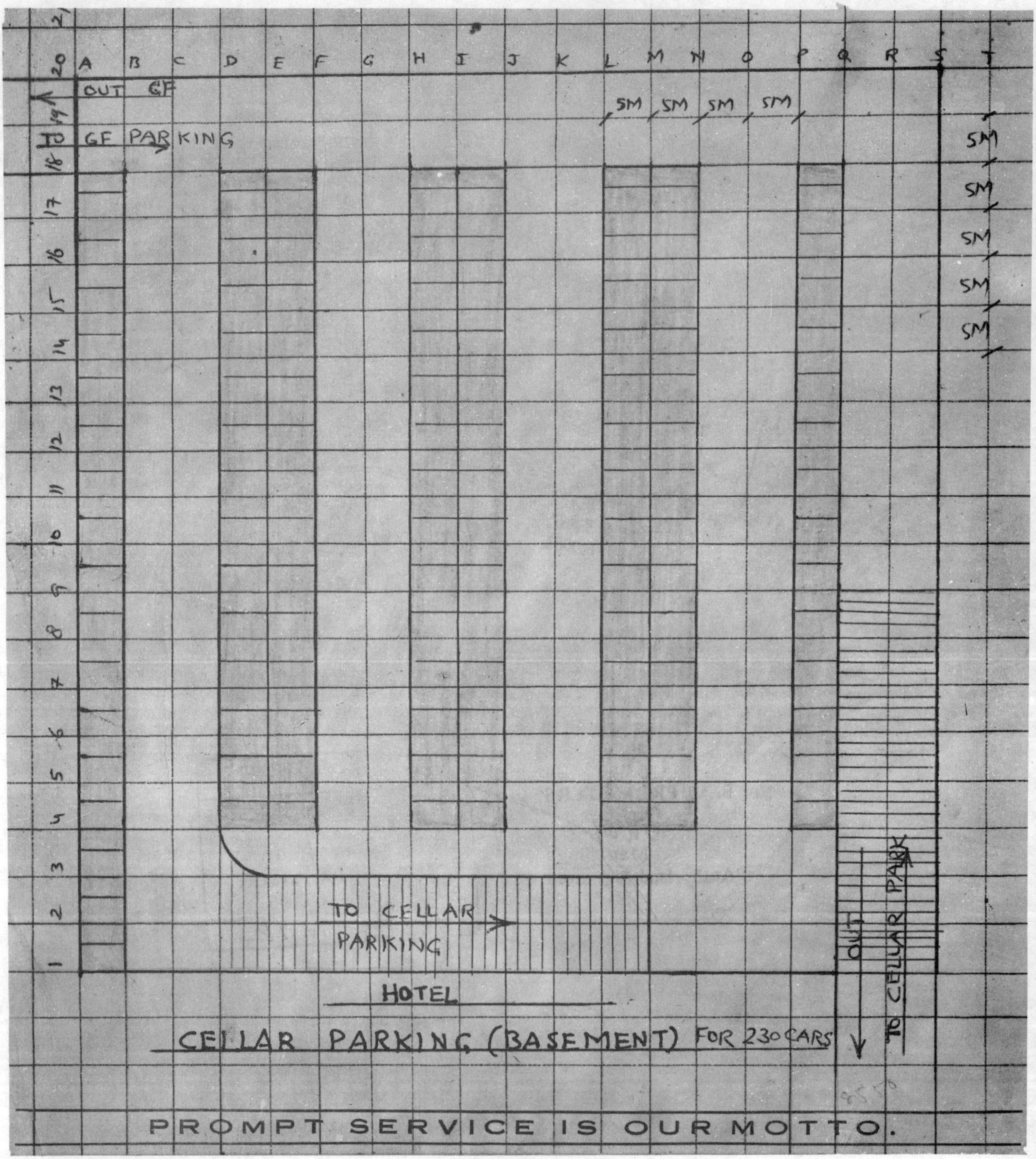

A B C D E F G H I J K L M N O P Q R S T
OUT GF
GF PARKING
5M 5M 5M 5M
5M
5M
5M
5M
5M
TO CELLAR PARKING
HOTEL
OUT
TO CELLAR PARK
CELLAR PARKING (BASEMENT) FOR 230 CARS
PROMPT SERVICE IS OUR MOTTO.

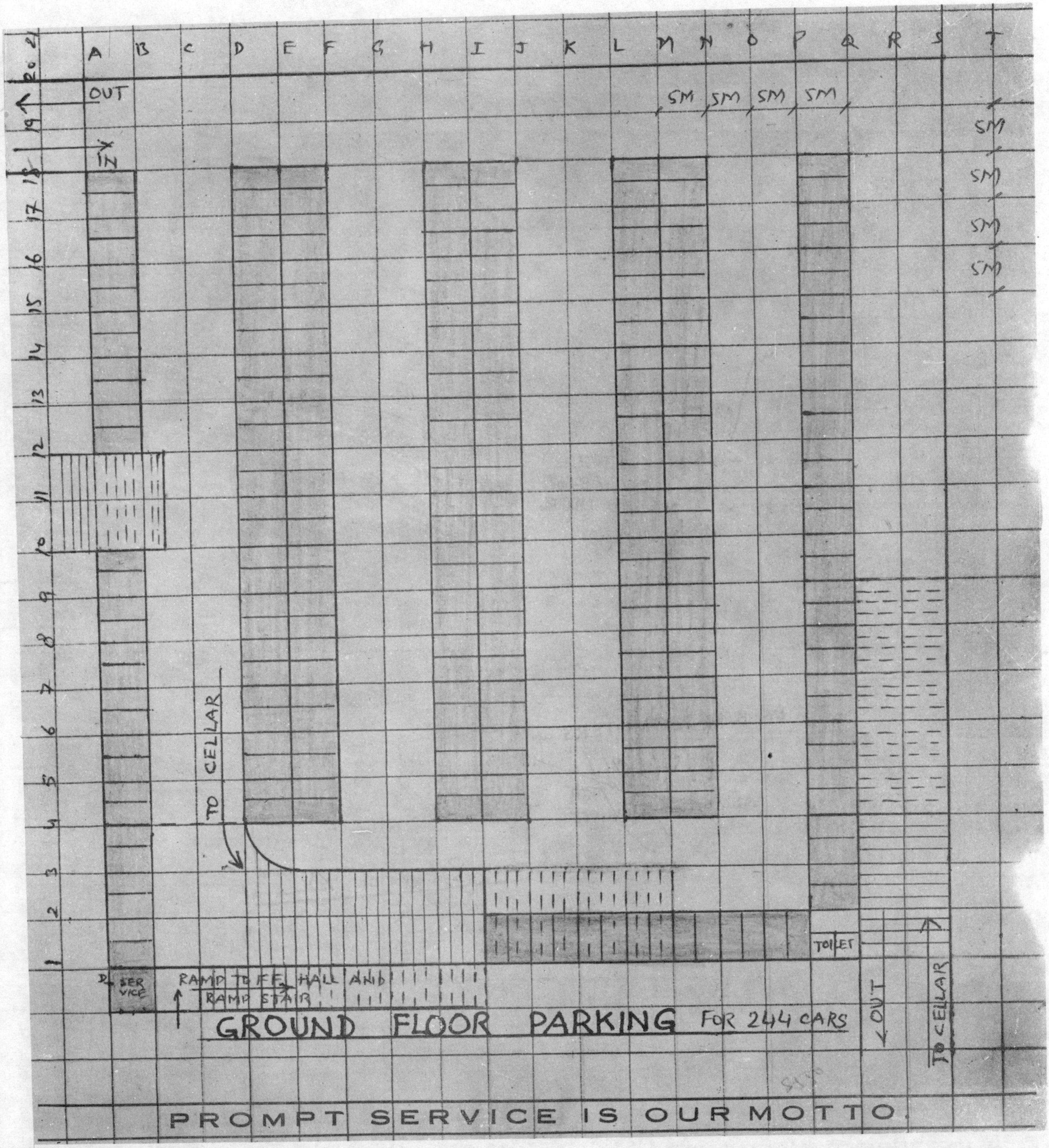

A B C D E F G H I J K L M N O P Q R S T

OUT

IN

SM SM SM SM

SM

SM

SM

SM

TO CELLAR

TOILET

OUT

TO CELLAR

SERVICE

RAMP TO F.F. HALL AND RAMP STAIR

GROUND FLOOR PARKING FOR 244 CARS

PROMPT SERVICE IS OUR MOTTO.

FIRST FLOOR PLAN

PROMPT SERVICE IS OUR MOTTO.

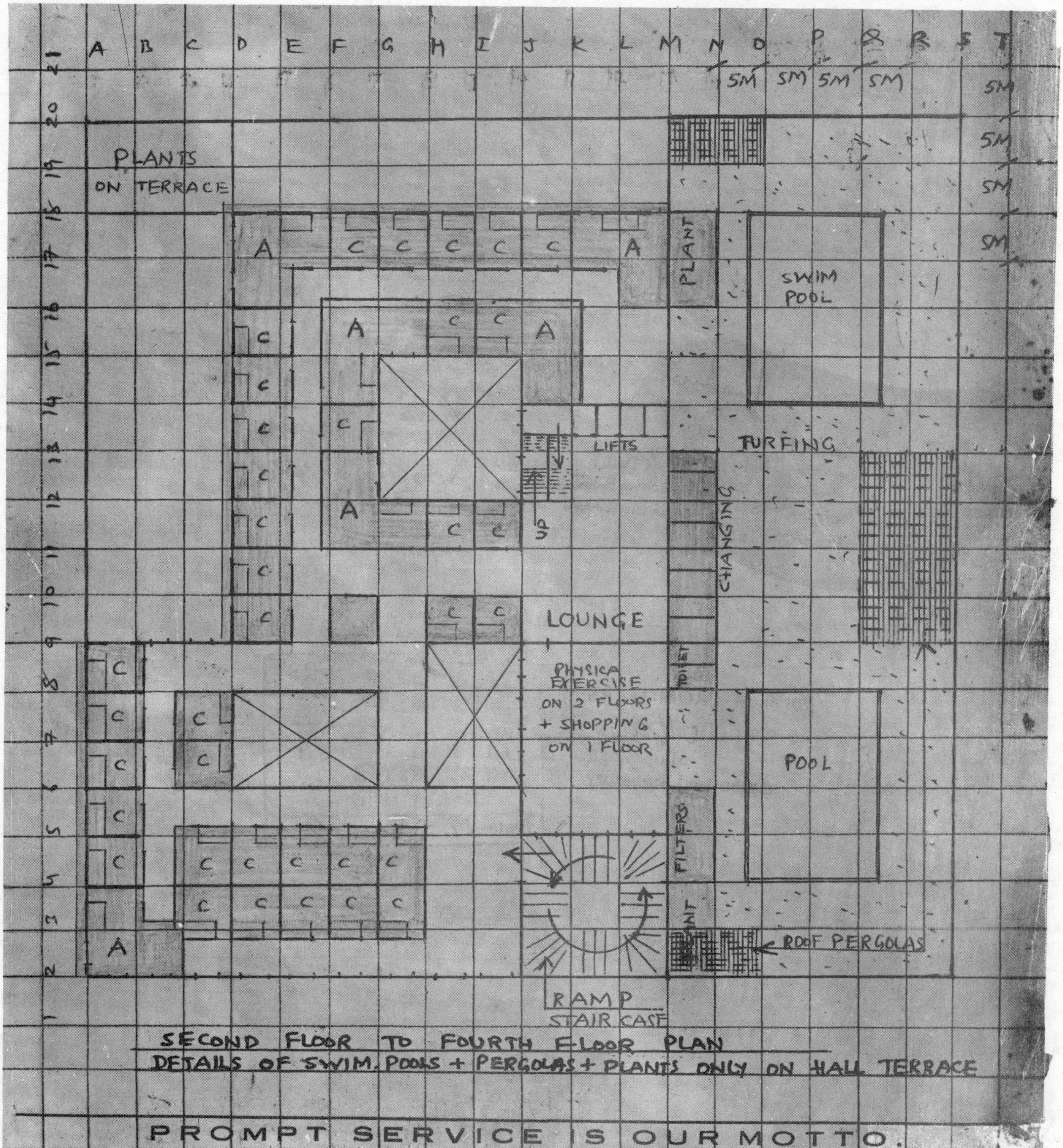
A B C D E F G H I J K L M N O P Q R S T
5M 5M 5M 5M
5M
5M
5M
5M
PLANTS
ON TERRACE
PLANT
SWIM
POOL
LIFTS
TURFING
CHANGING
UP
LOUNGE
PHYSICA
EXERCISE
ON 2 FLOORS
+ SHOPPING
ON 1 FLOOR
TOILET
POOL
FILTERS
ROOF PERGOLAS
RAMP
STAIR CASE
SECOND FLOOR TO FOURTH FLOOR PLAN
DETAILS OF SWIM. POOLS + PERGOLAS + PLANTS ONLY ON HALL TERRACE
PROMPT SERVICE IS OUR MOTTO.

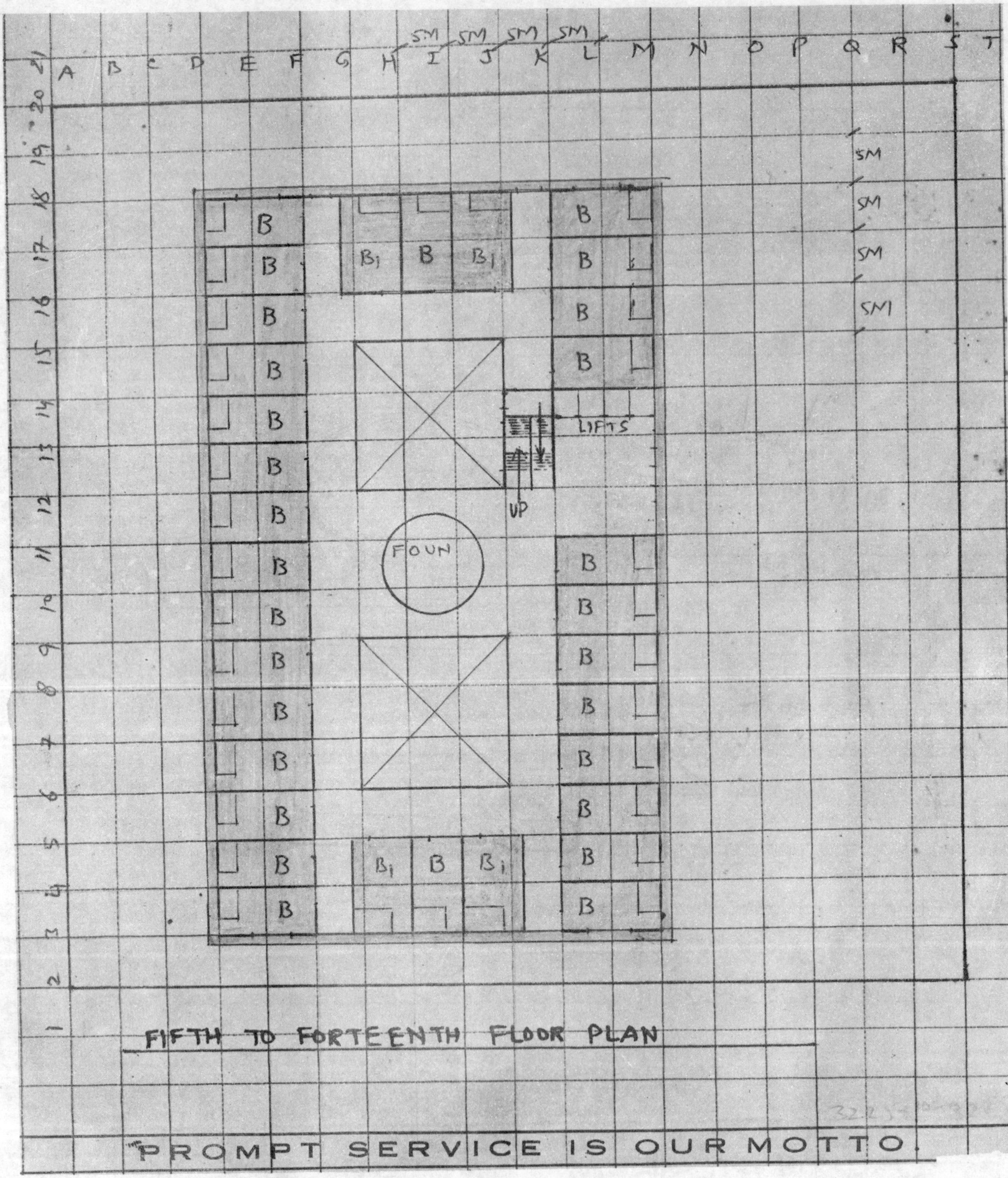
A B C D E F G H I J K L M N O P Q R S T
SM SM SM SM
SM SM SM SM
B
B₁ B B₁
LIFTS
UP
FOUN
FIFTH TO FORTEENTH FLOOR PLAN
PROMPT SERVICE IS OUR MOTTO.

FIFTEENTH FLOOR PLAN

PROMPT SERVICE IS OUR MOTTO.

SIXTEENTH FLOOR PLAN

PROMPT SERVICE IS OUR MOTTO.

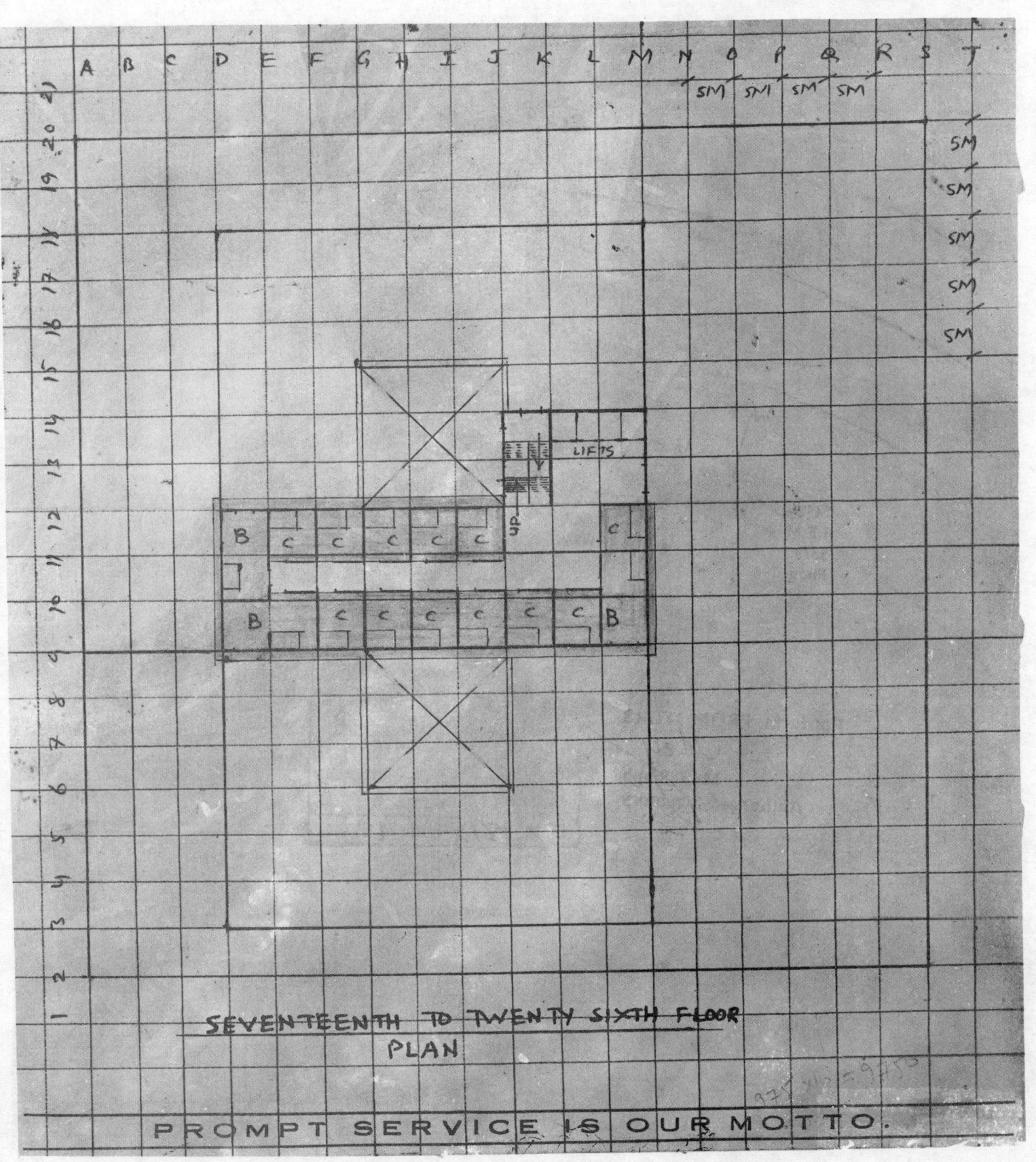

SEVENTEENTH TO TWENTY SIXTH FLOOR
PLAN

PROMPT SERVICE IS OUR MOTTO.

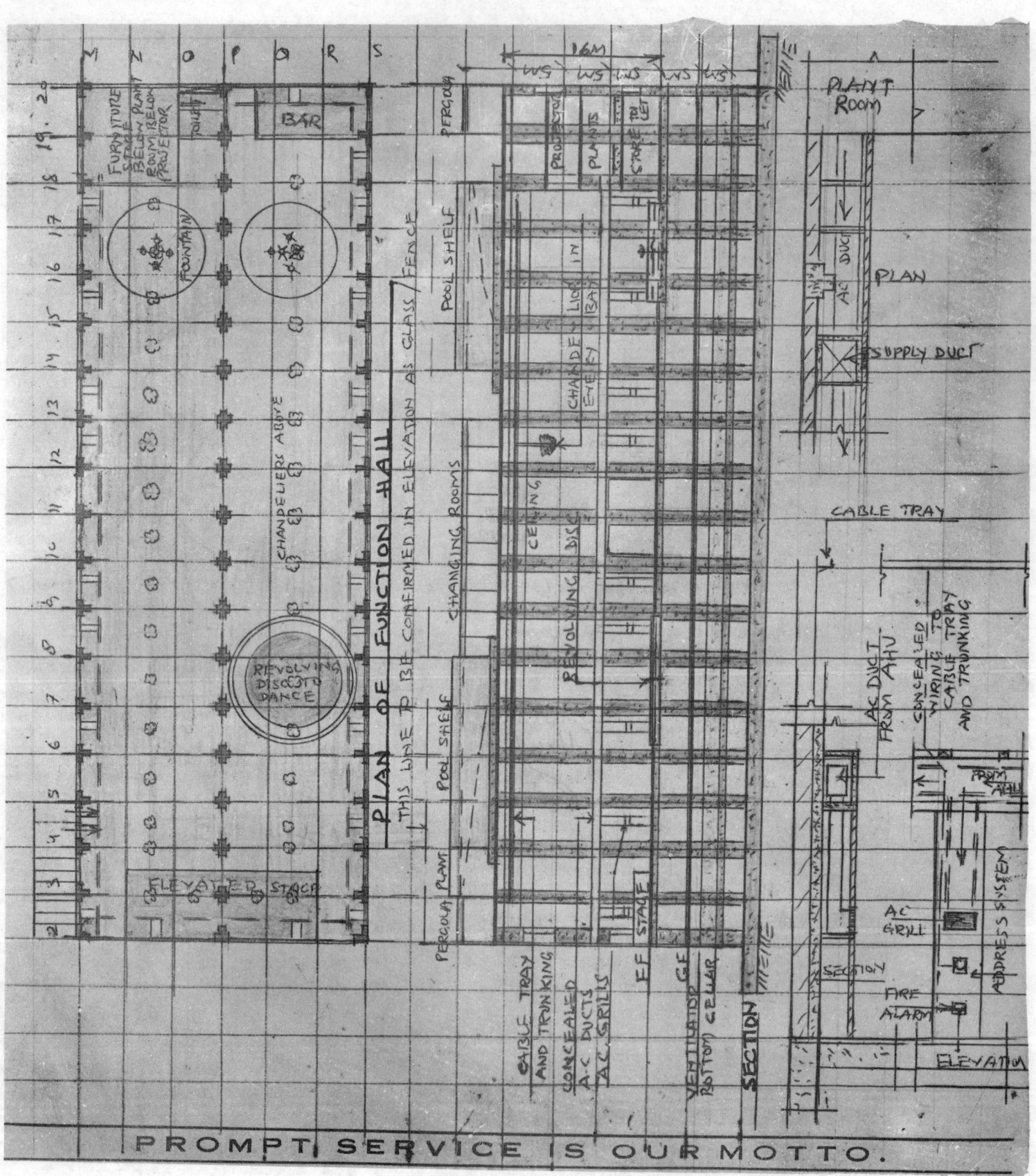

PLAN OF FUNCTION HALL

THIS LINE TO BE CONFIRMED IN ELEVATION AS GLASS/FENCE

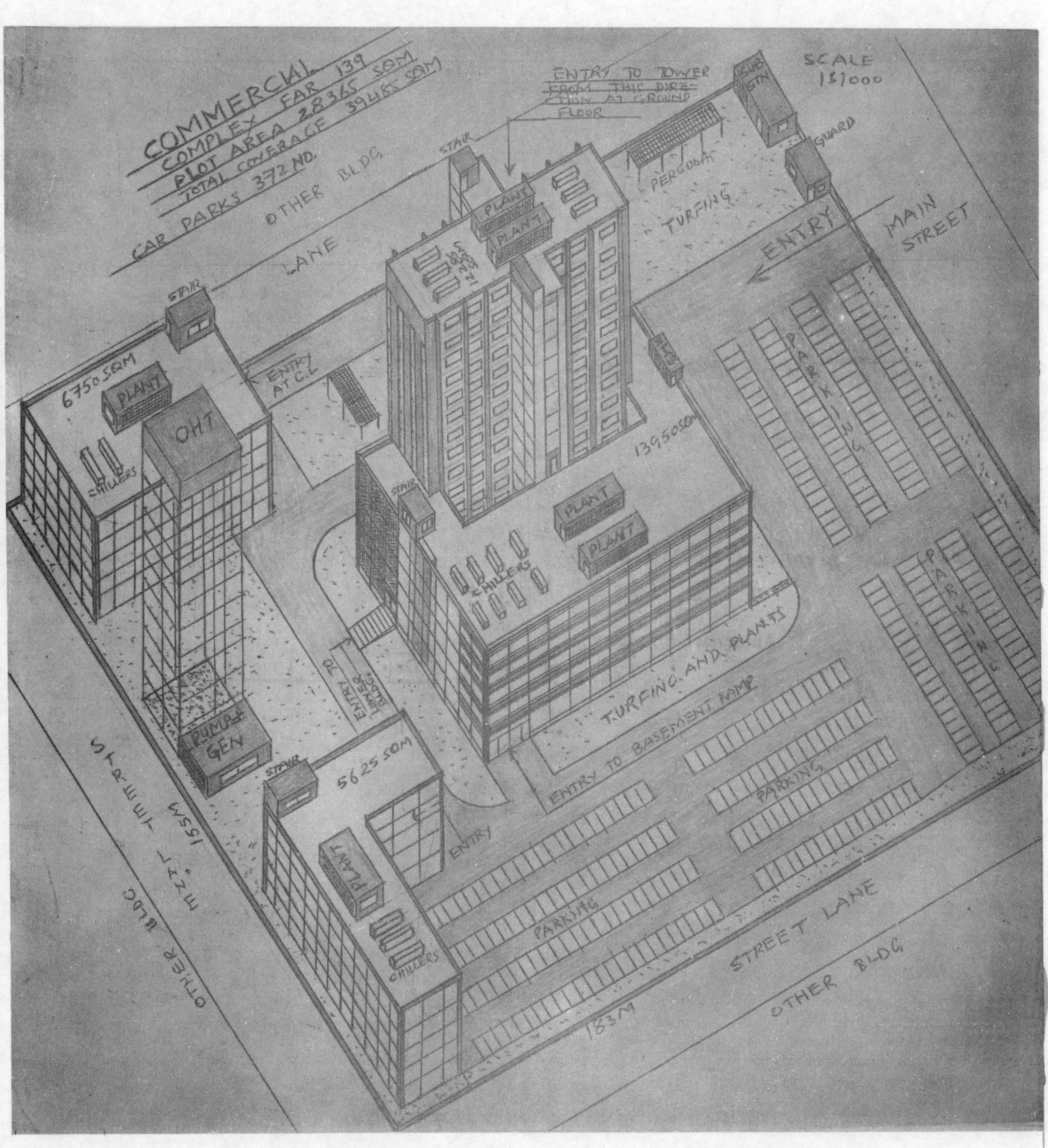

COMMERCIAL
COMPLEX FAR 139
PLOT AREA 28365 SQM
TOTAL COVERAGE 39465 SQM
CAR PARKS 372 NO.

OTHER BLDG.

LANE

ENTRY TO TOWER
FROM THIS DIRE-
CTION AT GROUND
FLOOR

SUB
STN

SCALE
1:1000

STAIR

PLANT
PLANT

PERGOLA

GUARD

TURFING

ENTRY

MAIN
STREET

STAIR

6750 SQM

PLANT

OHT

CHILLERS

ENTRY
AT G.L.

STAIR

13950 SQM

PARKING

PUMP &
GEN.

STAIR

ENTRY TO
TOWER BLDG.

PLANT
PLANT

CHILLERS

PARKING

STREET LINE 155M

5625 SQM

PLANT

ENTRY

TURFING AND PLANTS

ENTRY TO BASEMENT RAMP

PARKING

OTHER BLDG

CHILLERS

PARKING

STREET LANE

183M

OTHER BLDG

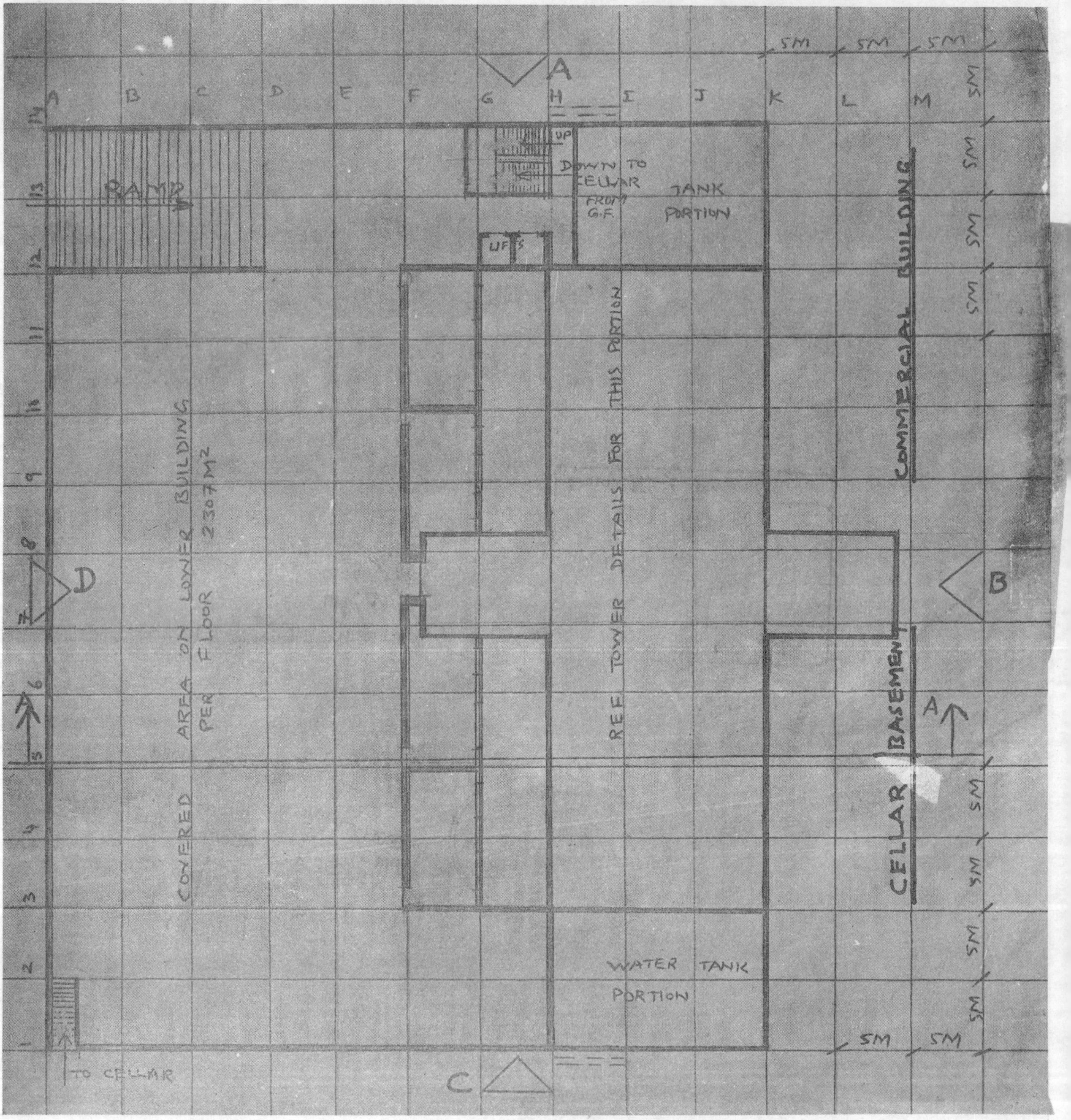

5M
5M
5M
A
A B C D E F G H I J K L M
RAMP
UP
DOWN TO CELLAR FROM G.F.
TANK PORTION
LIFTS
COMMERCIAL BUILDING
COVERED AREA ON LOWER BUILDING PER FLOOR 2307 M²
REF TOWER DETAILS FOR THIS PORTION
D
B
A
CELLAR BASEMENT
WATER TANK PORTION
TO CELLAR
C
1 2 3 4 5 6 7 8 9 10 11 12 13 14

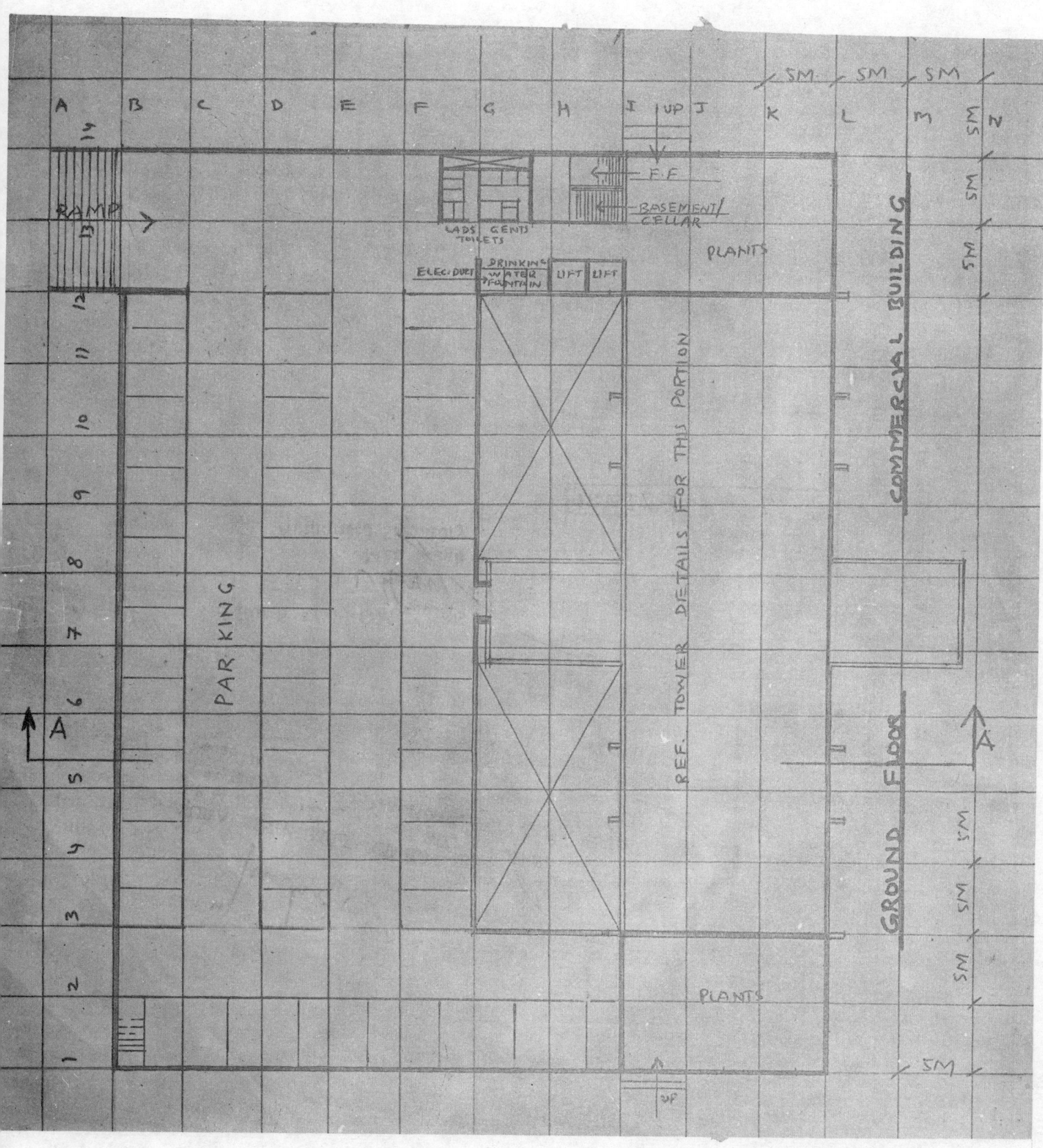

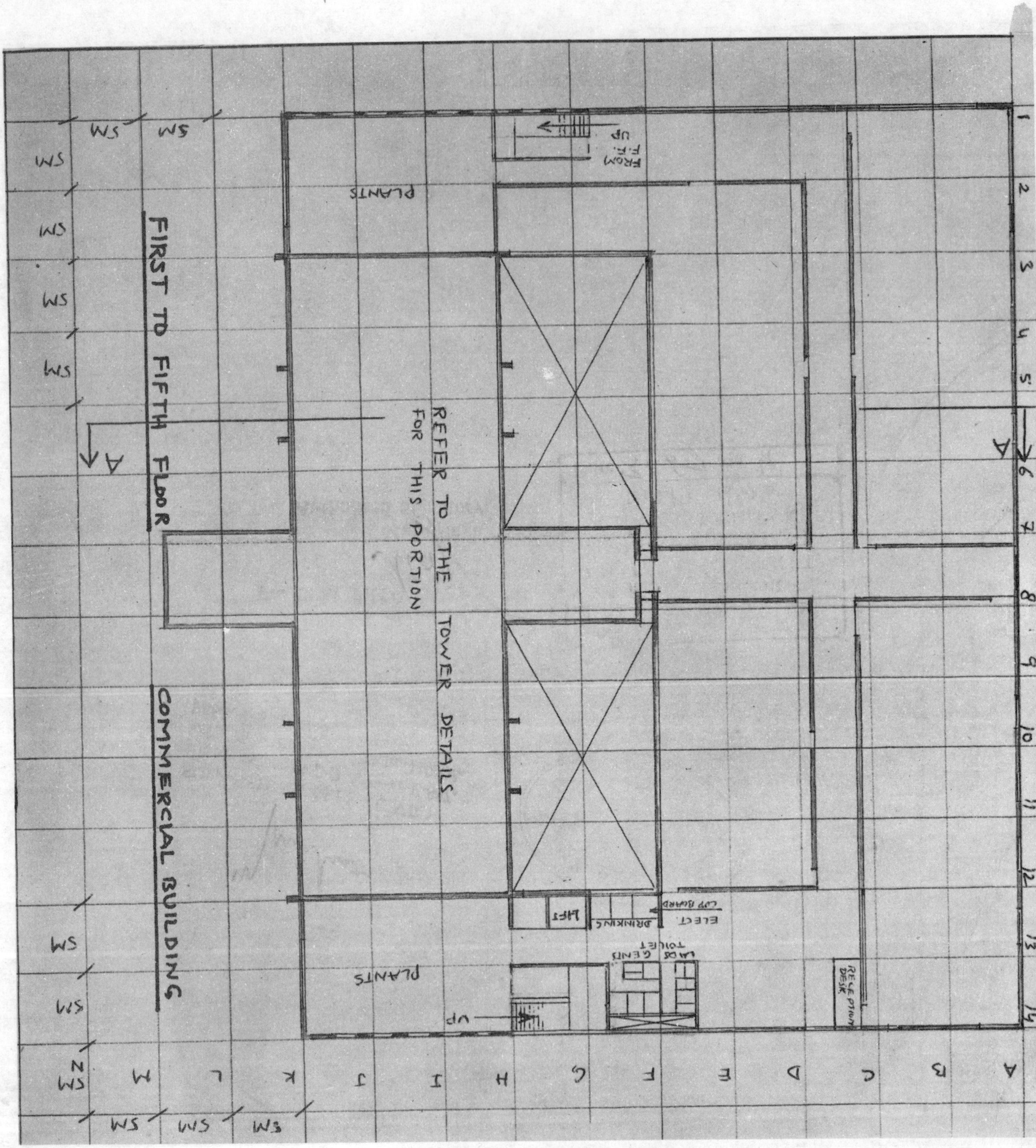

FIRST TO FIFTH FLOOR
COMMERCIAL BUILDING
REFER TO THE TOWER DETAILS
FOR THIS PORTION
PLANTS
PLANTS
FROM F.F. UP
UP
LIFT
DRINKING
ELECT. CUP BOARD
LADS GENTS TOILET
RECEPTION DESK
A
A
WS
SM
A B C D E F G H I J K L M N
1 2 3 4 5 6 7 8 9 10 11 12 13 14

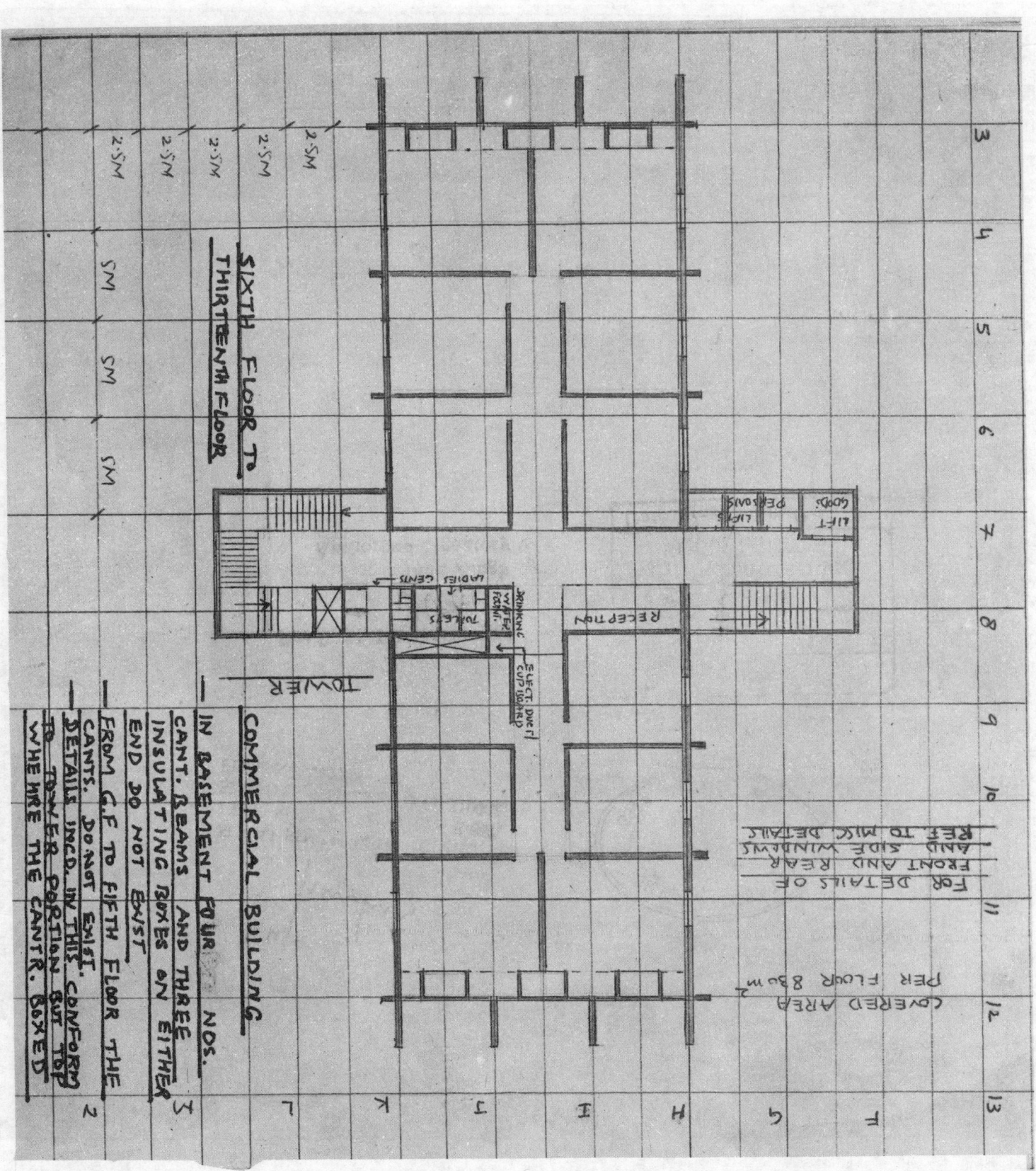

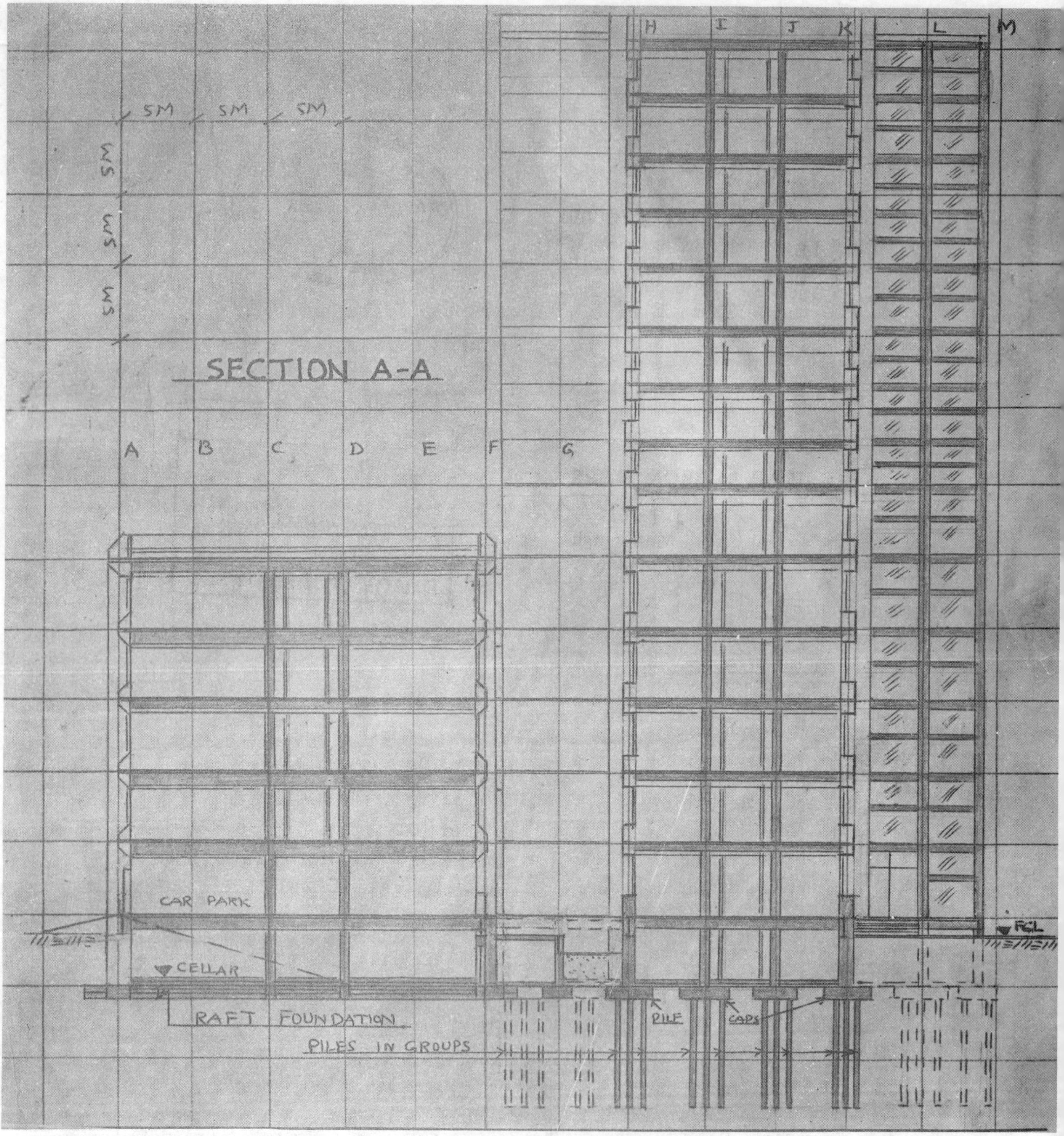

SM SM SM

SM

SM

SM

SM

SECTION A-A

A   B   C   D   E   F   G   H   I   J   K   L   M

CAR PARK

CELLAR

RAFT FOUNDATION

PILES IN GROUPS

PILE CAPS

RCL

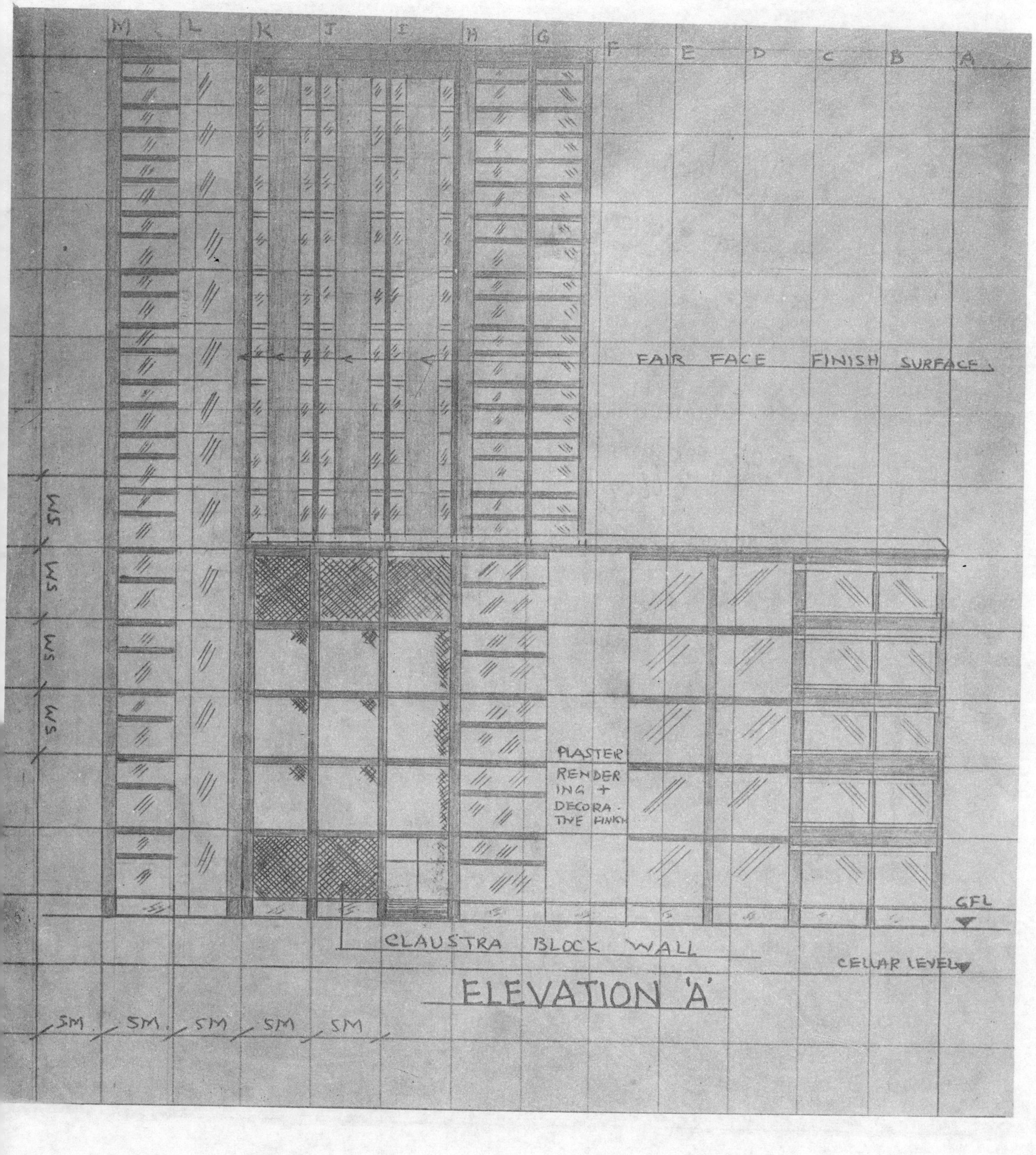

M L K J I H G
F E D C B A
FAIR FACE FINISH SURFACE
PLASTER RENDER ING + DECORA-TIVE FINISH
CLAUSTRA BLOCK WALL
GFL
CELLAR LEVEL
5M 5M 5M 5M
5M 5M 5M 5M 5M

ELEVATION 'A'

ELEVATION 'B'

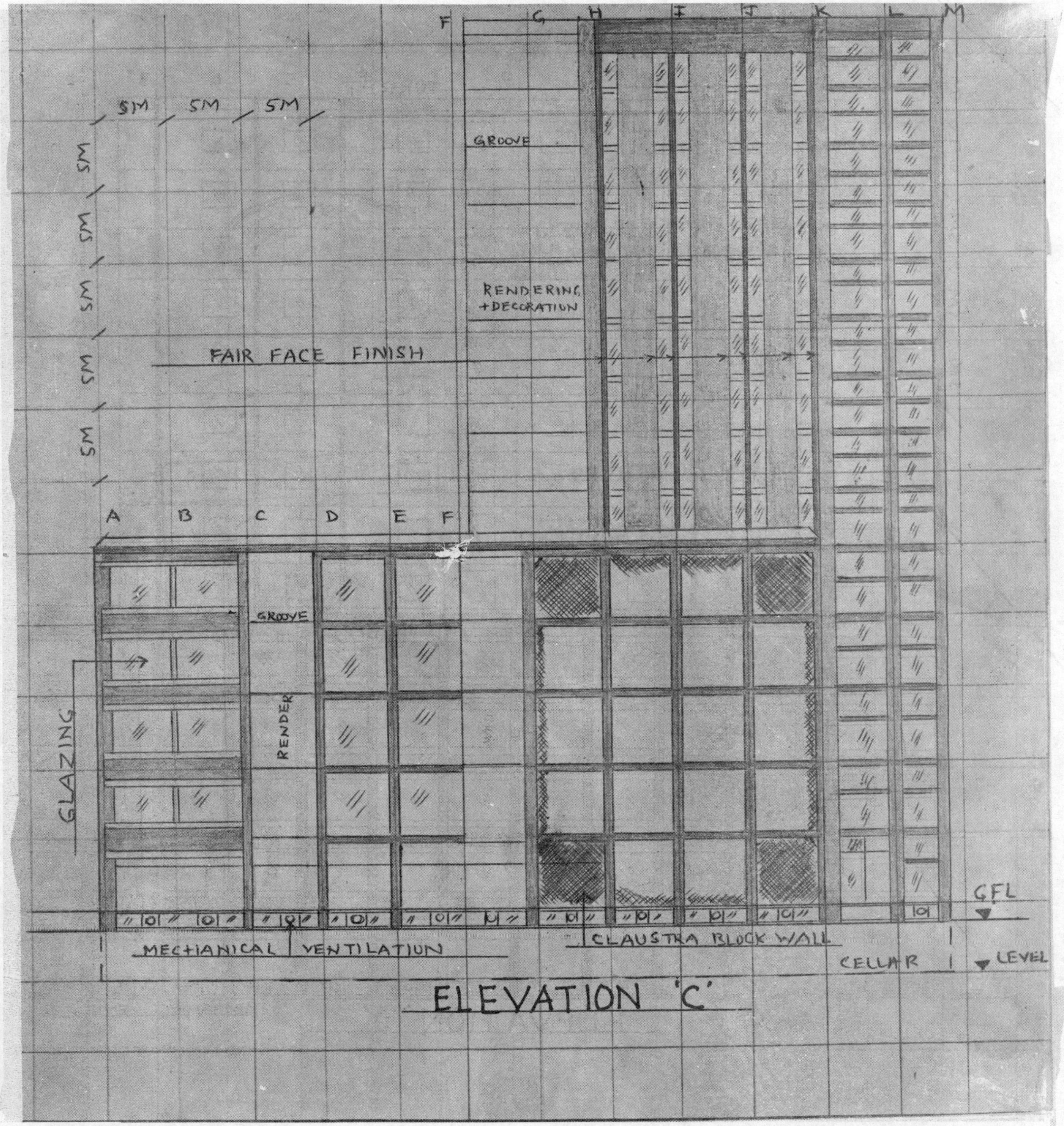
F
G
H
I
J
K
L
M
5M
5M
5M
5M
5M
5M
5M
5M
GROOVE
RENDERING
+DECORATION
FAIR FACE FINISH
A
B
C
D
E
F
GROOVE
GLAZING
RENDER
GFL
MECHANICAL VENTILATION
CLAUSTRA BLOCK WALL
CELLAR
LEVEL
ELEVATION 'C'

ELEVATION 'D'

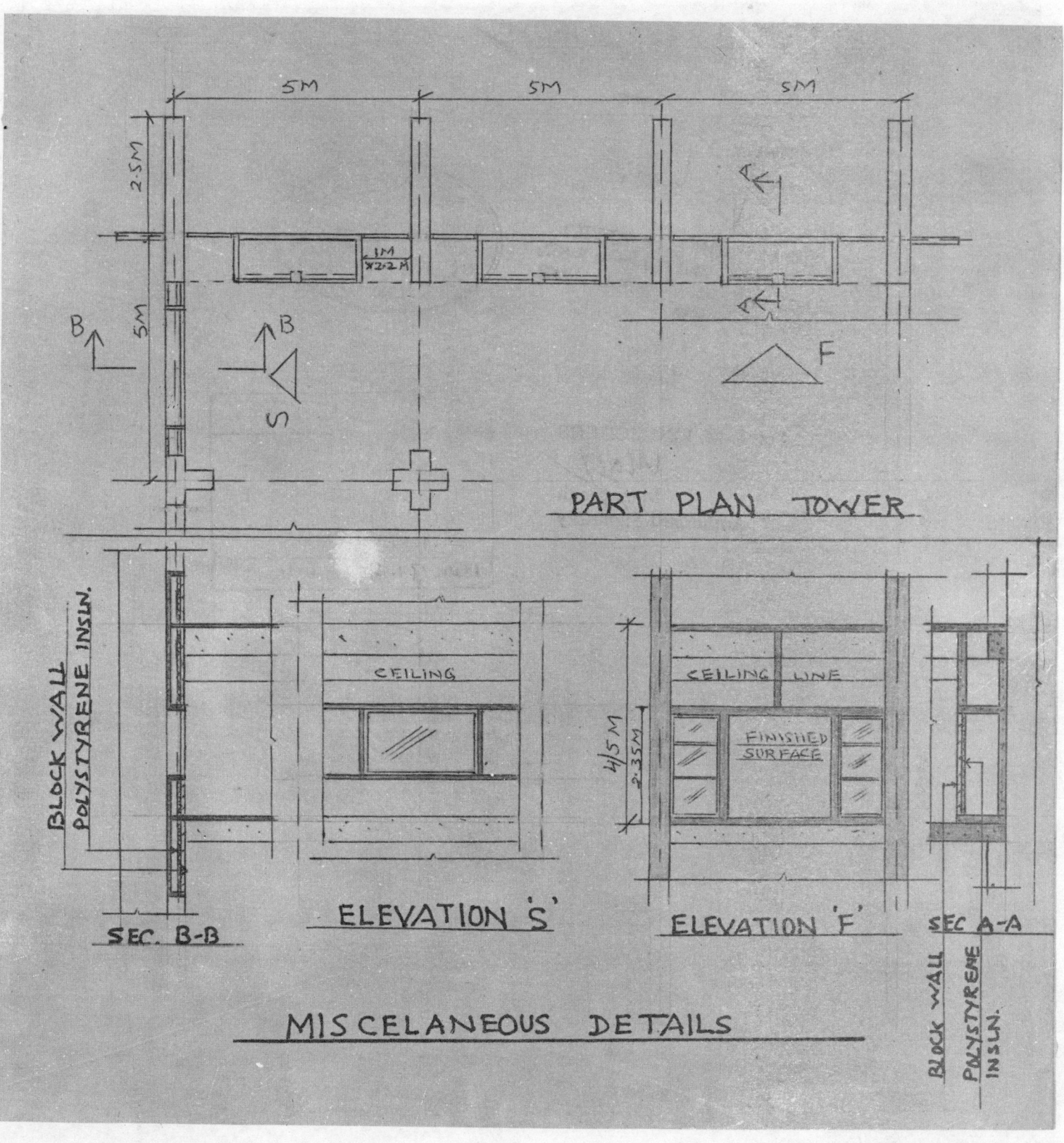

PART PLAN TOWER

SEC. B-B

ELEVATION 'S'

ELEVATION 'F'

SEC A-A

MISCELANEOUS DETAILS

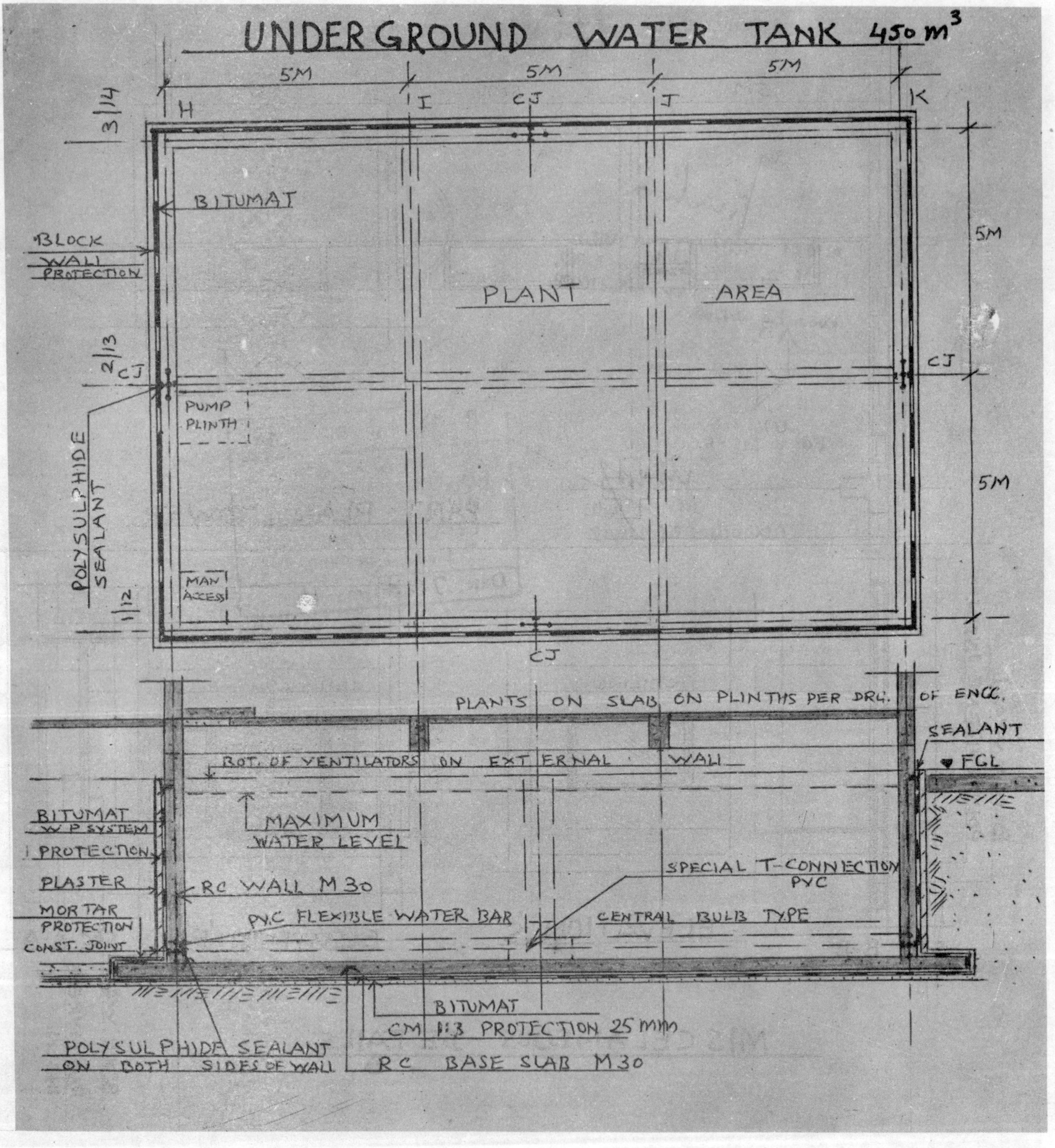

UNDERGROUND WATER TANK 450 m³

PLANT   AREA

BITUMAT

BLOCK WALL PROTECTION

POLYSULPHIDE SEALANT

PUMP PLINTH

MAN ACCESS

CJ

CJ

CJ

CJ

5M   5M   5M

3/14

2/13

1/12

5M

5M

H   I   CJ   J   K

PLANTS ON SLAB ON PLINTHS PER DRG. OF ENCC.

SEALANT

FGL

BOT. OF VENTILATORS ON EXTERNAL WALL

MAXIMUM WATER LEVEL

BITUMAT W P SYSTEM

PROTECTION

PLASTER

MORTAR PROTECTION

CONST. JOINT

RC WALL M 30

PVC FLEXIBLE WATER BAR

SPECIAL T-CONNECTION PVC

CENTRAL BULB TYPE

POLYSULPHIDE SEALANT ON BOTH SIDES OF WALL

BITUMAT CM 1:3 PROTECTION 25 mm

RC BASE SLAB M30

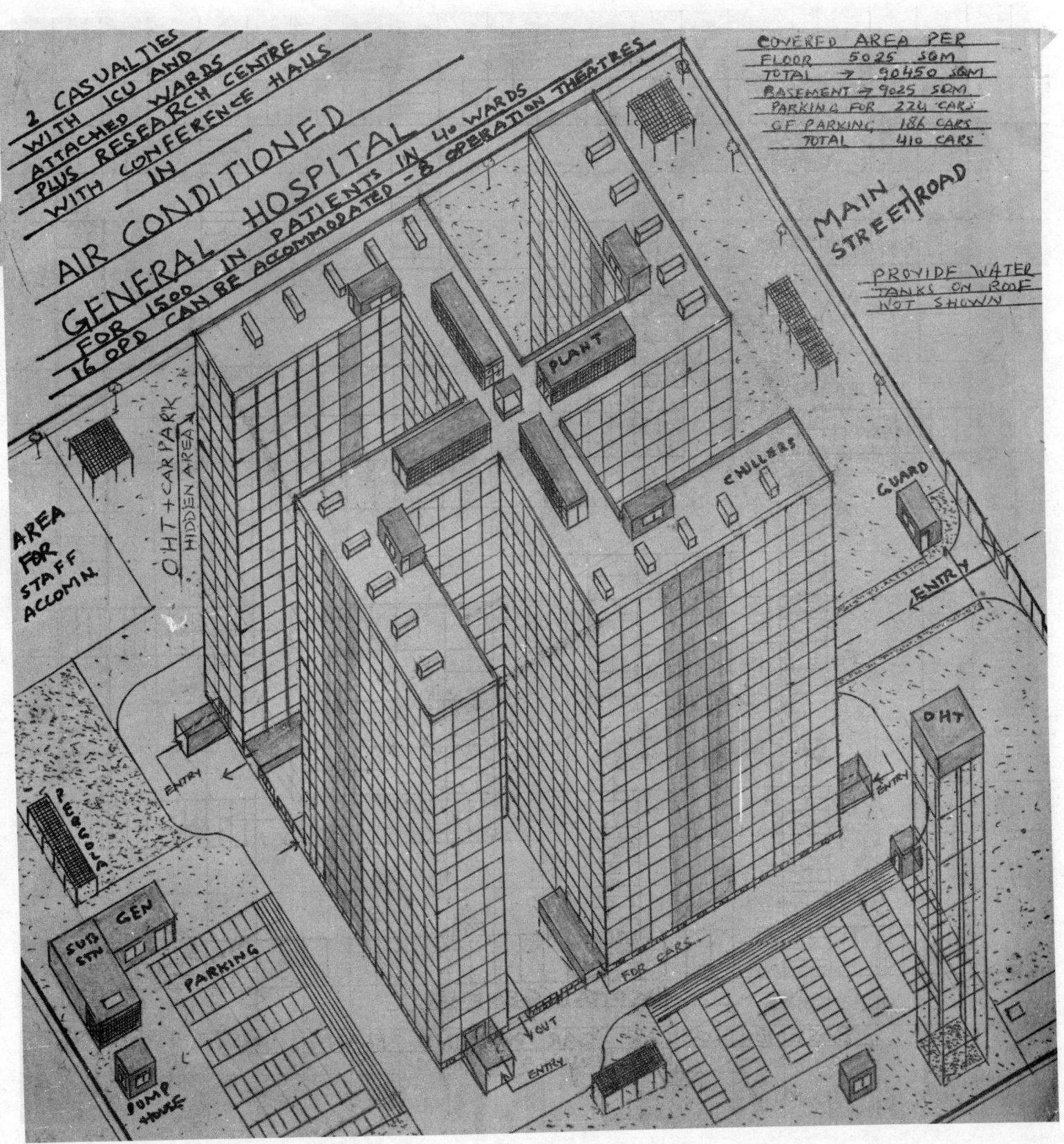

2 CASUALTIES WITH ICU AND ATTACHED WARDS PLUS RESEARCH CENTRE WITH CONFERENCE HALLS IN AIR CONDITIONED GENERAL HOSPITAL FOR 1500 IN PATIENTS IN 40 WARDS 16 OPD CAN BE ACCOMMODATED - 8 OPERATION THEATRES

COVERED AREA PER
FLOOR 5025 SQM
TOTAL → 90450 SQM
BASEMENT → 9025 SQM
PARKING FOR 224 CARS
GF PARKING 186 CARS
TOTAL 410 CARS

MAIN STREET/ROAD

PROVIDE WATER TANKS ON ROOF NOT SHOWN

AREA FOR STAFF ACCOMN.

OHT + CAR PARK HIDDEN AREA

PLANT

CHILLERS

GUARD

ENTRY

ENTRY

OHT

PERGOLA

SUB STN

GEN

PARKING

PUMP HOUSE

ENTRY

OUT

ENTRY

FOR CARS

HOSPITAL (GENERAL)

BASEMENT/CELLAR CAR PARKING PLAN — 9025 SQM COVERED AREA
FOR 224 CARS MINM.

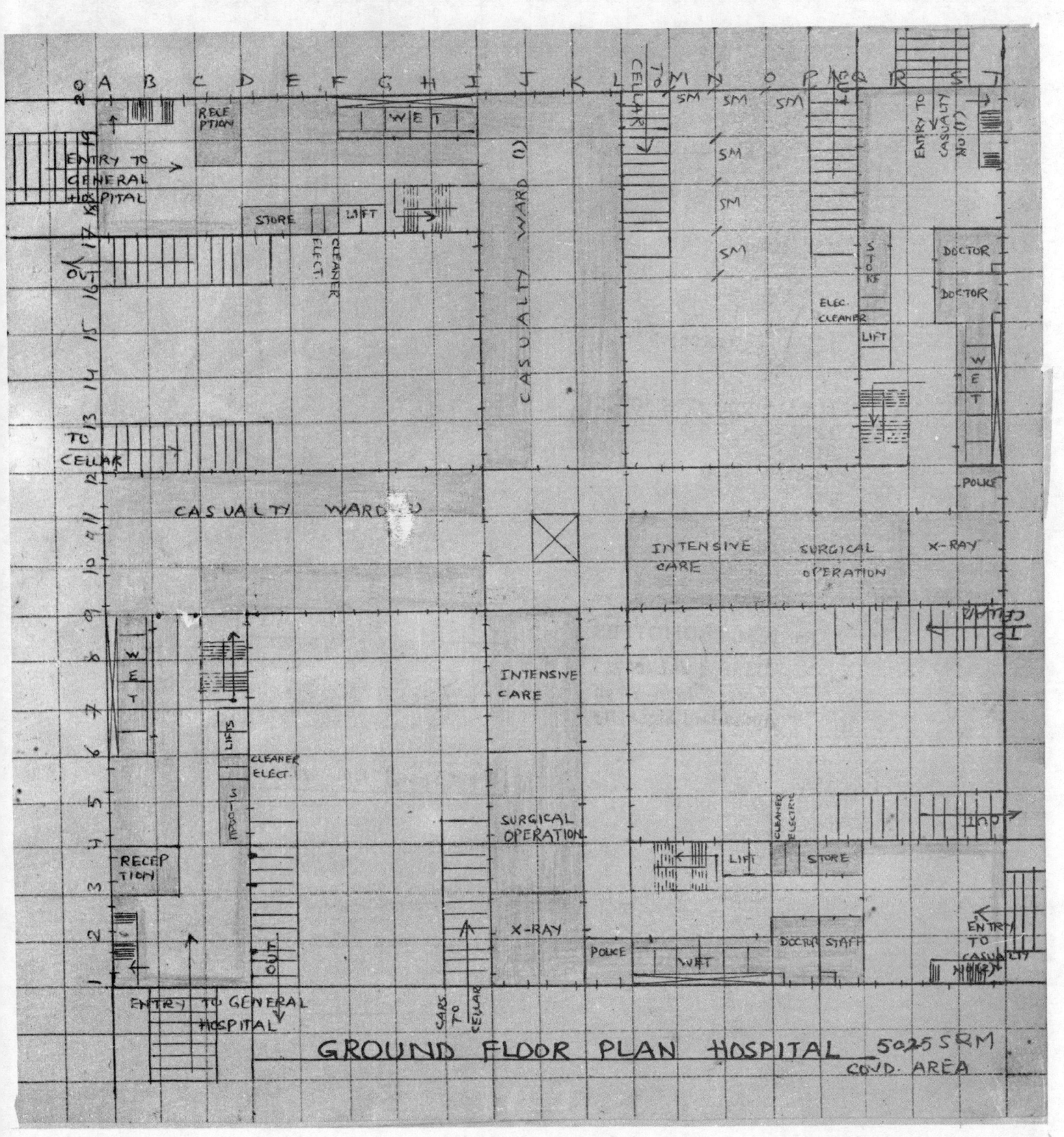

GROUND FLOOR PLAN HOSPITAL 5025 SRM COVD. AREA

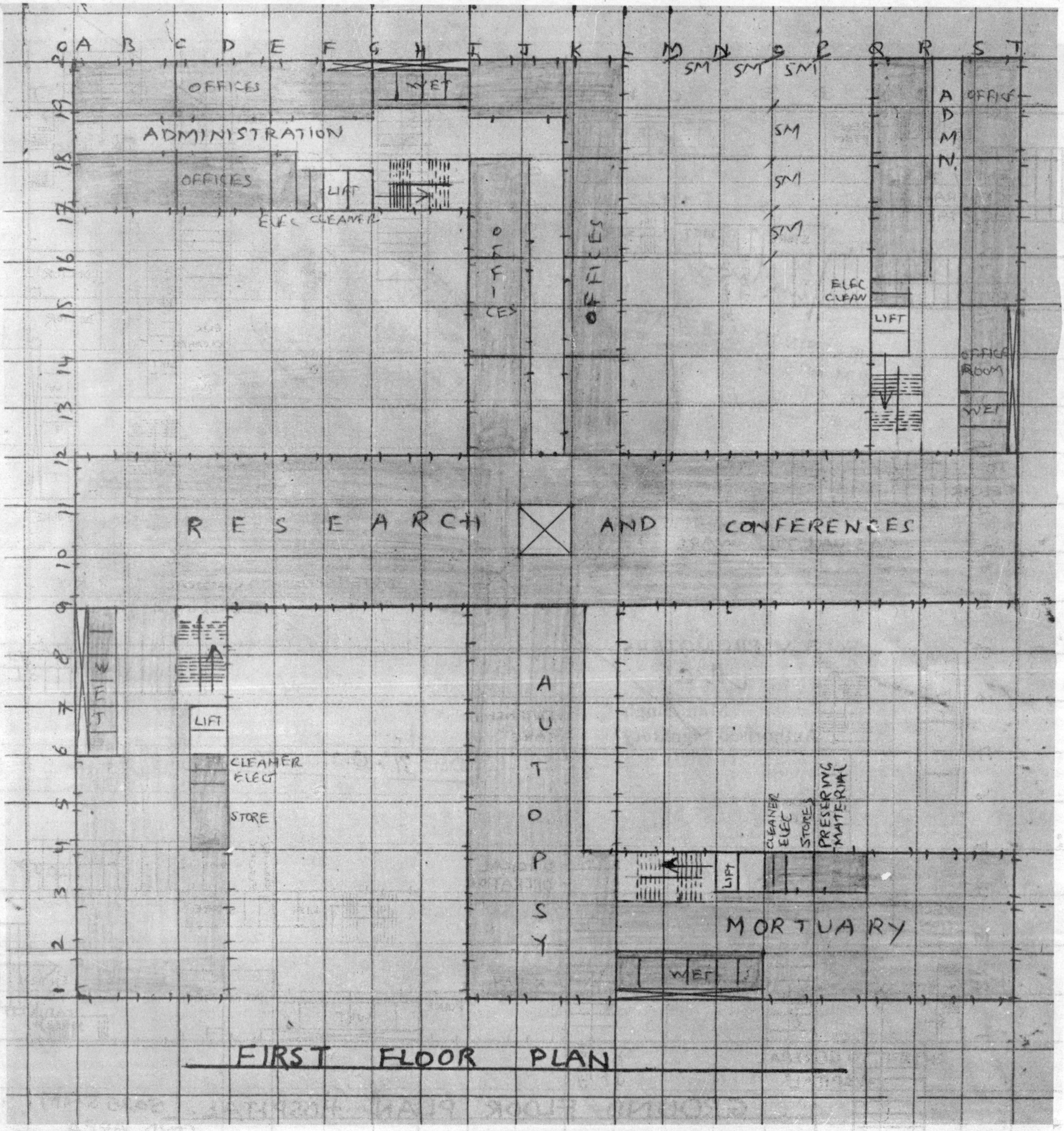

OFFICES

WET

ADMINISTRATION

OFFICES

LIFT

ELEC CLEANER

SM   SM   SM

SM

SM

SM

ADMIN.

OFFICE

OFFICES

OFFICES

ELEC CLEAN

LIFT

OFFICE ROOM

WET

RESEARCH   AND   CONFERENCES

WET

LIFT

CLEANER ELEC

STORE

AUTOPSY

LIFT

CLEANER ELEC

STORES

PRESERVING MATERIAL

MORTUARY

WET

FIRST FLOOR PLAN

TYPICAL OUT PATIENT DEPTT. PLAN (4 FLOORS)

DEPTTS — GYNA - ORTHO - PADEAT - OBSTER - ENT - DORMIT - TB - HEART - CANCER
PHY. HANDD - PARALYSIS - DENTRY - OPTHA

PLAN FOR SURGICAL OPERATION THEATRES

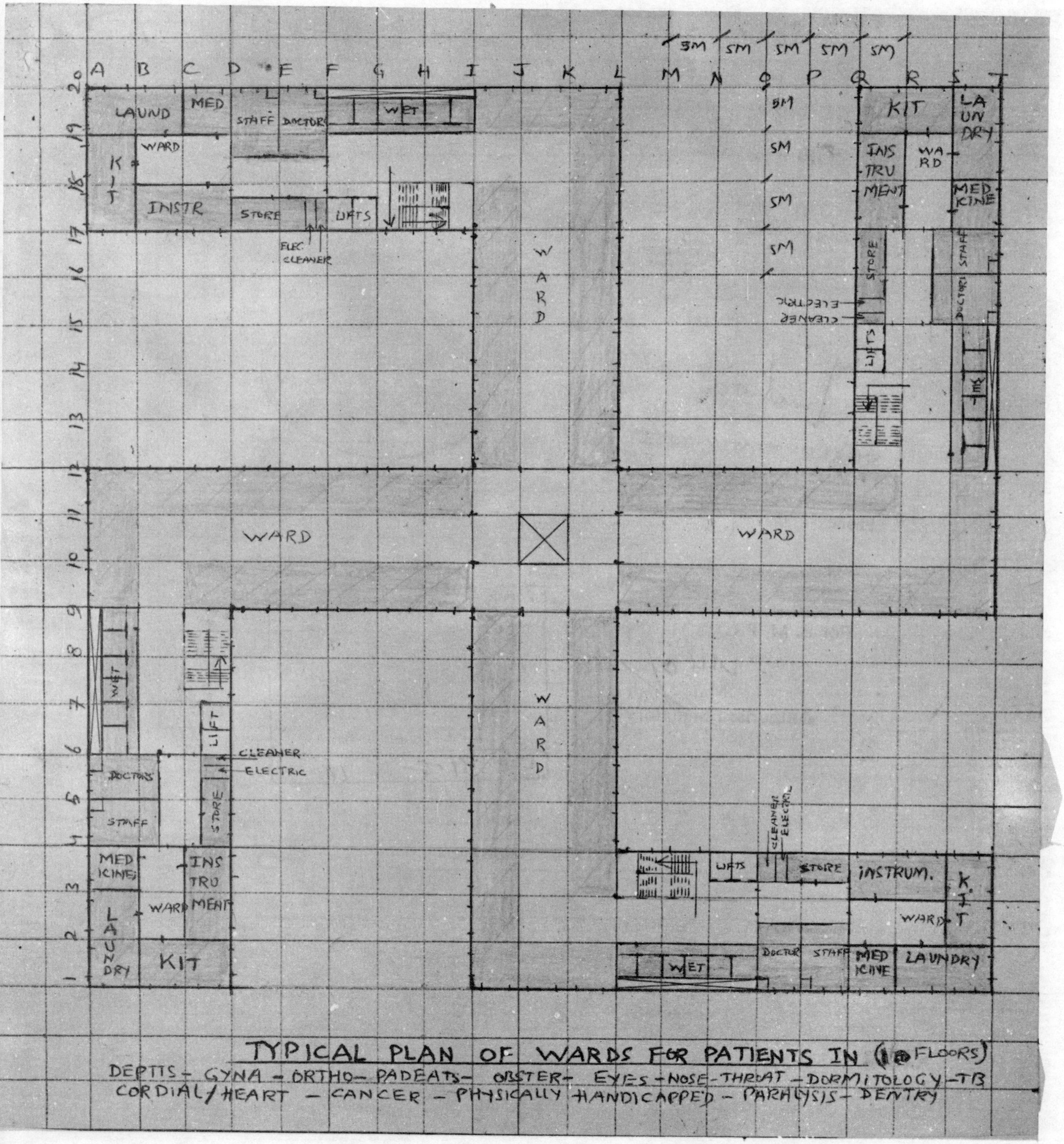

A B C D E F G H I J K L M N O P Q R S T

3M 5M 5M 5M 5M

LAUND MED KIT LA UN DRY

STAFF DOCTOR WET

WARD 5M INS TRU MENT WA RD

KIT 5M

INSTR STORE LIFTS 5M MED ICINE

ELEC. CLEANER 5M

W A R D STORE STAFF

CLEANER ELECTRIC DOCTORS

LIFTS

WARD WARD

WET

LIFT

CLEANER ELECTRIC

STORE

CLEANER ELECTRIC

LIFTS STORE INSTRUM. K I T

W A R D

INS TRU MENT WARD

MED ICINES

DOCTORS

STAFF WARD

LA UN DRY KIT DOCTOR STAFF MED ICINE LAUNDRY

WET

TYPICAL PLAN OF WARDS FOR PATIENTS IN (10 FLOORS)
DEPTTS- GYNA - ORTHO- PADEATS- OBSTER- EYES -NOSE- THROAT -DORMITOLOGY -TB
CORDIAL/HEART - CANCER - PHYSICALLY HANDICAPPED - PARALYSIS- DENTRY

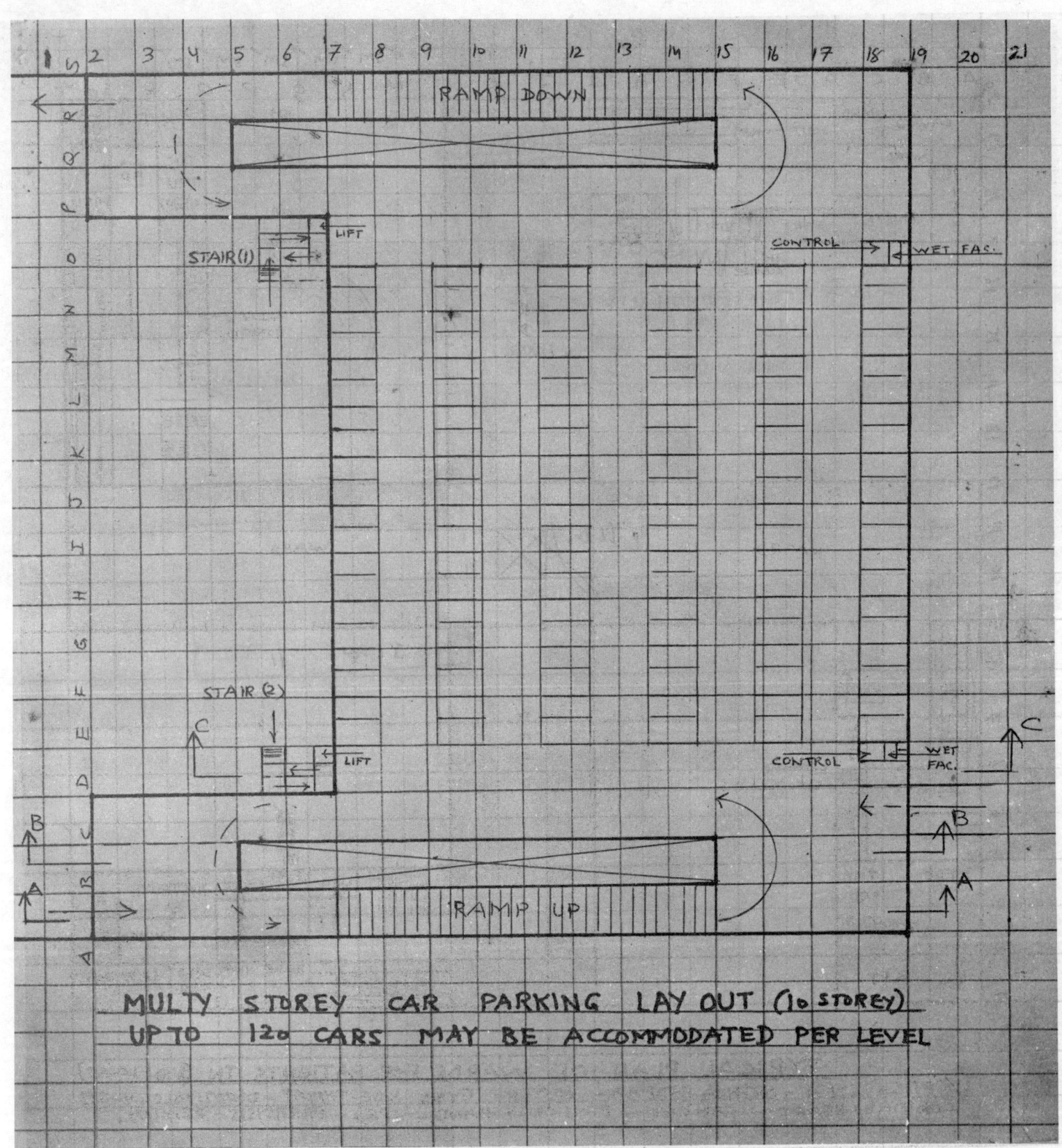

MULTY STOREY CAR PARKING LAY OUT (10 STOREY)
UPTO 120 CARS MAY BE ACCOMMODATED PER LEVEL

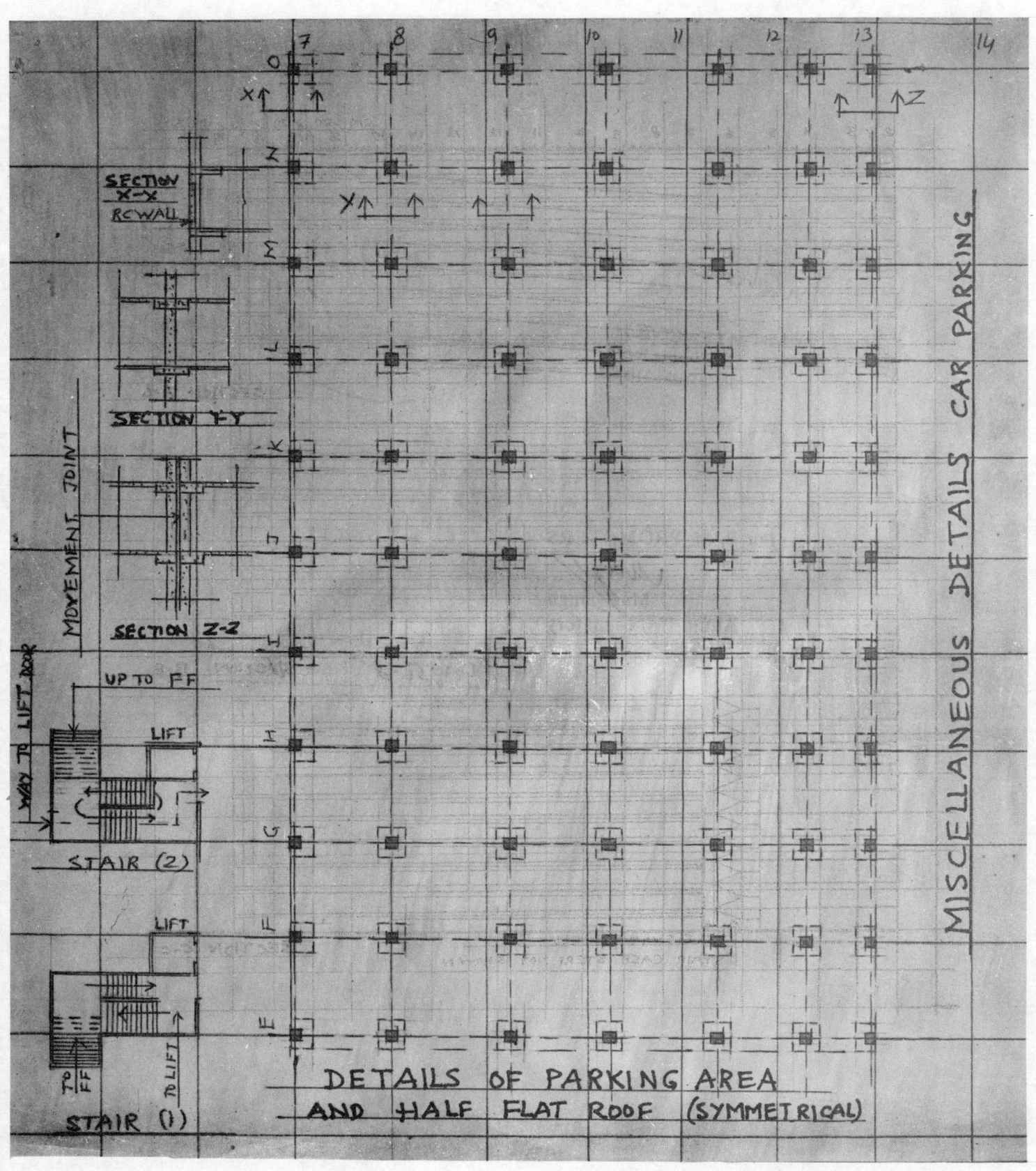

SECTION X-X
RC WALL

SECTION Y-Y

SECTION Z-Z

MOVEMENT JOINT

UP TO FF

LIFT

WAY TO LIFT ROOM

STAIR (2)

LIFT

TO FF

TO LIFT

STAIR (1)

DETAILS OF PARKING AREA
AND HALF FLAT ROOF (SYMMETRICAL)

MISCELLANEOUS DETAILS CAR PARKING

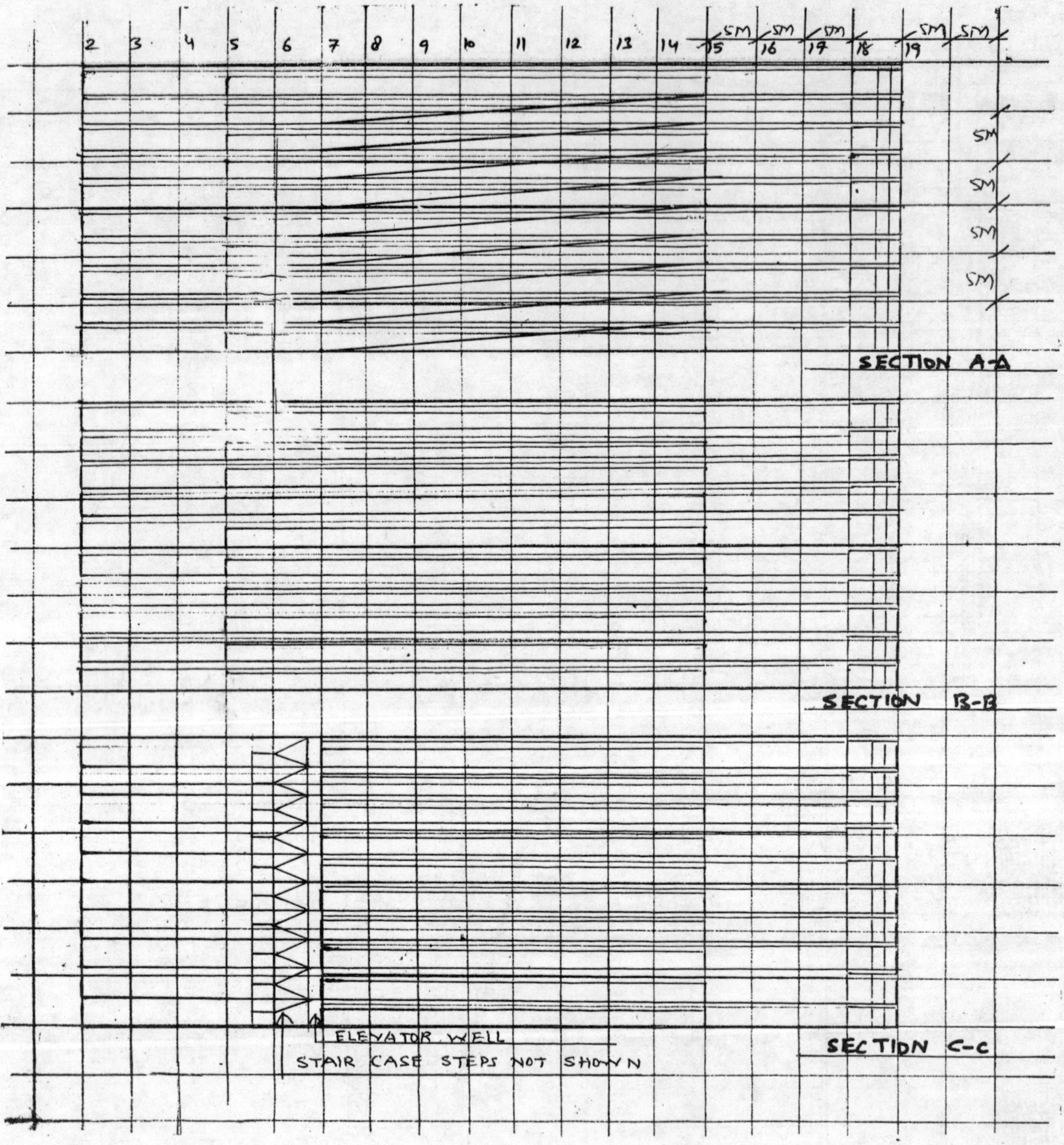
2 3 4 5 6 7 8 9 10 11 12 13 14 15 16 17 18 19
5M 5M 5M 5M 5M
5M
5M
5M
5M
SECTION A-A
SECTION B-B
ELEVATOR WELL
STAIR CASE STEPS NOT SHOWN
SECTION C-C

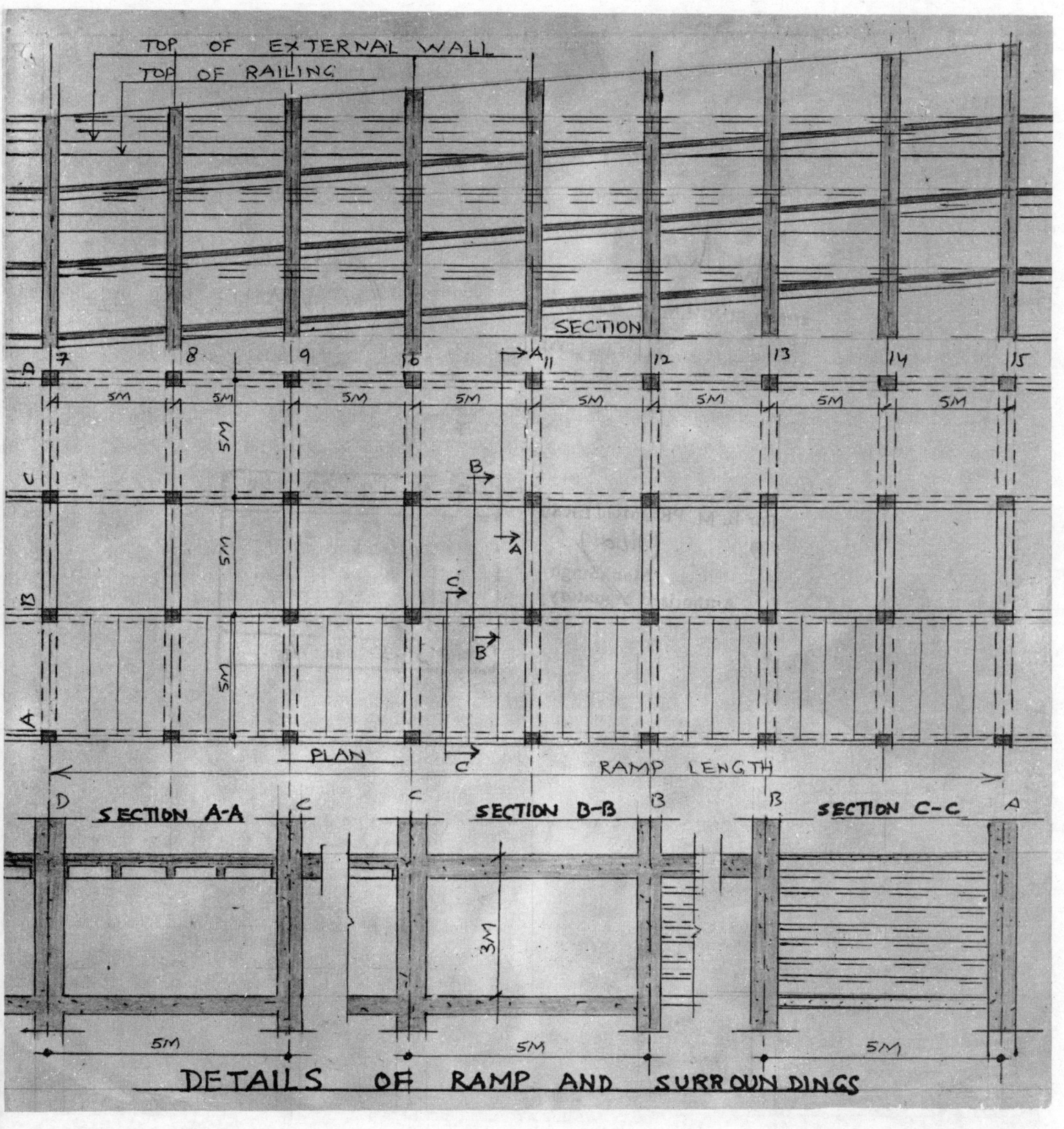

TOP OF EXTERNAL WALL

TOP OF RAILING

SECTION

D | 7 | 8 | 9 | 10 | A 11 | 12 | 13 | 14 | 15

5M   5M   5M   5M   5M   5M   5M   5M

C

B

B

C

A

PLAN

RAMP LENGTH

D | SECTION A-A | C   C | SECTION B-B | B   B | SECTION C-C | A

3M

5M   5M   5M

DETAILS OF RAMP AND SURROUNDINGS

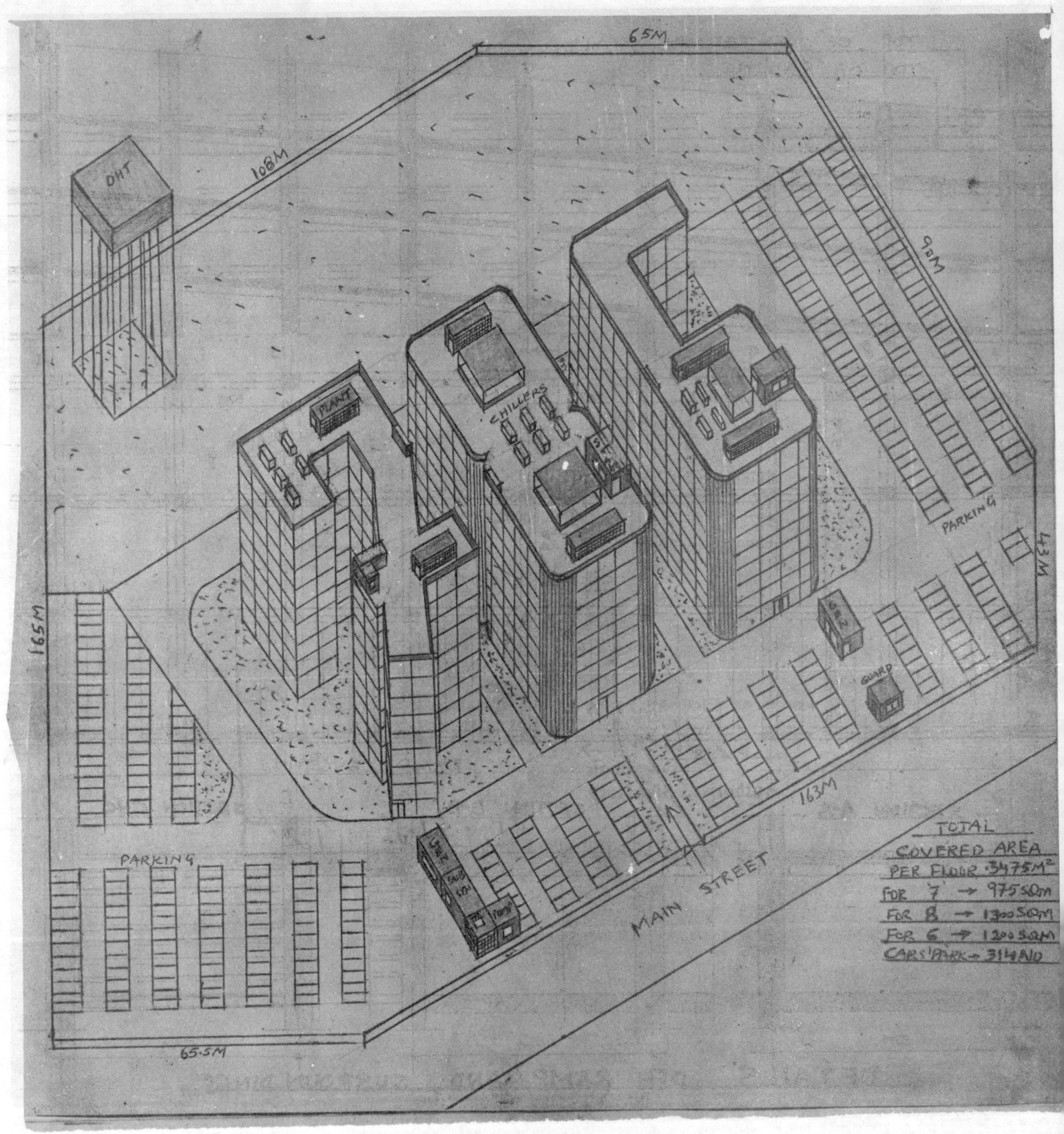

65M

108M

DHT

169M

PLANT

CHILLERS

9M

PARKING

43M

65.5M

PARKING

163M

MAIN STREET

GUARD

TOTAL
COVERED AREA
PER FLOOR 3475M²
FOR 7 → 975 SQM
FOR 8 → 1300 SQM
FOR 6 → 1200 SQM
CARS PARK → 314 NO

TYPICAL PLAN OF A MULTY FLOOR COMML. BLDG.
DETAILS OF PARKING, U/G TANK, SUB STN, GEN ONLY FOR GROUND
PLOT AREA 9975 SQM, OPEN 6275 SQM, COVD/FLOOR 3475 SQM

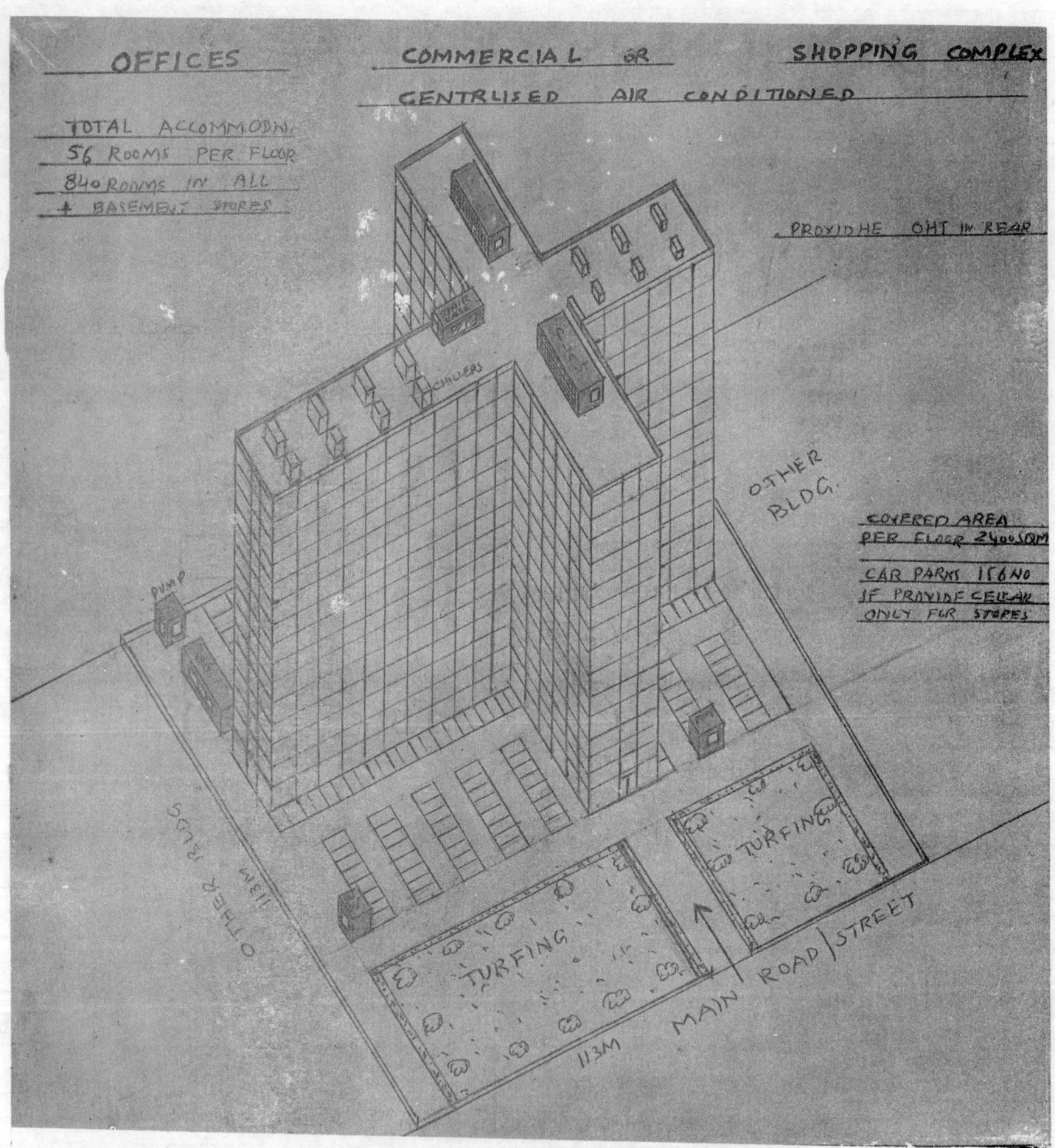

OFFICES   COMMERCIAL   OR   SHOPPING COMPLEX

                          CENTRLISED   AIR   CONDITIONED

TOTAL ACCOMMODN.
56 ROOMS PER FLOOR
840 ROOMS IN ALL   PROVIDHE OHT IN REAR
+ BASEMENT STORES

                                                    OTHER
                                                    BLDG.

                                                    COVERED AREA
                                                    PER FLOOR 2400 SQM

                                                    CAR PARKS 156NO
                                                    IF PROVIDE CELLAR
                                                    ONLY FOR STORES

DUMP

OTHER BLDG.

113M

TURFING   TURFING

113M   MAIN ROAD|STREET

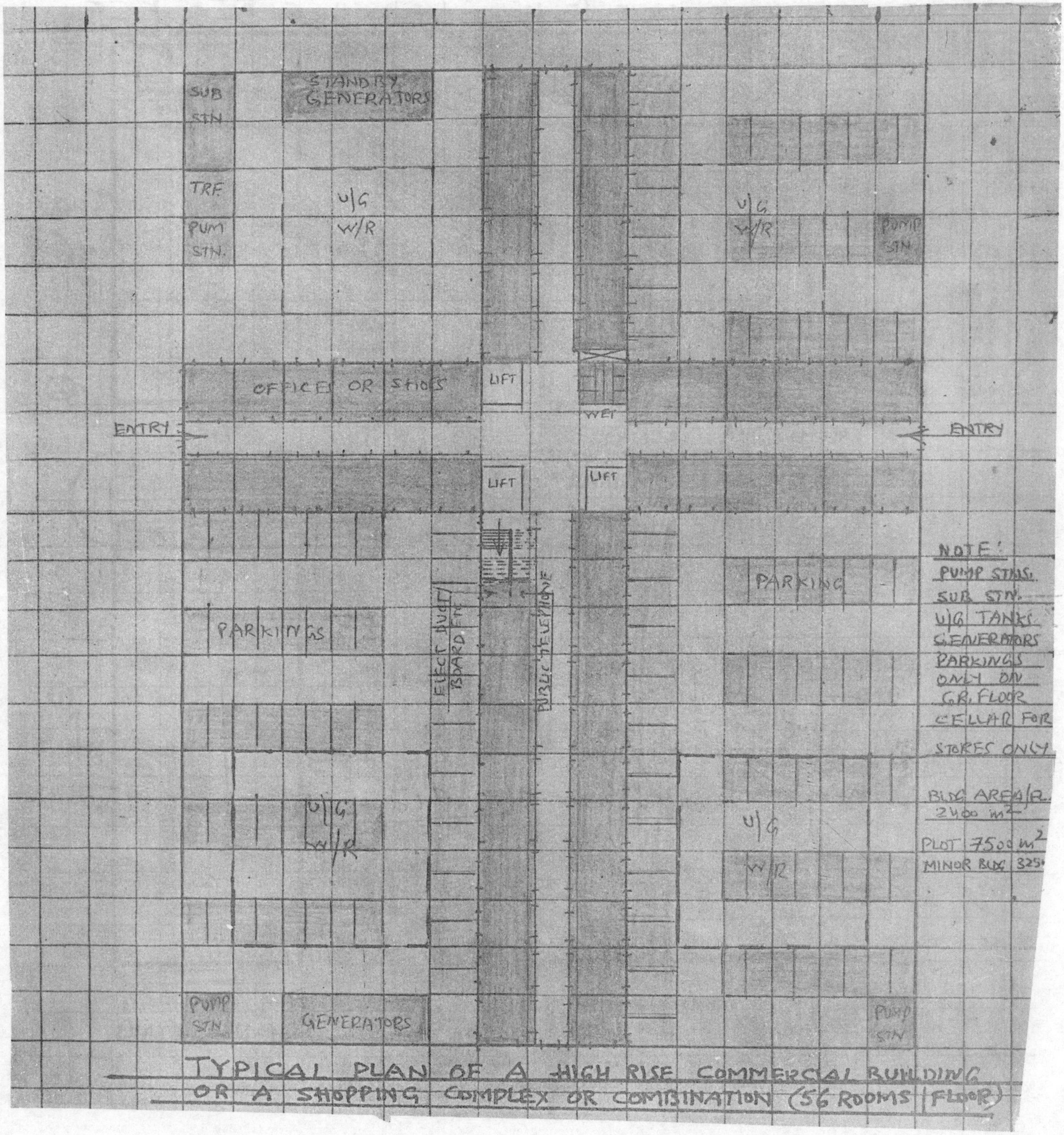

SUB STN

STANDBY GENERATORS

TRF.
PUM STN.

U/G W/R

U/G W/R

PUMP STN

OFFICES OR STORES

LIFT

WE!

ENTRY

ENTRY

LIFT

LIFT

PARKING

NOTE:
PUMP STNS.
SUB STN.
U/G TANKS
GENERATORS
PARKINGS
ONLY ON
GR. FLOOR
CELLAR FOR
STORES ONLY

PARKINGS

ELECT DUCT BOARD ETC

PUBLIC TELEPHONE

U/G W/R

BLDG AREA/R.
2400 m²

PLOT 7500 m²
MINOR BLDG 325

U/G W/R

PUMP STN

GENERATORS

PUMP STN

TYPICAL PLAN OF A HIGH RISE COMMERCIAL BUILDING
OR A SHOPPING COMPLEX OR COMBINATION (56 ROOMS/FLOOR)

SPORTS HALLS DETAILS

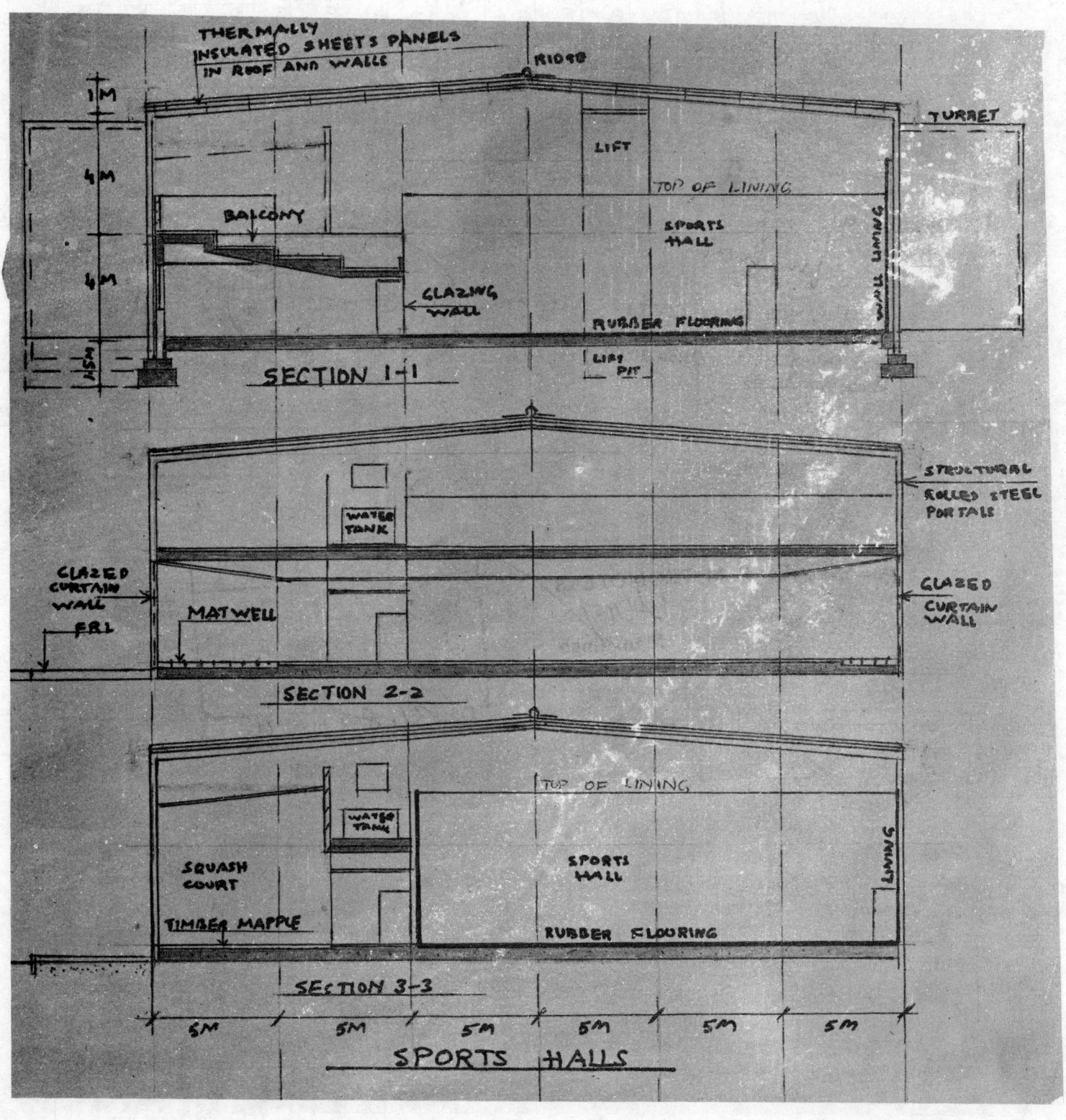

THERMALLY INSULATED SHEETS PANELS IN ROOF AND WALLS

RIDGE

TURRET

1M

4M

LIFT

TOP OF LINING

BALCONY

SPORTS HALL

WALL LINING

4M

GLAZING WALL

RUBBER FLOORING

SECTION 1-1

LIFT PIT

STRUCTURAL ROLLED STEEL PORTALS

CLAZED CURTAIN WALL FR1

WATER TANK

MATWELL

GLAZED CURTAIN WALL

SECTION 2-2

TOP OF LINING

WATER TANK

LINING

SQUASH COURT

SPORTS HALL

TIMBER MAPPLE

RUBBER FLOORING

SECTION 3-3

5M   5M   5M   5M   5M   5M

SPORTS HALLS

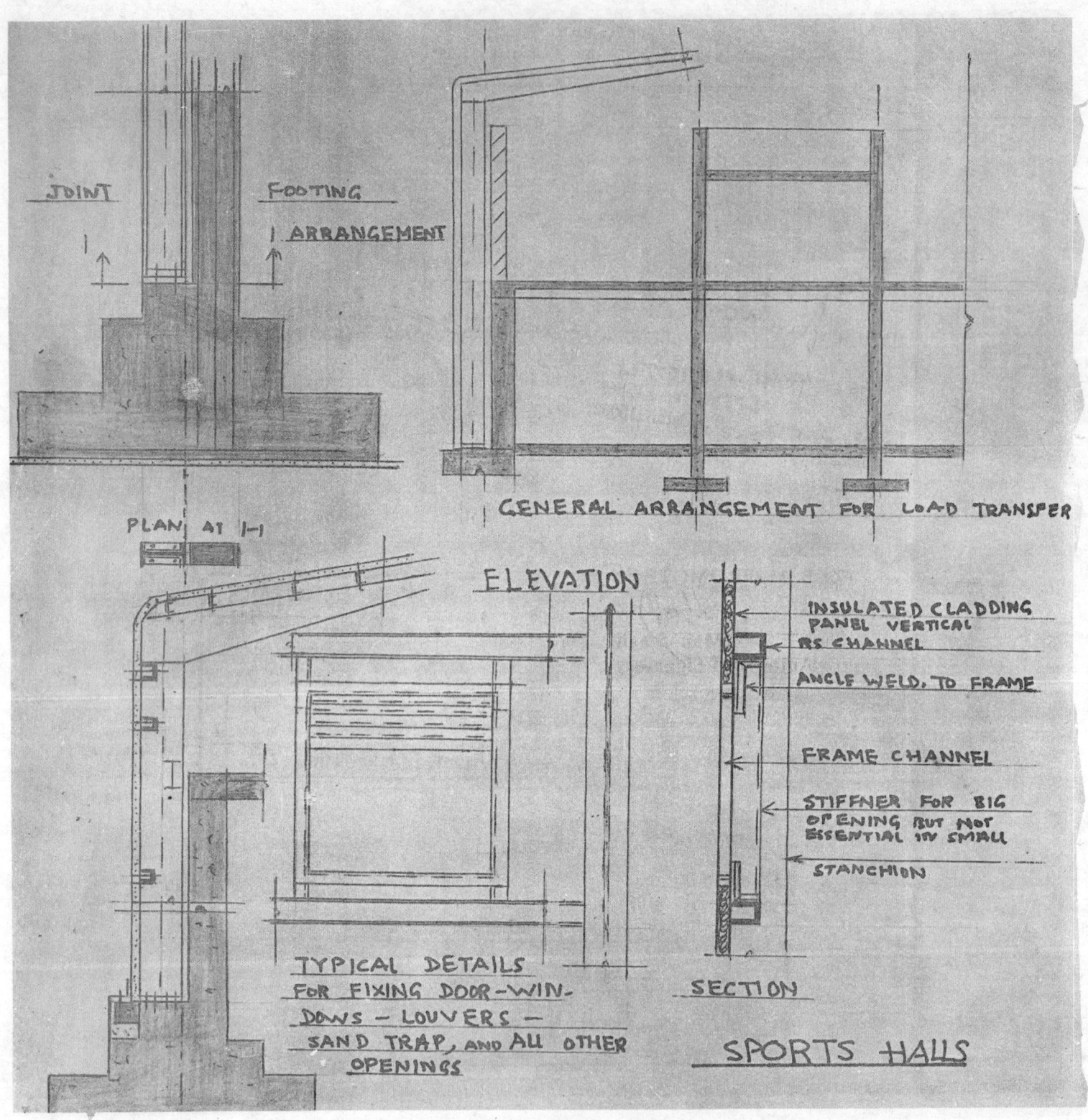

JOINT

FOOTING ARRANGEMENT

PLAN AT 1-1

GENERAL ARRANGEMENT FOR LOAD TRANSFER

ELEVATION

INSULATED CLADDING PANEL VERTICAL RS CHANNEL

ANGLE WELD, TO FRAME

FRAME CHANNEL

STIFFNER FOR BIG OPENING BUT NOT ESSENTIAL IN SMALL

STANCHION

TYPICAL DETAILS FOR FIXING DOOR-WIN-DOWS — LOUVERS — SAND TRAP, AND ALL OTHER OPENINGS

SECTION

SPORTS HALLS

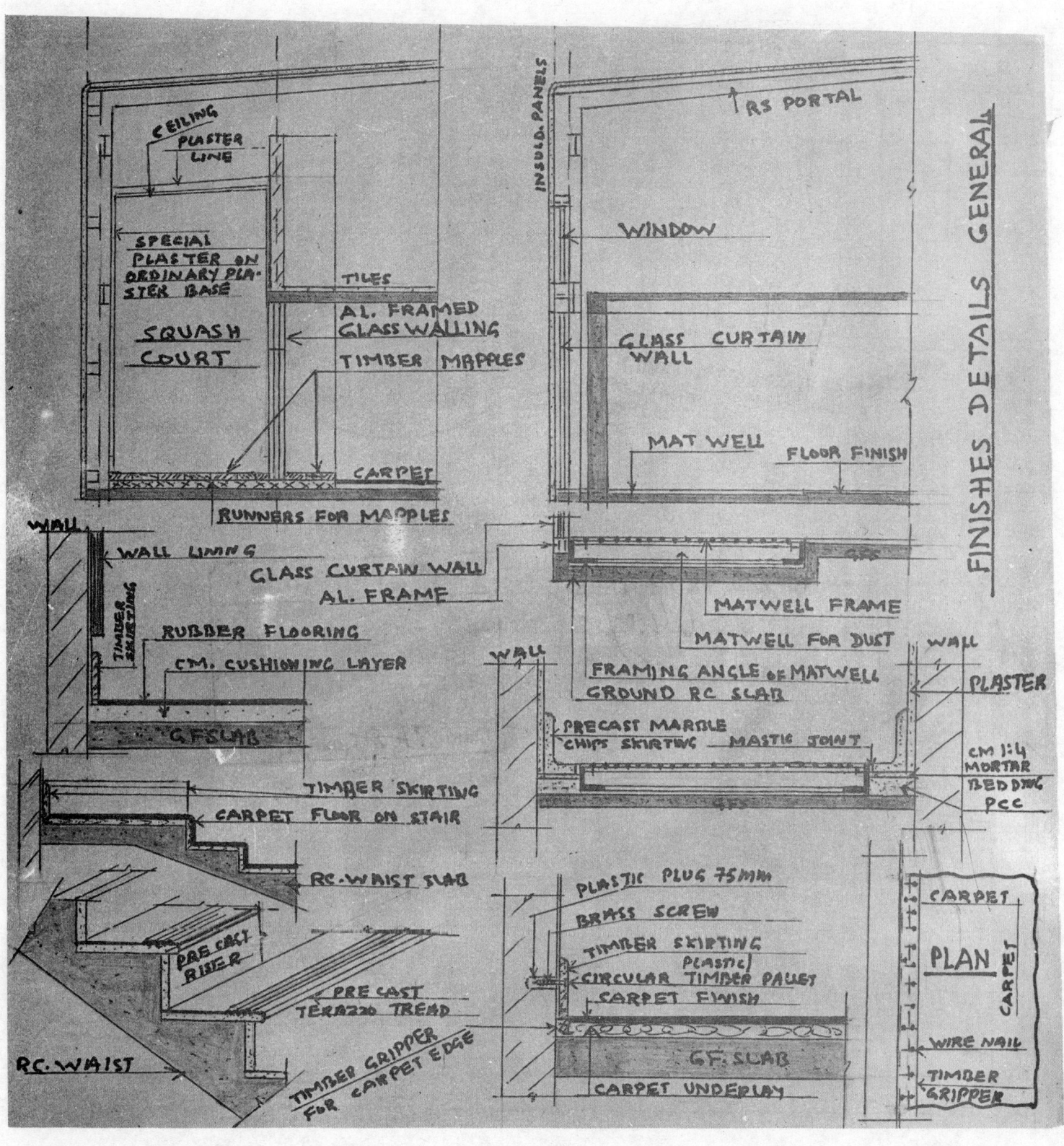

CEILING PLASTER LINE

SPECIAL PLASTER ON ORDINARY PLASTER BASE

SQUASH COURT

TILES

AL. FRAMED GLASS WALLING

TIMBER MAPPLES

CARPET

RUNNERS FOR MAPPLES

WALL

WALL LINING

TIMBER SKIRTING

GLASS CURTAIN WALL AL. FRAME

RUBBER FLOORING

CM. CUSHIONING LAYER

G.F. SLAB

TIMBER SKIRTING

CARPET FLOOR ON STAIR

RC. WAIST SLAB

PRE CAST RISER

PRE CAST TERAZZO TREAD

RC. WAIST

TIMBER GRIPPER FOR CARPET EDGE

INSULO. PANELS

RS PORTAL

WINDOW

GLASS CURTAIN WALL

MAT WELL

FLOOR FINISH

MATWELL FRAME

MATWELL FOR DUST

FRAMING ANGLE OF MATWELL GROUND RC SLAB

PRECAST MARBLE CHIPS SKIRTING   MASTIC JOINT

WALL

PLASTIC PLUG 75mm

BRASS SCREW

TIMBER SKIRTING (PLASTIC)

CIRCULAR TIMBER PAUET

CARPET FINISH

G.F. SLAB

CARPET UNDERLAY

FINISHES DETAILS GENERAL

WALL

PLASTER

CM 1:4 MORTAR BEDDING PCC

CARPET

PLAN

CARPET

WIRE NAIL

TIMBER GRIPPER

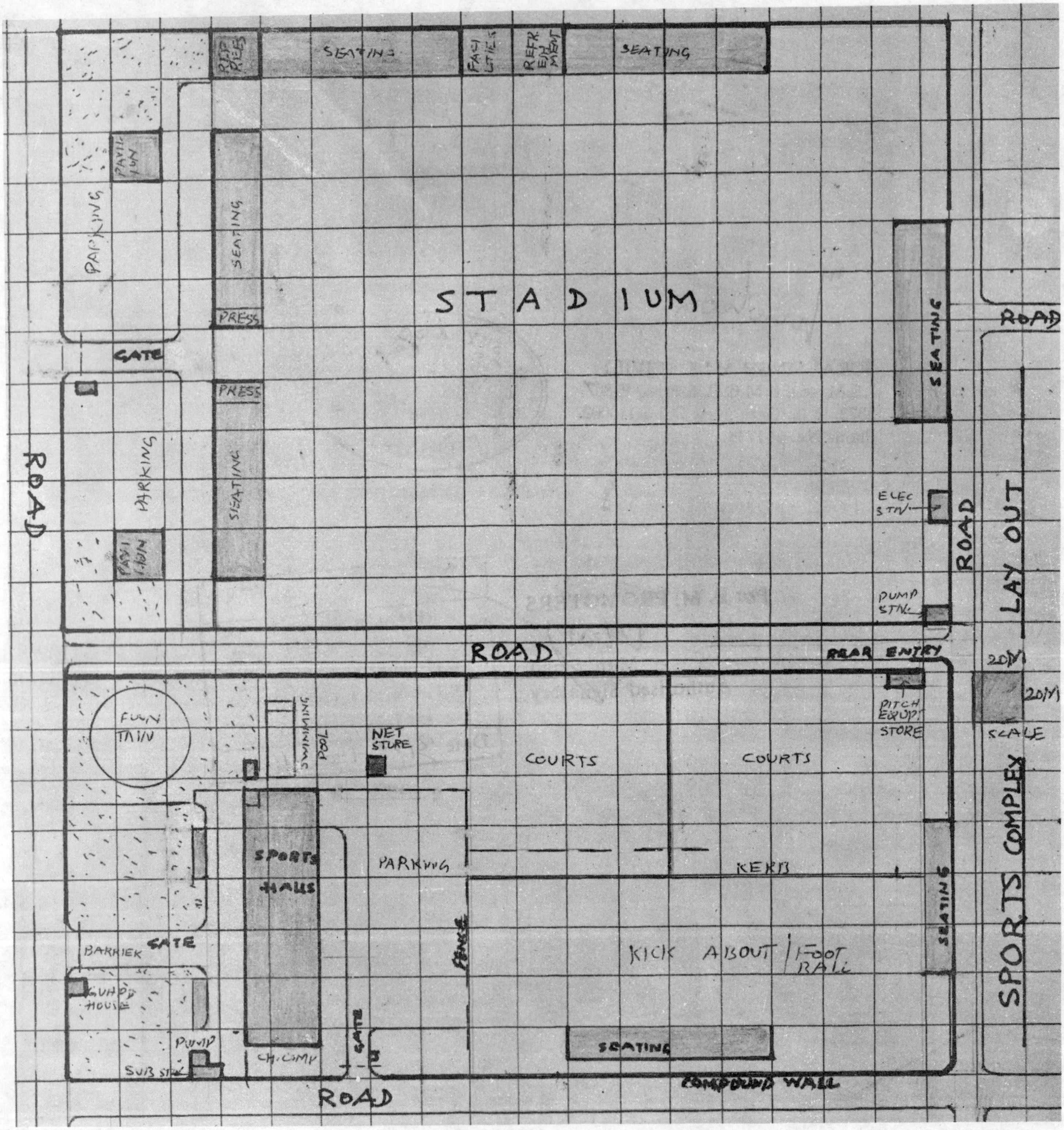

STADIUM

SPORTS COMPLEX LAY OUT

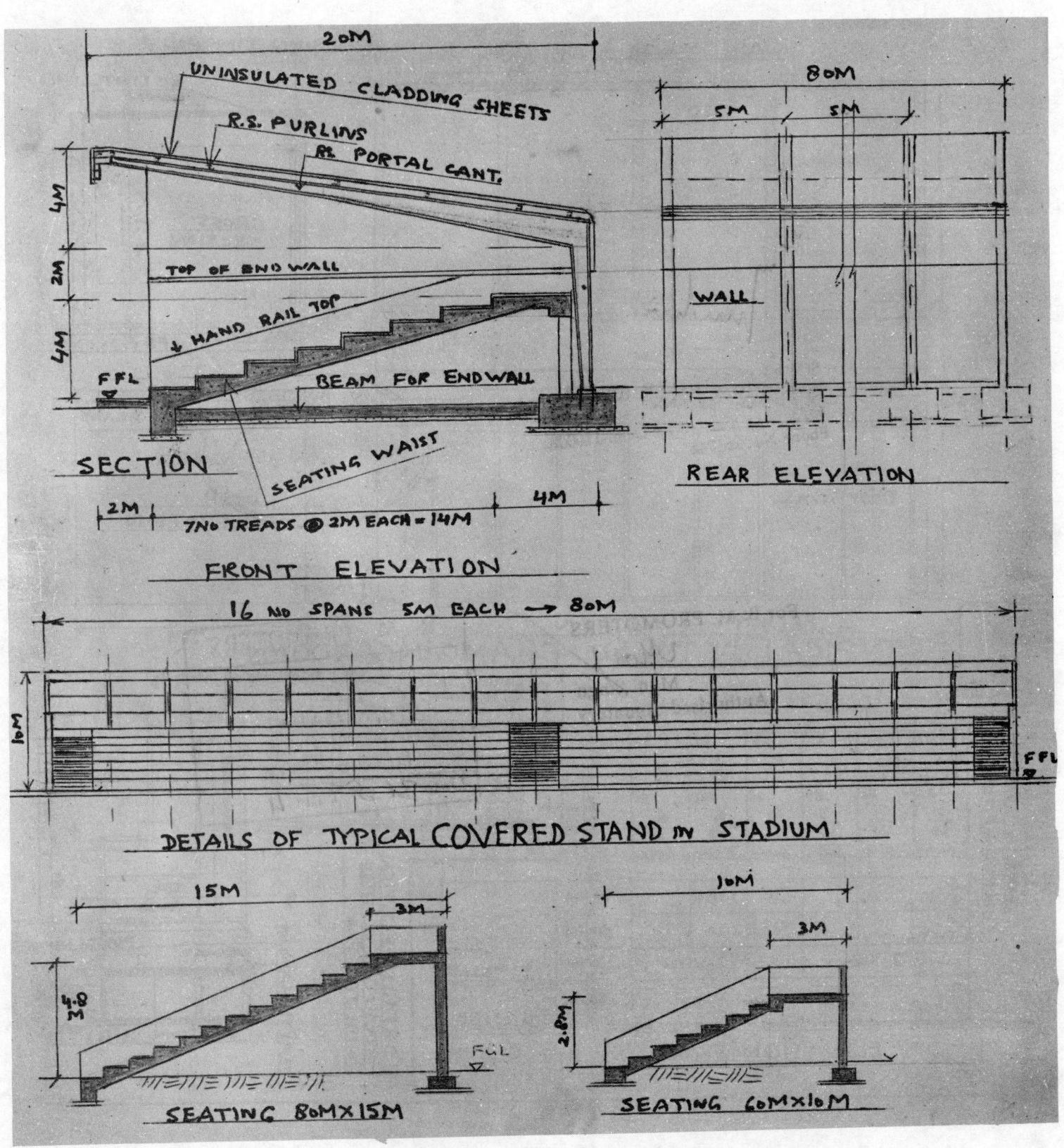

20M

UNINSULATED CLADDING SHEETS

R.S. PURLINS

R.S. PORTAL CANT.

4M

2M

TOP OF END WALL

↓ HAND RAIL TOP

4M

FFL

BEAM FOR ENDWALL

SECTION

SEATING WAIST

2M

7No TREADS @ 2M EACH = 14M

4M

FRONT ELEVATION

8OM

5M   5M

WALL

REAR ELEVATION

16 No SPANS 5M EACH → 80M

10M

FFL

DETAILS OF TYPICAL COVERED STAND IN STADIUM

15M

3M

4.8 M

FGL

SEATING 80M X 15M

10M

3M

2.8M

SEATING 60M X 10M

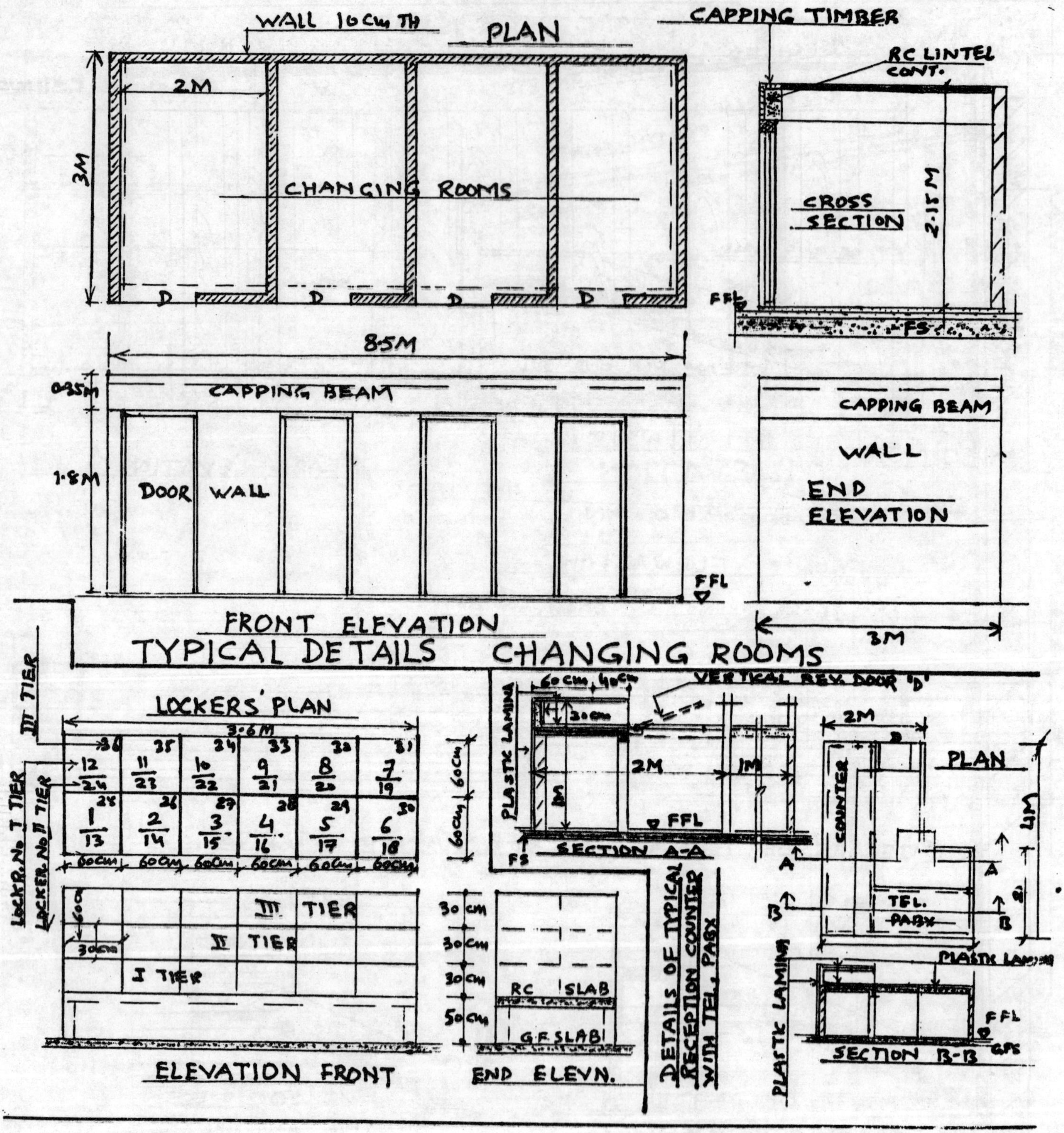

PLAN

WALL 10 CM TH

2M

3M

CHANGING ROOMS

D   D   D   D

CAPPING TIMBER

RC LINTEL CONT.

CROSS SECTION

2.15 M

FFL

FS

8.5M

0.25M

CAPPING BEAM

1.8M

DOOR WALL

FFL

FRONT ELEVATION

CAPPING BEAM

WALL

END ELEVATION

3M

TYPICAL DETAILS   CHANGING ROOMS

LOCKERS' PLAN

3.6 M

30	35	34	33	32	31
12 / 24	11 / 23	10 / 22	9 / 21	8 / 20	7 / 19
1 / 13	2 / 14	3 / 15	4 / 16	5 / 17	6 / 18

III TIER

Locker No I Tier
Locker No II Tier

28 26 27 28 29 30

60CM 60CM 60CM 60CM 60CM 60CM

60
30CM

III TIER

II TIER

I TIER

ELEVATION FRONT

60 CM, 40 CM

PLASTIC LAMINA

30 CM

VERTICAL REV. DOOR 'D'

2M   1M

3M

FFL

FS

SECTION A-A

60 CM   60 CM

30 CM
30 CM
30 CM

RC SLAB
50 CM

G.F. SLAB

END ELEVN.

DETAILS OF TYPICAL RECEPTION COUNTER WITH TEL, PABX

PLAN

2M

COUNTER

4M

A   A

B   B

TEL. PABX

PLASTIC LAMINA

PLASTIC LAMINA

FFL

GFS

SECTION B-B

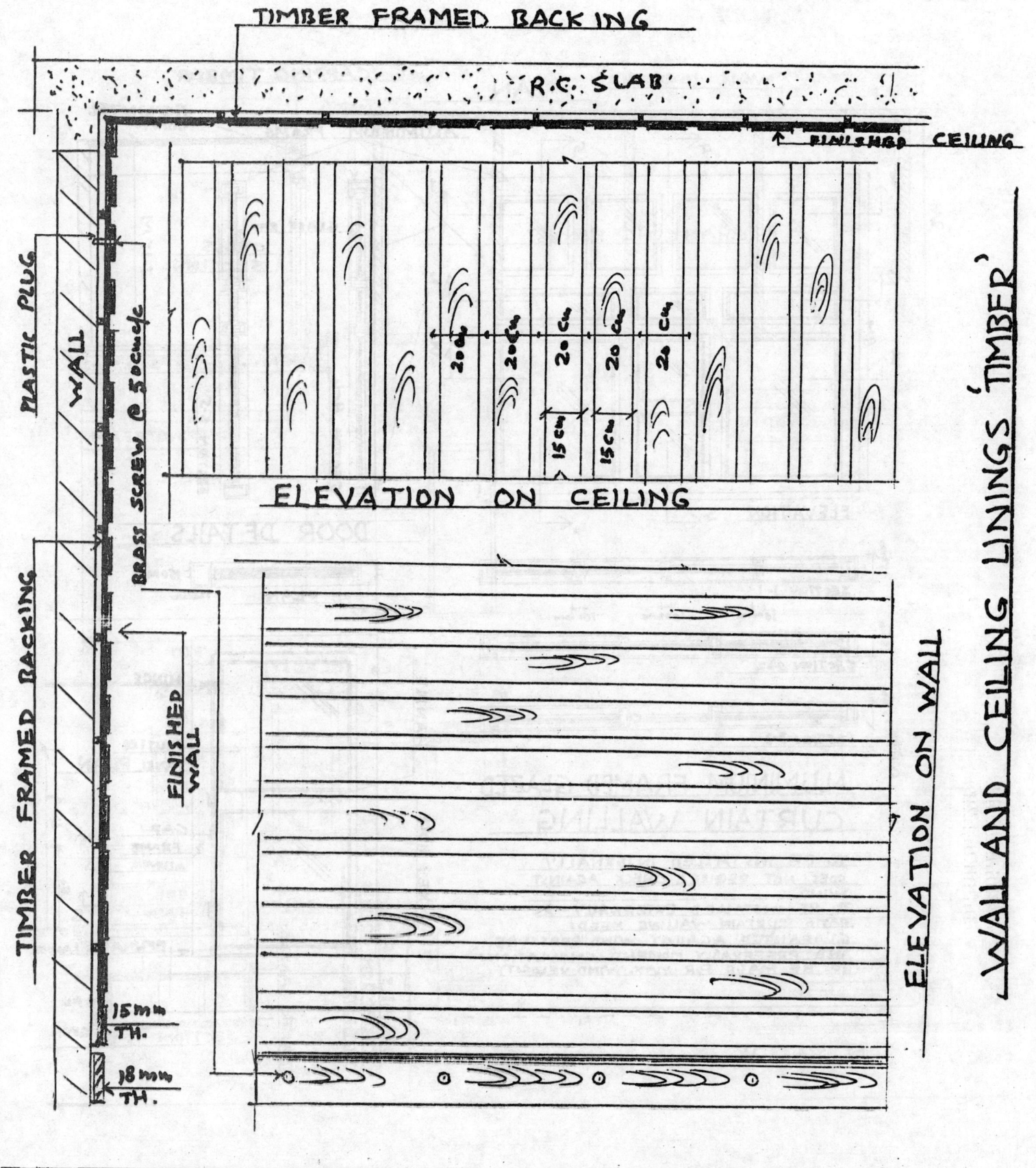

TIMBER FRAMED BACKING

R.C. SLAB

FINISHED CEILING

ELEVATION ON CEILING

PLASTIC PLUG

WALL

BRASS SCREW @ 50cm c/c

TIMBER FRAMED BACKING

FINISHED WALL

15mm TH.

18mm TH.

20 Cm   20 Cm   20 Cm   20 Cm   20 Cm

15 Cm   15 Cm

ELEVATION ON WALL

WALL AND CEILING LININGS 'TIMBER'

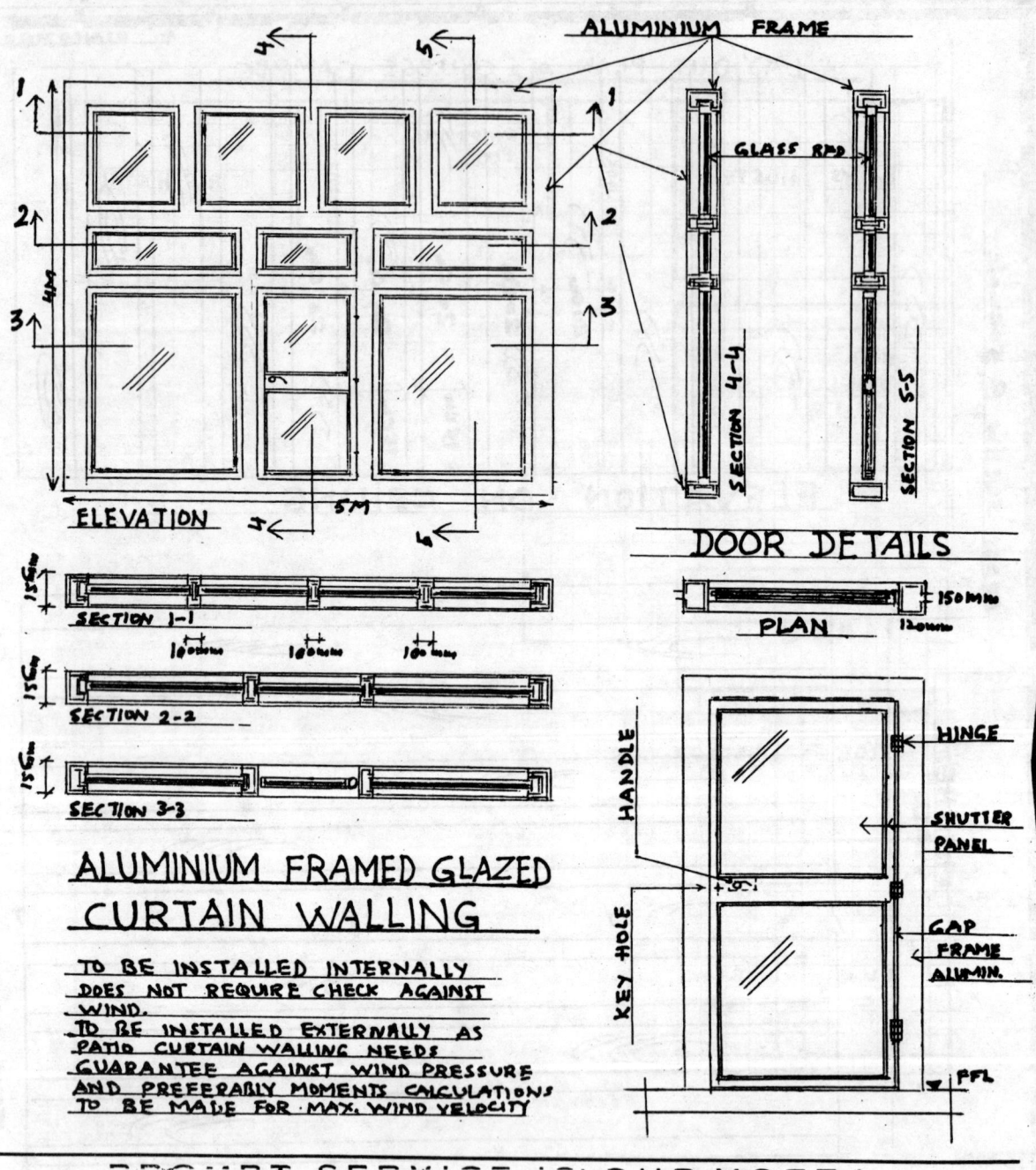

ALUMINIUM FRAME

GLASS RPD

SECTION 4-4

SECTION 5-5

1

ELEVATION

5M

4M

2

3

SECTION 1-1

15Cm

100mm 100mm 100mm

SECTION 2-2

15Cm

SECTION 3-3

15Cm

DOOR DETAILS

PLAN

150mm

120mm

HINGE

SHUTTER PANEL

GAP FRAME ALUMIN.

HANDLE

KEY HOLE

PFL

ALUMINIUM FRAMED GLAZED CURTAIN WALLING

TO BE INSTALLED INTERNALLY DOES NOT REQUIRE CHECK AGAINST WIND.

TO BE INSTALLED EXTERNALLY AS PATIO CURTAIN WALLING NEEDS GUARANTEE AGAINST WIND PRESSURE AND PREFERABLY MOMENTS CALCULATIONS TO BE MADE FOR MAX. WIND VELOCITY

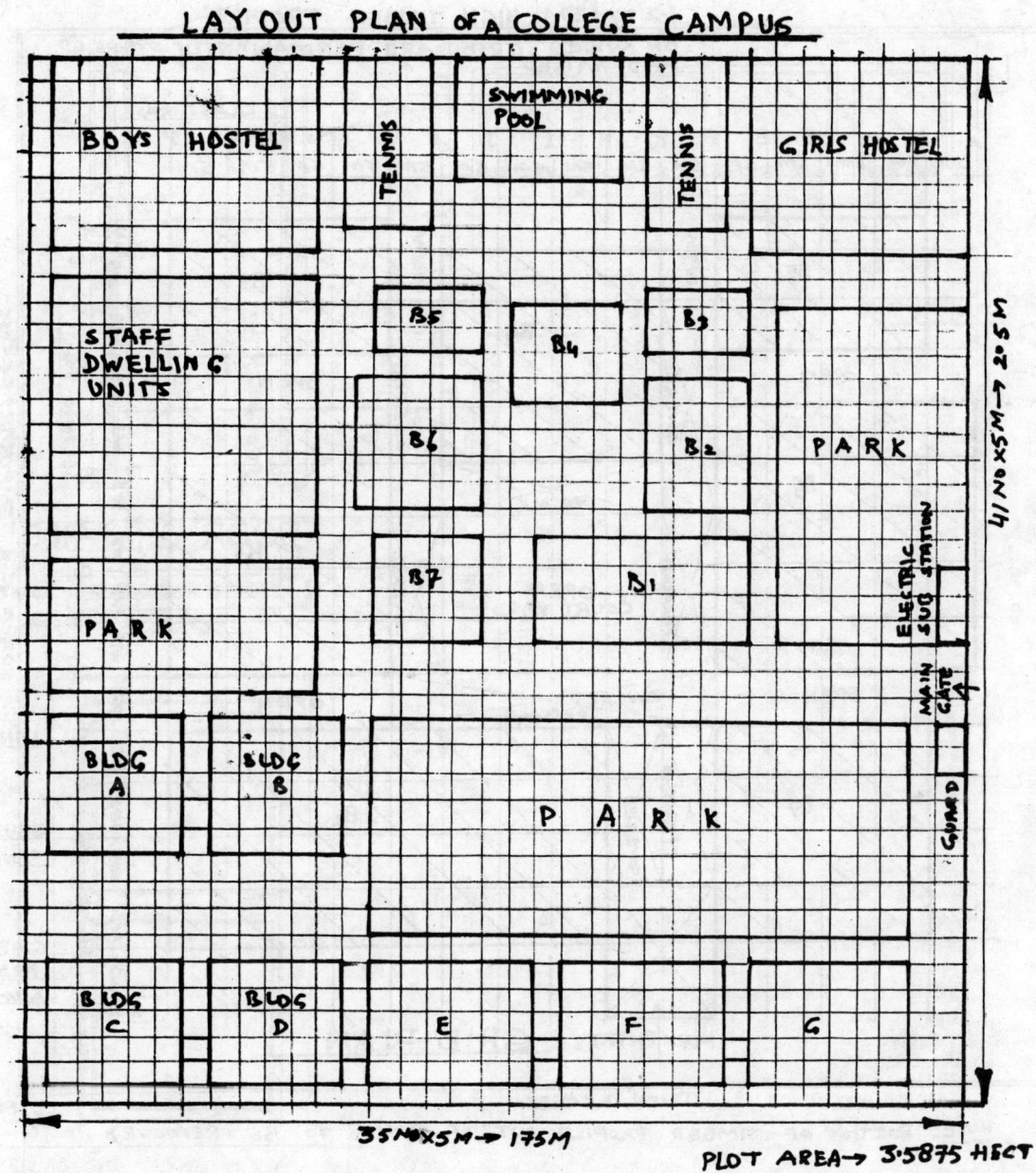

LAYOUT PLAN OF A COLLEGE CAMPUS

PLOT AREA → 3·5875 HECT

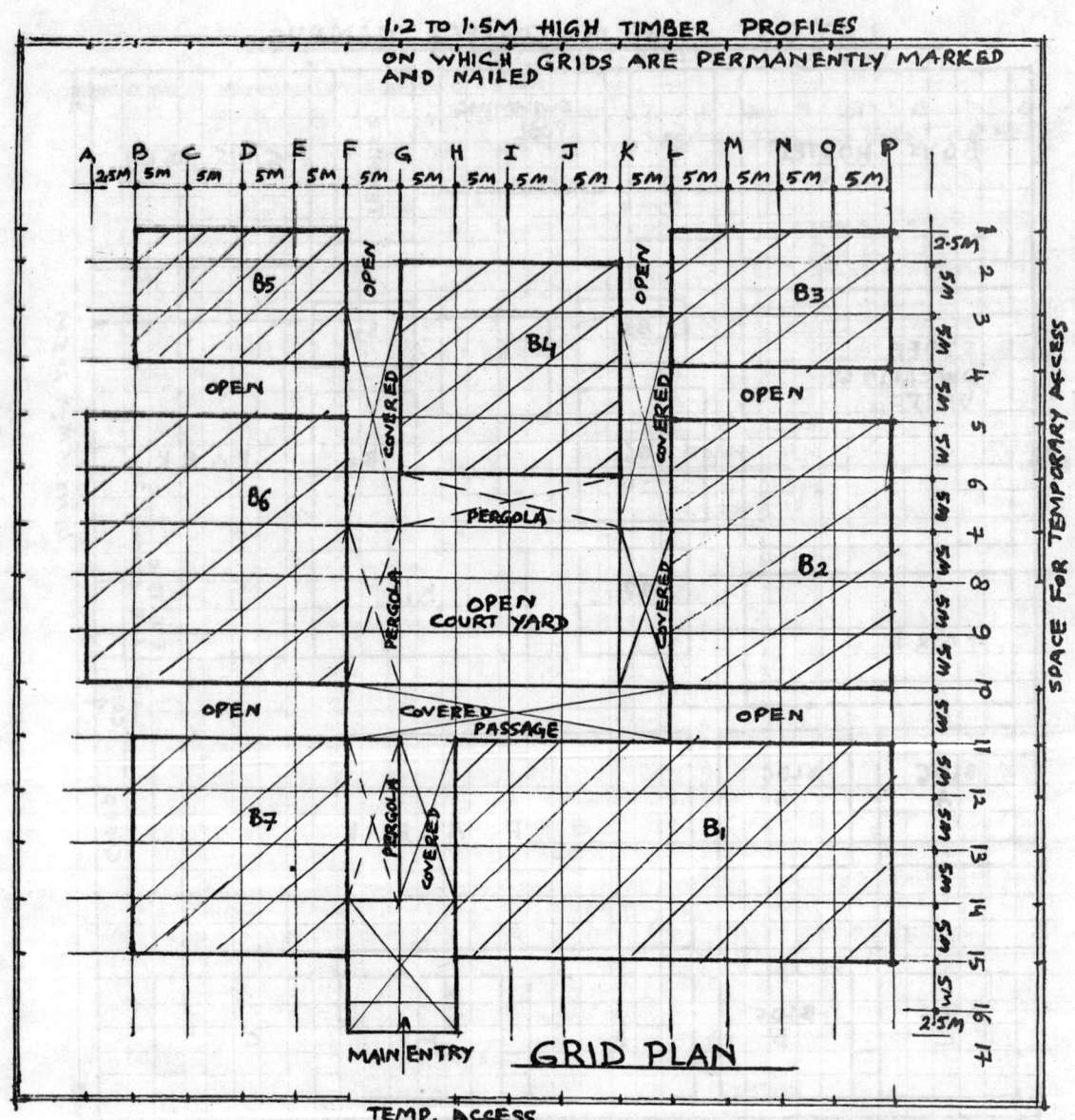

1.2 TO 1.5M HIGH TIMBER PROFILES
ON WHICH GRIDS ARE PERMANENTLY MARKED
AND NAILED

GRID PLAN

TEMP. ACCESS

NOTE: PORTION OF TIMBER PROFILE OUT OF GRIDS TO BE REMOVED

GROUND FLOOR LAYOUT

NOTE- OPENINGS NOT SHOWN FULLY IN WALLS

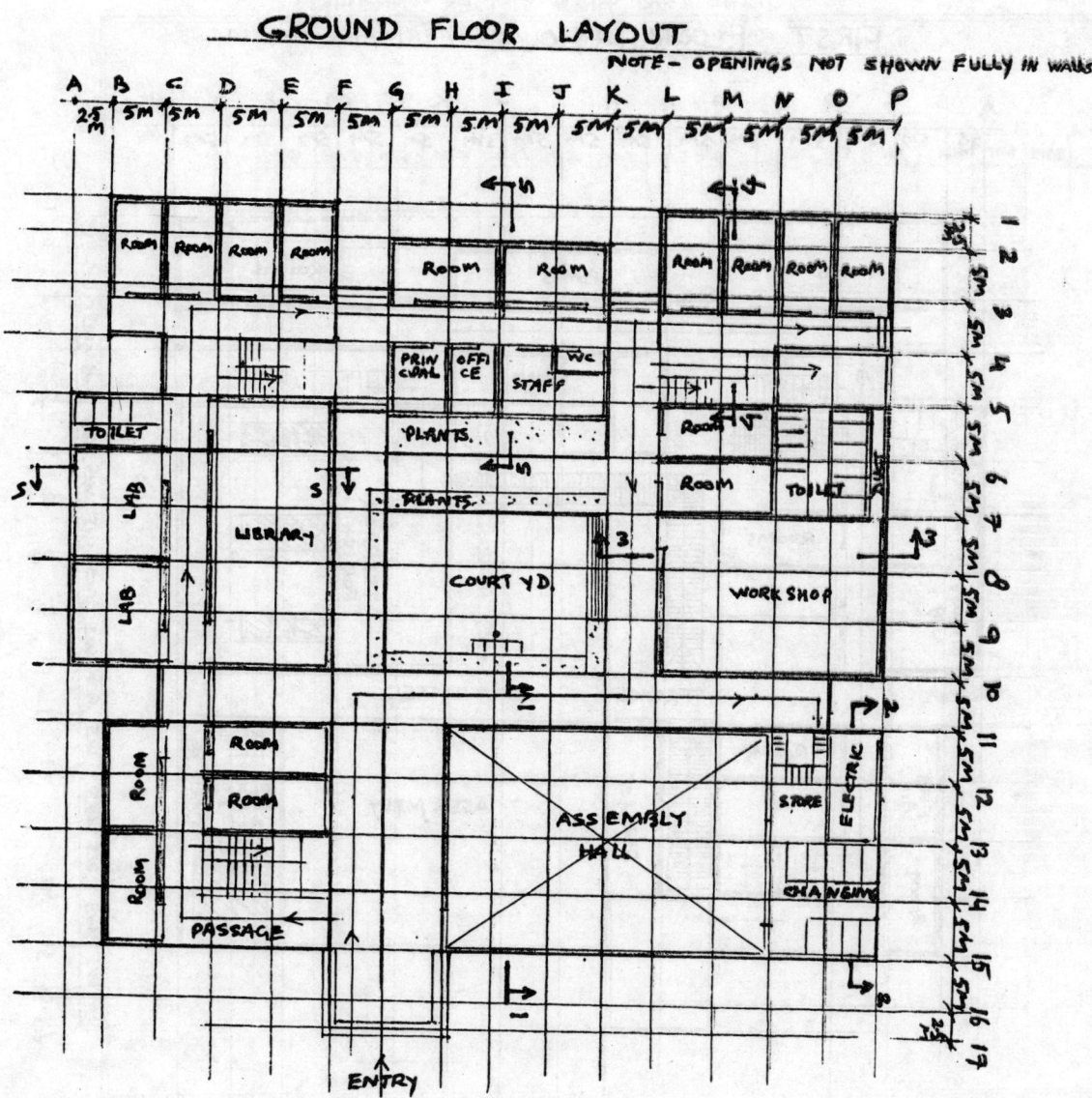

FIRST FLOOR LAY OUT

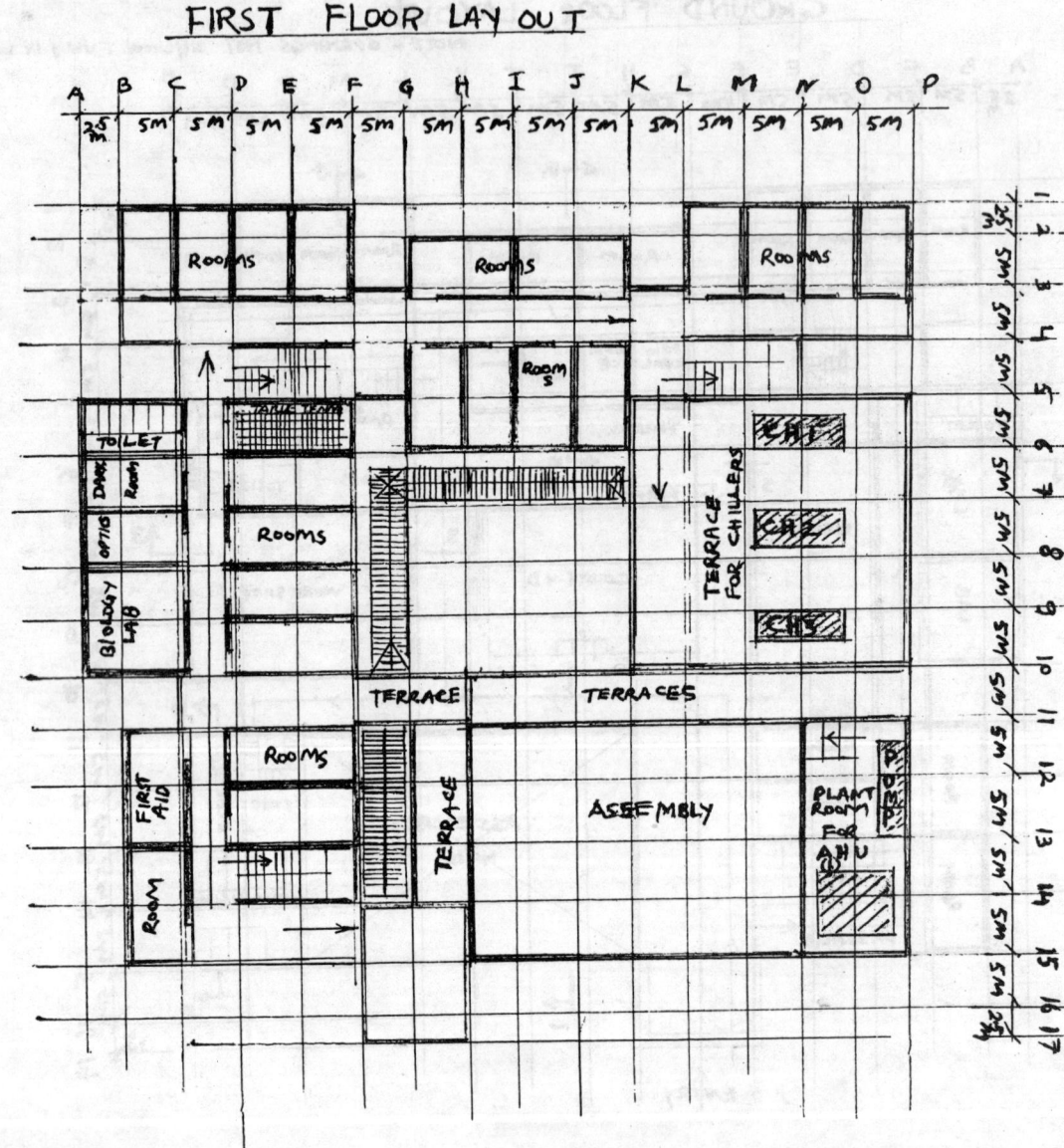

ROOF LAYOUT

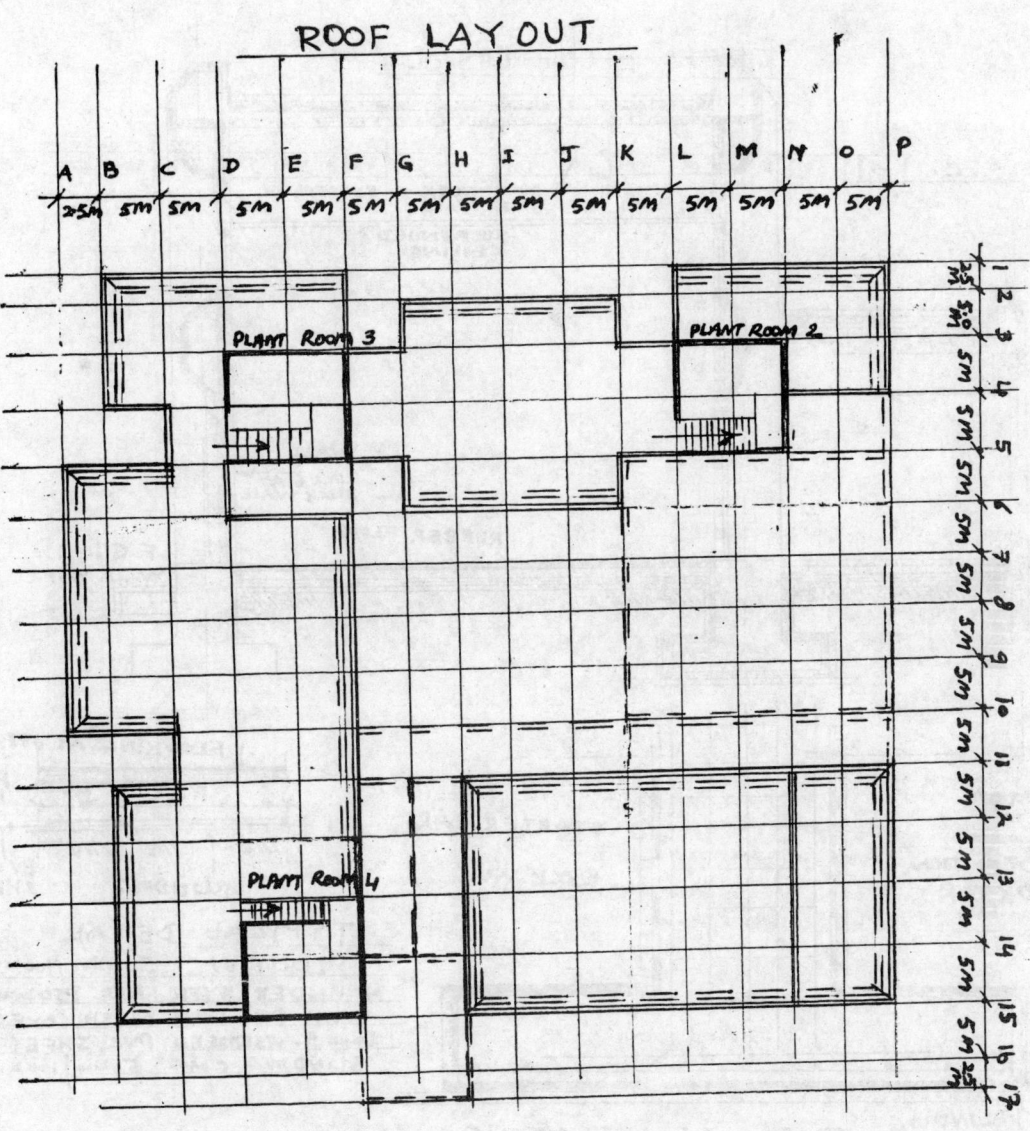

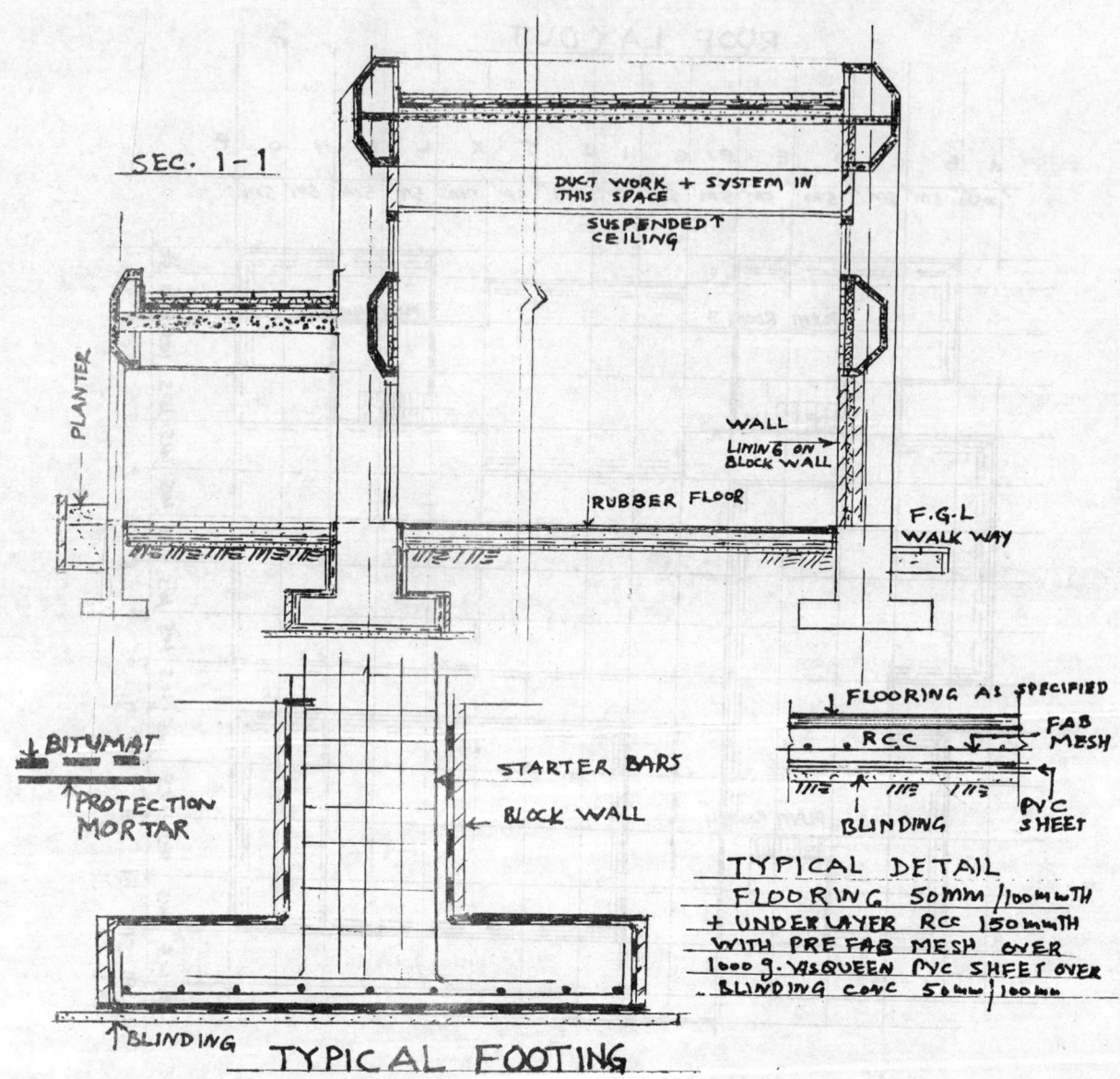

SEC. 1-1

DUCT WORK + SYSTEM IN
THIS SPACE

SUSPENDED
CEILING

PLANTER

WALL
LINING ON
BLOCK WALL

RUBBER FLOOR

F.G.L
WALK WAY

BITUMAT

PROTECTION
MORTAR

STARTER BARS

BLOCK WALL

BLINDING

FLOORING AS SPECIFIED

RCC

FAB
MESH

PVC
SHEET

BLINDING

TYPICAL DETAIL
FLOORING 50mm/100mm TH
+ UNDERLAYER RCC 150mm TH
WITH PREFAB MESH OVER
1000 g. VISQUEEN PVC SHEET OVER
BLINDING CONC 50mm/100mm

TYPICAL FOOTING

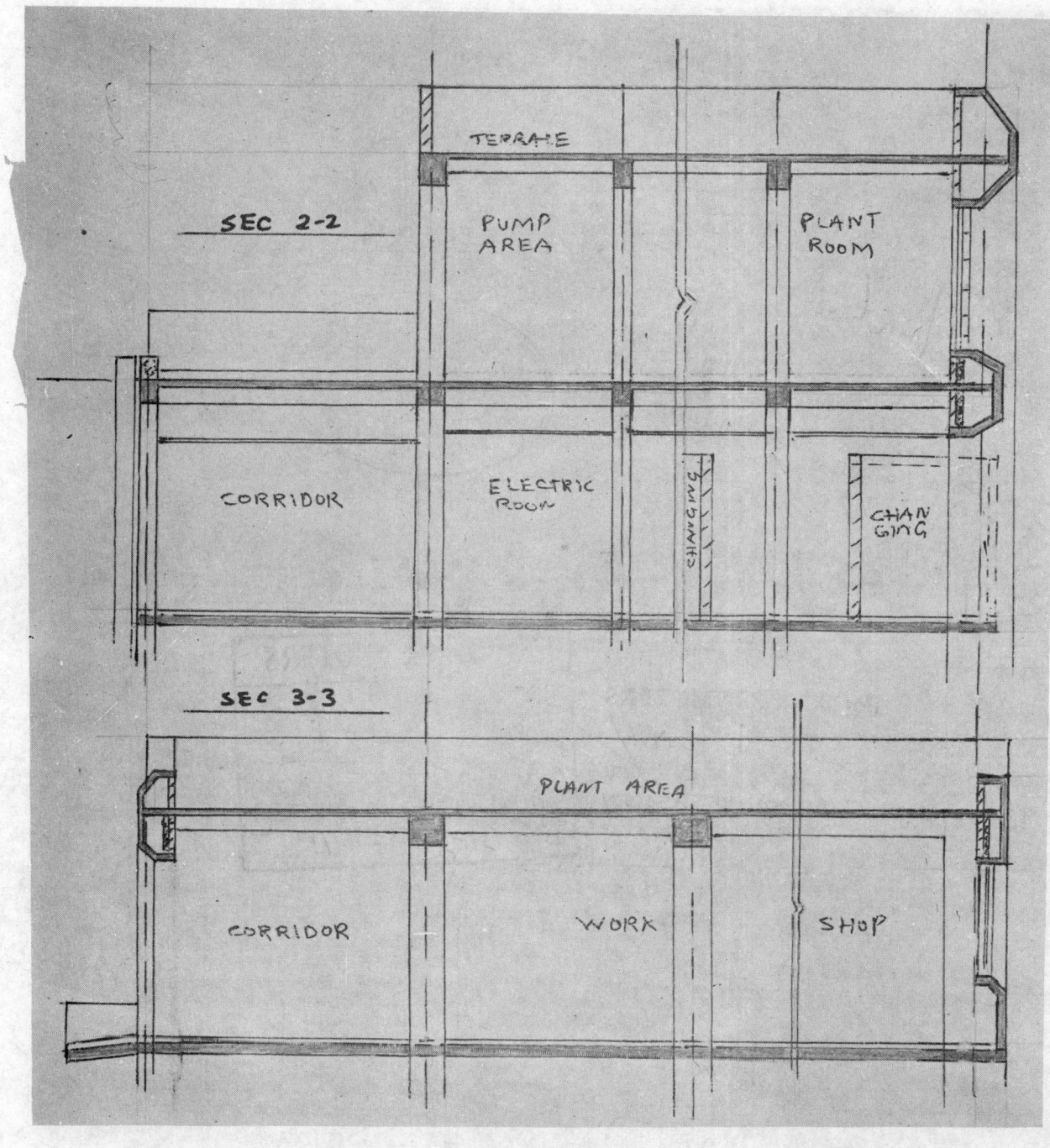

SEC 2-2

TERRACE

PUMP AREA

PLANT ROOM

CORRIDOR

ELECTRIC ROOM

CHANGING

CHANGING

SEC 3-3

PLANT AREA

CORRIDOR

WORK

SHOP

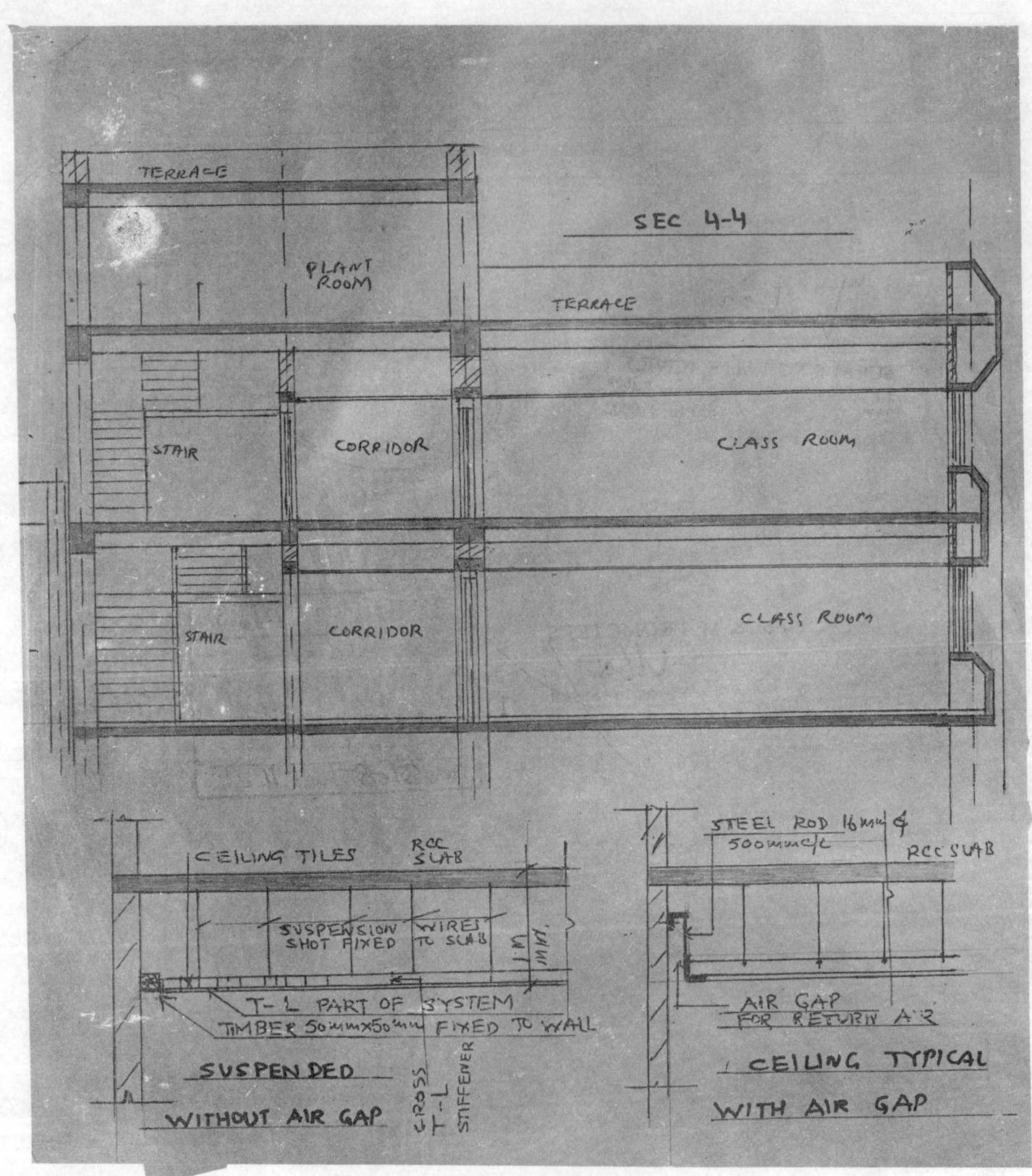

SEC 4-4

TERRACE

PLANT ROOM

TERRACE

STAIR

CORRIDOR

CLASS ROOM

STAIR

CORRIDOR

CLASS ROOM

CEILING TILES   RCC SLAB

SUSPENSION WIRES
SHOT FIXED TO SLAB

T-L PART OF SYSTEM
TIMBER 50mm×50mm FIXED TO WALL

CROSS T-L STIFFENER

SUSPENDED

WITHOUT AIR GAP

STEEL ROD 16mm φ
500mm c/c   RCC SLAB

AIR GAP
FOR RETURN AIR

CEILING TYPICAL

WITH AIR GAP

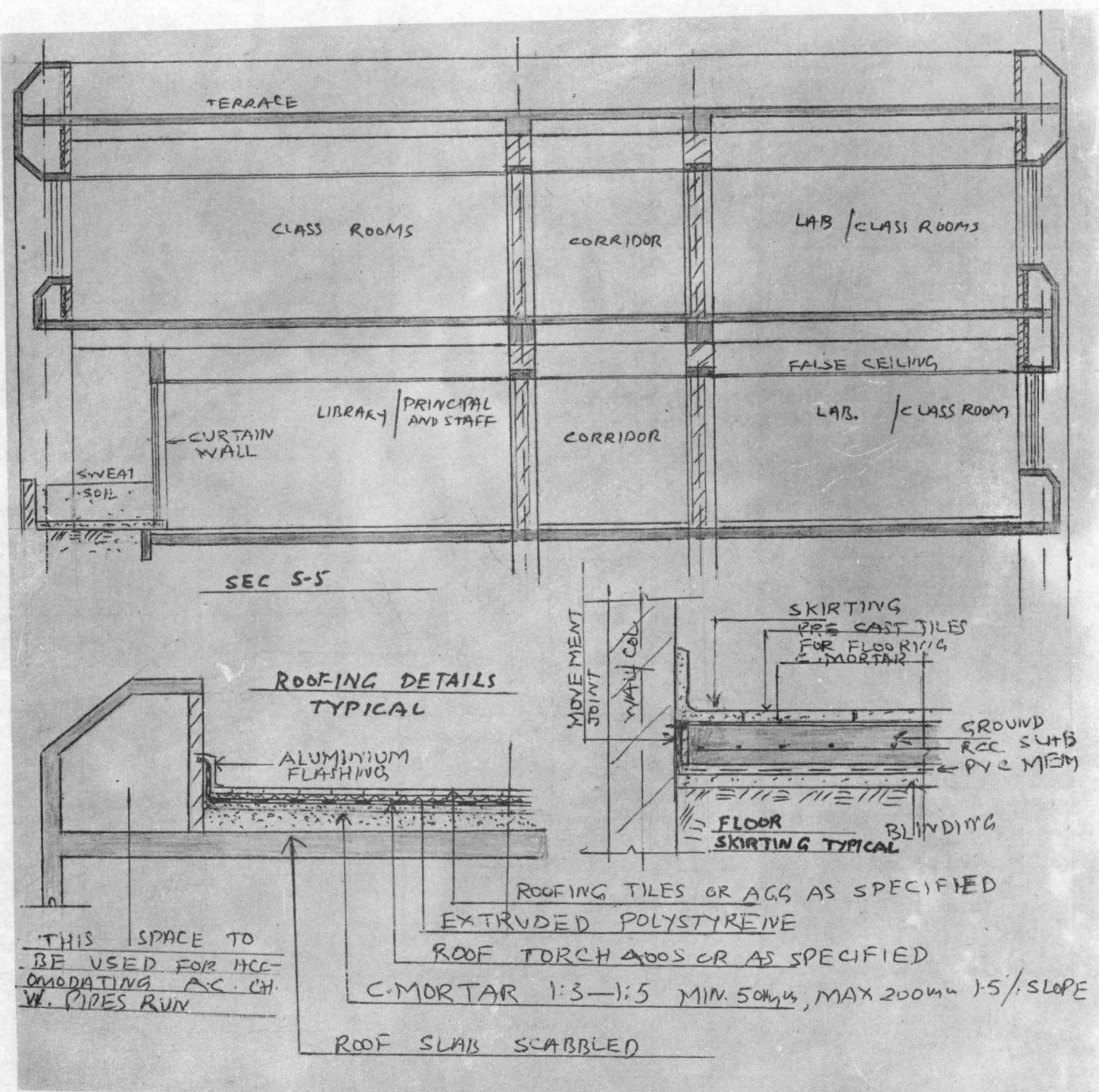

TERRACE

CLASS ROOMS

CORRIDOR

LAB / CLASS ROOMS

FALSE CEILING

LIBRARY | PRINCIPAL AND STAFF

CORRIDOR

LAB. | CLASS ROOM

CURTAIN WALL

SWEAT SOIL

SEC S-S

ROOFING DETAILS TYPICAL

ALUMINIUM FLASHING

MOVEMENT JOINT

WALL COL

SKIRTING

PRE CAST TILES FOR FLOORING C MORTAR

GROUND RCC SLAB PVC MBM

FLOOR SKIRTING TYPICAL

BLINDING

THIS SPACE TO BE USED FOR ACCOMODATING A.C. CH. W. PIPES RUN

ROOFING TILES OR AGG AS SPECIFIED

EXTRUDED POLYSTYRENE

ROOF TORCH 400S OR AS SPECIFIED

C. MORTAR 1:3—1:5 MIN. 50mm, MAX 200mm 1.5% SLOPE

ROOF SLAB SCABBLED

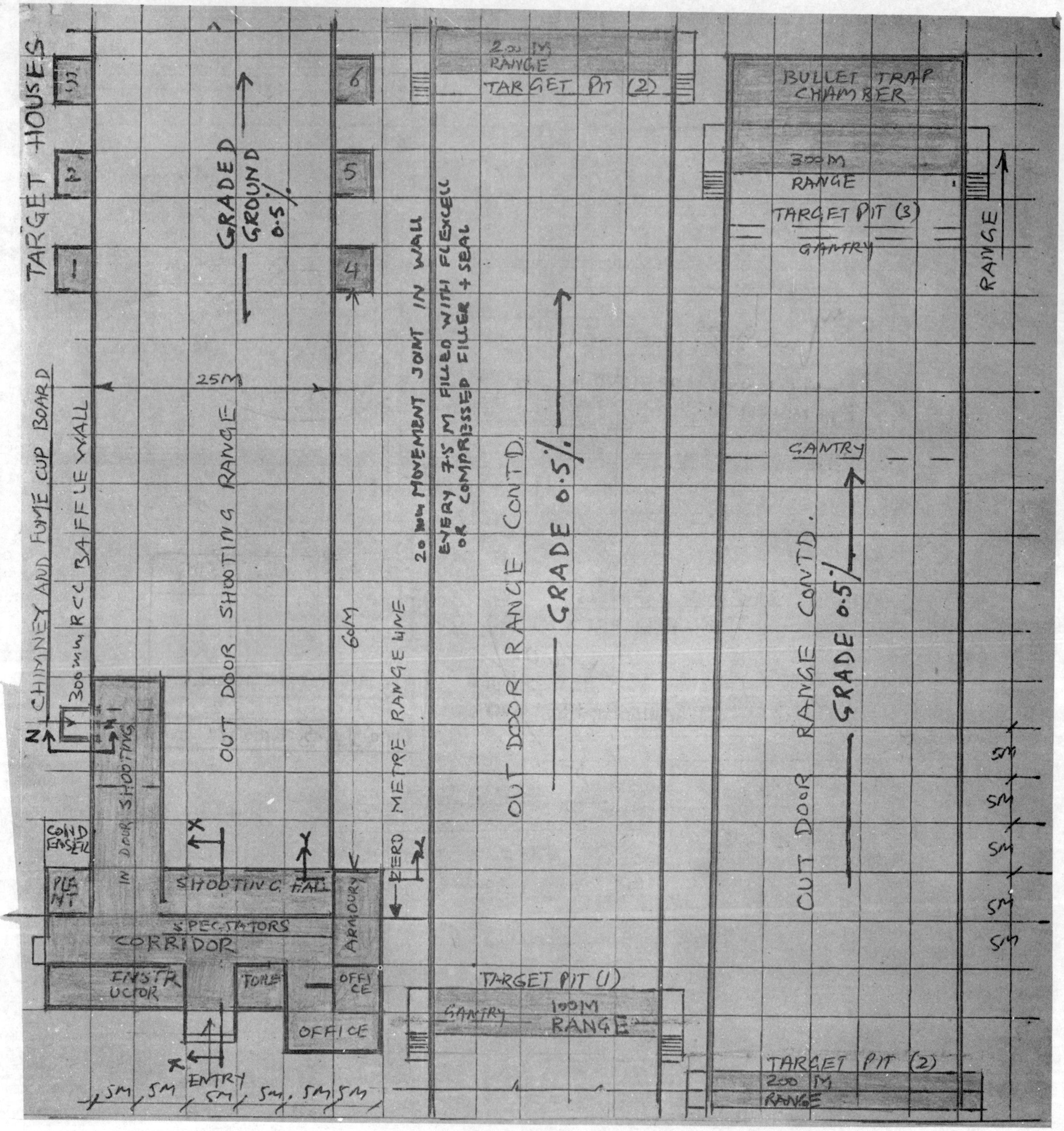

TARGET HOUSES

'3' '2' '1'

CHIMNEY AND FUME CUP BOARD

300mm RCC BAFFLE WALL

GRADED GROUND 0.5%

25M

6 5 4

OUT DOOR SHOOTING RANGE

20 mm MOVEMENT JOINT IN WALL

EVERY 7.5 M FILLED WITH FLEXCELL OR COMPRESSED FILLER + SEAL

2.00 M RANGE
TARGET PIT (2)

OUT DOOR RANGE CONTD.

GRADE 0.5%

BULLET TRAP CHAMBER

300 M RANGE
TARGET PIT (3)
GANTRY

RANGE

GANTRY

60M

METRE RANGE LINE

OUT DOOR RANGE CONTD.

GRADE 0.5%

CUT DOOR RANGE CONTD.

GRADE 0.5%

5M
5M
5M
5M
5M

N

IN DOOR SHOOTING

COND
EASEL

PLS
M.

SHOOTING HALL

SPECTATORS

CORRIDOR

ARMOURY

INSTRUCTOR

TOILE

OFFICE

OFFICE

TARGET PIT (1)

GANTRY 100M RANGE

TARGET PIT (2)
2.00 M
RANGE

ENTRY

5M 5M 5M 5M 5M 5M

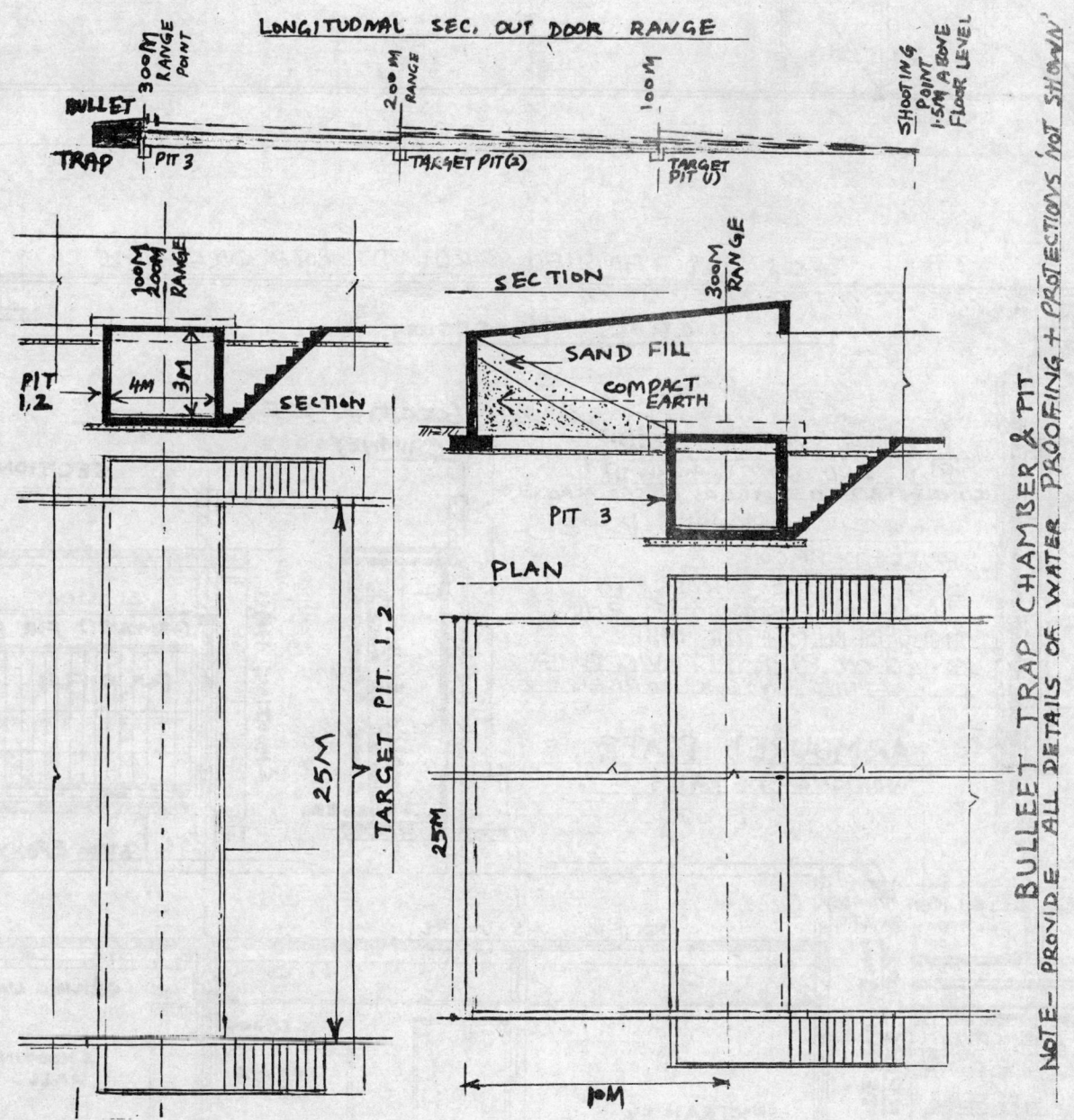

LONGITUDINAL SEC. OUT DOOR RANGE

BULLEET TRAP CHAMBER & PIT

NOTE— PROVIDE ALL DETAILS OF WATER PROOFING + PROTECTIONS 'NOT SHOWN'

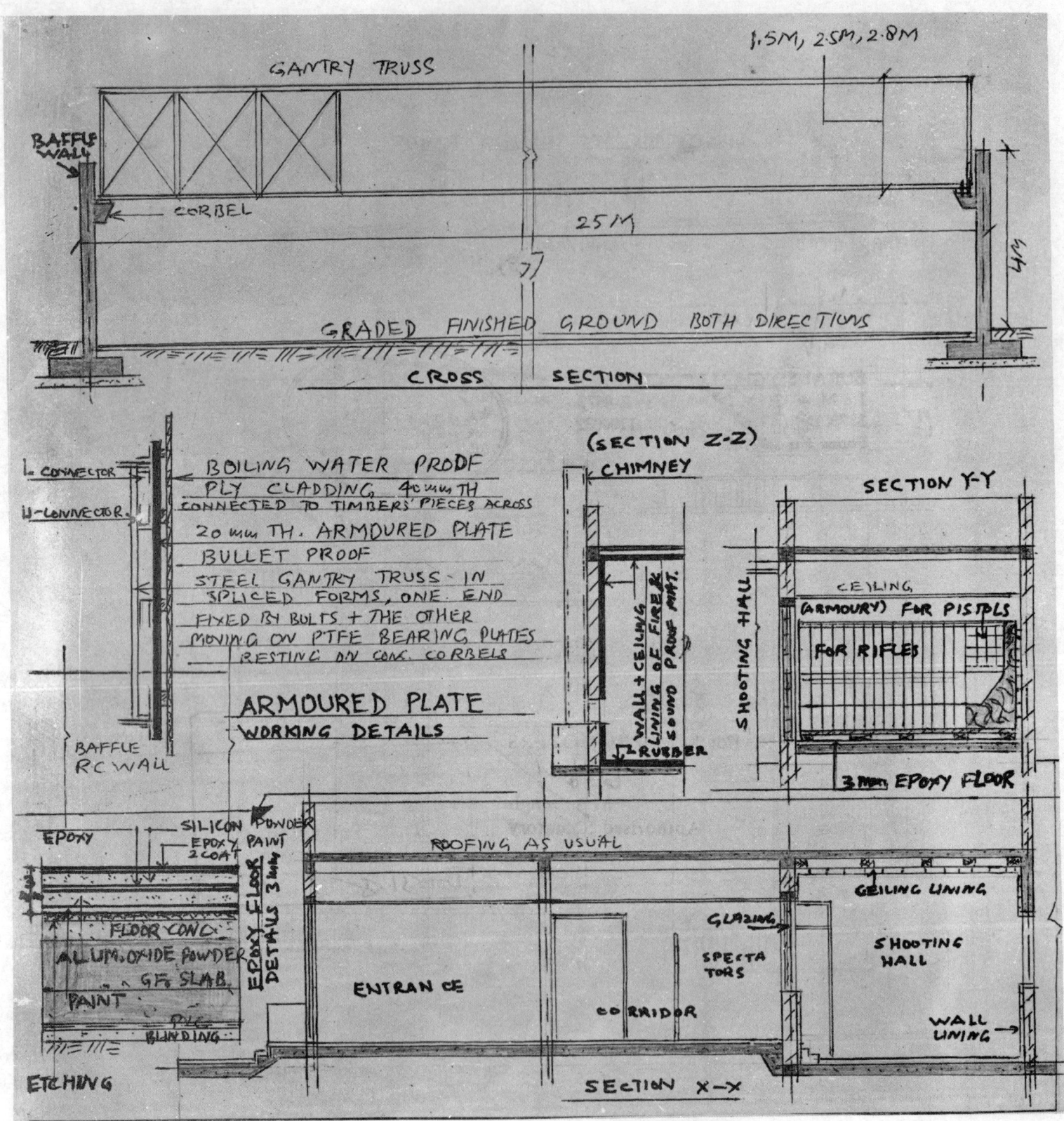

GANTRY TRUSS

1.5M, 2.5M, 2.8M

BAFFLE WALL

CORBEL

25M

GRADED FINISHED GROUND BOTH DIRECTIONS

CROSS SECTION

L CONNECTOR

U CONNECTOR

BOILING WATER PROOF
PLY CLADDING 40mm TH
CONNECTED TO TIMBERS PIECES ACROSS

20mm TH. ARMOURED PLATE
BULLET PROOF

STEEL GANTRY TRUSS IN
SPLICED FORMS, ONE END
FIXED BY BOLTS + THE OTHER
MOVING ON PTFE BEARING PLATES
RESTING ON CONC. CORBELS

ARMOURED PLATE
WORKING DETAILS

(SECTION Z-Z)
CHIMNEY

WALL+CEILING LINING OF FIRE & SOUND PROOF MAT.

RUBBER

BAFFLE RC WALL

SECTION Y-Y

CEILING
(ARMOURY) FOR PISTOLS

FOR RIFLES

SHOOTING HALL

3mm EPOXY FLOOR

EPOXY

SILICON POWDER
EPOXY PAINT
2 COAT

PAINT

FLOOR CONC.
ALLUM. OXIDE POWDER
GF. SLAB.

PAINT

PCC BLINDING

ETCHING

EPOXY FLOOR DETAILS 3mm

ROOFING AS USUAL

CEILING LINING

GLAZING

SPECTATORS

SHOOTING HALL

ENTRANCE

CORRIDOR

WALL LINING

SECTION X-X

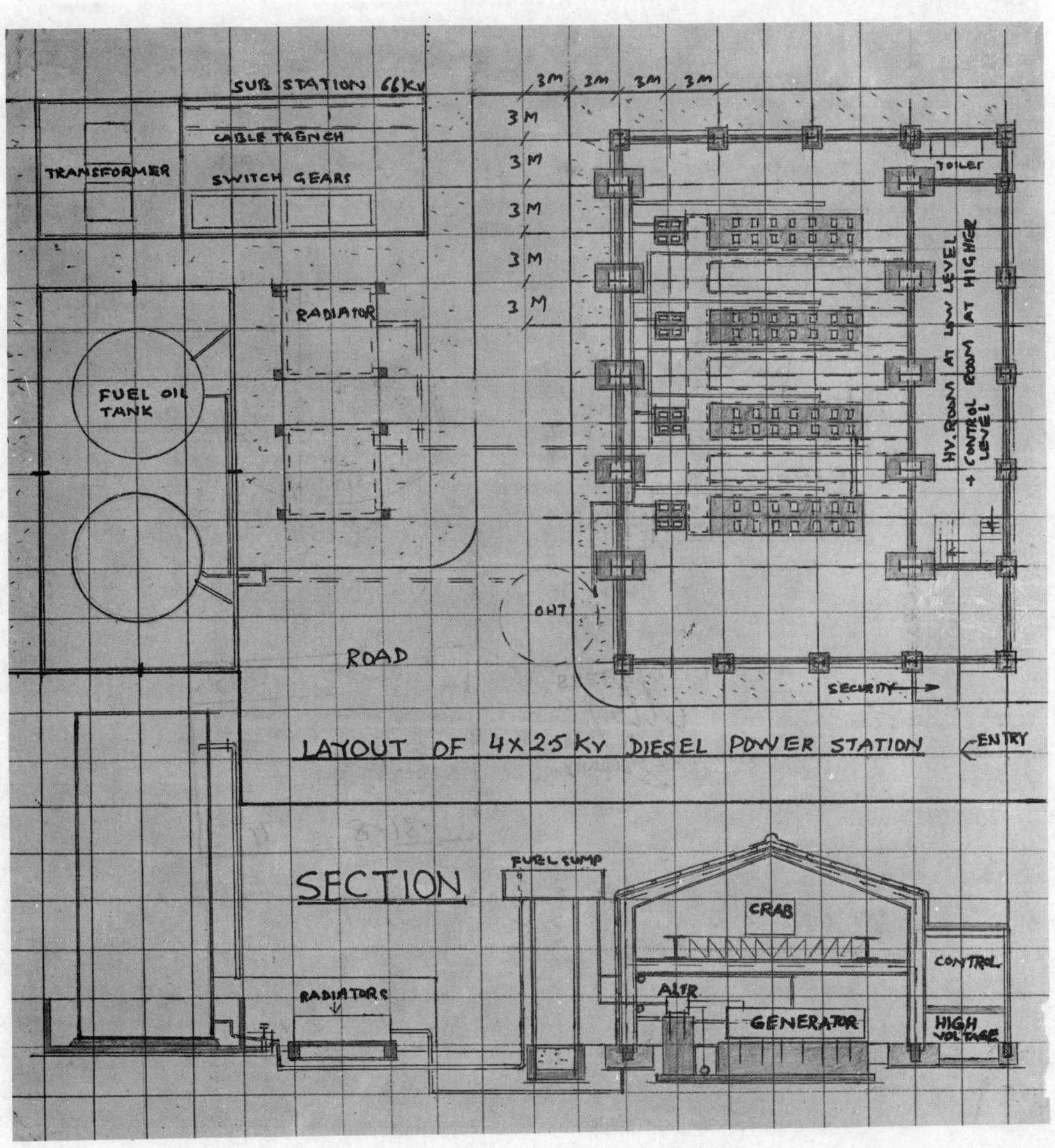

SUB STATION 66Kv

TRANSFORMER

CABLE TRENCH

SWITCH GEARS

3M 3M 3M 3M

3 M

3 M

3 M

3 M

3 M

RADIATOR

FUEL OIL
TANK

TOILET

HV. ROOM AT LOW LEVEL
+ CONTROL ROOM AT HIGHER
LEVEL

OHT

ROAD

SECURITY

LAYOUT OF 4×25 KV DIESEL POWER STATION

ENTRY

SECTION

FUEL SUMP

CRAB

RADIATORS

AIR

GENERATOR

CONTROL

HIGH
VOLTAGE

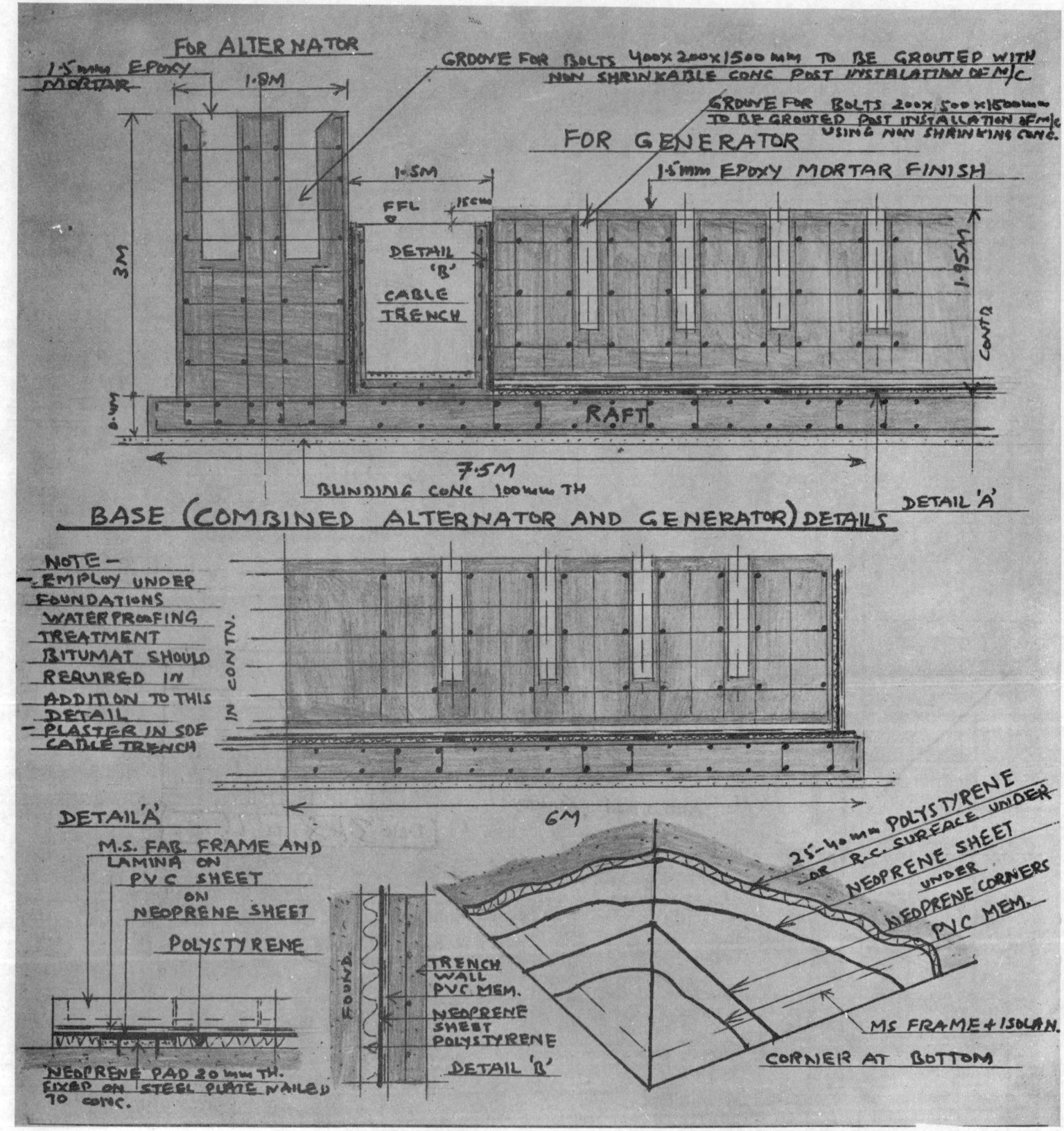

FOR ALTERNATOR

1·5 mm EPOXY MORTAR

1·8M

3M

GROOVE FOR BOLTS 400×200×1500 MM TO BE GROUTED WITH NON SHRINKABLE CONC POST INSTALATION OF M/C

GROOVE FOR BOLTS 200×500×1500mm TO BE GROUTED POST INSTALLATION OF M/C USING NON SHRINKING CONC.

FOR GENERATOR

1·5mm EPOXY MORTAR FINISH

1·5M

FFL

15cm

DETAIL 'B' CABLE TRENCH

1·95M

CONTR

0·4M

RAFT

DETAIL 'A'

7·5M

BLINDING CONC 100mm TH

BASE (COMBINED ALTERNATOR AND GENERATOR) DETAILS

NOTE –
– EMPLOY UNDER FOUNDATIONS WATERPROOFING TREATMENT BITUMAT SHOULD REQUIRED IN ADDITION TO THIS DETAIL
– PLASTER IN SDE CABLE TRENCH

IN CONTN.

6M

DETAIL 'A'

M.S. FAB. FRAME AND LAMINA ON PVC SHEET ON NEOPRENE SHEET POLYSTYRENE

FOUND.

NEOPRENE PAD 20 mm TH. FIXED ON STEEL PLATE NAILED TO CONC.

TRENCH WALL PVC MEM. NEOPRENE SHEET POLYSTYRENE

DETAIL 'B'

25-40 mm POLYSTYRENE OR R.C. SURFACE UNDER NEOPRENE SHEET UNDER NEOPRENE CORNERS PVC MEM.

MS FRAME & ISOLAN.

CORNER AT BOTTOM

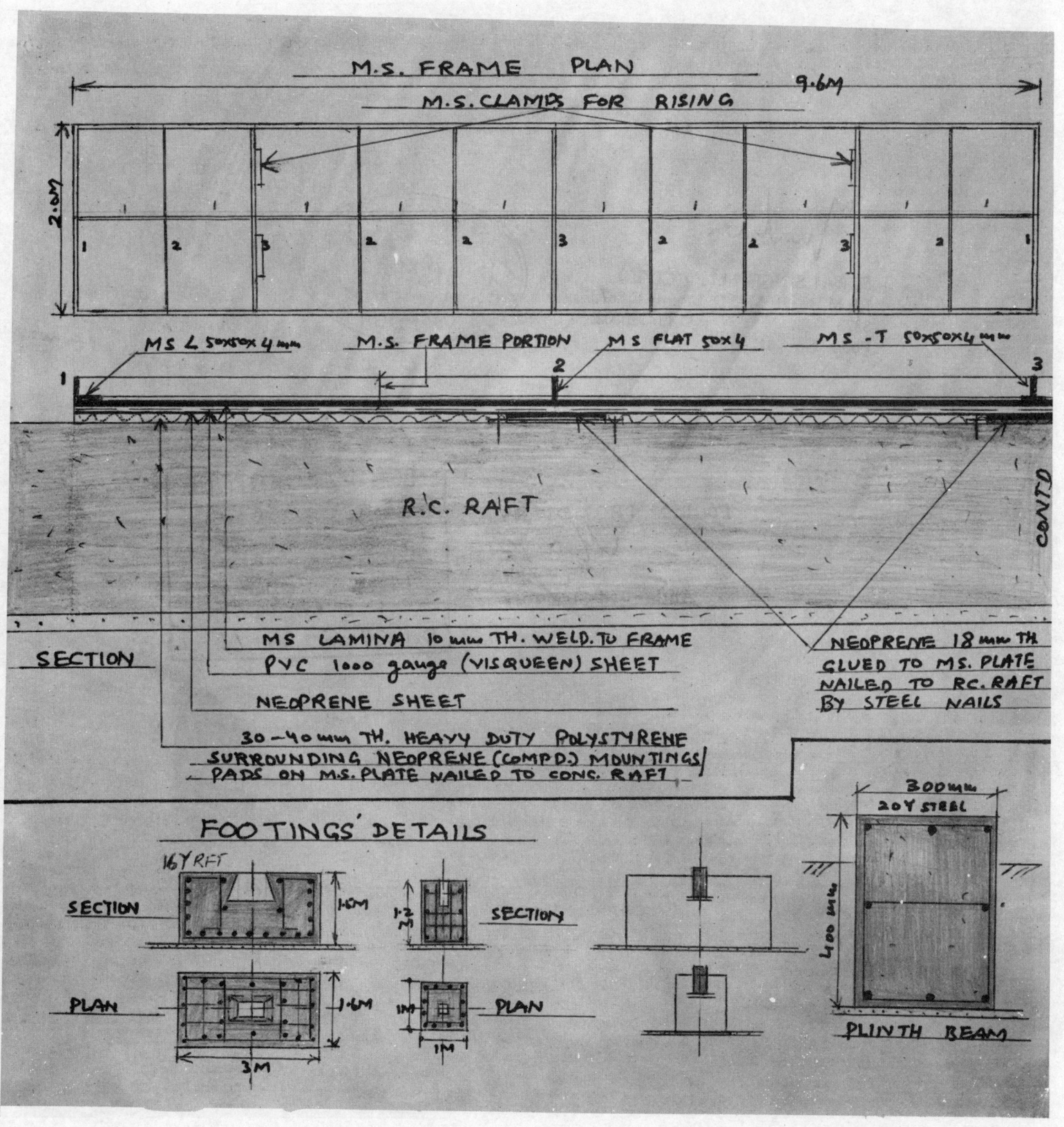

M.S. FRAME   PLAN

M.S. CLAMPS FOR RISING

9.6M

2.0M

MS L 50x50x4mm   M.S. FRAME PORTION   MS FLAT 50x4   MS - T 50x50x4mm

SECTION

R.C. RAFT

CONTD

MS LAMINA 10mm TH. WELD. TO FRAME

PVC 1000 gauge (VISQUEEN) SHEET

NEOPRENE SHEET

30 - 40mm TH. HEAVY DUTY POLYSTYRENE
SURROUNDING NEOPRENE (COMPD.) MOUNTINGS/
PADS ON M.S. PLATE NAILED TO CONC. RAFT

NEOPRENE 18mm TH.
GLUED TO M.S. PLATE
NAILED TO R.C. RAFT
BY STEEL NAILS

FOOTINGS DETAILS

16Y RFT

SECTION   1.5M

1.2M   SECTION

PLAN   1.6M

1M   PLAN

1M

300mm

20Y STEEL

400mm

PLAN   3M

PLINTH BEAM

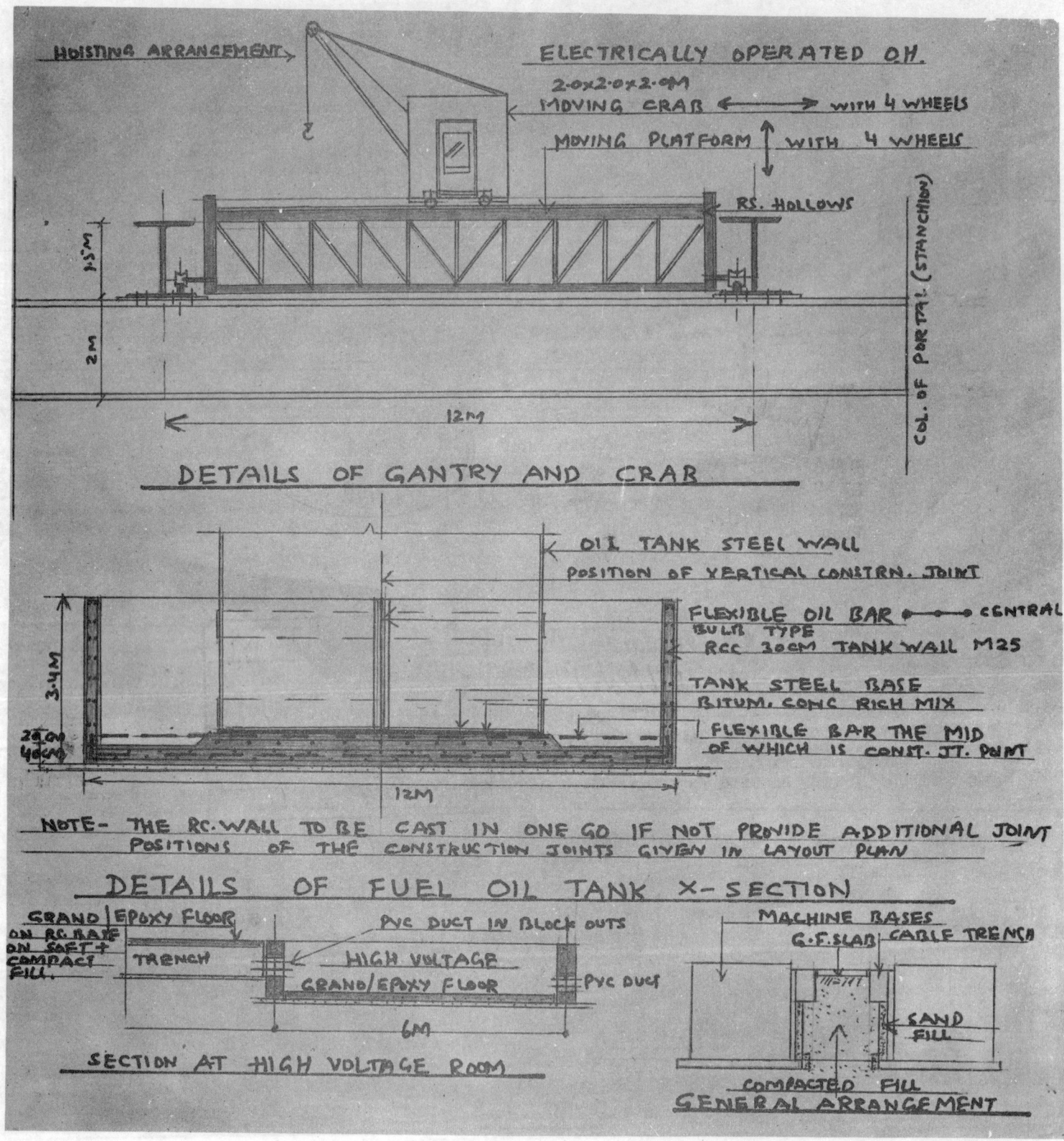

HOISTING ARRANGEMENT

ELECTRICALLY OPERATED OH.
2·0×2·0×2·0M
MOVING CRAB ← → WITH 4 WHEELS
MOVING PLATFORM ↕ WITH 4 WHEELS

RS. HOLLOWS

COL. OF PORTAL (STANCHION)

1·5M

2M

12M

DETAILS OF GANTRY AND CRAB

OIL TANK STEEL WALL
POSITION OF VERTICAL CONSTRN. JOINT

FLEXIBLE OIL BAR ← → CENTRAL
BULB TYPE
RCC 30CM TANK WALL M25

TANK STEEL BASE
BITUM. CONC RICH MIX

FLEXIBLE BAR THE MID
OF WHICH IS CONST. JT. POINT

3·4M

25CM
40CM

12M

NOTE- THE RC. WALL TO BE CAST IN ONE GO IF NOT PROVIDE ADDITIONAL JOINT
POSITIONS OF THE CONSTRUCTION JOINTS GIVEN IN LAYOUT PLAN

DETAILS OF FUEL OIL TANK X-SECTION

GRANO/EPOXY FLOOR
ON RC. BASE
ON SOFT &
COMPACT
FILL.

PVC DUCT IN BLOCK OUTS

TRENCH

HIGH VOLTAGE
GRANO/EPOXY FLOOR

PVC DUCT

6M

MACHINE BASES
G.F. SLAB   CABLE TRENCH

SAND
FILL

COMPACTED FILL

SECTION AT HIGH VOLTAGE ROOM

GENERAL ARRANGEMENT

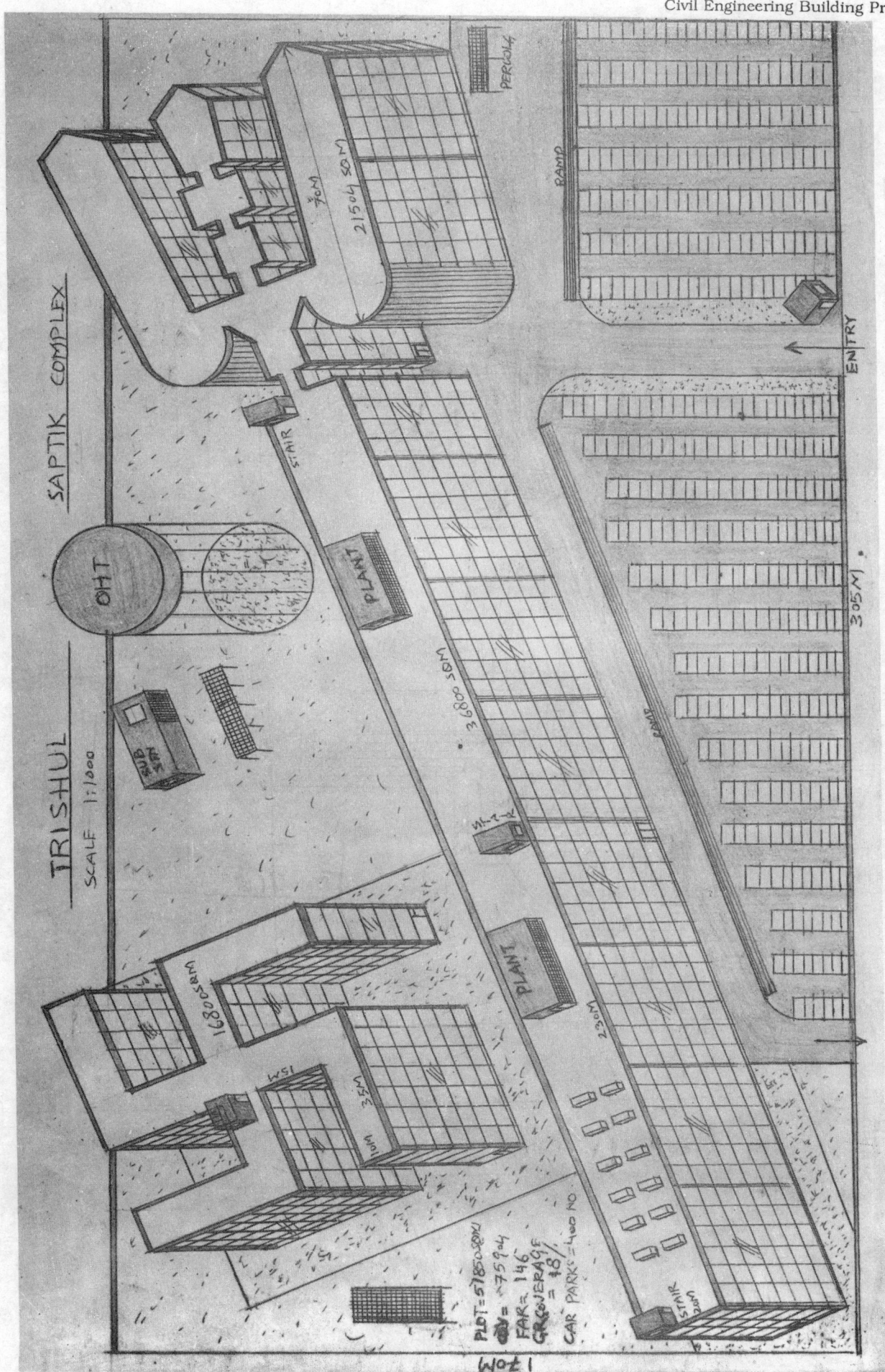

TRISHUL

SAPTIK COMPLEX

SCALE 1:1000

OHT

PLANT

PLANT

21504 SQM

36800 SQM

16800 SQM

70M

15M

3.5M

10M

23.0M

PERGOLA

STAIR

STAIR 30M

SUB STN

RAMP

RAMP

ENTRY

305 M.

170M

PLOT = 57850 SQM
=75904
FAR = 1.46
GROUND COVERAGE = 18.1
CAR PARKS = 400 NO

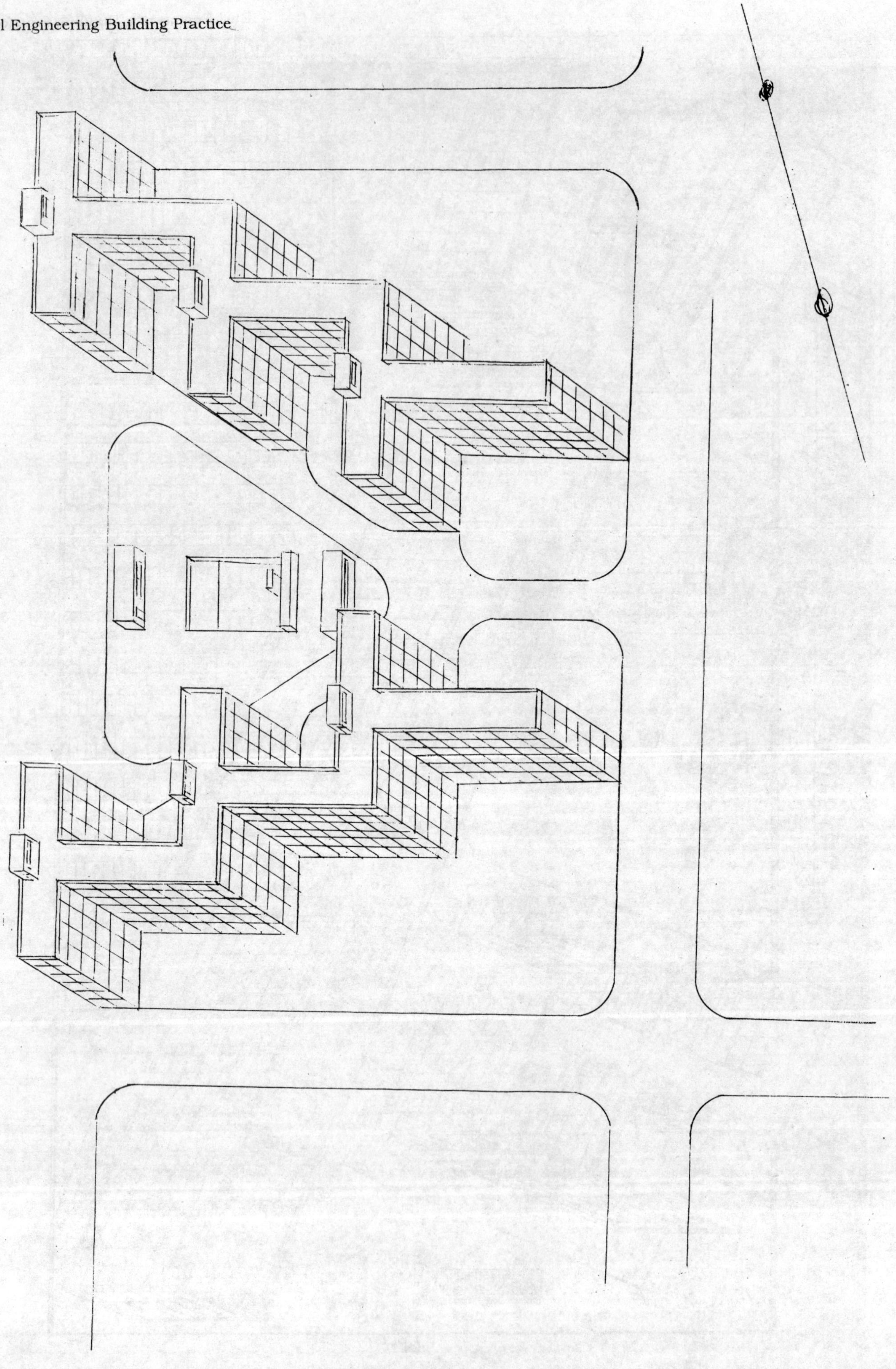

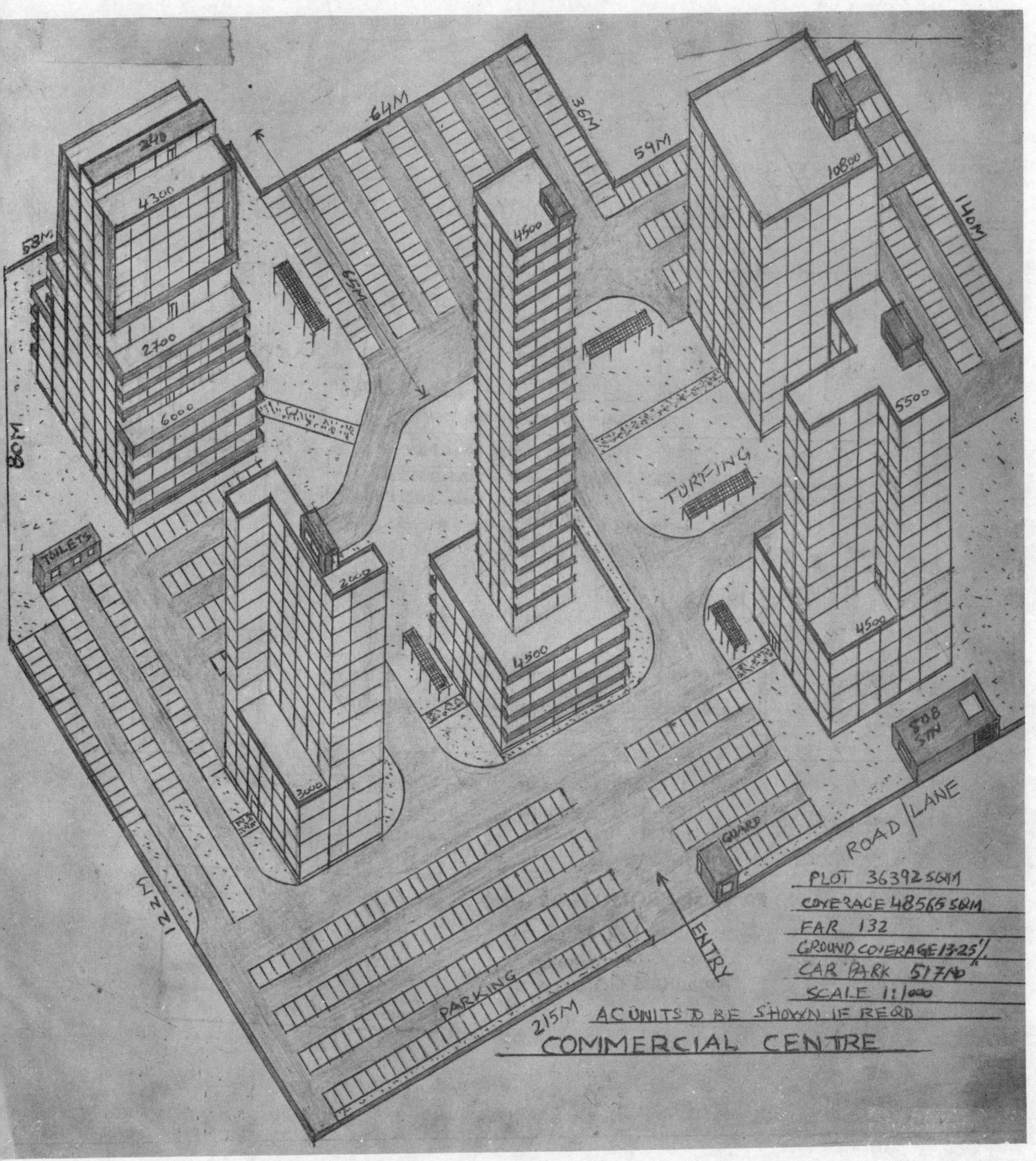

COMMERCIAL CENTRE

25M×13M FINISHED SIZE, 3·2M MAX. AND 1·2M MIN.

SWIMMING POOL LAYOUT AND SERVICES
TYPICAL

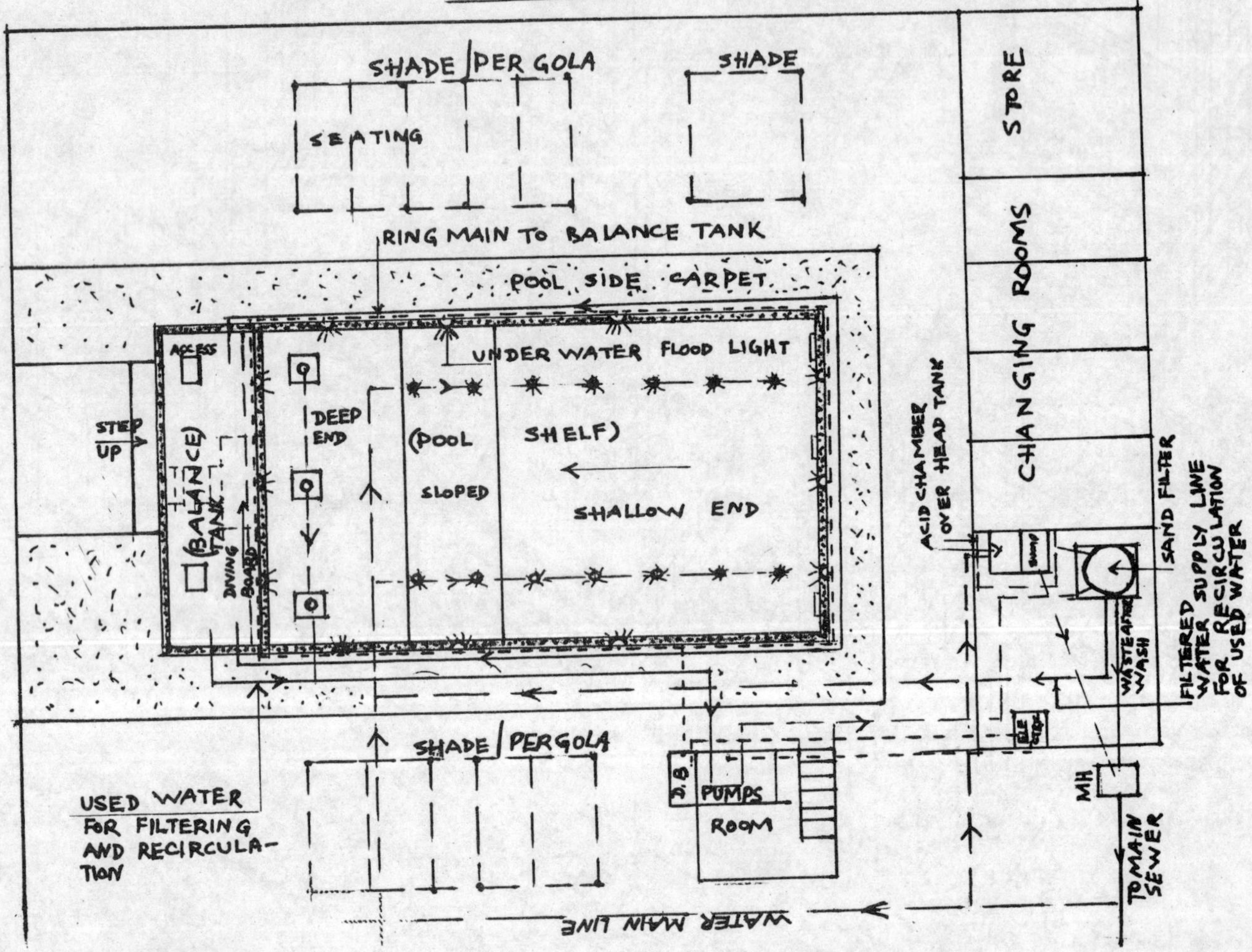

SHADE PERGOLA   SHADE

STORE

SEATING

CHANGING ROOMS

RING MAIN TO BALANCE TANK

POOL SIDE CARPET

ACCESS

UNDER WATER FLOOD LIGHT

STEP UP

(BALANCE TANK)

DEEP END

(POOL SLOPED

SHELF)

SHALLOW END

DIVING BOARD

ACID CHAMBER OVER HEAD TANK

PUMP

SAND FILTER

WASTE AFTER WASH

FILTERED SUPPLY LINE FOR RECIRCULATION OF USED WATER

USED WATER FOR FILTERING AND RECIRCULATION

SHADE PERGOLA

PUMPS ROOM

MH

TO MAIN SEWER

WATER MAIN LINE

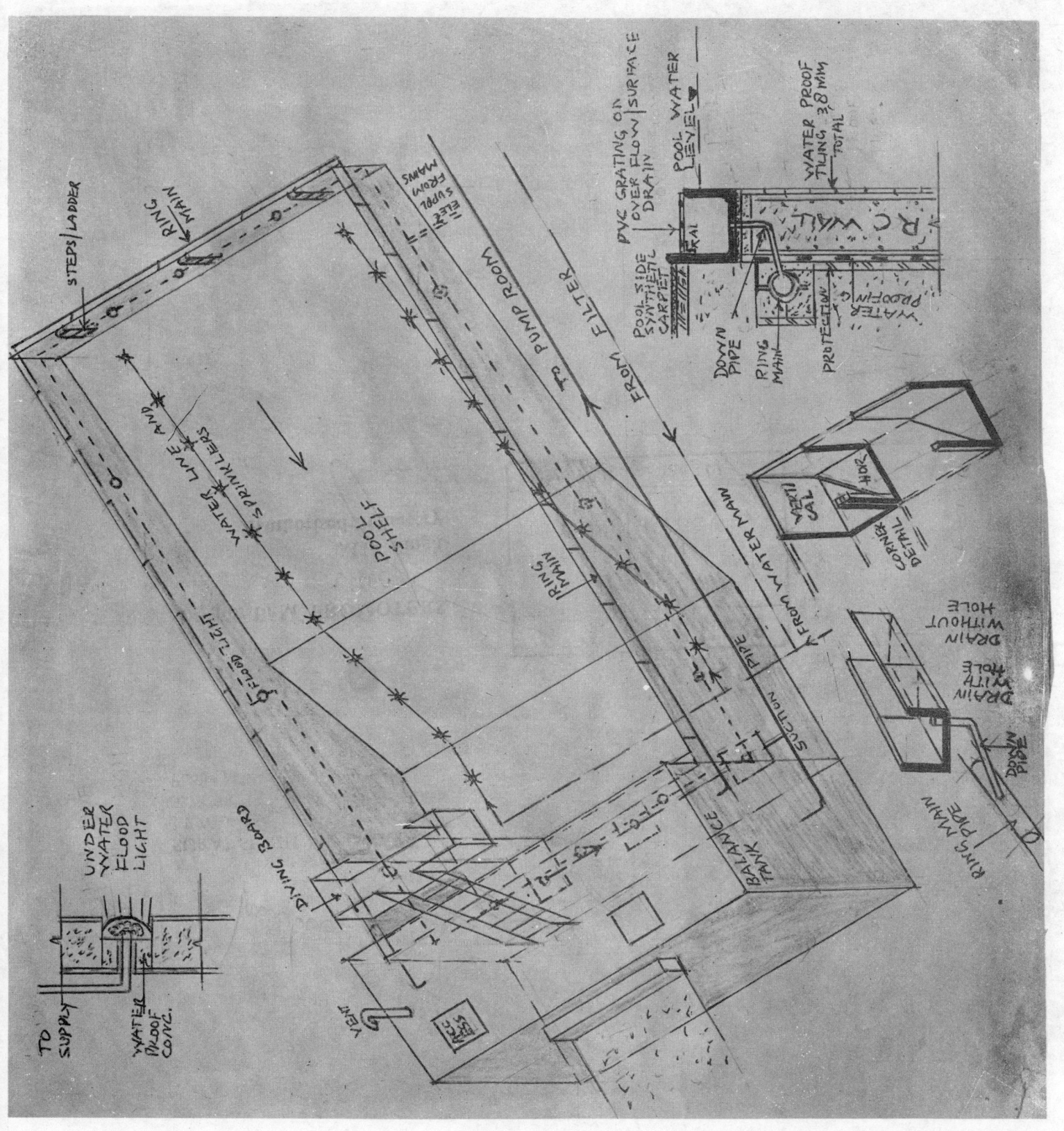

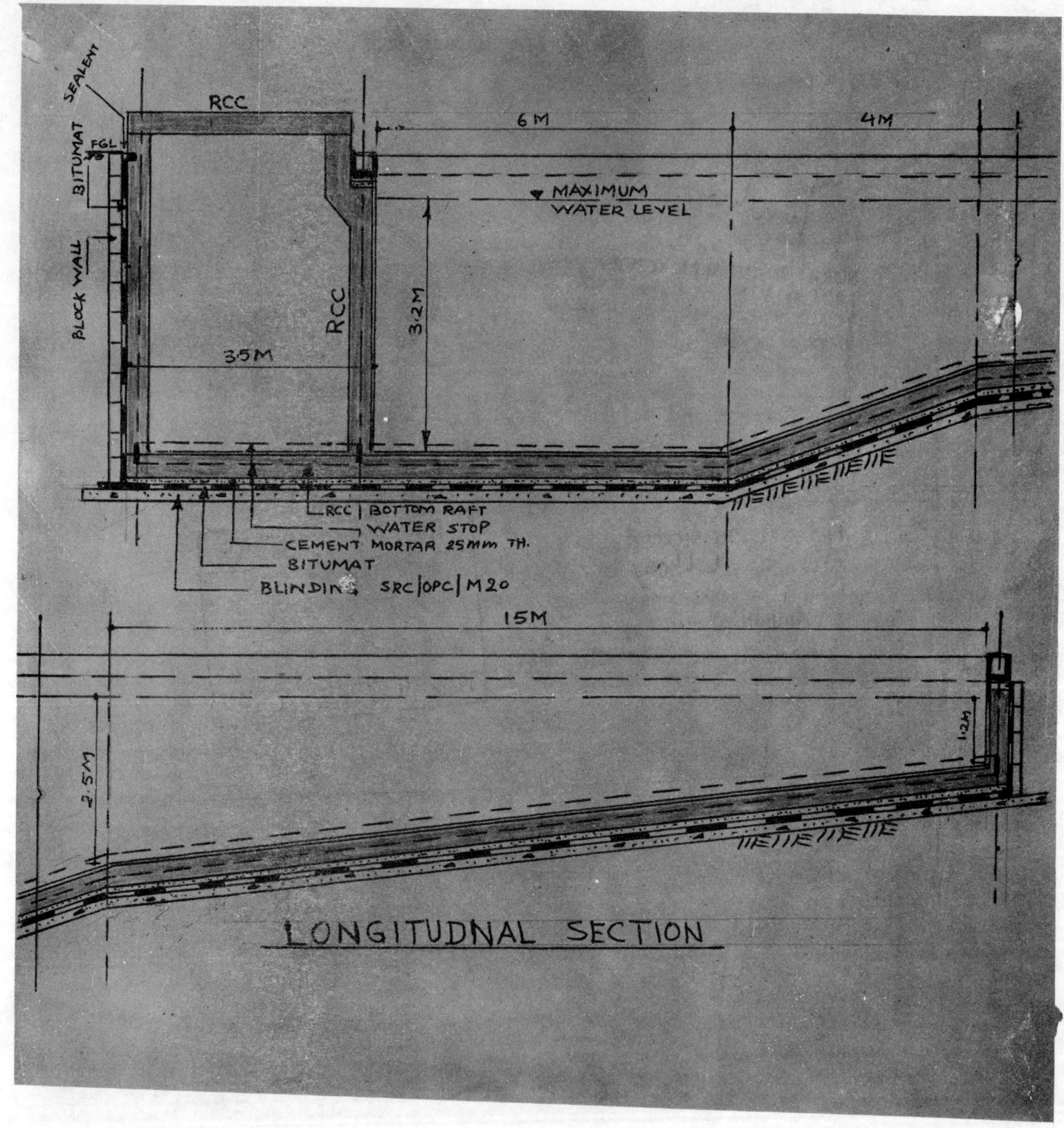

SEALENT

BITUMAT

FGL

BLOCK WALL

RCC

RCC

6 M

4M

MAXIMUM
WATER LEVEL

3·2M

3·5M

RCC BOTTOM RAFT
WATER STOP
CEMENT MORTAR 25mm TH.
BITUMAT
BLINDING SRC/OPC/M20

15M

2·5M

1·2M

LONGITUDNAL SECTION

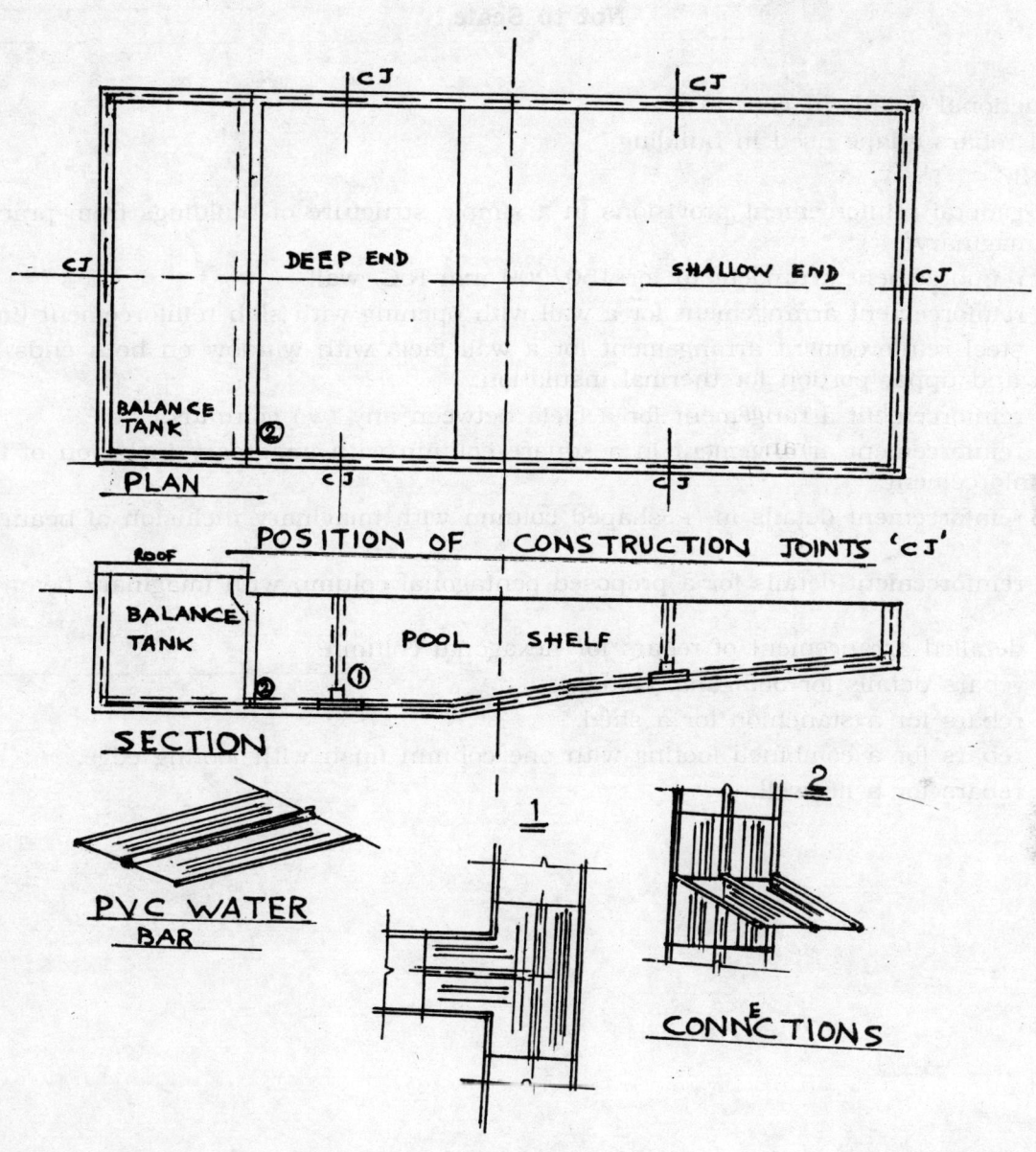

CJ   CJ

CJ   DEEP END   SHALLOW END   CJ

BALANCE
TANK   ②

PLAN   CJ   CJ

POSITION OF CONSTRUCTION JOINTS 'CJ'

ROOF
BALANCE
TANK   POOL SHELF
         ②   ①

SECTION

PVC WATER
BAR

1   2

CONNCTIONS

Sketches 135 to 151

Constructional Site Details for Information

First Hand Written

Not to Scale

135. Constructional details in general.
136. General rebars shape used in building.
137. As in 136.
138. Typical general reinforcement provisions in a simple structure of buildings from practical view point (imaginary).
139. Typical reinforcement arrangement for 150/200 mm R.C. wall.
140. Typical reinforcement arrangement for a wall with opening with slab reinforcement (imaginary).
141. Typical steel reinforcement arrangement for a wall facia with window on both ends with R.C. column and upper portion for thermal insulation.
142. Typical reinforcement arrangement for a facia between any two columns.
143. Typical reinforcement arrangement in a square column with imaginary inclusion of beam and slab reinforcement.
144. Typical reinforcement details in '+' shaped column with imaginary inclusion of beam and slab rebars.
145. Typical reinforcement details for a proposed pentagonal column with imaginary beam and slab rebars.
146. Typical detailed arrangement of rebars for hexagonal column.
147. Typical rebars details for octagonal column.
148. Typical rebars for a stanchion for a shed.
149. Typical rebars for a combined footing with one column flush with footing edge.
150. Typical rebars for a lift well.

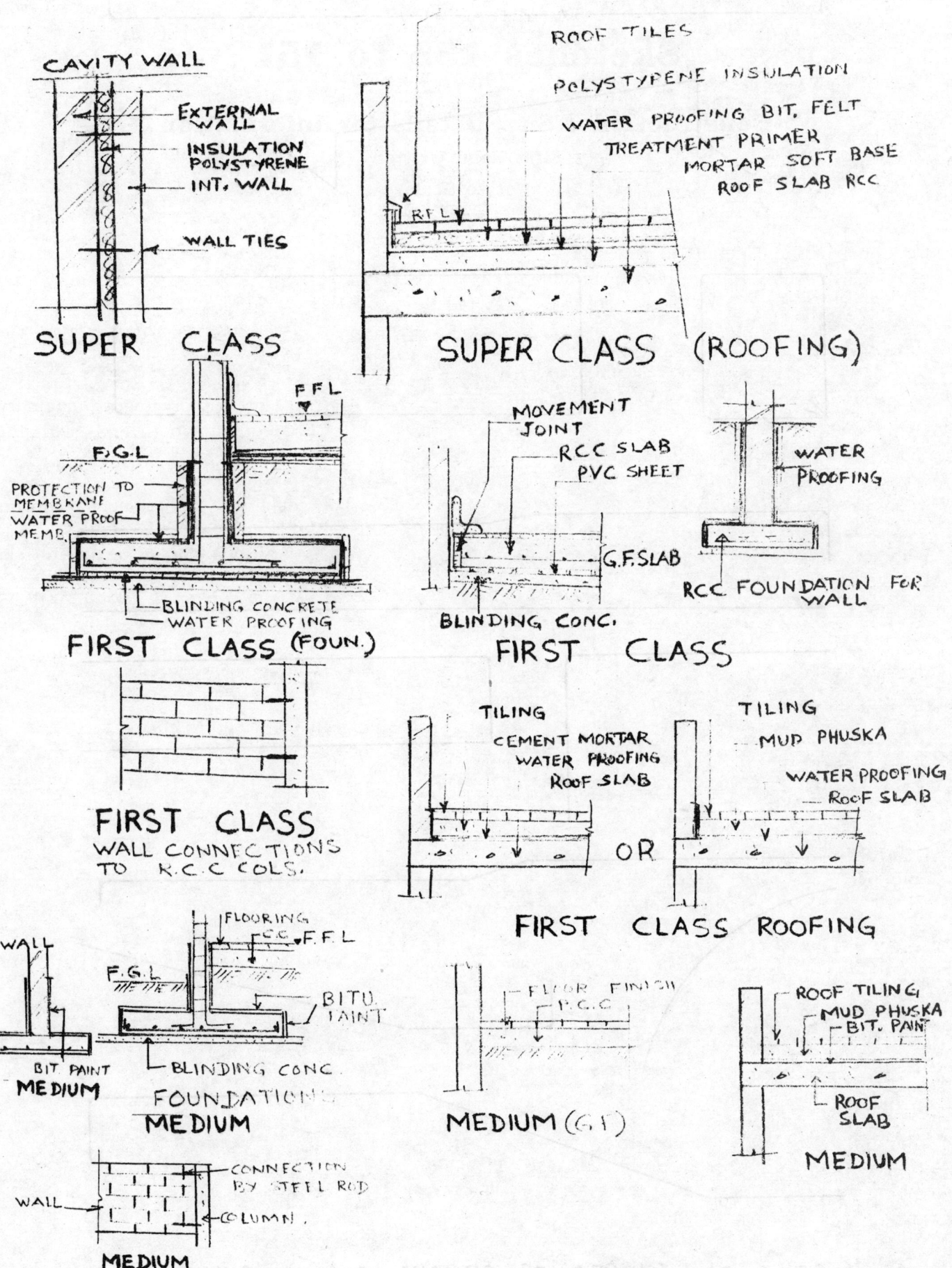

ALUMINIUM SOLAR REFLECTOR

CAVITY WALL

EXTERNAL WALL
INSULATION POLYSTYRENE
INT. WALL
WALL TIES

SUPER CLASS

ROOF TILES

POLYSTYRENE INSULATION

WATER PROOFING BIT. FELT
TREATMENT PRIMER
MORTAR SOFT BASE
ROOF SLAB RCC

R.F.L

SUPER CLASS (ROOFING)

F.F.L
F.G.L
PROTECTION TO MEMBRANE
WATER PROOF MEMB.
BLINDING CONCRETE
WATER PROOFING

FIRST CLASS (FOUN.)

MOVEMENT JOINT
RCC SLAB
PVC SHEET
G.F. SLAB
BLINDING CONC.

FIRST CLASS

WATER PROOFING

RCC FOUNDATION FOR WALL

FIRST CLASS
WALL CONNECTIONS TO R.C.C COLS.

TILING
CEMENT MORTAR
WATER PROOFING
ROOF SLAB

TILING
MUD PHUSKA
WATER PROOFING
ROOF SLAB

OR

FIRST CLASS ROOFING

WALL
FLOORING
C.C. F.F.L
F.G.L
BITU. PAINT
BIT. PAINT
BLINDING CONC.

MEDIUM

FOUNDATIONS
MEDIUM

FLOOR FINISH
P.C.C

MEDIUM (G.F)

ROOF TILING
MUD PHUSKA
BIT. PAINT
ROOF SLAB

MEDIUM

WALL
CONNECTION BY STEEL ROD
COLUMN

MEDIUM

GENERAL SHAPES OF REBARS USED IN BUILDINGS

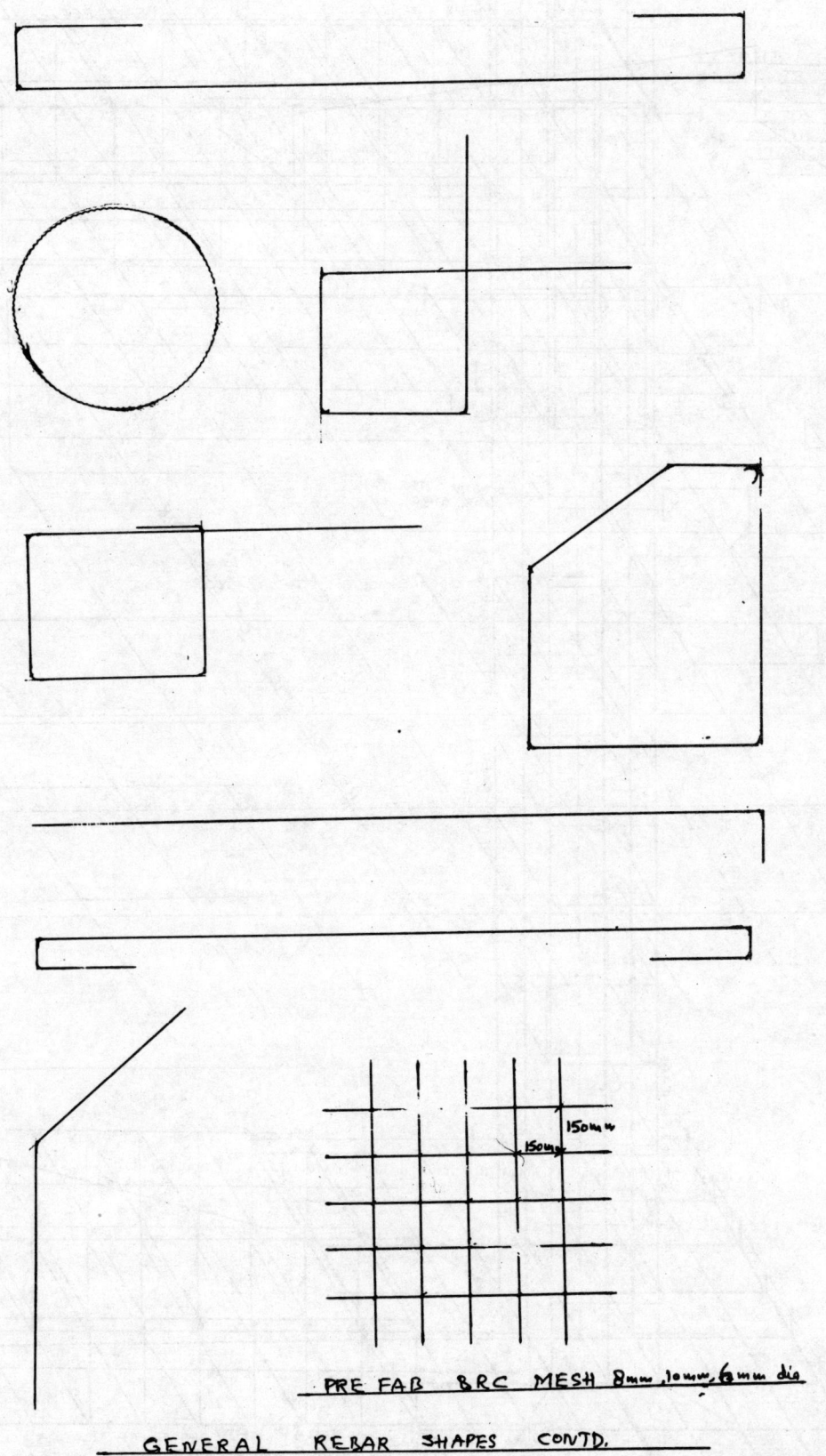

PRE FAB BRC MESH 8mm, 10mm, 6mm dia

GENERAL REBAR SHAPES CONTD.

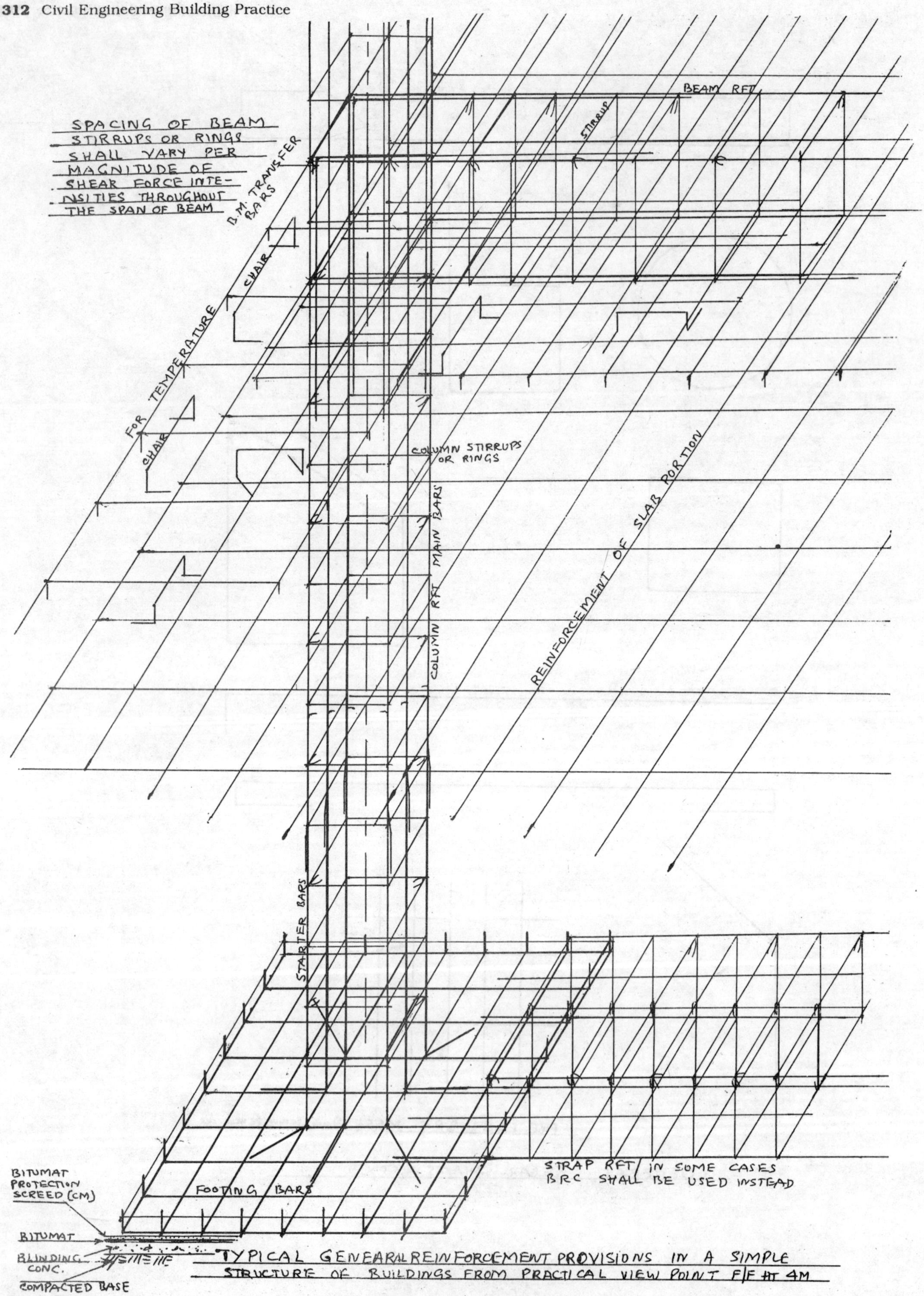

SPACING OF BEAM STIRRUPS OR RINGS SHALL VARY PER MAGNITUDE OF SHEAR FORCE INTENSITIES THROUGHOUT THE SPAN OF BEAM

D.M. TRANSFER BARS

FOR TEMPERATURE CHAIR

CHAIR

BEAM RFT

STIRRUP

COLUMN STIRRUPS OR RINGS

COLUMN RFT MAIN BARS

REINFORCEMENT OF SLAB PORTION

STARTER BARS

BITUMAT PROTECTION SCREED (CM)

BITUMAT

BLINDING CONC.

COMPACTED BASE

FOOTING BARS

STRAP RFT IN SOME CASES BRC SHALL BE USED INSTEAD

TYPICAL GENEARAL REINFORCEMENT PROVISIONS IN A SIMPLE STRUCTURE OF BUILDINGS FROM PRACTICAL VIEW POINT F/F HT 4M

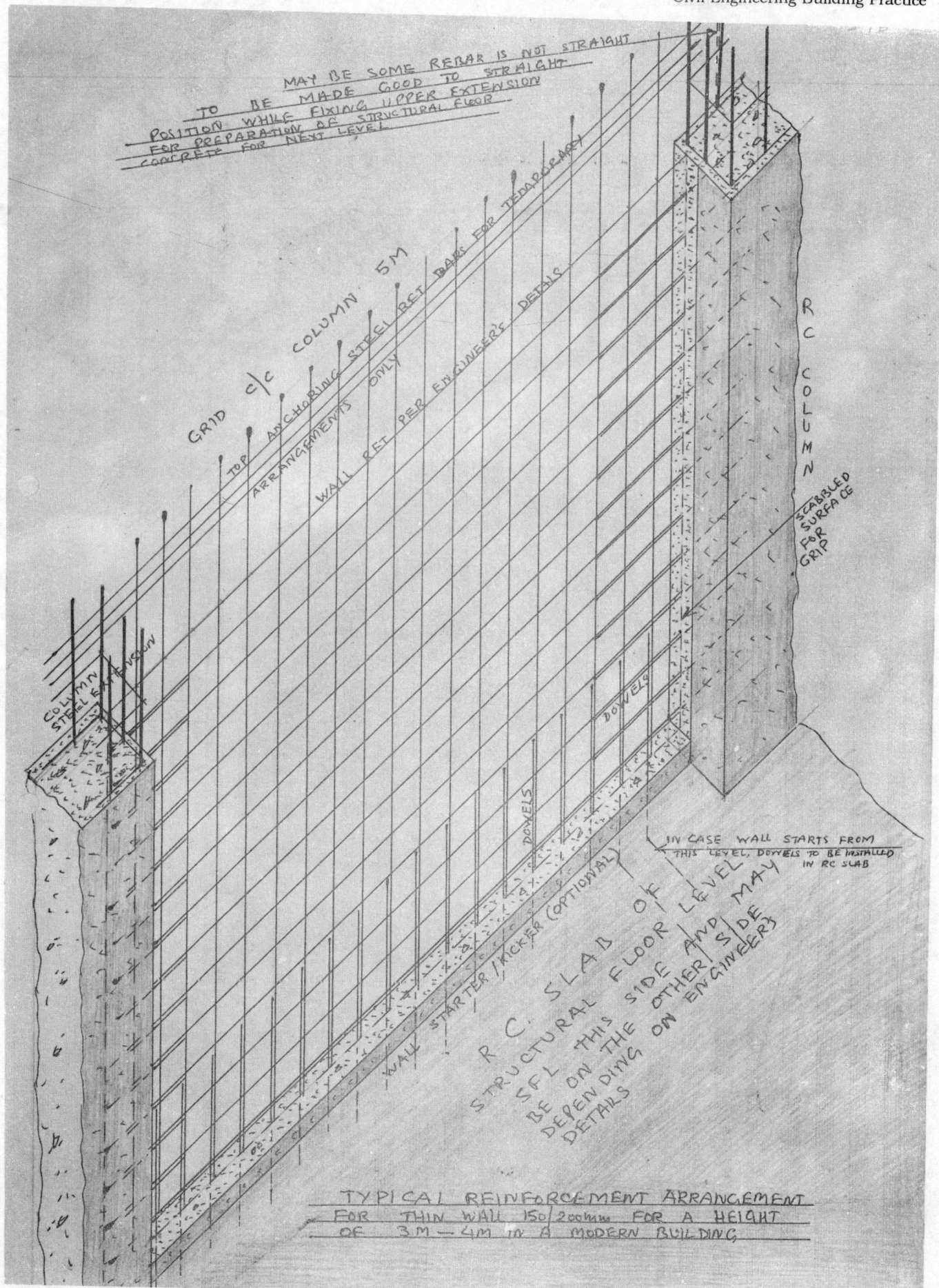

MAY BE SOME REBAR IS NOT STRAIGHT TO BE MADE GOOD TO STRAIGHT POSITION WHILE FIXING UPPER EXTENSION FOR PREPARATION OF STRUCTURAL FLOOR CONCRETE FOR NEXT LEVEL

GRID c/c COLUMN 5M

TOP ANCHORING STEEL REF BARS FOR TEMPORARY ARRANGEMENTS ONLY

WALL REF PER ENGINEER'S DETAILS

RC COLUMN

SCABBLED SURFACE FOR GRIP

COLUMN STEEL EXTENSION

DOWELS

DOWELS

IN CASE WALL STARTS FROM THIS LEVEL, DOWELS TO BE INSTALLED IN RC SLAB

WALL STARTER / KICKER (OPTIONAL)

RC SLAB OF STRUCTURAL FLOOR LEVEL SFL THIS SIDE AND MAY BE ON THE OTHER SIDE DEPENDING ON ENGINEER'S DETAILS

TYPICAL REINFORCEMENT ARRANGEMENT FOR THIN WALL 150/200mm FOR A HEIGHT OF 3M—4M IN A MODERN BUILDING.

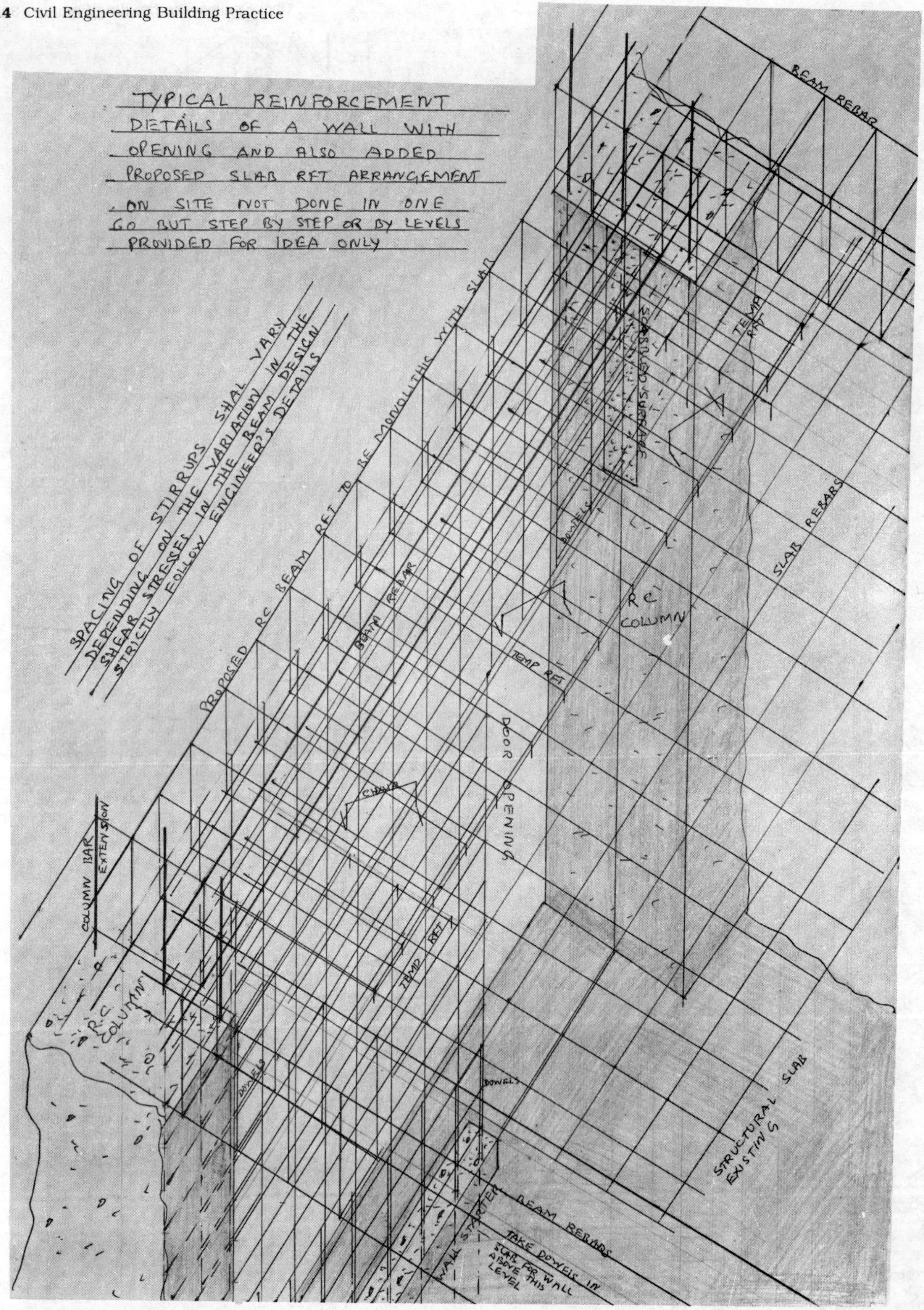

TYPICAL REINFORCEMENT DETAILS OF A WALL WITH OPENING AND ALSO ADDED PROPOSED SLAB RFT ARRANGEMENT. ON SITE NOT DONE IN ONE GO BUT STEP BY STEP OR BY LEVELS PROVIDED FOR IDEA ONLY

SPACING OF STIRRUPS SHALL VARY DEPENDING ON THE VARIATION IN THE SHEAR STRESSES IN THE BEAM DESIGN STRICTLY FOLLOW ENGINEER'S DETAILS

PROPOSED RC BEAM RFT TO BE MONOLITHIC WITH SLAB

BEAM REBAR

COLUMN BAR EXTENSION

R C COLUMN

DOWELS

WALL STARTER

BEAM REBARS

TAKE DOWELS FOR WALL IN SLAB ABOVE THIS LEVEL

TEMP RFT

CHAIR

DOOR OPENING

TEMP RFT

DOWELS

BEAM REBARS

STRUCTURAL EXISTING SLAB

SLAB REBARS

R C COLUMN

TEMP RFT

SCABBLED SURFACE

DOWELS

BEAM REBAR

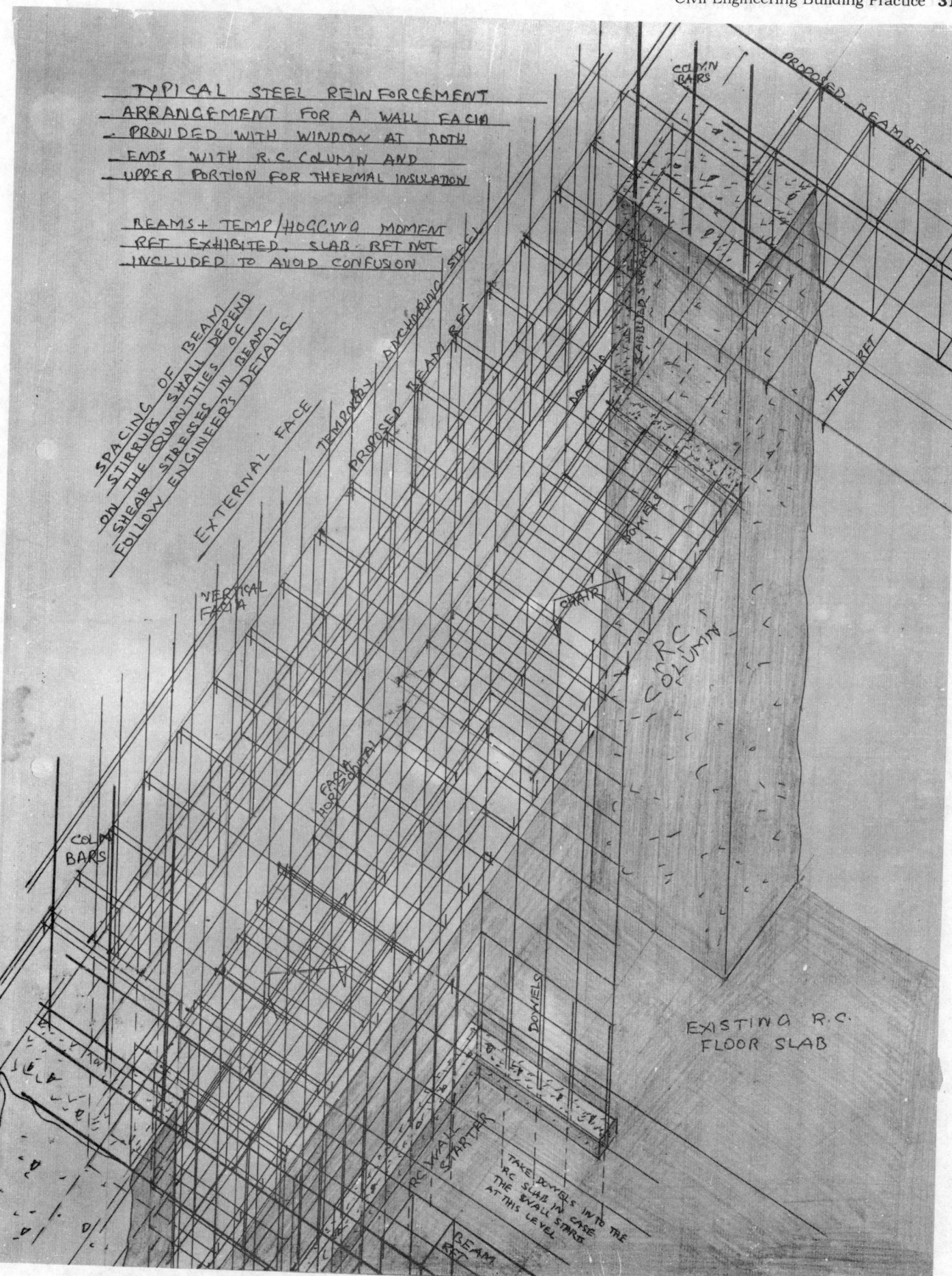

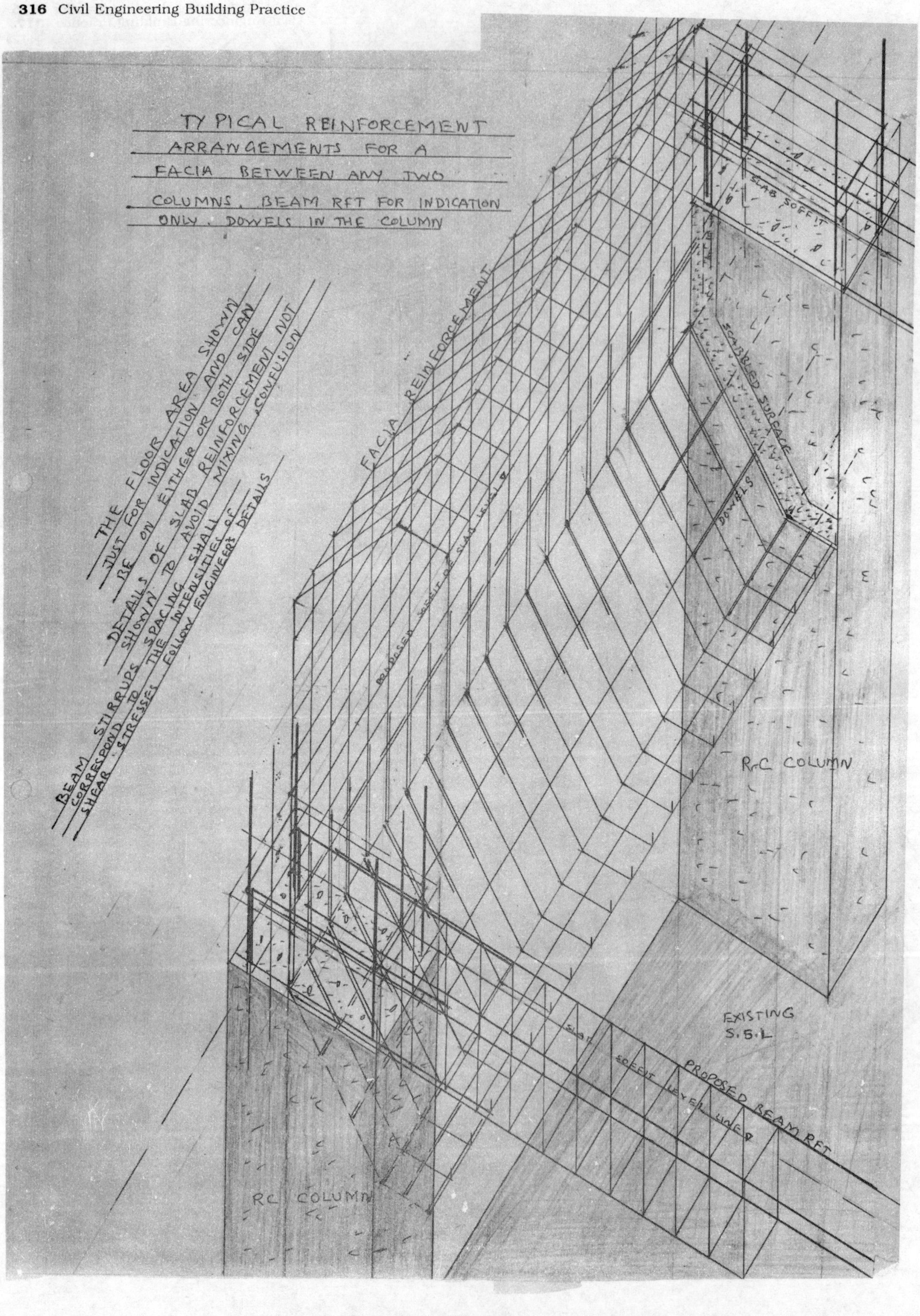

TYPICAL REINFORCEMENT
ARRANGEMENTS FOR A
FACIA BETWEEN ANY TWO
COLUMNS. BEAM RFT FOR INDICATION
ONLY. DOWELS IN THE COLUMN

THE FLOOR AREA SHOWN
JUST FOR INDICATION AND CAN
BE ON EITHER OR BOTH SIDE

DETAILS OF SLAB REINFORCEMENT NOT
SHOWN TO AVOID MIXING CONFUSION

BEAM STIRRUPS SPACING SHALL
CORRESPOND TO THE INTENSITIES OF
SHEAR STRESSES FOLLOW ENGINEERS DETAILS

FACIA REINFORCEMENT

SLAB SOFFIT

SCABBLED SURFACE

DOWELS

REMAIN SOFFIT OF SLAB LEVEL

RrC COLUMN

EXISTING
S.S.L

PROPOSED BEAM RFT

SCREET LEVER WED

SLAB R

RC COLUMN

CHAIR

PROPOSED BEAM RFT

PROPOSED BEAM RFT

SLAB REINFORCEMENT TWO WAY

DOWELS/OVERLAP

TYPICAL RFT ARRANGEMENTS IN A' SQUARE COLUMN

EXISTING STRUCTURAL FLOOR SLAB AROUND THE PROPOSED COLUMN + TO THE OTHER COLUMNS

KICKER/STARTER

BEAMS ARE FROM COLUMN/COLUMN
NOTE - DO NOT CONFUSE FOR THE
BEAMS TO BE CANTILEVER

PROPOSED FLOOR BEAM REFT

TEMP. REINFORCEMENT

SLAB REINFORCEMENT

DOWELS OR STARTER OVERLAP

KICKER STARTER

EXISTING STRUCTURAL FLOOR SLAB AROUND UPTO THE NEXT OR SURROUNDING COLUMNS NOT SHOWN

TYPICAL REINFORCEMENT DETAILED ARRANGEMENT OF + SHAPE COLUMN AND THE FLOOR LEVEL BEAMS AND THE SLAB RFT. FOR GUIDANCE ONLY

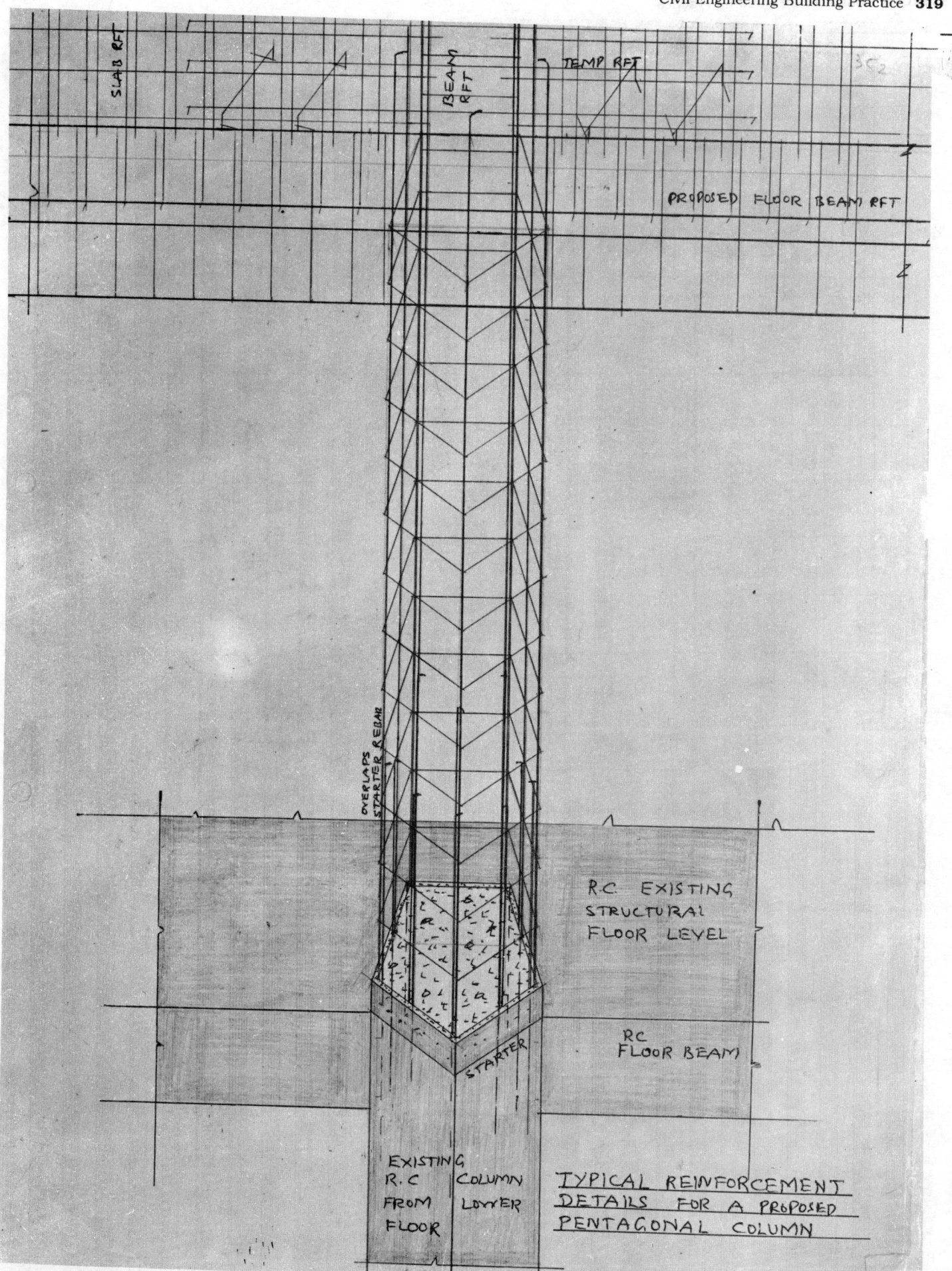

SLAB RFT

BEAM RFT

TEMP RFT

PROPOSED FLOOR BEAM RFT

OVERLAPS STARTER REBAR

R.C EXISTING
STRUCTURAL
FLOOR LEVEL

RC
FLOOR BEAM

STARTER

EXISTING
R.C COLUMN
FROM LOWER
FLOOR

TYPICAL REINFORCEMENT
DETAILS FOR A PROPOSED
PENTAGONAL COLUMN

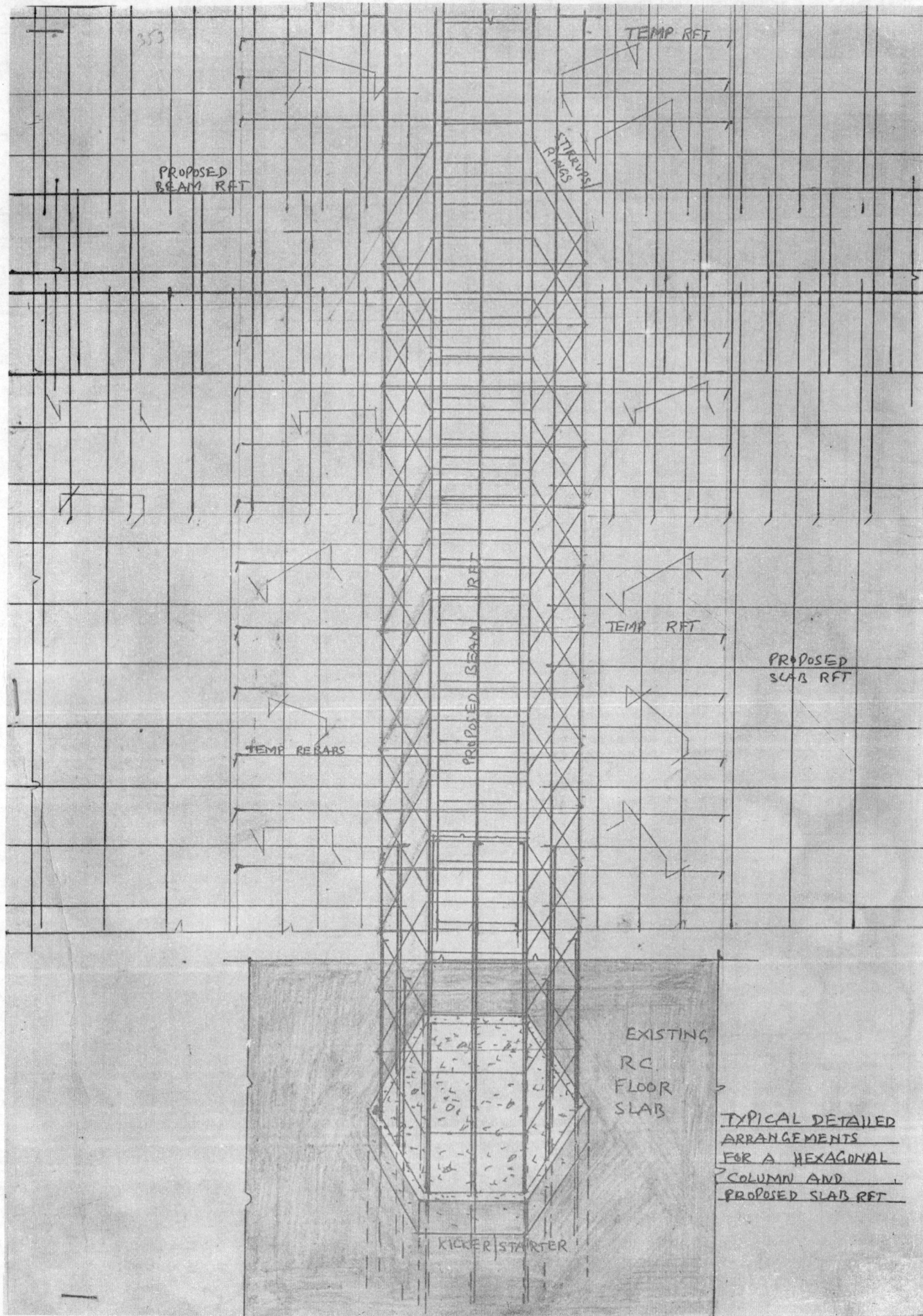

TEMP RFT

STIRRUPS/
RINGS

PROPOSED
BEAM RFT

TEMP RFT

PROPOSED
SLAB RFT

PROPOSED BEAM RFT

TEMP REBARS

EXISTING
R.C.
FLOOR
SLAB

TYPICAL DETAILED
ARRANGEMENTS
FOR A HEXAGONAL
COLUMN AND
PROPOSED SLAB RFT

KICKER STARTER

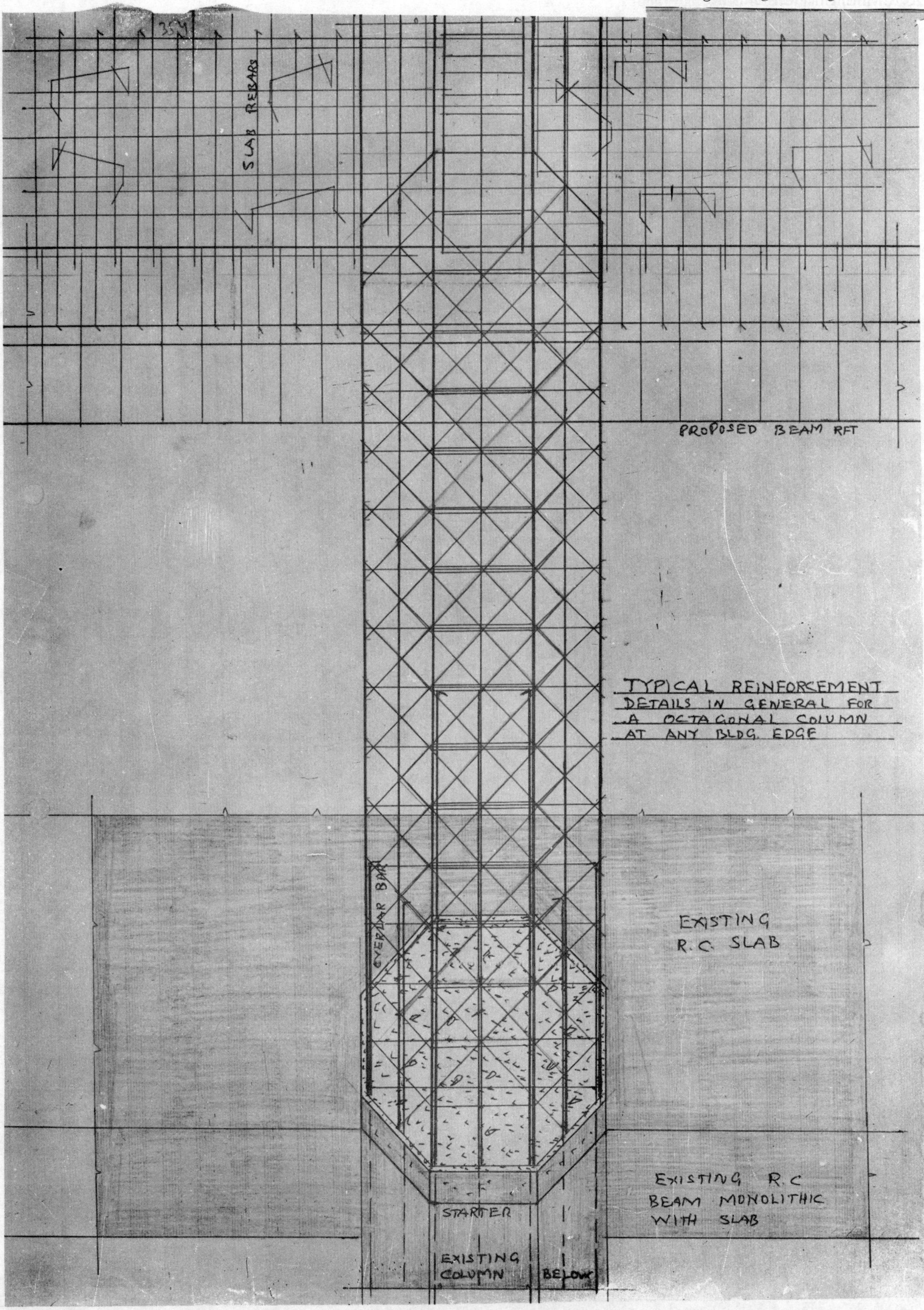

SLAB REBARS

PROPOSED BEAM RFT

TYPICAL REINFORCEMENT
DETAILS IN GENERAL FOR
A OCTAGONAL COLUMN
AT ANY BLDG. EDGE

OVERBAR BAR

EXISTING
R.C SLAB

STARTER

EXISTING R.C
BEAM MONOLITHIC
WITH SLAB

EXISTING
COLUMN BELOW

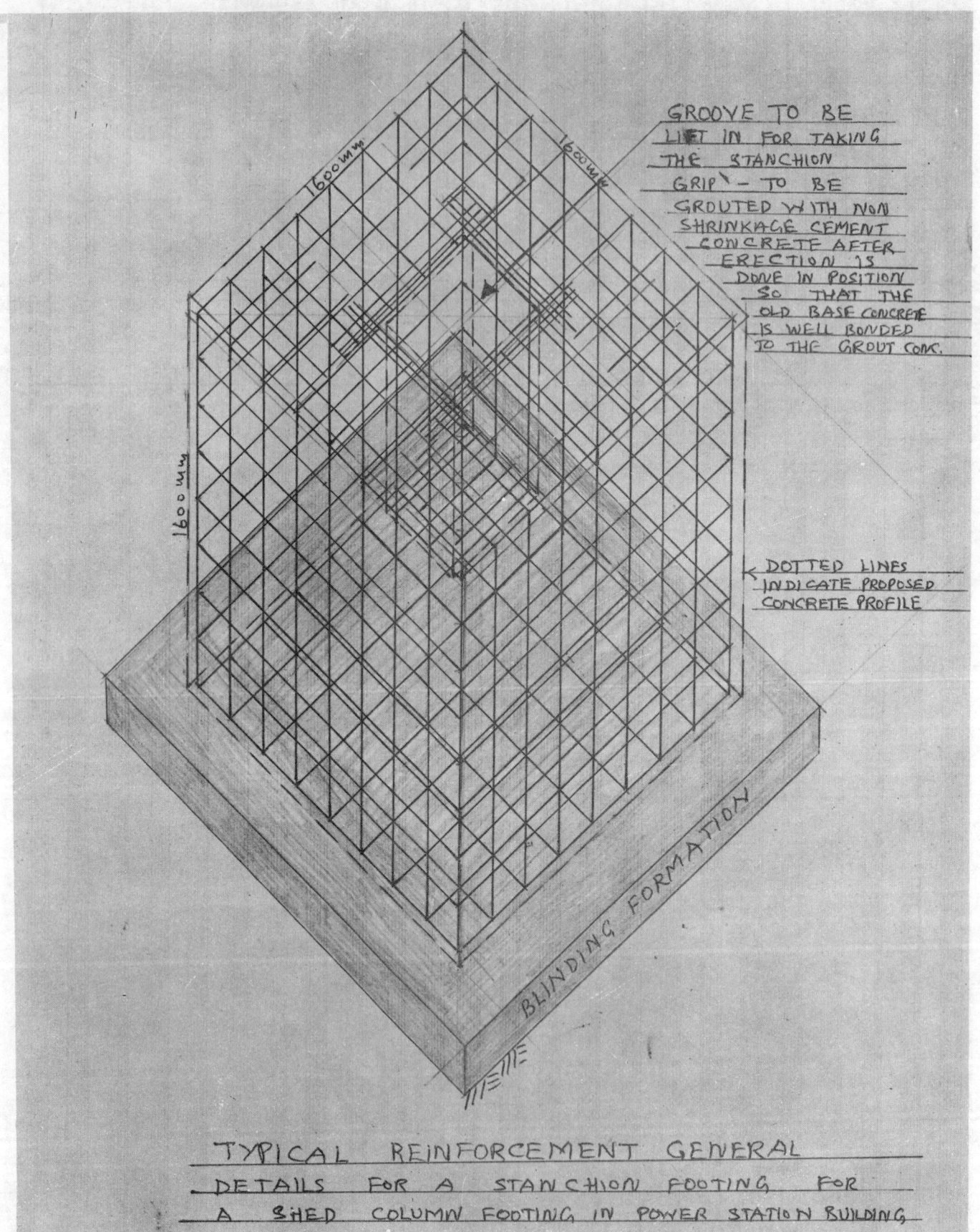

GROOVE TO BE LEFT IN FOR TAKING THE STANCHION GRIP - TO BE GROUTED WITH NON SHRINKAGE CEMENT CONCRETE AFTER ERECTION IS DONE IN POSITION SO THAT THE OLD BASE CONCRETE IS WELL BONDED TO THE GROUT CONC.

DOTTED LINES INDICATE PROPOSED CONCRETE PROFILE

1600 mm

1600 mm

1600 mm

BLINDING FORMATION

TYPICAL REINFORCEMENT GENERAL DETAILS FOR A STANCHION FOOTING FOR A SHED COLUMN FOOTING IN POWER STATION BUILDING

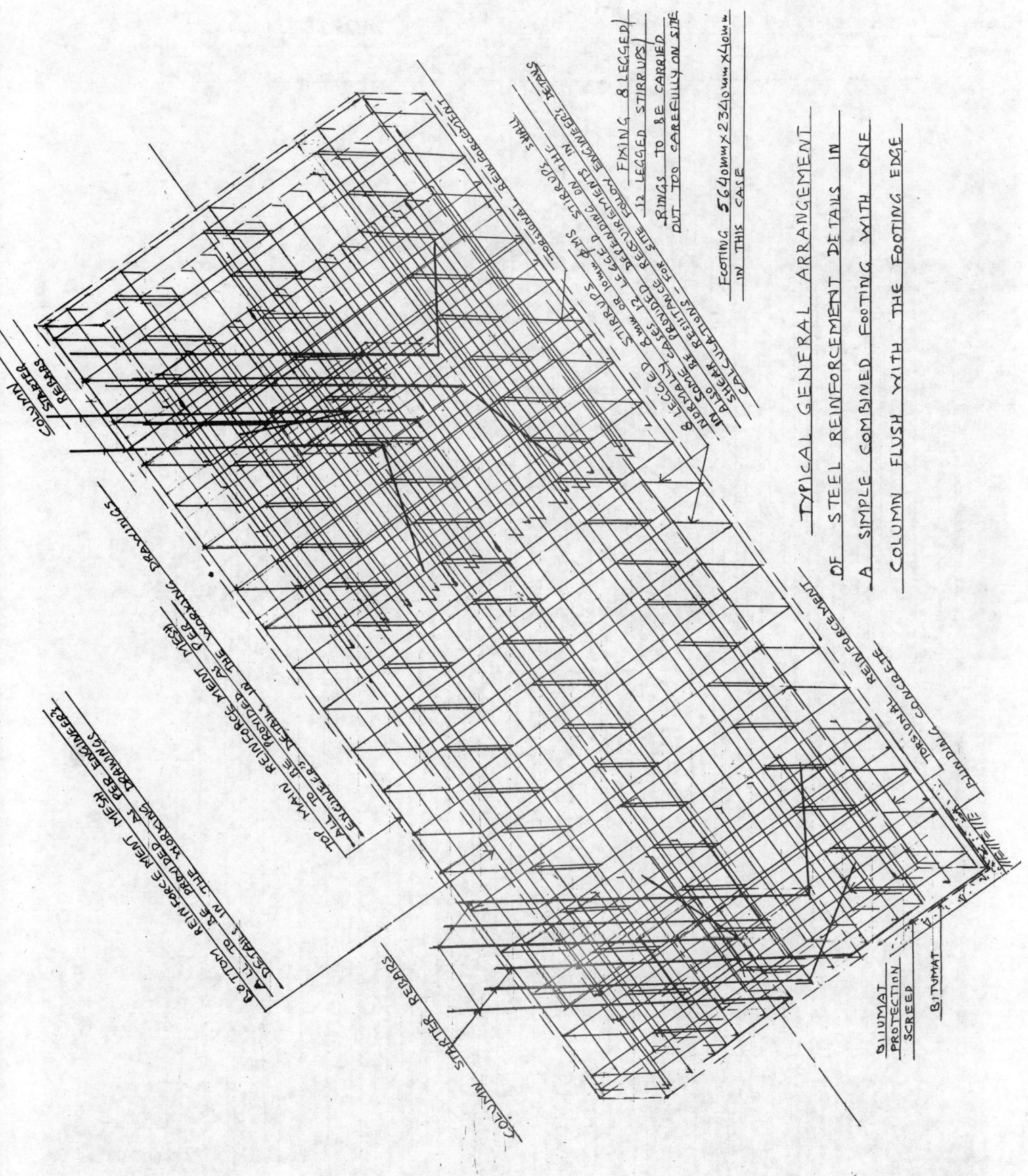

TYPICAL GENERAL ARRANGEMENT OF STEEL REINFORCEMENT DETAILS IN A SIMPLE COMBINED FOOTING WITH ONE COLUMN FLUSH WITH THE FOOTING EDGE

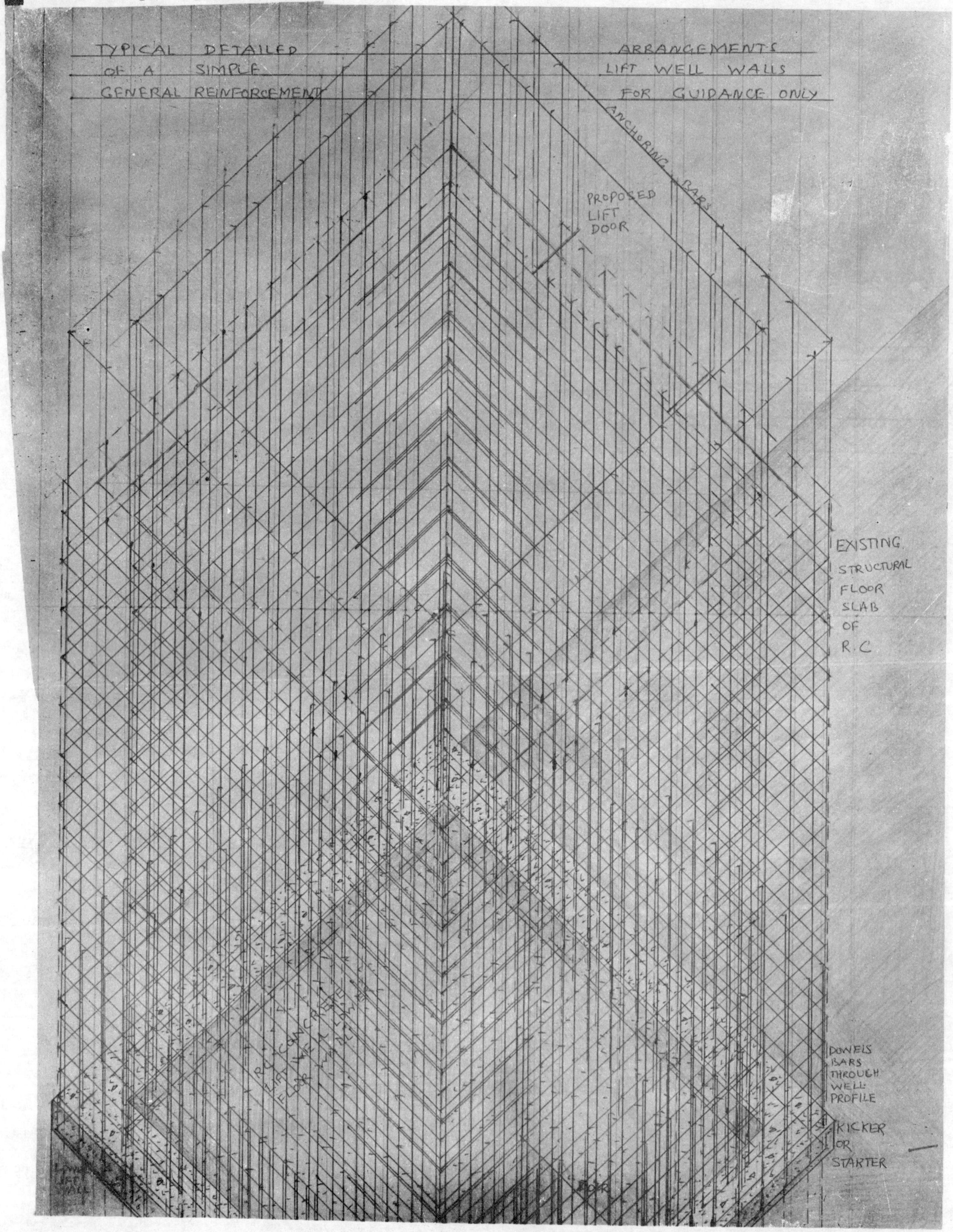

TYPICAL DETAILED ARRANGEMENTS OF A SIMPLE LIFT WELL WALLS GENERAL REINFORCEMENT FOR GUIDANCE ONLY

ANCHORING BARS

PROPOSED LIFT DOOR

EXISTING STRUCTURAL FLOOR SLAB OF R.C

DOWELS BARS THROUGH WELL PROFILE

KICKER OR STARTER

TYPICAL GENERAL ARRANGEMENT OF A CANTILEVER SLAB PROJECTIONS
RESISTED BY A CONTINUOUS R.C. BEAM WITH TORSIONAL RESISTANCE

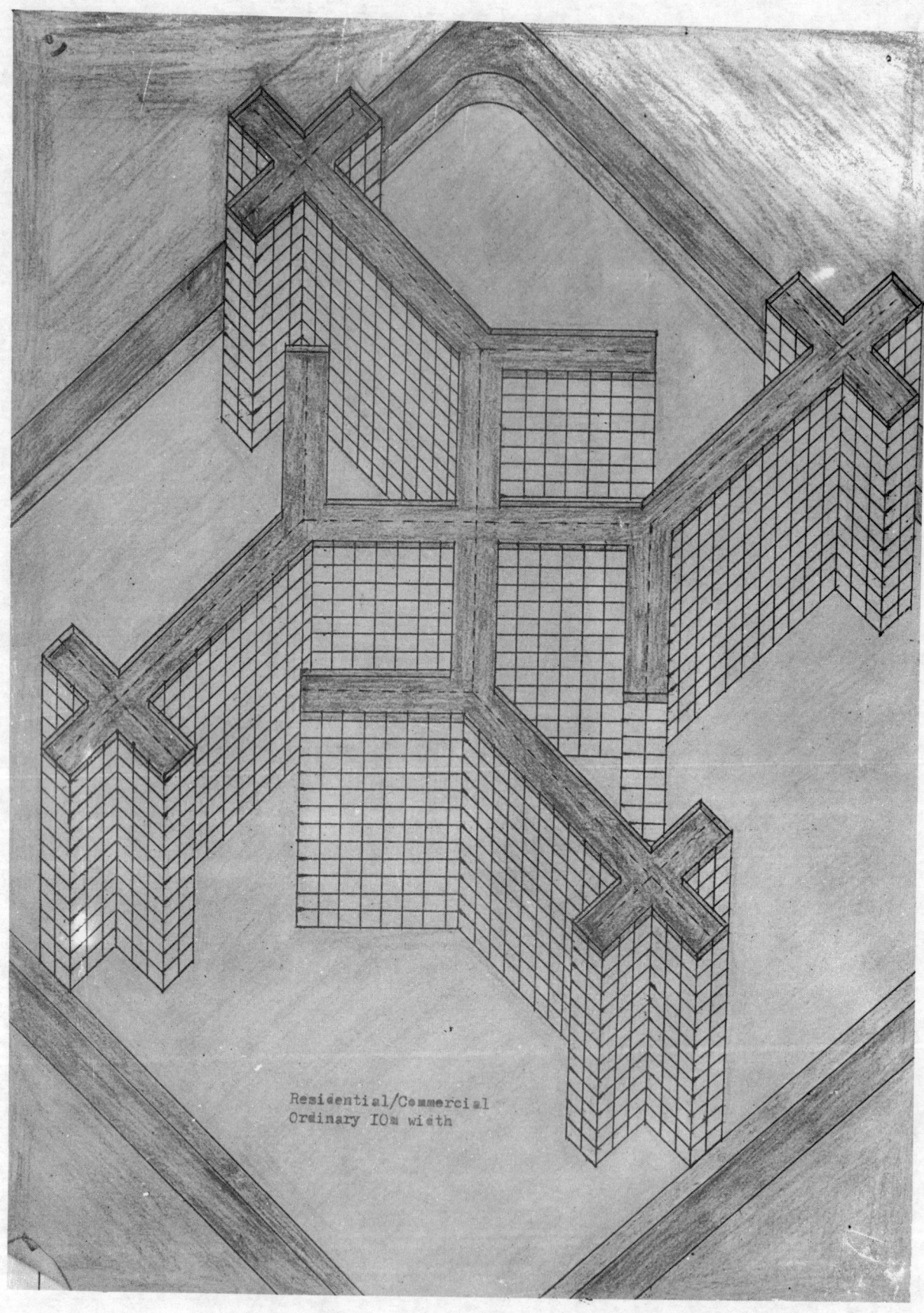

Residential/Commercial
Ordinary 10m width

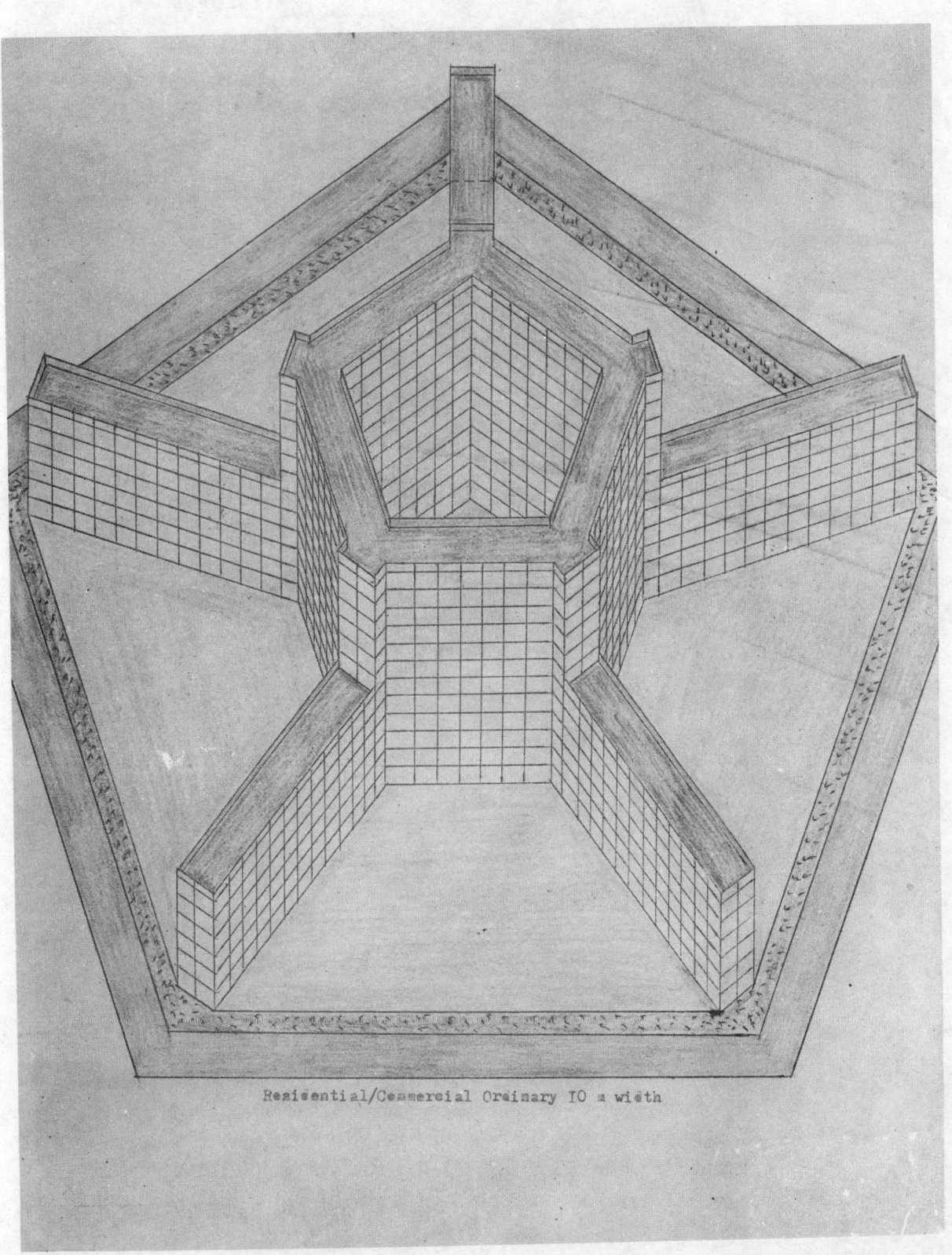

Residential/Commercial Ordinary IO a width

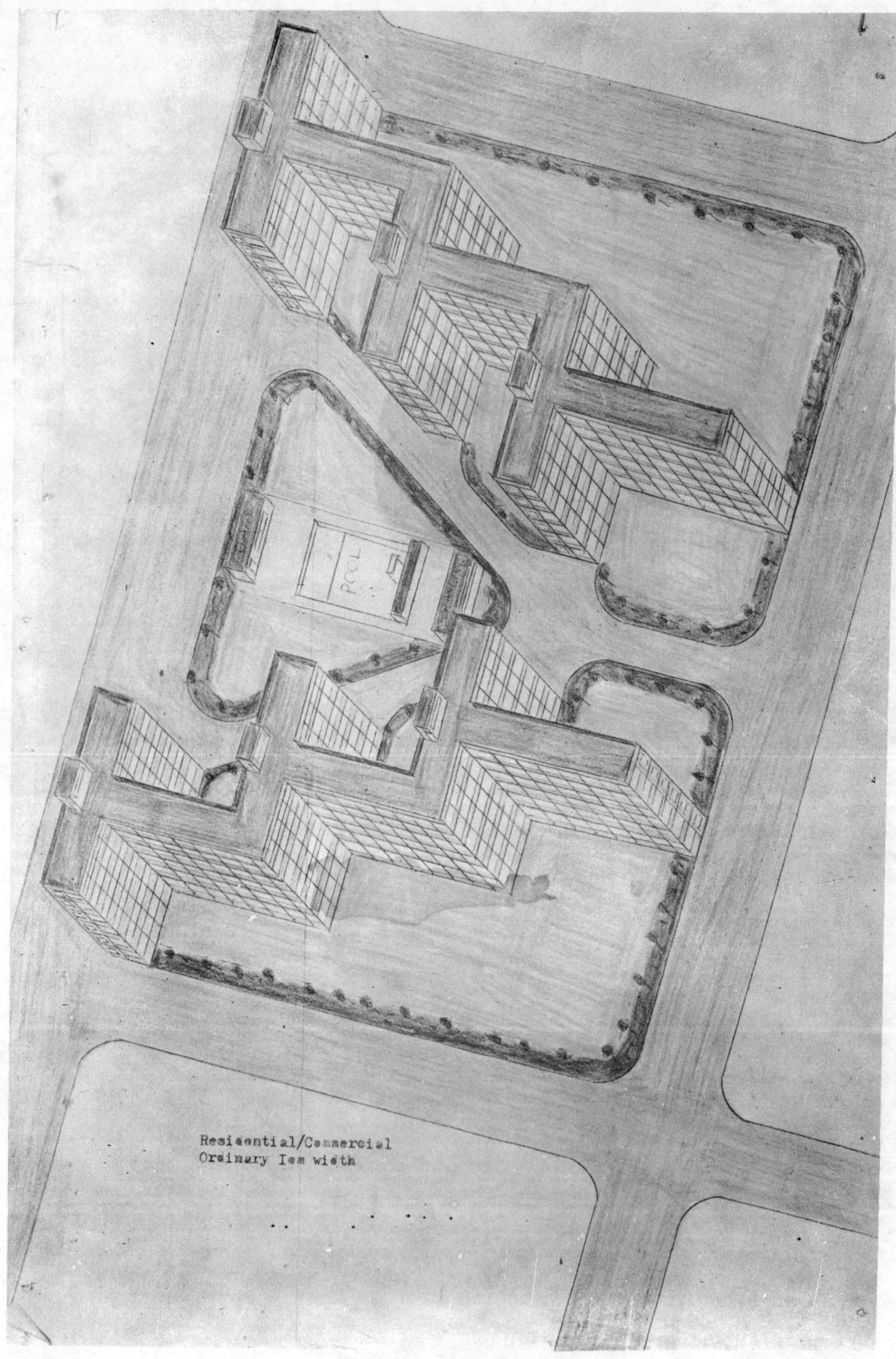

Residential/Commercial
Ordinary Ion width

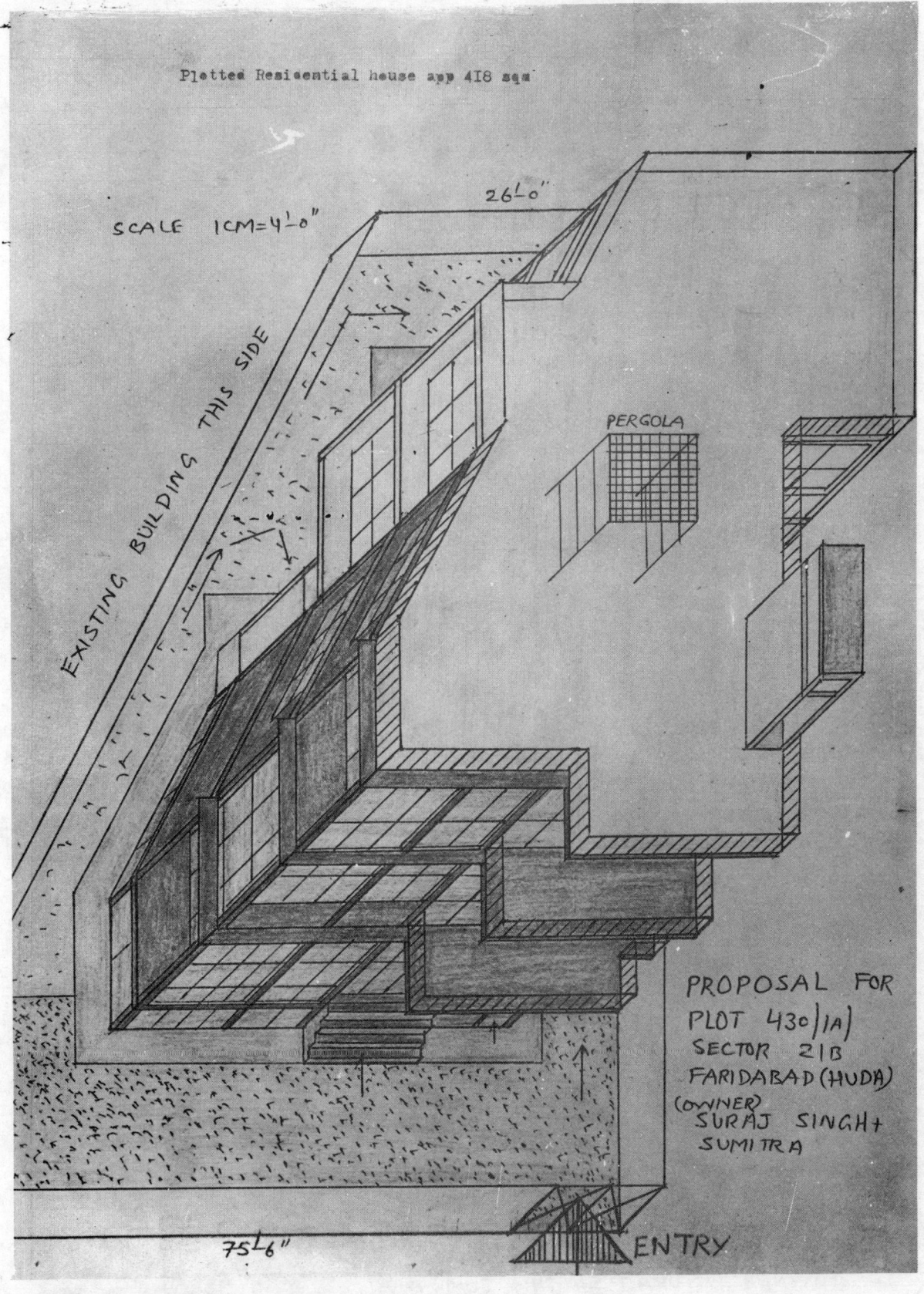

Plotted Residential house app 418 sqm

SCALE 1CM=4'-0"

26'-0"

EXISTING BUILDING THIS SIDE

PERGOLA

PROPOSAL FOR
PLOT 430/1A)
SECTOR 21B
FARIDABAD (HUDA)
(OWNER)
SURAJ SINGH+
SUMITRA

75'-6"

ENTRY

Plotted Residential house app 418 sqm

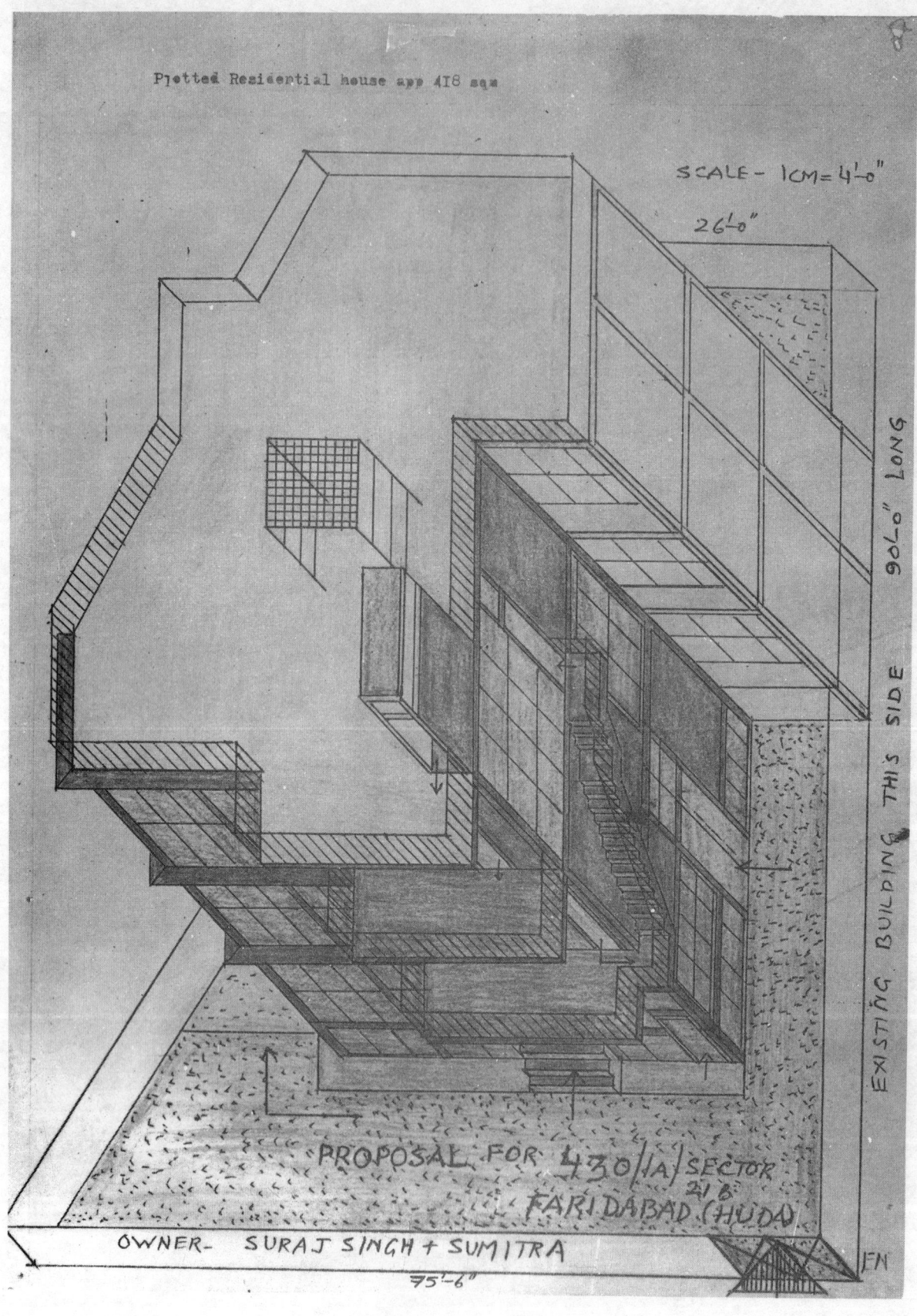

SCALE- 1CM=4'-0"

26'-0"

90'-0" LONG

EXISTING BUILDING THIS SIDE

PROPOSAL FOR 430/1A/SECTOR 21 B
FARIDABAD (HUDA)

OWNER- SURAJ SINGH + SUMITRA

75'-6"

FN

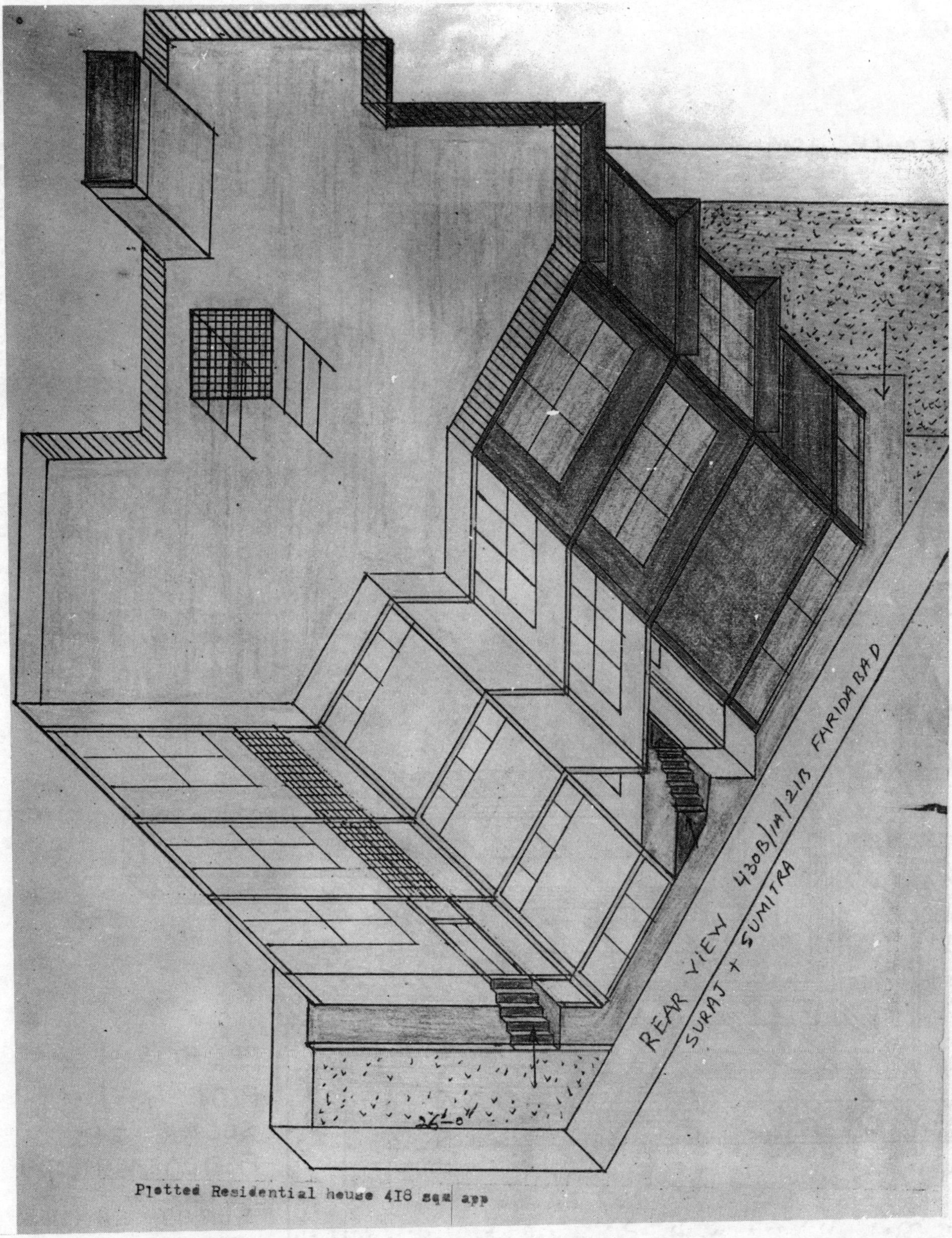
REAR VIEW 430B/1A/21B FARIDABAD
SURAJ + SUMITRA
26'-0"

Plotted Residential house 418 sqm app

Plotted Residential house app 4I8 sqm

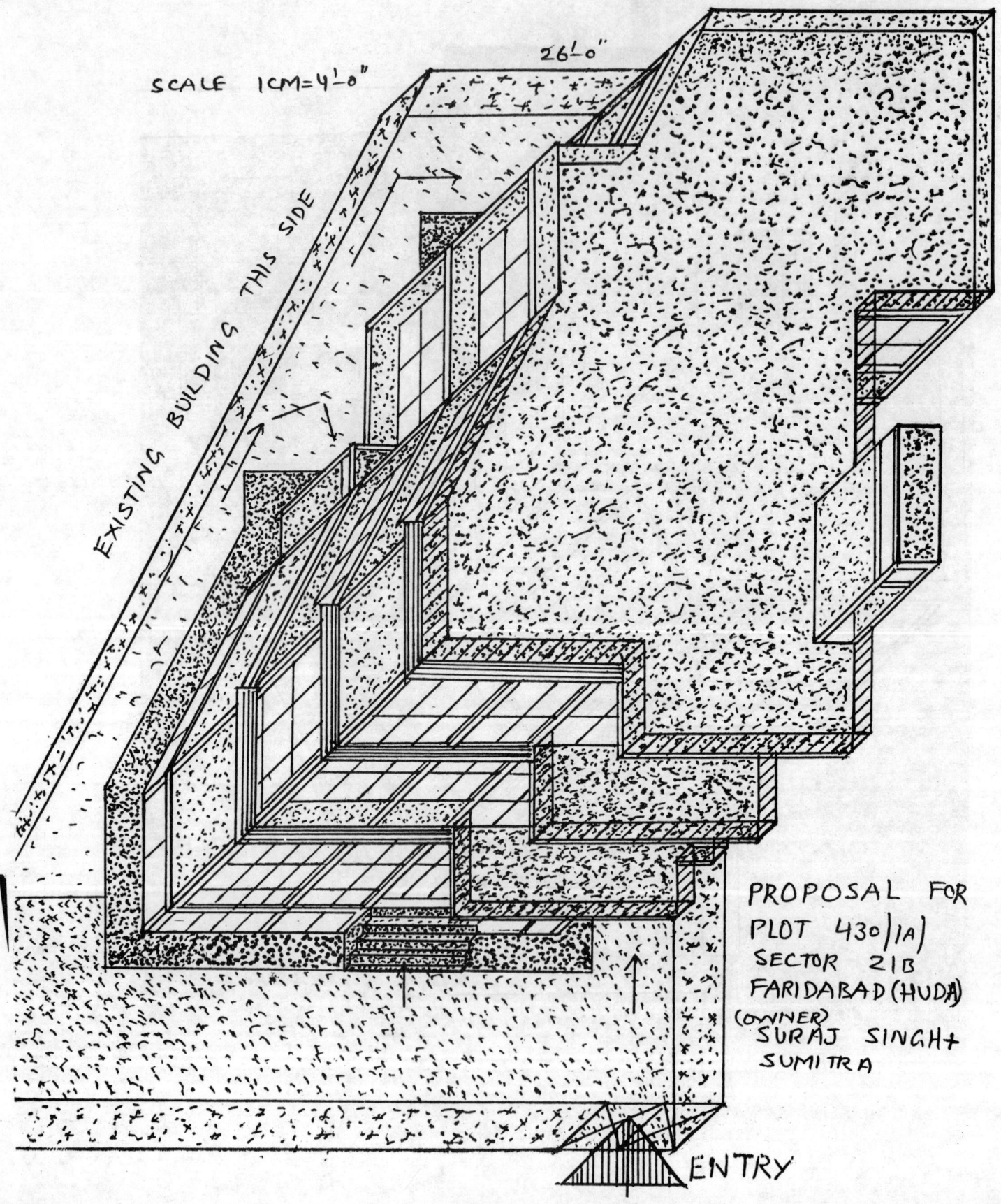

SCALE 1CM=4'-0"

26'-0"

EXISTING BUILDING THIS SIDE

PROPOSAL FOR
PLOT 430/1A)
SECTOR 21B
FARIDABAD (HUDA)
(OWNER)
SURAJ SINGH+
SUMITRA

ENTRY

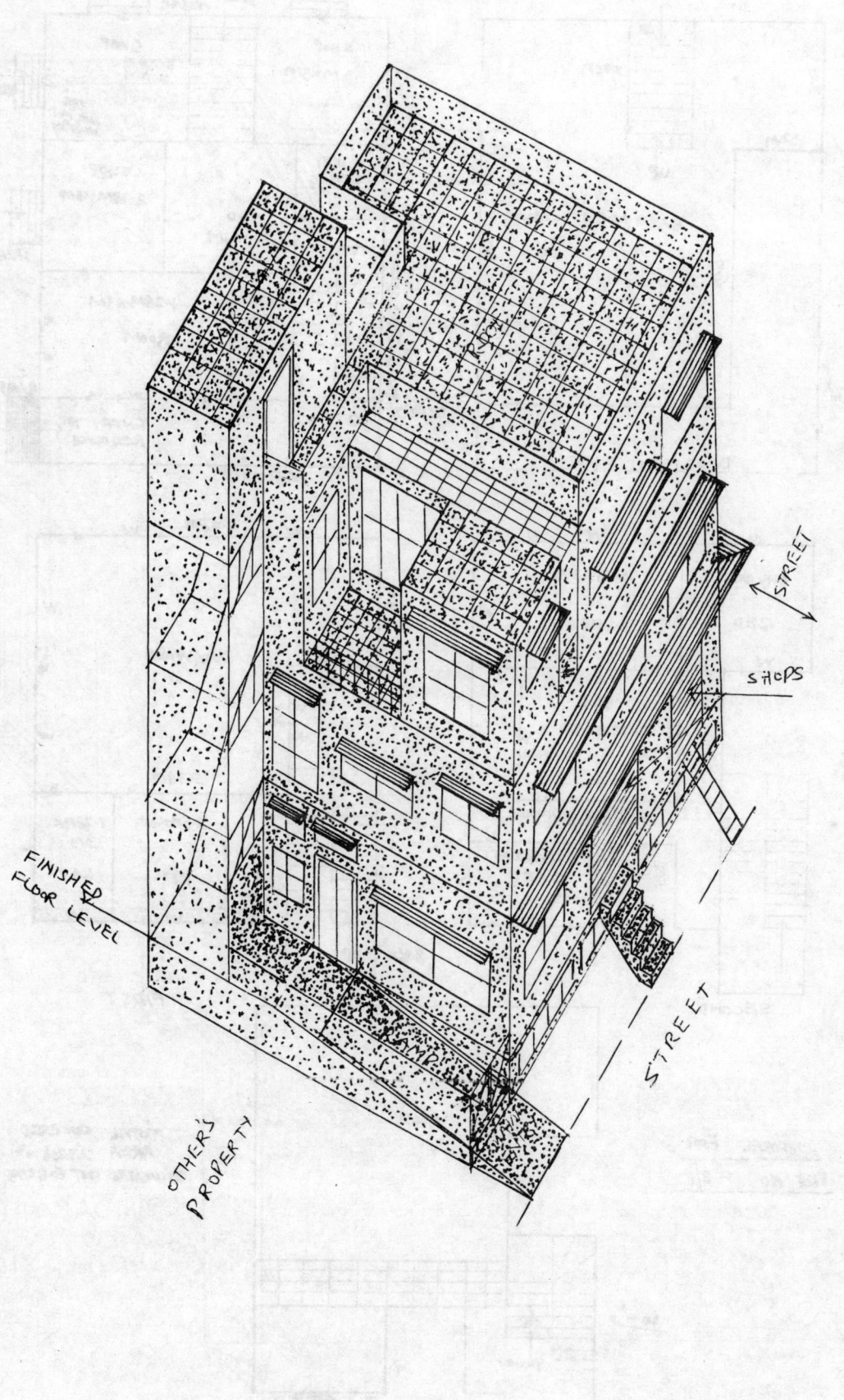

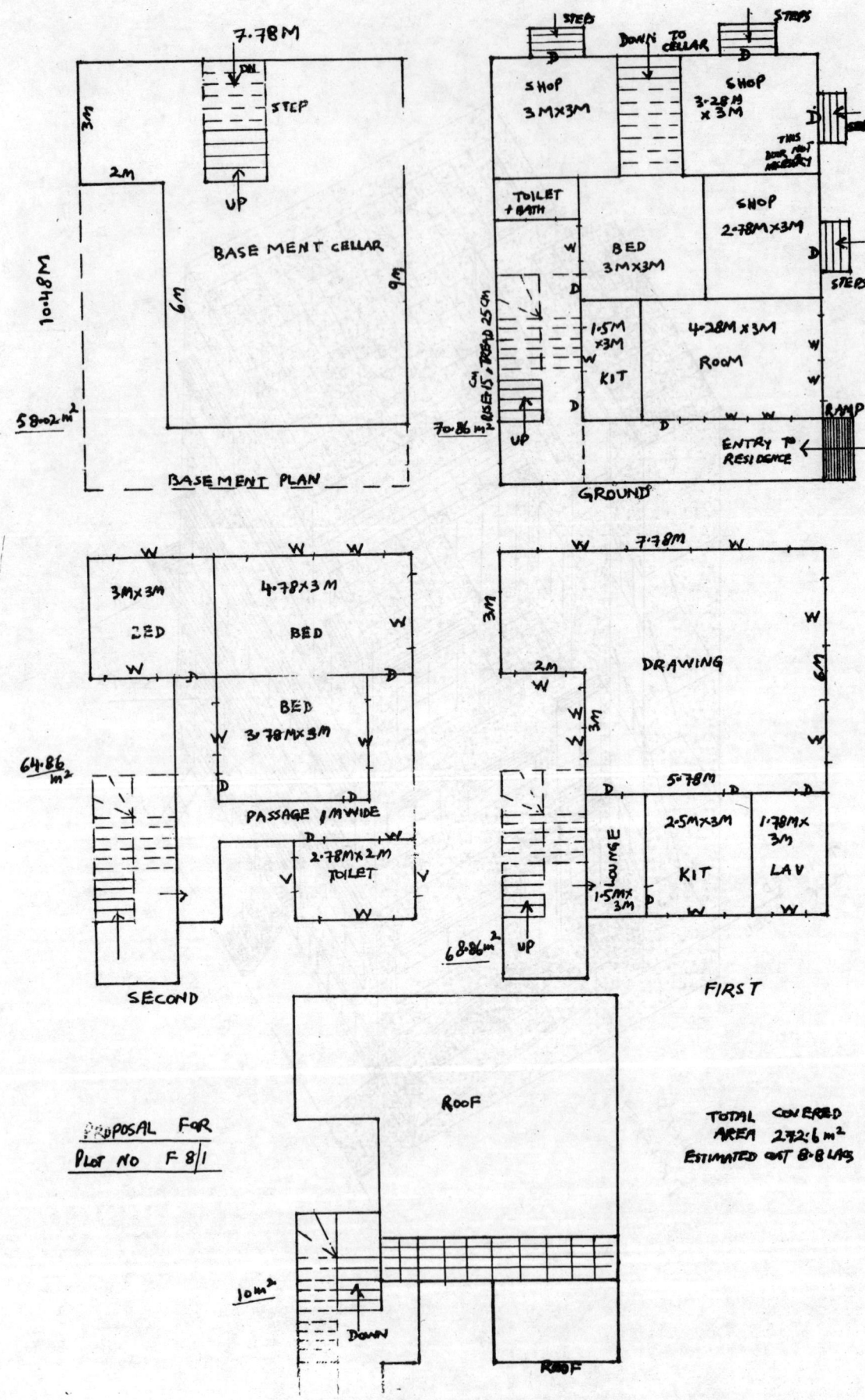

7·78 M

3M

2M

DN

STEP

UP

BASEMENT CELLAR

10·48 M

6M

9M

58·02 m²

BASEMENT PLAN

STEPS

DOWN TO CELLAR

STEPS

D

D

SHOP
3 M X 3M

SHOP
3·28 M
X 3M

THIS
DOOR NOT
NECESSARY

D

STEPS

TOILET
+ BATH

BED
3M X 3M

SHOP
2·78M X 3M

D

STEPS

C·I 1·05 & 1·5, 1·75 & 2·25 CM

70·86 m²

W

D

W

1·5M
X3M
KIT

4·28M X 3M
ROOM

W

UP

D

W

W

W

RAMP

ENTRY TO
RESIDENCE

GROUND

W   W   W

3MX3M
BED

4·78 X 3 M
BED

W

D

D

BED
3·78M X 3M

W

W

64·86 m²

D

D

PASSAGE 1 M WIDE

D

2·78M X 2M
TOILET

V   V

W

SECOND

W   7·78 M   W

3M

DRAWING

2M

W

6M

W

3M

68·86 m²

UP

D   5·78M   D   D

LOUNGE

1·5MX
3M

2·5M X 3M
KIT

1·78M X
3M
LAV

W   W

FIRST

ROOF

PROPOSAL FOR
PLOT NO F 8/1

TOTAL COVERED
AREA 272·6 m²
ESTIMATED COST 8·8 LACS

10 m²

DOWN

ROOF

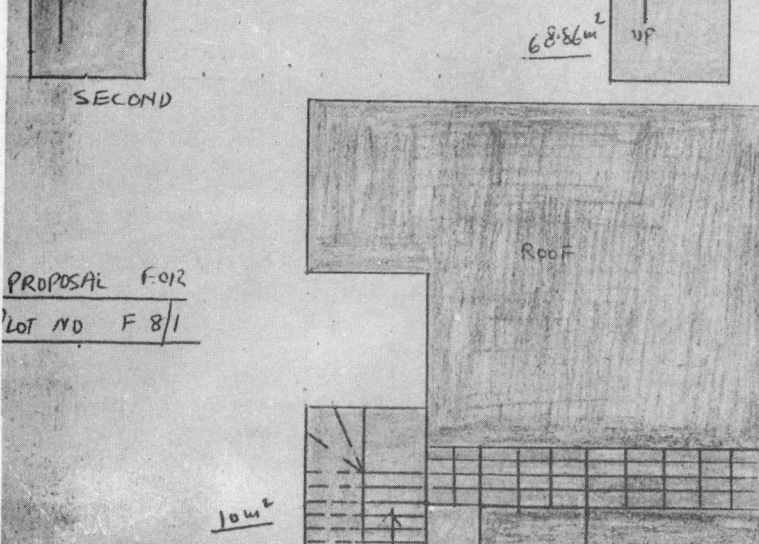

7.78 M

STEP 1½M WIDE

UP

3M

6M

BASEMENT/CELLAR

9M

BASEMENT PLAN

STEPS

DOWN TO CELLAR

STEPS

SHOP
3M×3M

SHOP
3.2M ×3M

STEPS

TOILET
+BATH

SHOP
2.75M×3M

BED
3M×3M

RISE 15 . TREAD 25

STEPS

70.86 M²

1.3M
×3M

4.28M×3M
ROOM

KIT

UP

W W

RAMP

ENTRY TO
RESIDENCE

GROUND

W W W

3M×3M
BED

4.78×3M
BED

W

BED
3.78M×3M

D D

PASSAGE 1M WIDE

2.78M×2M
TOILET

W

SECOND

W 7.78M W

3M

2M

3M

DRAWING

5.78M

LOUNGE

2.5M×3M
KIT

1.78M×
3M
LAV

1.5M×
3M

68.86 M²

UP

FIRST

ROOF

PROPOSAL F-012

PLOT NO F 8/1

TOTAL COVERED
AREA 272.6 M²
ESTIMATED COST 3.5 LAKHS

Residential cum shopping
on small plots commonly
built unauthorisedly/una-
pproved, by the people

10 M²

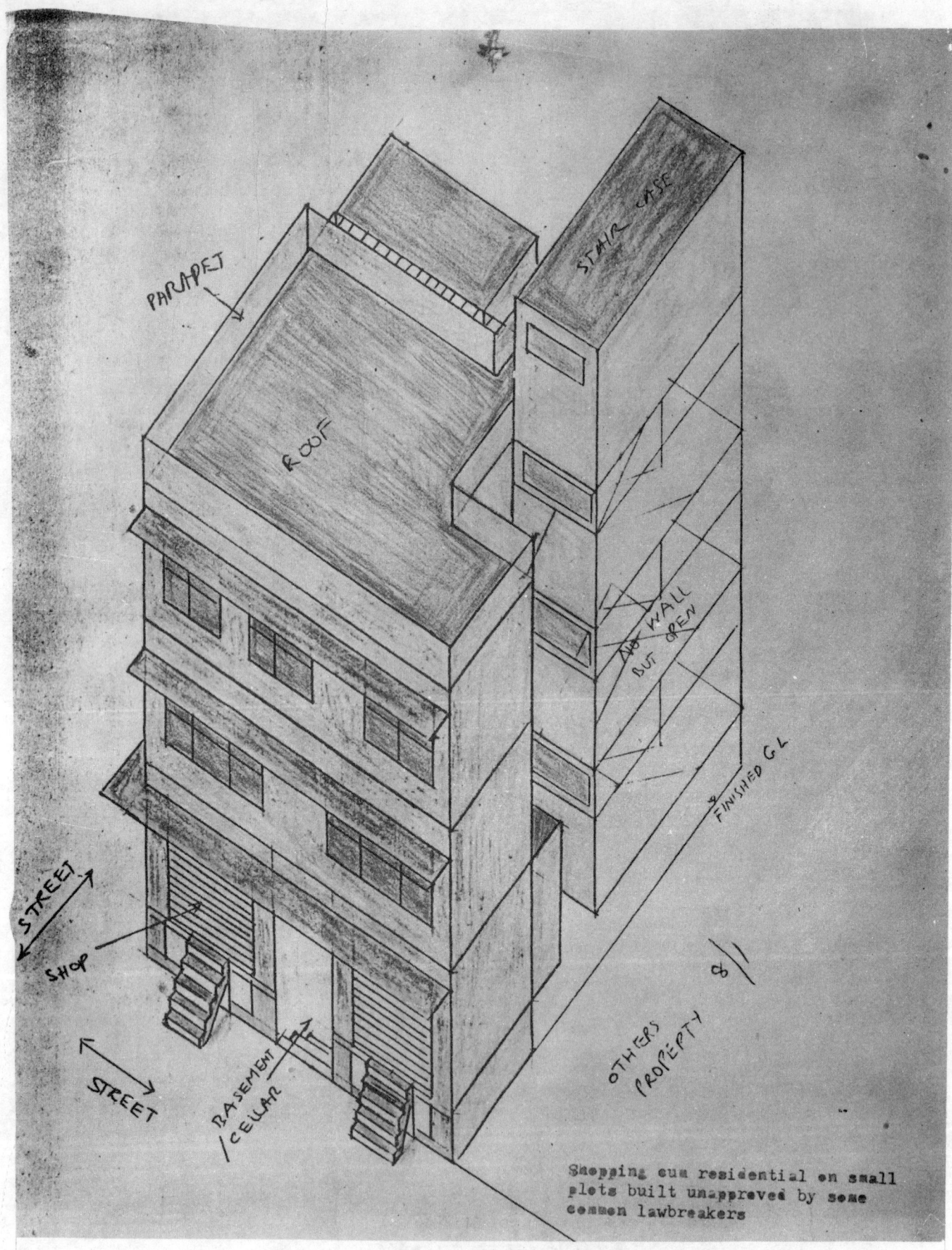

Shopping cum residential on small
plots built unapproved by some
common lawbreakers

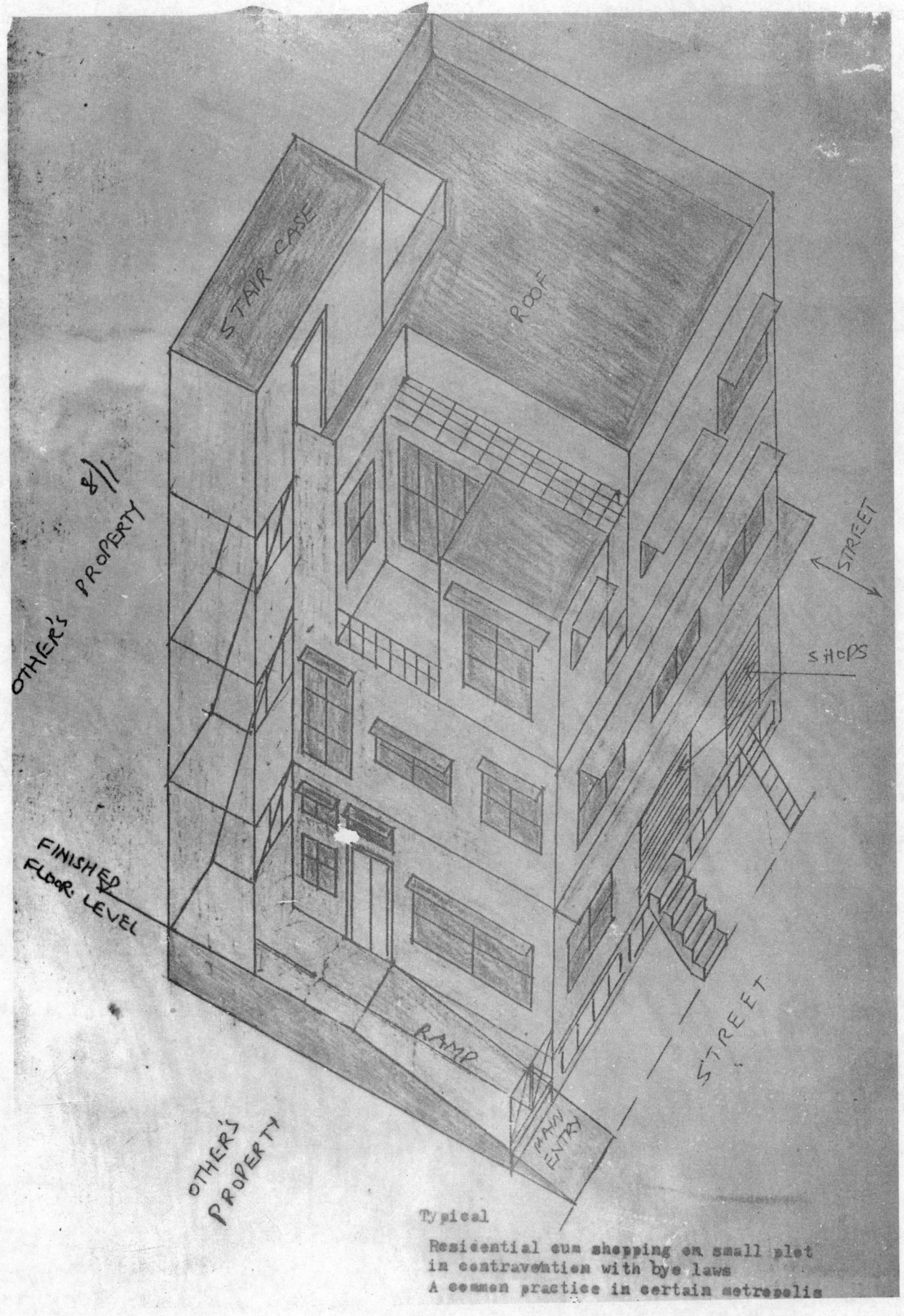

STAIR CASE

ROOF

OTHER'S PROPERTY 8/1

STREET

SHOPS

FINISHED
FLOOR LEVEL

OTHER'S
PROPERTY

RAMP

STREET

MAIN
ENTRY

Typical

Residential cum shopping on small plot
in contravention with bye laws
A common practice in certain metropolis

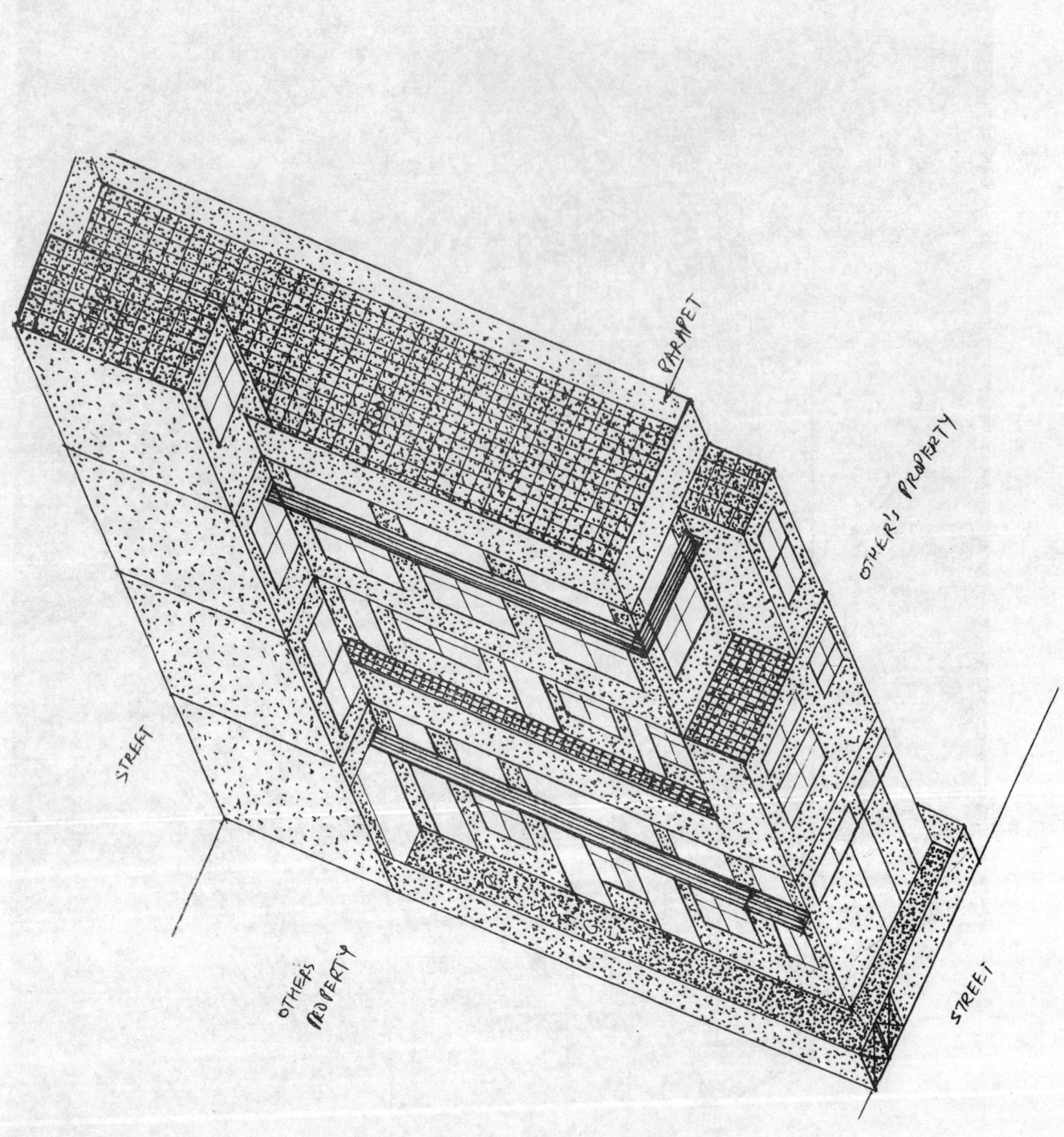

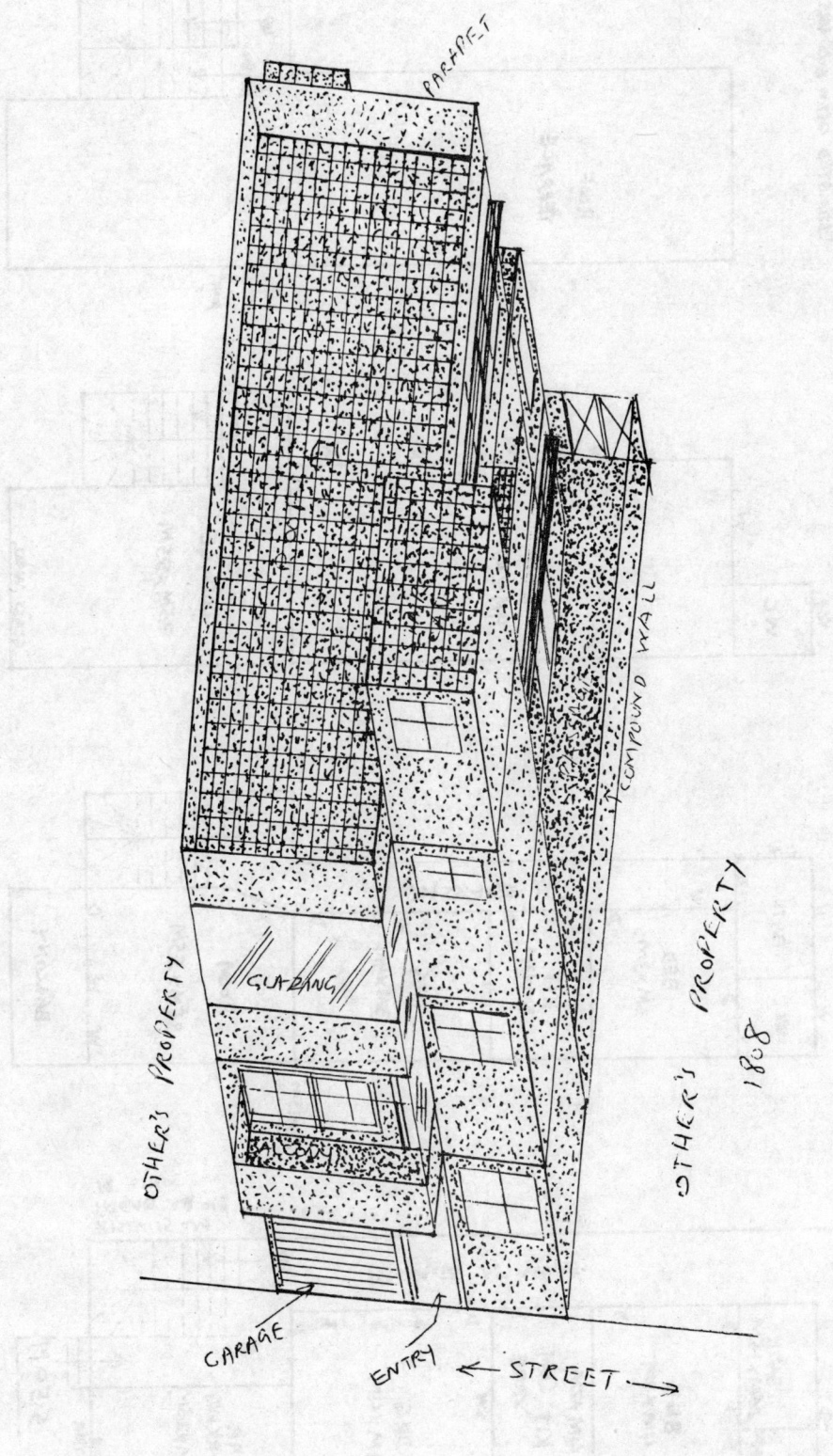

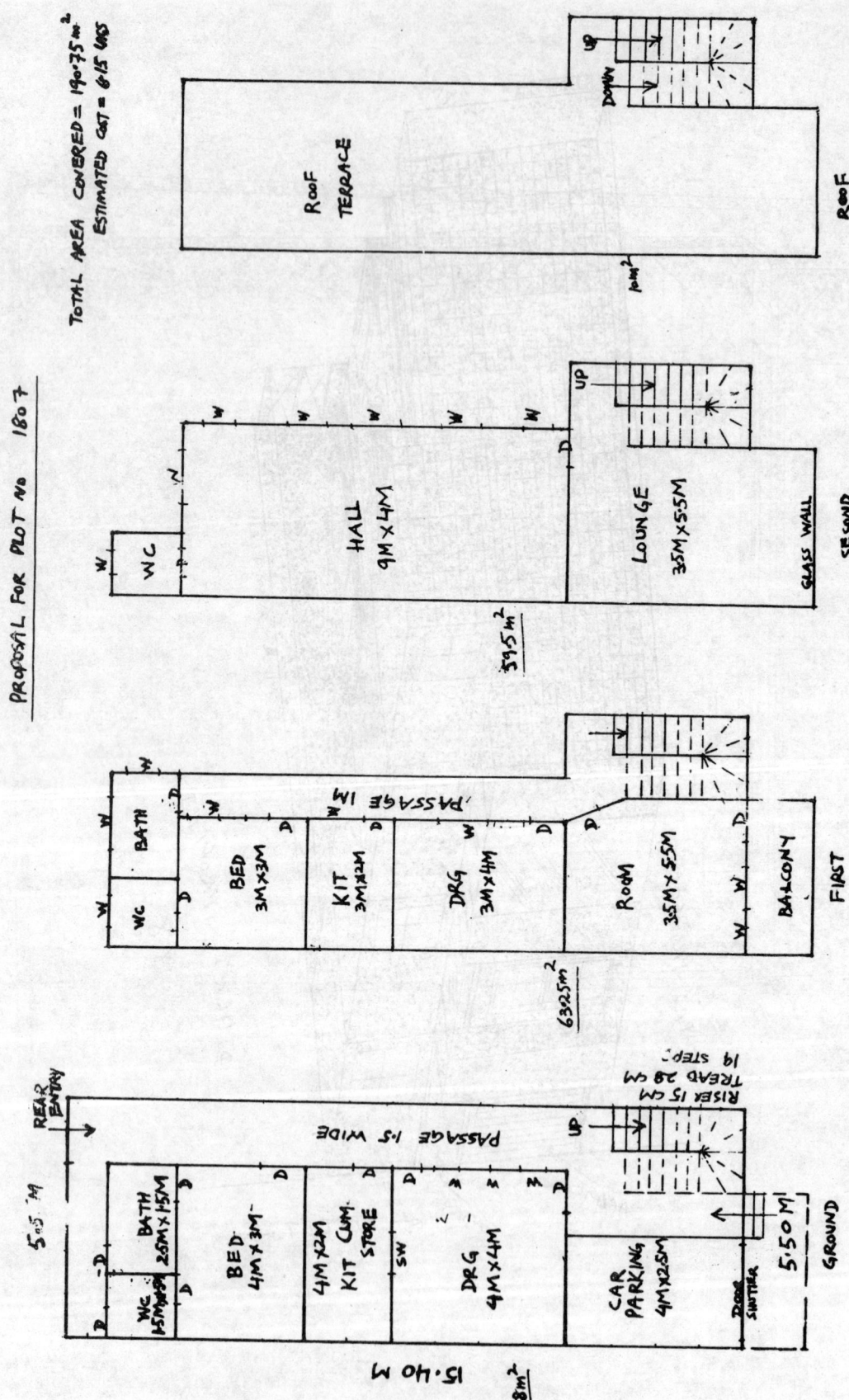

PROPOSAL FOR PLOT No 1807

TOTAL AREA COVERED = 190·75 m²
ESTIMATED COST = 6·15 LAKS

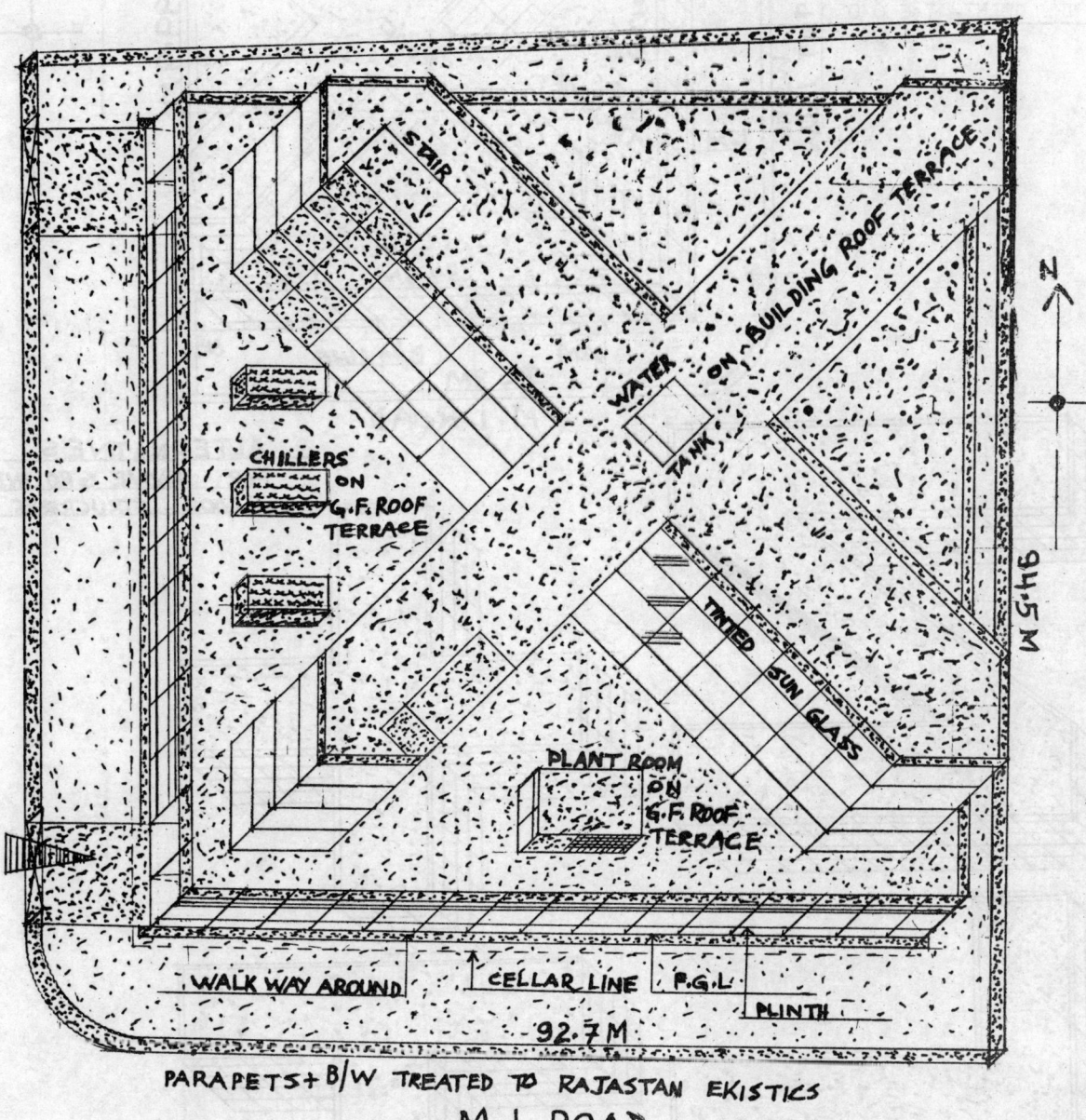

STAIR

WATER ON BUILDING ROOF TERRACE

TANK

CHILLERS ON G.F. ROOF TERRACE

TINTED SUN GLASS

PLANT ROOM ON G.F. ROOF TERRACE

N

94.5 M

WALK WAY AROUND   CELLAR LINE   P.G.L   PLINTH

92.7 M

PARAPETS + B/W TREATED TO RAJASTAN EKISTICS

M I ROAD

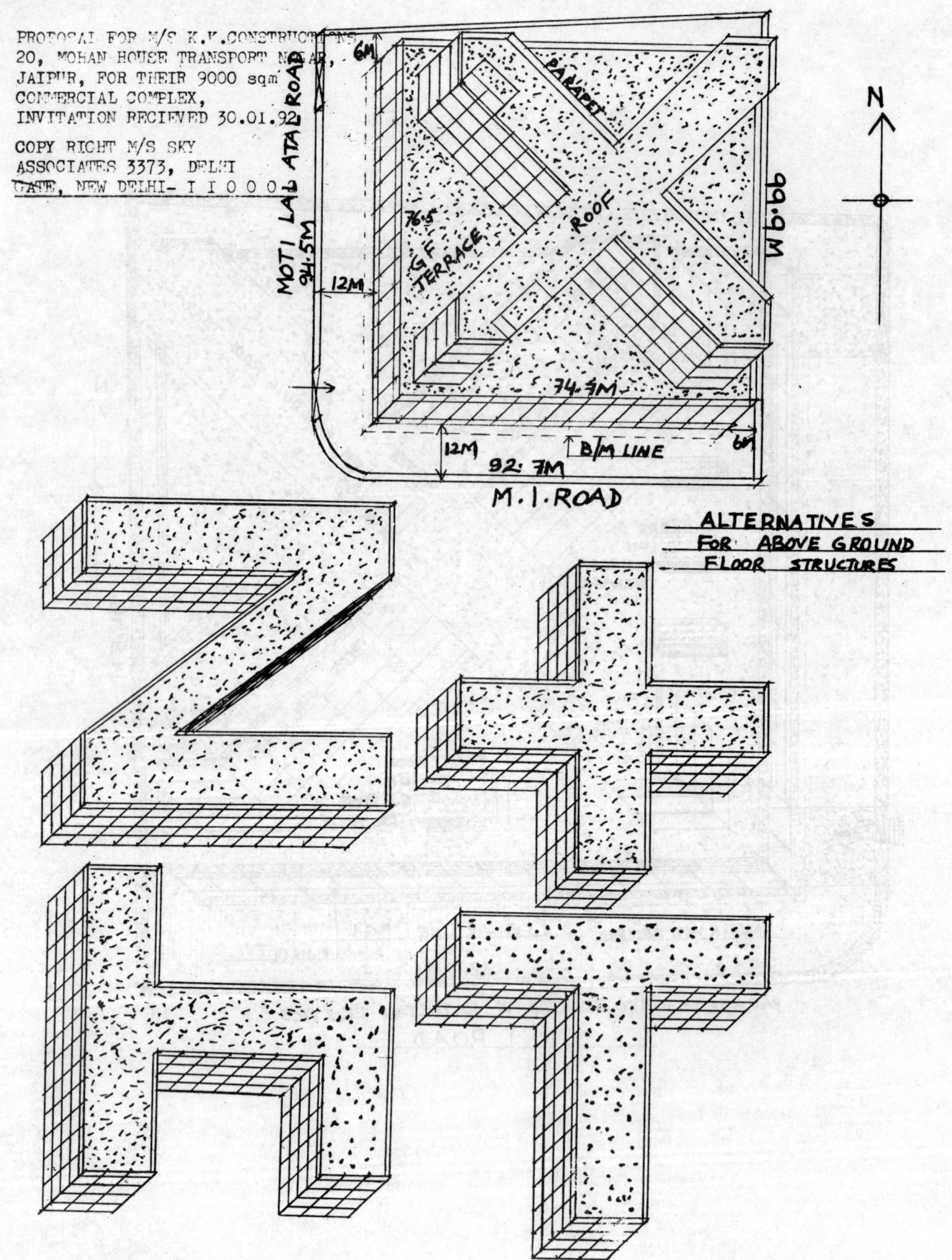

PROPOSAL FOR M/S K.K.CONSTRUCTIONS,
20, MOHAN HOUSE TRANSPORT NAGAR,
JAIPUR, FOR THEIR 9000 sqm
COMMERCIAL COMPLEX,
INVITATION RECIEVED 30.01.92

COPY RIGHT M/S SKY
ASSOCIATES 3373, DELHI
GATE, NEW DELHI- 110000

MOTI LAL ATAL ROAD

6M

24.5M

12M

76.5

G F
TERRACE

PARAPET

ROOF

99.9M

N

74.5M

12M

B/M LINE

6M

92.7M

M.I.ROAD

ALTERNATIVES
FOR ABOVE GROUND
FLOOR STRUCTURES

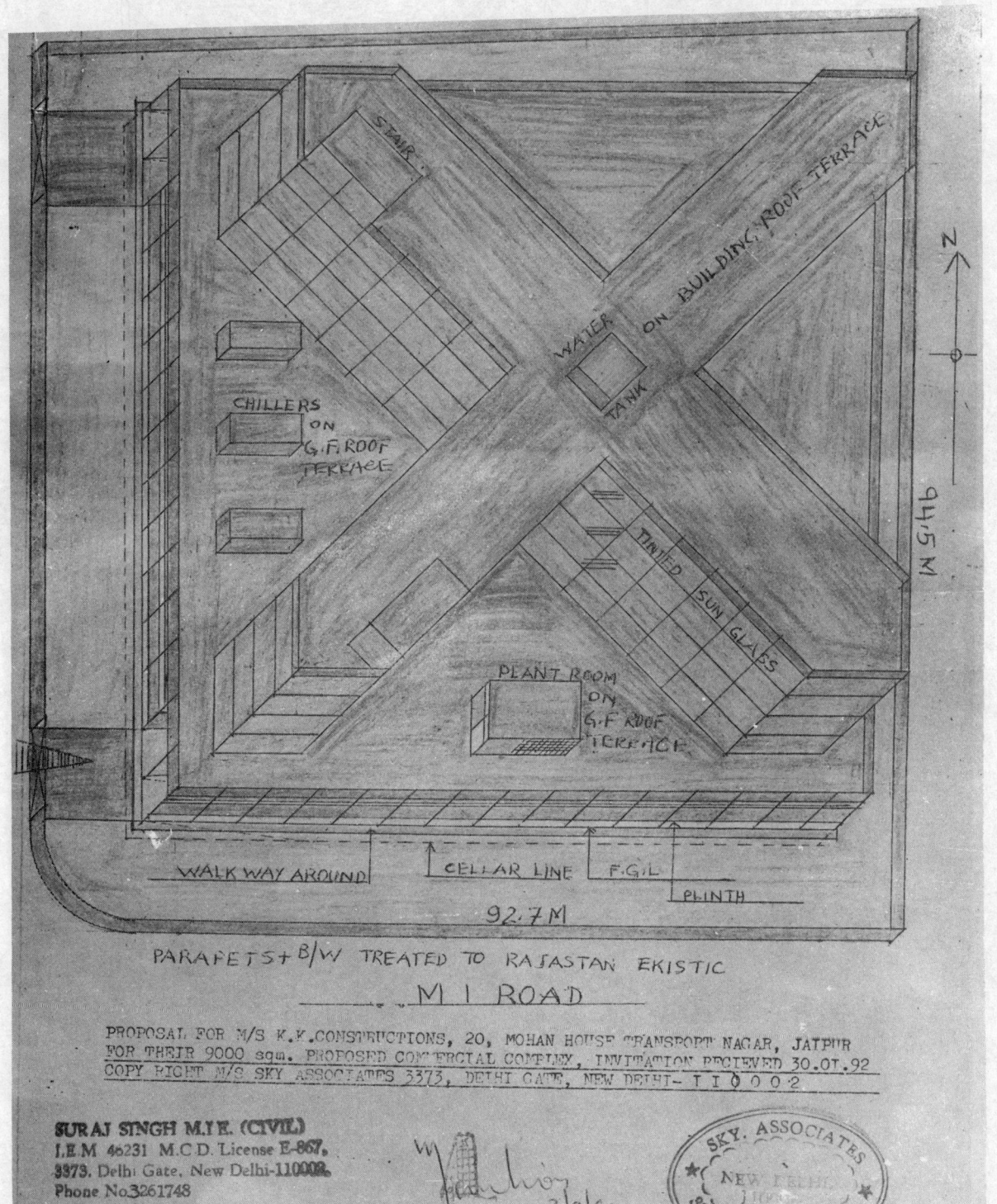

PARAFETS+B/W TREATED TO RAJASTAN EKISTIC

M I ROAD

PROPOSAL FOR M/S K.K. CONSTRUCTIONS, 20, MOHAN HOUSE TRANSPORT NAGAR, JAIPUR
FOR THEIR 9000 sqm. PROPOSED COMMERCIAL COMPLEX, INVITATION RECIEVED 30.01.92
COPY RIGHT M/S SKY ASSOCIATES 3373, DELHI GATE, NEW DELHI- 1 1 0 0 0 2

SURAJ SINGH M.I.E. (CIVIL)
I.E.M 46231 M.C.D. License E-867,
3373, Delhi Gate, New Delhi-110002.
Phone No.3261748

3/2/92

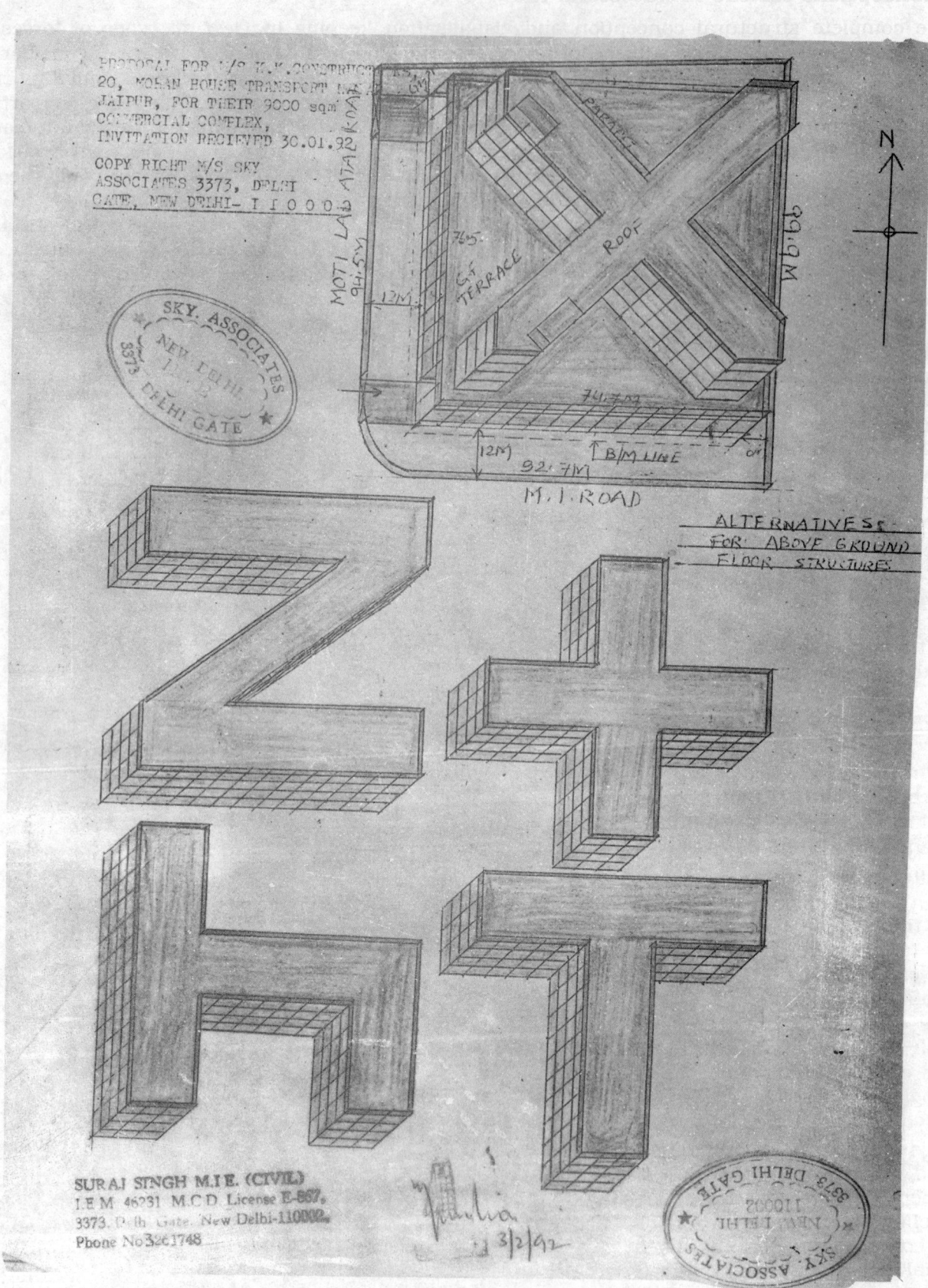

ALTERNATIVES
FOR ABOVE GROUND
FLOOR STRUCTURES

SURAJ SINGH M.I.E. (CIVIL)
I.E.M 46231 M.C.D. License E-867,
3373, Delhi Gate, New Delhi-110002.
Phone No.3261748

Structural Conceptions for the Architectural Aid

Based on the complete structural conception and visualisation keeping in view all types of forces, a three dimensional structure shall be subject to during the intended normal service life time, considering various uses, architectural requisites, keeping the economic consideration in 1st place, assimilating new ides which so far has been neglected though available freely in globe society, literature, imparting rhythm, vertical and horizontal expressions with uniformity and harmony satisfying all the requirements of the principles of architecture of modern and post modern era, the structures' profiles of various kinds, shapes, cultures, etc., have been expressed in terms of simple designs which could be by application of efforts of civil and structural engineers to imagine and conceive to extend a base to the struggling and searching architects about new idea or the dream structures who always feel short of these conceptions of structures to accommodate the required spaces for the proposed use criteria, as always building design flows from architecture desk to structure desk and vice versa and back to the structure desk for the incorporation of many alterations and changes with the whole conceptual intuitive profile being far different from what the architect originally dreamt with a fact that the final results or shape shall be dictated by the structural provisions.

Thousands of structures may be dreamt and designed preliminarily and space would be set in but for this outstanding objective or target, an unified approach and cooperation between the two designers is certainly a pre-requisite which lacks practically all over the world or globally due to the professional egos and valuation of the professional personality measurements while the fact remains that the building shall be likely to collapse if the structure miscarries functionally according to the architectural estimates. Amendments can be applied at the architectural faults if an excellent, durable and sound structure has been erected.

The owner or the client is not concerned about the professional differences between the two designers but certainly very much concerned about the ultimate results of the investments made on the building project and in terms of an efficient and good delivery of the goods with profitable returns as expected in the owner's initial estimates of the project promotion in relation to the overall costs. This expectation is a right of the owner/client/society and the designers must respect and respond to this fact without any regard to architectural and structural engineering differences among the building engineering professionals.

The beauty of the skin and excellent make up shall not be much useful for long if the bones, the muscles and the complete mechanism of the body are defective. The structural designer should be strong enough, dynamic, efficient in all professional approaches, the client must know the facts of the structural responsibilities, importance and significances. If an excellent structure is designed preliminarily, it shall virtually generate architecture in certain cases where imagination of architecture steps down and the preliminary design of structure shall provide a track for the architectural train.

The architectural professional is morally and professionally required to consult in every respect before taking up the assignment from the owner or client and must involve the structural professional in generating the scheme, tell the client about the levels of the responsibilities in the project management team. This problem shall not appear in case the owner appoints a civil engineer for the project management of the promotions. Even though no structural professional can guarantee cent per cent accuracy of buildings of all sorts but something is always better than nothing. The architect neglecting the structural or the civil engineer shall always produce the defective and sub standard buildings and this is not fair in profession for dealing client as such as most clients do not know about this matter. A reasonably good building shall be produced by a combined management of the team of civil or structural engineer, architectural engineer, building services engineers, specialist engineers, expert construction contractor and intelligent and the prudent financer or the promoter or the owner.

An unified approach in preliminary design consideration has to be adopted in respect of the estimated and anticipated forces due to the dead load, imposed or live load, weather effects, impact load, sonic boom effects, explosion effect, seismic or earthquake effect, wind effect, vibration effect, temperature changes effect, three-dimensional movement effects, two dimensional movement effect, shear walls effect, cores effect, perforations effect, material hysteresis effect, creep effect, redistribution of forces, thermal stresses, torsional effects, strains in concrete, secondary stresses effects, effects due to the foundation behaviour, settlements, stability, instability effects and estimates and not the least.

Too much of these factors' considerations depend on the experience, the attainments and understanding of an engineer about the overall judgement regarding the structure. Every designer may adopt different approach in making the structure. The complete planning, detailed design analysis, design checks shall depend upon the preliminary design decisions and the other disciplines should follow that to make a sensible rationale structure. This part of the job shall effect the construction cost during life time of the structure. To keep the maintenance cost low, the designer must think at lengths and come out with many alternatives, fix one as best of the various brain children. This approach shall provide a bank of generalised structures with the designer which shall enable the designer to count on the future projects to minimise the cost and attempt research on the subject.

Once a project is completed, the designer and all the related managers leave the place and may never go back to that structure for years together to observe or check the predicted performance. The factual loads the structure takes may differ from what were assumed during designs and the use of the structure be entirely different. The architectural aspect shall definitely alter its look after years and probably the architect shall not like to see the defaced elevations. Practically, desired maintenance shall never be carried out. After many decades the behaviour of the original structures change, redistribution of forces and stresses shall take place and vary by many percentages of the original assumed or calculated quantities or magnitudes rendering unwanted movements and torsional stresses. Therefore, it is stressed that at the planning stage the structural and the architectural professional/s must work together sincerely and seriously for the betterment of the project and the society keeping in view all the foregoing aspects and produce as far as possible a good erected structure by real supervision with adherence to the strict quality controls.

The profiles are based upon adoption of 11 m c/c of the columns of the frame in general maintaining a floor height of 3.5 m spacing grids 5 m c/c but shall vary at the intersections or the change in the direction of the centre line where the lifts and the staircases have been arranged.

Occupants shall escape to the fire free zones in case of the fire break out. Wind shall effect only on one side in some cases. Community feeling shall develop by living in the same building, the complex with security arrangements shall cost comparatively less and affordable. Set of such buildings shall extend to the formation of the self contained complexes or the integrated plans of the colonies by extending the layout/s to include all the modern facilities needed in such modern complexes. The construction of such structures shall be repetitive, simple and better control can be effected easily.

Structure/s can be applied to the residential, commercial and the office accommodations. In general a height for the 13 number of levels is suggested and a raft or the normal piles' foundation shall work in the areas of the normal bearing capacities and shall in fact depend on the soil investigation results adopted by the concerned engineer. An attempt has been made to eliminate the cantilever provision in all respects to make the structure more resistant to the seismic effects and unnecessary stresses which are expected by virtue of the inclusion of the cantilevers in general. No claim is made for the profiles but hoped that the proposals shall definitely assist in the formation of further new ideas by the very senior engineers distinguished in the building engineering field of design and construction and also the architects not to keep themselves in the darkness of the personality clashes with the civil and the structural engineers but to think for the betterment of the society as a whole.

Sketches 172 to 289

(Line Work Aids)
Concepts of Structural Profiles in General
Not to Scale

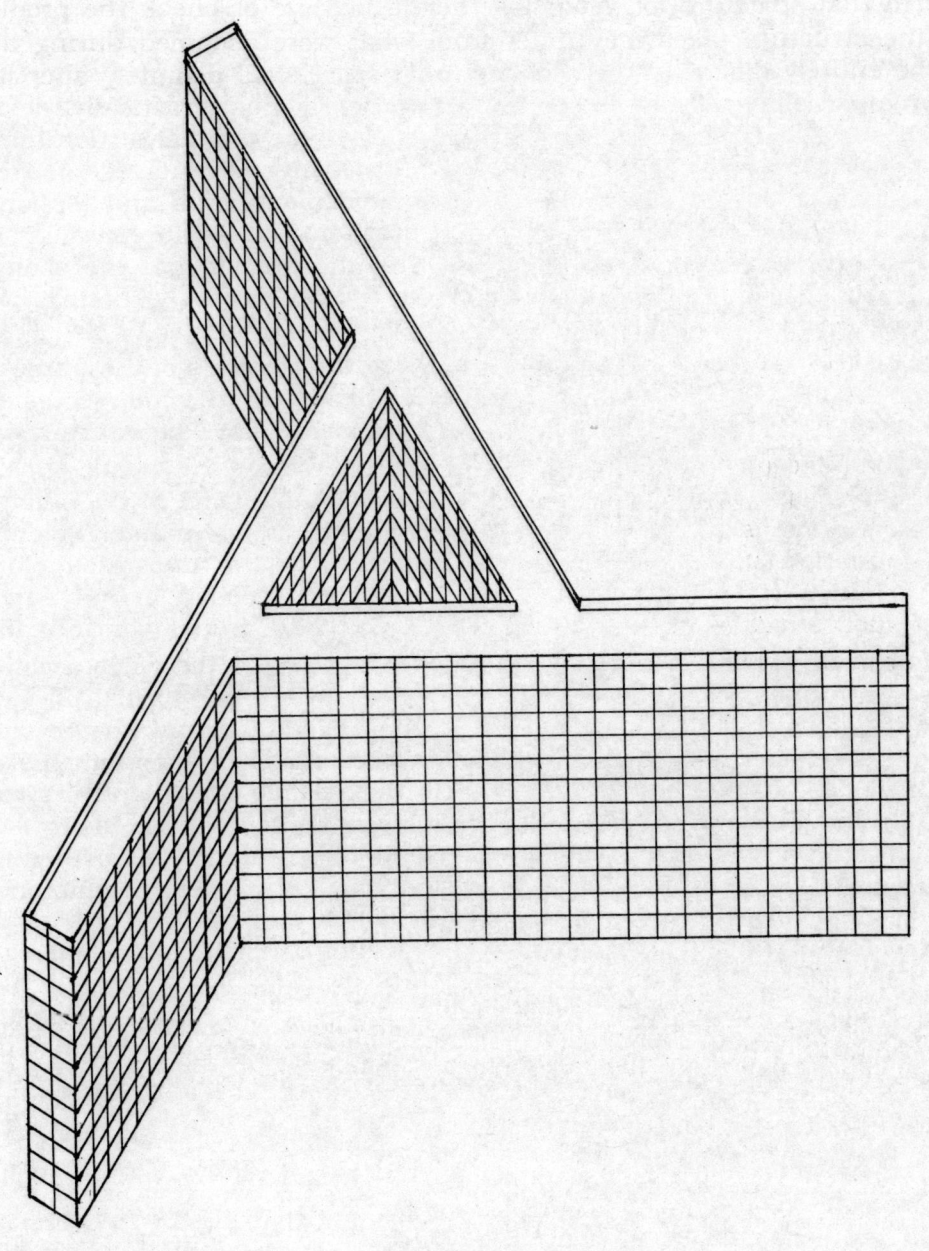

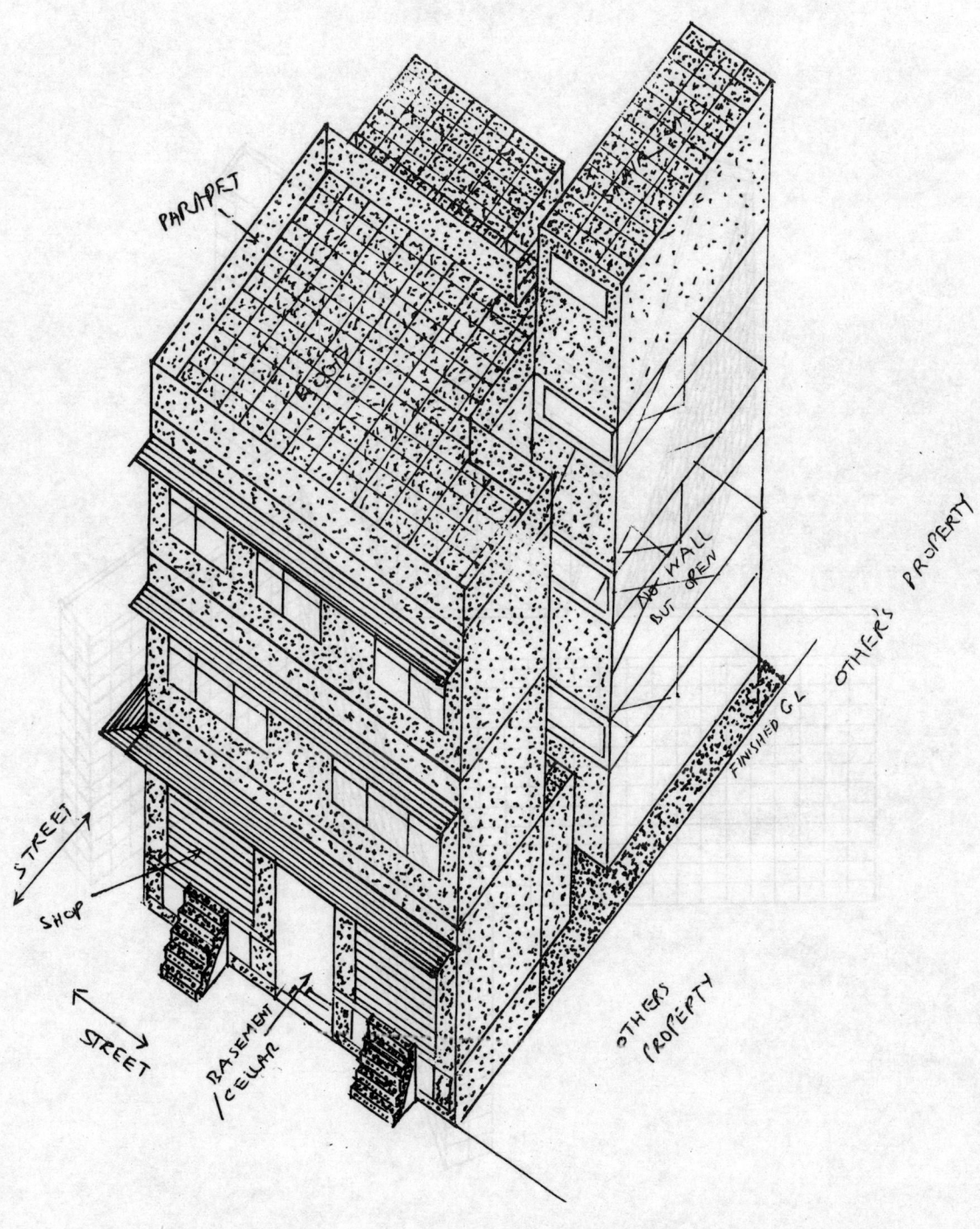

PARAPET

NOT WALL
BUT OPEN

OTHER'S PROPERTY

FINISHED G L

STREET

SHOP

STREET

BASEMENT
/ CELLAR

OTHERS
PROPERTY

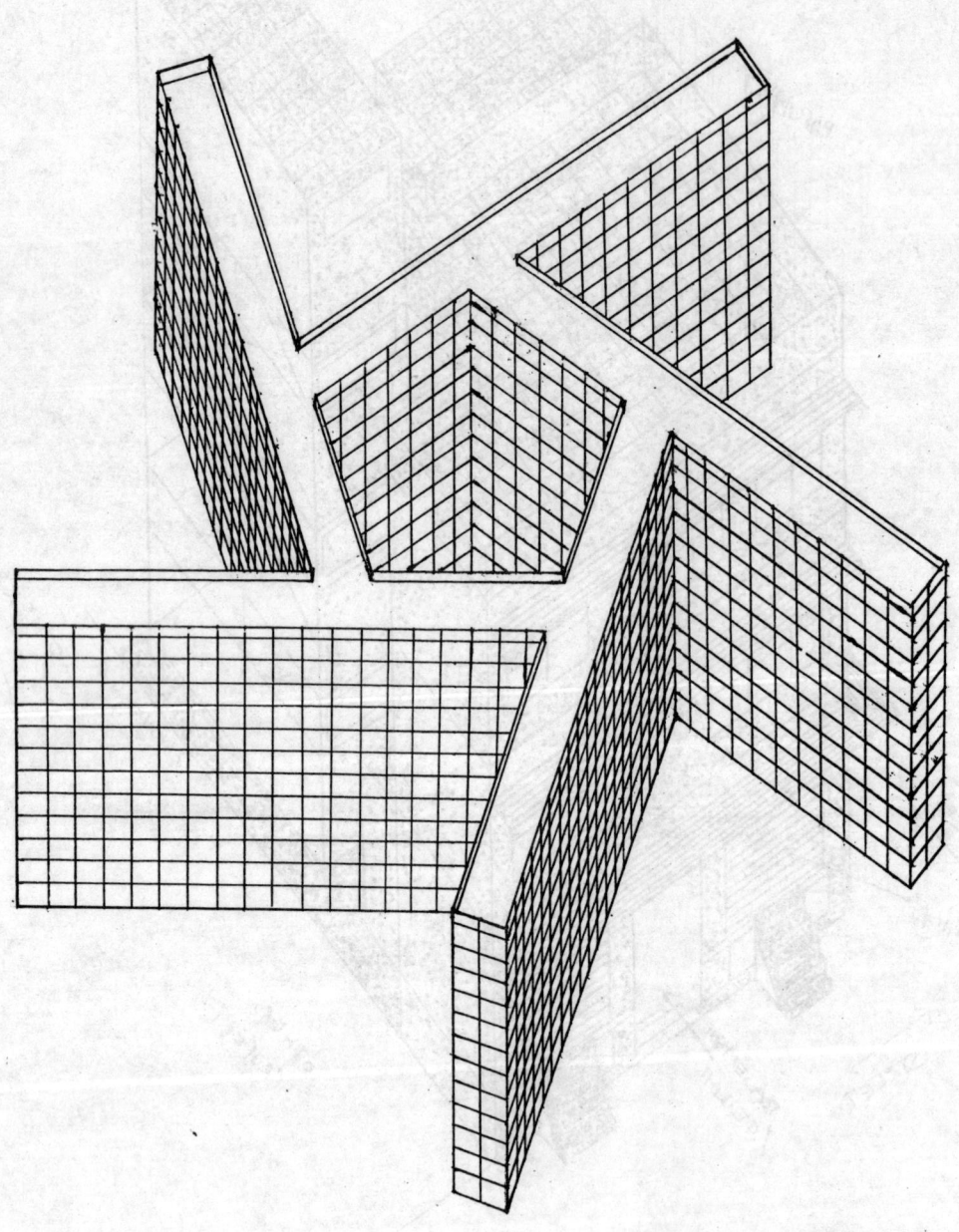

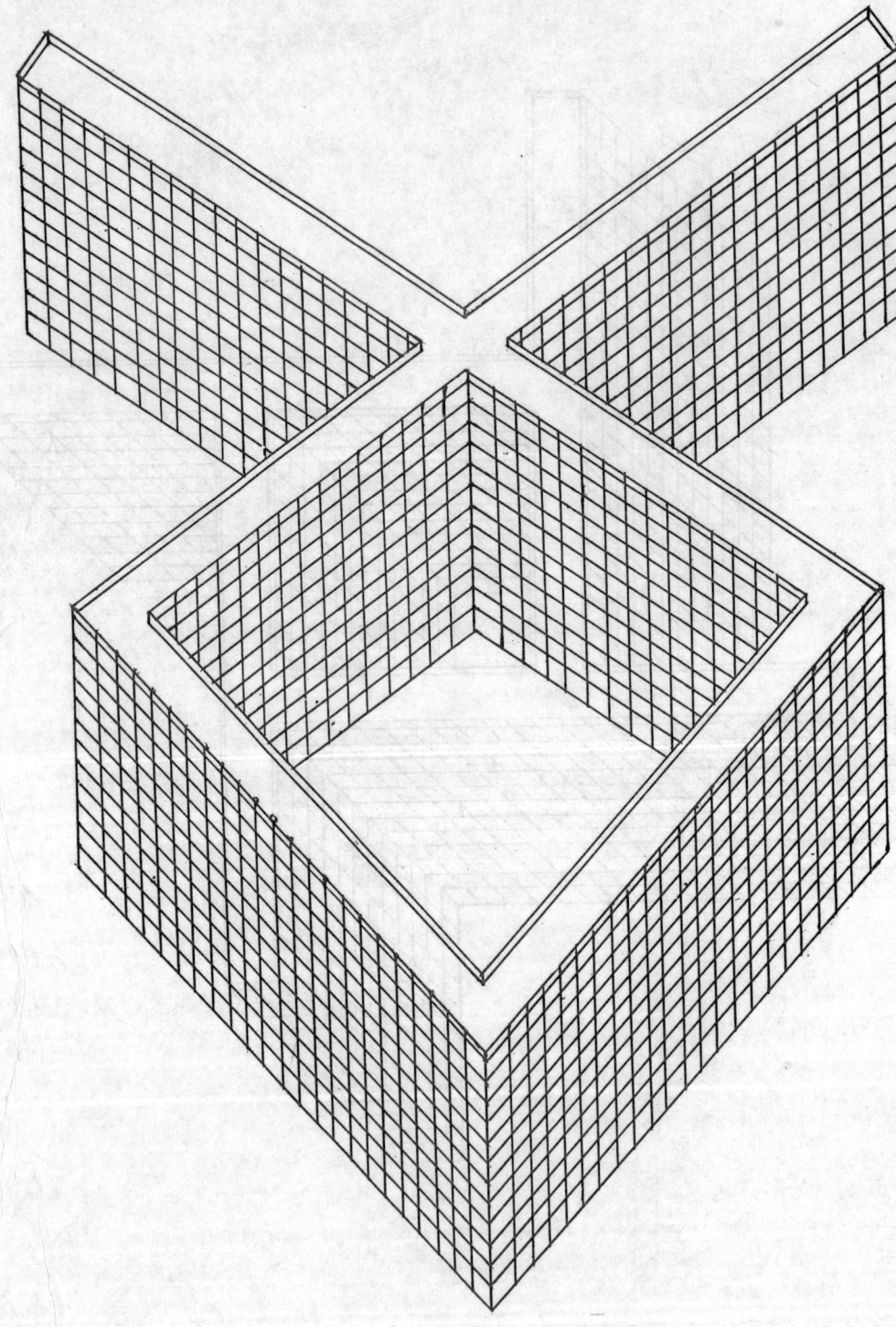

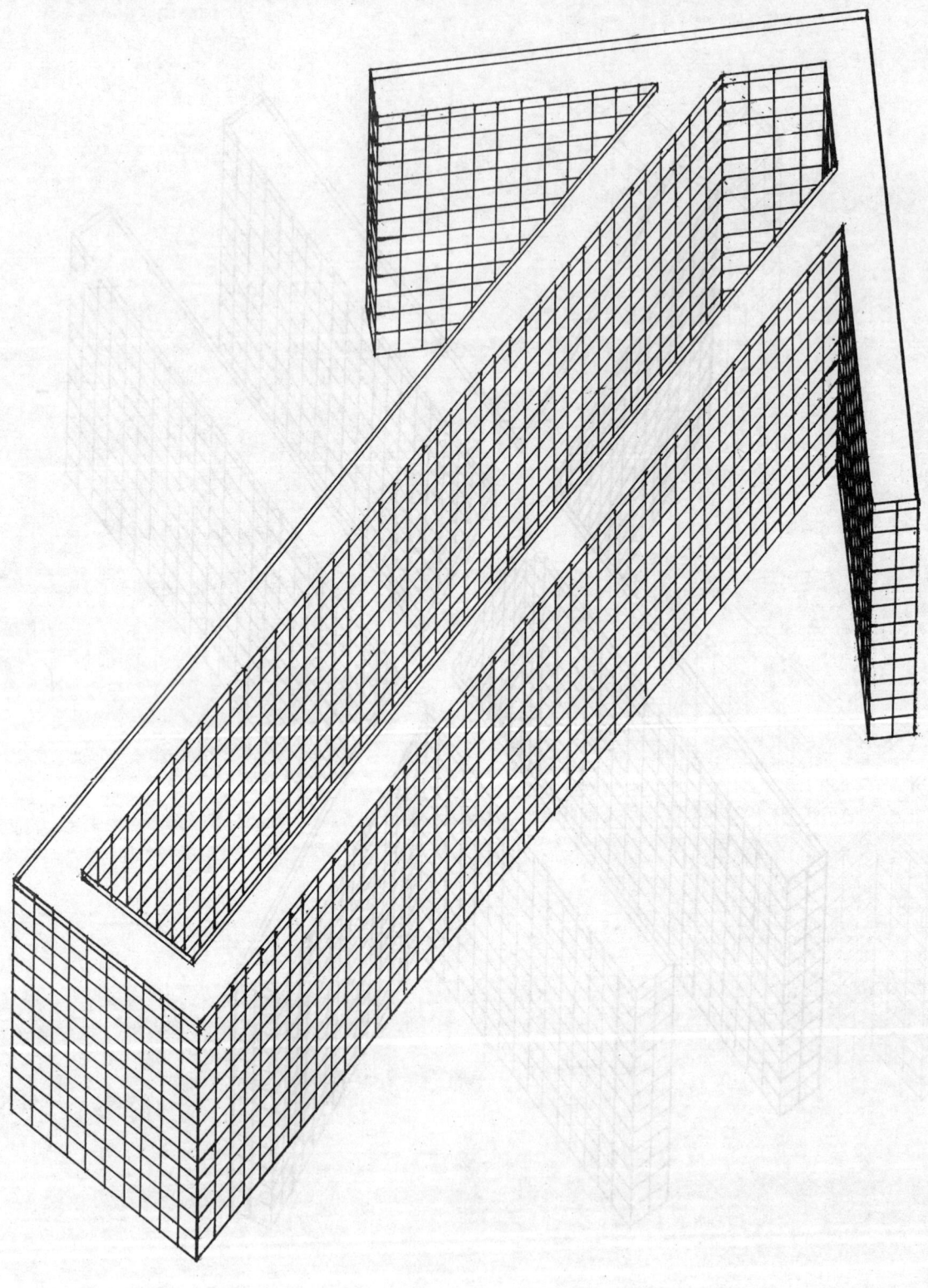

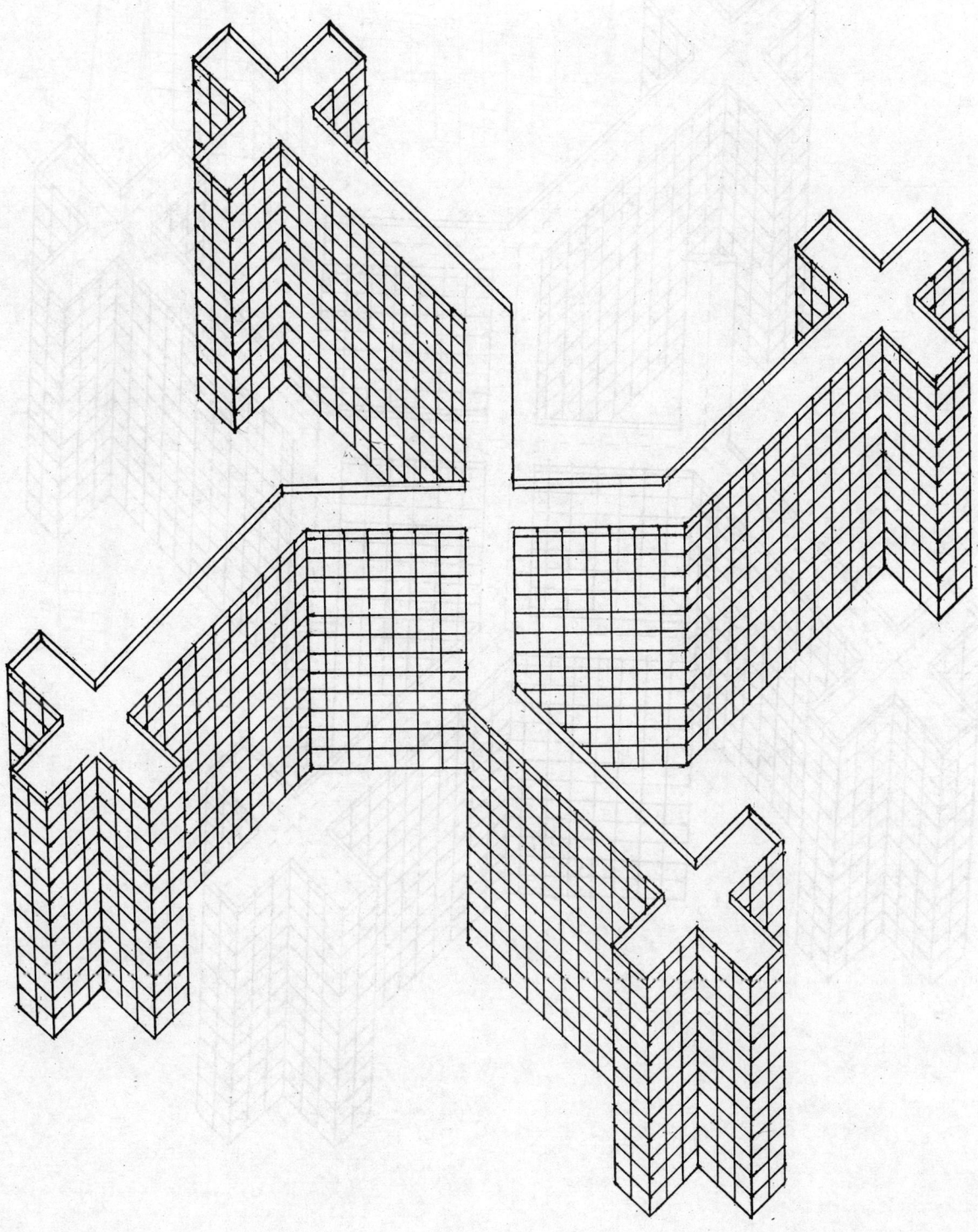

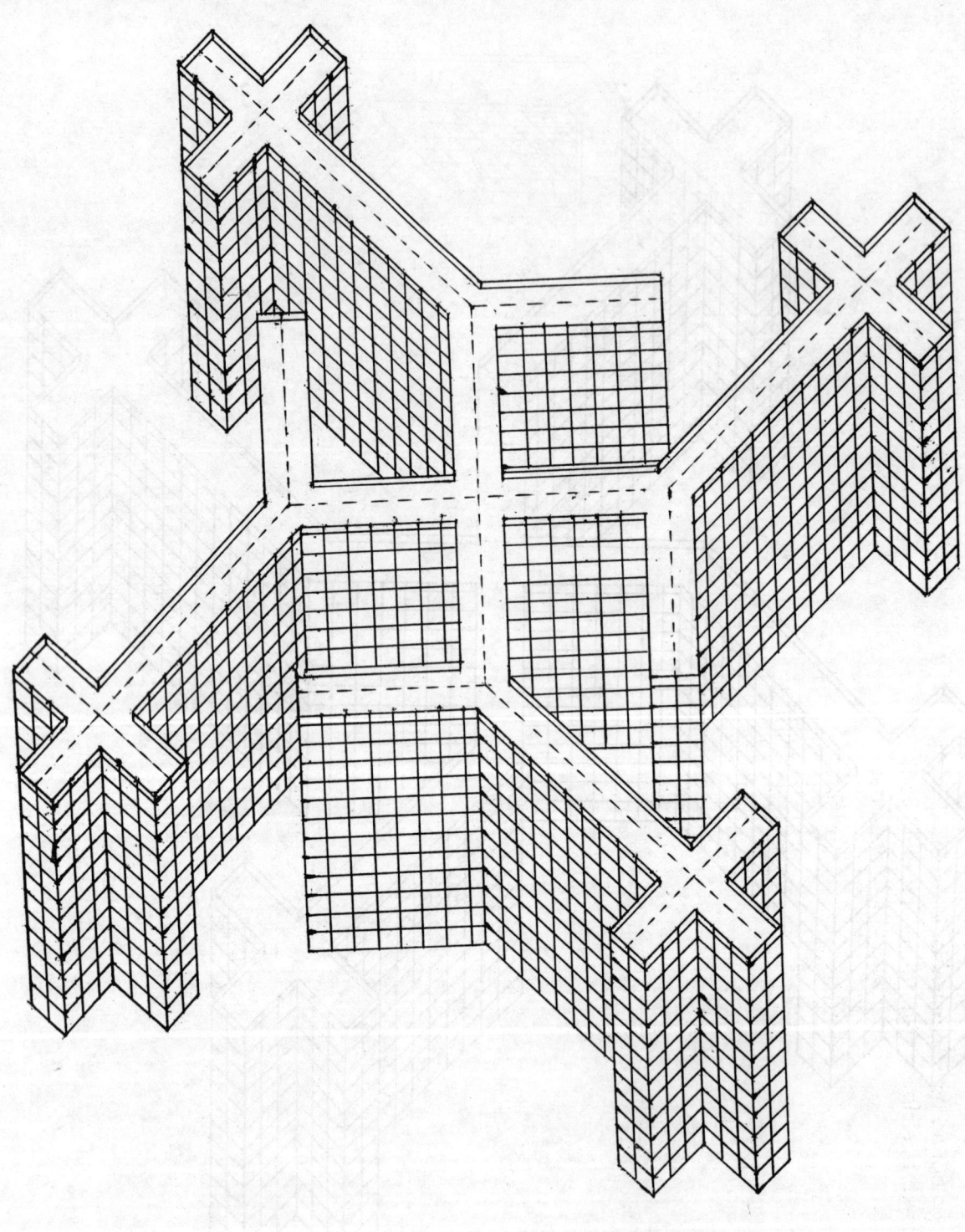

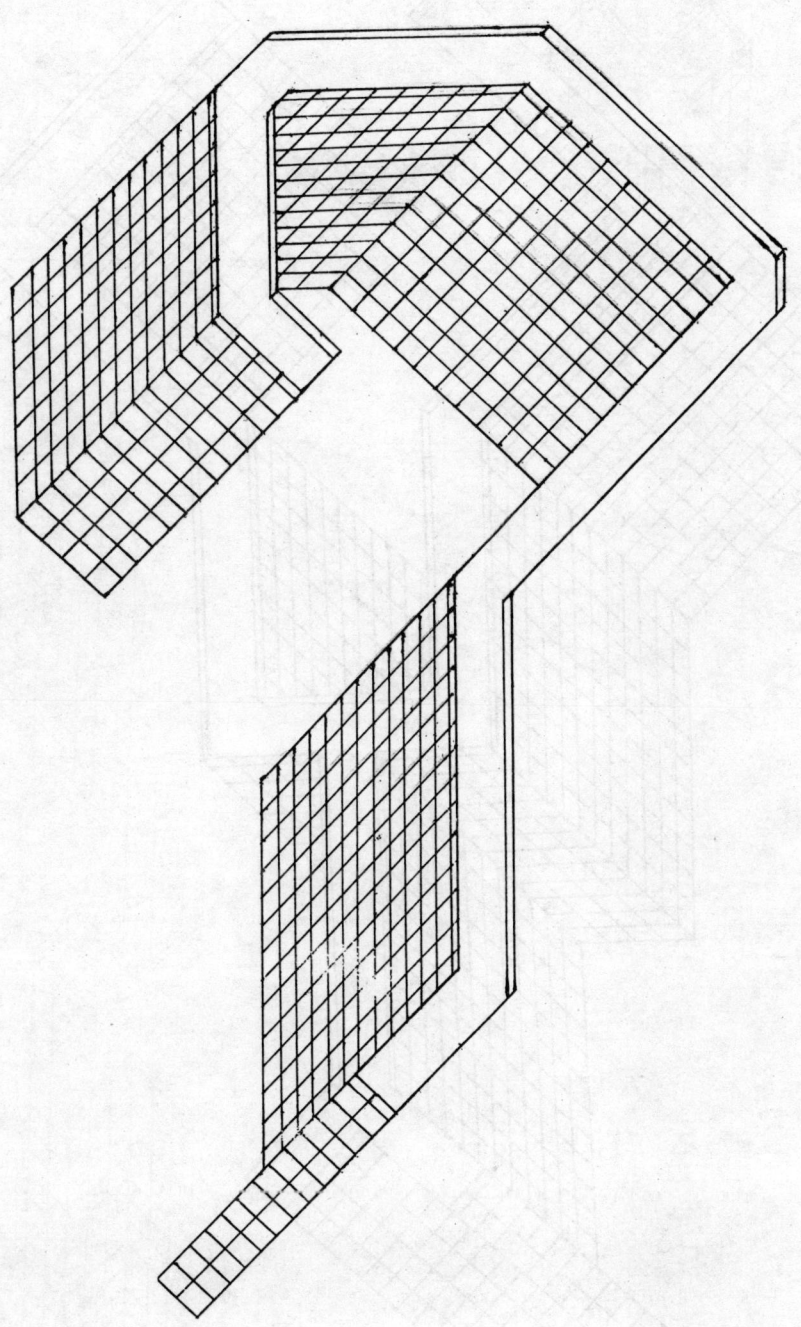

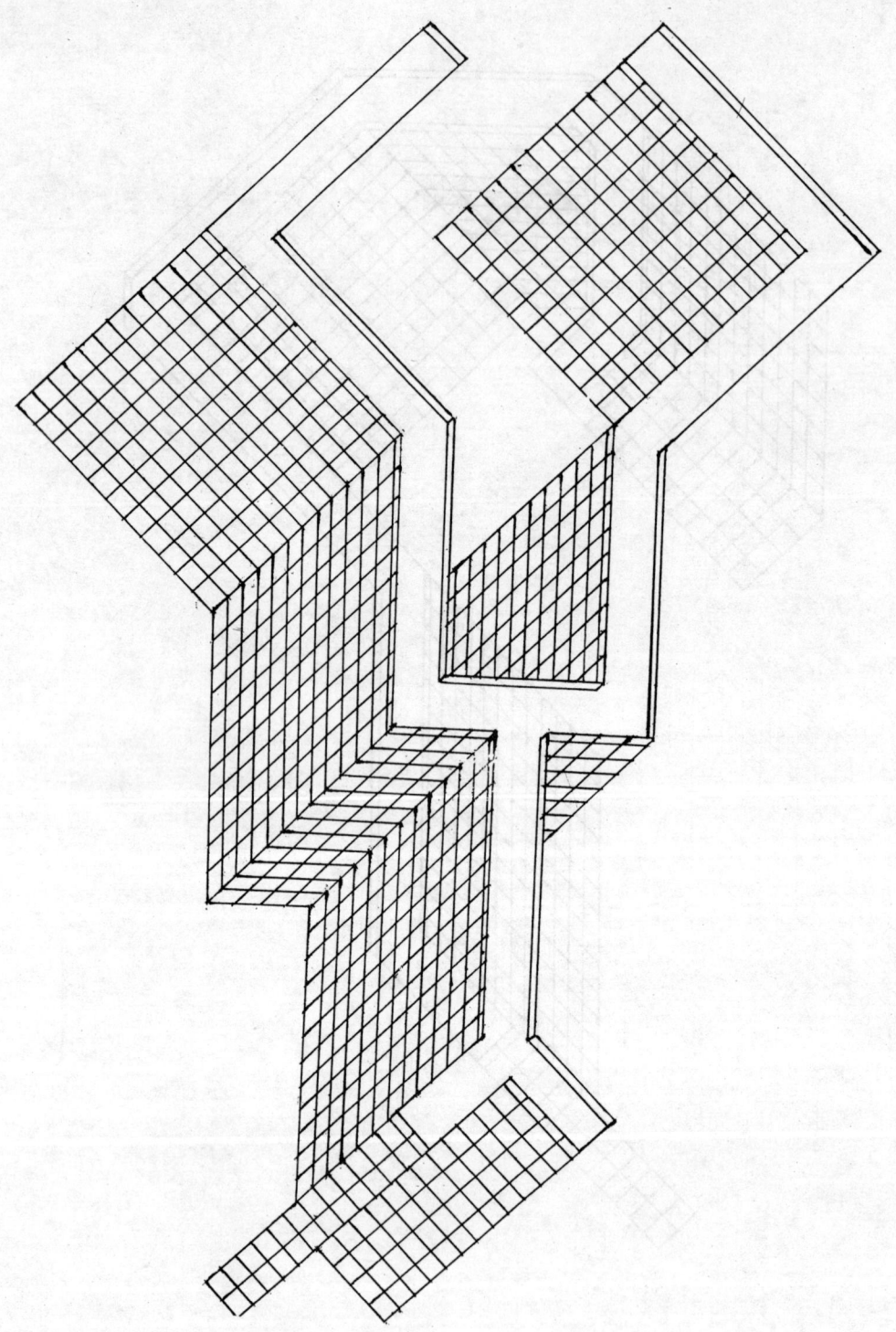

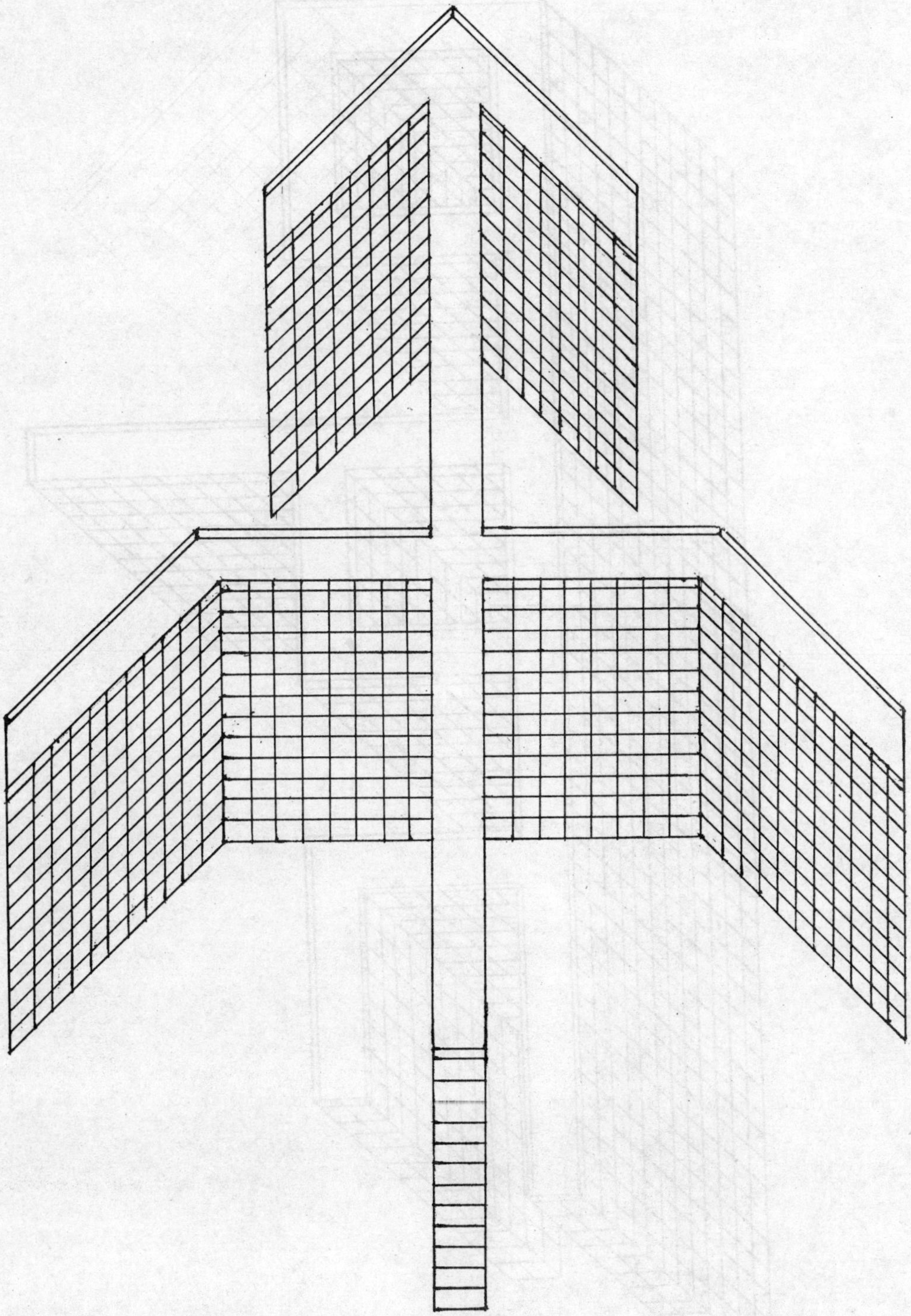

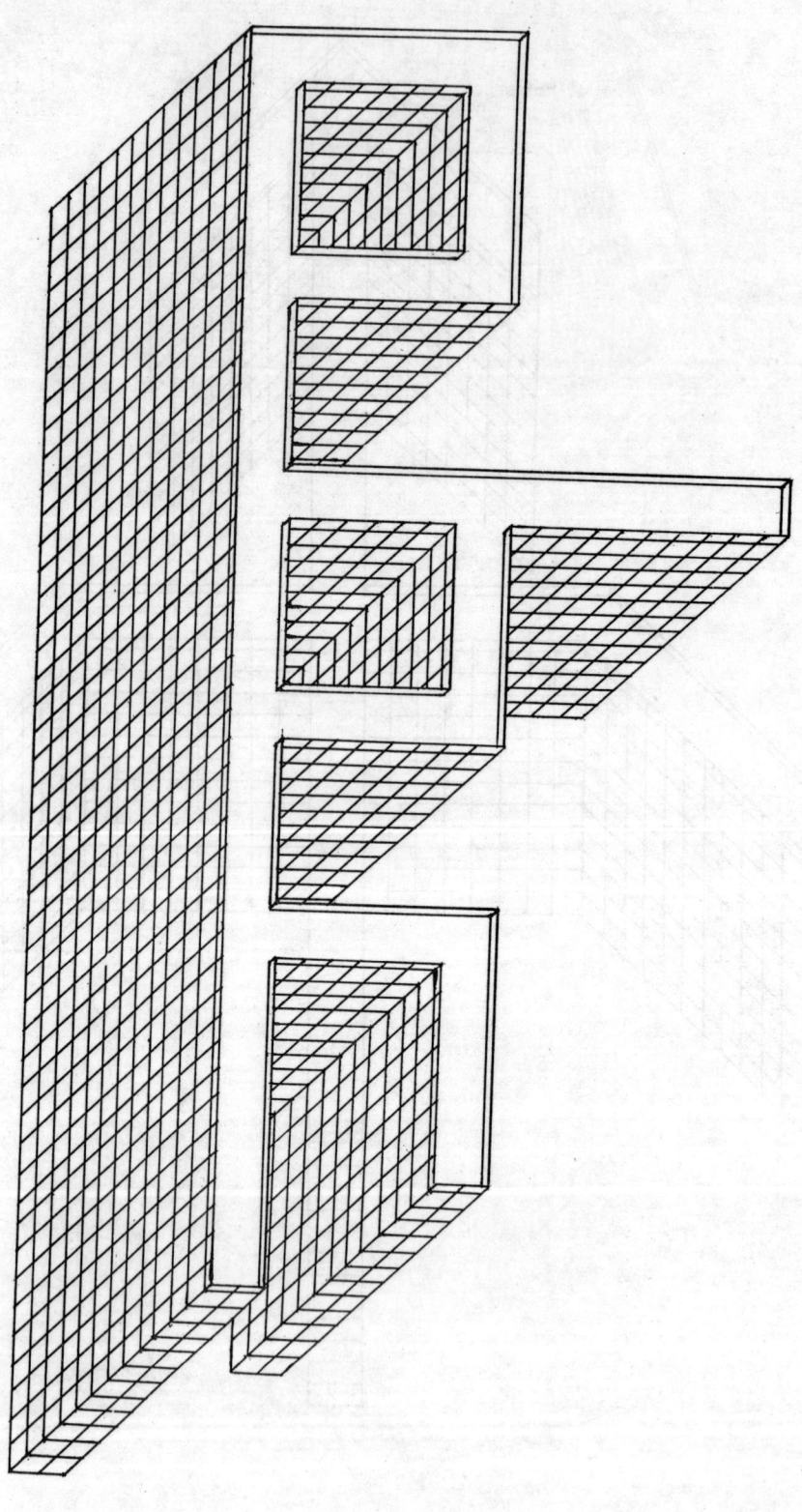

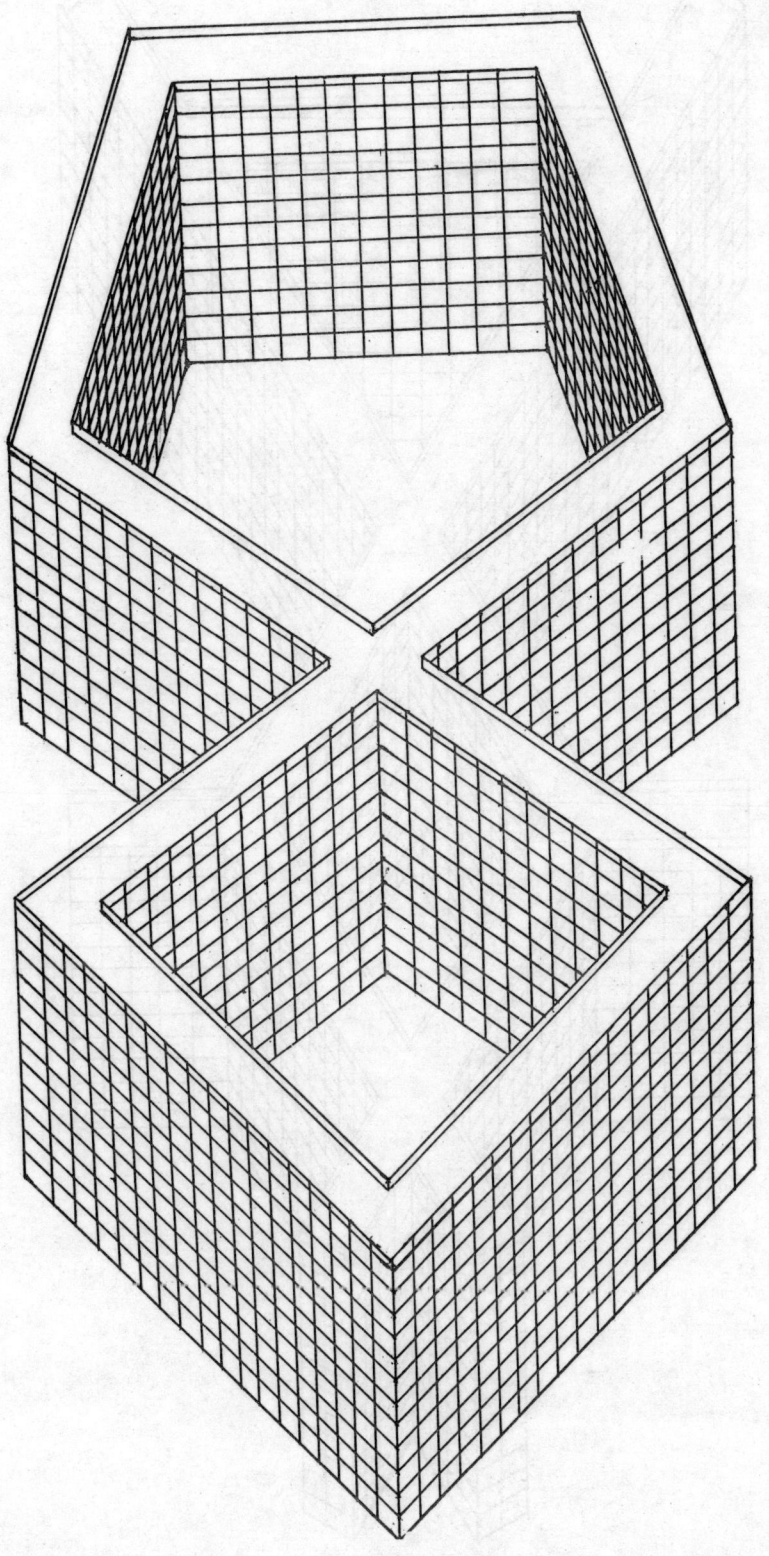

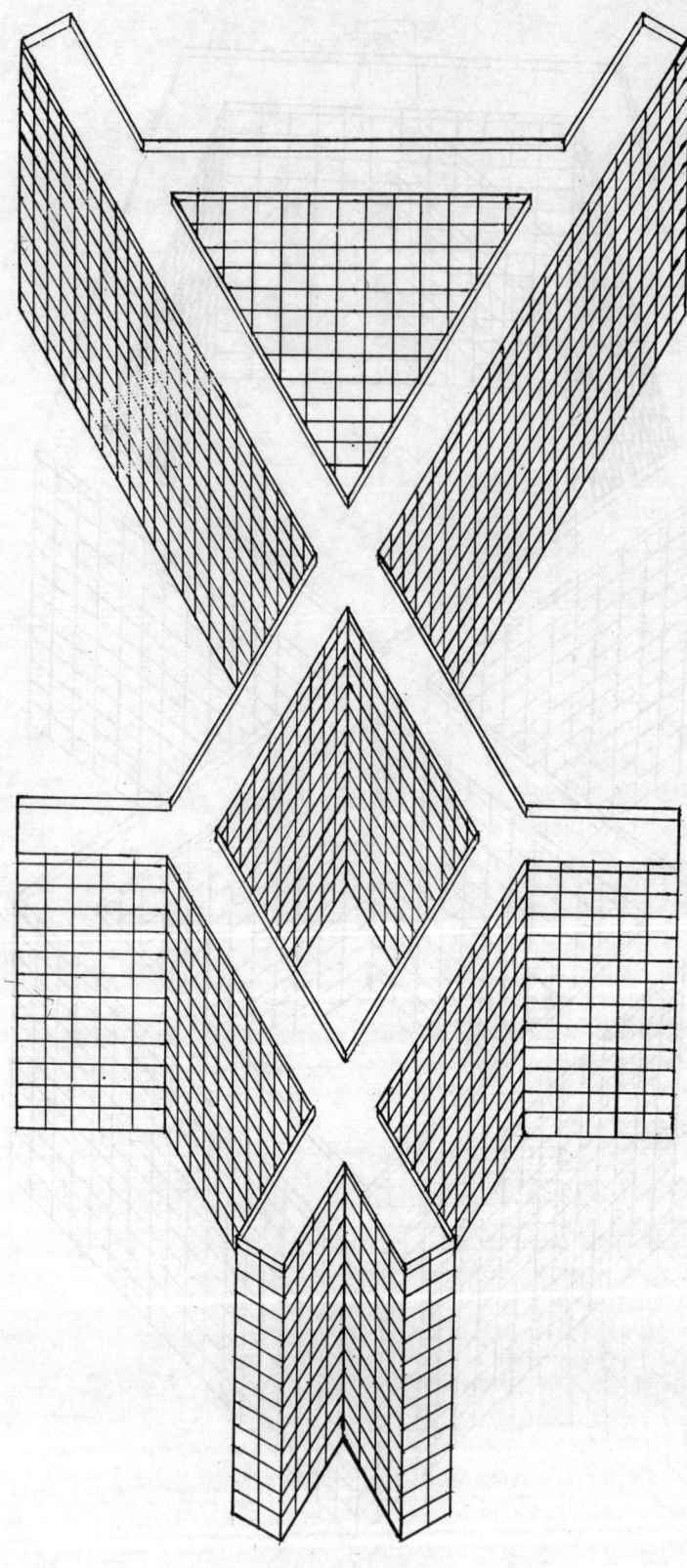

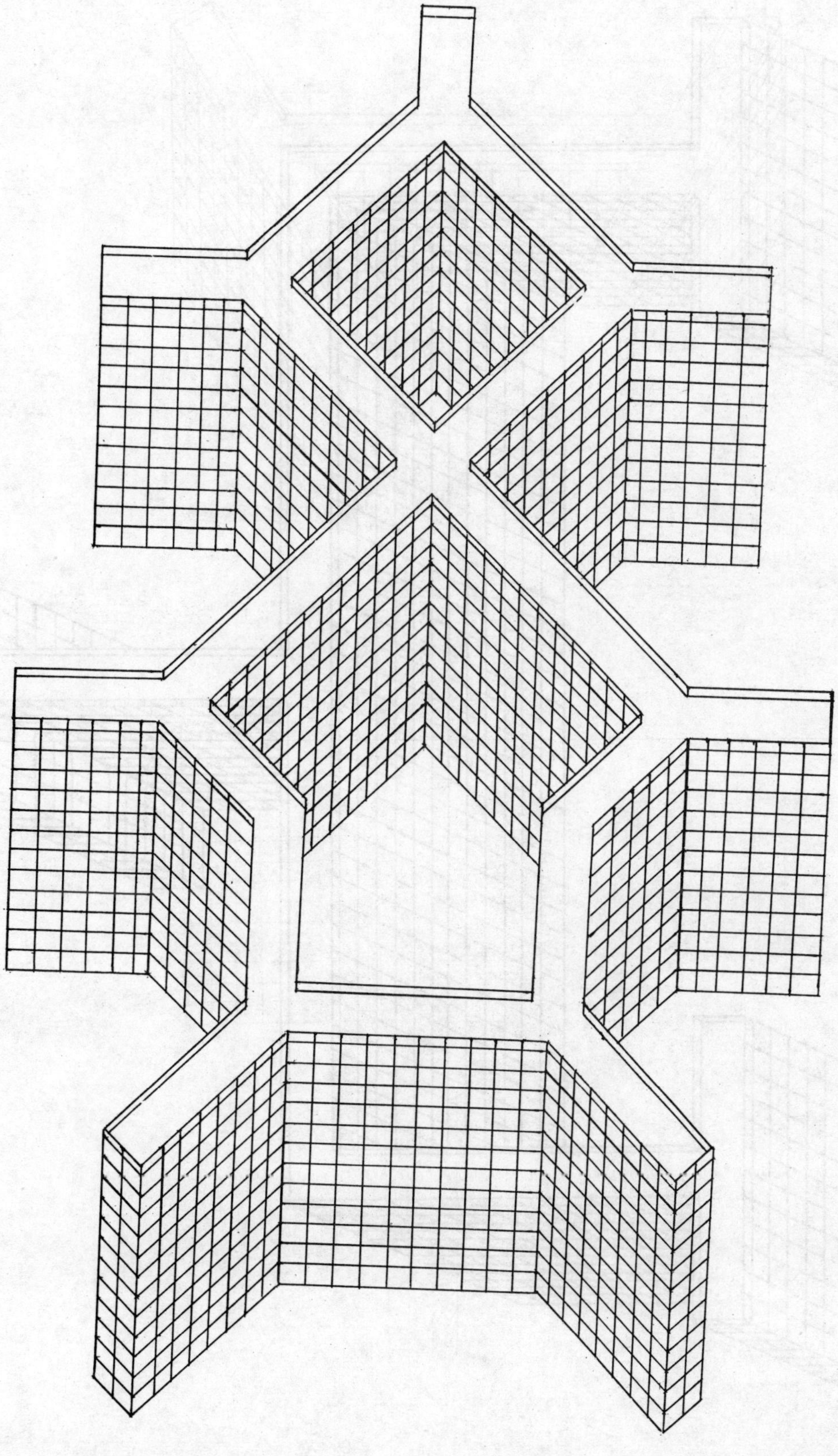

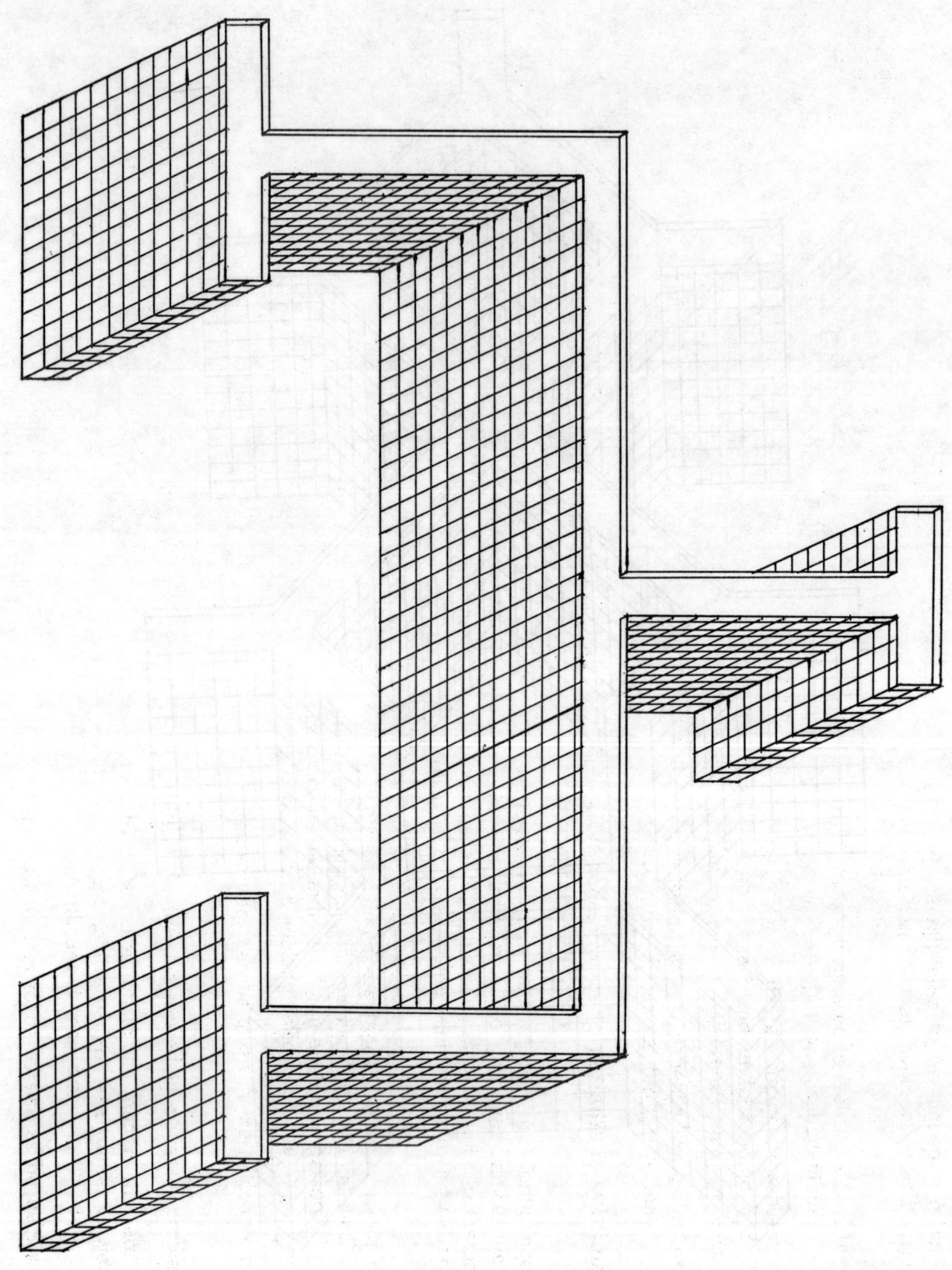

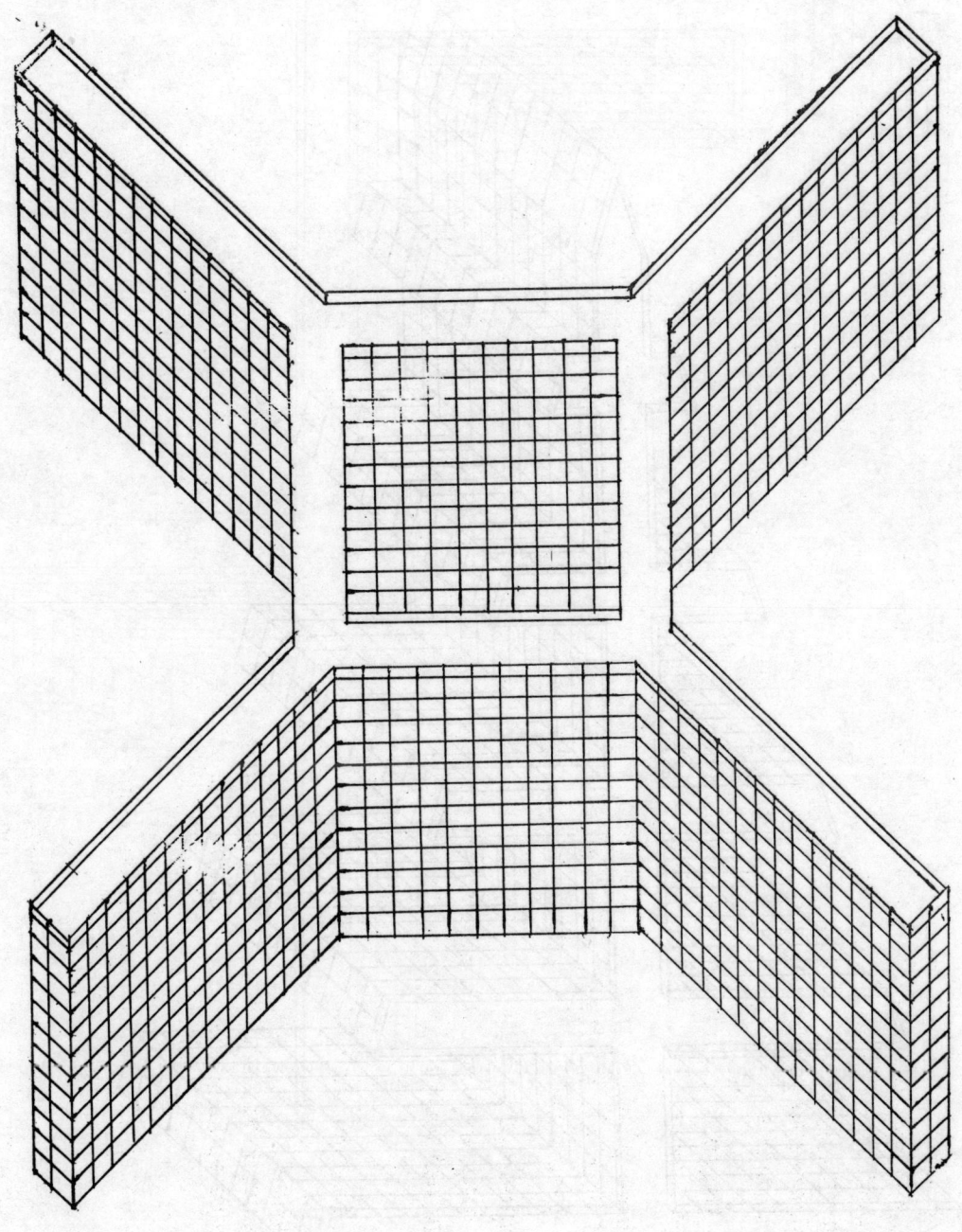

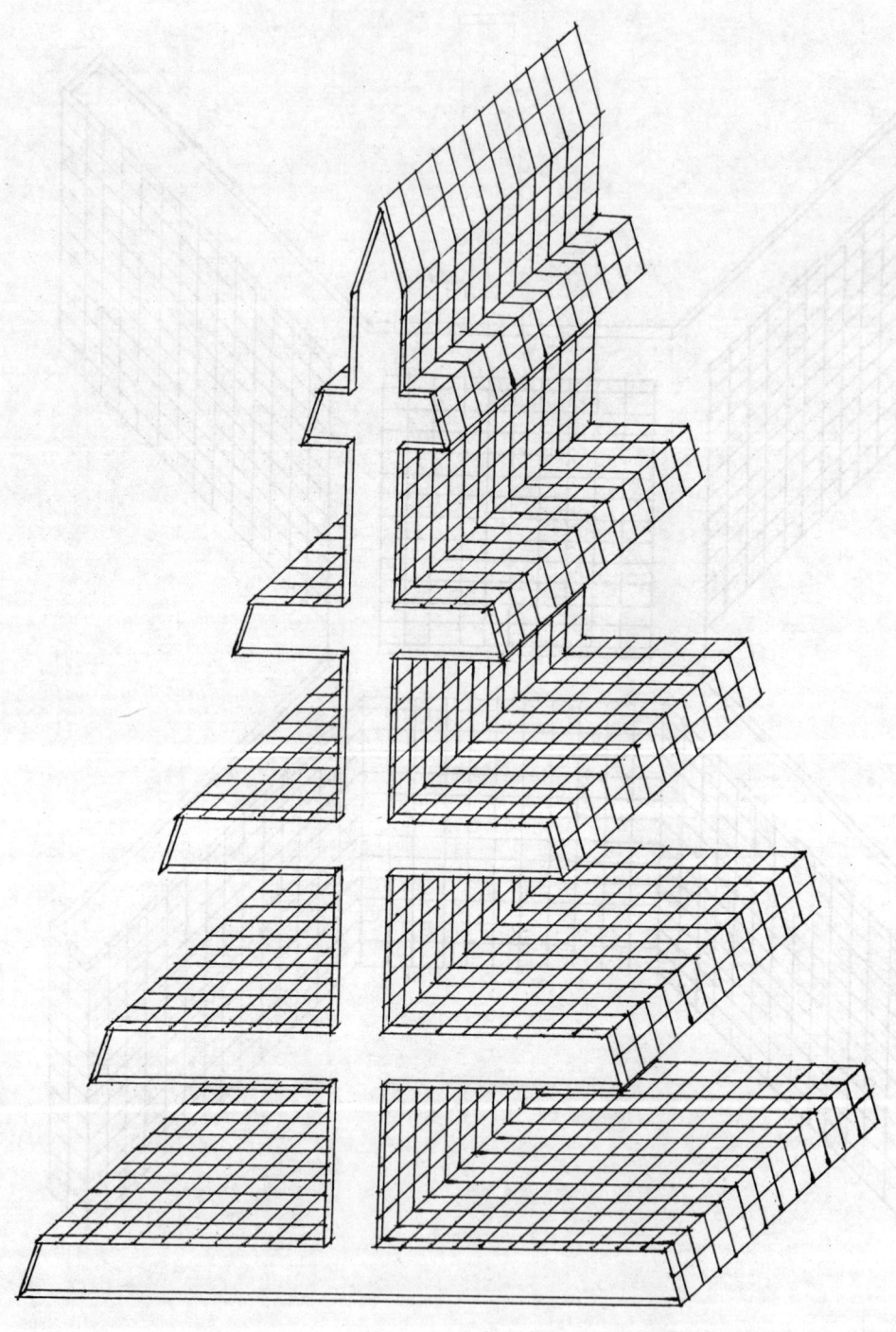

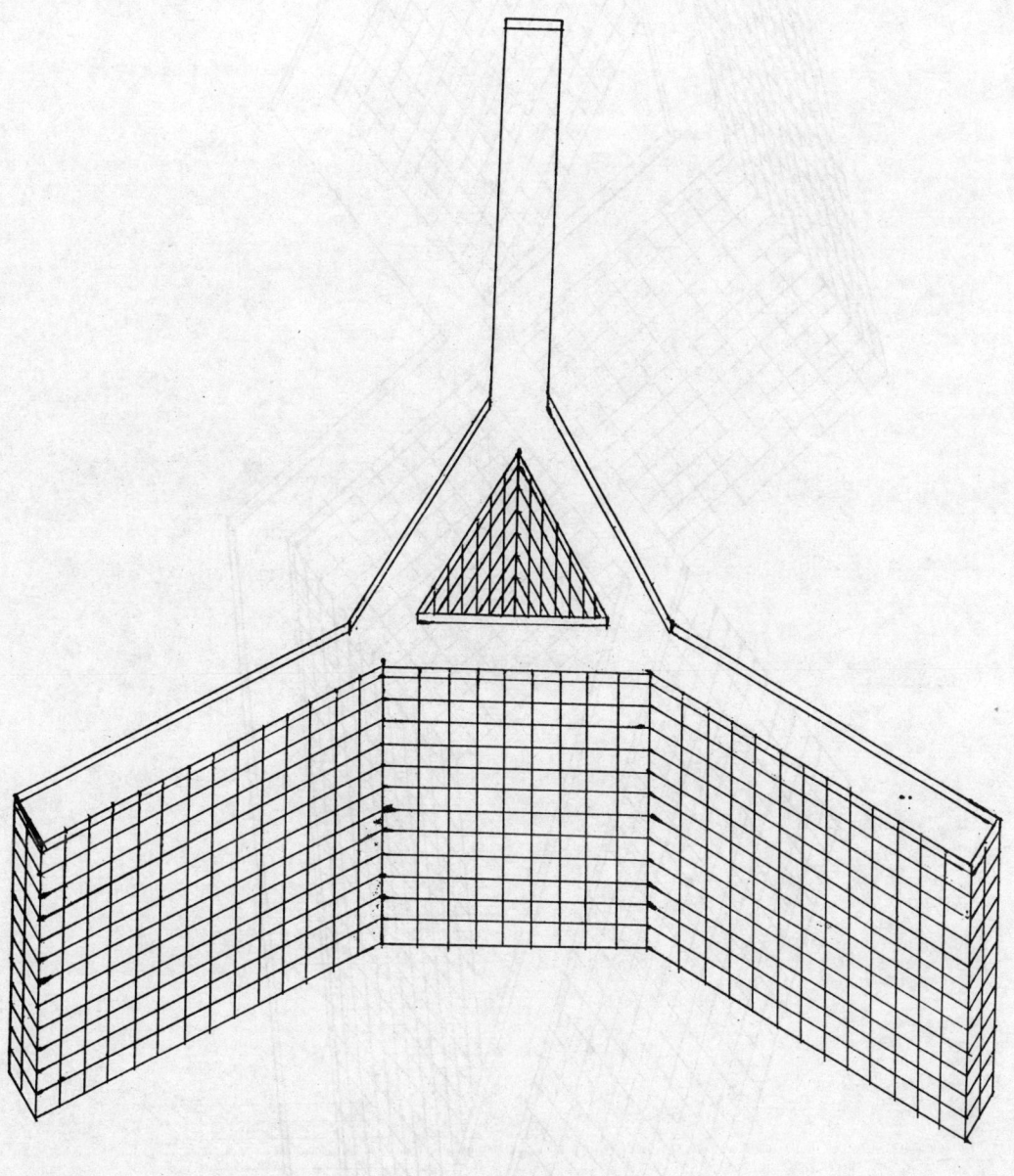

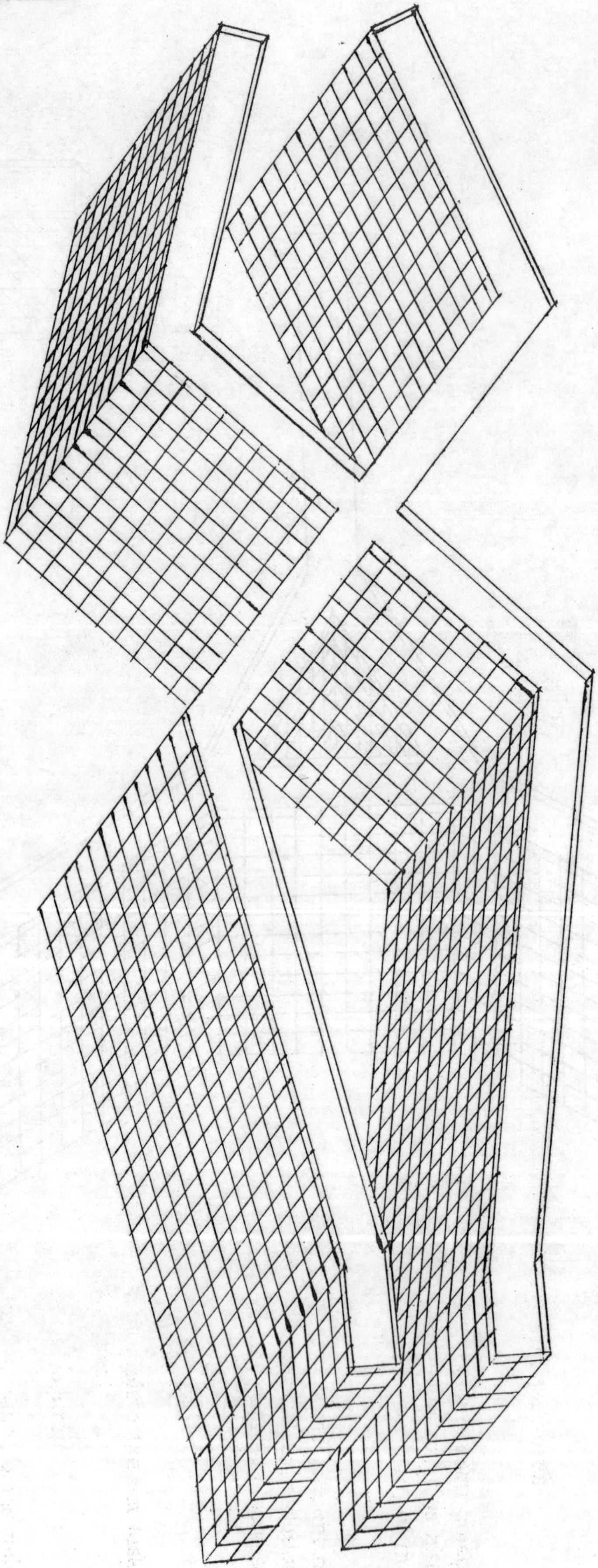

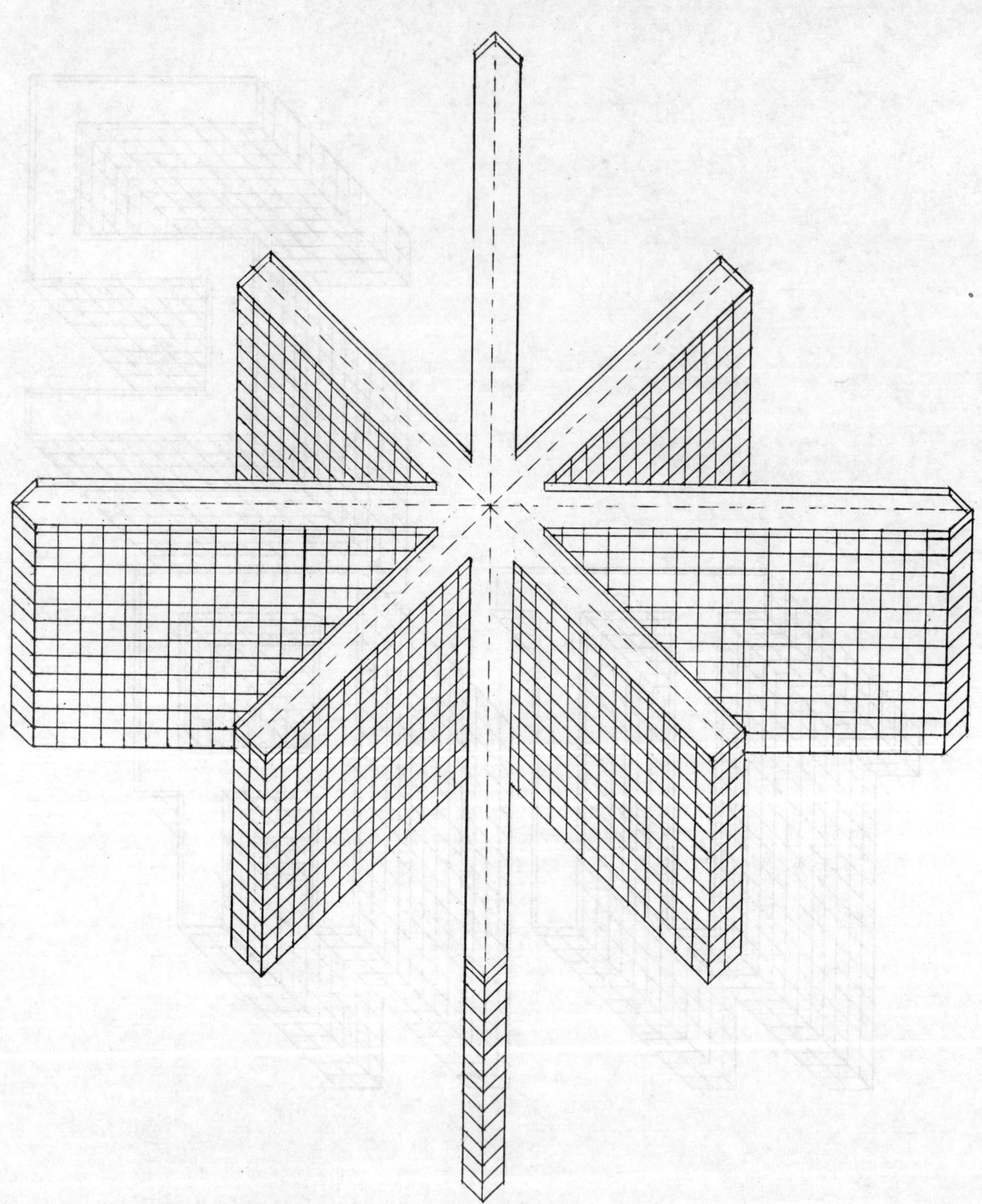

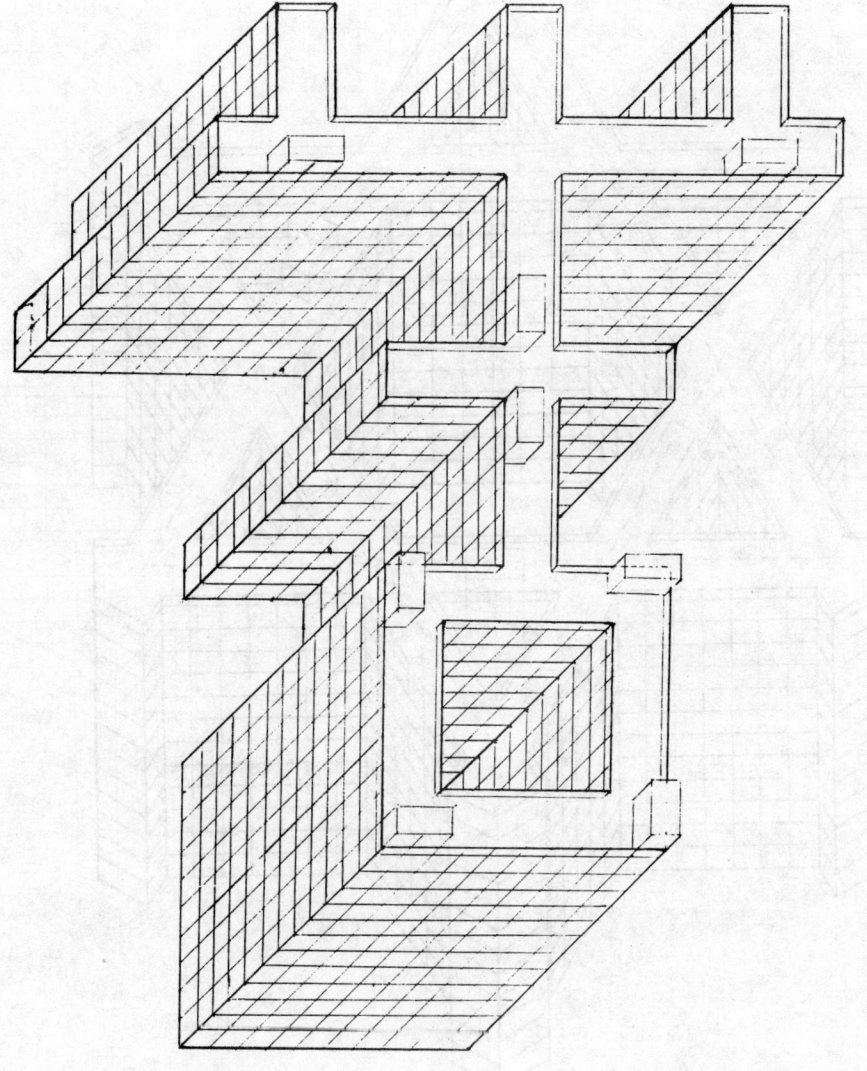

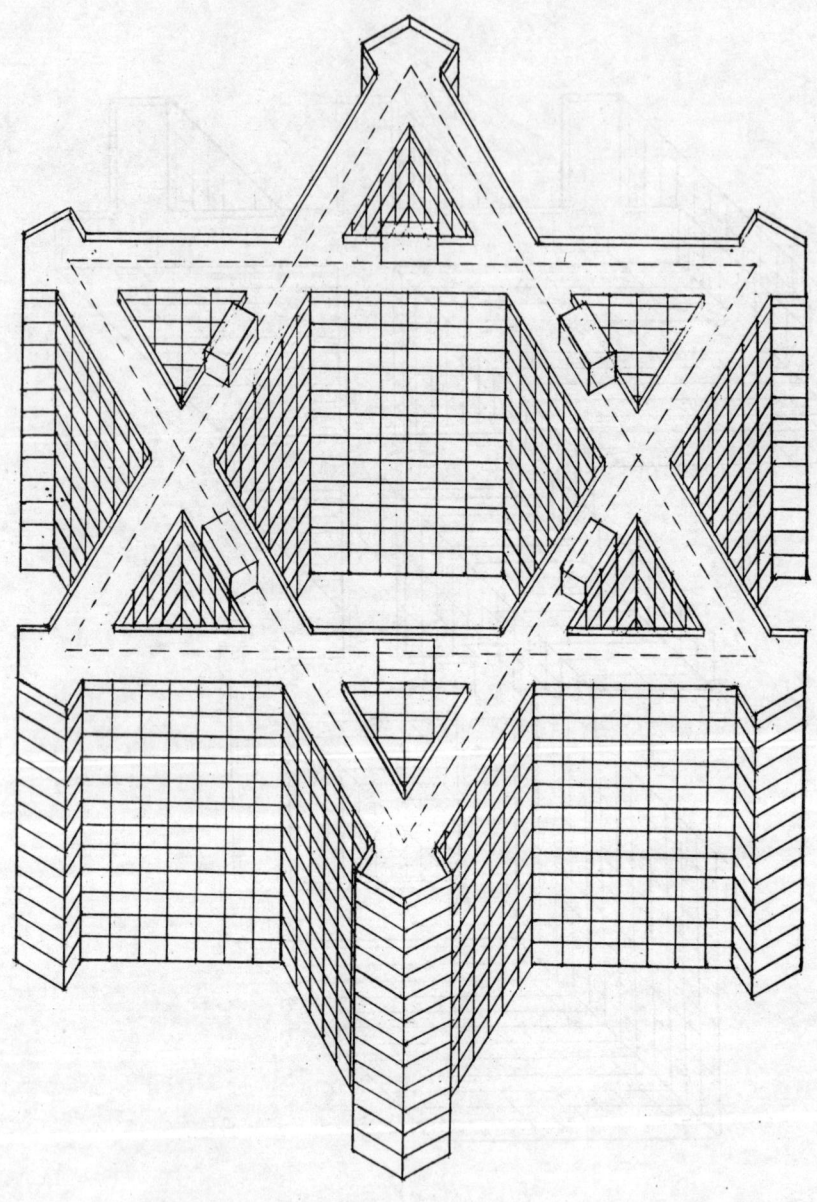

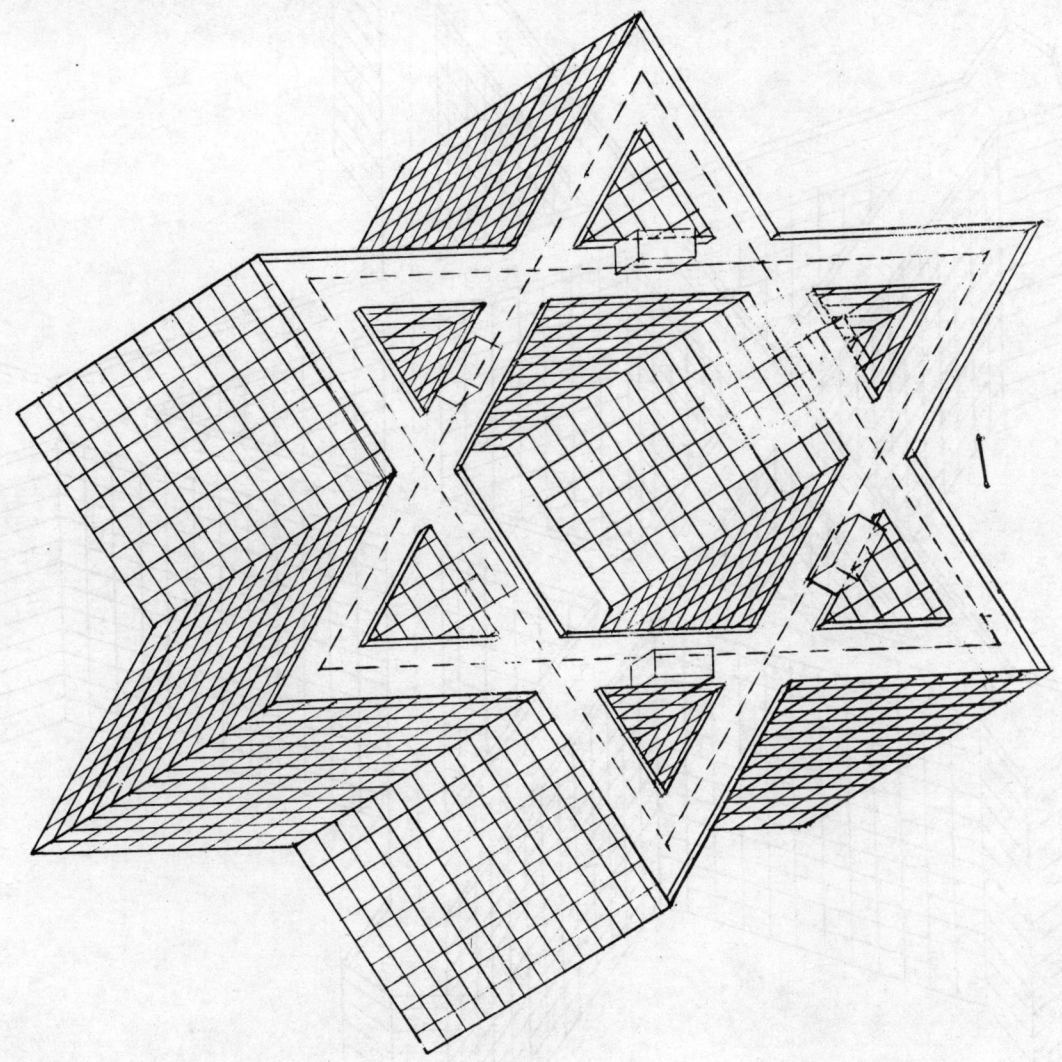

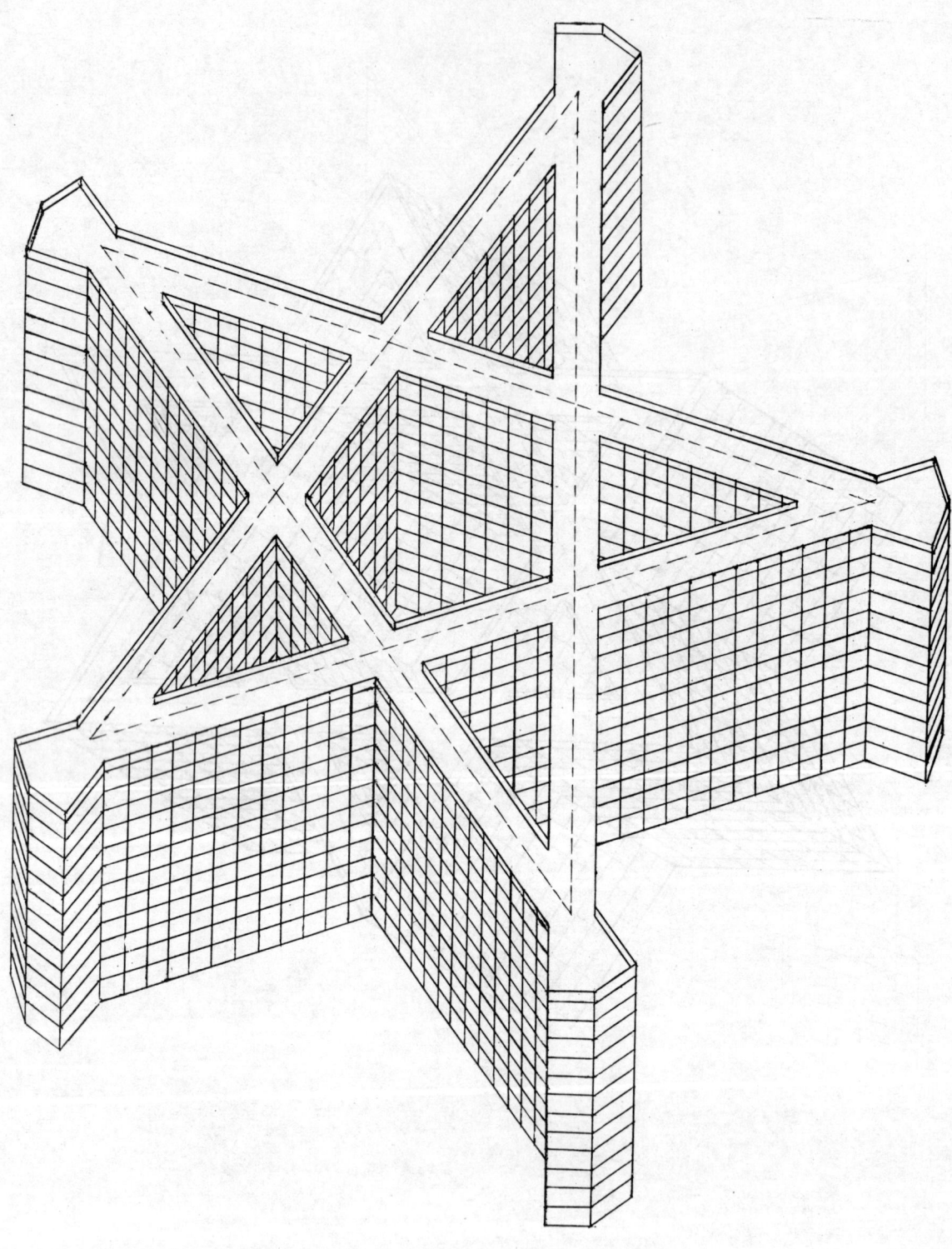

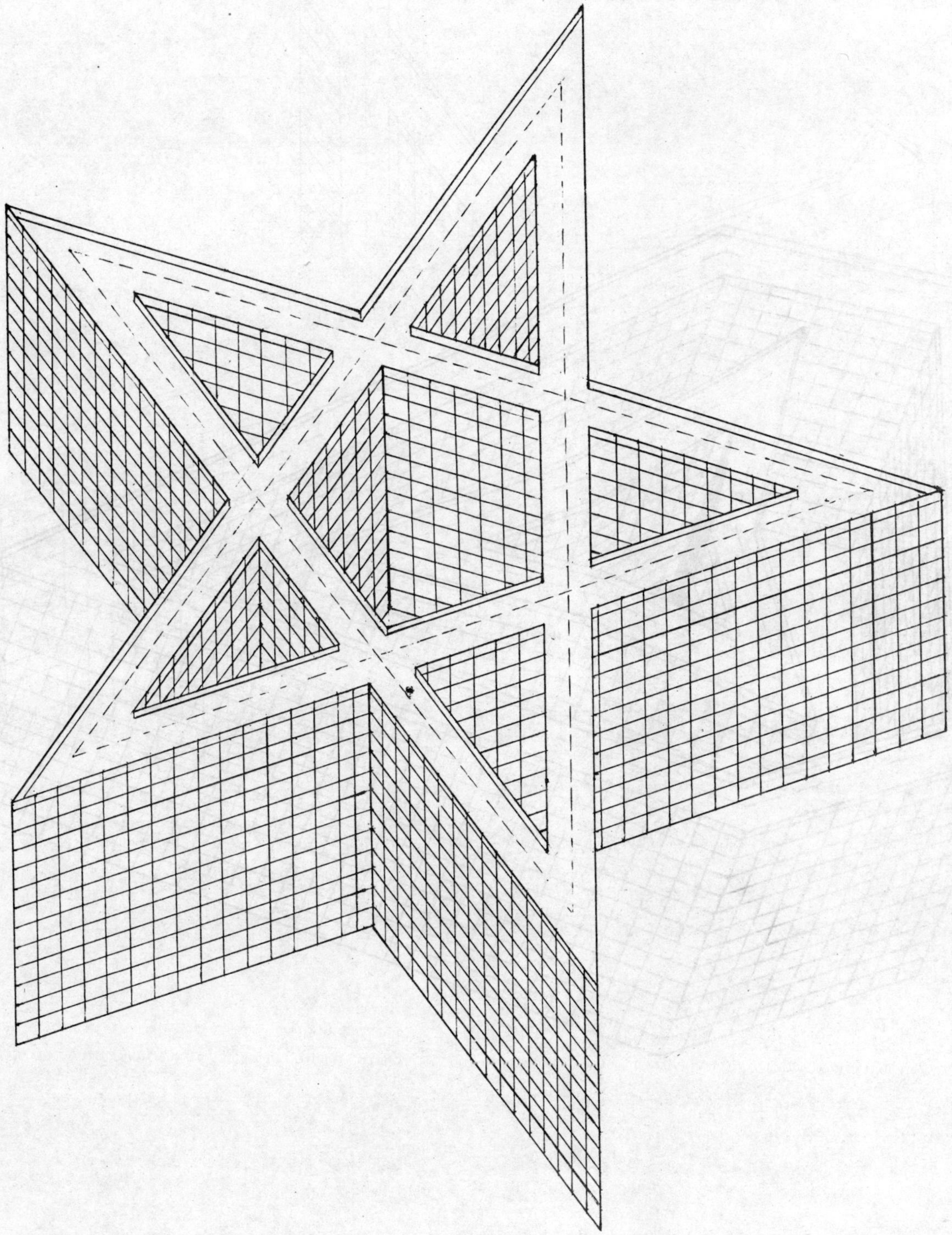

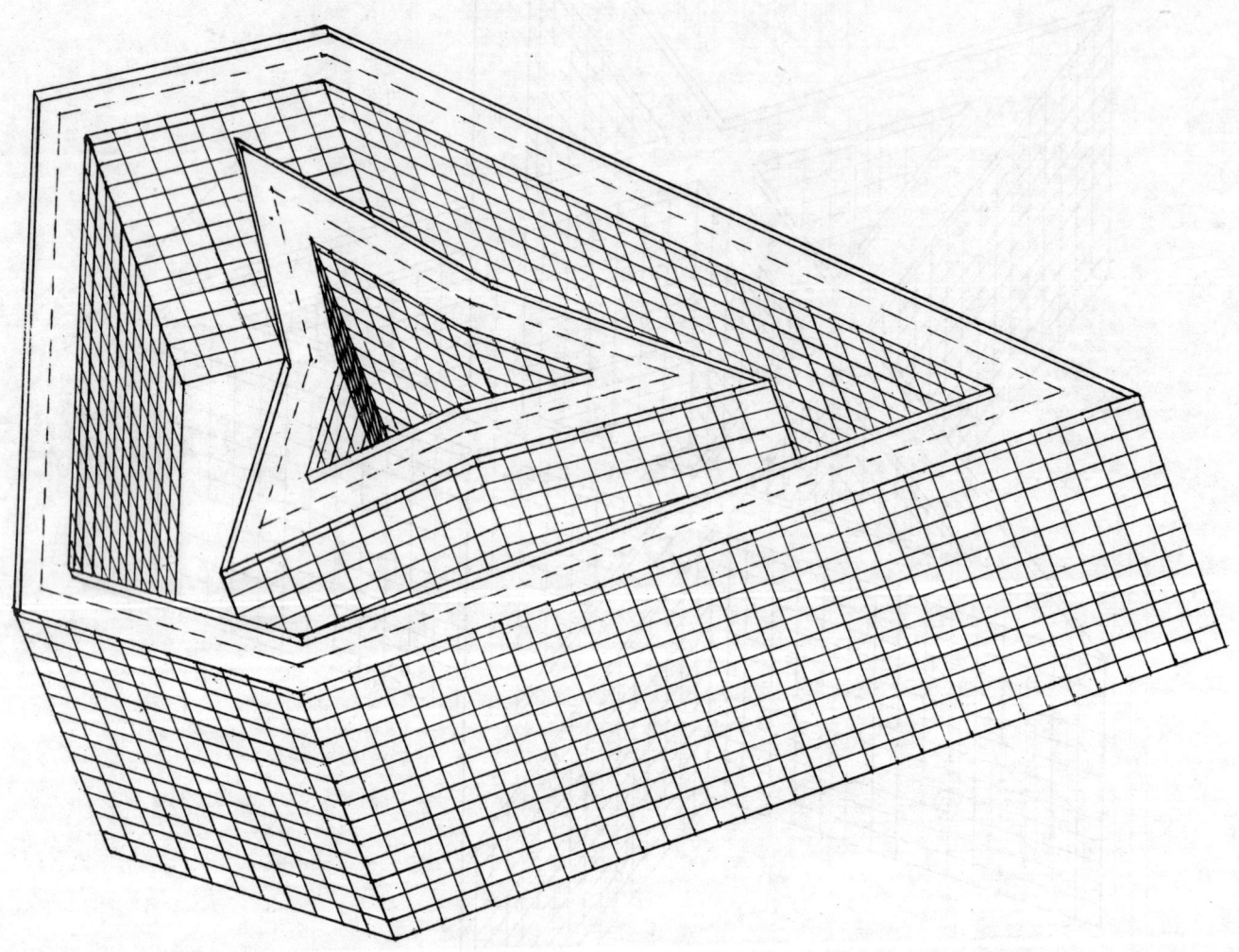

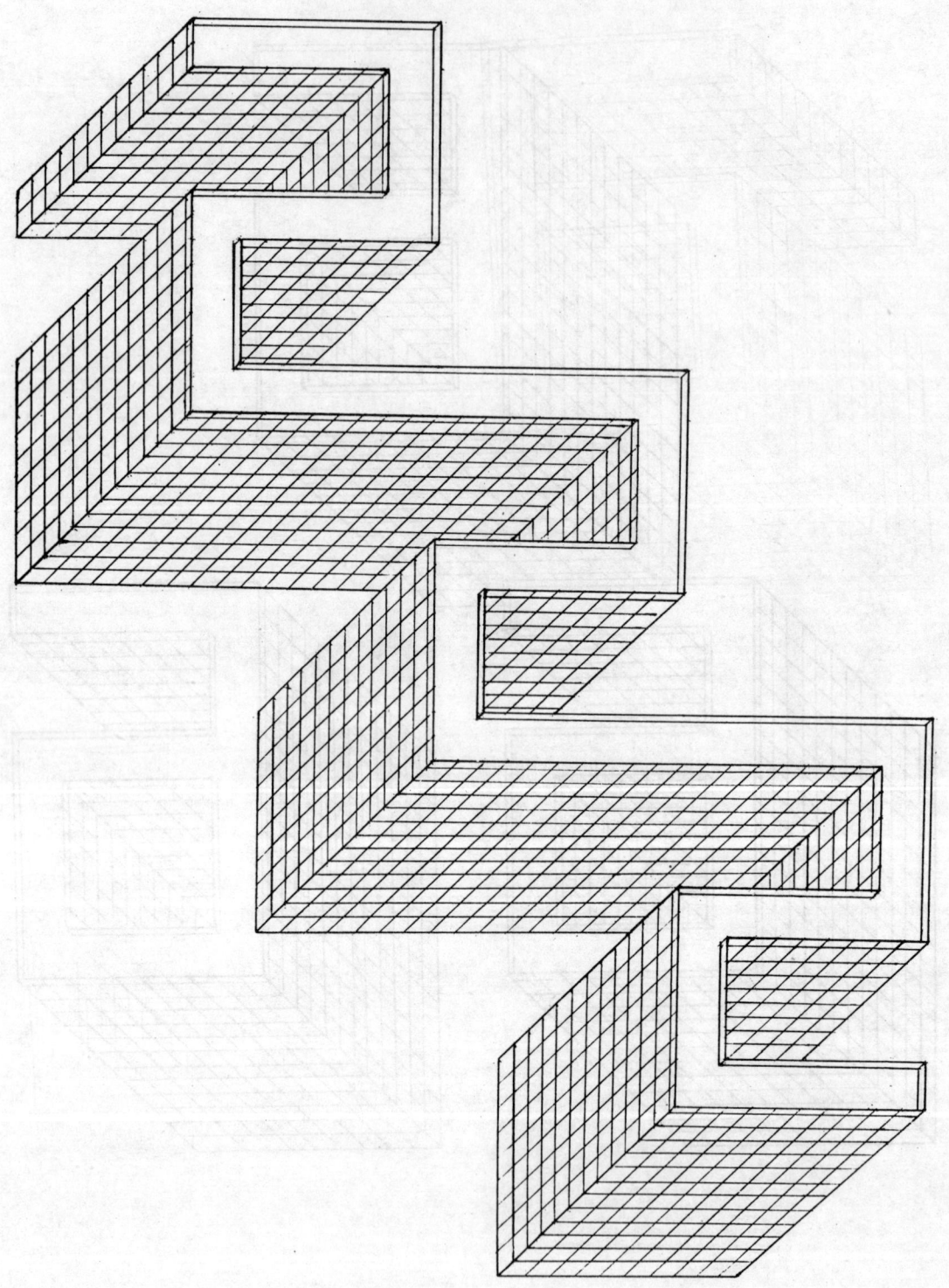

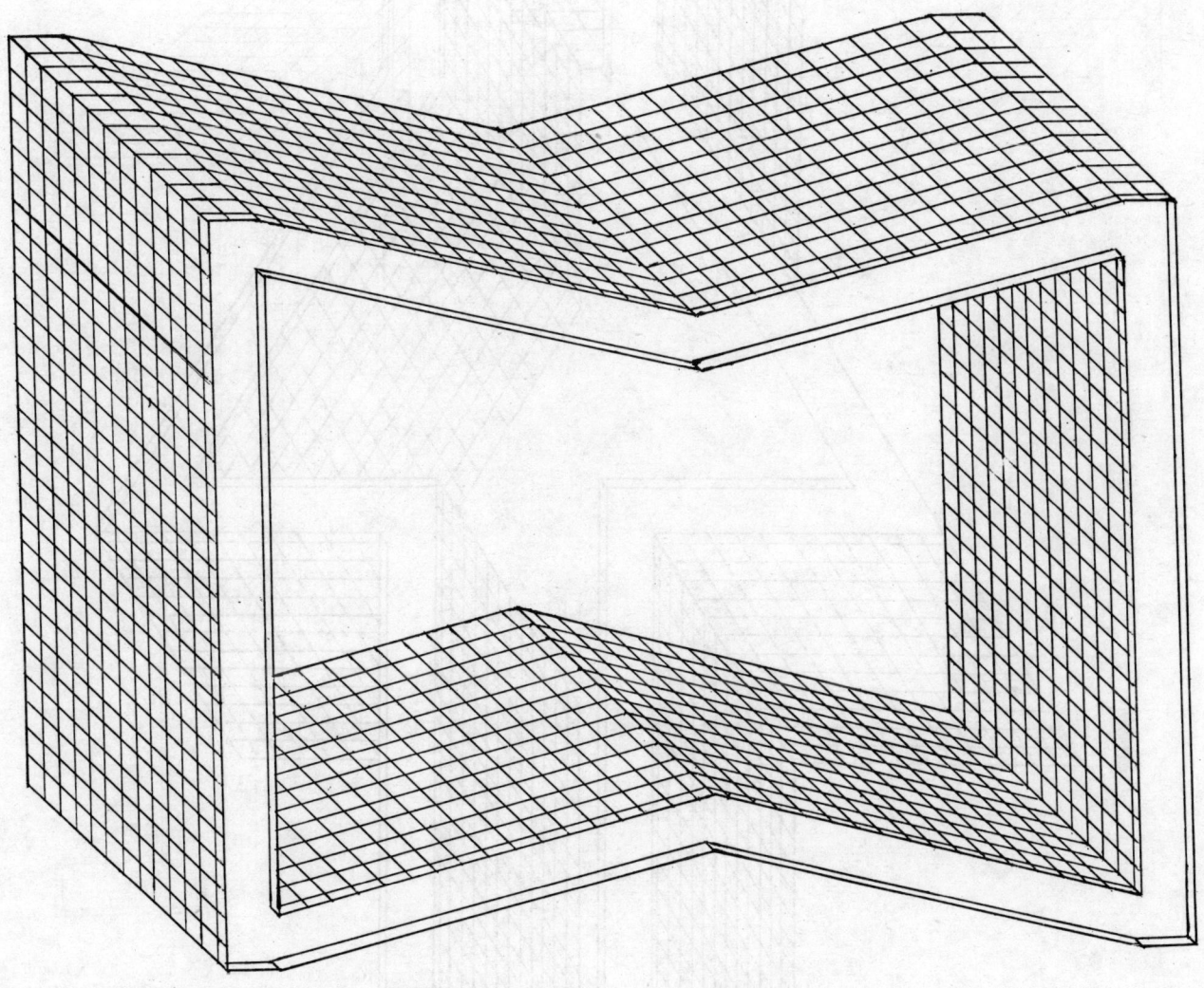

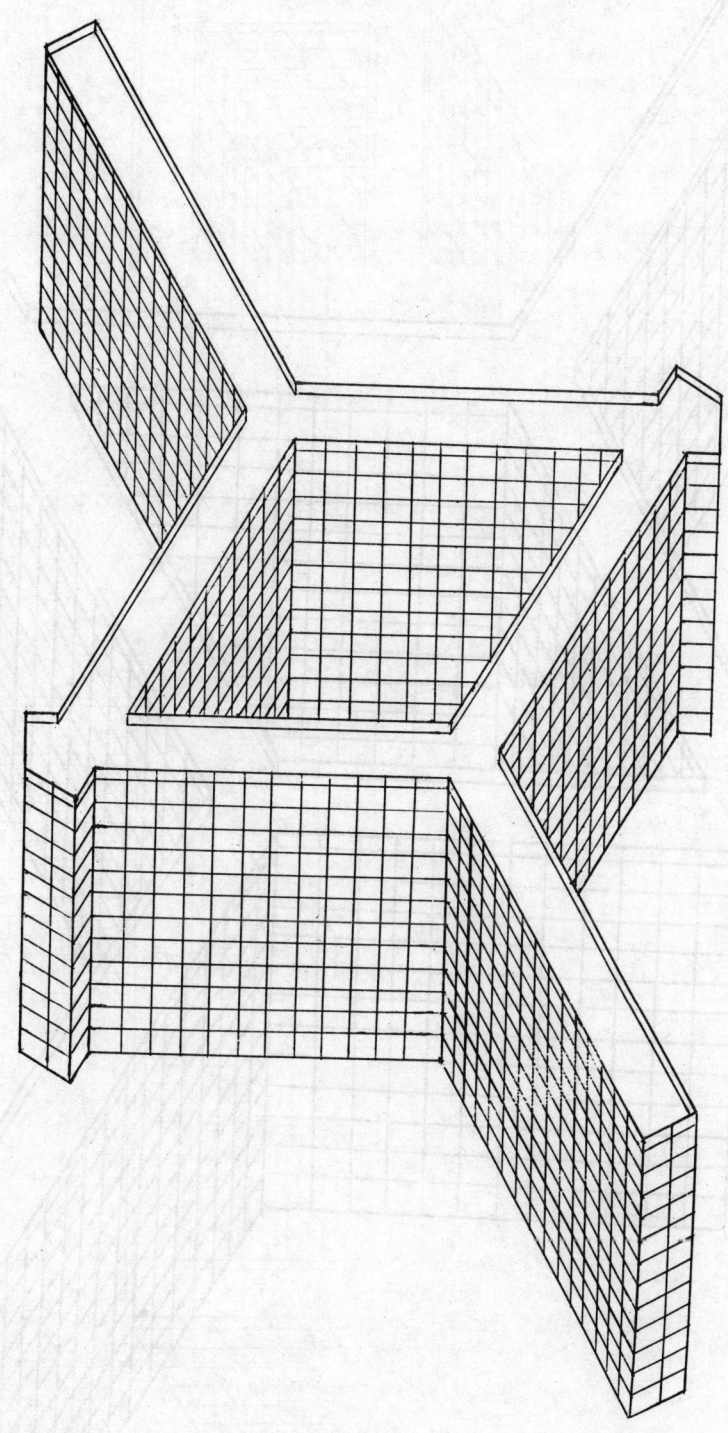

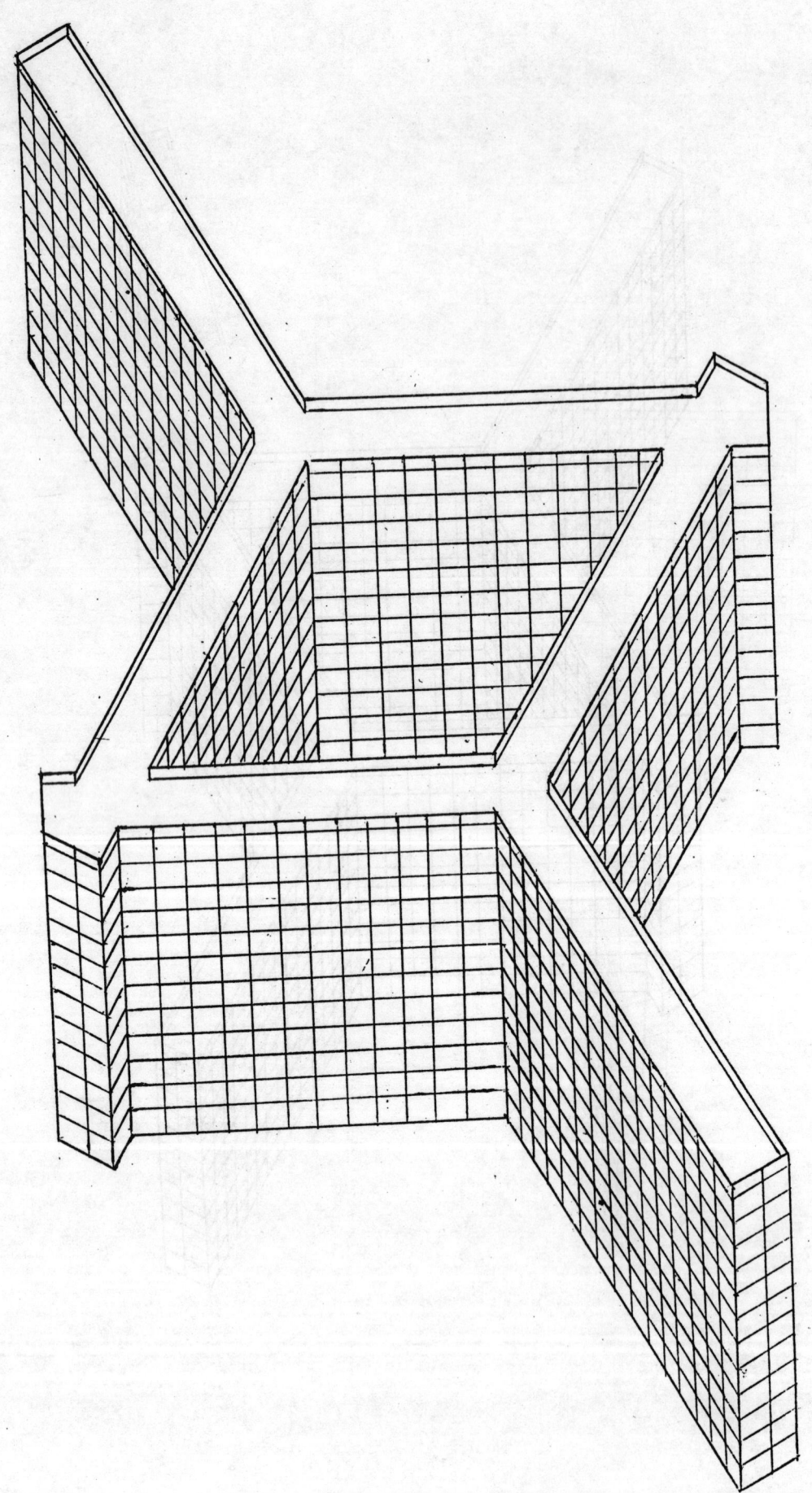

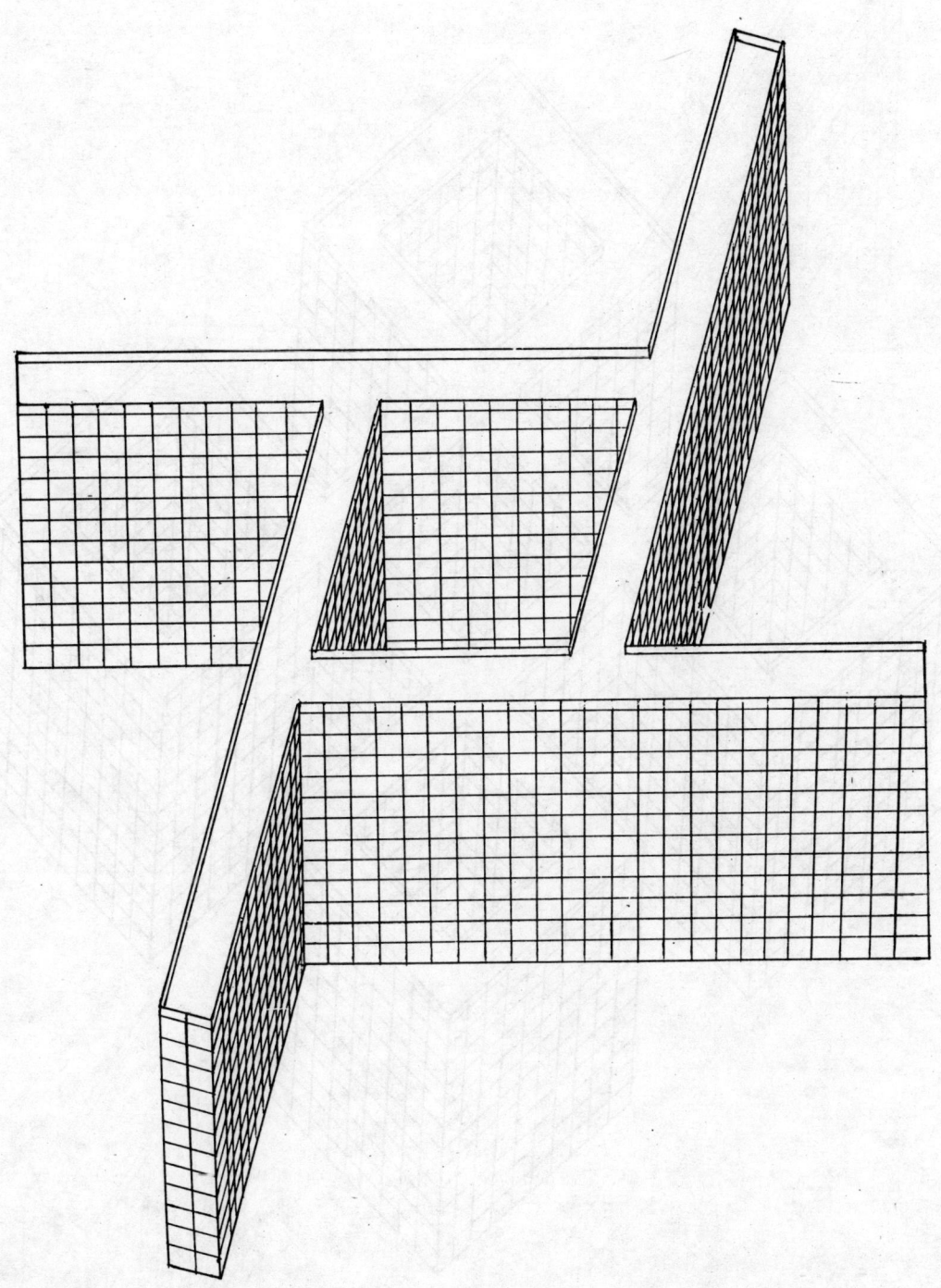

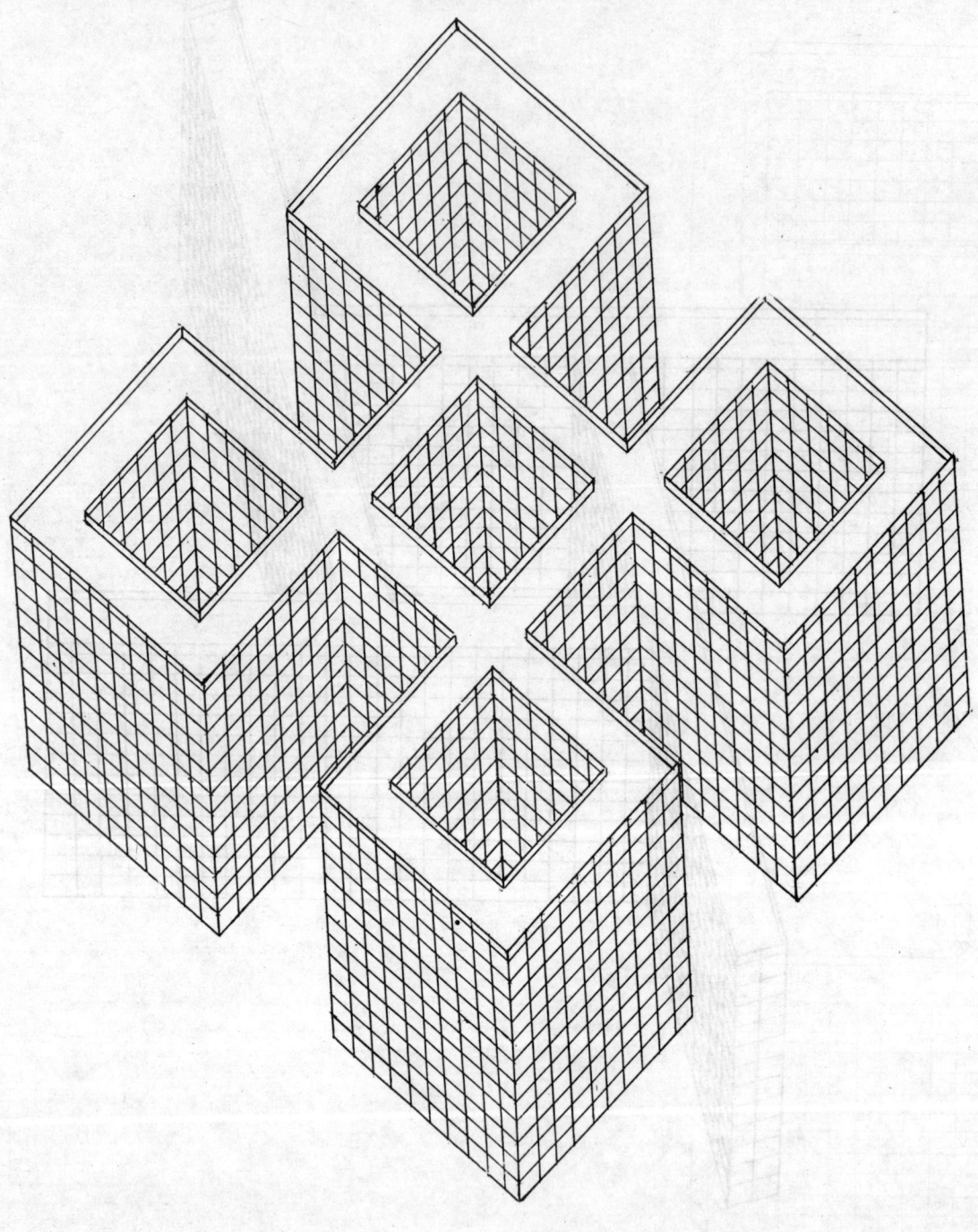

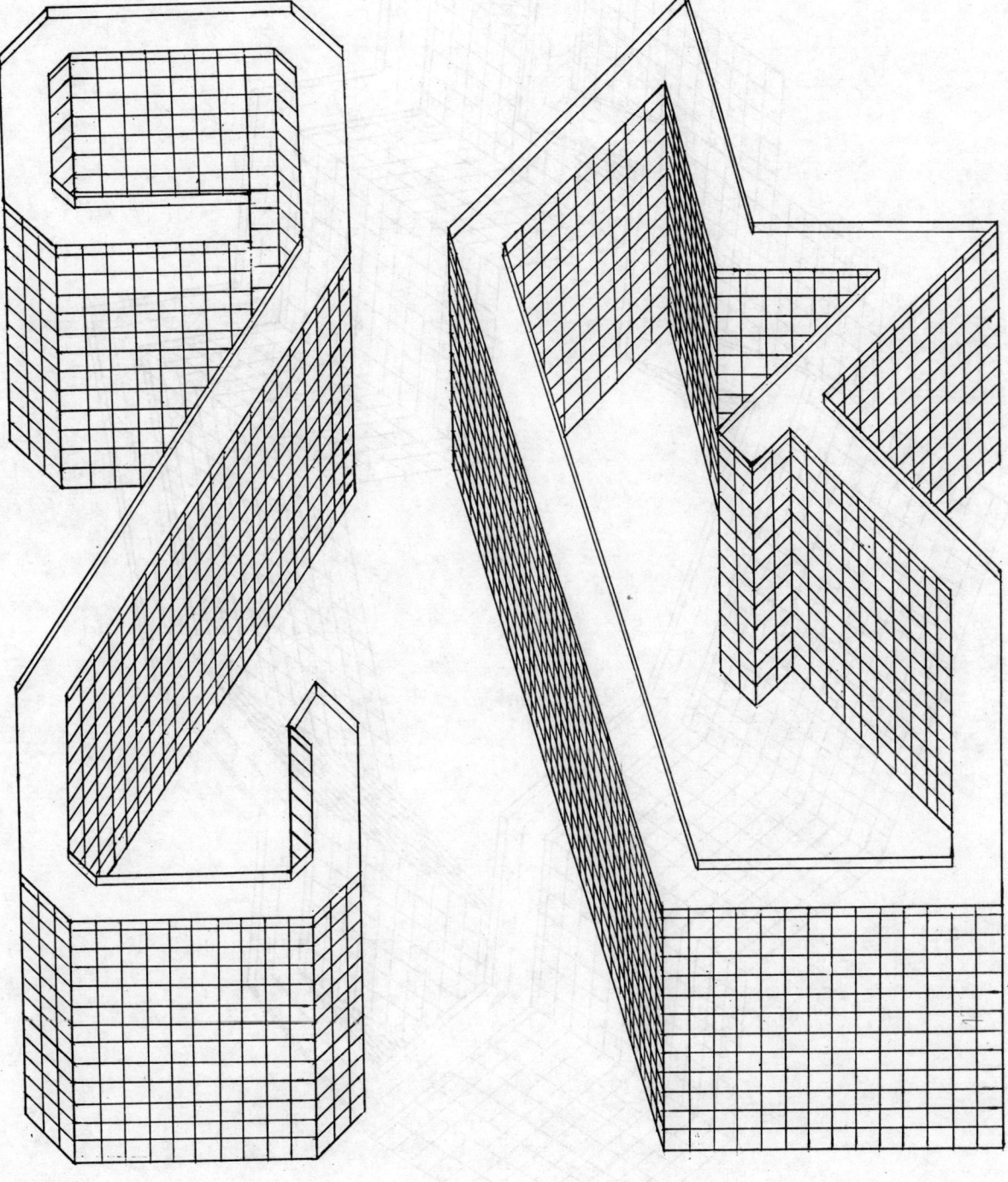

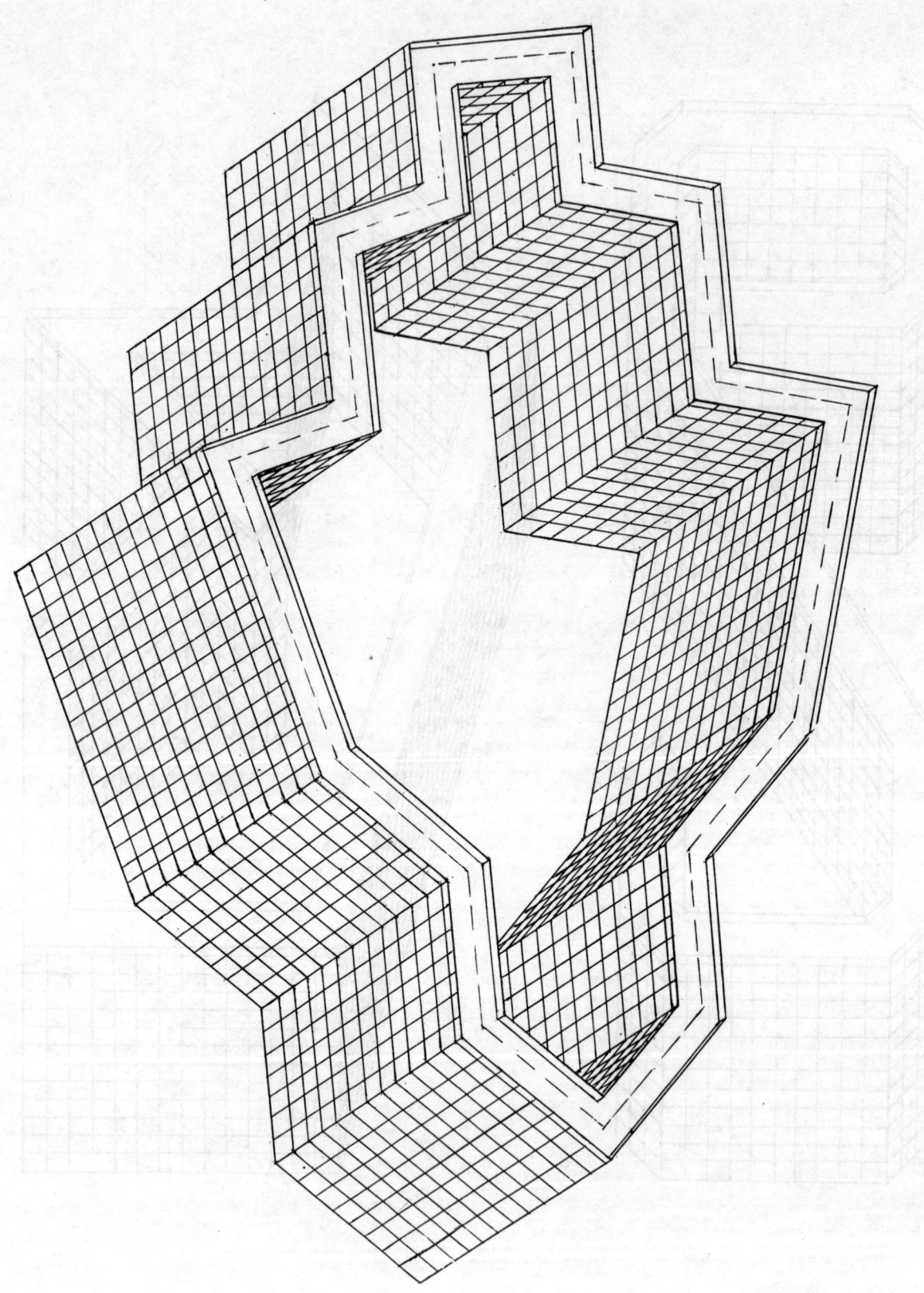

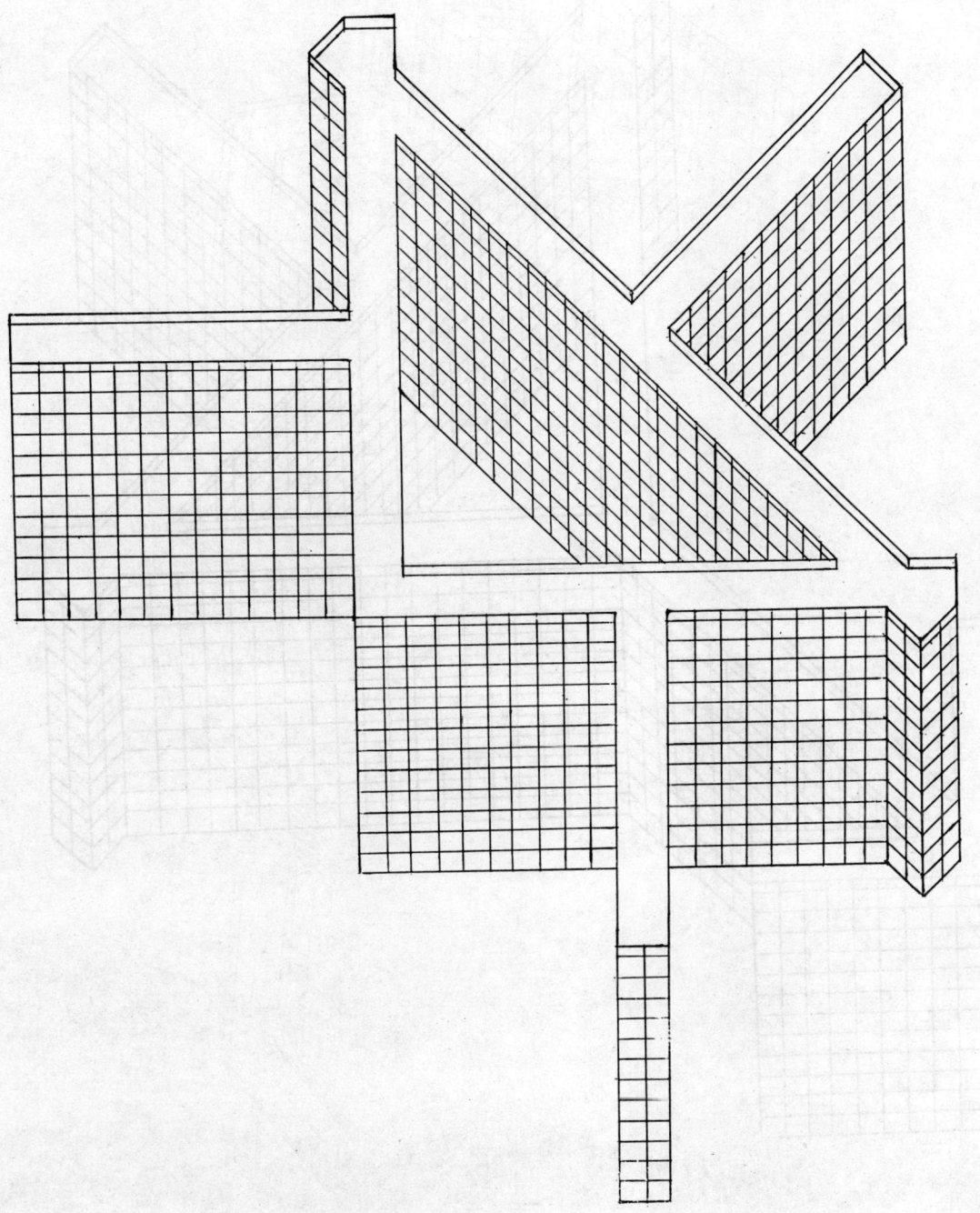

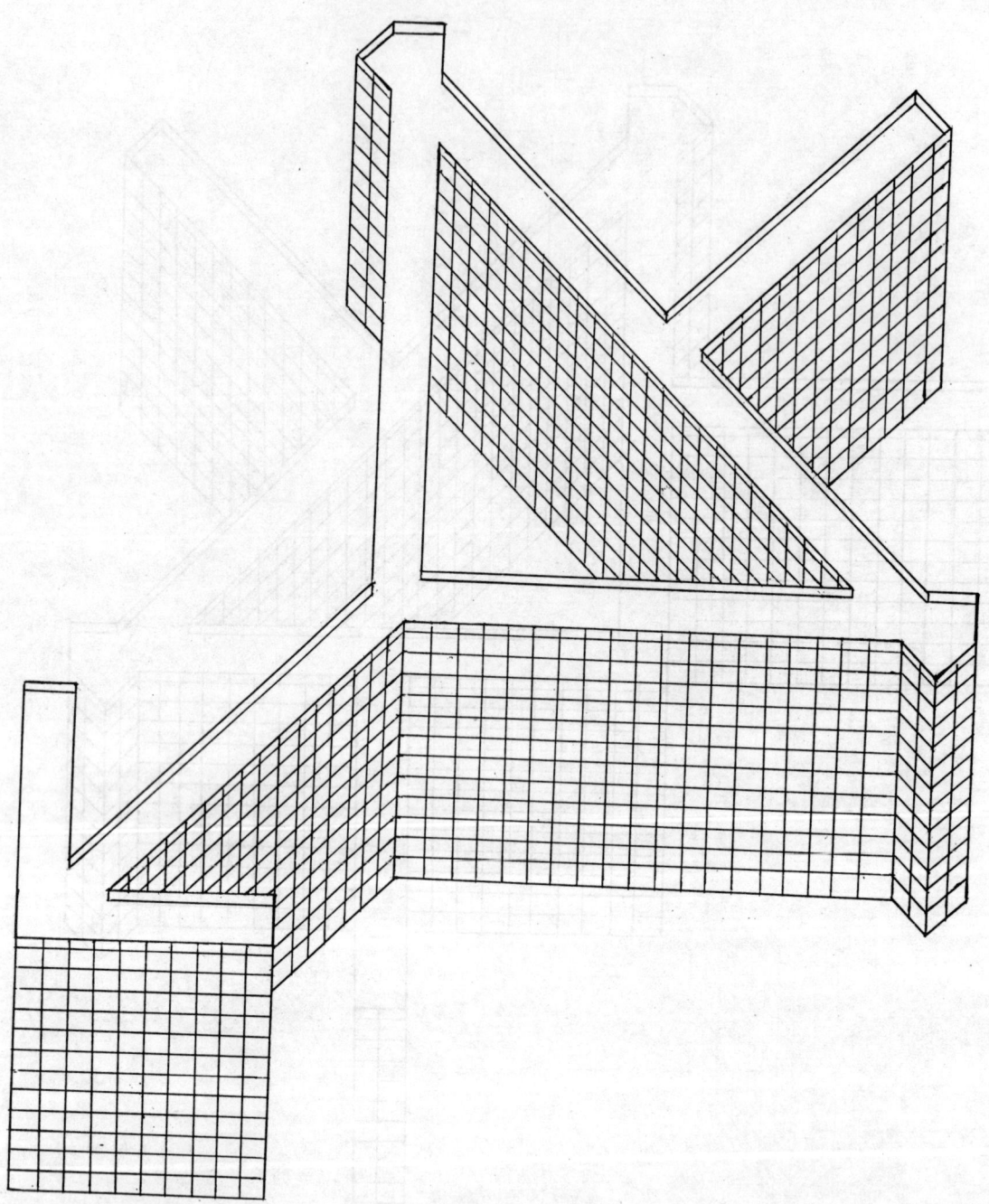

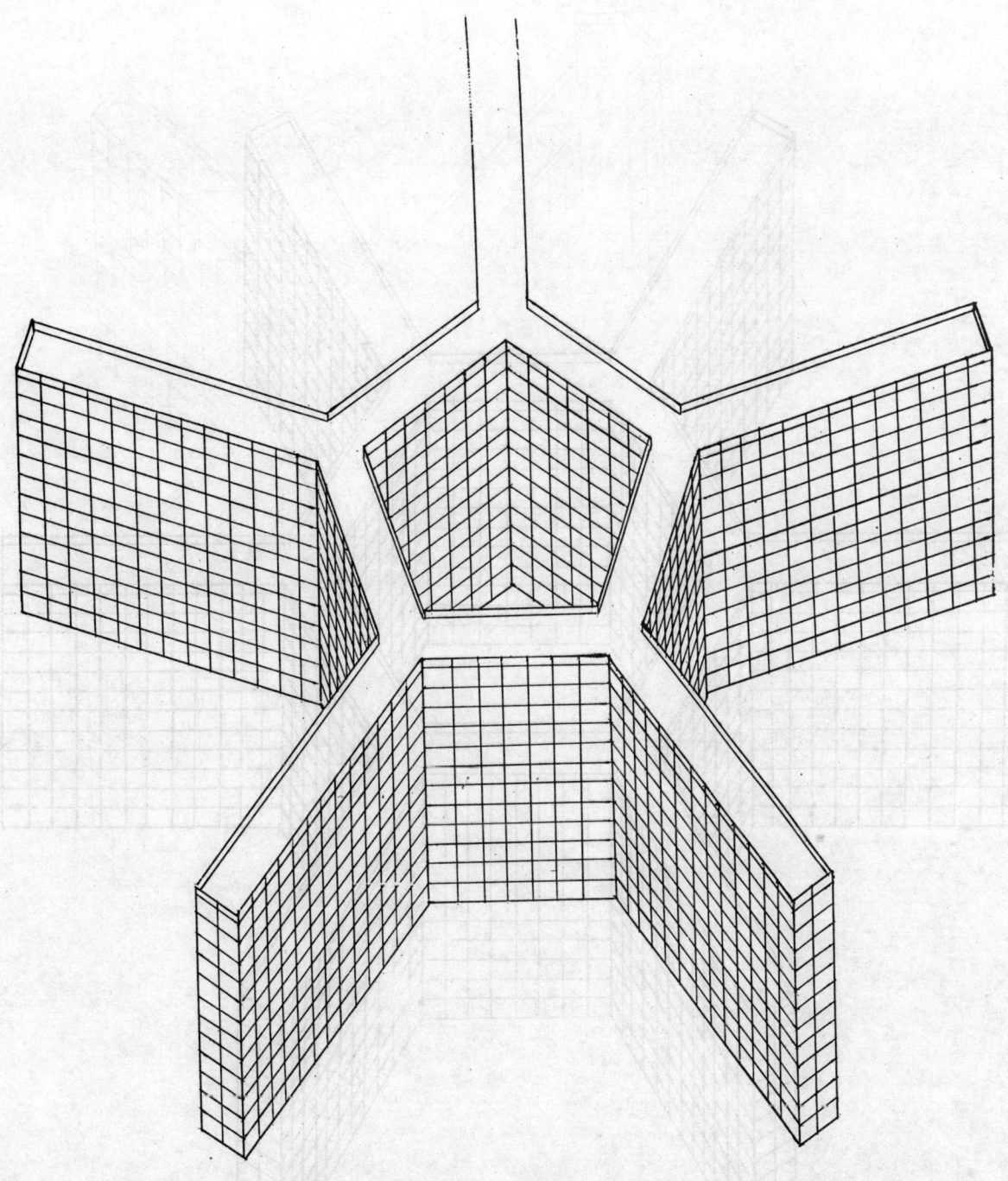

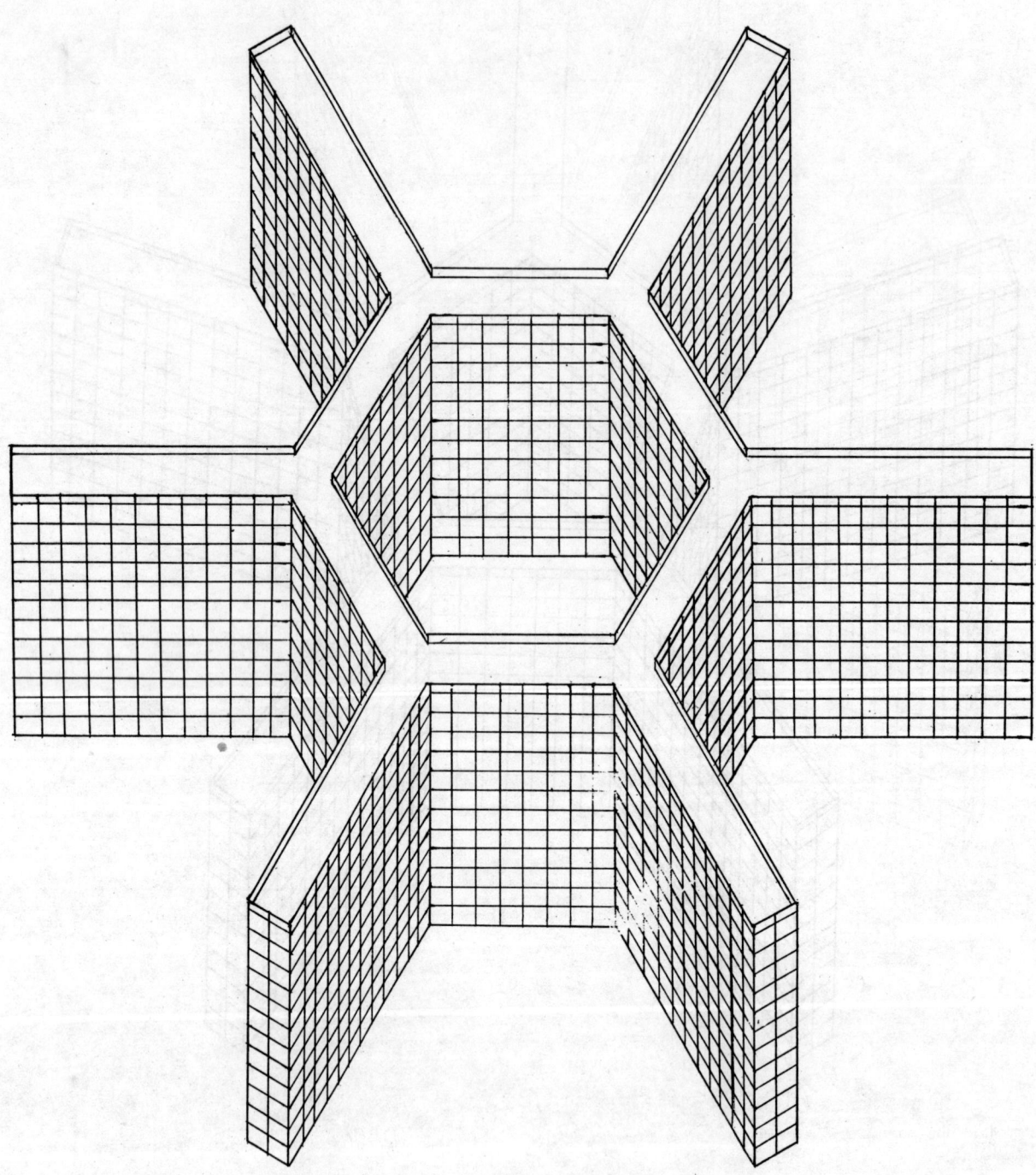

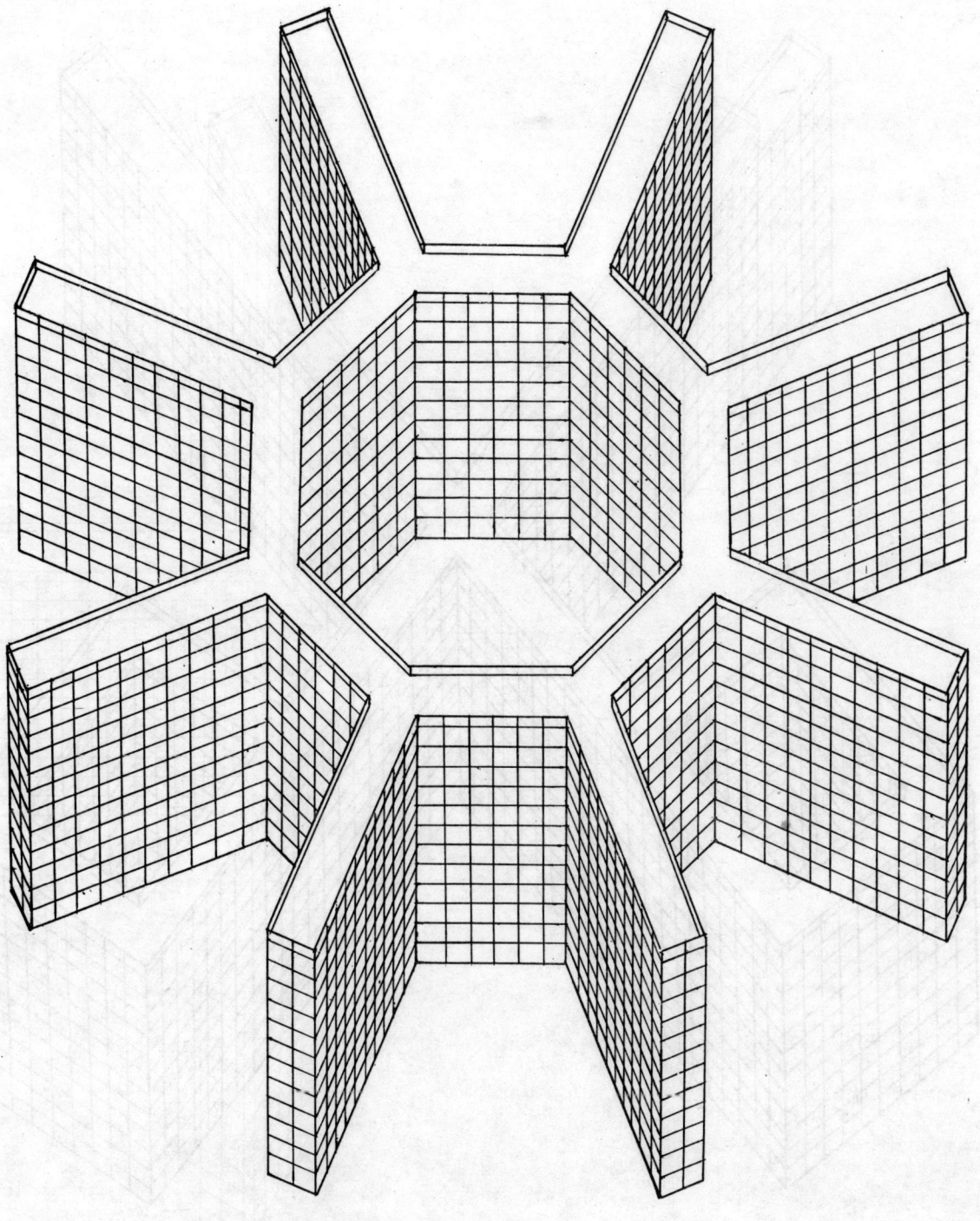

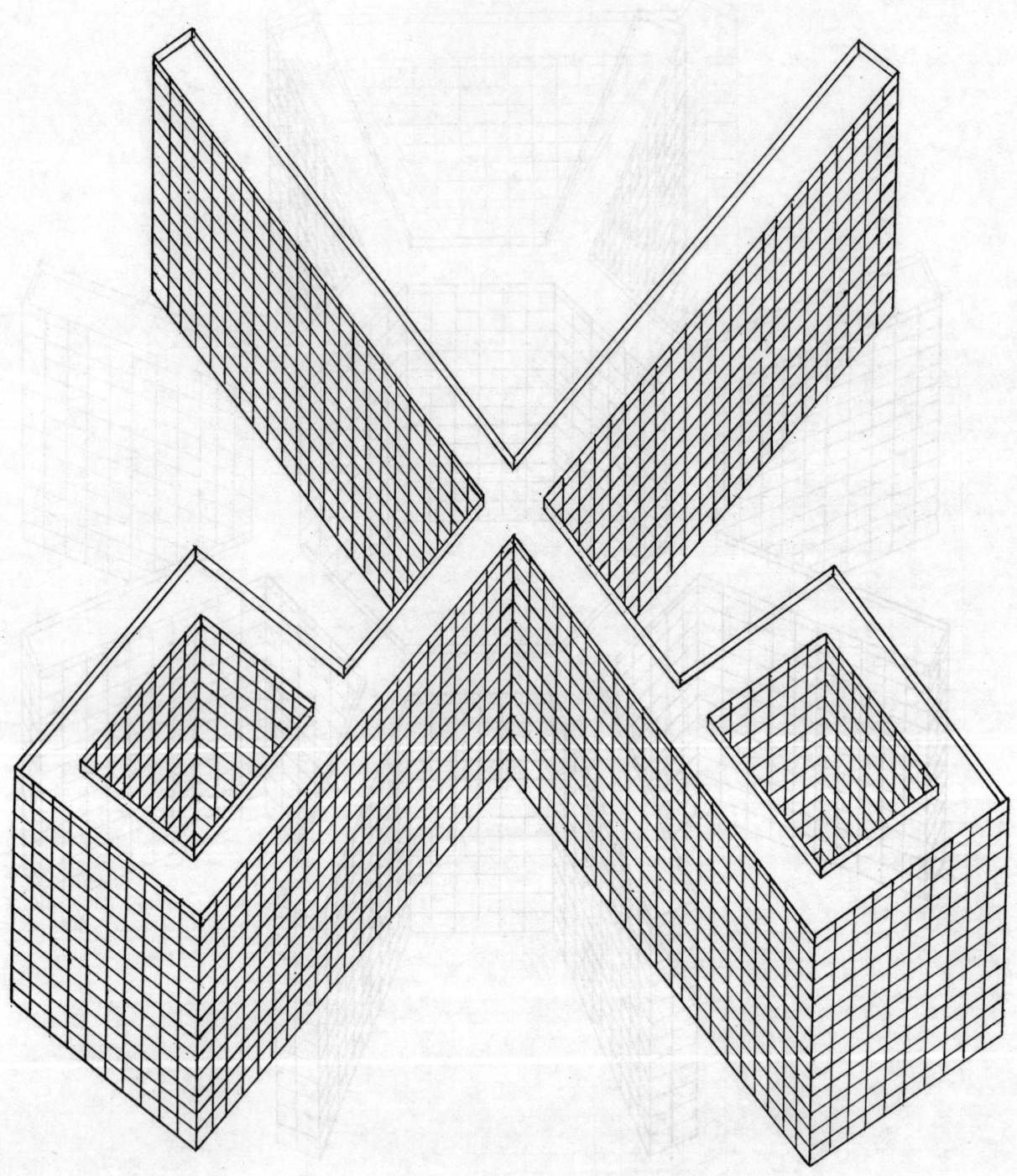

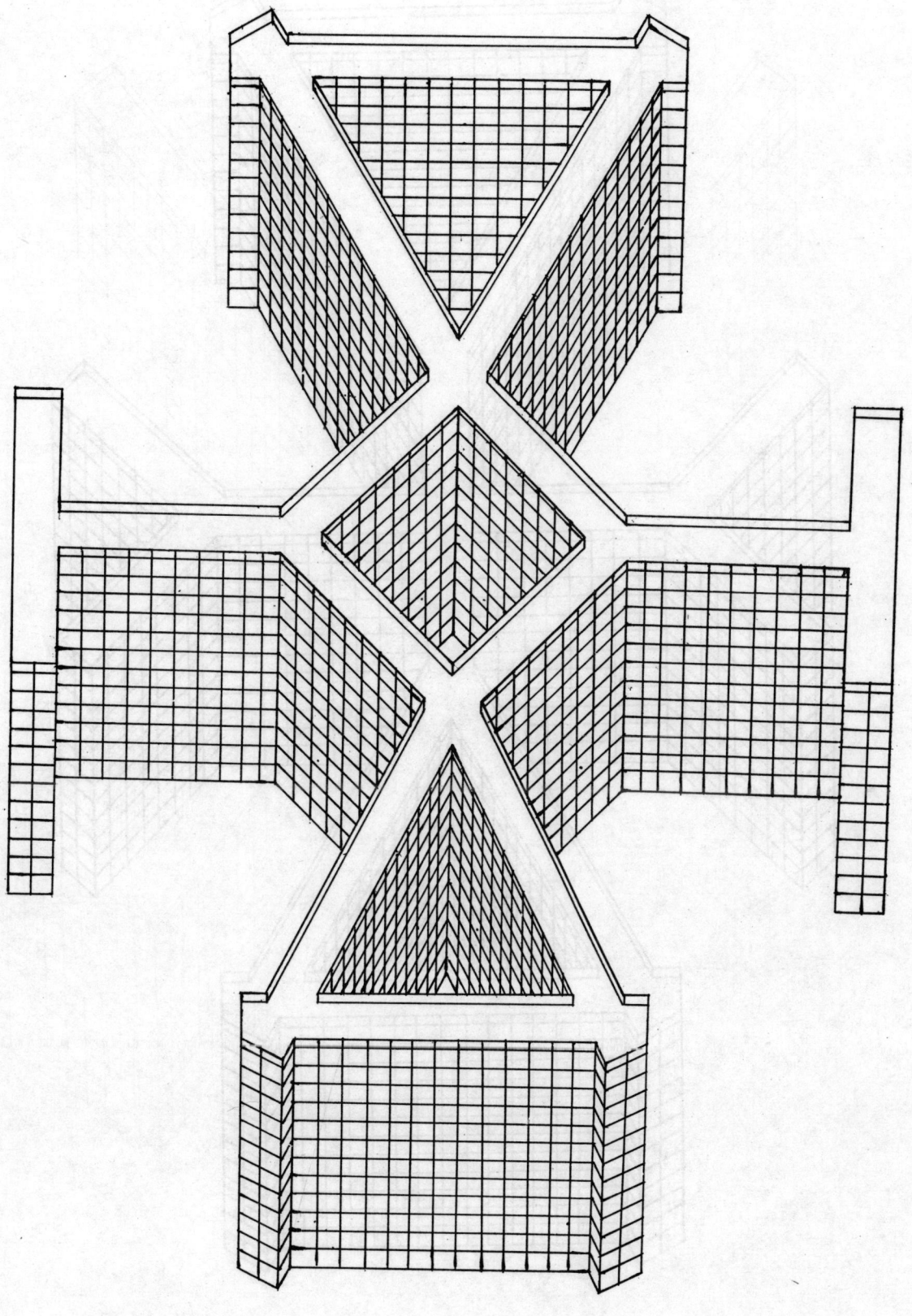

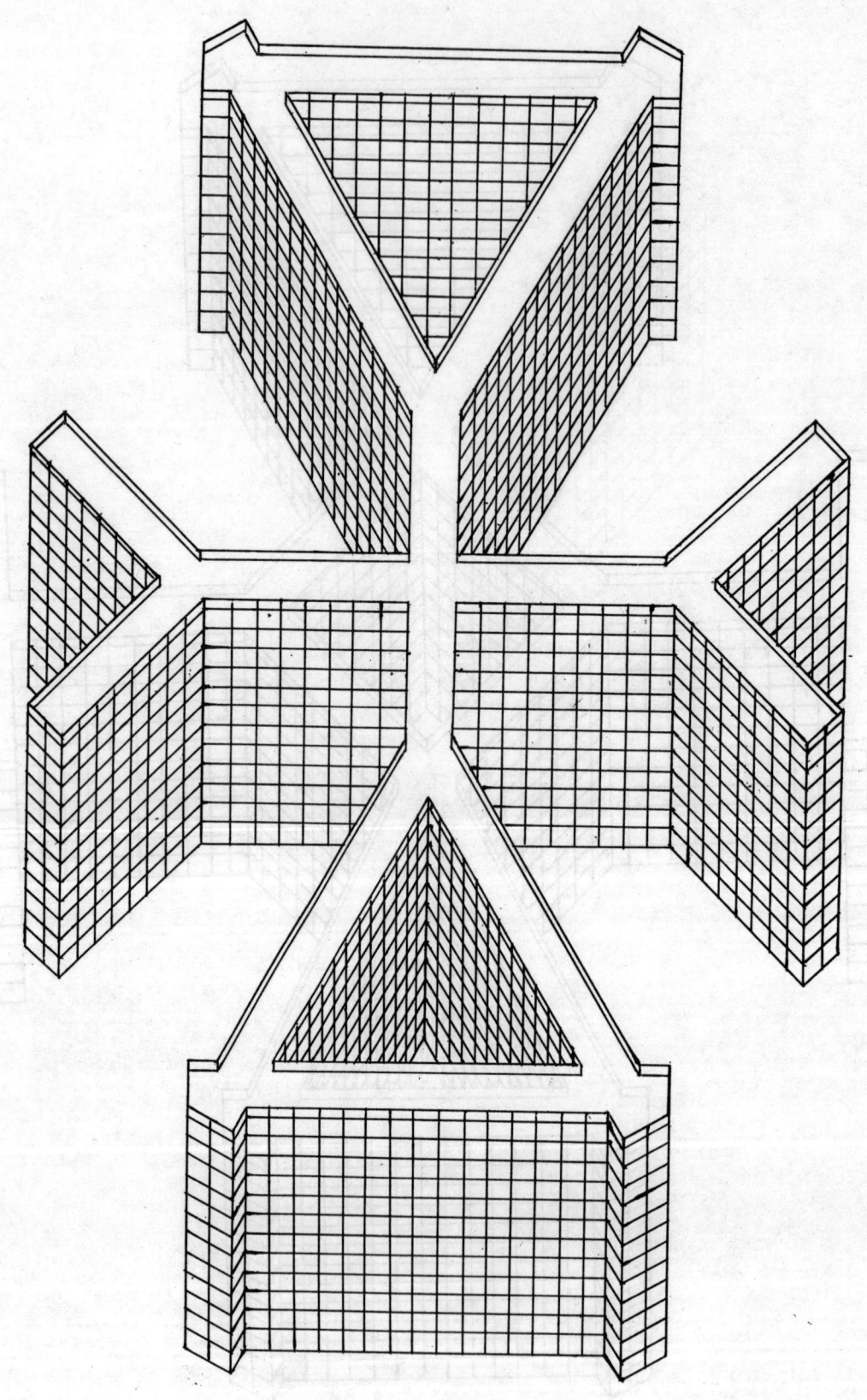

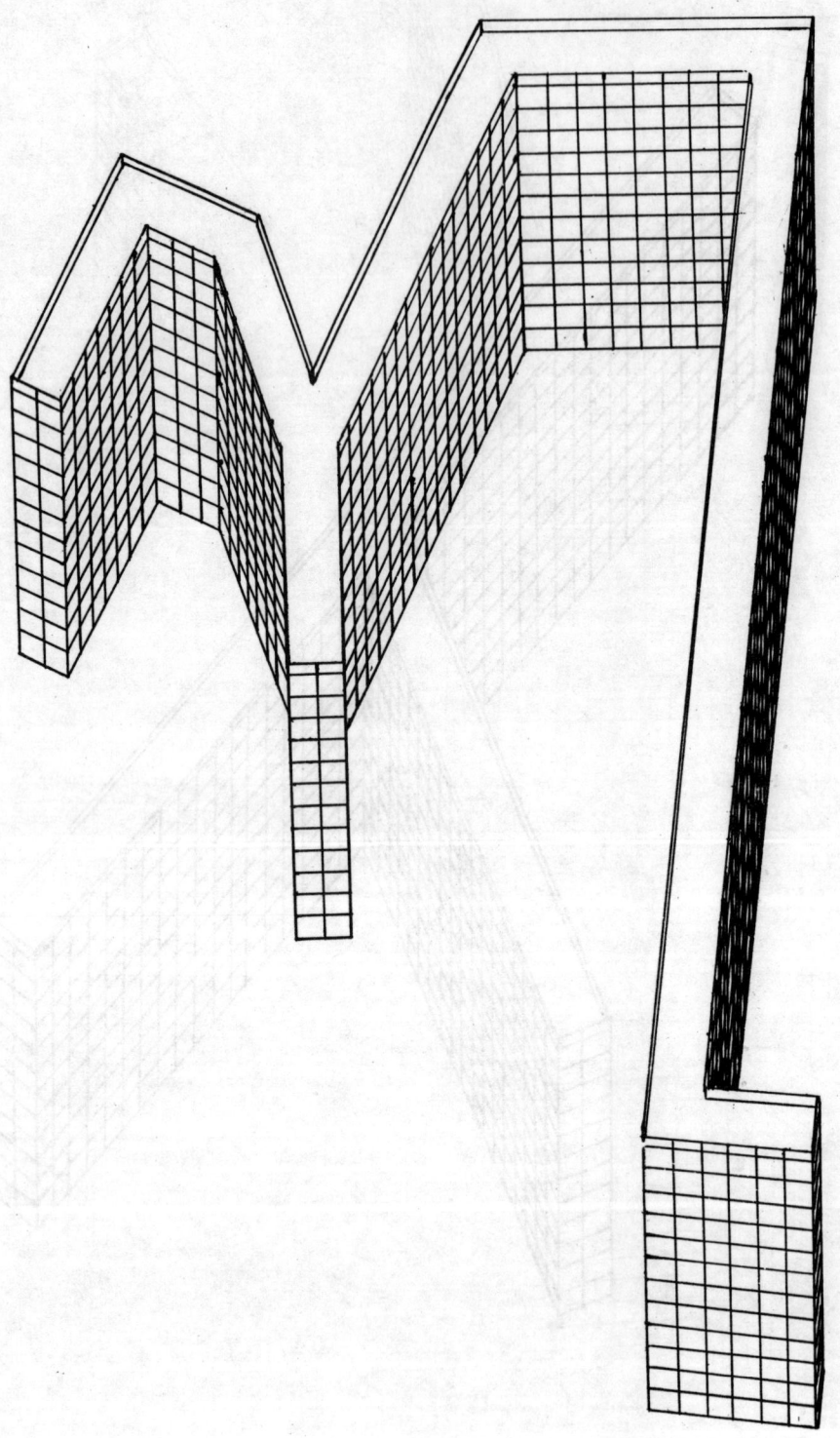

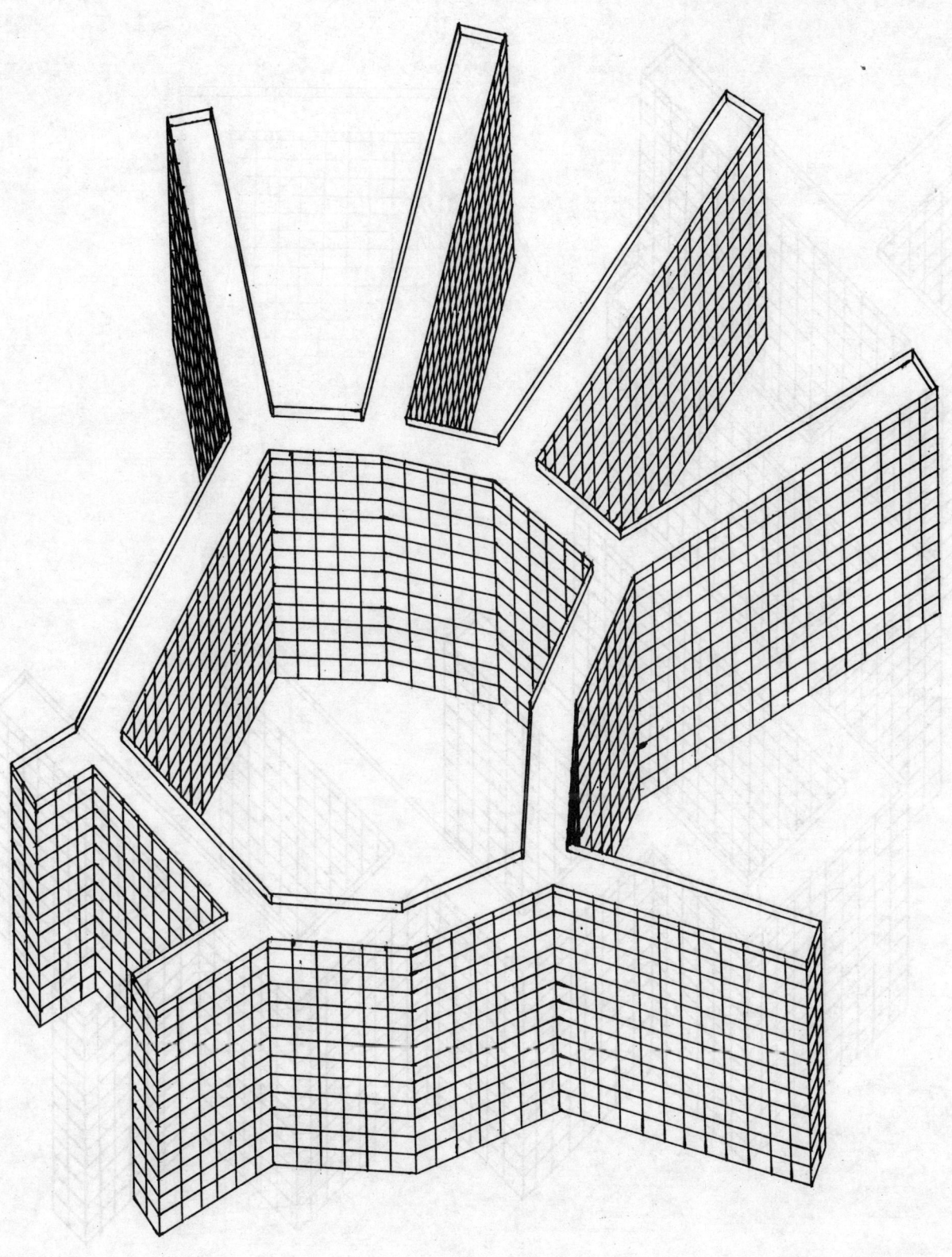

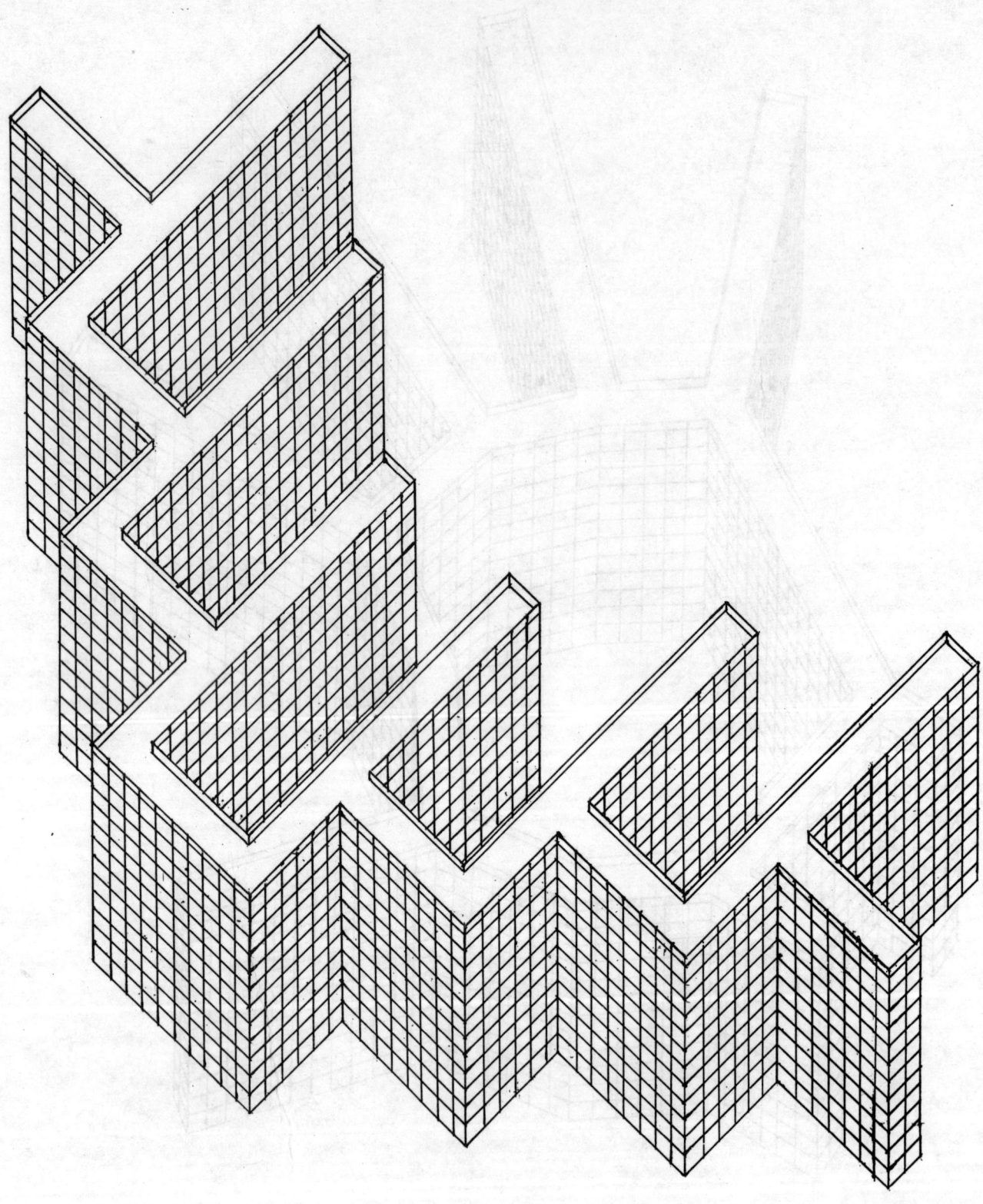

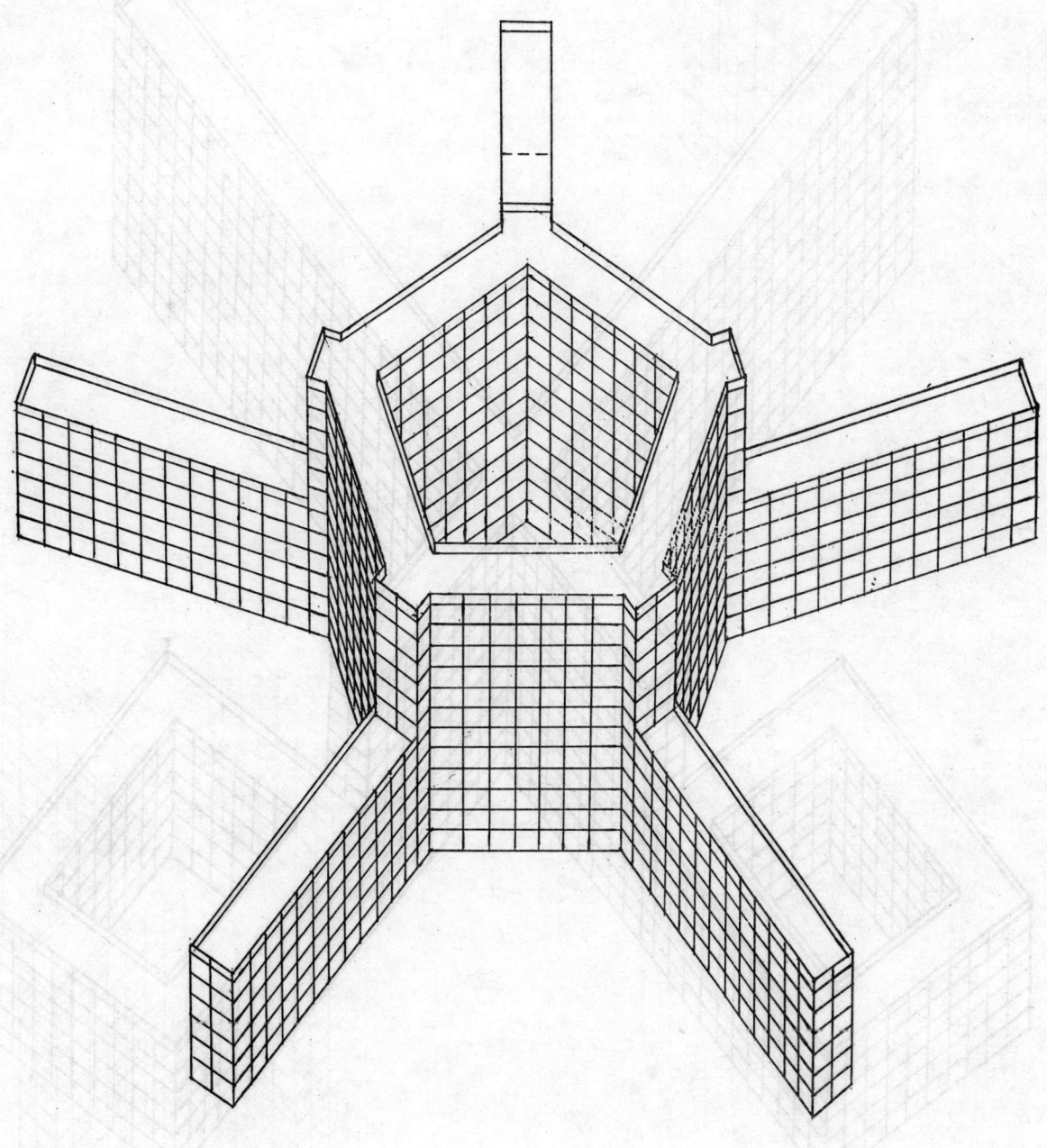

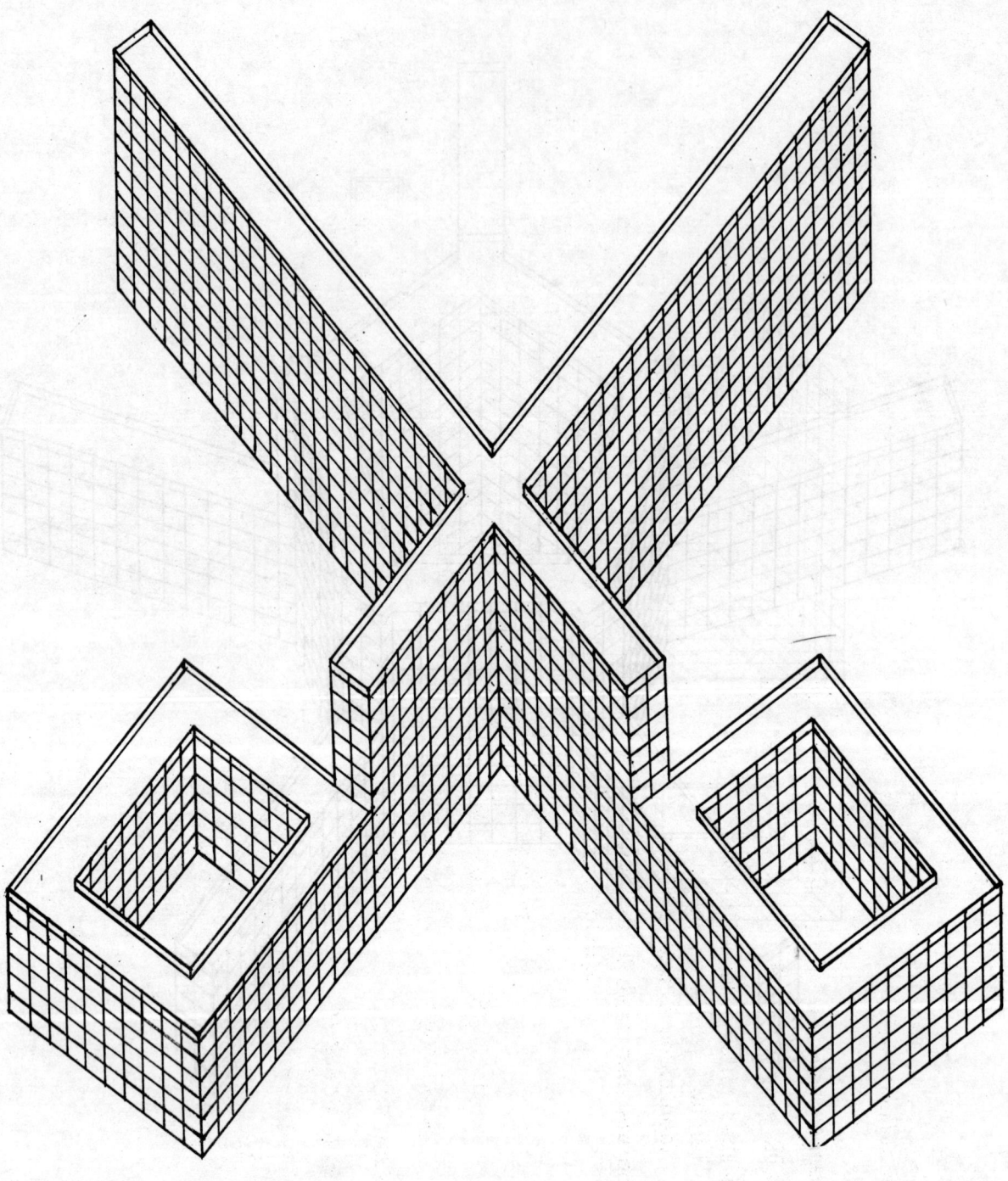

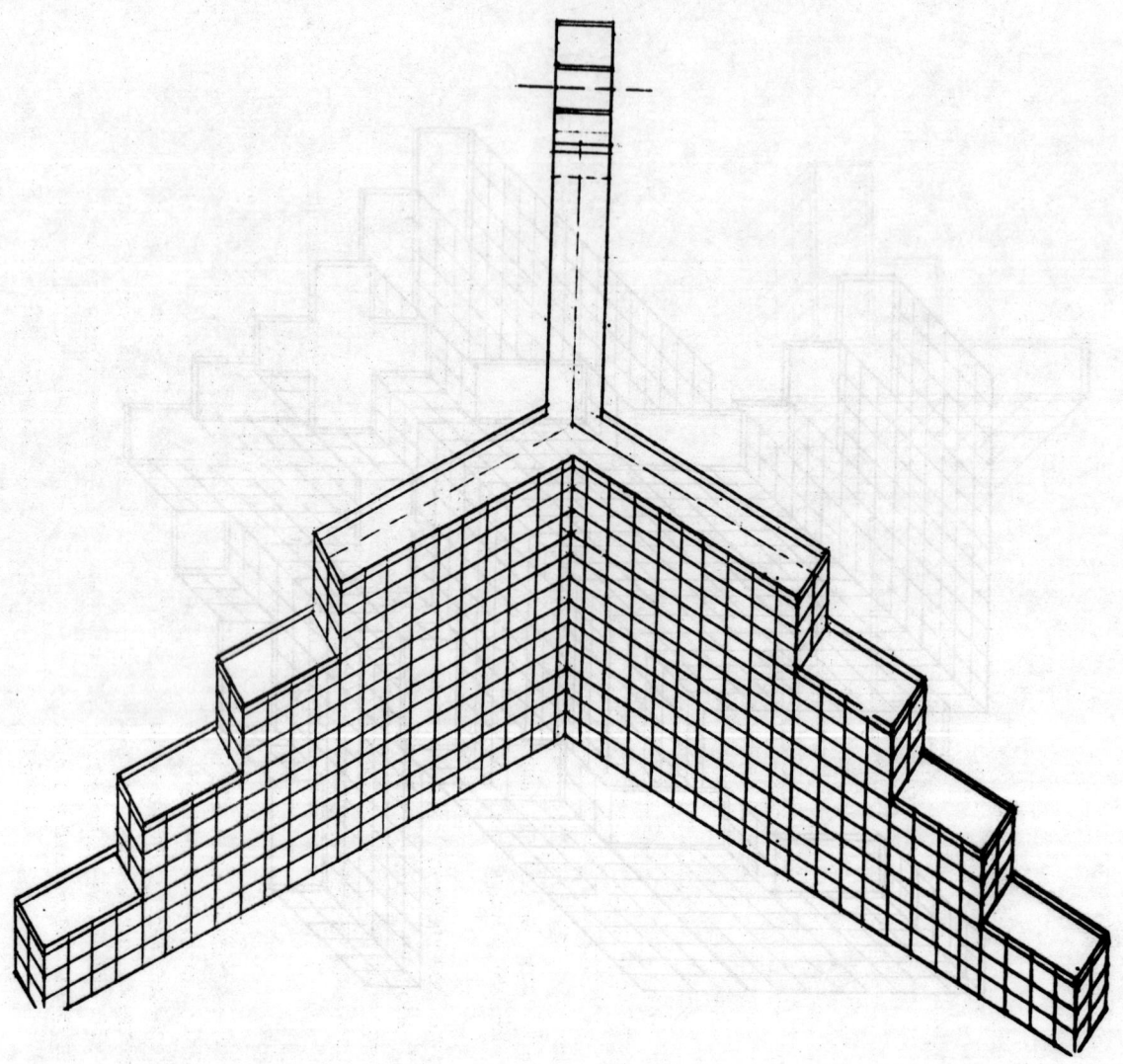

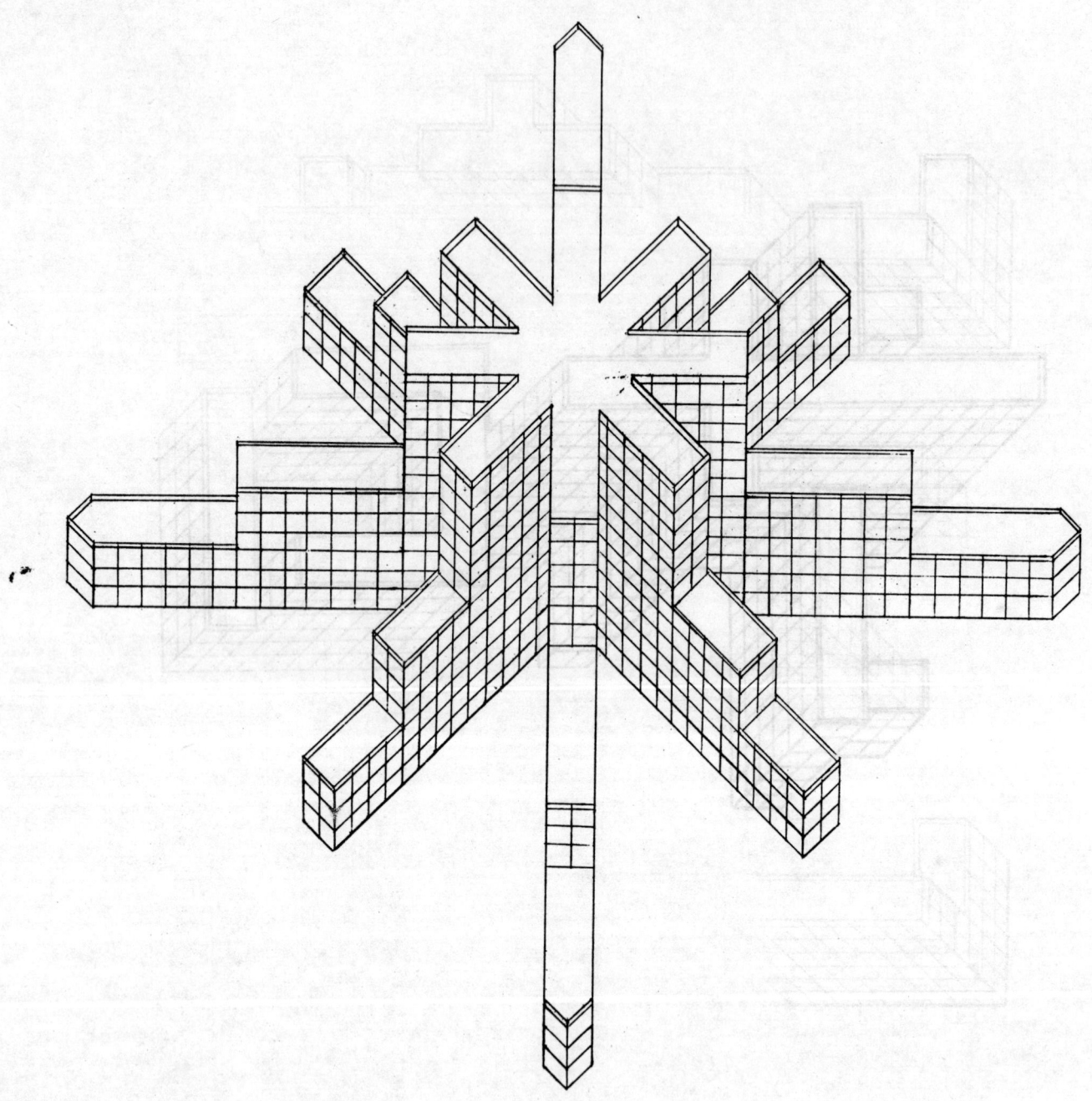

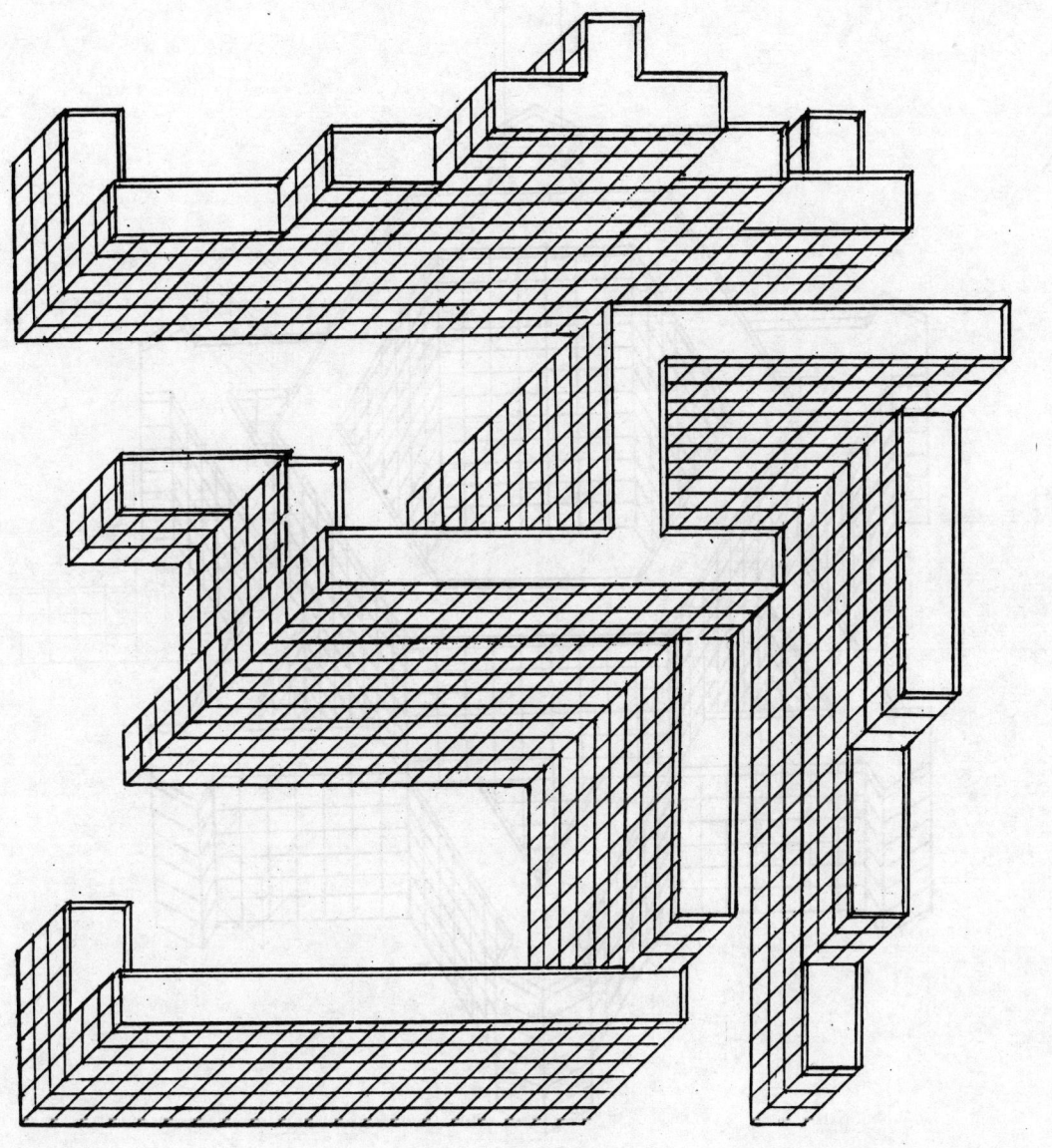

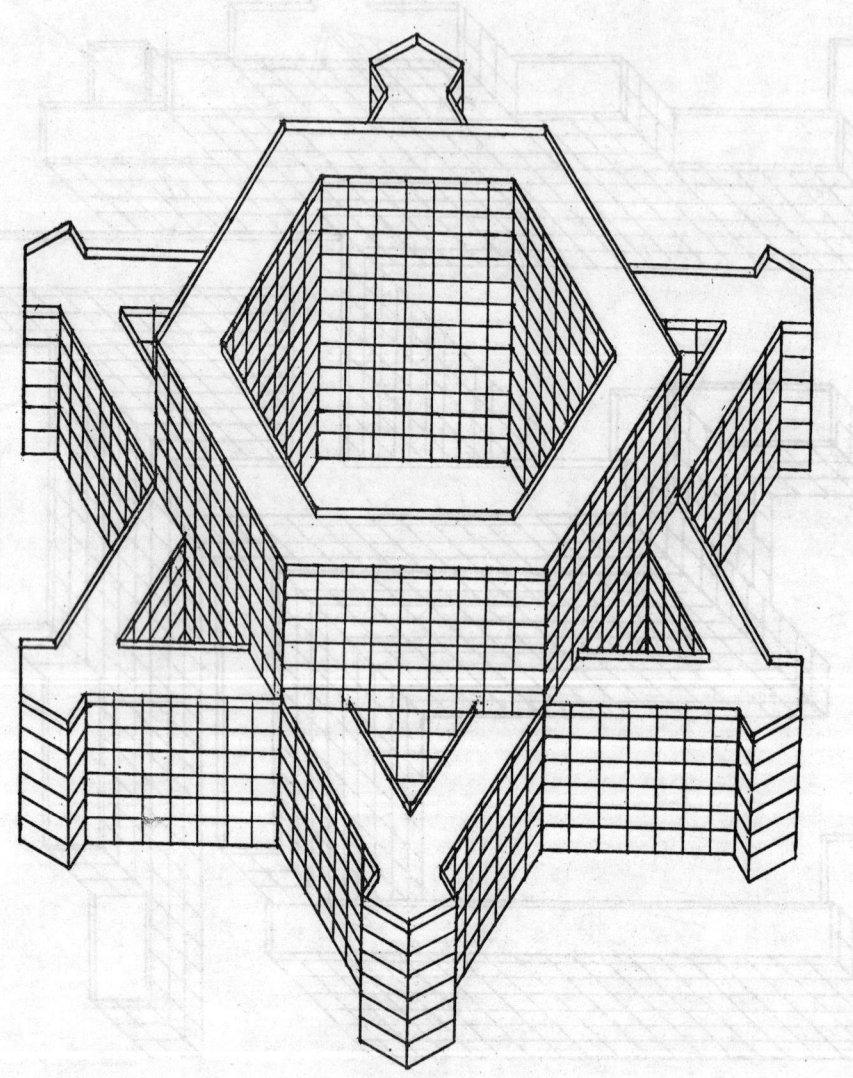

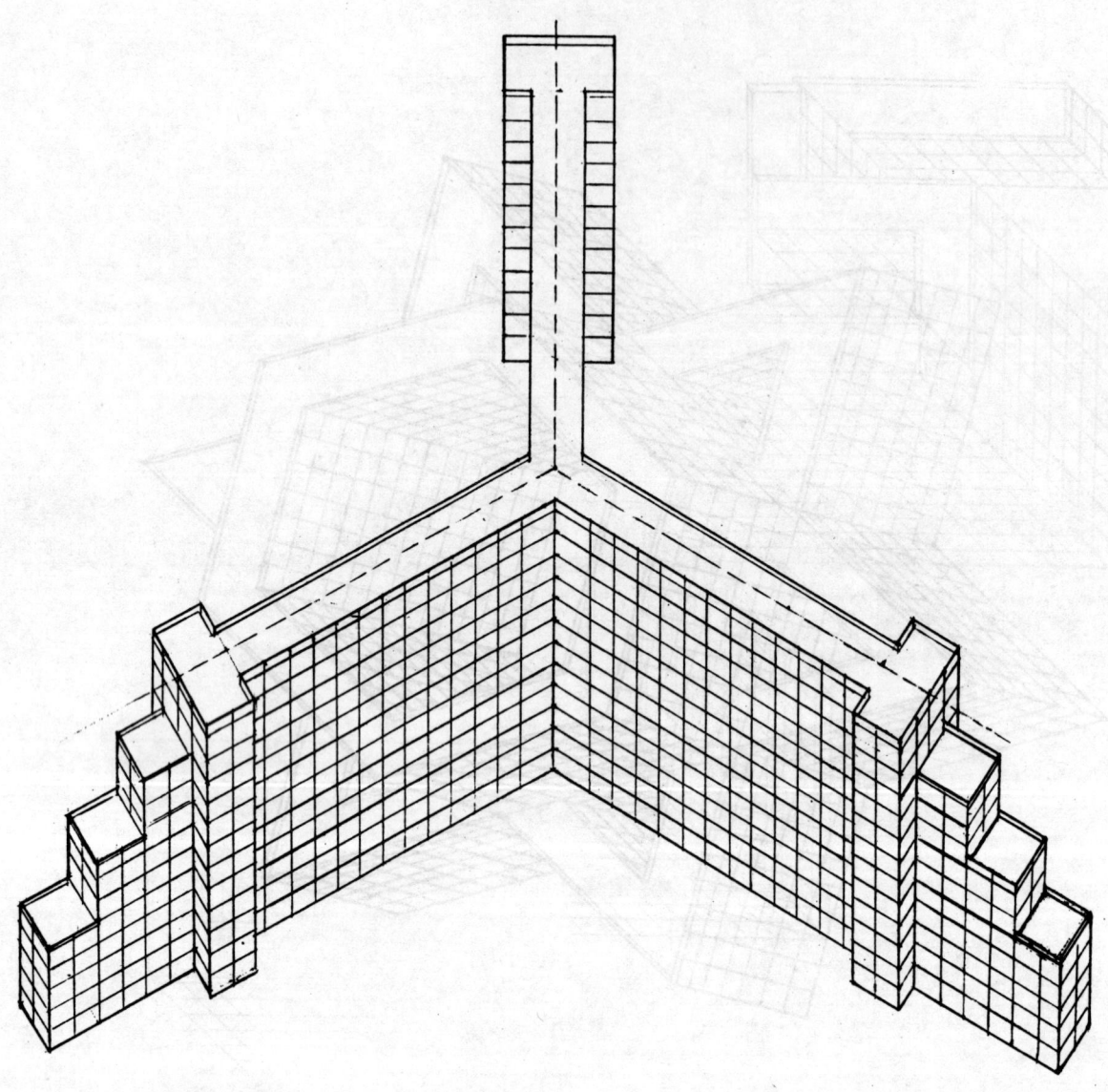

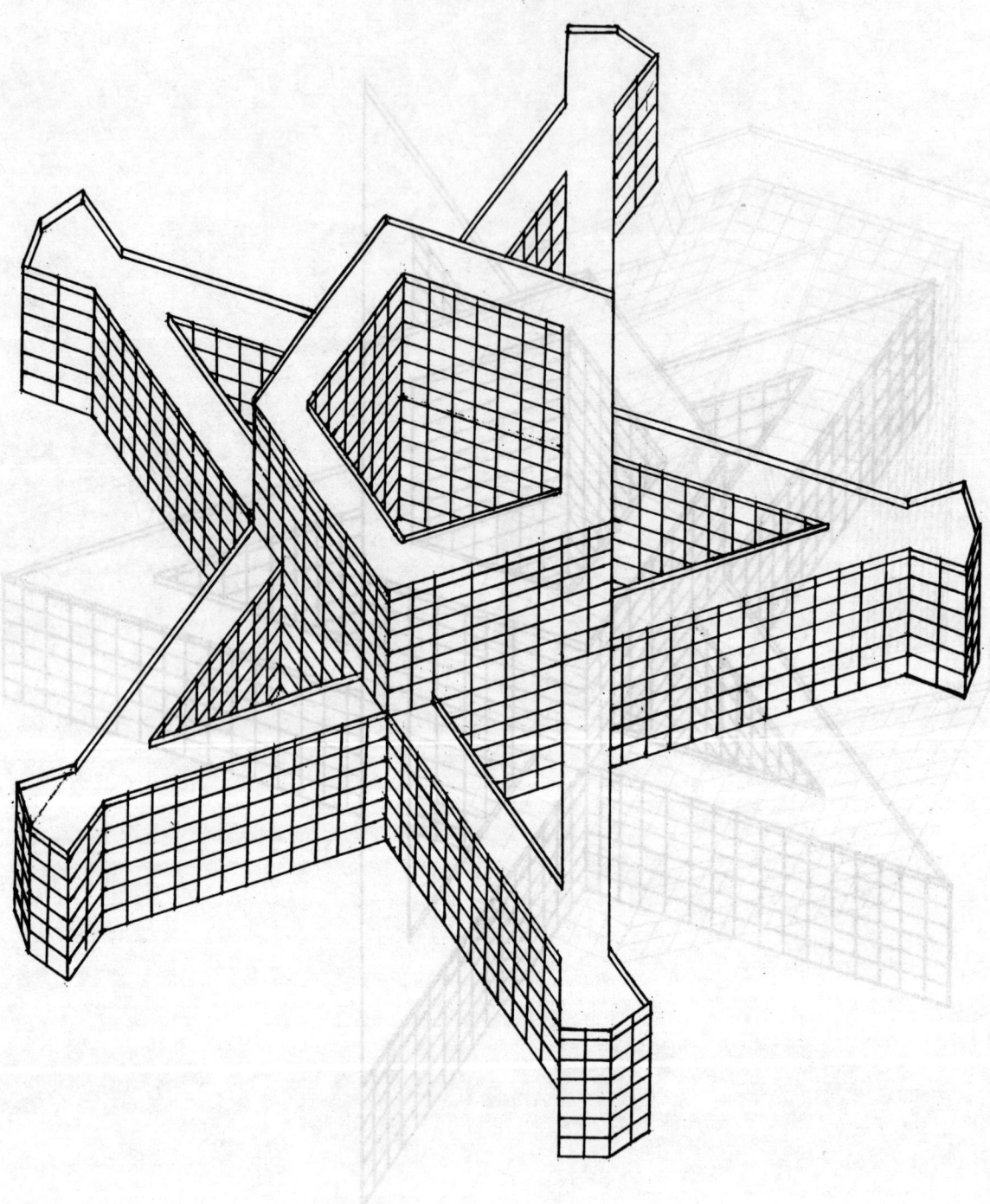

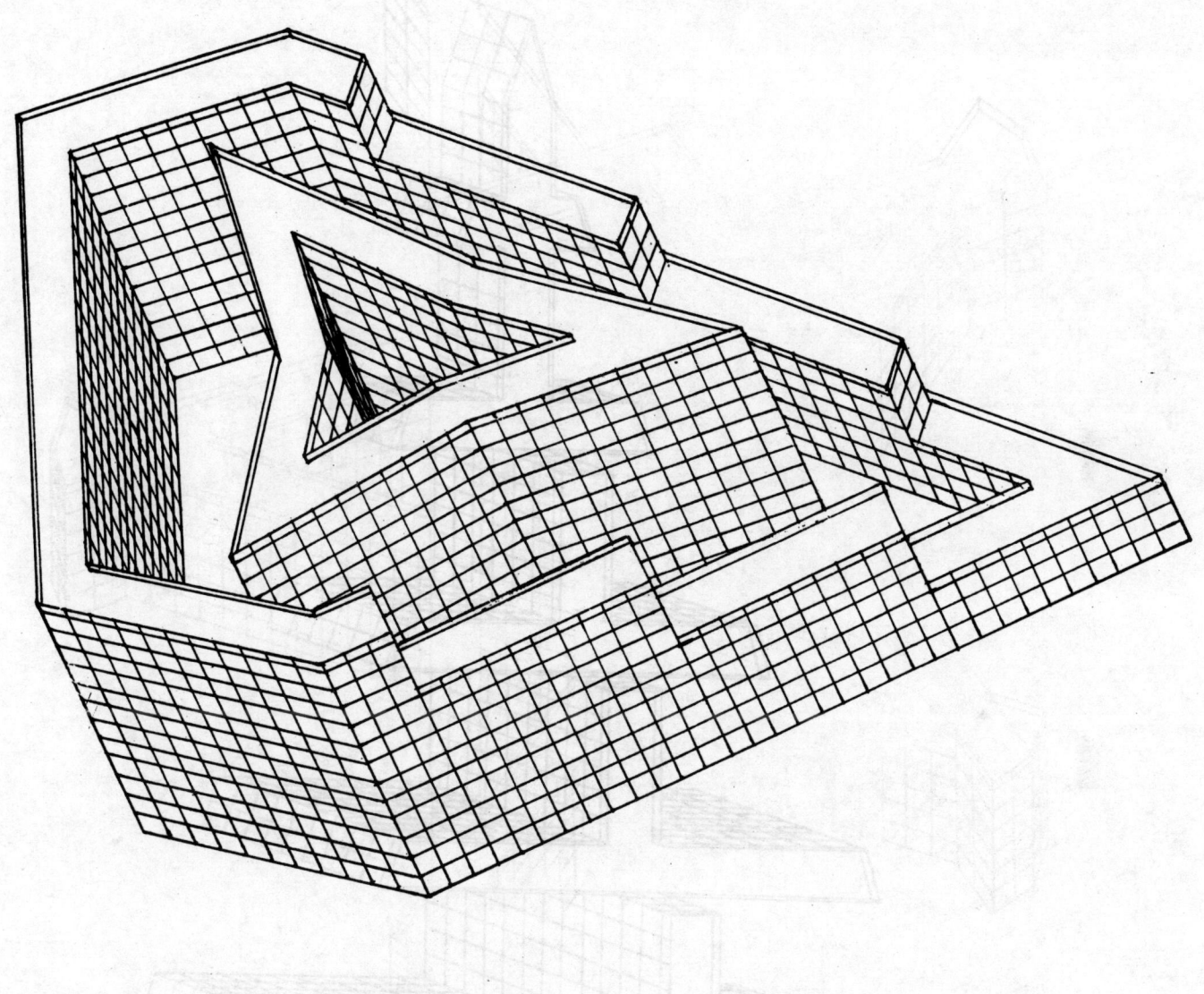

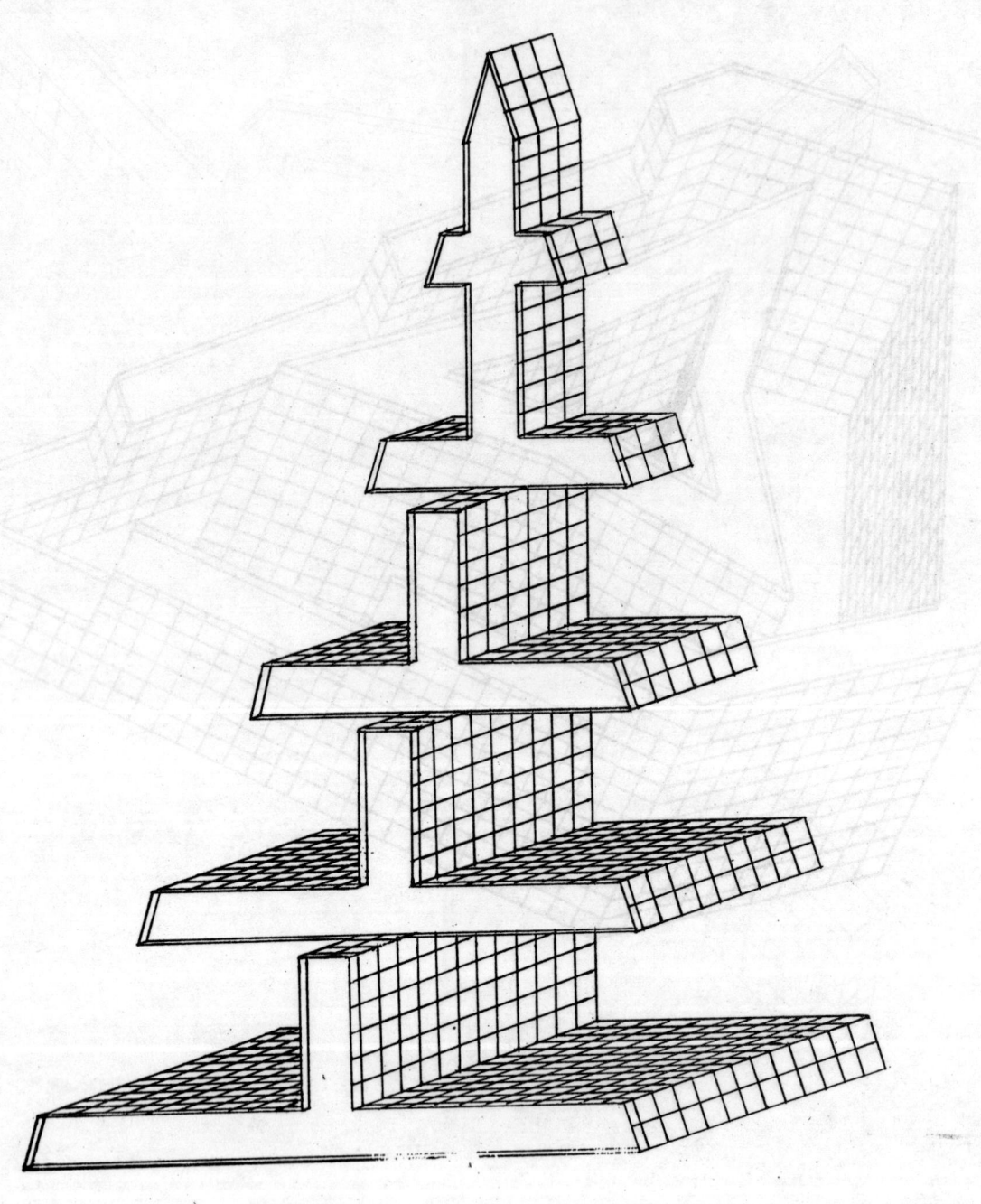

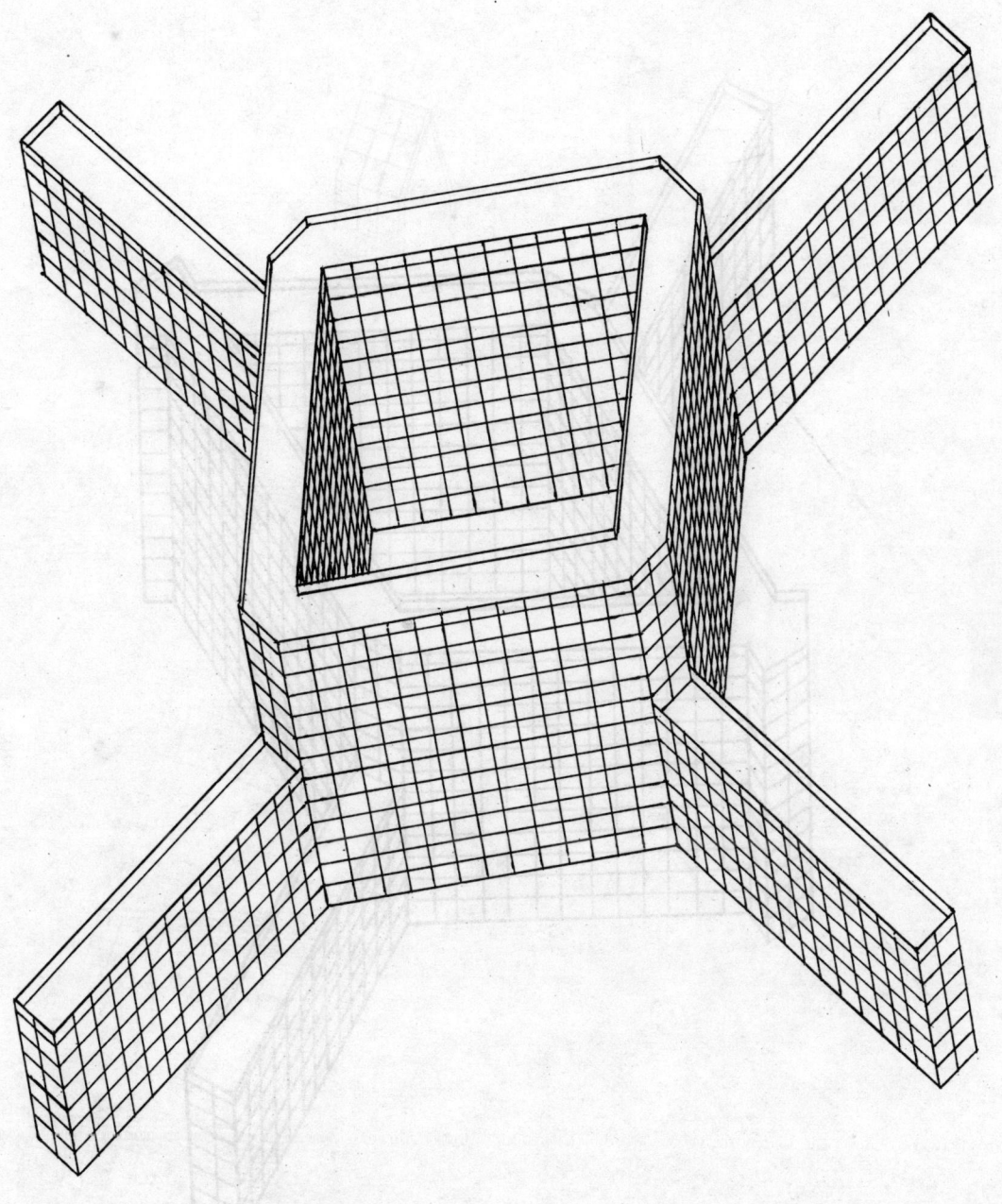

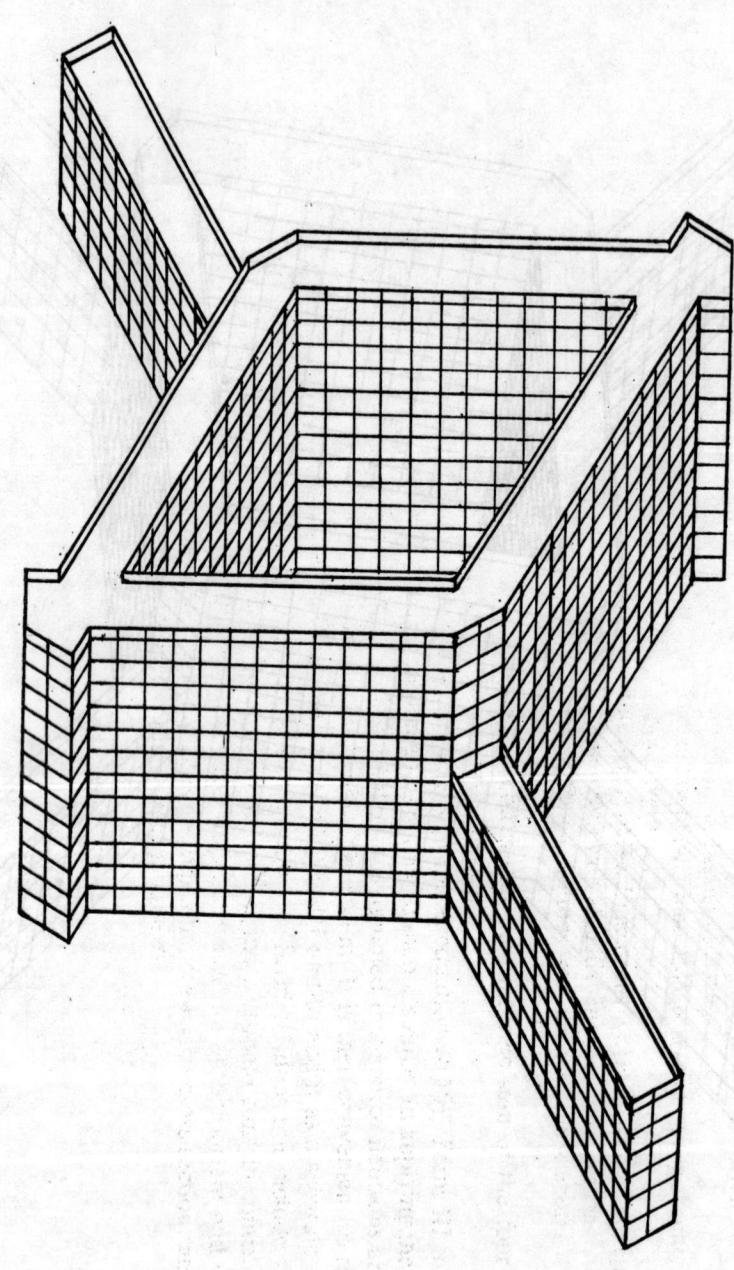

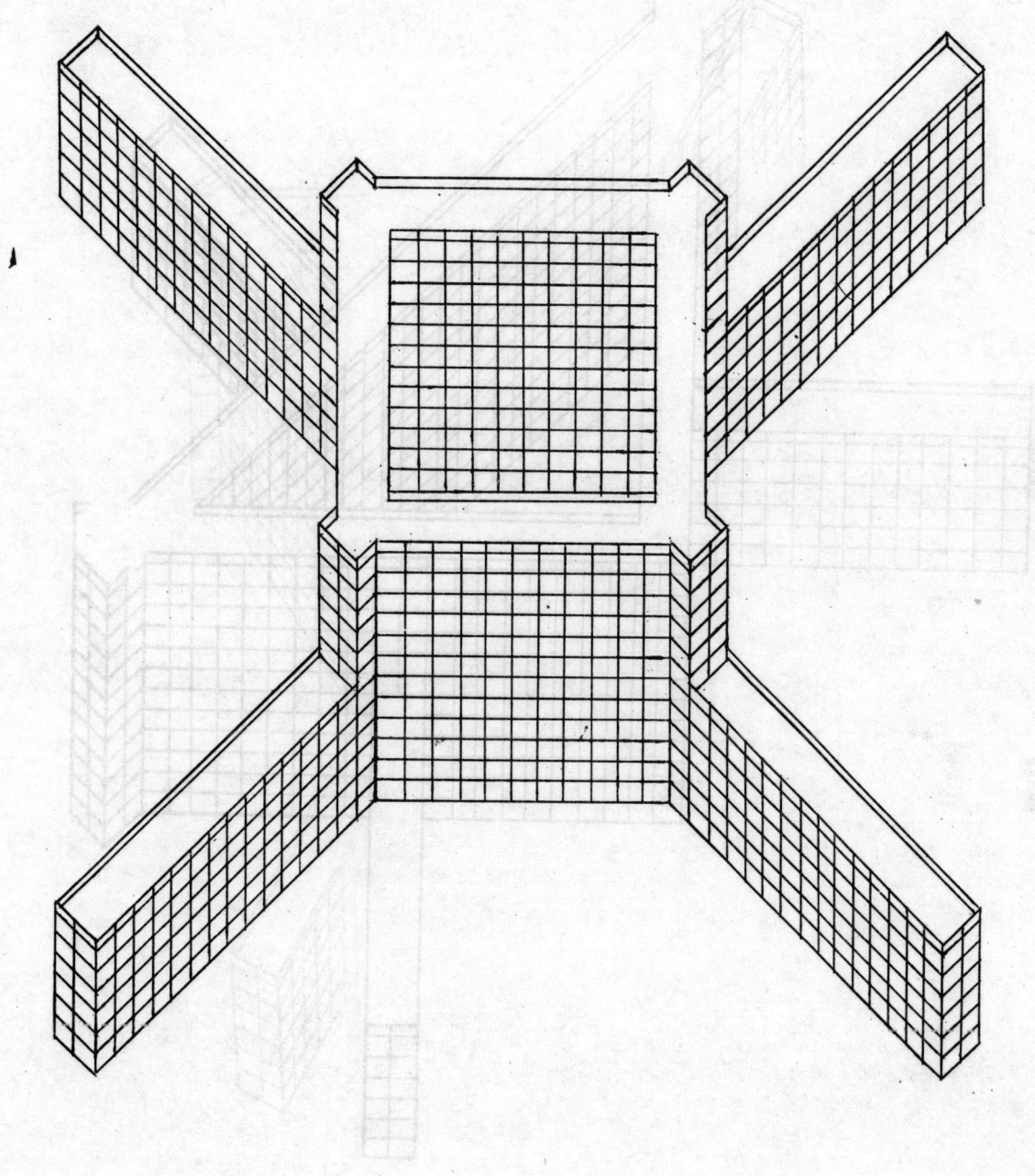

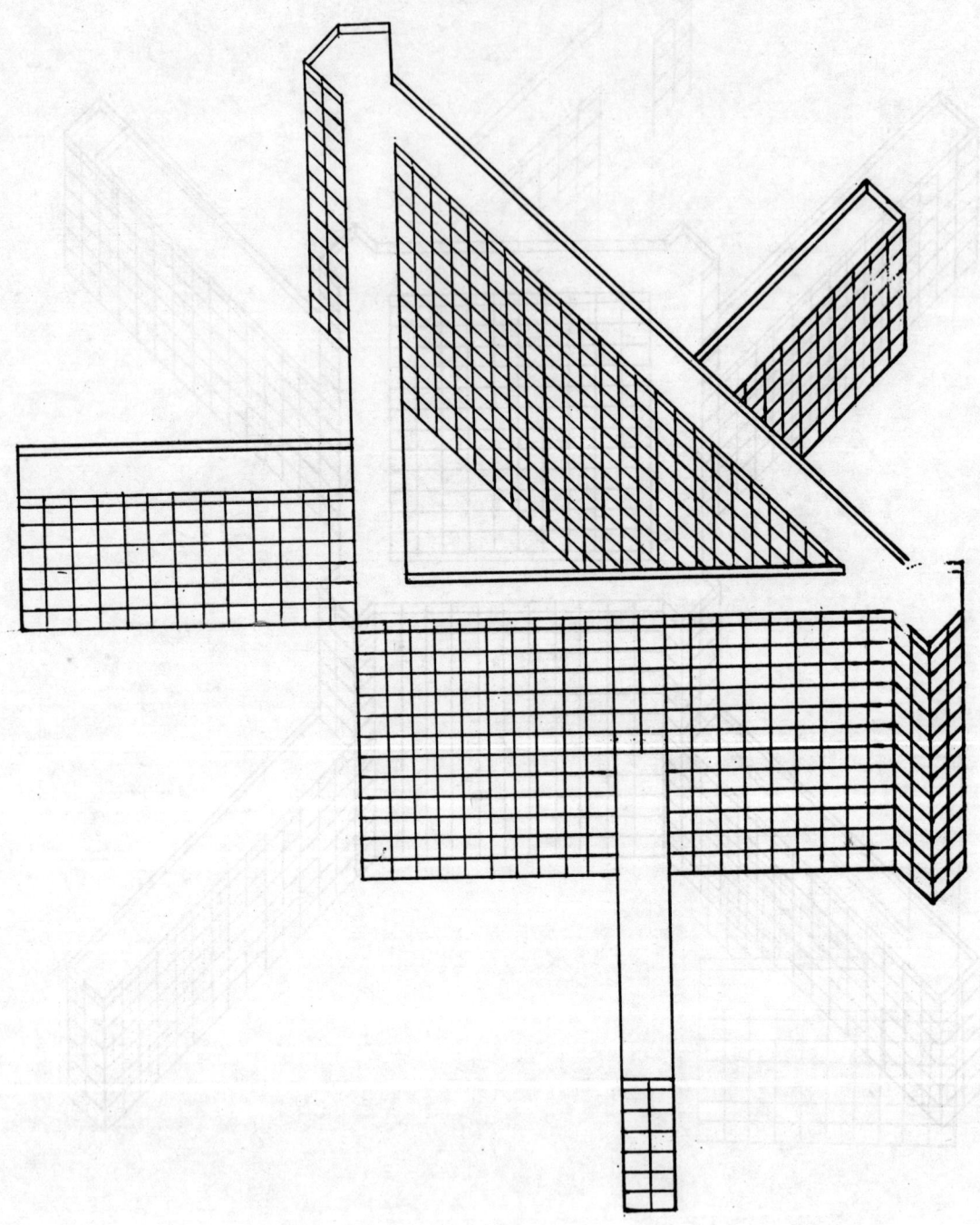

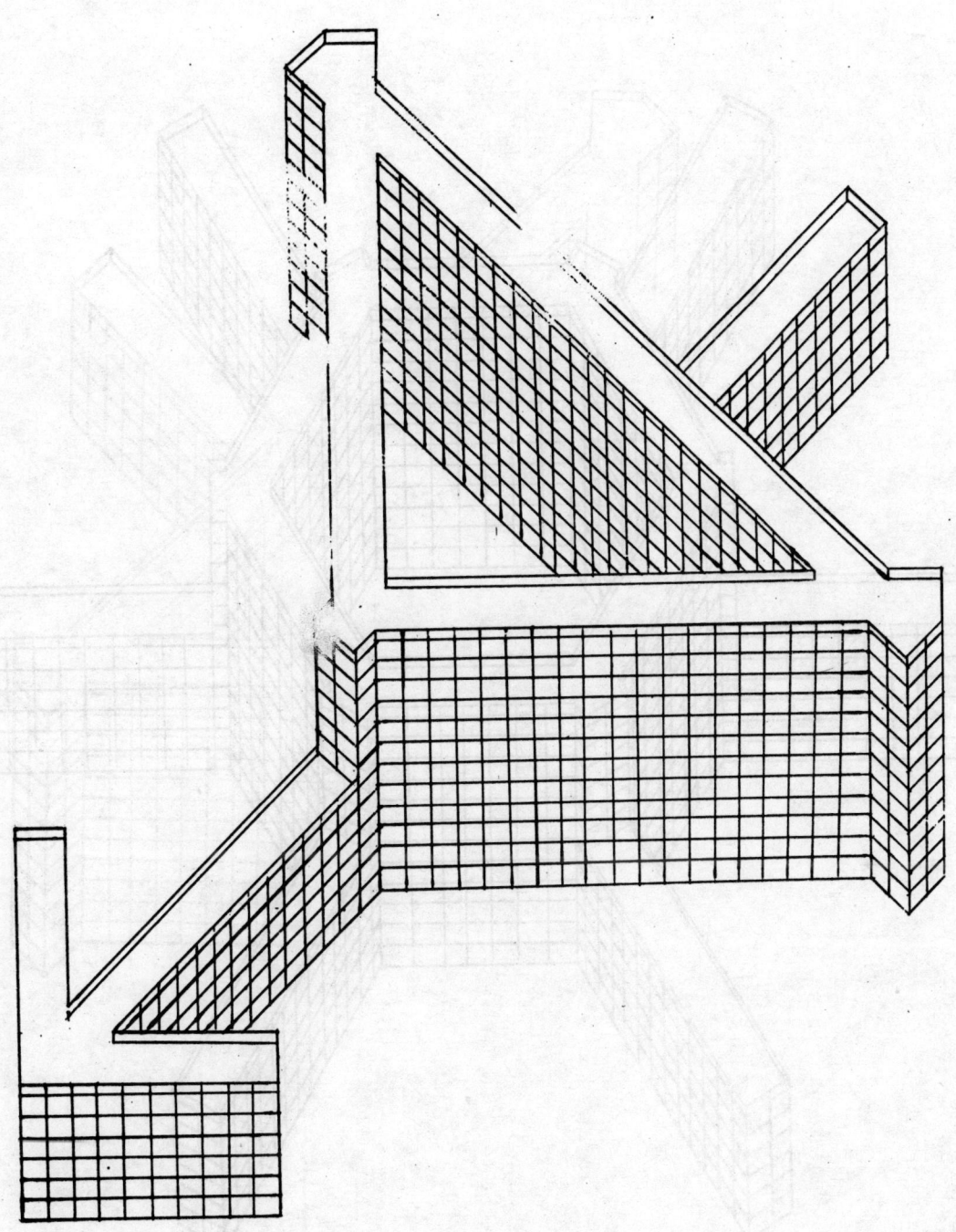

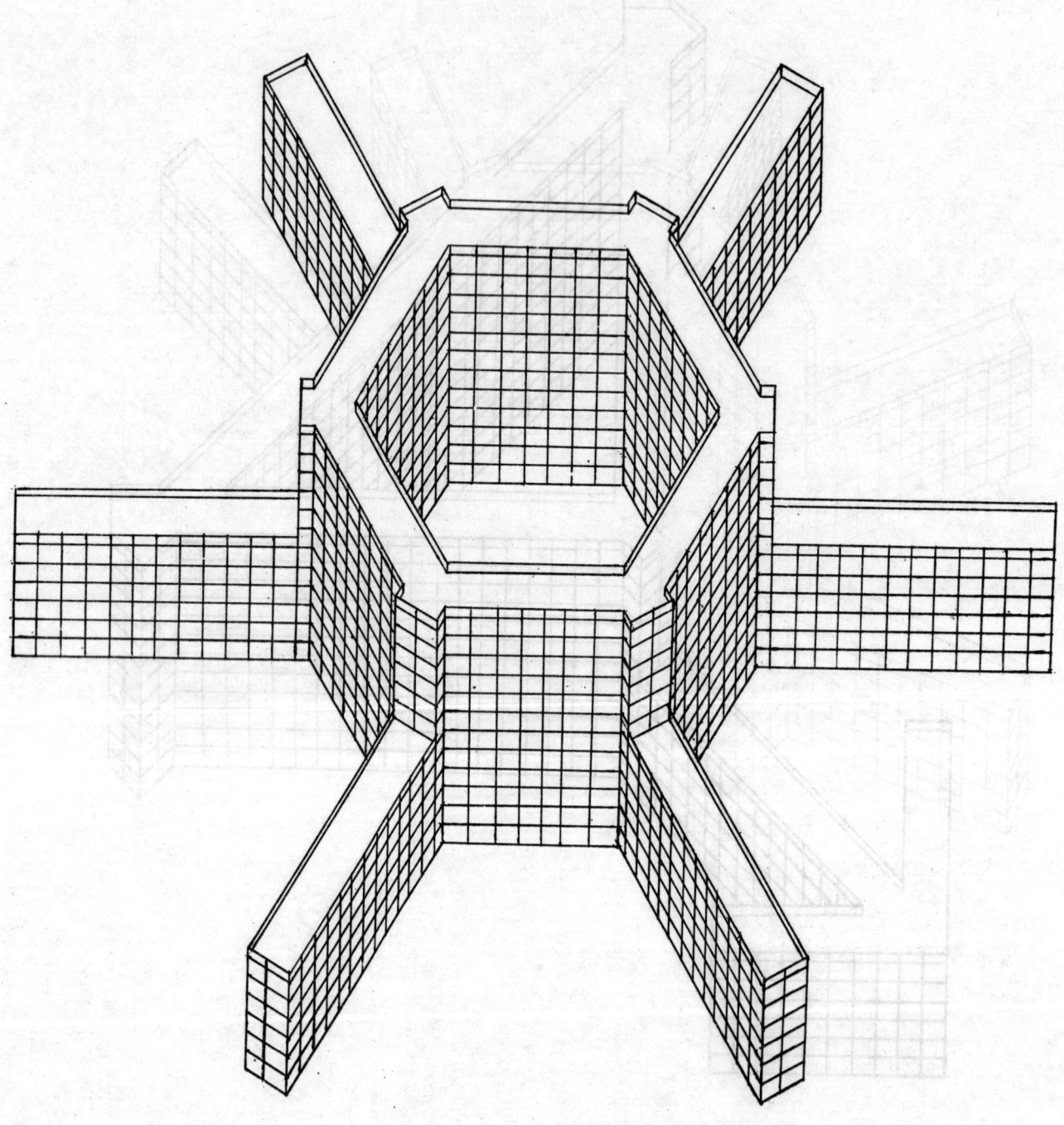

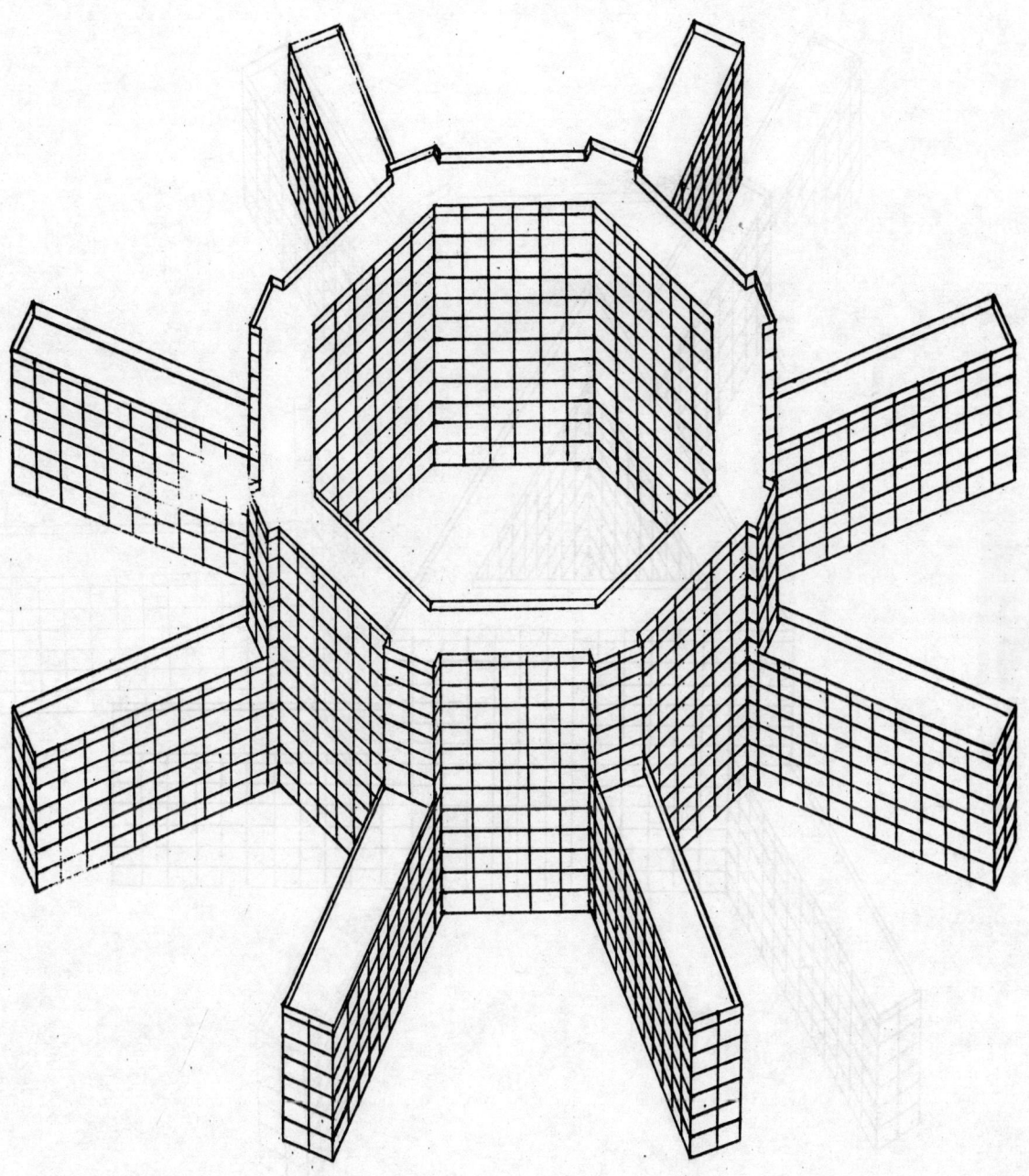

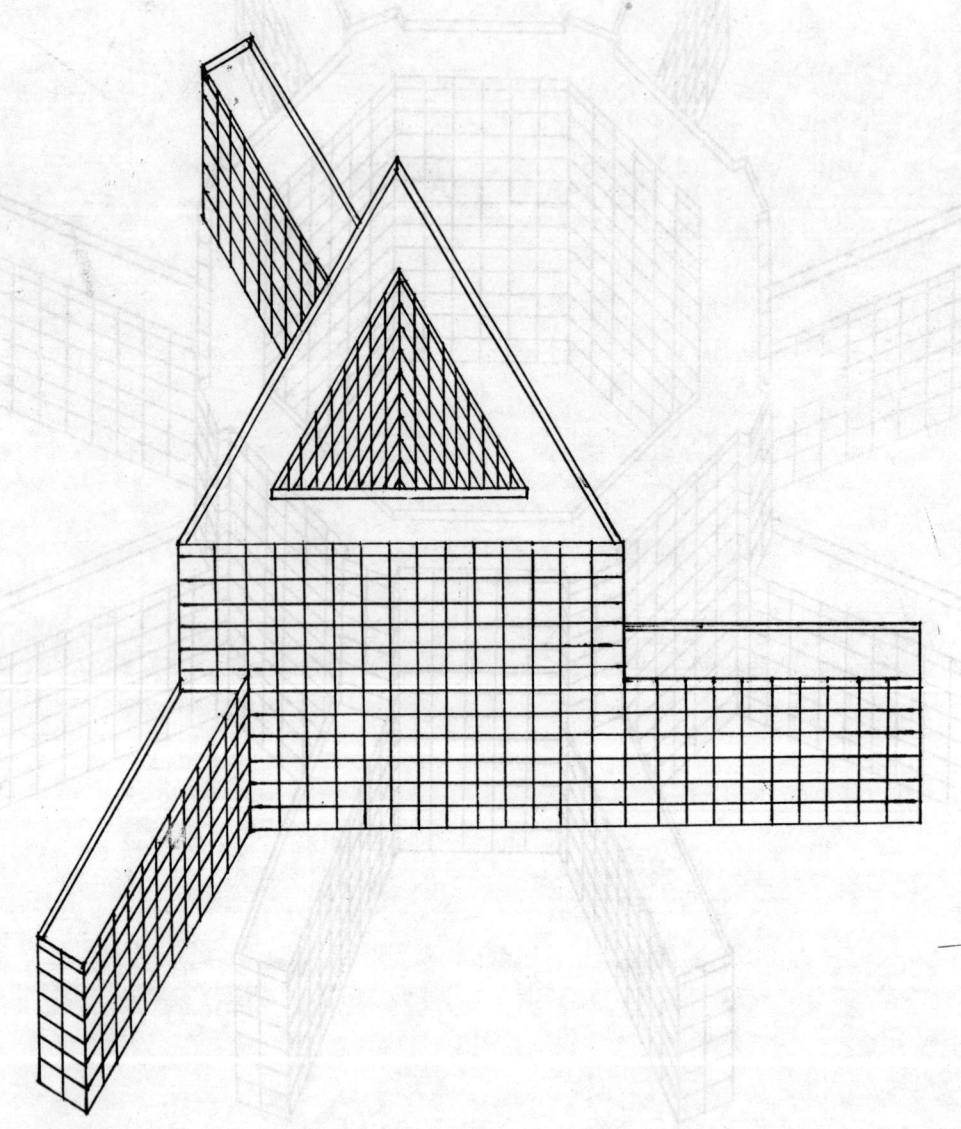

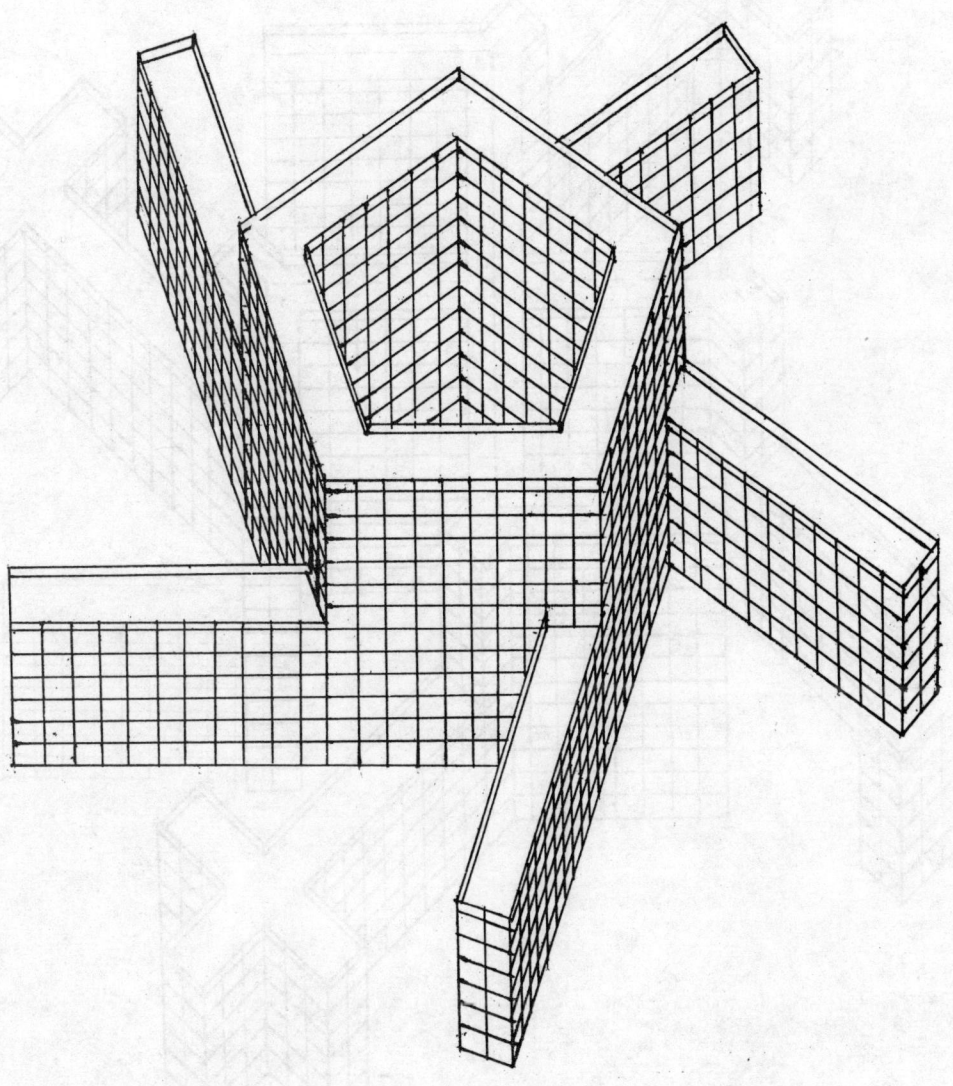

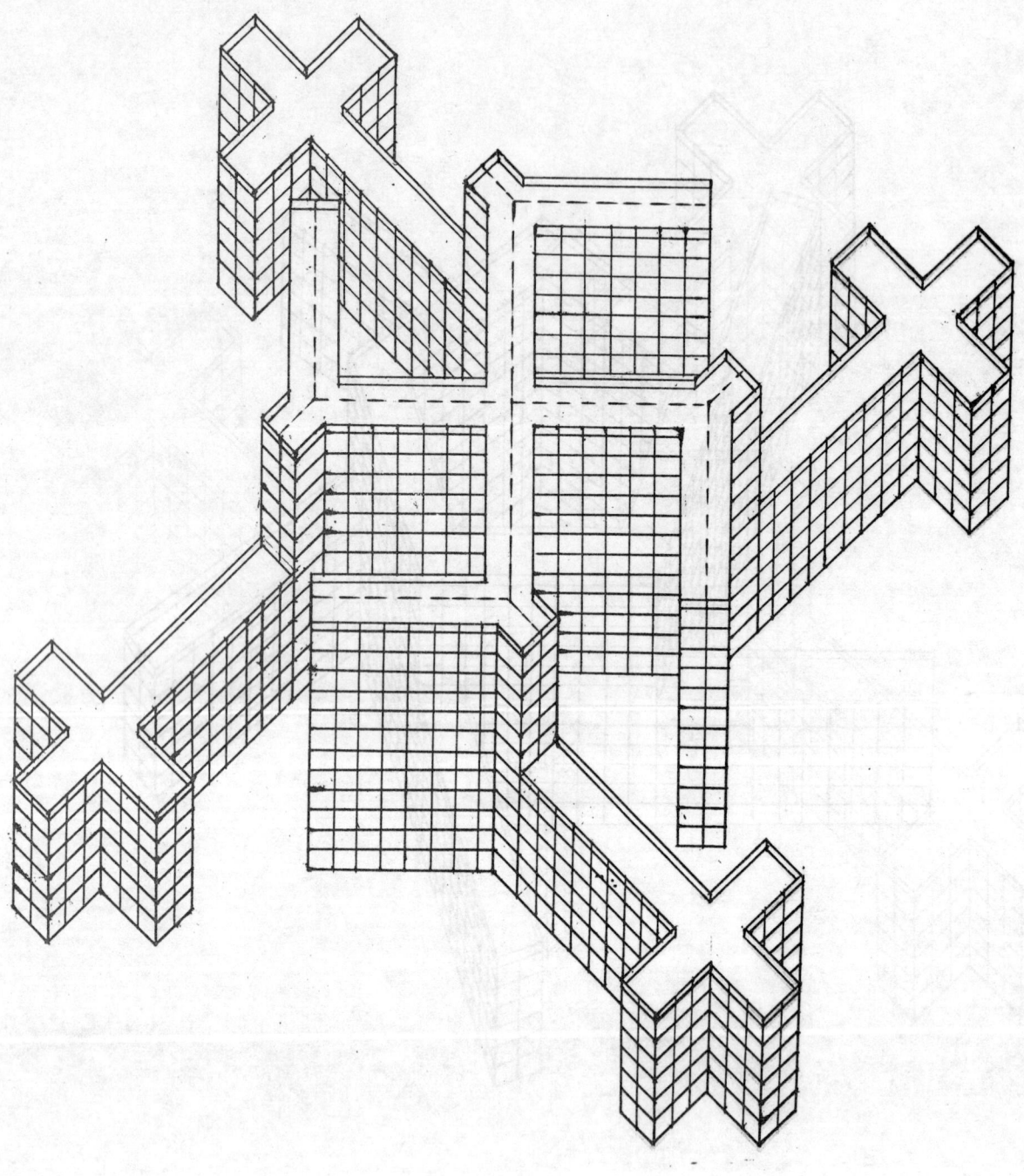

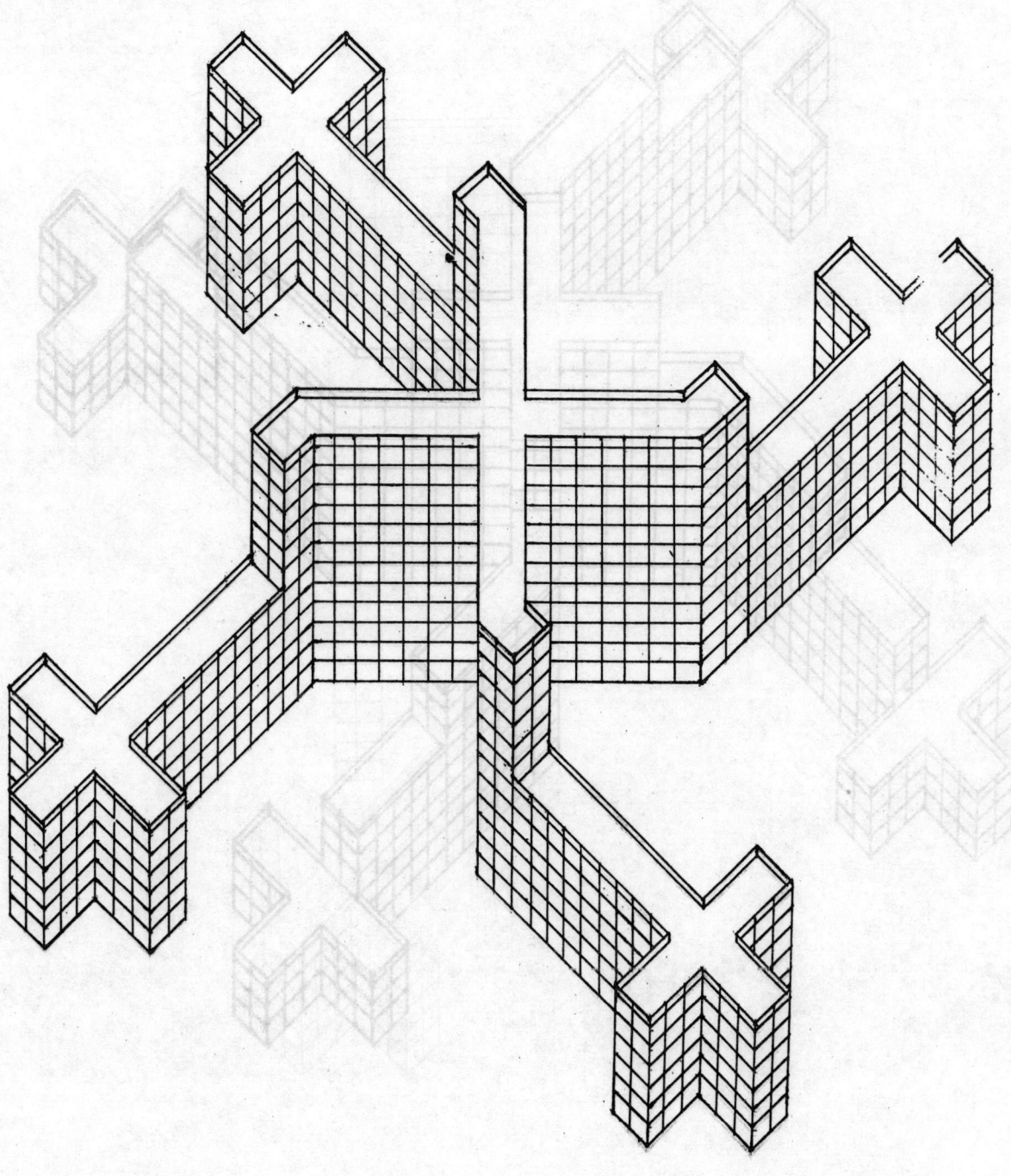

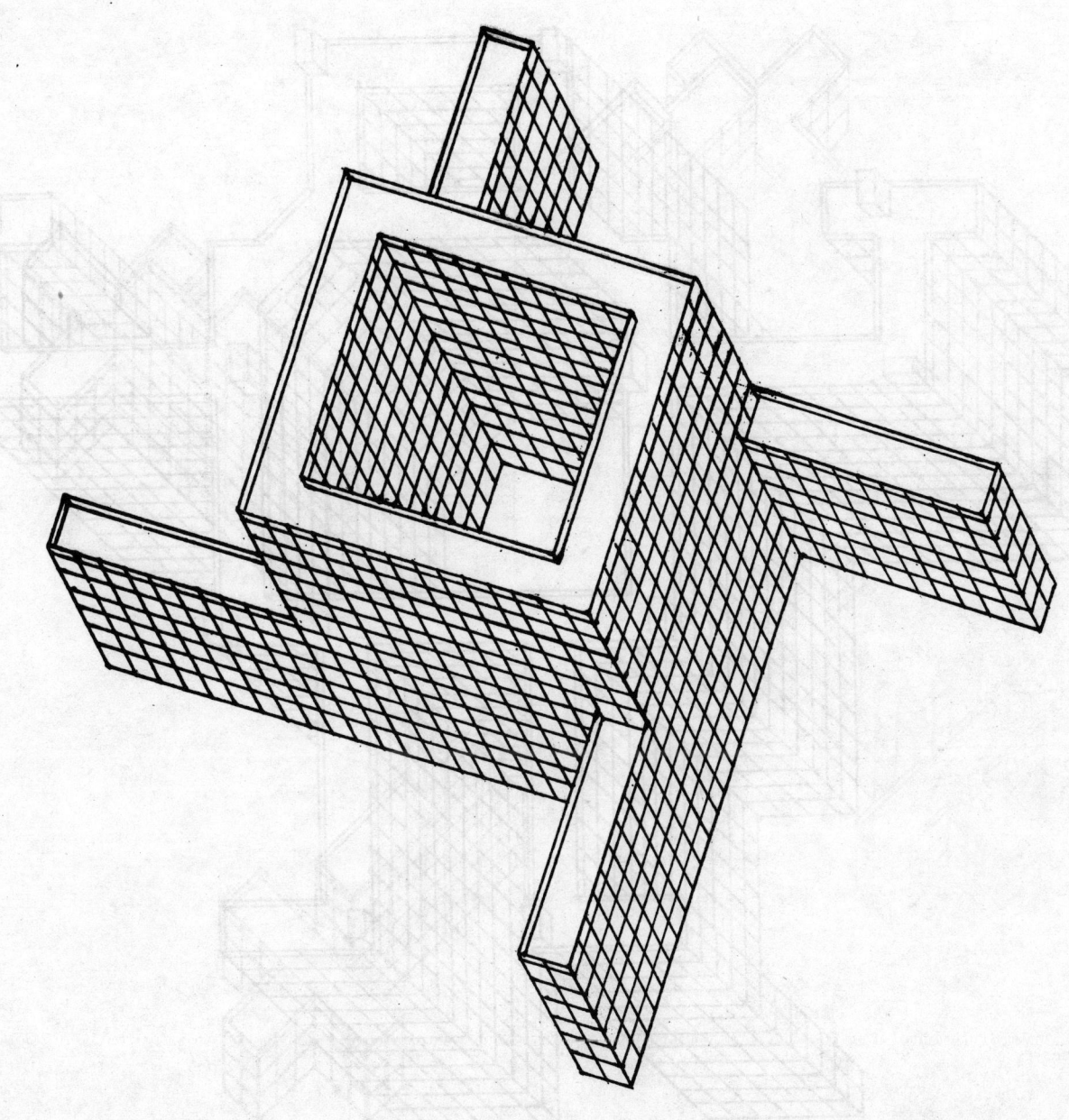

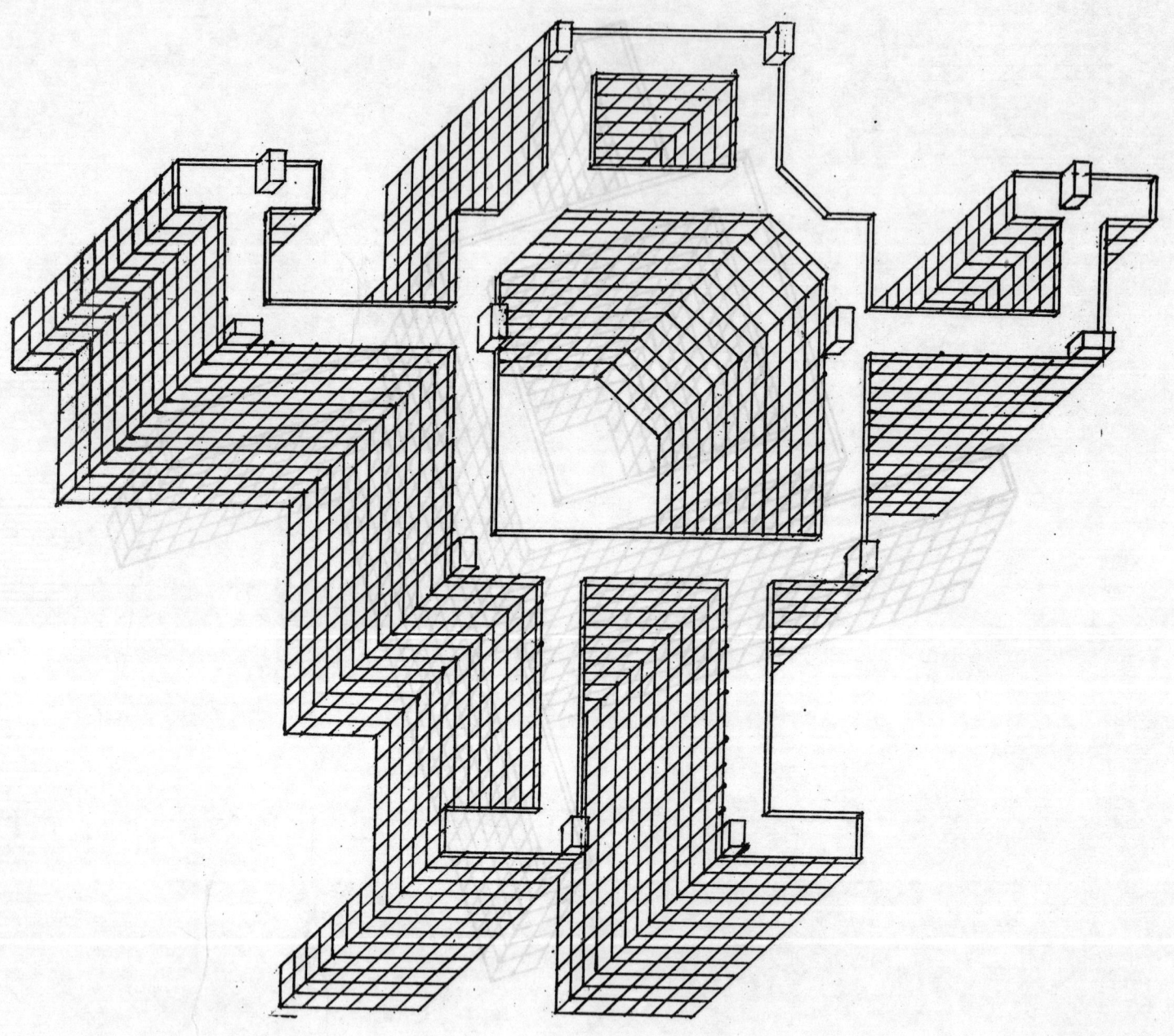

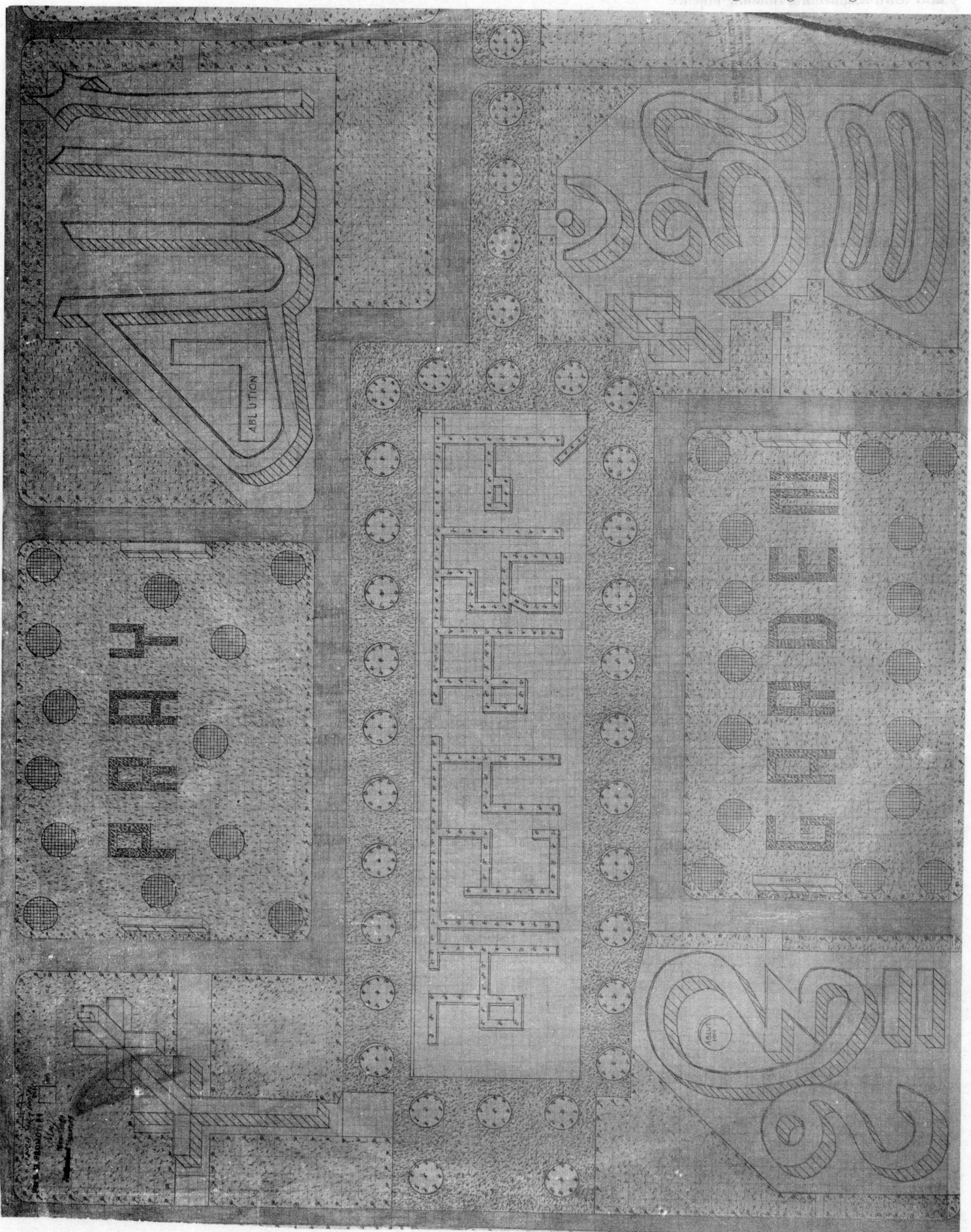

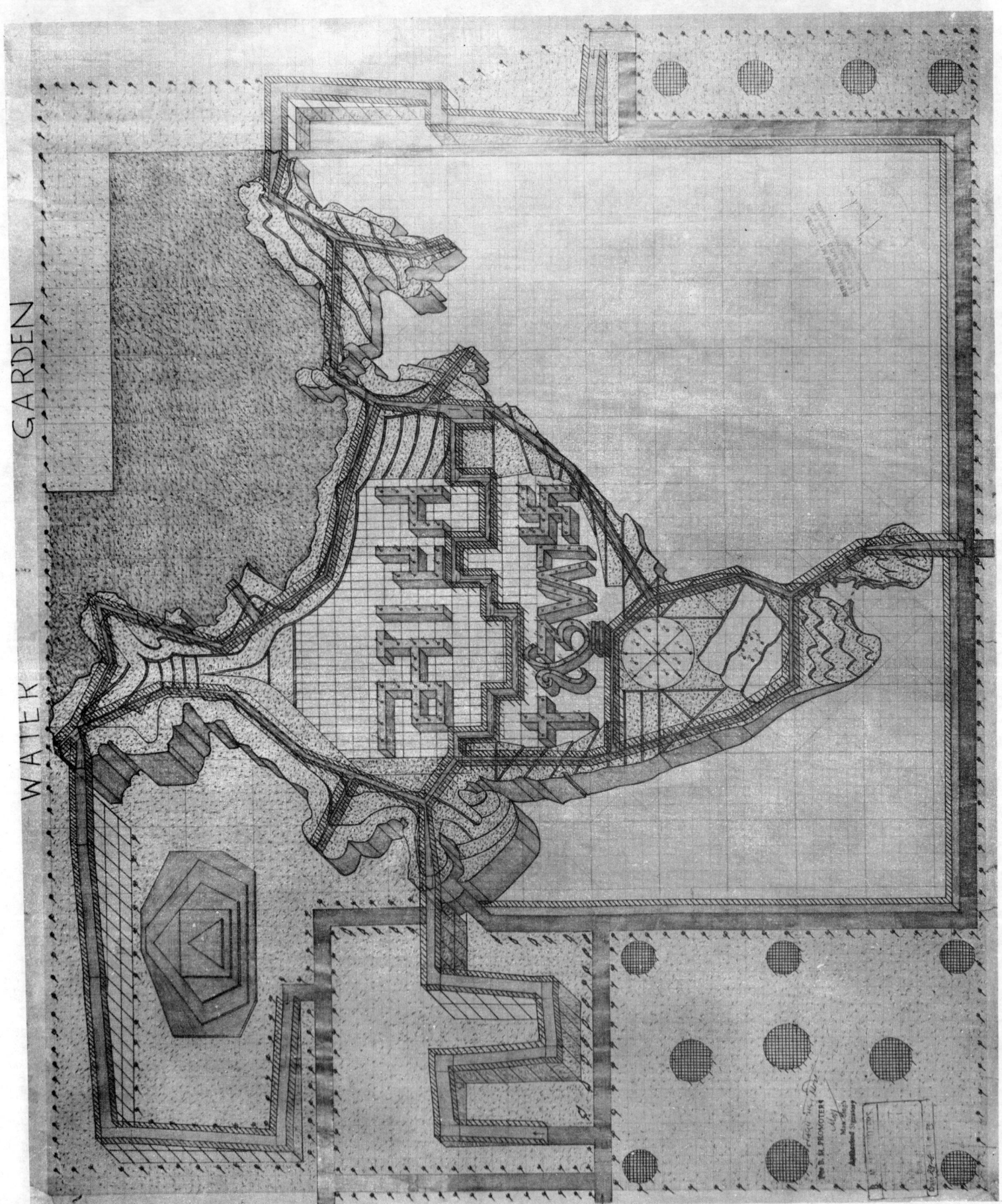

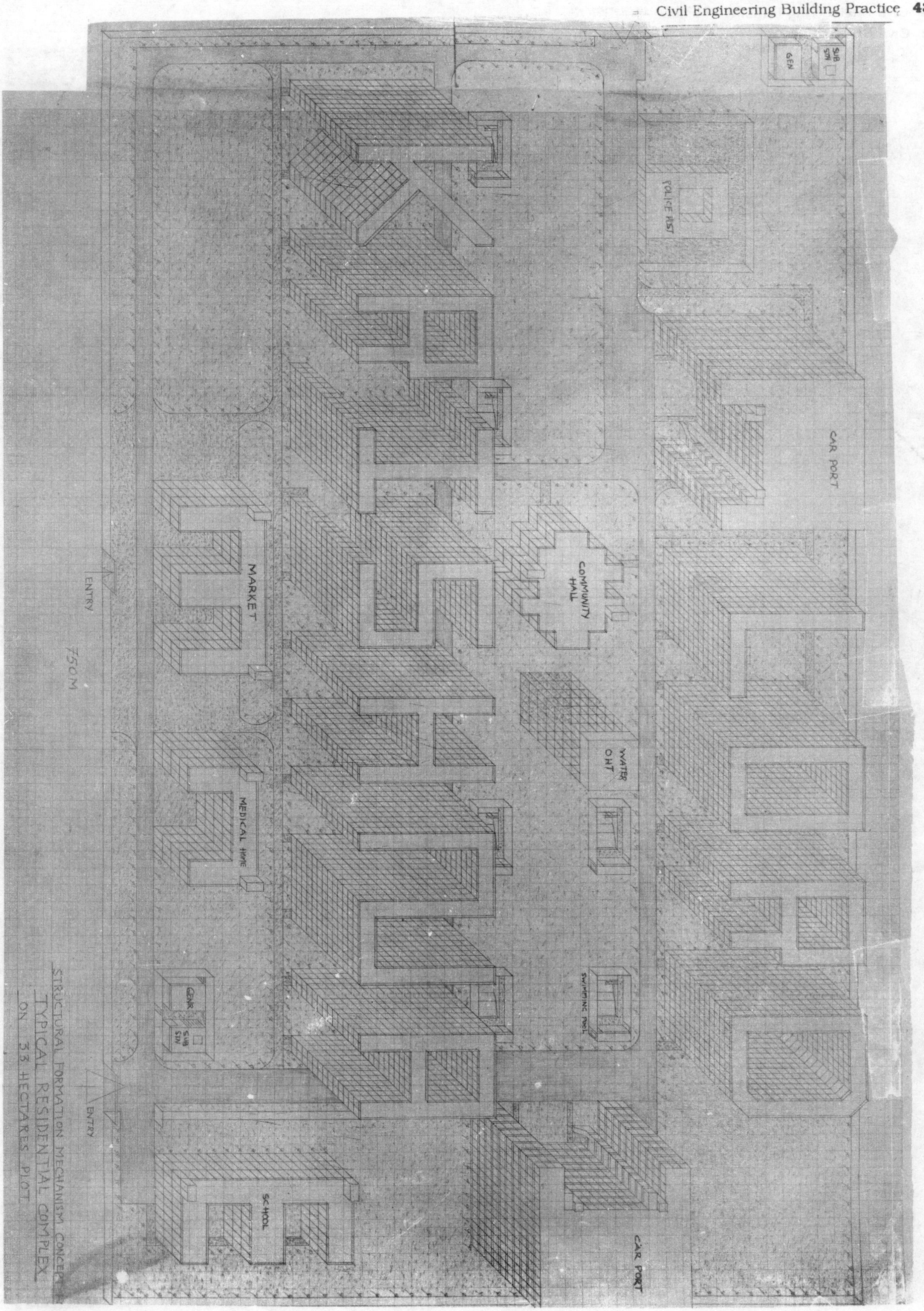

STRUCTURAL FORMATION MECHANISM CONCEPT
TYPICAL RESIDENTIAL COMPLEX
ON 33 HECTARES PLOT

ENTRY

750M

MARKET

MEDICAL HOME

SCHOOL

GEN · SUB STN

ENTRY

COMMUNITY HALL

WATER OHT

PARKING AREA

CAR PORT

POLICE PST

SUB STN · GEN

CAR PORT

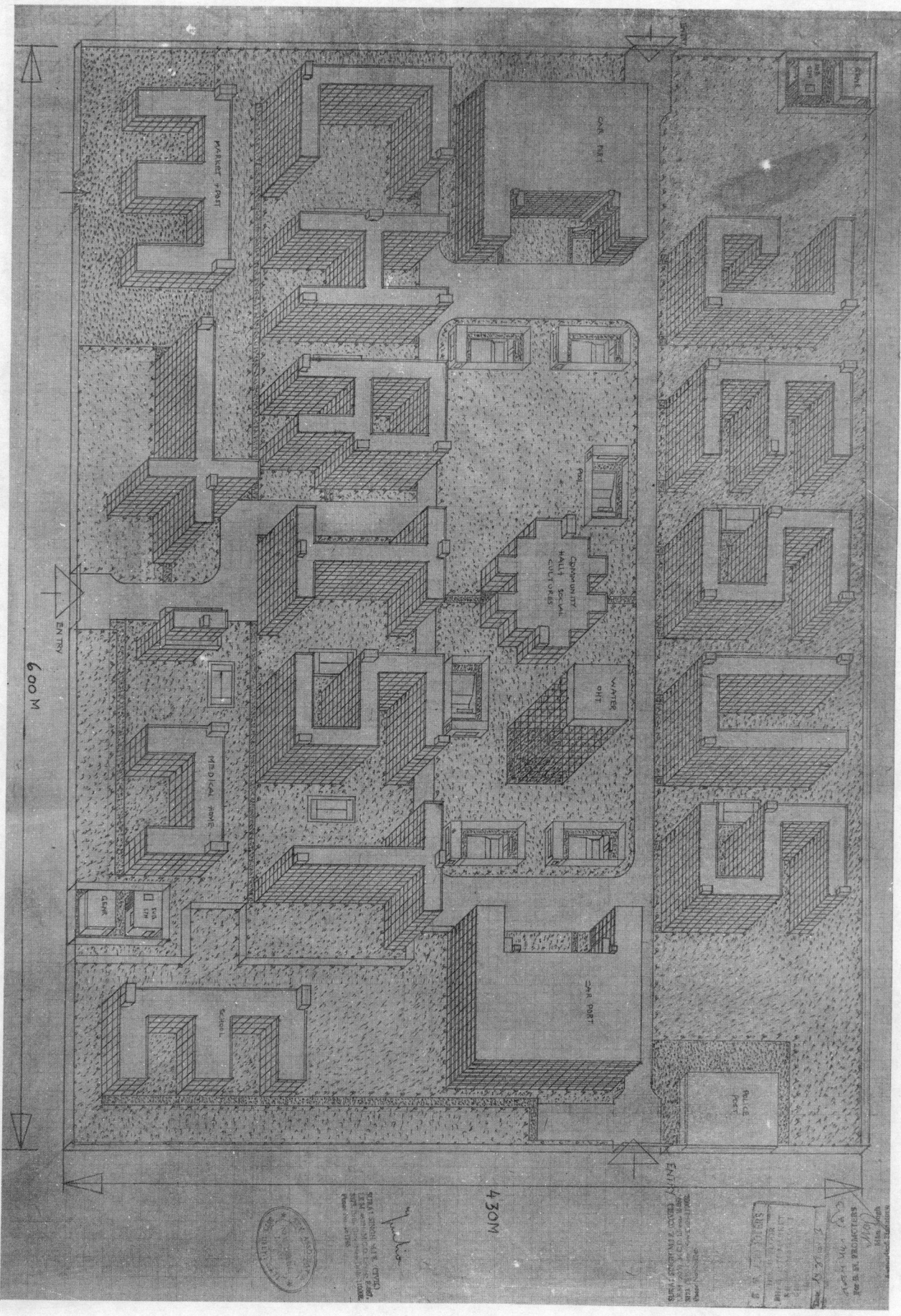

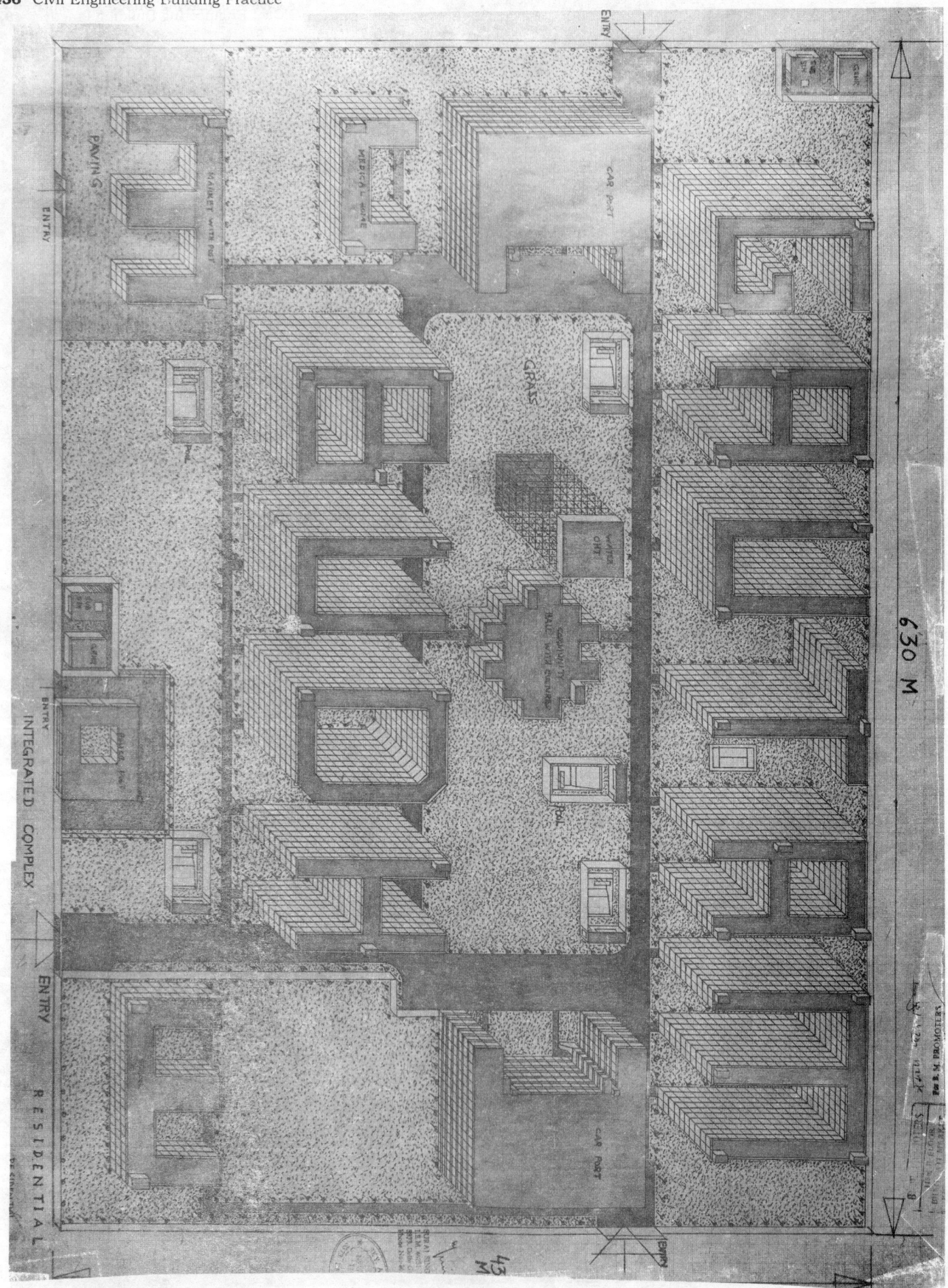

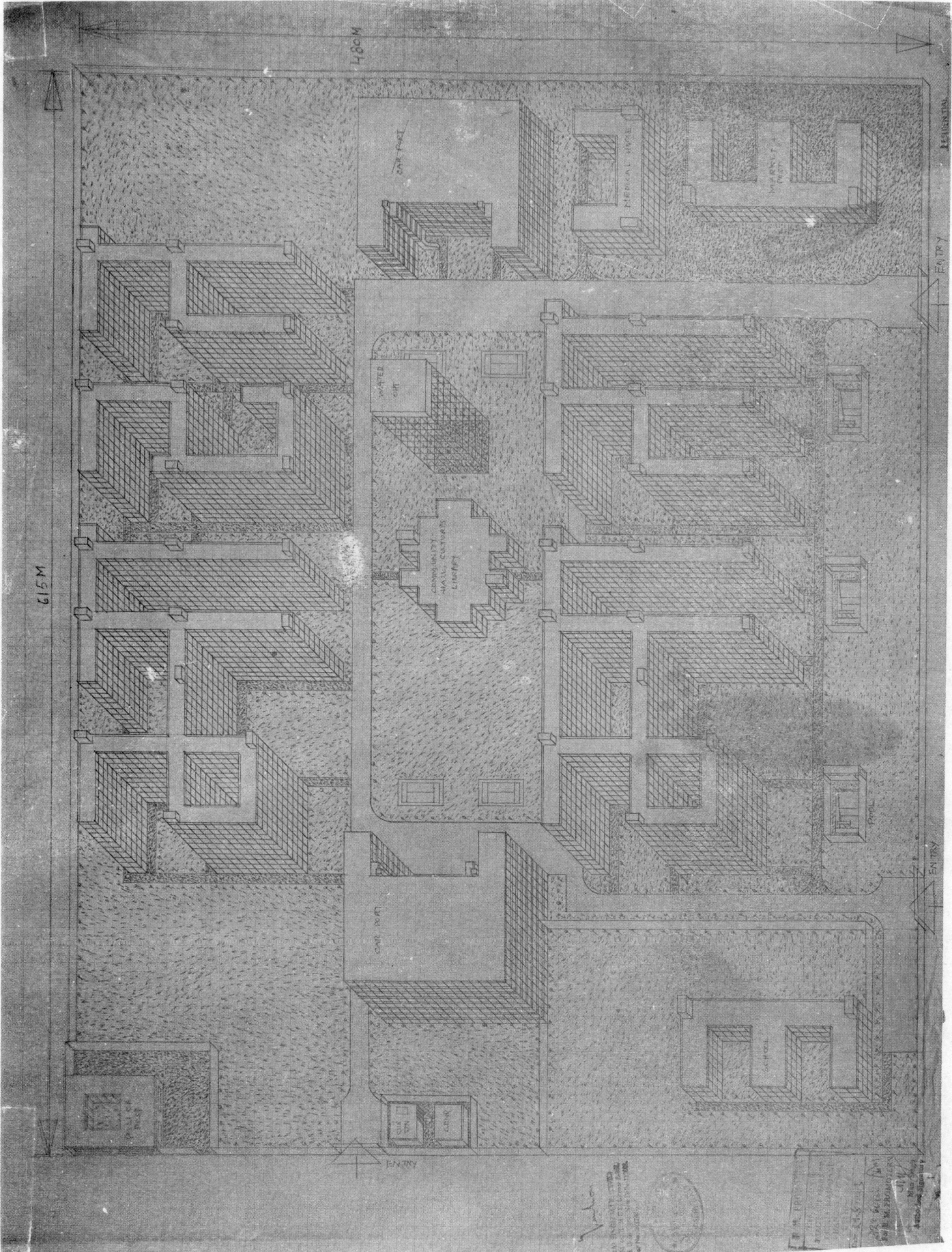

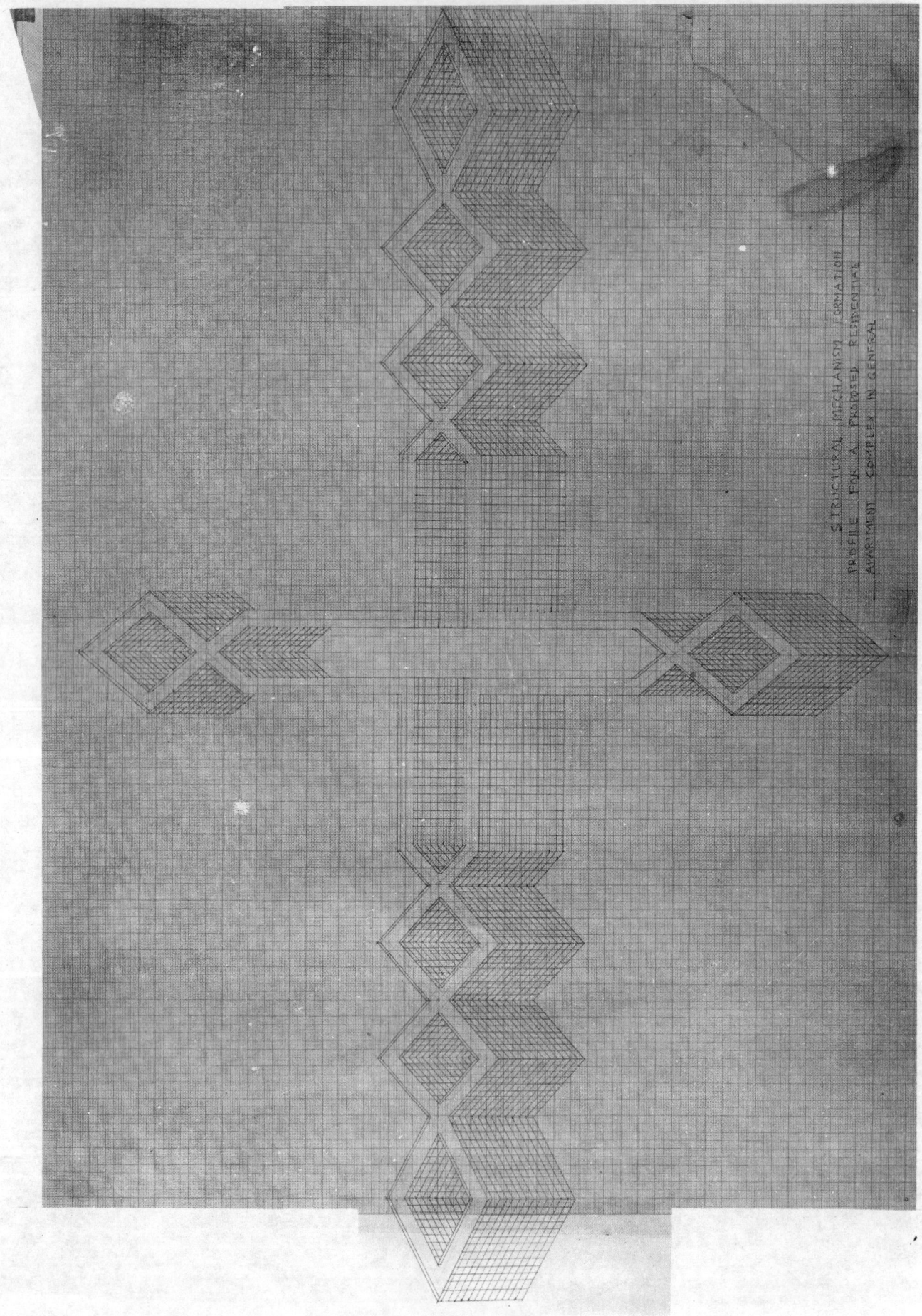

STRUCTURAL MECHANISM FORMATION
PROFILE FOR A PROPOSED RESIDENTIAL
APARTMENT COMPLEX IN GENERAL

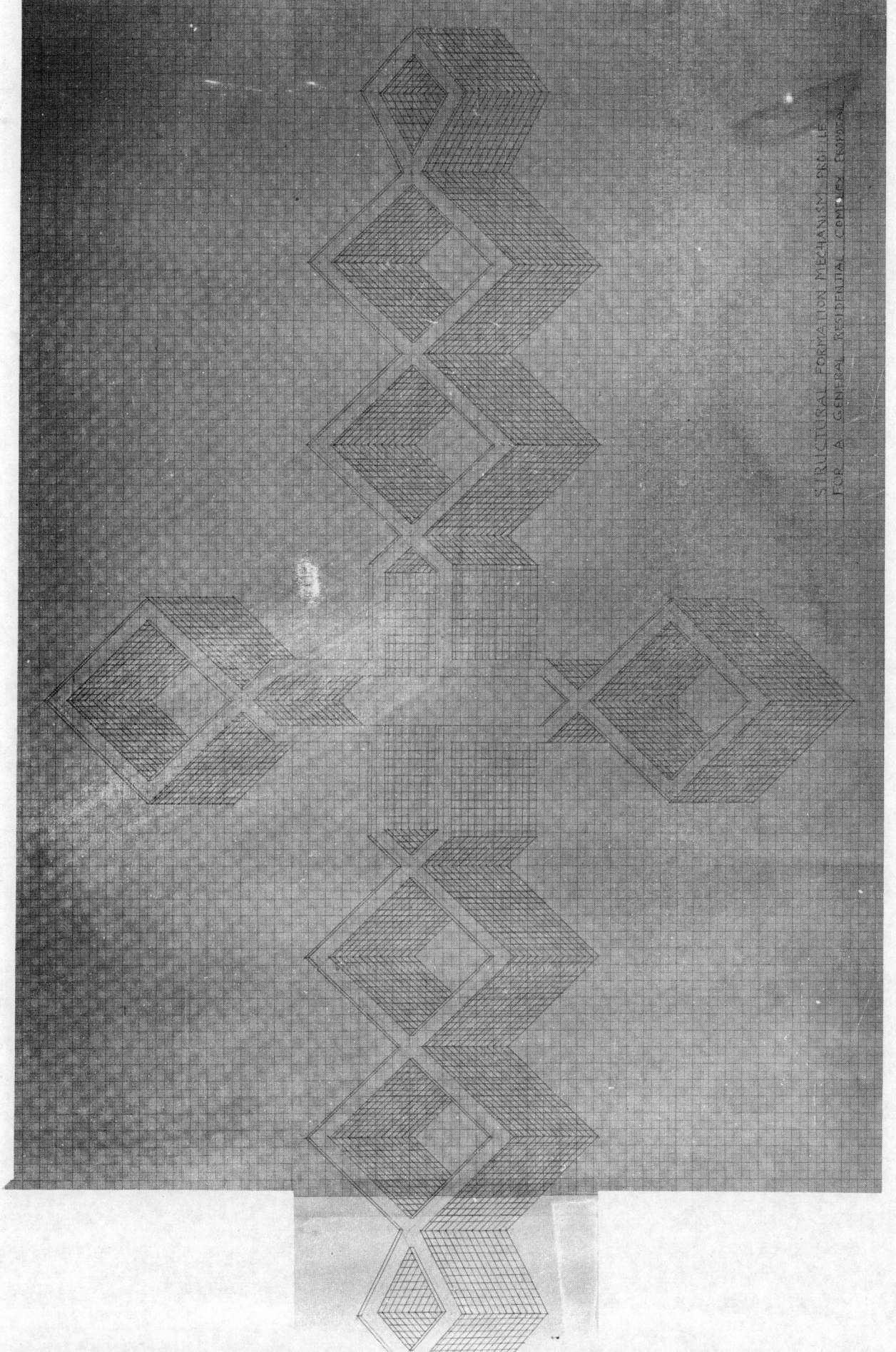

STRUCTURAL FORMATION MECHANISM PROFILE
FOR A GENERAL RESIDENTIAL COMPLEX PROPOSAL

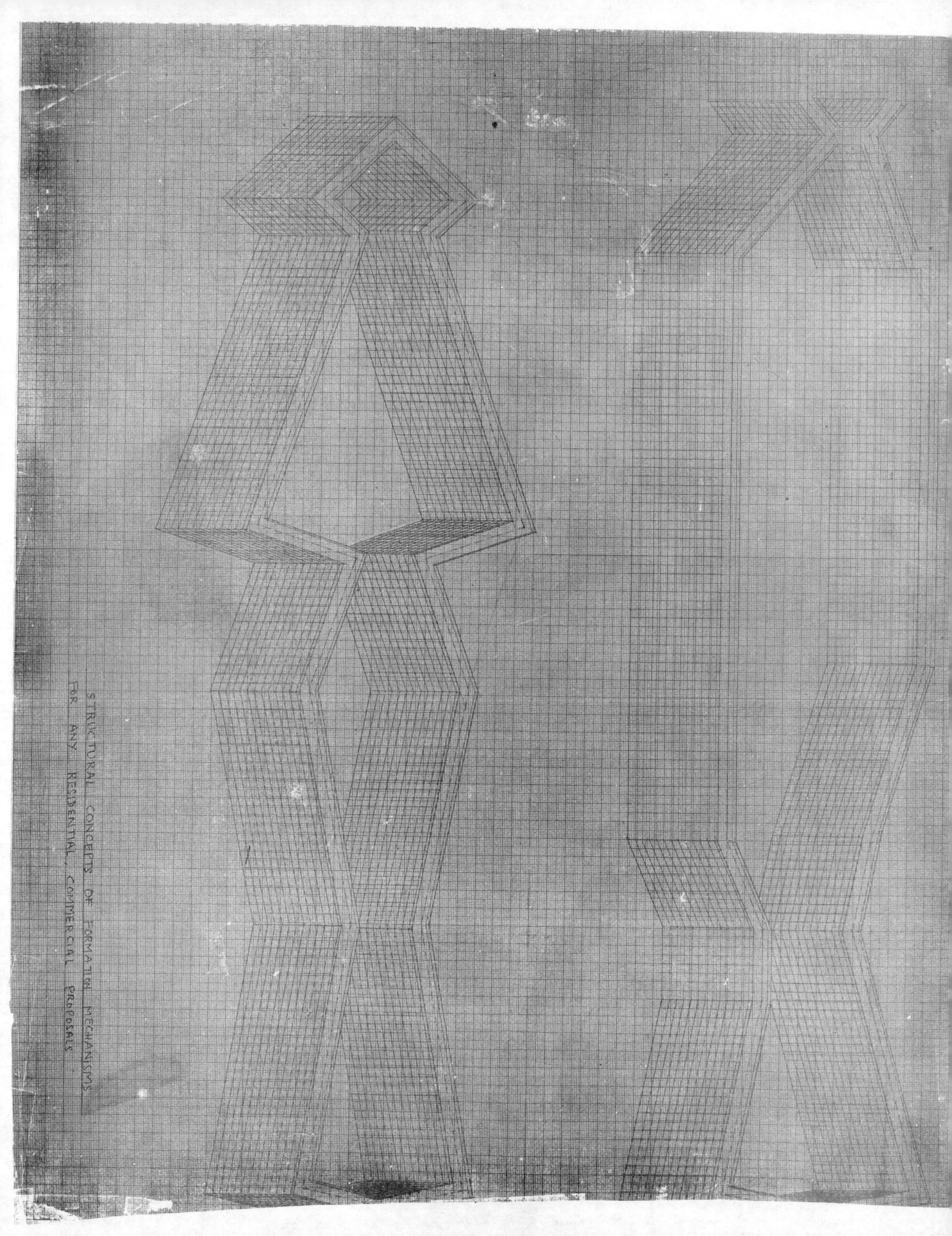

STRUCTURAL CONCEPTS OF FORMATION MECHANISMS
FOR ANY RESIDENTIAL, COMMERCIAL PROPOSALS

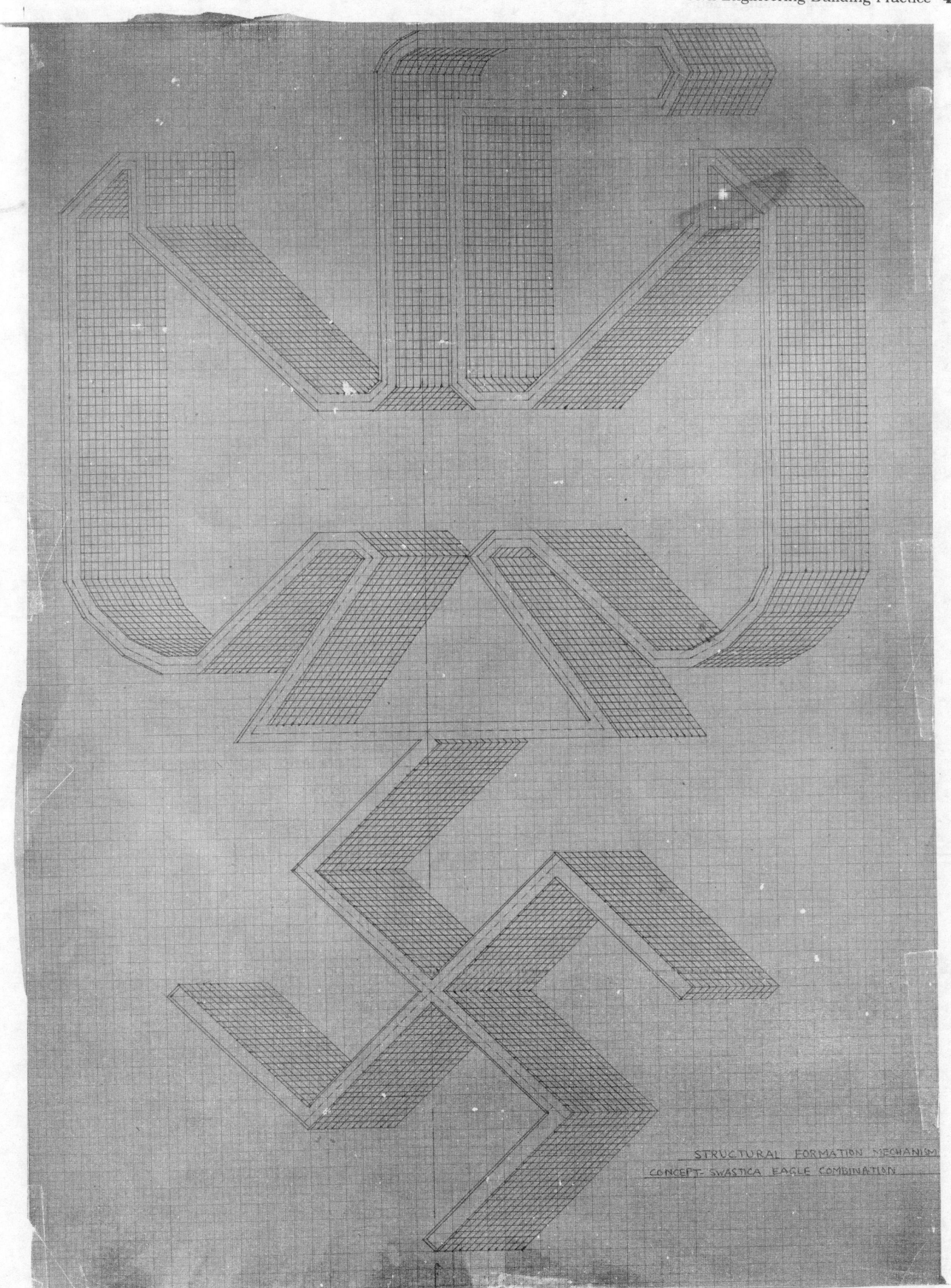

STRUCTURAL FORMATION MECHANISM
CONCEPT: SWASTICA EAGLE COMBINATION

STRUCTURAL FORMATION MECHANISM
CONCEPT ON A DOLL PROFILE

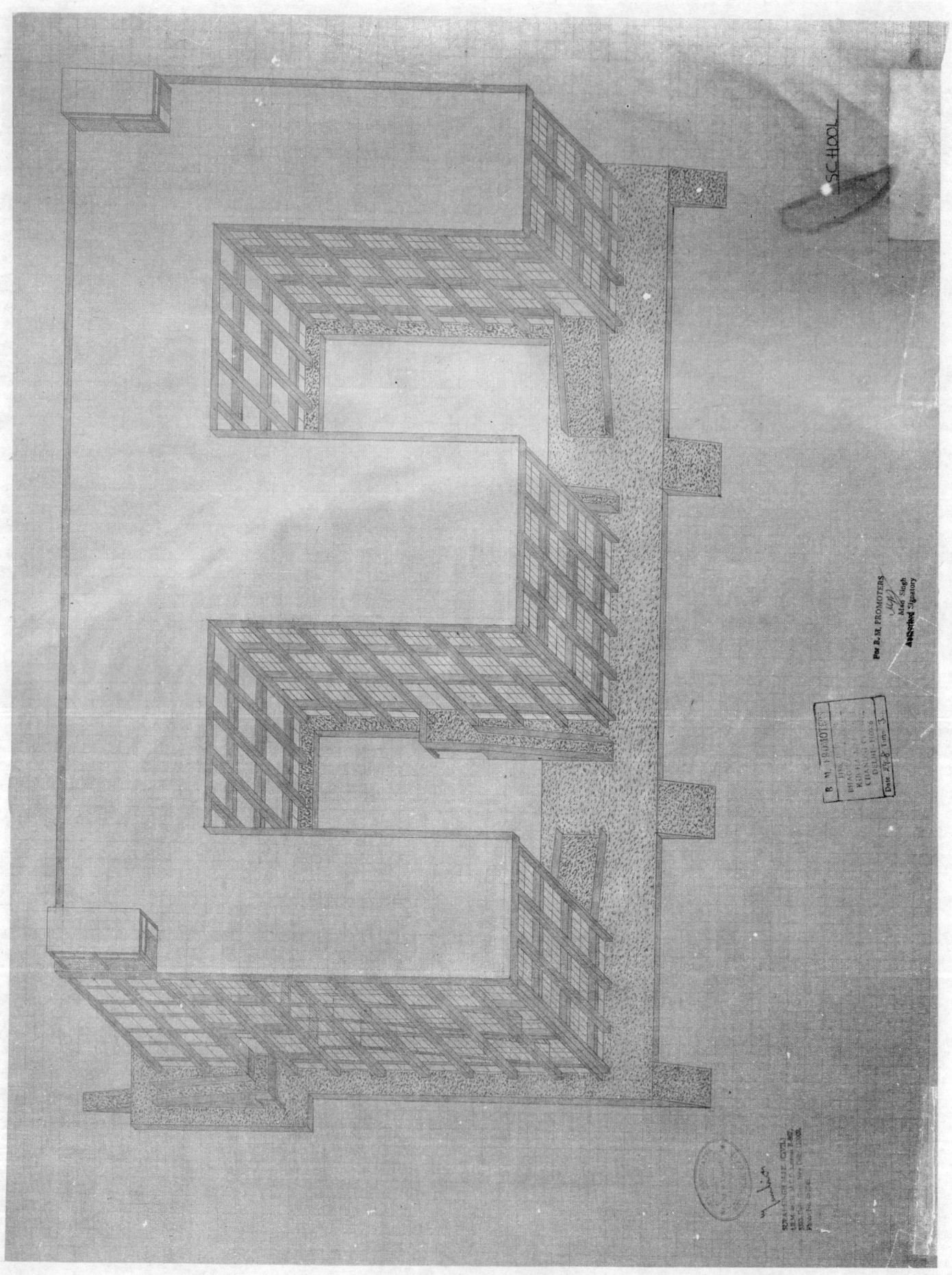

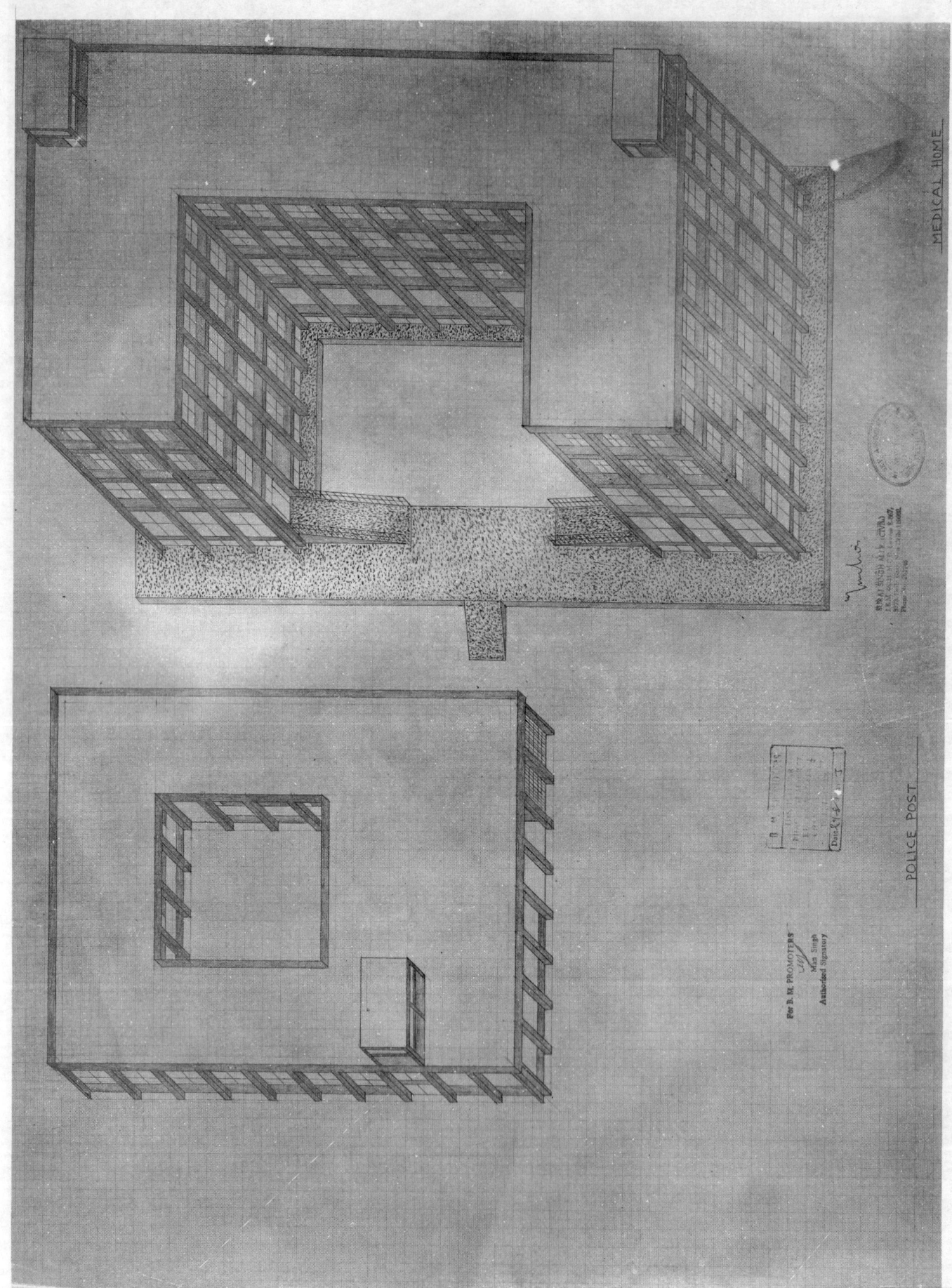

MEDICAL HOME

POLICE POST

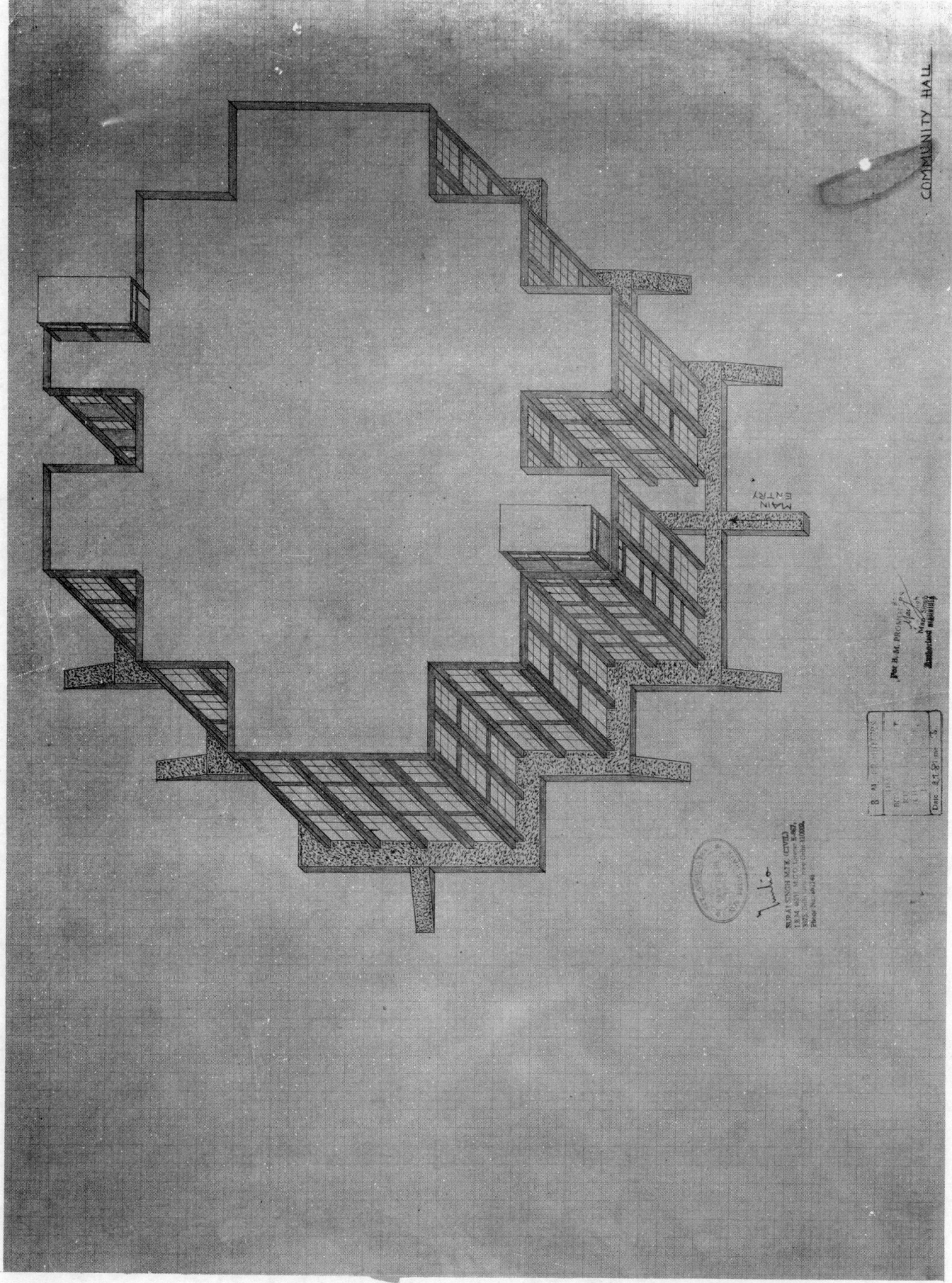

COMMUNITY HALL

MAIN ENTRY

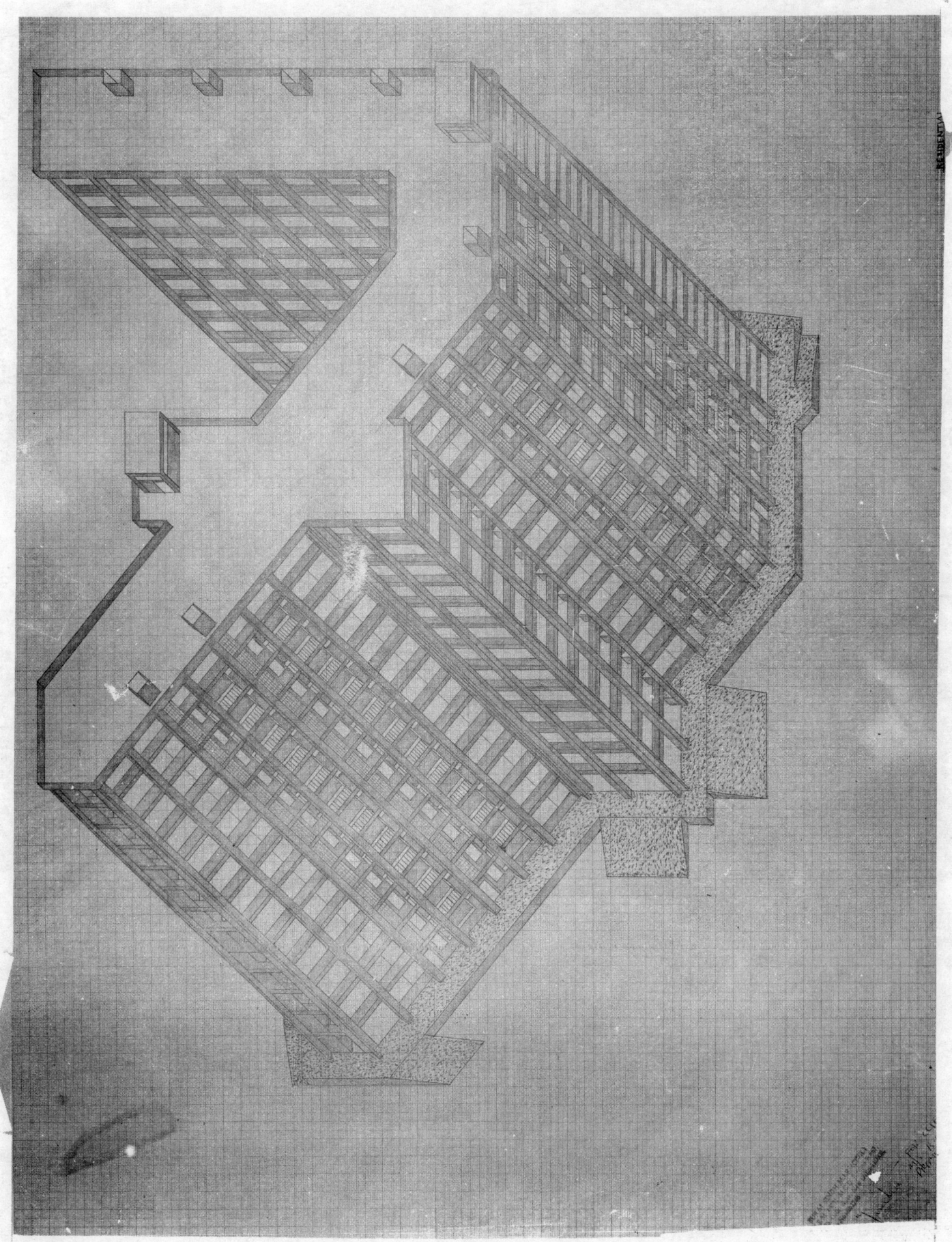

RESIDENTIAL
mansion

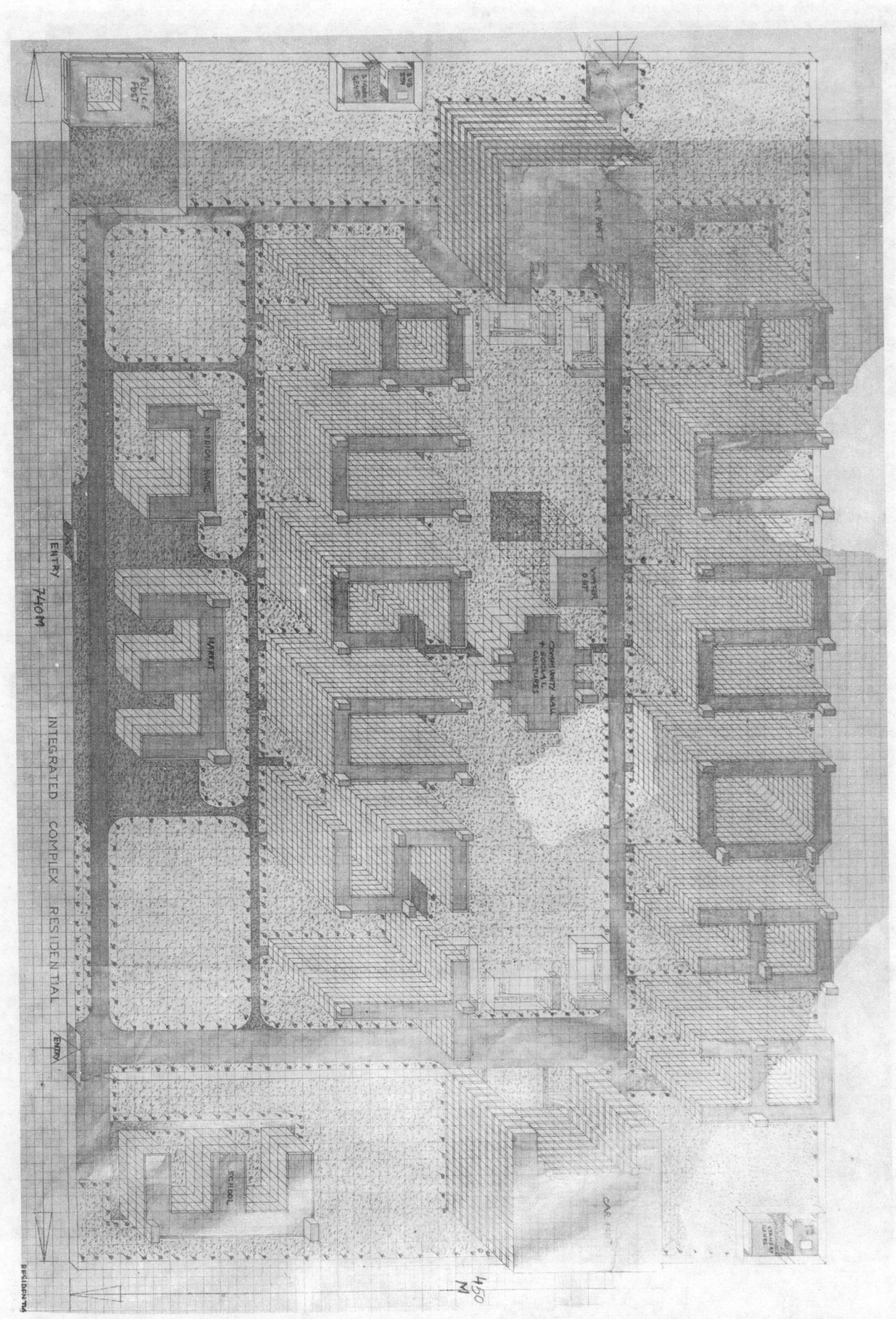

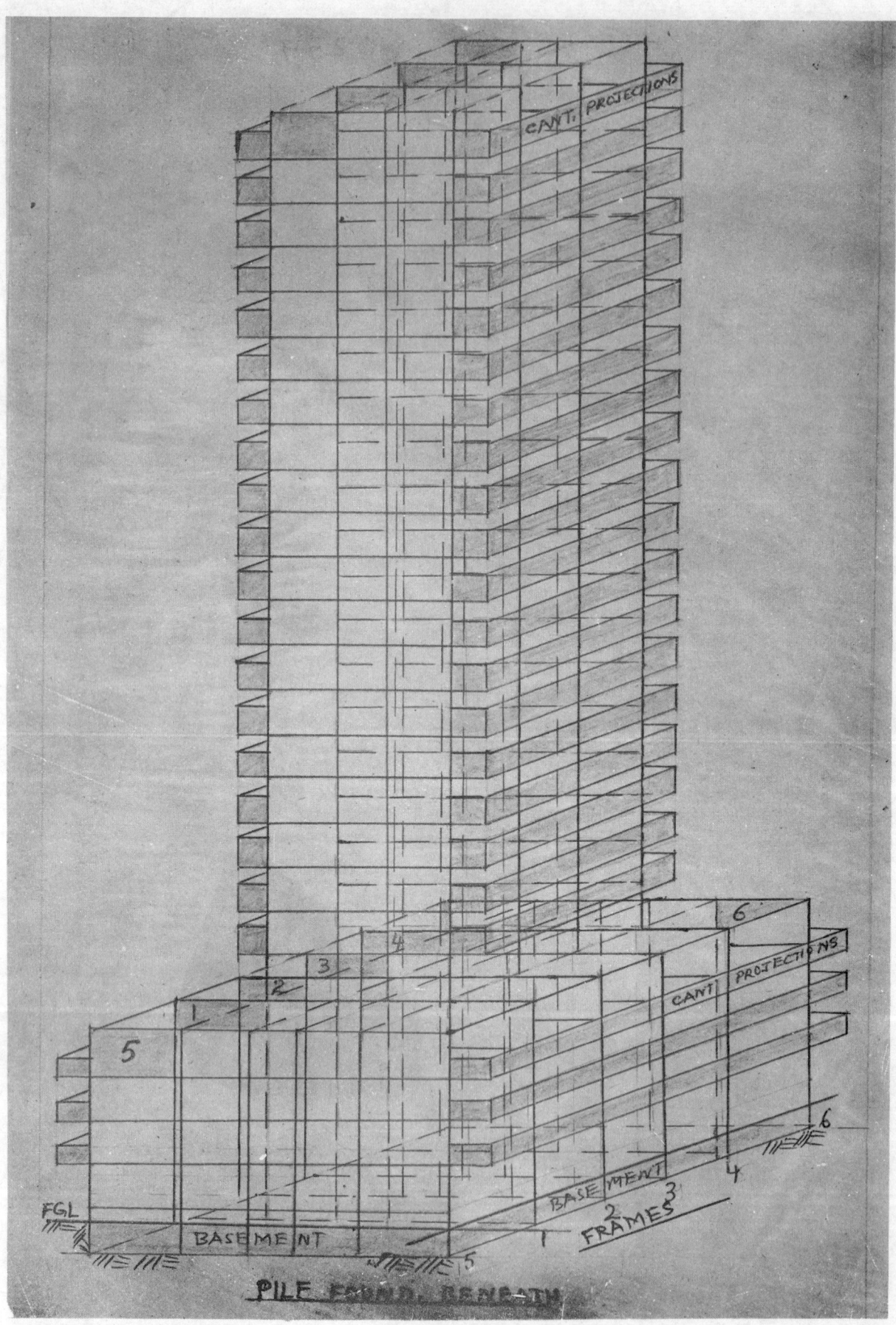

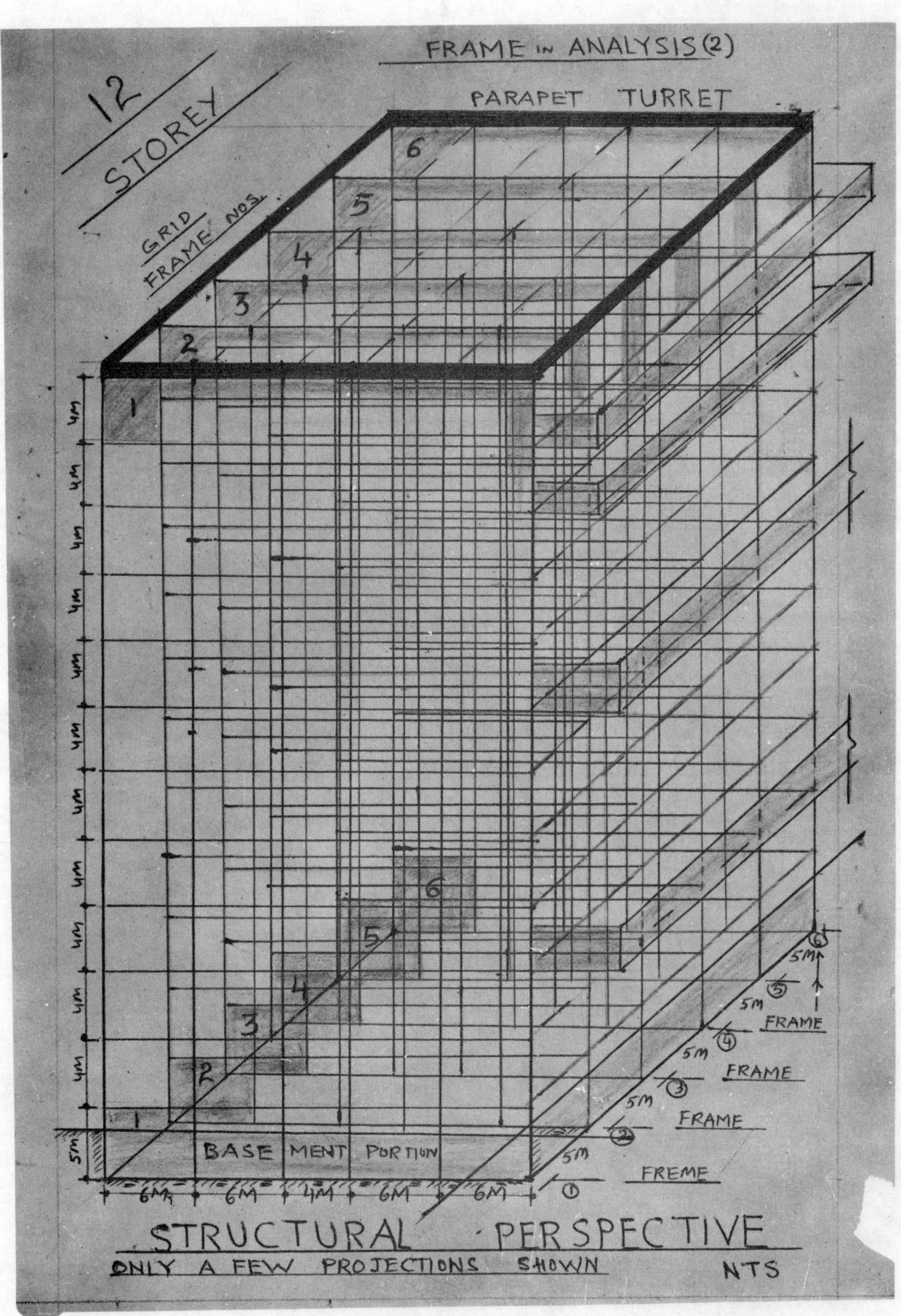

FRAME IN ANALYSIS (2)

12 STOREY

PARAPET TURRET

GRID

FRAME NOS.

BASEMENT PORTION

STRUCTURAL · PERSPECTIVE

ONLY A FEW PROJECTIONS SHOWN   NTS

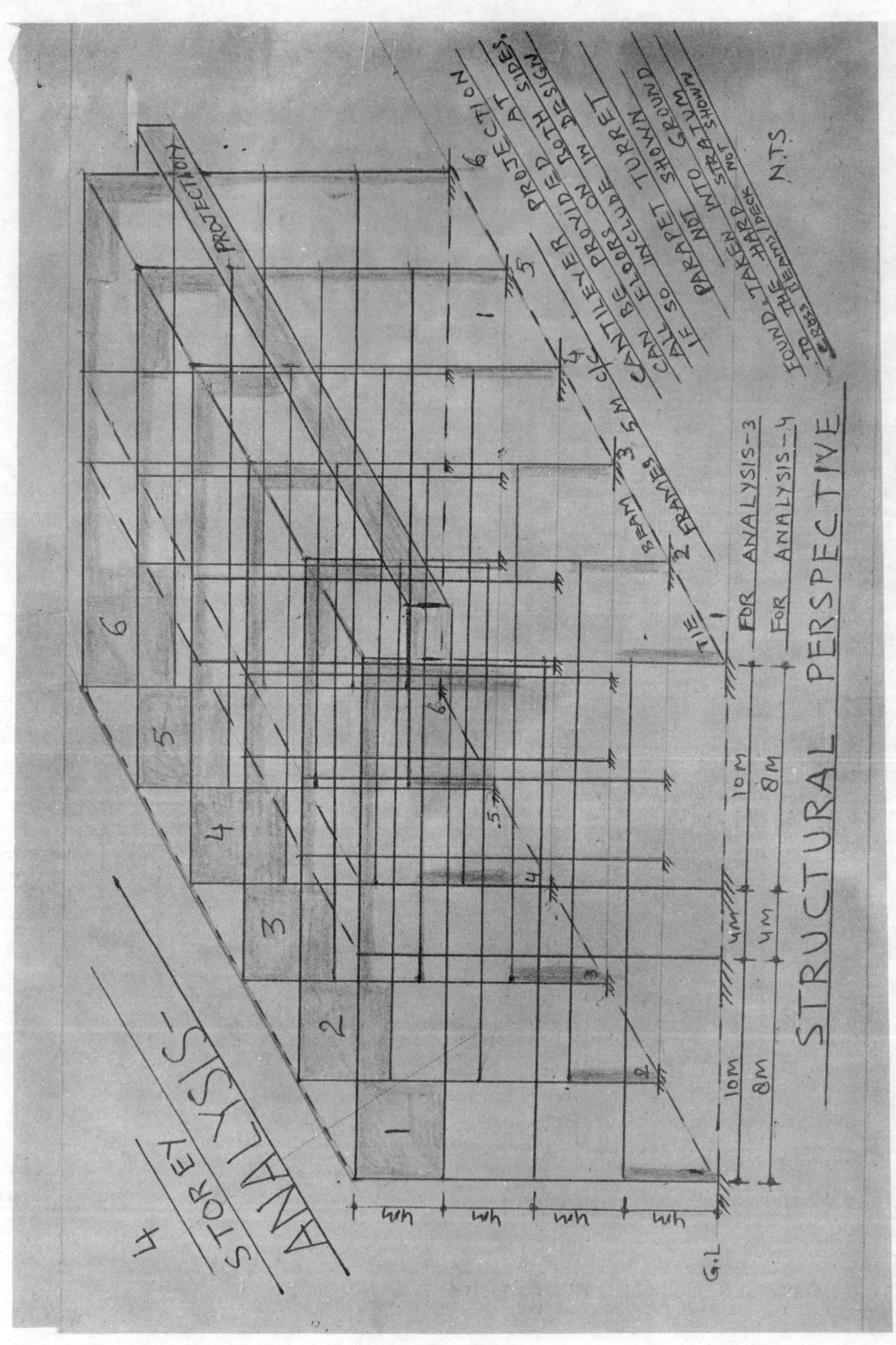

STRUCTURAL PERSPECTIVE

ANALYSIS-4

STOREY

6
5
4
3
2
1

PROJECTION

6 SECTION

5 PROJECTION AT SIDES

CANTILEVER PROVIDE BOTH SIDES

CAN BE FLOORS OR INCLUDE TURRET

ALSO PROVD IN DESIGN

IF PARAPET PROVIDED

PARAPET SHOWN

NOT INTO GROUND

FOUND. TAKEN HARD NOT SHOWN

TO THE BEAM/DECK STRATUM

CROSS STRATUM SHOWN

N.T.S.

BEAM 3.5 M c/c

FRAMES 3 M c/c

TIE

FOR ANALYSIS-3

FOR ANALYSIS-4

10M
8M

10M
8M

4M
4M

G.L

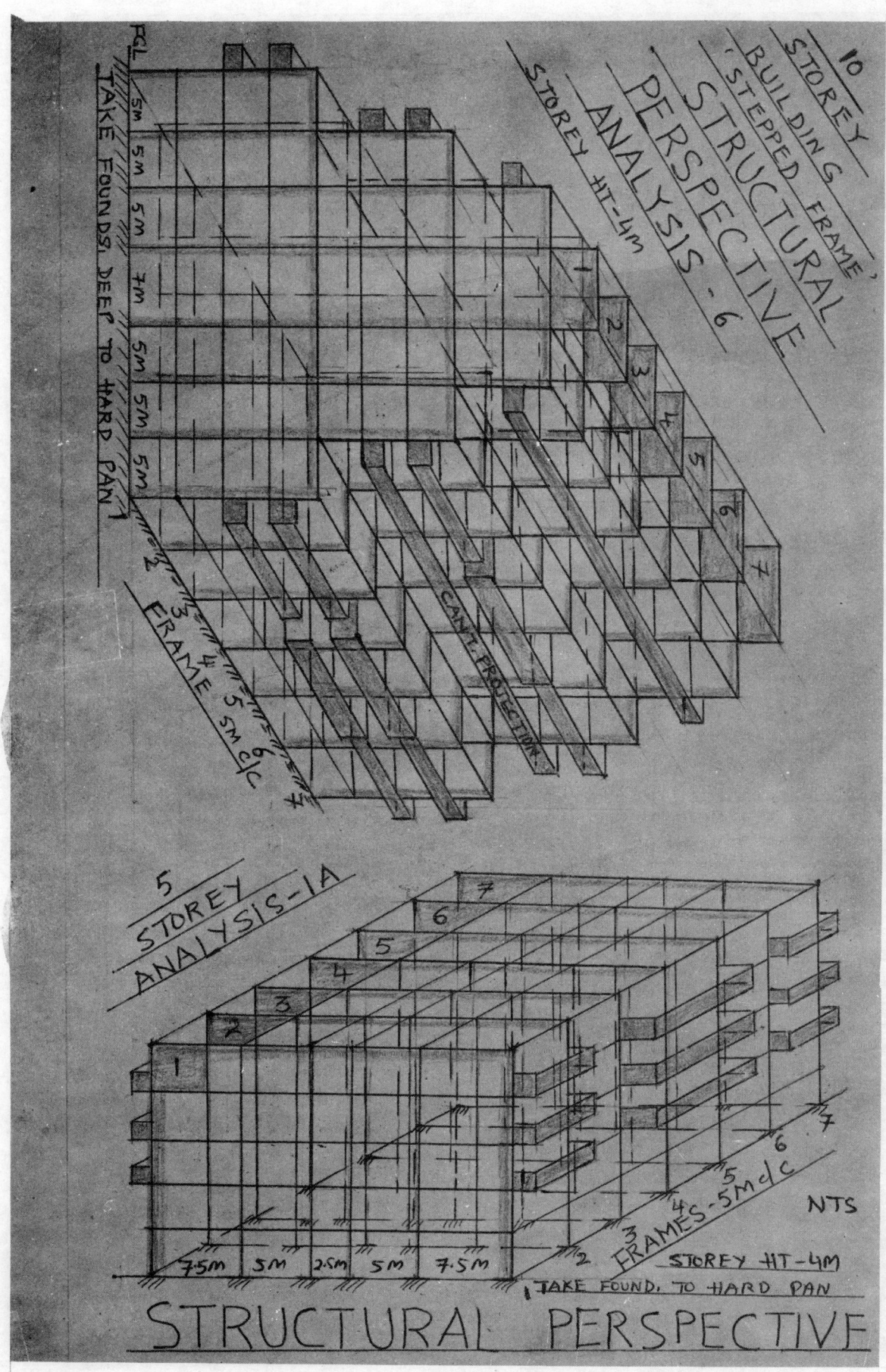

STRUCTURAL PERSPECTIVE

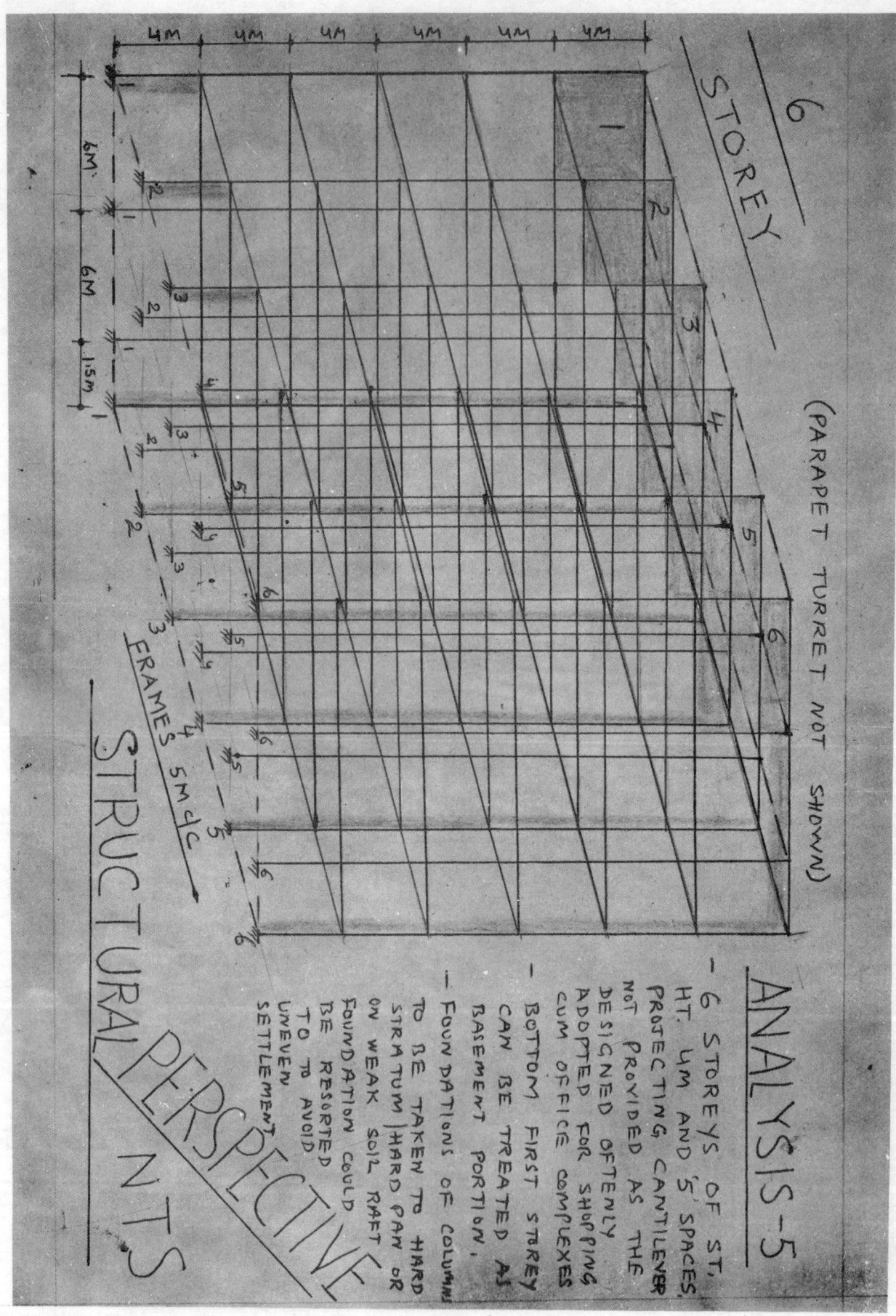

6 STOREY (PARAPET TURRET NOT SHOWN)

STRUCTURAL PERSPECTIVE NTS

FRAMES 5m c/c

ANALYSIS-5

- 6 STOREYS OF ST. HT. 4M AND 5' SPACES, PROJECTING, CANTILEVER NOT PROVIDED AS THE DESIGNED OFTENLY ADOPTED FOR SHOPPING CUM OFFICE COMPLEXES

- BOTTOM FIRST STOREY CAN BE TREATED AS BASEMENT PORTION.

i FOUNDATIONS OF COLUMNS TO BE TAKEN TO HARD STRATUM/HARD PAN OR ON WEAK SOIL, RAFT FOUNDATION COULD BE RESORTED TO TO AVOID UNEVEN SETTLEMENT

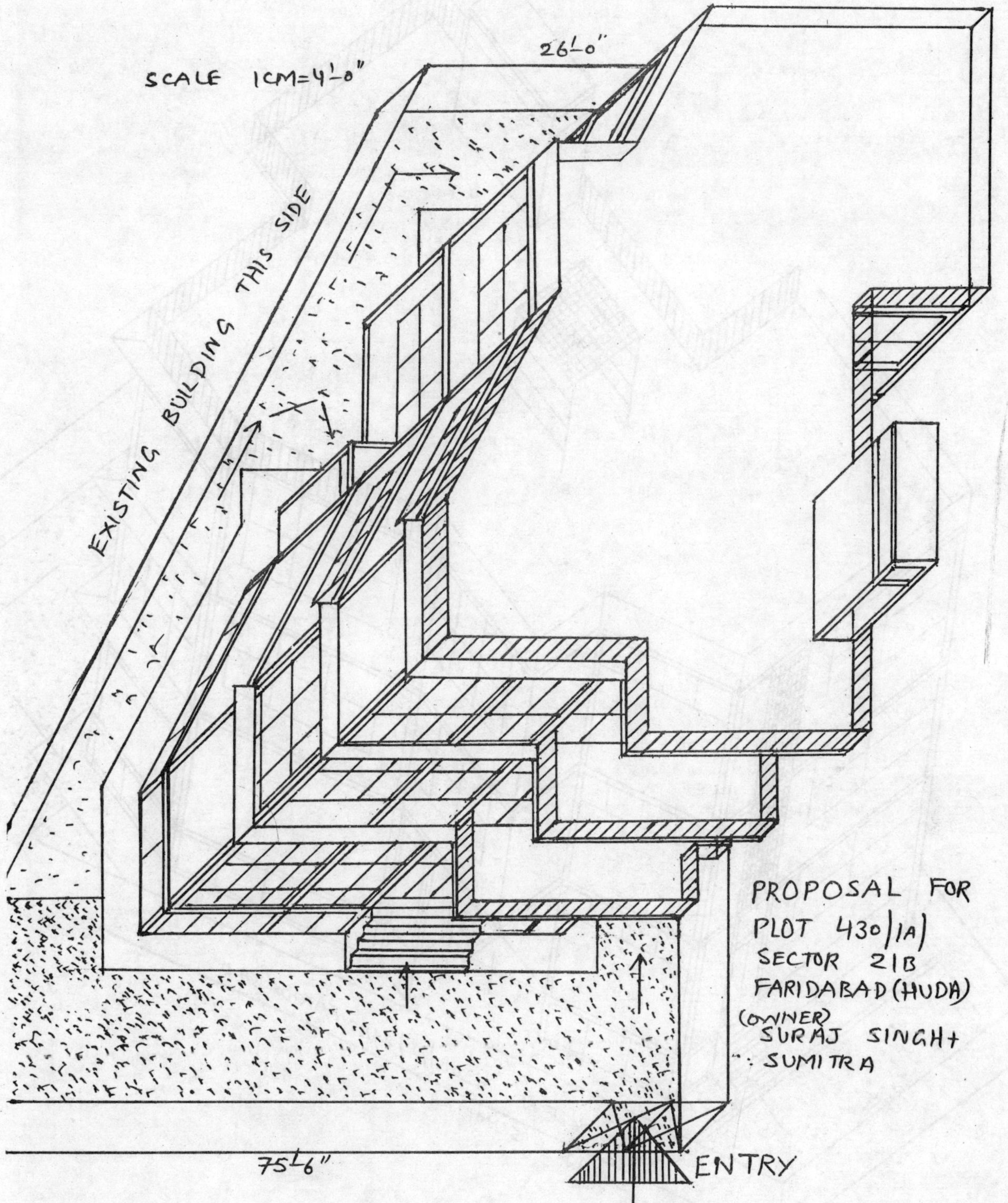

SCALE 1CM=4'-0"

26'-0"

EXISTING BUILDING THIS SIDE

PROPOSAL FOR
PLOT 430/1A)
SECTOR 21B
FARIDABAD (HUDA)
(OWNER)
SURAJ SINGH+
SUMITRA

75'-6"

ENTRY

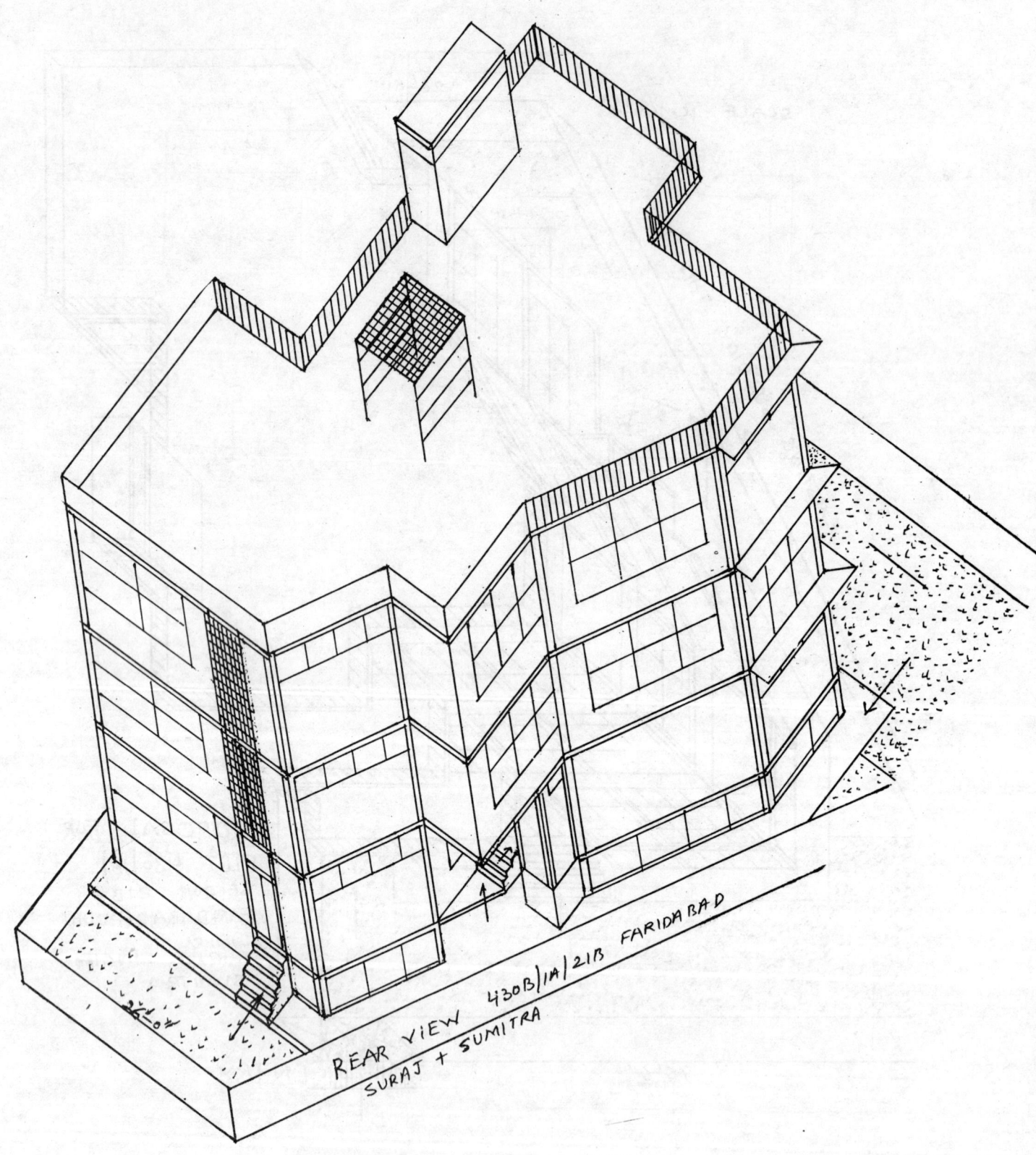

REAR VIEW 430B/1A/21B FARIDABAD
SURAJ + SUMITRA

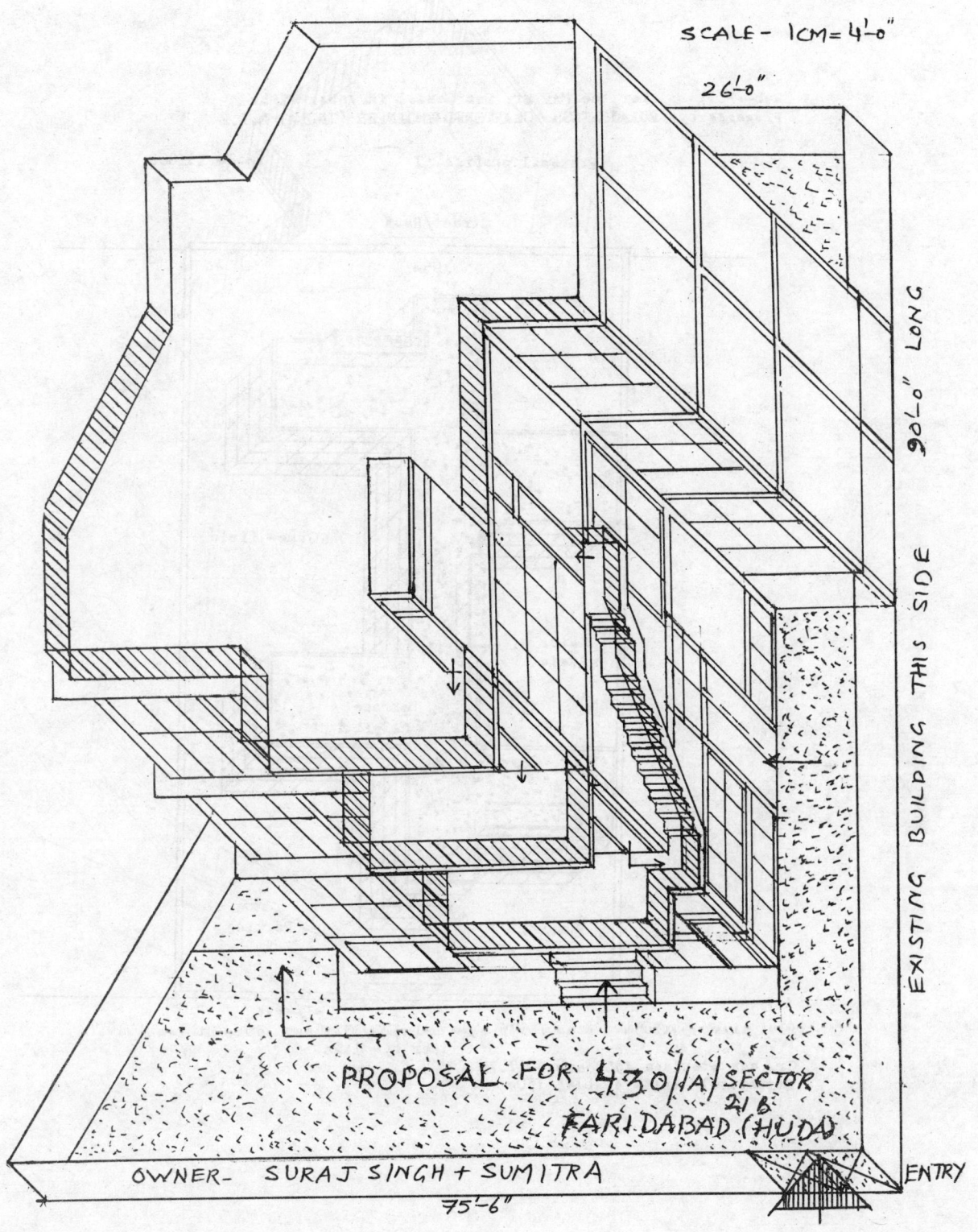

SCALE- 1CM=4'-0"

26'-0"

90'-0" LONG

EXISTING BUILDING THIS SIDE

PROPOSAL FOR 430/1A/SECTOR
21 B
FARIDABAD (HUDA)

OWNER- SURAJ SINGH + SUMITRA

75'-6"

ENTRY

Sub- Proposal for the MAP Sr. Sec School in Ashok Vihar
Proposed by- SURAJ SINGH CHARTERED ENGINEER (INDIA) M.I.E

Proposal profile I   Scale I:I000

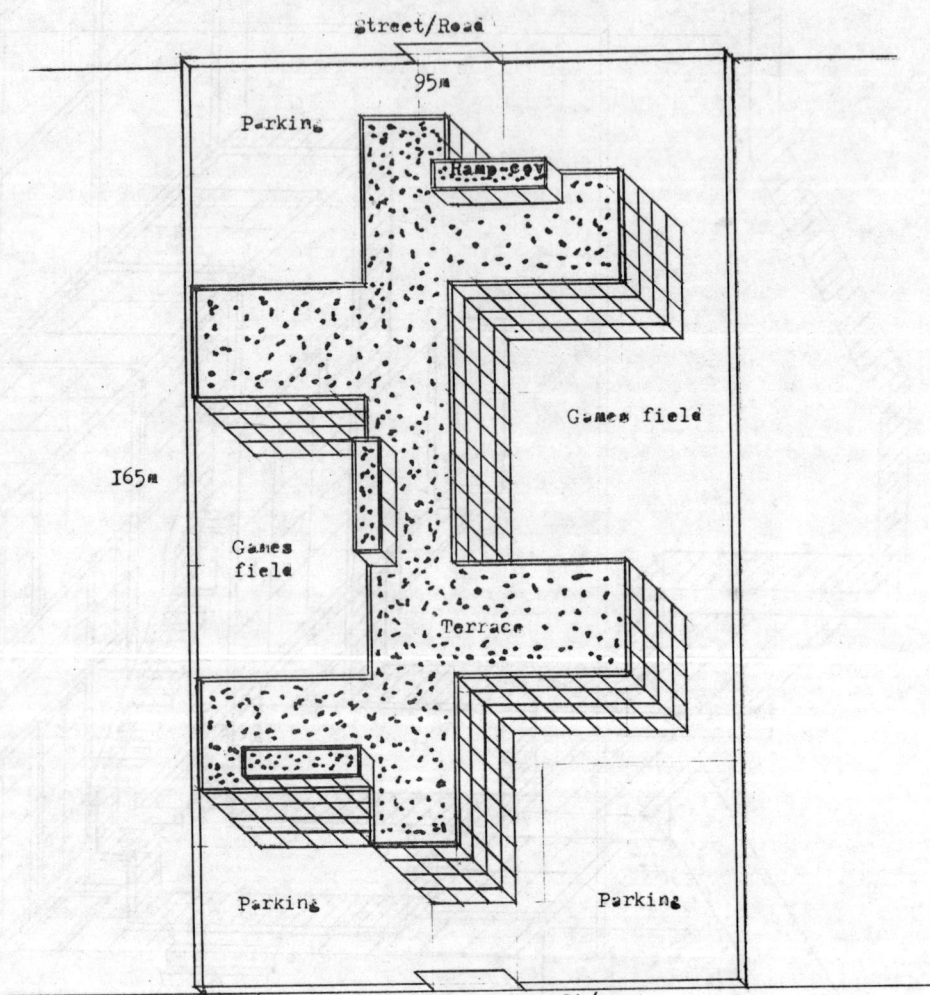

Plot size 95mx165m   Area I5675 sqm   FAR I20, Gr. cov 30%, Ht I4m
                                       Per bye laws
Gr. cov provides 4350 sqm - 27.75%
Structural length of bldg I30m

Proposal profile II   Scale 1:1000

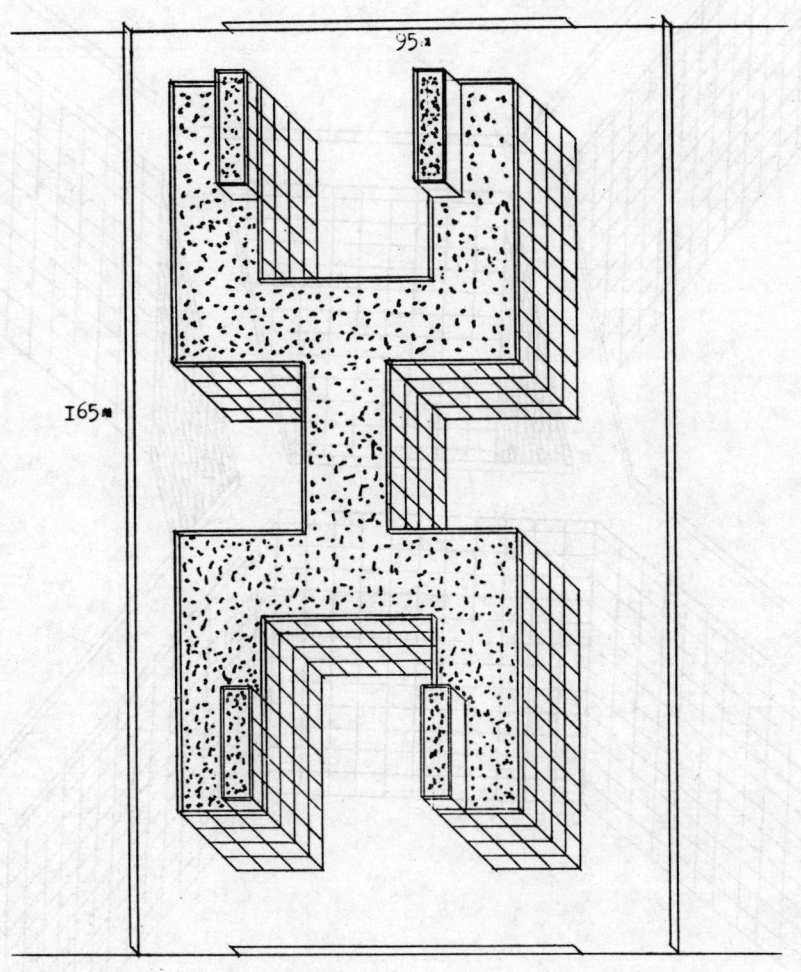

GENERAL INFORMATIONS

Part's Introduction

This part deals with the general exchange of available informations collected from various sources for the augmentation of the reader's available information. I want to disseminate these informations which by virtue of my being abroad from India for many years collected, to these who are in fact in the requisition of this short chapter.

I inquired personally regarding the matter with many colleagues of the engineering discipline about the contents included but could not discover even a few who were well informed about these facts. To suit the purpose I decide to extend the comprehensive matters for the benefit of the readers particularly those in the higher classes of the degree programmes, the technicians, the budding engineers and shall also impart benefit to the professional engineers to some extent to my anticipation.

The inclusions are :

1. General planning and design of swimming pools (Specialist work)
2. Hydromassage and baths (Specialist work)
3. The designer and the profession in Building engineering
4. Professional bodies involved in the building Industry and laws, Requirements
   a. Bye law The Institution of Engineers (India)
   b. The Architects Act 1972
   c. The Institution of Civil Engineers, London — Routes to Membership Chartered Engineer
   d. The Chartered Institute of Buildings, U.K.

Now I am giving an extract of personally conducted interviews/discussions with the engineering students, technicians, senior technicians, Bachelor Engineers, working engineers, local statutory authorities, municipality, etc.

The inquiry revealed that most of the technicians, and the engineers in various levels/classes are not clear of the **title of profession 'Chartered Engineer'**. The technicians and the bachelor engineers could not reply the meaning of the title and some replied that for the first time it is heard. Some technicians answered regarding the AMIE degree that it is a private course/programme which is not a degree but also a degree in engineering. Some even asked me whether it was true that Sec A and B pass shall make a bachelor degree in engineering. Some very confidently answered that after passing sec A and B a certificate shall be issued that the candidate has become AMIE. Some told that just to appear in the UPSC conducted IES and IAS examinations they are doing the course. Some replied because they could not hunt a seat in the university or regular formal colleges of engineering they are studying this programme. Some told that AMIE is not an engineering degree but equivalent to BE. Some bachelor engineers replied that civil engineer got no authority to sign the plans and drawings and only architect could sign. Some of the technicians were not sure about the official recognition of sec A and B as engineering degree. None of these interviewed could reply about the **Institution of Engineers status**. None of the interviewed replied about the meaning of the **designation/title Chartered Engineer (India)**.

I discussed the matter with many architects and none could reply the real meaning. I discussed the matter with many other educated people none replied. I faced the similar situation in some countries of the Middle East and that too posed by the Indian engineers and the technicians. I was effected negatively by an inquiry made by the govt. of Iraq engineer from the Indian project engineer hired by a Jordanian contracting company. The Indian project engineer informed the resident engineer of Iraq that the AMIE was not a degree, the diploma of membership was merely a diploma and not the B.Sc. engineering. The consequence was that the curriculum vitae were not approved till the clarification was done by the Indian authorities by the attestations' endorsement on the instrument. On an inquiry made by the head of buildings, a Lebanese engineer from an American university, about the AMIE from any assistant engineer of Indian nationality in connection with the application moved by one AMIE candidate from India, the Indian technician responded that this was merely a private study and not BE/B.Sc. I also experienced a problem with the civil services officer in the Indian Embassy at Qatar/Doha while applying for the change of profession from the 'service' to the 'Chartered Engineer' endorsement. The guy was not clear about the title of the Chartered Engineer and replied abruptly that he was not all concerned who chartered you but who appointed you. The guy was also not sure about the equivalence/at par of the Sections A and B to the Bachelor degree in engg. Consequently the ICS did not accept the diploma of membership and stressed for the appointment letter only. The endorsement was done 'Engineer (Civil) based on the appointment letter only. The officer was not fully informed about CE. I in fact did not find any difficulty in this connection whenever dealt with UK engineers or the similarly qualified engineers from the countries other than India.

The people in India in general do not know **what is the Institution of Engineers India** and **what is the official status** and the **relevant functions** and **the objectives the Institution has been performing**. The people in general do not advice their wards or relatives to join the Institution for qualifying the Sec A and B for obtaining the 1st degree in engineering due to the fear that the ward shall not be able to get through the examinations even in ten years while there shall not be much difficulty if joined any college of engineering of formal education level or any affiliated college to any recognised university even at the cost of spending money for the capitation amounts of the magnitude of hundreds of thousands. I told many inquiring persons that the pass of Sec A and B of Institution of Engineers India is recognised qualification and at par with the bachelor of engineering degree from any recognised university and the **AMIE/MIE/FIE are the classes of the corporate membership grades for the designation Chartered Engineer (India) an official authorisation for the profession and the business of an Engineer. This title is a mark of competence and excellence in the field of practice and titular professional can serve the public as Engineer according to the provisions of the law**.

Even the statutory municipality and the authorities of India do not care for the lawful validity of the title 'Chartered Engineer' and the rights to practice of the incumbent as 'Licensed Engineer.' I have experienced the MCD (Municipal Corporation of Delhi) where I too was registered as Licensed Engineer despite holding the title 'Chartered Engineer.' I corresponded with the commissioner many times and supported my claims but the municipality full of civil engineers never cared to send me a reply that the Chartered Engineer was acceptable to the department of MCD to practice as Engineer without obtaining local license. I also brought the problem to the notice of the Delhi High Court about the matter and the international practice about the profession of Engineer, without obtaining local license. The High Court ordered the complete change in the bye laws in a case against the MCD, for not controlling the encroachment and illegal constructions commercially on the residential area which were made a news in the paper. I poured my suggestions into the High Court under copy to the Institution of Engineers, MCD, various politicians, etc. Even after the amendment of the bye laws, the MCD, mentioned in the draft that MCD shall license the corporate member (Civil) to practice as Engineer MCD. The architects snatched the authority to practice on the registration with the council of architecture under the Architects Acts 1972 and they use the title Architect as the council is directly under the control of the central government of India on similar grounds as the Medical Council of India but the engineers are not aware of their rights to obtain the statutory professional designations/title which is a requirement of the law in all countries 'Chartered Engineer' awarded after completion of Chartered Professional Review or the written tests depending on the regulations of the different Institutions and once registered as Chartered Engineer of the related discipline, that is considered an international regulated qualification comprising of academic degree and the attainment to the specified standards by the candidate in the field and that reference of the registration is employed for the practice but in this country the practice is entirely different. The engineers in the government are appointed even without the official regulated qualifications by the official national council of India, i.e., the National Council of the Institution of Engineers. In the governments or the private sectors or anywhere, where the law needs the properly qualified engineer, the regulated qualifications should be mentioned to formulate the uniform standards throughout the country. The constitution has provided the Engineers,

Institution but the people and the government shall have to make use of that professional lawful body for the benefit of the society by obtaining the services of the excellent engineers on the similar ground that MCI authorises a medical practitioner to serve the people in the professional capacity and the Institute of Chartered Accountants and others authorise the professionals to serve the people/public as consultants. The powers are being misused by the authorities and it is the only engineer who should realise the legal rights to act as public adviser/consultant instead of always thinking to be public servant where the engineer can share only delegated responsibility and authority. As a public consultant the engineer is absolutely responsible and liable while in level as public servant, the engineer is not liable but likely to be out of job. In the capacity as public consultant or councellor, the society shall be served with higher authority. Rest the engineer shall refer to the law enclosed in last. (ENGINEER IN LAW)

SWIMMING POOLS GENERAL INFORMATION

Planning design and operation of swimming pools in general consideration subject to the domination or dictation by the local codes, formula or bye laws.

Types Residential (homes), Semi-public or semi-commercial (flats, clubs, institutional), Public or commercial (recreational centres).

Residential pool Above ground, part in ground, part above ground. All are subject to regulations regarding electrical supply, sanitation, fencing or enclosure and water usage and vary with the situation.

Above ground pools Usually sold in packaged form fully equipped. The main hydraulic difference between the equipment for the above and in ground pools is the lower filter pressures on the above ground installations due to the absence of the extended pipelines which results in the use of the filters and accessories designed for the above ground application.

Part in ground/part above ground Usually packed pools requiring some excavation on site and transitional between the above two types. Generally resemble the above ground pools in construction but shall be deep enough for diving and large enough to require conventional filters.

In ground pools Many factors influence the design of the permanent in ground pools which include available space, position in relation to sun and wind, deck areas and dividing facilities, safety privacy and security, landscaping and aesthetics, location of equipment room or plant room, total cost of construction and maintenance, Orientation is important with regard to sun, wind, and trees. The combination of direct sunlight and shelter from the wind increases the pleasure derived from the pool and keeps the water warm while the trees add to the maintenance. However, both the sun and shade should be available for the complete enjoyment. The minimum deck width around the pool should be 1 m. The addition of compound walls or fences alongwith the required level of landscape imparts comforts and privacy, promotes safety and does not permit the unauthorised use. Sheltered pools need less maintenance due to the reduction in the air borne dirt, etc.

A pool may be covered for safety reasons when not in use as well as to reduce maintenance, evaporation and loss of chemicals. Pool cover may be simple plastics or fabric tarpaulins to the power driven folding structures.

Indoor pools The construction around an indoor pool should incorporate a vapour barrier to isolate the pool area from the rest of the building. The enclosure should be well ventilated to reduce condensation on walls, windows and ceiling with sufficient space heating to keep the temperature of the air at least 10 degrees F above that of the pool water. Access to outdoor patio is important in good weather. Various types of enclosures are available which permit year long use of outdoor pools.

Construction of residential pools The choice between the several methods of construction for small swimming pools must take into account :

The purpose of the pool (swimming only or diving included), the site conditions, material for construction, cost and maintenance cost, regulations, etc.

Size of pool A pool may be as small as 3.6 m wide and 7.2 m long with a sloping floor giving 1 m depth at one end and 2 m at the other. Fig shows dimensions of rectangular pools used for both swimming and diving but actual shape may be any other free form or irregular to suit the site and purpose of the pool. A safety ledge of 1 m depth may be constructed around the perimeter. Bathing load is not usually a factor in the design of a small private pool but a typical 4.8 m by 10.2 m pool can accommodate as many as 15 persons at one time allowing an area of 3 sqm per person.

Pool size width × length	Dimensions			Capacity
	A	B	C	
4.5 m × 9 m	9 m	3.6 m	2.4 m	73430 litres
4.8 × 10.2	10.2	3.6	3.6	84780
5.4 × 11.4	11.4	3.6	4.8	103330
6.0 × 12.0	12.0	3.6	5.4	124905
7.5 × 15.0	15.0	4.8	7.2	194925

The depth of 2.55 m recommended for diving from a board not more than 0.6 m above deck level and could be reduced to 2 m only for swimming.

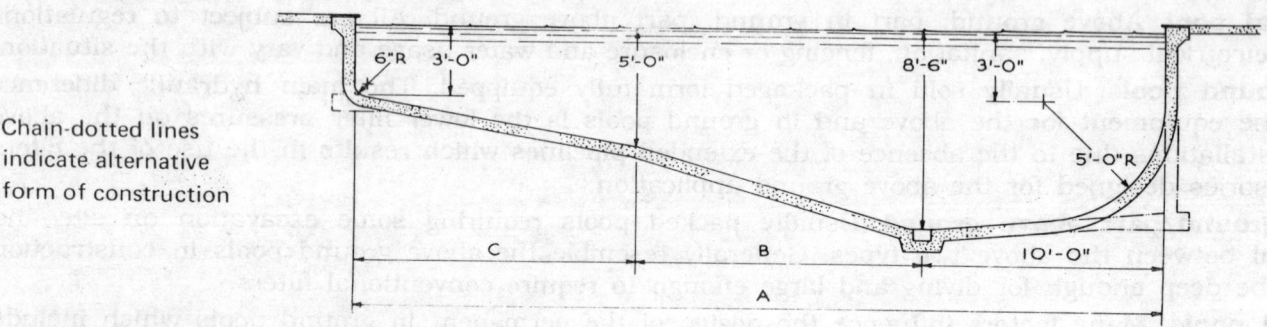

Chain-dotted lines indicate alternative form of construction

SEMI COMMERCIAL AND SEMI PUBIC POOLS

Commercial and public pools While many of the considerations for private and residential pools including several methods of construction also apply to semi commercial and semi public pools. The majority of large pools shall be concrete construction. Fig gives dimensions of rectangular pools which serve as a guide for the design.

It is in the public interest that the regulations governing the design, construction and operations of various types should be much more comprehensive for public, institutional and commercial than for private and residential pools. Consequently these should be taken into considerations from the initiation of the planning. Pools for competitive swimming, diving and water polo should incorporate the requirements of the various athletic organisations.

Size of the pool The size of a small semi commercial or semi public pool can be determined on the basis of 3.4 sqm per person at the maximum capacity after the probable usage has been estimated from the population of the building or the size of organisation.

The attendance at a public or commercial pool depends on such factors as — size of community, weather and pool temperature, convenience and adjacent facilities, competition from beaches and other pools, admission fee.

The following examples, which give the size of the community and the maximum attendance serves as a guide to pool size although local regulations vary considerably and shall be subject to different interpretations.

Population	Maximum Attendance	Pool size
4000	500 persons per day	12 m × 18 m
6000	800	12.6 × 24.6
10000	1100	15 × 30
20000	2000	19.2 × 49.2
40000	3100	22.5 × 49.2

The proportionate attendance at an outdoor public pool is usually higher in a small community than in a large one.

Studies showing the following for a sample of cities with population below 30000

Average daily attendance 2-3% of population

Maximum daily attendance 5-10% of population

Maximum at any time about 1/3 of daily attendance

In the absence of the regional and local regulations, a method of sizing an outdoor pool generously for average days and adequately for busy shall be to allow 1.86 sq m of combined pool and immediate deck area per person at the maximum planned capacity of the entire pool enclosure. The final dimensions should include the length, depth and areas required for athletic competitions. Deep water platforms and other high diving can be provided in bays away from straight away swimming courses.

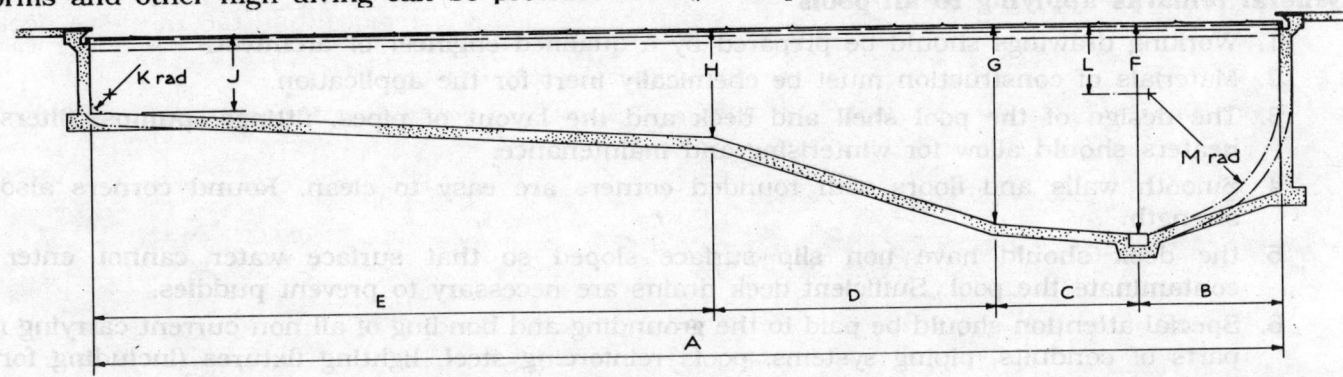

Pool size width × length	Dimensions													Approximate Capacity
	A	B	C	D	E	F	G	H	J	K	L	M	
6 m × 12 m	12 m	2 m	2 m	3.6 m	4.8 m	2.6 m	2.4 m	1.4 m	1 m	2.7 m	1 m	1.5 m	124900 Lt
7.5 × 15	15	2	2	3.6	7.8	2.7	2.6	1.4	1	2.7	1	1.68	195300
9 × 18	18	2.4	3.6	3.2	9	2.7	2.6	1.5	1	2.7	1	1.68	281225
12.6 × 24.6	25	3.3	4.2	4.8	12.3	3.51	3.2	1.5	1.1	0.3	1	2.3	639665
15 × 24.6	25	3.3	4.2	4.8	12.3	3.51	3.2	1.5	1.1	0.3	1	2.3	757000
15 × 30	30	3.3	4.2	4.8	18	3.6	3.2	1.5	1.1	0.3	1	2.3	946250
19.2 × 49.2	49.2	3.3	5.7	4.2	36	3.6	3.2	1.5	1.2	0.3	1	2.3	1816800
22.5 × 49.2	49.2	3.3	5.7	4.2	36	3.6	3.2	1.5	1.2	0.3	1	2.3	2119600

1. Dimensions are nominal only and do not provide for electronic finishing and timing devices on pools used for the competitions. The actual dimensions of such pools should be in accordance with the specifications of the athletic organisations. An olympic size pool is 50 m long and 16 m wide for 6 lanes and at least 21 m for 8 lanes.

2. The following items should be arranged in accordance with the requirements of the athletic organisations

   The extent of deep water in front and on both sides of the board or platform. The spacing of multiple diving boards and platforms. In addition, the width of the deck should be sufficient to accommodate towers of standard construction so that the expense of special towers with side ladders is avoided. If a board with a quick adjustable fulcrum is mounted on a platform, there should be a safety handrail on the side where fulcrum adjustments are made. The maximum gradient of the pool bottom in the diving well must not exceed 1 V : 2 H.

3. The depth at the shallow end depends on the purpose of the pool. Bathing pools used by children can be shallower than the minimum of 1 m. The minimum depth of water for competitive swimming is 1.07 m.

Minimum requirements for diving

Dimension F	Suitable for purpose
2.6 m	Diving board not more than 0.6 m above water level
3.6	1 m diving tower
3.9	3, m diving tower, 5 m diving platform
4.2	6.5 m diving platform
4.5	7.5 m diving platform
4.8	10 m diving platform

General remarks applying to all pools

1. Working drawings should be prepared by a qualified engineer or architect.
2. Materials of construction must be chemically inert for the application.
3. The design of the pool shell and deck and the layout of pipes, fittings, pumps. filters and heaters should allow for winterising and maintenance.
4. Smooth walls and floors with rounded corners are easy to clean. Round corners also add strength.
5. the deck should have non slip surface sloped so that surface water cannot enter and contaminate the pool. Sufficient deck drains are necessary to prevent puddles.
6. Special attention should be paid to the grounding and bonding of all non current carrying metal parts of conduits, piping systems, pools reinforcing steel, lighting fixtures (including forming shells), junction boxes, fences within 3 m of the pool, ladders, diving board supports, and the like, which must be bonded together and connected to a common ground. This and other important electrical requirements for swimming pools are dealt with in article incorporated with the National Electrical Code.

A complete pool installation comprises :

1. The pool shell as described previously.
2. Functional equipment in and around the pool such as ladders, underwater lights, life rings, and diving boards.
3. A recirculation system for filtering, heating (if necessary) disinfecting the pool water.
4. Maintenance equipment for removing dirt and debris from the pool surface, walls and the floor.

Functional equipments

Access — 2 tread ladder for water depth of less than 1.2 m
3 tread ladder for water depth of 1.2 to 1.5 m
4 tread ladder for water depth of more than 1.5 m
Stair rail for use with built in stairs
Grab rails for use with recessed steps.

Diving — Residential spring boards, 1.2 to 2.4 m long at deck level
Residential diving boards, 2.4 to 3.6 m long on standards 0.45-0.6 m above water level
Pool slides
Heavy duty diving boards, 3 and 3.6 m long. on stands 0.6 m and 1 m above water level
Diving boards, 4.2m and 4.8 m long on towers with adjustable fulcrum
1 and 3 above water level
Diving platforms at 1.5, 2 and 3 m above water level

Refer to the figures regarding depths of water required for diving

Competition   Rope, floats, cup anchors and rope hooks for racing lanes starting platforms

Stanchions posts

Safety   Life ring and life hook

Life guard chair (one chair per 185 sqm of pool area/surface)

Rope, floats, cup anchors and rope hooks for safety lines

Gas mask (where gas chlorinator is used)

Under water lights and ground fault detectors

(A minimum of 7.2 watt per sqm of pool surface is necessary to provide safe, attractive underwater lighting. Lights should be directed away from the house or deck area to prevent glare and the best illumination should be at the deep end.)

Recirculation System The recirculation system includes all the pipes, fittings and equipment concerned with filtering, heating and disinfecting the pool water.

With pool water treatment, the same water recirculates continually, thus providing multiple opportunity for the removal of suspended dirt, the destruction of organisms, and the maintenance of an equable temperature. At the same time, the water is subject to continual recontamination by leaves, seeds algae, bacteria, hair dead skin, lint, cosmetic oils, food, clay, sand and such like, Nevertheless, the net result is that the water retained in a swimming pool represents valuable savings compared with the filtering, heating and disinfecting raw make up water. The principal components of the recirculation system perform as follows :

The main outlet is located at the deepest point of the pool on the end of the suction piping from the filter pump drawing water from the bottom of the pool where dirt accumulates and permits complete drainage of the pool for maintenance purposes. It often incorporates a **hydro relief valve** for admitting ground water to an empty pool to prevent the shell from floating should the ground water rise above the bottom of the pool. The installation of a hydro relief valve is strongly recommended for all concrete and fibre glass pools.

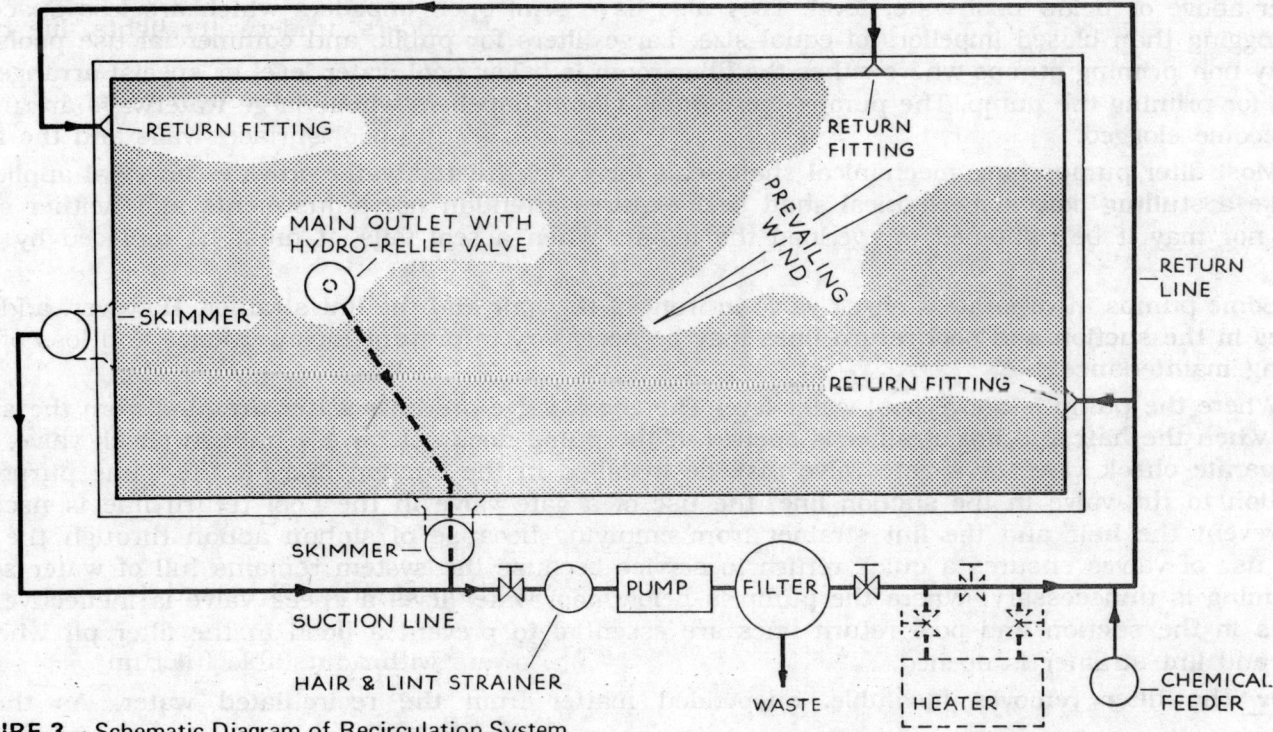

FIGURE 3 — Schematic Diagram of Recirculation System

An **automatic skimmer** is mounted on the pool wall at water level and connects to a branch of the suction piping from the **filter pump**. Its purpose is to promote the surface flow across the pool. Besides drawing in water which requires most treatment because of exposure to sun, wind, and rain, it also removes floating dust and debris, leaves and oil films before they can accumulate on floors and walls. A combination **skimmer** connects directly into the line between the main outlet and the filter pump. It incorporates a trimmer valve which divides the total flow through the filter between that across the surface of the pool and that through the main outlet. It makes a further contribution to simple pipework if it also provides a connection for the hose of a **vacuum cleaner**. The number of skimmers required depends on the size and purpose of the pool. Their arrangement on a small outdoor pool should promote surface flows across the pool in connection with the return fittings and the prevailing wind. **Skimmers** are essential for a pool which receives only a periodic attention, whereas a large pool with full time attendants may have an overflow gutter instead of skimmers. The over flow gutter extends around the pool at water level and incorporates drains at frequent intervals. These feed into a **recirculation tank** connected to the filter pump. The principal function of the **recirculation tank** is to salvage treated water which would otherwise go to the sewer. It is arranged to prevent air being drawn into the filter pump allows convenient introduction of raw water ahead of the filter and other apparatus. The recirculation tank should have a capacity of 40.75 litres for each sqm of pool surface.

Hair and lint strainers are provided in the skimmer(s) and at the inlet of the filter pump. They catch hair and large particles which would otherwise accumulate inside the system to clog pump, piping or filter. Emptying the hair and lint strainer is part of daily maintenance to keep the recirculation system working at top efficiency.

Filter pump (recirculation pump) The centrifugal type filter pump develops capacity and pressure to suit the size of the filter and the resistance of the recirculation system. It may be of either self priming or non priming design. A self priming pump mounted at or above deck level is capable of exhausting air from the suction system and raising water from the pool to establish and maintain operation. It can also operate the vacuum cleaner.

A non-priming pump operates satisfactorily where mounted below deck level, because the pool water keeps the pump and suction system full of water at all times. It is unsuitable for operating a vacuum cleaner which requires a separate vacuum pump in this case.

In practice many small filter systems use self priming pumps because they adopt to installation either above or below pool water level. They also have semi open impellers which are less susceptible to clogging than closed impellers of equal size. Large filters for public and commercial use pools often empty non priming pumps where either the filter room is below pool water level or special arrangements exist for priming the pump. The pump incorporates closed impellers whose large waterways are unlikely to become clogged.

Most filter pumps have mechanical shaft seals for which treated water provides an ideal application. Unlike a stuffing box, a mechanical shaft seal requires attention nor adjustments and neither shall it leak nor may it be tightened to overload the motor. When a seal fails, it must be replaced by a new one.

Some pumps incorporate a check valve in front of the hair and the lint strainer. However, additional valves in the suction and pool return lines may be necessary to prevent inconvenience and loss of water during maintenance.

Where the pump is above pool water level The check valve prevents water draining from the suction pipe when the hair and lint strainer is opened. If the pump does not have a built in check valve, either a separate check valve or a gate valve may be installed in the suction line for the same purpose. In addition to the valve in the suction line, the use of a gate valve in the pool return line is necessary to prevent the hair and the lint strainer from emptying because of siphon action through the filter. This use of valves ensures a quick return to service because the system remains full of water so that repriming is unnecessary. Where the pump is below pool water level a check valve is ineffective. Gate valves in the suction and pool return lines are essential to prevent a flood in the filter pit when the hair and lint strainer is opened.

Filter The filter removes insoluble, suspended matter from the recirculated water. As the dirt accumulates within the filter, the filter pump cannot maintain the required rate of flow against the increased resistance. Consequently, the filter requires regular cleaning or backwashing to maintain its efficiency. There are four principal types of filters for swimming pools.

1. Vacuum diatomaceous earth
2. Pressure diatomaceous earth
3. Rapid sand (pressure)
4. High rate sand (pressure)

SCHEMATIC DIAGRAM FOR RECIRCULATION SYSTEM ABOVE

Pressure diatomaceous earth filter A typical pressure d.e. filter contains an arrangement of hollow elements covered with filter cloth. The pump forces water through the elements from the outside to a central manifold. The filter cloth serves as a support for a coating of diatomaceous earth which forms the filter medium and traps the dirt. By reversing the flow, both dirt and the diatomaceous earth are flushed to waste during backwashing so that the coating procedure must be repeated when the filter is returned to service. The operating cycles can be prolonged by adding diatomaceous earth at intervals continually on large filters and by using mechanical means periodically to disturb the coating. Diatomaceous earth filters generally require more maintenance in time and materials than sand filters.

Rapid sand filters A rapid sand filter contains a layer of fine silica sand supported on a graded bed of gravel. The filter pump forces water slowly 122 ltr to 204 ltr pm per sq m down through the filter bed where the top layer of sand traps the dirt. By using an upward flow of water for backwashing at 4 or 5 times the filtering rate the sand can be agitated to release the dirt which is flushed to waste.

High rate sand filter A high rate filter contains a deep bed of very fine silica sand supported on mechanical under drain. The filter pump forces water down through the filter bed at a high rate 612-816 ltr pm per sqm where the dirt penetrates the sand to a considerable depth.

This gives the high rate sand filter a much greater dirt holding capacity than a rapid sand filter of equal size. Backwashing is carried out at about the same rate as filtering 122 ltr pm per sqm which makes pump selection much easier than for a raid sand filter.

Filter area This is the surface area of the filter medium exposed to the flow of unfiltered water. For a sand filter it equals the cross sectional area of the tank. For a diatomaceous earth filter it equals the total useful area of all the elements.

Flow rate This is basic to the type of filter with values shown below :

Type of filter	Flow rate/sqm of filter area
Diatomaceous earth	40.72 ltr pm
Rapid sand	122-204 ltr pm
High rate sand	612-816 ltr pm

Filter rate = Flow rate × filter area

Turn over time represents the number of hours required to recirculate the total capacity of the pool through the filter = Pool capacity/Filter rate and shall be subject to the local regulations but for guidance the following values could be used.

Class of pool	Contamination	Turn over time
Residential	Light	8-12 hours
Apartments, hotels, clubs	Moderate	8
Public, Institutional	Heavy	6-8
Institutional (children)	Very heavy	4
Wading pools	Extra heavy	1.5-3
Health spas, roman baths	Hot water, Extra heavy	1-2

Wading pools and roman baths are often adjacent to larger swimming pools but they should always have separate recirculation systems.

Filter size Specifications for filters usually relate filter models to pool capacities and turn over times determined on the following basis.

1. Determine the filter rate from the pool capacity and turn over time required from :

    Flow rate = Pool capacity/Turn over time

2. Calculate the filter area from the flow rate for the type of filter to be used from :

    Filter area = Filter rate/Flow rate

3. Select the size and number of filters to match the filter area required.

Pool heater The need for a pool heater depends on the purpose of the pool and the general climate. A heated pool is more enjoyable at all times and extends the season at both ends by at least a month. The heater is connected into a loop in the return line from the filter and may either be gas fired, oil fired or electric or consist of a heat exchanger where stream or hot water service is available. The heater is more vulnerable to lime deposits and corrosion than pool equipment. As a result it should be checked frequently during the early weeks of operation to ensure that it suffers no adverse effects from the pool water chemistry.

Chemical feeders Chemical for disinfection and other purposes are usually added after the water has passed through the filter. There are so many varieties of feeding equipments that no attempt is made to deal with them except to emphasize the importance of applying any equipment in strict accordance with the instructions which accompany it as manufacturer's instructions.

Return fittings The return fittings in the pool wall below water level control the flow rate and direction of the filtered water entering the pool. Their arrangement on a small outdoor pool is usually to promote surface flows in conjunction with the skimmers and the prevailing winds. Directional return fittings have the advantage of adjustments to set the entire body of water circulating in the same direction as the water entering the pool.

Hydro air fittings One or more of the return fittings on a residential pool may be of the hydro therapy variety. To obtain the maximum effect for the hydromassage, the filter is by passed and alternative return fittings are closed thus diverting the entire pump output directly to the hydro air fittings. While the foregoing is an economical arrangement, most filter pumps are too small to give effective hydro message action in a swimming pool. In addition, users often forget to put the filter back into operation afterward or fail to open the other return fittings thus upsetting the condition and circulation of the water. The best arrangement is a separate hydro air system having its own pump and recirculation pipework.

Maintenance equipment Which require consideration during the design and construction of pool.

Vacuum fittings Vacuum fittings are necessary where a permanent vacuum line is installed for maintenance purposes. The vacuum line may connect either to the suction system of the filter pump on small pools or to a separate vacuum pump on large pools. The vacuum fitting is mounted on the pool wall below water level and is kept plugged except when the hose of the vacuum cleaner is connected to it by means of a hose to wall adapter.

The vacuum cleaner should discharge to waste only when handling heavy dirt which would impose a heavy load on the filter. During normal maintenance it should discharge to either the filter or the recirculation tank to avoid the loss of treated and heated water.

Vacuum fittings are not essential equipment. Many small pools need no separate vacuum line as the vacuum cleaner can be connected into the skimmer(s) to reach all parts of the pool. On large pools, a portable vacuum cleaner pump with a long power cord is used in preference to a stationary vacuum pump with a permanent vacuum line.

Automatic cleaning devices are available for performing most of the day to day maintenance within the pool. These require special provision for electric or hydraulic power.

Hydromassage A relaxing therapy obtained with a unique process of mixing air and water giving a wide range of beneficial effects the result of which is combined action of three factors, water pressure, air agitation and temperature. An incorrect balance of these factors cancels the desired effect. Only well proven and scientifically tested processes guarantee a proper and scientific hydro massage action. The flow of small bubbles of predetermined size is actually mixed with the water flow, not merely added to it. At a 35-36 degree C temperature, the blood circulation and body muscles structure is stimulated as with a penetrating and thorough massage carried out by experienced fingers.

Pool water chemistry As untreated water provides an environment which encourages the rapid growth of organisms, the water in a swimming pool must be continually filtered and treated to prevent the creation of a health hazard. Its quality must equal or exceed the standard prescribed for drinking water.

While filtration makes the water clear and attractive by removing suspended matter, chemical treatment of invisible contamination is vital to the safe operation and proper maintenance of the pool. The purpose of chemical treatment is to disinfect the pool water by killing algae, bacteria and viruses. As the process upsets the chemical balance, secondary treatment is necessary to prevent discomfort to bathers and damage to the pool and its equipments. Additional treatment may be necessary to precipitate dissolved material/minerals for removal by either the filter or the vacuum cleaner. The following discussion relates principally to the application of chlorine and its compounds for disinfecting swimming pools because these are the chemicals in the common use. The initial dose of chemical must exceed the chlorine demand of the pool which is the amount required 1. to kill all bacteria present, 2. to oxidize all other organic matter 3. to satisfy chemical reactions with inorganic substances. This phase of treatment is complete when tests show the existence of a stable Chlorine residual for attacking the continual contamination from bathers and other sources. This residual comprises very active uncombined chlorine (free residual) and less active chlorine in chemical combination with other substances (combined residual). Chlorine must be added continually to maintain the residual of chlorine but the amount varies considerably from day to day depending on the weather and the bathing load. The addition of the chlorine gas to water results in the formation of hypochlorous and hydrochloric acids. Hypochlorous acid is very unstable and breaks down readily for disinfection purposes. Hydrochloric acid contributes nothing to the treatment process and so increases the acidity of water. This change in acidity is a typical side effect of any process of disinfection and requires additional chemicals to counteract it. Excessive counteraction in this case would result in too much alkalinity. Acidity and alkalinity are measured by the pH concentration. This is a logarithmic system with values below 7.0 for acidity above 7.0 for alkalinity, 7.0 being neutral (neither acid nor alkaline). Acid water fosters the corrosion of metal parts which can lead to the expense of replacing heaters, filters and other equipments. Alkaline water causes mineral deposits and interferes with disinfection, cloudy water which irritates bathers' skins, eyes and breathing passages is typical of a pool where the pH concentration is too high or too low.

The foregoing shows that correct levels of chlorine residual and pH concentration are essential to satisfactory pool operation. Pool owners use simple test kits to check chlorine residual and pH concentration daily. Acceptable values fall within the limits given below.

Chemical factor	Swimming pools	Drinking water
Free chlorine residual with chlorine and hypochlorites	0.3 to 0.7 ppm	0.2 to 0.6 ppm
Cyanurates	1.0 to 2.0	
pH concentration	7.2 to 7.6	7.2 to 8.0

(ppm — parts per million)

A third factor in a pool chemistry is the total alkalinity. This refers to the quantity of alkaline minerals in water, which should be from 80 to 100 ppm. It becomes significant in a residential pool only when a chronic condition persists in spite of normal levels of chlorine residual and pH concentration. Total alkalinity must not be confused with pH concentration which refer to the balance between alkaline and acid substances.

A steady reading for chlorine residual is only possible by the continual addition of chemicals to replace the chlorine expended to disinfect contamination from raw water, bathers and air borne sources and that dissipated by evaporation and sunlight. The methods of feeding chlorine into the water to meet the chlorine demand vary with the chemicals used and the forms in which they are supplied. Several systems of chlorination and pH control are described in the following discussions.

Chlorine gas Cl_2 As it is elemental, chlorine gas is the most concentrated and inexpensive form of chlorine but such reasons as safety, cost of equipment and the type of secondary treatment limit the use of that to the large pools. A gas chlorinator allows the chlorine to flow at a constant rate from a high pressure cylinder into the return pipe from the filter. The chemical reaction between the chlorine gas and the water produces hypochlorous and hydrochloric acids. Hypochlorous acid is the useful disinfectant. Hydrochloric acid which is an unwanted product requires secondary treatment because of its adverse effect on the pH concentration. This treatment consists of a chemical pot feeding a solution of soda ash (sodium carbonate Na_2CO_3) into the water entering the filter. The alkaline soda ash neutralises the acid which would otherwise reduce the pH concentration. Soda ash is required at the rate of 1.25-1.5 lb for 1 lb of chlorine consumed. The gas cylinder is usually mounted on a platform scale for checking the weight of gas used each day.

Calcium Hypochlorite $Ca(ClO)_2$ It is a dry white compound available in granular and tablet forms under various trade names. It releases 70% of its weight in free chlorine when dissolved in water but with a different effect on pH concentration than chlorine gas. As a result of the alkalinity of the calcium by product the pool water will become alkaline so that it requires the addition of muriatic (20 deg Baume hydrochloric) acid or sodium bisulphate to restore the balance. The granular form is prepared for use by dissolving it in water. As this results in a precipitation of calcium carbonate, an extra container is necessary for mixing purposes so that only the clear liquid is allowed to pass through the chemical feed pump (hypochlorinator) for discharge into the return line. The importance of using only the clear liquid is that residue of calcium carbonate would clog the feeder system and form lime scale on the walls and floor of the pool. Another method which reduces calcium deposits in the pool is to add soda ash to the same container as the calcium hypochlorite. This converts the calcium hypochlorite into sodium hypochlorite. The same precaution is necessary to ensure that only the clear liquid passes through the chemical feed pump (hypochlorinator). The heavy residue of calcium carbonate should be discarded. Tablets of calcium hypochlorite dissolve more slowly than the granules with the advantage of providing a more constant and even supply of chlorine when thrown into the pool or used in a dissolving basket. These slow dissolving tablets may also be placed in the skimmer basket.

Sodium Hypochlorite NaOCl It is the principal constituent of liquid household bleaches which usually contain 10-15% of available chlorine. It can be either fed though a hypochlorinator (chemical feed pump) in commercial strength or diluted or poured directly into the pool.

Commercial solutions of sodium hypochlorite contain sodium hydroxide to prevent loss of chlorine during storage. As sodium hydroxide is alkaline, it reacts with acids in the pool to make the use of soda ash unnecessary. Should the pool be already alkaline, the pH concentration will increase which will require concentration by muriatic (20 deg Baume hydrochloric) acid or sodium bisulphate.

Cyanurates These are complex compounds of cyanuric acid which are readily soluble in water. They have a slightly acidic effect so that the addition of acid for pH control is seldom necessary. As cyanurates are more expensive than conventional chemicals, initial treatment of the pool with 30 ppm of cyanuric acid is essential to prevent the dissipation of chlorine residual due to exposure to the ultraviolet rays of sunlight. This dissipation with an untreated pool may amount to as much as 70% of the chlorine on hot sunny days. When using cyanurates the free chlorine residual is maintained between 1.0 - 2.0 ppm. This level of stable chlorine residual is very effective against algae. The advantage of cyanurates make them ideal for the residential pools. The pool owner should use the test kit which shows values of free chlorine residual up to 3.0 ppm. The pool serviceman should also have a test kit for cyanuric acid up to 100 ppm.

Super chlorination It is necessary when the pool water becomes cloudy or discoloured when a false residual test is discovered or when signs of algae appear. It involves closing the pool, applying heavy doses of chemicals to raise the chlorine residual to 5 ppm or more and allowing the excess residual to dissipate overnight.

Break down chlorination Usually applies to outdoor pools with a high organic contents. When the initial dose of chemicals react with the pool water to satisfy the original chlorine demand, most of it is absorbed by organic matter to form chloramines and other chlorine compounds. These substances are responsible for chlorine odours and tastes. The addition of small amounts of excess chlorine are necessary for normal pool, operation, but in order to eliminate tastes and odours it is sometimes necessary to raise the free chlorine (say to 5 ppm). The free chlorine then starts to oxidise the chloramines and other chlorine compounds. As these reactions absorb much of the excess free chlorine, the residual falls suddenly to a lower value and the chlorine compounds disappear together with their tastes and odours. This is called the break point at which most of the chlorine in the water is in the free state. Other chemicals for the disinfection of swimming pools includes iodine and certain of their compounds. Other methods of disinfecting water are the use of silver slats, the electrolytic production of silver ion and ultraviolet radiation but all are either impractical or unsatisfactory for swimming pools.

Other pool chemicals : Alum Alum sulphate $Al_2(SO_4)_3 . 18 H_2O$ is used as a coagulant with sand filters to give improved clarity to the water. It dissolves slowly to make an acid solution which neutralises in reaction with the pool water to form aluminium hydroxide. Under the right conditions, aluminium hydroxide precipitates in white flocs which accumulates as a gelatinous mat on the filter bed. This mat will catch particles which would otherwise pass through the sand. The successful use of alum for flocculation requires critical alkaline conditions with a pH of 7.4 to 7.6. Use 4-5 oz of lump alum per sq ft of the filter area. Put the lump alum in the basket of the skimmer or the hair and lint strainer after backwashing the filter.

During the initial disinfection of a pool containing water with a high iron content, the chlorine will cause the iron to precipitate as red iron oxide which will discolour and cloud the water. If powdered alum or a saturated alum solution is sprinkled on the surface of water and left overnight, it will coagulate the turbidity and hasten the settling process, so that the precipitate can be vacuumed from the pool bottom in the morning. The pool should not be used during the treatment.

Algicide The safest chemicals for controlling algae in swimming pools are proprietary preparations made for the purpose. The ideal algicide kills existing algae, prevents new algal growth, creates no problems with filter or pool operation and is harmless to pool users.

Chemicals for cleaning filters When sand is returned to the pool an inspection of the filter bed of a sand and gravel filter may reveal the formation of mud balls. This condition can be corrected by purging the bed with lye (sodium hydroxide NaOH) or calcium hypochlorite. Open the top of the filter and drain the water below the top of the sand. Break up the sand and mud balls and allow the top layer to dry before spreading chemical evenly over the bed in the following amounts

Lye 1 lb. per sq. ft. of filter area

Calcium hypochlorite 0.5 lb. per sq. ft. of filter area.

Be sure that the top of sand is dry before applying the chemical otherwise the chemical will be emitting dangerous fumes. Close the filter and raise the water level to about 50 mm above the sand. Allow the chemical to stand in the filter from 4-6 hours, then backwash thoroughly continuing for at least 10 minutes after the water runs clear.

When the sand bed is partly calcified due to hard water the use of muriatic (20 deg Baume hydrochloric) acid or sodium bisulphate as an acid wash may correct the condition. The acid is applied like the purging chemicals just described at a rate of about 2 lb per sq ft area. Proprietary preparations are also available for this purpose and may include detergents and sequesting agents for removing oils and cosmetics. If calcination is so bad that the bed is no longer porous, it is better to change the sand than to use an acid wash which would corrode the tank and fittings.

Heat loss retardants These are proprietary products which spread across the pool surface in molecular thickness to retard evaporation and conserve the heat in the pool water. They serve to reduce steaming and condensation around indoor pools. The ideal retardants should be colourless, odourless, non-toxic and non-corrosive and should have no adverse effect on disinfection.

Sequesting agents are usually proprietary preparations which separate the insoluble scale and scum forming minerals in pool water and convert them into harmless soluble forms. An important application is to prevent the accelerated corrosion, scale formation and staining which may occur with heated pools.

Muriatic acid is the commercial name for 20 deg Baume hydrochloric acid HCl. It is used for reducing the pH concentration of pool water but requires careful handling by trained people. It is also used as an acid wash for calcified sand filters and for scrubbing the pools walls to get rid of established algae. Sulphuric acid H_2SO_4 is not recommended for use around swimming pools.

Sodium Bisulphate NaH$_2$SO$_4$ is an acid salt used for pH control and cleaning purposes. It is known as dry acid because it serves as a safe alternative to muriatic acid. It is available in the pure form and in combination with other chemicals under various trade names.

Always exercise greater care when mixing chemicals Add chemicals to water and never water to chemicals. Never add acids and alkalise to the same container. Follow the instructions given on the package. Keep children away from the pool chemicals.

pH concentration Salt water pools with water of natural origin have a high pH concentration due to the mineral content (approximately 3.5% by weight). As a result the reduction of pH to recommended values may not be economical.

Reasons why it may be difficult to adjust the pH concentration in a new pool.

1. the raw water may be low or high pH concentration by nature
2. new concrete is very alkaline
3. many of the alkaline salts on the walls of an unpainted concrete pool will dissolve in the first fill of water.

Mineral deposits Scale formation by hardness minerals is a common problem among pool owners. The accumulation of scale on the pool structure and on pipes, heat exchangers, filters and ladders include the insoluble carbonates, aluminates, silicates and sulphites of calcium, iron and magnesium. Calcium carbonate (lime) is the largest component. Water with high values of temperature, pH, dissolved solids, alkalinity and calcium hardness encourages scale formation. pH concentration is the easiest factor to control. A pool heater should handle a large flow of water with a low temperature rise to avoid mineral deposits within the heat exchanger. Ten gallons of water heated through 5 deg F will add the same amount of heat as five gallons hated through 10 deg. F.

Red water Red water is usually caused by iron oxide (rust). It is sometimes due to a low pH concentration which fosters corrosion of steel pipes, fittings and tanks. Where concentrations of iron in swimming pool water exceeds 0.3 ppm, damage to the equipment and finish of the pool is likely to occur. Most iron bearing waters are clear as they flow from the source and the discolouration occurs when the iron is oxidized either by exposure to the air or by chlorination. The method of removing this discolouration by using alum has already been described in Alum. Where the local water supply stains sinks and bathtubs, either fills the pool from another source which is iron free or instal an iron removal filter (A portable iron removal filter). A portable iron removal filter may be used during the filling operation and the removal of iron from the subsequent make up water could be effected by precipitation in a surge tank connected to the suction of the filter pump. Alum and a heavy dosage of chlorine are fed continually into the surge tank to cause precipitation of the iron which is retained on the sand bed as the water passes through the filter. If it is clear, the water from the source may be used for back washing, provided that chlorine and alum are not supplied to the surge tank during the process. Where a gas chlorinator is in use, the raw water used to operate the chlorinator may be sufficient for make up purposes.

Brown water Brown water is usually caused by tannin, an organic acid found in swamp water. Tests show low chlorine residual. The remedy is superchlorination followed by adjustments of pH concentration as necessary.

Green water It is usually caused by algae and tests show low chlorine residual. The remedy is superchlorination followed by adjustment of the pH concentration and the use of an effective algicide.

Blue water Blue or clear green water is due to the presence of copper salts formed by the acid corrosion of copper pipes and fittings. The actual colour varies with the pH concentration and the type of ions in the water.

POOL WINTERIZING

In areas where freezing temperatures occur, it is usual to close an outdoor pool for the winter. This is known as winterizing, and should be taken into account during the design of the pool. In general, winterizing involves :

1. emptying the system of all water above the frost line;
2. closing pool openings against the entry of rain, snow, surface water, and debris;
3. removal or protection of mechanical equipment to prevent damage by vandalism and exposure to the elements.

Winterizing practices vary with the locality and the severity of the climate, so that the experience of local pool builders forms the best qualified source of information. The action required at the end of the season is summarized below.

1. SAND FILTER : backwash for at least 30 minutes to clean the filter bed thoroughly. DIATOMACEOUS EARTH FILTER : Backwash until the waste water runs clear.
2. Drain or pump the pool water level below the inlet fittings. Drain the water from the filter tank (refer to Owner's Manual) and all vulnerable piping. Check that the dial valve on a sand filter is set to WINTERIZE; this will speed the draining process by allowing air to enter the tank.
3. SAND FILTER : Open the filter tank and inspect the condition of the sand. DIATOMACEOUS EARTH FILTER : dismantle the filter, and scrub the elements with a detergent solution; if the elements are encrusted with lime, wash them in a dilute solution of muriatic acid, and rinse them thoroughly; replace elements which have defects in the cloth. Protect the filters from the elements and extreme temperatures.
4. Drain the FILTER PUMP (refer to Owner's Manual). This is the time to make repairs in readiness for next season. Store the pump and motor indoors, or protect them from the elements and extreme temperatures.
5. Remove the covers and empty the HAIR AND LINT STRAINER and CHEMICAL FEED POTS.
6. Remove the orifice plates or directional balls from the INLET FITTINGS, and use pipe plugs or rubber expansion plugs to make these wall openings watertight.
7. Remove and store the strainer basket and valve plate from the SKIMMER(S). Plug the wall fitting for the equalizer line (where used). Drain, bail, or pump as much water as possible from the skimmer body and piping. Take an 8 ft length of one-half or five- eighths inch rubber garden hose; plug and seal both ends to prevent leakage from it. Insert the ends into the skimmer piping as far as possible, so that the air-filled hose will absorb the expansion of any ice which forms in the pipes. Stuff a burlap bag into the weir throat from the inside of the skimmer to hold the weir closed. Seal the wall and deck openings to the skimmer against rain and snow by taping plastic sheet or waterproof building paper over them.

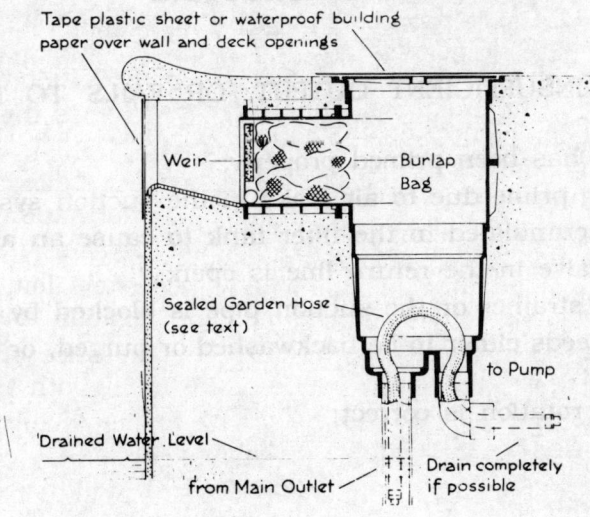

8. HEATERS. GAS-FIRED; close the main gas valve; drain the heater; remove the safety pressure switch and store it indoors so that it cannot be damaged by frost. OIL-FIRED : open the main disconnect switch and remove the fuses; close the valve in the oil supply line; drain the heater; protect the burner and motor from the elements and extreme temperatures. ELECTRIC : open the main disconnect switch and remove the fuses; drain the heater. HEAT EXCHANGER : close the valves to all vulnerable pipes, and drain both primary and secondary circuits.

After draining the heater, replace the plugs if the heaters are of cast iron or steel. This precaution will prevent deterioration of the ferrous threads. If the heater has pockets which will not drain completely, pour in about a quart of antifreeze, but remember to flush it out before putting the heater back into service. Use a tarpaulin or plastic sheet to protect an outdoor heater from the elements and extreme temperatures. Drain indoor heater if there is any risk of freezing by outside air, such as that drawn down the stack by the combustion draft of adjacent furnaces or boilers.

9. UNDERWATER LIGHTS. Either disconnect the light at the deck box and remove it from the pool for storage indoors, or take it from its niche and store it in a box on deck, or lower it carefully below water level to the extent of its cord where the water will not freeze.

10. Remove and store indoors such items as ladders, stair rails, and diving boards. Cover wedge and flange anchors, or seal them with special plugs. Store diving boards on a flat, level surface in a dry place. This precaution will prevent cracks and warping. Never lean a diving board against a wall.

11. Use petroleum jelly, or other suitable rust preventative, to coat any unpainted or unprotected metal parts subject to corrosion.

12. With all piping drained and pool openings plugged, fill the pool to about two feet below the normal level. The weight of water will reduce the chances of the pool floating out the ground in the event of excessive ground water after the spring thaw or heavy rain. A series of logs, 6 to 8 inches in diameter, is sometimes installed around the pool to relieve pressure from the ice which forms before the ground freezes. This practice is no longer general, but if used, logs should be anchored to prevent them damaging the pool walls.

A method of relieving the water pressure below the ice is to use a special electric heating probe to prevent the pool freezing over completely.

13. Protect all equipment left outdoors against the elements and extreme temperatures.

14. Dirt and debris can be kept out of the pool by either installing a proprietary pool cover, or covering the pool and adjacent deck with polyethylene sheeting held down by sand bags. Before covering the pool, dose the water heavily with chlorine and algicide, so that the water remains in good condition until the next season. If the winterized pool forms a potential hazard to children or others, surround it by a snow fence or similar temporary barrier.

TROUBLESHOOTING

FILTER PUMP

1. IF THE PUMP HAS INSUFFICIENT OUTPUT, OR FAILS TO MOVE WATER, CHECK THE FOLLOWING :

a whether the pump has been primed properly;

b whether it is losing prime due to air leaks in the suction system;

c whether air has accumulated in the filter tank to cause an air lock;

d whether the gate valve in the return line is open;

e whether either the strainer or the suction pipe is blocked by debris;

f whether the filter needs either to be backwashed or purged, or to have the sand bed replaced due to calcification;

g whether the pump rotation is correct;

   h whether either incorrect or low voltage is causing the motor to run at reduced speed;

   i whether the impeller passages are obstructed by debris;

   j whether the total suction lift is excessive (use a vacuum gauge at the pump inlet) total suction lift includes the effects of the total height of pump above pool, the distance of pump from pool, and the size of suction pipe;

   k whether the filter system is too small for the size of pool;

   l whether the impeller is properly adjusted.

2. IF THE PUMP DOES NOT PRIME, OR LOSES PRIME WHILE RUNNING, CHECK THE FOLLOWING :

   a whether the check valve (on a self-priming pump) is closing;

   b whether there are leaks in the suction system (including valve stems and glands);

   c whether the mechanical shaft seal is leaking;

   d whether either the strainer or the suction pipe is blocked by debris;

   e whether the impeller passages are obstructed by debris;

   f whether the pool water level is below the skimmer opening;

   g whether the flow rate is too great for the suction conditions (partly close the gate valve in the return line to correct this).

3. IF THE PUMP IS NOISY OR VIBRATES, CHECK THE FOLLOWING :

   a whether the bearings are worn;

   b whether the impeller is rubbing on the pump case;

   c whether the impeller passages are partly obstructed;

   d whether the pump unit is properly supported;

   e whether the total suction lift is excessive, thus causing cavitation within the pump (cavitation makes rumbling or gravelly noises, and can be destructive).

ELECTRIC MOTOR-FILTER PUMP

Open the disconnect switch and remove the fuses before disconnecting or repairing an electric motor.

Check that the motor is grounded properly, and that there are no couplings, gaskets, or non-metallic parts which break the continuity of the ground connection.

References to capacitor, starting winding, or internal starting switch apply only to a single-phase motor.

1. IF MOTOR RUNS AT RATED SPEED BUT OVER, HEATS, TRIPS OVERLOAD, OR BLOWS FUSES, CHECK THE FOLLOWING :

   a whether the current is excessive (use an ammeter to compare the actual current with the nameplate rating);

   b whether the supply voltages is low;

   c whether the motor rating corresponds to the supply voltage;

   d whether the motor is correctly connected for the supply voltage

   e whether the pump operating pressure is too low, causing an overload on the motor (partly close the gate valve in the return line to correct this);

   f whether the ventilation to the motor is adequate;

   g whether the impeller is rubbing inside the pump case;

   h whether there is a short circuit to ground;

   i whether there is a short circuit between the windings.

2. IF MOTOR FAILS TO START, CHECK THE FOLLOWING :

   a whether the motor is plugged in;

   b whether there is an open switch, blown fuse, or loose or incorrect wiring in the supply line;

   c whether the overload has tripped and failed to reset;

d whether there is a loose connection either to the starting winding or to the capacitor

e whether the internal starting switch is stuck in the open position;

f whether the capacitor is defective (internal open or short circuit).

3. IF MOTOR RUNS AT ABOUT HALF SPEED, OVERHEATS, TRIPS OVERLOAD, OR BLOWS FUSES, CHECK THE FOLLOWING :

a whether the motor rating corresponds to the supply voltage;

b whether the motor is correctly connected for the supply voltage;

c whether the internal starting switch fails to open;

d whether there is a short to ground;

e whether there is a short between the windings.

JACUZZI

TECHNICAL BULLETIN

CARTRIDGE FILTERS SERIES CFT

Description

Jacuzzi filter systems are cartridge type polyester cloth filters designed for fresh water swimming pools and large whirlpool baths. The cartridge element is composed of fine interwoven fibers which form a net work of open pores. As the water passes through the pores, particles are retained (16 micron average). As filtering proceeds a thickness is built up and progressively smaller particles are retained, thus increasing the effectiveness of the filtration.

To clean or replace the filter element it is only necessary to remove the cover and lift the cartridge assembly out of the cell.

Cleaning of filter element is done manually with clear water. There is no need for selector valve or backwashing.

A complete filter includes

Filter cell : constructed of thermoplastic with 1½" threaded inlet and outlet. The series CFT features quick opening Belt-Lock™ fastener, automatic air release valve, drain plug and pressure gauge. Operating pressure is 3.5 Atm.

Filter Cartridge : a cloth of spun-bounded polyester fibers. Patented sediment catcher and internal support prevents dirt from falling into tank when element is removed.

FILTER AND PUMP PERFORMANCES

Model	Filter area	Pump power		Pump capacity	Filter rate	Turn-over rate m^3		
		HP	kW	m^3h	m^3h/m^2	6 h	8 h	10 h
CFT 50	4.6	½	0.37	4.2	0.92	25	33	42
		¾	0.55	7	1.53	42	56	70
		1	0.74	11.3	2.45	68	90	113
CFT 100	9.2	¾	0.55	8.4	0.92	50	67	84
		1	0.74	14	1.53	84	112	140
		1½	1.10	19	2.06	114	152	190
		2	1.47	22.6	2.45	135	180	226

Application

To use the filter performance table for size selection, follow these three steps :

1. Calculate pool capacity by measuring the volume (m^3) of the pool up to the normal water line.
2. Determine turn-over rate which is defined as the number of hours required to pump the entire pool volume through the filter system once. The correct turn-over rate depends on the use of the pool.
3. Under desired turn-over rate, find approximate pool capacity in m^3. To the left will be recommended filter and pump size.

Type of pool	Turn-over rate/hour	Usage
Residential, single family	8-10	light
Semi-commercial (hotels, condominium, etc.)	6-8	moderate
Commercial, public	4-6	heavy
Wading pool, children	2-4	very heavy
Health spas (small)	0.5	very heavy

Filter rates

Different qualities of filtration can be obtained depending on filter rate of the cartridge.

Excellent filtration : with flows of 0.92 m^3h/m^2, for heavy use.

Good filtration : with flows of 1.53 m^3h/m^2, for moderate use.

Fair filtration : with flows of 2.45 m^3h/m^2, for light use.

Jacuzzi recommends a maximum filter rate of 2.45 m^3h/m^2 as higher flow rates will decrease filtration performance and cycle duration.

The filter performance table indicates the filter rate of each system so you may select one to meet your requirements.

All pipe outlet are 1½".

In case of salt water application contact factory.

MENSIONI D'INGOMBRO in cm   **OVER-ALL DIMENSIONS in cm**

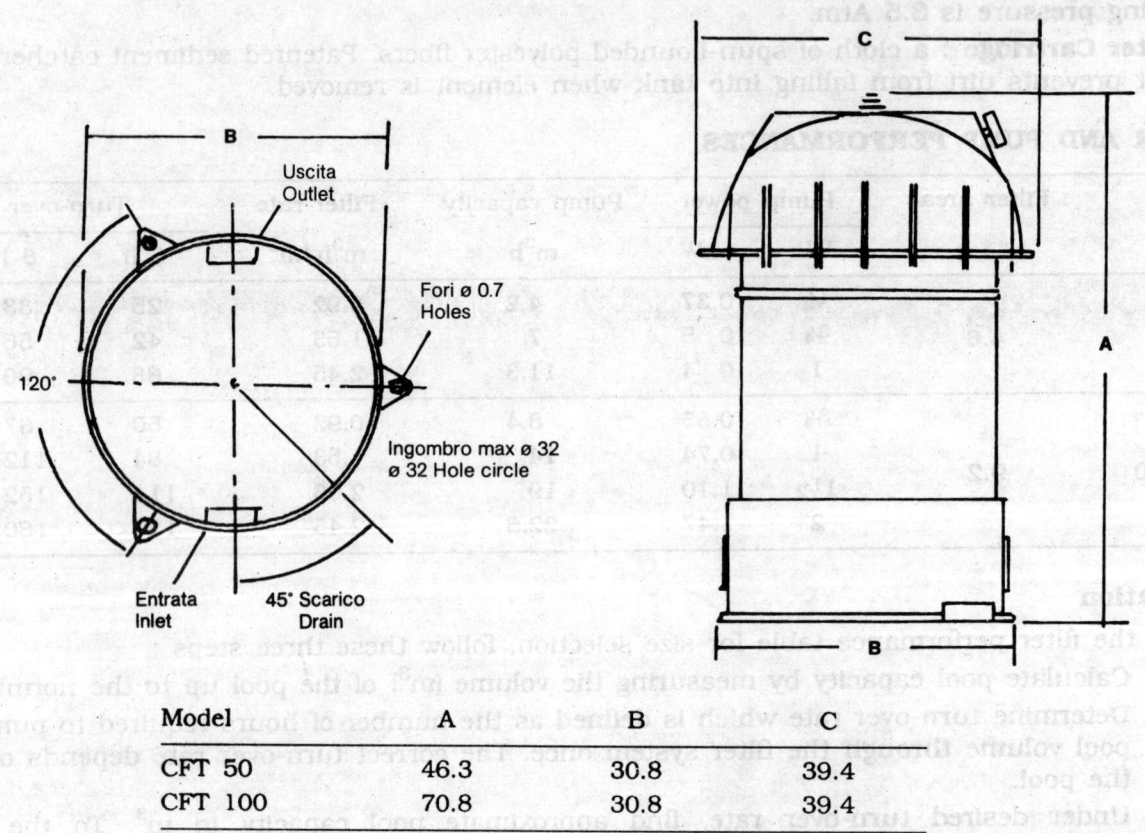

Model	A	B	C
CFT 50	46.3	30.8	39.4
CFT 100	70.8	30.8	39.4

The above data are indicative only.

SAND FILTERS SERIES FM-SS-ST

Description

Jacuzzi filter systems are high rate sand media filters, designed for fresh water swimming pools.

The natural filtering action of sand has proved best through 70 years of experience. The filtering quality is such that the sand bed retains 50% of the floating debris and lets pass those less than 15 micron. Floculation can be made by inserting a quality of alum (approximately kg. 0.800 for each sq. metre of filtering surface) : in this manner, a filtering characteristic of 70% is obtained with a filtering **rate** of 5-7 micron. The sand stays in the tank permanently and is cleaned by reversing the water **through** the sand bed.

Backwash rates are shown on performance table. The backwash is carried out when the difference in pressure between the two pressure gauges installed on the dial valve reaches a value of about 1 Atm. Sufficient waste disposal must be provided for up to 5 minutes operation.

FM 17-25-24

FM 30-36-42

SS

ST

A complete filter includes

Filter tank with a maximum working pressure of 3.0 Atm, base and drain plug.

FM Series tank is molded plastic with glass fiber reinforcing.

SS Series tank is of AISI304 stainless steel with 150 × 110 mm access opening.

ST Series tank is of carbon steel with the inside coated with special submarine paint and magnesium anode mounted internally to neutralize potential corrosion from electrolysis. Access openings : 150 × 110 mm for filters 24″ and 30″ ; 150 × 200 mm for filters 36″ and 42″ ; 220 × 320 mm for filters 48″ and 54″.

Filter media not included. The special sand and gravel used must be purchased separately.

Distribution system : diffuser for incoming water and underdrain system within the media that eliminates potential disruption of the media bed by uniform direction of flow throughout the tank.

This exclusive system is constructed of non-corrosive molded material and incorporates an internal return pipe. All components are threaded together for easy replacement.

Dial valve : operation is by simple rotation of the handle. Two pressure gauges are mounted to indicate pressure drop across the sand bed and proper time for backwashing of media.

Filters 17″, 20″, 24″ and 30″ use a 1½″ size molded plastic valve having seven positions : Filter; Test; Backwash; Rinse; Whirlpool or Recirculate; Drain; Winterize. Filters 36″ and 42″ use a 2″ size cast iron (Series ST, FM) or bronze (Series SS, FM) valve having five positions : Filter; Backwash; Recirculate; Drain; Winterize. Filters 48″ and 54″ use a 3″ size cast iron valve having five positions same as 2″ size.

FILTER AND PUMP PERFORMANCES

Model	Tank diameter		Filter area	Pump power		Pump capacity	Filter rate	Minimum backwash rate	Turn-over rate — hours			
									4 h	6 h	8 h	10 h
	Inches	Meters	m^2	HP	kW	m^{3}h	m^{3}h/m^2	m^{3}h	Pool capacity m^3			
17	17″	0.430	0.145	½	0.37	6	41	6	—	36	48	60
20	20″	0.510	0.204	¾	0.55	9	44	7	—	54	72	90
24	24″	0.610	0.292	1	0.74	12	41	10	—	72	96	120
				1½	1.10	14	48	10	—	84	112	140
30	30″	0.750	0.442	1½	1.10	16	35	16	—	96	128	160
				2	1.47	21	46	16	—	126	168	210
36	36″	0.900	0.636	2	1.47	27	42	24	108	162	216	270
				3	2.20	30	48	24	120	180	240	300
42	42″	1.100	0.950	3	2.20	36	38	32	144	216	288	360
				5	3.68	44	47	32	176	264	352	440
48	48″	1.250	1.226	5	3.68	49	40	42	196	294	392	490
				5	3.68	58	47	42	232	348	464	580
54	54″	1.400	1.539	5	3.68	61	40	53	244	366	488	610
				7½	5.52	72	47	53	288	432	576	720

Notes : For larger pool sizes select "Multiple tanks" or "Commercial Sand Filters".

Application

To use the filter performance table for size selection, follow these three steps :

1. Calculate pool capacity by measuring the volume (m^3) of the pool up to the normal water line.

2. Determine turn-over rate which is defined as the number of hours required to pump the entire pool volume through the filter system once. The correct turn-over rate depends on the use of the pool.
3. Under desired turn-over rate, find approximate pool capacity in m³. To the left will be recommended filter and pump size.

Type of pool	Turn-over rate/hour	Usage
Residential, single family	8-10	light
Semi-commercial (hotels, condominium, etc.)	6-8	moderate
Commercial, public	4-6	heavy
Wading pool, children	2-4	very heavy
Health spas (small)	0.5	very heavy

Filter rates

Different qualities of filtration can be obtained depending on the filter rate through the sand bend.

Excellent filtration : with flows between 35 to 40 m^3h/m^2.

Good filtration : with flows between 41 to 45 m^3h/m^2.

Fair filtration : with flows between 46 to 50 m^3h/m^2.

A filter rate beyond 50 m^3h/m^2 will disrupt the sand bed and no filtering action will occur.

The filter performance table indicates the filter rate of each system so you may select one to meet your requirements.

Friction losses in suction and discharge pipes can reduce filter rates. If filter is located away from pool, increase pipe sizes and/or pump size to accommodate these losses.

when the difference in pressure between the two pressure gauges installed in the face piping reaches a value of 1 Atm.

Sufficient waste disposal must be provided for up to 5 minutes operation.

Filter media

It is recommended that two grades of media be placed in the filter.

1. Gravel supporting base of hard rounded stones, 3 to 6 mm in diameter, with less than 2% by weight of thin, flat or elongated pieces. Material should be washed clean. It should be installed to completely cover the arms of the underdrain system.
2. Filter media of quartz sand with oval shaped grains or with rounded edges, free from clay and other organic impurities, with grain size range from 0.4 to 0.7 mm and a uniformity index of 1.45 ÷ 1.70.

Jacuzzi filters were designed for permanent use of this filter media and is not normally necessary to replace it for the life of the equipment.

Tank diameter	Gravel quantity kg	Gravel quantity m³	Sand quantity kg	Sand quantity m³
17"	25	0.019	50	0.027
20"	25	0.019	75	0.040
24"	50	0.038	125	0.069
30"	100	0.076	350	0.190
36"	175	0.130	400	0.220
42"	225	0.170	525	0.290
48"	375	0.280	900	0.500
54"	425	0.320	1150	0.630

DIMENSIONI D'INGOMBRO in cm

OVER-ALL DIMENSIONS in cm

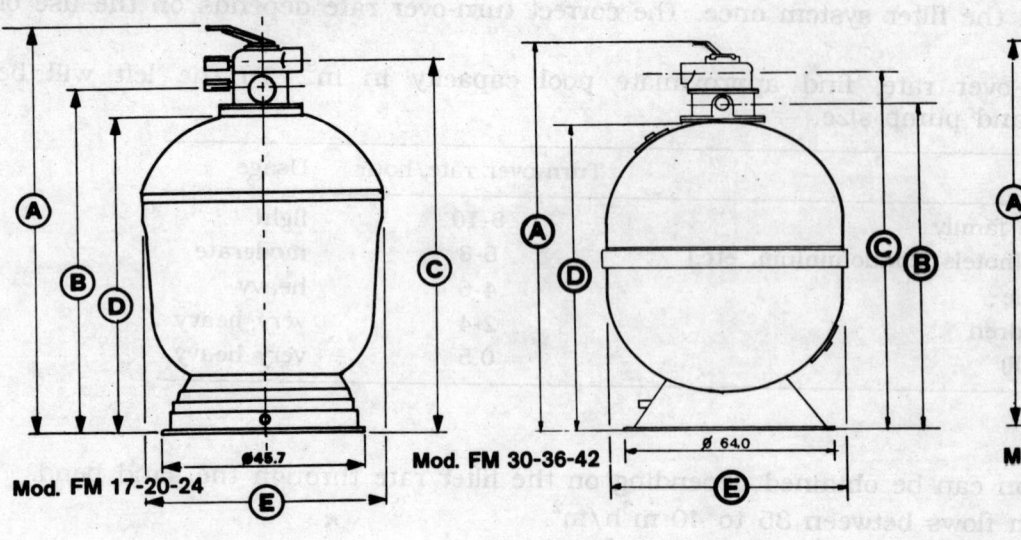
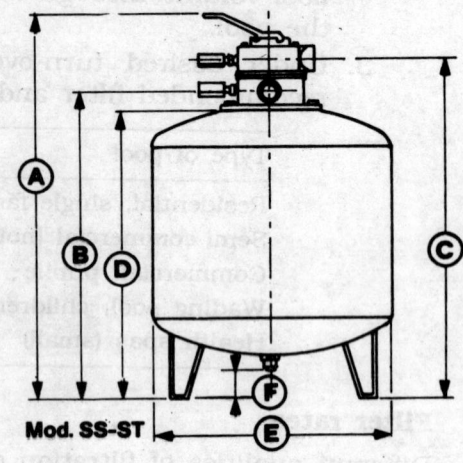

Mod. FM 17-20-24   Mod. FM 30-36-42   Mod. SS-ST

Tank Model	A	B	C	D	E	F
FM 17	83.8	65.4	74.8	60.6	46	–
FM 20	93.9	77.1	84.9	72.3	55.2	–
FM 24	96.5	80	87.5	75.2	62.2	–
FM 30	118	102	108	98	76	–
FM 36	123	104	115	99	91	–
FM 42	145	126	137	121	111	–
ST 24 - SS 24	93.6	78.1	85.1	67.8	61	3.5
ST 30 - SS 30	122.7	107.2	114.2	96.9	76	6.2
ST 36 - SS 36	118.5	100.5	111	95	91	5
ST 42	116.9	98.9	109.4	93.4	111	5
ST 48	163	144.3	152.8	137.3	126	6.5
ST 54	163	144.3	152.8	137.3	141	6.5

The above data are indicative only.

COMMERCIAL SAND FILTERS SERIES SM

Description

The series SM Jacuzzi Commercial Filter Systems are high rate sand media filters, designed for large public type fresh water swimming pools or other heavy duty applications.

Special hydraulic design and media selection allow maximum flow rates of 36 m³/m² (15 gpm per sq. ft.). This flow rate results in compact tank sizes and more economical systems. A deep sand bed is used to provide three-dimensional filtration, high dirt-holding capacity and long filter cycles.

The natural filtering action of sand has proved best through 70 years of experience. The filtering quality is such that the sand bed retains 50% of the floating debris and lets pass those less than 15 micron. Flocculation can be made by inserting a quantity of alum (approximately kg 0.800 for each sq. metre of filtering surface) : in this manner, a filtering characteristic of 70% is obtained with a filtering rate of 5-7 micron. The sand stays in the tank permanently and is cleaned by reversing the water through the sand bed.

DESCRIPTION

A complete system includes

Filter tank : constructed of carbon steel, for working pressure of 3.0 Atm (50 PSI) and test pressure of 5 Atm (75 PSI). Side wall height is 1.2 meters (60″). The inside has two coats of corrosion resisting epoxy paint. The tank is mounted on three legs and equipped with to 40 cm dia. (ϕ 15″) access openings; one on top and one on the side for media removal. An automatic air relief valve is mounted on top.

Filter media not included. The special sand and gravel used must be purchased separately.

Distribution system : diffuser for incoming water and underdrain system within the media that eliminates potential disruption of the media bed by uniform direction of flow throughout the tank. This exclusive system is constructed of a steel main header with laterals of molded plastic pipes. Each lateral has orifices designed for even distribution. All components are threaded together for easy replacement.

Backwash control : supply and discharge headers are directly flanged to backwash face piping system. The flanged fittings include four quick acting butterfly valves type PN10-UNI2277. Valves can be positioned to have following operations : Filter - Backwash - Drain - Recirculate. Face piping system includes pressure gauge panel to indicate pressure drop across sand bed and proper time for backwashing media, sight glass mounted on waste outlet to inspect backwash water, and drain valve.

Recirculation pumps : not included with filter tank. Select recommended self-priming filter pumps series ULS as indicated or centrifugal pumps (*) to match filter flow capacity at 15 meters (50 ft.) head.
(*) Contact factory for price quotation and installation recommendations.

FILTER PERFORMANCES

Model	Tank diameter		Filter area	Filter flow rate	Face pipe	Recommended pumps		Turn-over rate — hours		
						Self-priming	Centrifugal	4 h	6 h	8 h
	Inches	Meters	m^3	m^{3}h	Inches	Model	HP	Pool capacity m^3		
SM 54″	54″	1.40	1.54	55	4″	5 ULS	5	220	330	440
SM 60″	60″	1.55	1.88	70	4″	7.5 ULS	7½	280	420	560
SM 66″	66″	1.70	2.27	80	4″	10 ULS	10	320	480	640
SM 72″	72″	1.85	2.68	95	5″	5 ULS-2	10	380	570	760
SM 78″	78″	2.00	3.14	115	5″	7.5 ULS-2	10	460	690	920
SM 84″	84″	2.15	3.63	130	5″	7.5 ULS-2	15	520	780	1040

Application

To use the filter performance table for size selection, follow these three steps :

1. Calculate pool capacity by measuring the volume (m^3) of the pool up to the normal water line.
2. Determine turn-over rate which is defined as the number of hours required to pump the entire pool volume through the filter system once. The correct turn-over rate depends on the use of the pool.
3. Under desired turn-over rate, find approximate pool capacity in m^3. To the left will be recommended filter and pump size.

Two smaller diameter tanks, which equal required filter area, may be selected.

Type of pool	Turn-over rate/hour	Usage
Residential, single family	8-10	light
Semi-commercial (hotels, condominium, etc.)	6-8	moderate
Commercial, public	4-6	heavy
Wading pool, children	2-4	very heavy
Health spas (small)	0.5	very heavy

Filter rates

The series SM units are sized for 36 m^{3}h/m^2 maximum. A filter rate beyond 50 m^{3}h/m^2 will disrupt the sand bed and no filtering action will occur. The recirculation pump and system must be designed not to exceed the recommended filter rates at any time. Friction losses in suction and discharge pipes can reduce filter rates. If filter is located away from pool, increase pipe sizes and/or pump size to accommodate these losses. Backwash rates are equal to the filtration rate. The backwash is carried out

Filter media

Filter media not included with filter. The special sand and gravel must be purchased separately.

A) The base of tanks should be filled with concrete to approximately 1 cm (½″) from bottom of header. Volume of concrete shown on chart is approximate and may vary in relation to curvature of tank head. Smooth and level concrete by hand. After concrete is hard, underdrain laterals are installed with orifices facing down.

B-C-D) Supporting gravel should consist of clean, hard, rounded stones with an average specific gravity of not less than 2.5 and should be free from shale, mica, clay, sandstone, sand loam or other impurities. Not more than 1% may be flat or micaceous particles.

Three grades of gravel are to be placed in even layers within the filter :

B — Large 1″ × 1½″
C — Medium ½″ × ¾″
D — Fine ⅛″ × ¼″

E — Filtering media should be hard durable grains of rounded or subangular silica sand free from clay, loam, dirt and organic matter. It should have an effective size of 0.4 ÷ 0.6 mm and an uniformity index of 1.45 ÷ 1.70.

F — The free area between sand bed and inlet pipes must be maintained to provide proper operation of filter and backwash cycles. **Orifices of inlet laterals must be facing up**.

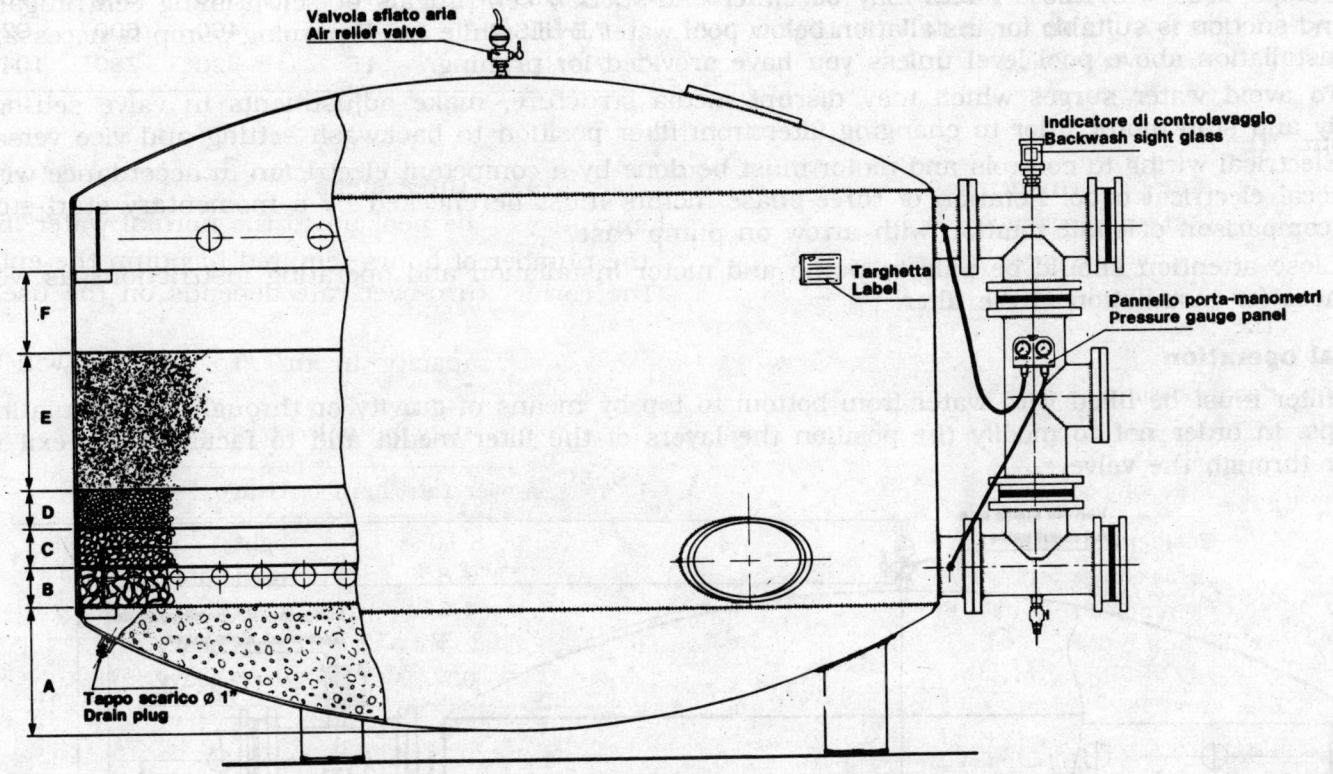

TABLE OF FILTER MATERIAL QUANTITIES

Model	Dimensions		Volumes in m³					Quantities in kg					Total Sand and Gravel
	Diam. (m)	Height (m)	A Concrete/ Gravel 20 × 30	B Gravel 20 × 30	C Gravel 10 × 20	D Gravel 5 × 9	E Sand 0.4 × 0.7	A Concrete/ Gravel 20 × 30	B Gravel 20 × 30	C Gravel 10 × 20	D Gravel 5 × 9	E Sand 0.4 × 0.7	
SM 54″	1.40	2.43	0.220	0.215	0.215	0.215	0.770	385	375	375	375	1230	2740
SM 60″	1.55	2.43	0.270	0.264	0.264	0.264	0.940	470	460	460	460	1510	3360
SM 66″	1.70	2.43	0.340	0.317	0.317	0.317	1.135	595	555	555	555	1830	4090
SM 72″	1.85	2.43	0.421	0.376	0.376	0.376	1.320	735	660	660	660	2140	4855
SM 78″	2.00	2.43	0.528	0.440	0.440	0.440	1.550	925	770	770	770	2510	5745
SM 84″	2.15	2.43	0.651	0.508	0.508	0.508	1.800	1140	890	890	890	2900	6710

The above data are indicative only.

Installation notes

In planning your filter installation, considerable attention should be directed to the foundation. Concrete should be of sufficient strength to support the weight of the installation. The system, of course, should be located as close to the pool as possible, at or below deck level.

The location preferably should be in a filter room, or if it is outside, a dry and shaded area is preferred. If you are using a filter room, it should be dry and well ventilated. Consideration should be given to the waste sump and proper drainage.

You should also consider location of the various types of other equipment, such as pump and motor, chemical solution tank, chemical feed pump, chlorinator, chlorine cylinders, exhaust fan, etc. All equipment should be easily accessible for servicing and winterizing.

Pumps used with these filters may be either end-suction centrifugals or self-priming centrifugals. An end-suction is suitable for installation below pool water level, while a self-priming pump is necessary for installation above pool level unless you have provided for priming.

To avoid water surges which may disrupt media structure, make adjustments in valve settings slowly and stop motor prior to changing filter from filter position to backwash setting and vice versa.

Electrical wiring to controls and motor must be done by a competent electrician in accordance with the local electrical code. Rotation of three-phase motors must be checked by a momentary start-stop and comparison of shaft rotation with arrow on pump case.

Close attention should be paid to pump and motor installation and operating instructions as well as those for installation of the filter.

Initial operation

The filter must be filled with water from bottom to top by means of gravity or through the circulating pumps, in order not to modify the position the layers of the filter media and to facilitate the exit of water through the valve.

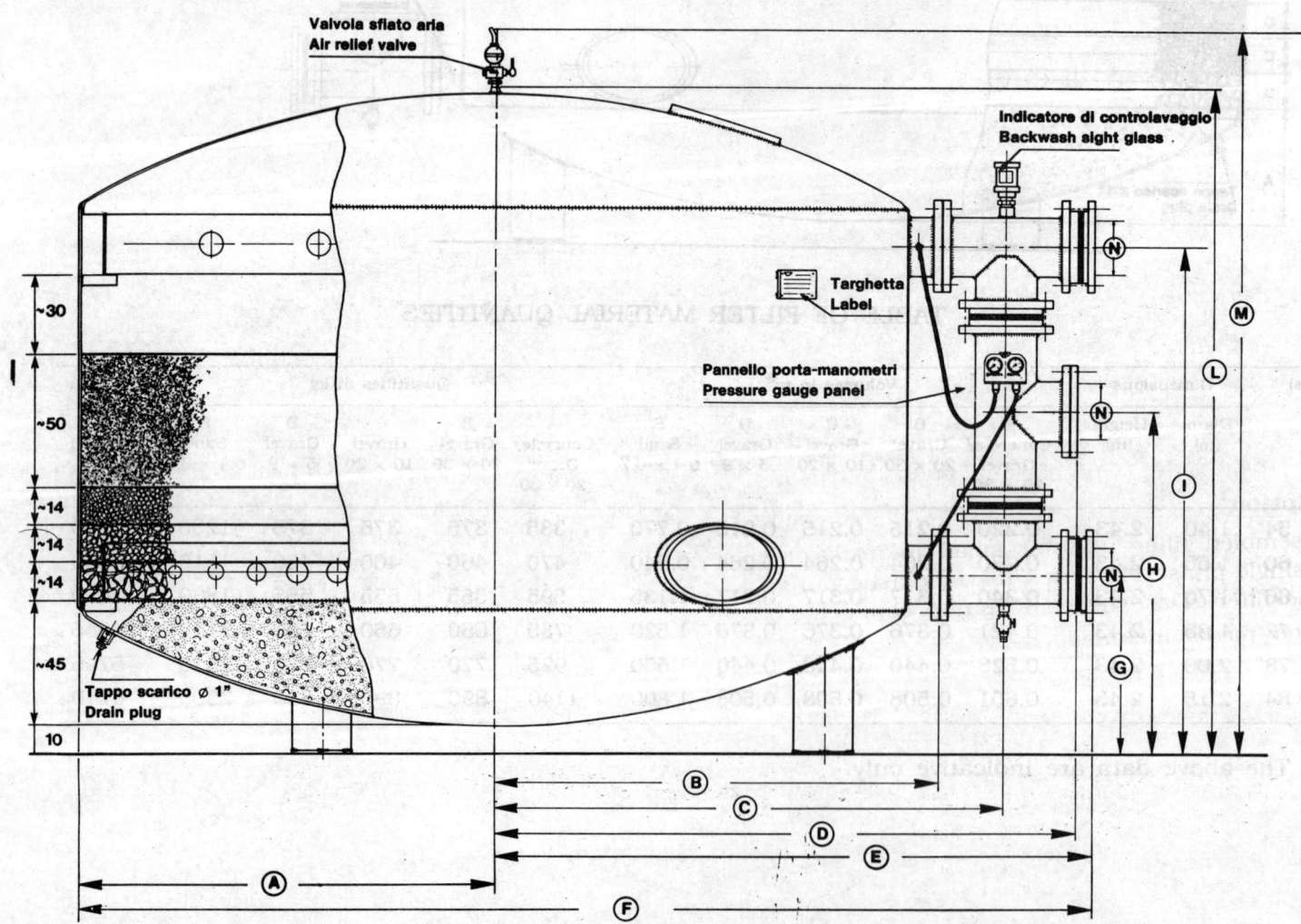

Overall dimensions in cm

Model	A	B	C	D	E	F	G	H	I	L	M	N
SM 54″	70.5	82.5	99.7	119.1	124	194.5	61.3	128.1	179.9	219.6	243.1	4″
SM 60″	78	90	1072 —	126.6	131.5	209.5	61.3	128.1	179.9	219.6	243.1	4″
SM 66″	85.5	97.5	114.7	134.1	139	224.5	61.3	128.1	179.9	219.6	243.1	4″
SM 72″	93.1	105.1	123.8	144.9	150	243.1	62.8	128.2	178.6	219.8	243.3	5″
SM 78″	100.6	112.6	131.3	152.4	157.5	258.1	62.8	128.2	178.6	219.8	243.3	5″
SM 84″	108.1	120.1	138.8	159.9	165	273.1	62.8	128.2	178.6	219.8	243.3	5″

The above data are indicative only.

SPA-PAK

Description

Pre-assembled filter unit for small to medium pools, consisting of : fibreglass resin base, FM-series filter, single-phase PH series thermoplastic pump, PVC fittings, quick-fit connections between the pump and the filter, 1½″ valve and pump connections.

Filter media included.

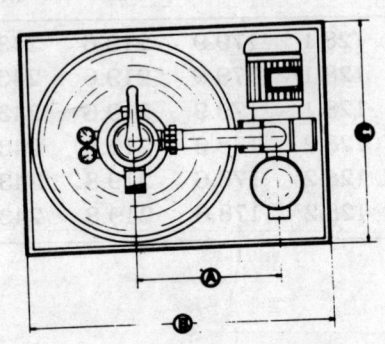

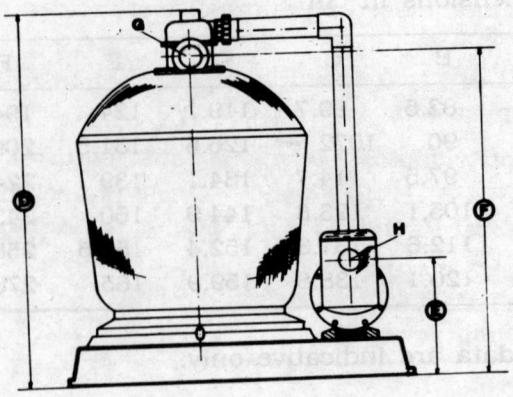

FILTER AND PUMP PERFORMANCES

Model	Pump	HP	KW	Filter	Sand 0.3 + 0.7 mm kg	Gravel 3 + 6 mm kg	Filter area m²	Filter rate m³h/m²	Pump capacity m³/h	Minimum Backwash rate m³/h
0.5 SP 1P 17 FM	5 PH	½	0.37	FM17	50	25	0.145	41	6	6
0.7 SP 1P 20 FM	7 PH	¾	0.55	FM20	75	25	0.204	44	9	7
1 SP 1P 24 FM	1 PH	1	0.74	FM24	125	50	0.292	41	12	10
1.5 SP 1P 24 FM	1.5 PH	1½	1.10	FM24	125	50	0.292	48	14	10

Overall dimensions in cm

Model	A	B	C	D	E	F	G	H
0.5 SP 1P 17 FM	34	82	72	96	36	77.5	1½"	1½"
0.7 SP 1P 20 FM	37	92	74	106	36	89	1½"	1½"
1 SP 1 P 24 FM	42	92	74	108.5	36	92	1½"	1½"
1.5 SP 1P 24 FM	42	92	74	108.5	36	92	1½"	1½"

RECIRCULATION PUMPS SERIES PH

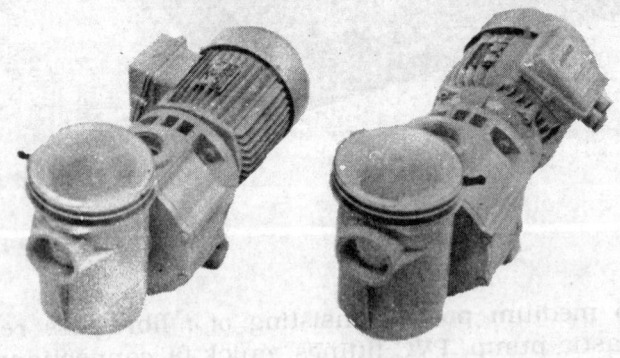

Description

The Jacuzzi series PH pumps are constructed of molded fiber-filled Noryl for corrosion free operation. The design maintains isolation of water from the metal part of the pump or motor.

Series PH : is a self-priming centrifugal with 6" prefilter, including transparent lid and molded strainer.

All pumps include closed type impeller of molded material, leakless mechanical seal and brass pump shaft.

Series PH2V : are two speed pumps (2900-1450 rpm, single-phase only) for energy saving filter or whirlpool spa application.

Standard electric motors, asynchronous, squirrel cage type, totally enclosed, fan cooled.

Enclosure 1P44 - B 5 Form - Single-phase 220 V with permanently connected capacity and three-phase 220-380 V and 415 V - 50 Hz - 2900 rpm.

PERFORMANCE TABLE

Model	Pump power		Discharge head in metres												
	HP	KW	2	3	4	5	6	8	10	12	14	16	18	20	22
			Capacity m^3/hour												
2900 n'/rpm															
5 PH	½	0.37					10.5	8.4	6.6	4.2	2.5				
7 PH	¾	0.55					14.1	12	9.9	7.5	3.9	1			
1 PH	1	0.74					19.8	17.1	14.1	10.5	7	3	1		
1.5 PH	1½	1.10					24	22	19.5	15.8	12	6			
2 PH	2	1.47					33.6	31.2	28.8	25.2	20.4	16	10.2		
3 PH	3	2.20					39	36	33.6	30	27.6	24	19.5	15	8
1450 n'/rpm															
1 PH2V	½	0.37	8.2	7.8	5.4	0.5									
1.5 PH2V	¾	0.55	12	11.4	8.4	3									

Application

1. Series PH filter pumps should be selected to provide the proper capacity at the total head developed by : a) suction lift; b) pressure drop across filter media, about 5 metres of water column; c) pipe friction in suction and discharge lines.
2. More than 4 metres lift and/or friction in suction line is not recommended.
3. Normal operating head of the pump on a sand filter is between 8 and 13 metres.
4. Maximum water temperature of installation is 60°C (140°F).
5. A control box with motor protection circuit breakers or fuses must be installed for each motor.

PERFORMANCE CURVES
Overall dimensions in cm

Model	A	B	C	D	E	F	G	H	I	L
5 PH	53	23	30.5	23	25	31	1½″	1½″	12.7	16.5
7 PH	53	23	30.5	23	25	31	1½″	1½″	12.7	16.5
1 PH	56	23	30.5	23	25	31	1½″	1½″	12.7	16.5
1.5 PH	56	23	30.5	23	25	31	2″	1½″	12.7	16.5
2 PH	62.5	23	33.1	23	25	31	2″	2″	12.7	16.5
3 PH	64.5	23	33.1	23	25	31	2″	2″	12.7	16.5

The above data are indicative only.

CURVE CARATTERISTICHE

PERFORMANCE CURVES

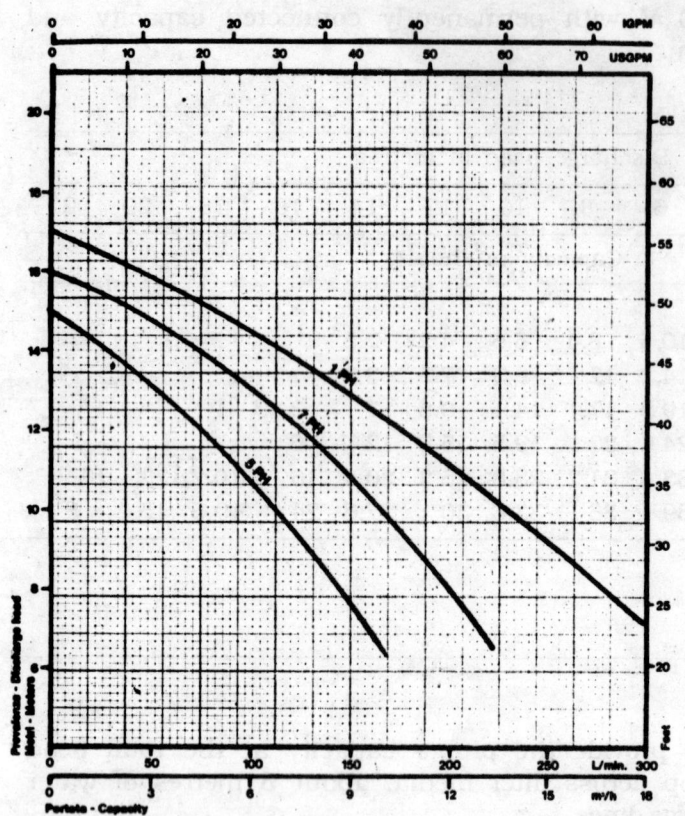

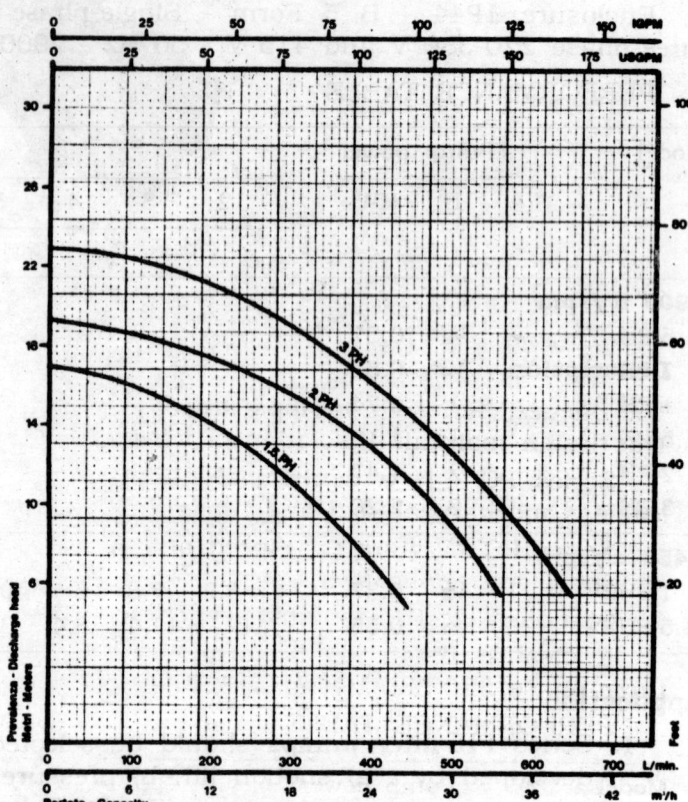

DIMENSIONI D'INGOMBRO in cm

OVER-ALL DIMENSIONS in cm

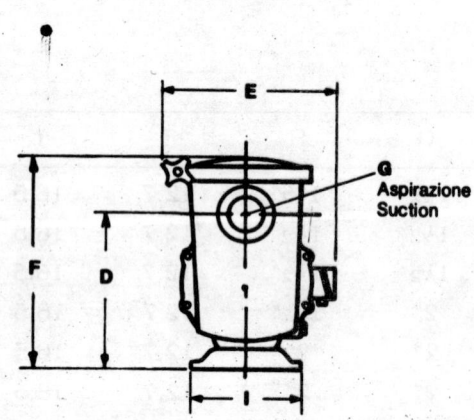

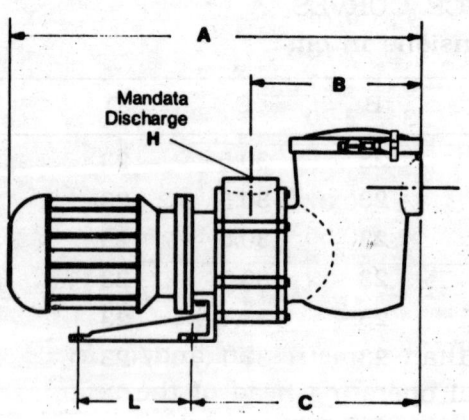

RECIRCULATION PUMPS SERIES PC

Description

The Jacuzzi series PC pumps are constructed of molded fiberfilled Noryl for corrosion free operation. The design maintains isolation of water from the metal part of the pump or motor.

Series PC : is a centrifugal pump. All pumps include closed type impeller of molded material, leakless mechanical seal and brass pump shaft.

Series PC 2V : are two speed pumps (2900-1450 rpm, single-phase only) for energy saving filter or whirlpool spa application.

Standard electric motors, asynchronous, squirrel cage type, totally enclosed, fan cooled.

Enclosure IP44 - B 5 Form - Single-phase 220 V with permanently connected capacitor and three-phase 220-380 V and 415 V - 50 Hz - 2900 rpm.

PERFORMANCE TABLE

Model	Pump power		Discharge head in metres											
	HP	KW	2	3	4	5	6	8	10	12	14	16	18	20	22
							Capacity m^3/hour								
2900 n'/rpm															
5 PC	½	0.37					10.5	8.4	6.6	4.2	2.5				
7 PC	¾	0.55					14.1	12	9.9	7.5	3.9	1			
1 PC	1	0.74					19.8	17.1	14.1	10.5	7	3	1		
1.5 PC	1½	1.10					24	22	19.5	15.8	12	6			
2 PC	2	1.47					33.6	31.2	28.8	25.2	20.4	16	10.2		
3 PC	3	2.20					39	36	33.6	30	27.6	24	19.5	15	8
1450 n'/rpm															
1 PC2V	½	0.37	8.2	7.8	5.4	0.5									
1.5 PC2V	¾	0.55	12	11.4	8.4	3									

Application

1. More than 4 metres lift and/or friction in suction line is not recommended.
2. Normal operating head of the pump on a sand filter is between 8 and 13 metres.
3. Series PC pumps must always be mounted below water level to maintain prime.
4. Maximum water temperature of installation is 60°C (140°F)
5. A control box with motor protection circuit breakers or fuses must be installed for each motor.

CURVE CARATTERISTICHE

PERFORMANCE CURVES

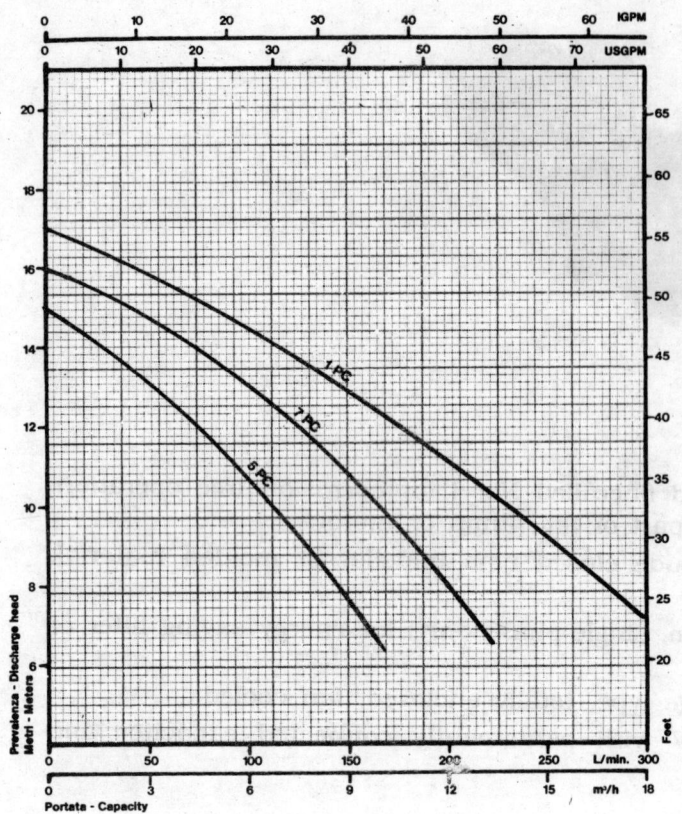

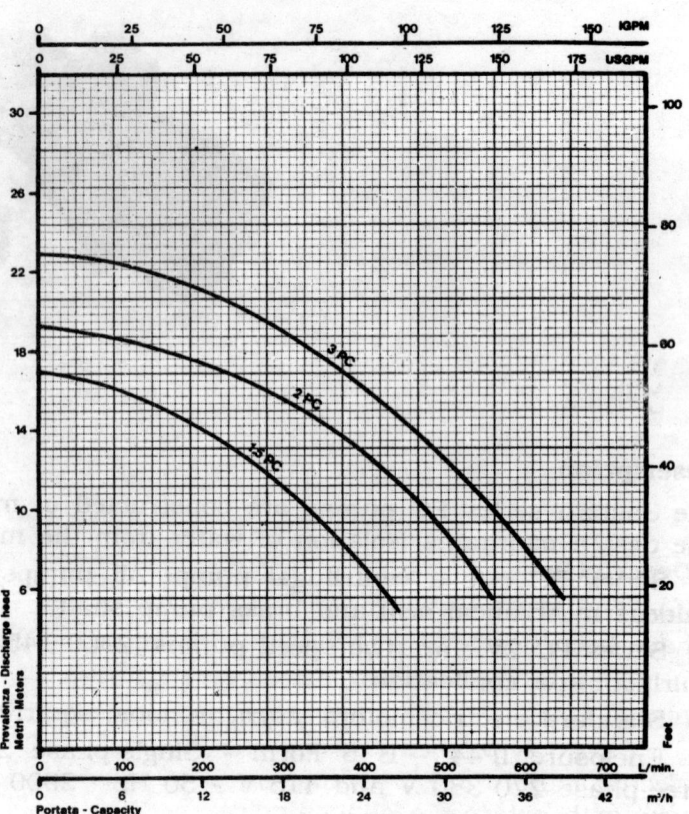

DIMENSIONI D'INGOMBRO in cm

OVER-ALL DIMENSIONS in cm

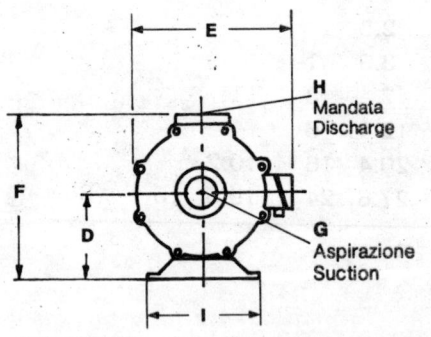

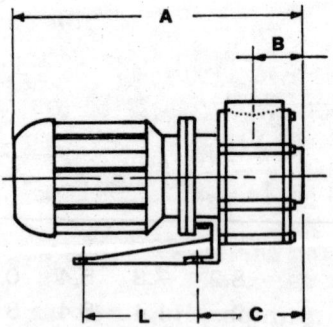

PERFORMANCE CURVES
Overall dimensions in cm

Model	A	B	C	D	E	F	G	H	I	L
5 PC	36.5	6	14.5	12.5	25	23.6	2″	1½″	12.7	16.5
7 PC	36.5	6	14.5	12.5	25	23.6	2″	1½″	12.7	16.5
1 PC	39.5	6	14.5	12.5	25	23.6	2″	1½″	12.7	16.5
1.5 PC	39.5	6	14.5	12.5	25	23.6	2″	1½″	12.7	16.5
2 PC	46	6	17.3	12.5	25	25	2″	2″	12.7	16.5
3 PC	46	6	17.3	12.5	25	25	2″	2″	12.7	16.5

The above data are indicative only.

RECIRCULATION PUMPS SERIES WPH

Description

Jacuzzi's WPH series pumps are made from corrosion-resistant fibreglass-reinforced ABS plastic. They have been designed so that no water comes into contact with any metallic part of the electric motor.

These are electrically driven centrifugal pumps with a built-in 6″ prefilter and basket, and a clear plastic cover.

Closed type plastic impeller; mechanical seal made of graphite and porcelain. Single phase flanged electric motor, 220/240 Volts, with permanently engaged condenser, 2900 RPM - 50 Hz. Squirrel cage type with external cooling. Enclosure IP-44-B 5 form.

Pumps are available in two versions :

a. with threaded 1½″ suction and pressure connections:

b. with threaded suction and adjustable union on pressure side (diameter 1½″).

PERFORMANCE TABLE

Model	Pump power		Discharge head in metres												
	HP	KW	2	3	4	5	6	8	10	12	14	16	18	20	22
							Capacity m^3/hour								
6 WPH	0.6	0.44					14.9	12.7	9.5	4.2	–				
8 WPH	0.8	0.59					19.0	16.5	13.1	7.8	–				
12 WPH	1.2	0.88					23.2	20.7	17.4	12.5	5.0				

Application

1. Series WPH filter pumps should be selected to provide the proper capacity at the total head developed by : a) suction lift; b) pressure drop across filter media, about 5 metres of water column; c) pipe friction in suction and discharge lines.
2. More than 4 metres lift including friction in suction line is not recommended.
3. Normal operating head of the pump on a sand filter is between 8 and 12 metres.
4. Maximum water temperature of installation is 60°C (140°F).
5. A control box with motor protection circuit breakers or fuses must be installed for each motor.

E CARATTERISTICHE

PERFORMANCE CURVES

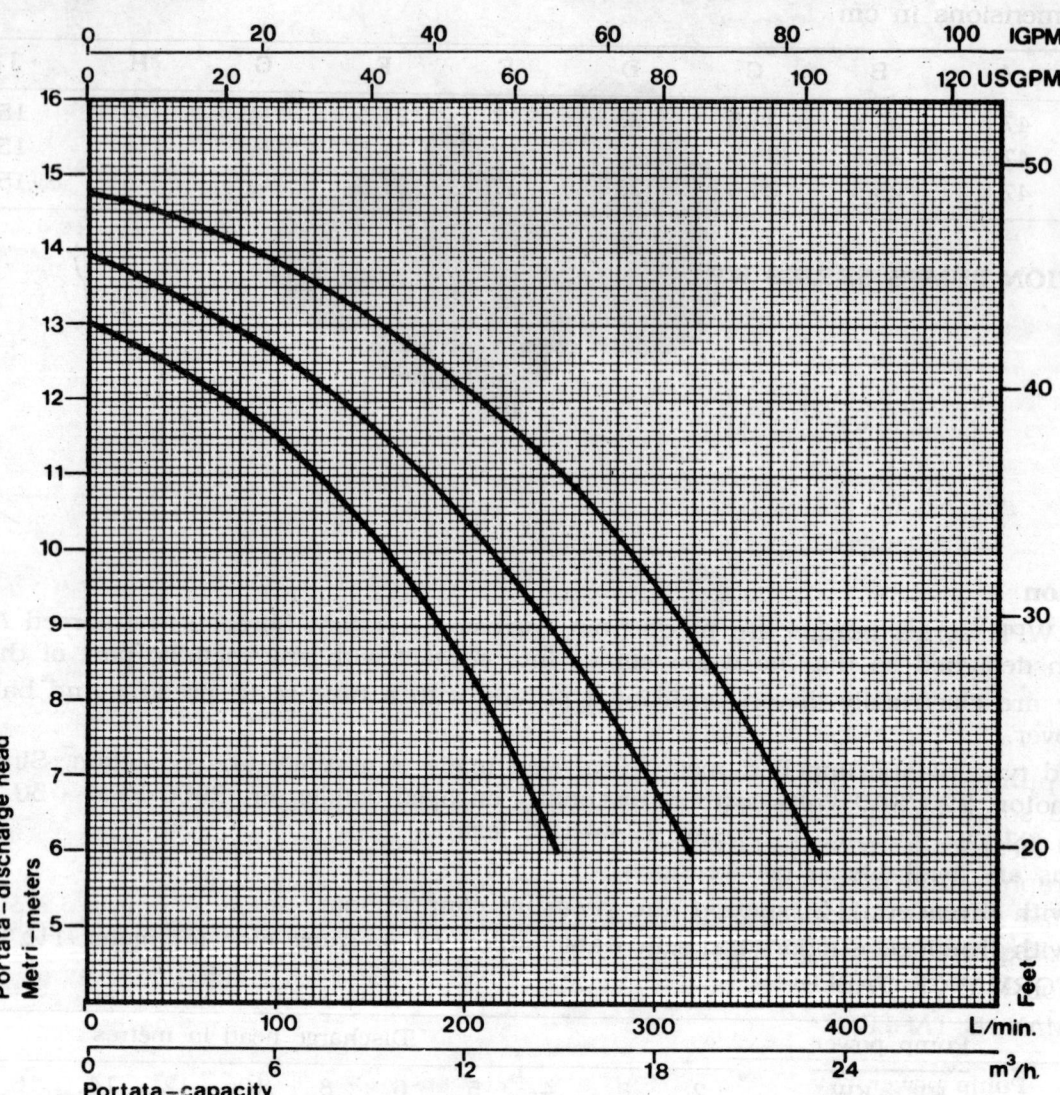

NSIONI D'INGOMBRO in cm

OVER-ALL DIMENSIONS in cm

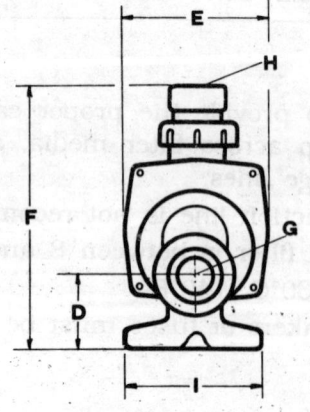

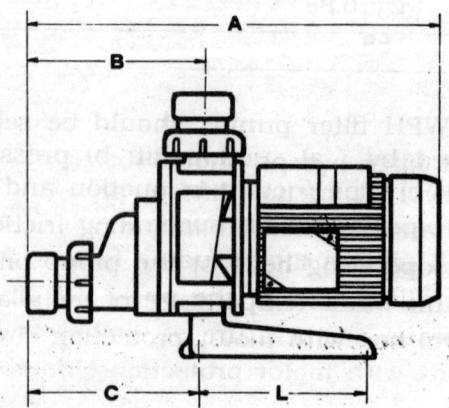

PERFORMANCE CURVES
Overall dimensions in cm

Model	A	B	C	D	E	F	G	H	I	L
6 WPH	47.5	25.5	24.5	23	17	31.5	1½"	1½"	15	17
8 WPH	47.5	25.5	24.5	23	17	31.5	1½"	1½"	15	17
1.2 WPH	47.5	25.5	24.5	23	17	31.5	1½"	1½"	15	17

RECIRCULATION PUMPS SERIES WP

Description

Recirculation pumps made from corrosion-resistant fibreglass- reinforced ABS plastic. These pumps have been designed so as to prevent water coming in contact with any metallic part of the electric motor.

These are electrically driven centrifugal pumps with a closed- type plastic impeller; mechanical seal is made of graphite and porcelain.

Single-phase flanged type electric motor, 220/240 Volts, with permanently engaged condenser; 2900 RPM - 50 Hz. Squirrel cage type with external cooling. Enclosure IP-44-B5 form.

The suction and pressure sides of the pumps are fitted with adjustable connections.

PERFORMANCE TABLE

Model	Pump power		Discharge head in metres												
	HP	KW	2	3	4	5	6	8	10	12	14	16	18	20	22
			Capacity m³/hour												
6 WP	0.6	0.44					14.9	12.7	9.5	4.2	–				
8 WP	0.8	0.59					19.0	16.5	13.1	7.8	–				
12 WP	1.2	0.88					23.2	20.7	17.4	12.5	5.0				

Application

1. Series WPH filter pumps should be selected to provide the proper capacity at the total head developed by : a) suction lift; b) pressure drop across filter media, about 5 metres of water column; c) pipe friction in suction and discharge lines.
2. More than 4 metres lift including friction in suction line is not recommended.
3. Normal operating head of the pump on a sand filter is between 8 and 12 metres.
4. Maximum water temperature of installation is 60°C (140°F).
5. A control box with motor protection circuit breakers or fuses must be installed for each motor.

VE CARATTERISTICHE

PERFORMANCE CURVES

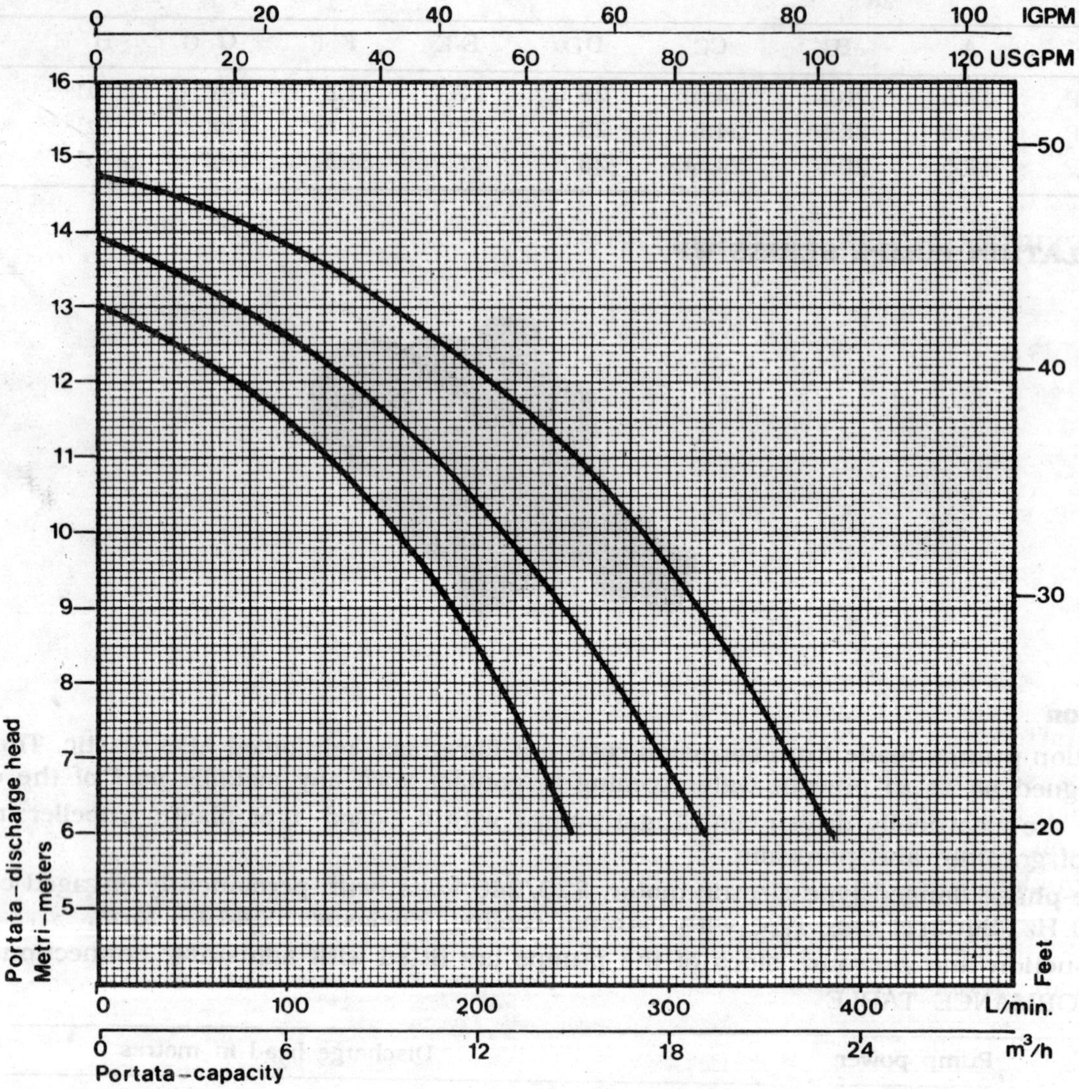

NSIONI D'INGOMBRO in cm

OVER-ALL DIMENSIONS in cm

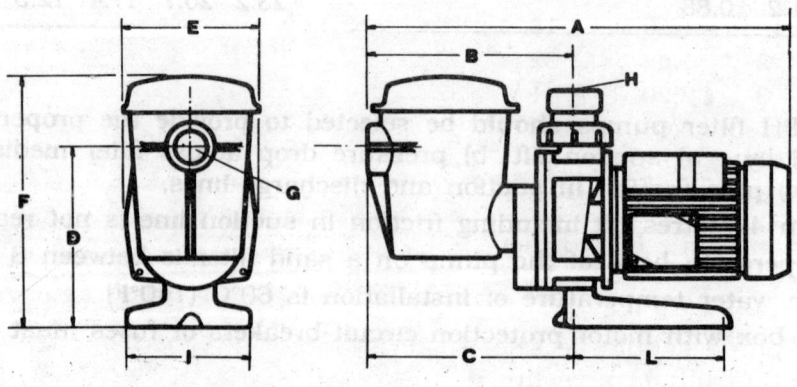

PERFORMANCE CURVES
Overall dimensions in cm

Model	A	B	C	D	E	F	G	H	I	L
6 WP	42	19.5	19.5	8.5	17	29.5	1½"	1½"	15	17
8 WP	42	19.5	19.5	8.5	17	29.5	1½"	1½"	15	17
1.2 WP	42	19.5	19.5	8.5	17	29.5	1½"	1½"	15	17

POOL FILTER PUMPS SERIES USF-ULS

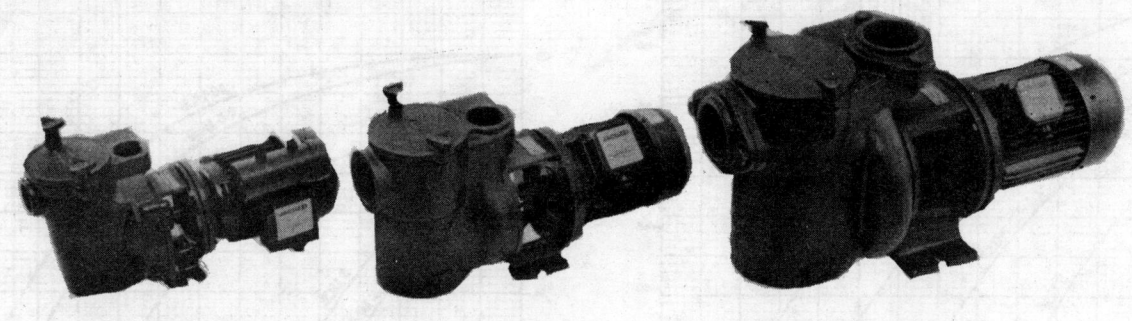

Description

Jacuzzi self-priming centrifugal units equipped with a prefilter and basket made of non-corrosive material. Impeller has open blades to pass small debris without clogging. Units are made of close grain cast iron with mechanical seal and brass pump shaft (USF) and stainless steel (ULS).

Standard electric motors, asynchronous, squirrel cage type, totally enclosed, fan cooled. Enclosure IP44 - B5 Form - Single- phase 220 V with permanently connected capacitor and three-phase 220-380 V and 415 V - 50 Hz - 2900 rpm.

PERFORMANCE TABLE

Model	Pump Power		Discharge head in metres										
	HP	KW	6	8	10	12	14	16	18	20	22	24	26
			Capacity m³/hour										
5 USF	½	0.37	9	8.2	7.2	5.7	4.1	2					
7 USF	¾	0.55	12.4	11.4	10.1	8.7	6.8	4.2					
1 USF	1	0.74	16	14.5	13.1	11.4	9.6	7.2	3				
1.5 USF	1½	1.10	18	16.9	15.5	14	12.2	10	6.6	1			
2 ULS	2	1.47	32	29	26	23	21	16	8				
3 ULS	3	2.20	41	37	33	30	26	22	16	8			
5 ULS*	5	3.68	68	64	61	57	53	48	42	32	15		
7.5 ULS*	7½	5.52	81	78	74	71	68	64	59	54	45	31	1
10 ULS*	10	7.36	88	83	80	78	74	70	66	61	54	42	1

* Available also in bronze construction

Application

1. Filter pump should be selected to provide the proper capacity at the total head developed by :
a) suction lift; b) pressure drop across filter media, about 5 metres of water column; c) pipe friction in suction and discharge lines.

2. More than 4 metres lift and/or friction in suction line is not recommended.
3. Normal operating head of the pump on a sand filter is between 10 and 20 metres.
4. A control box with motor protection circuit breakers of fuses must be installed for each motor.
5. All units are designed for fresh water, in case of salt water application contact factory.

CURVE CARATTERISTICHE

PERFORMANCE CURVES

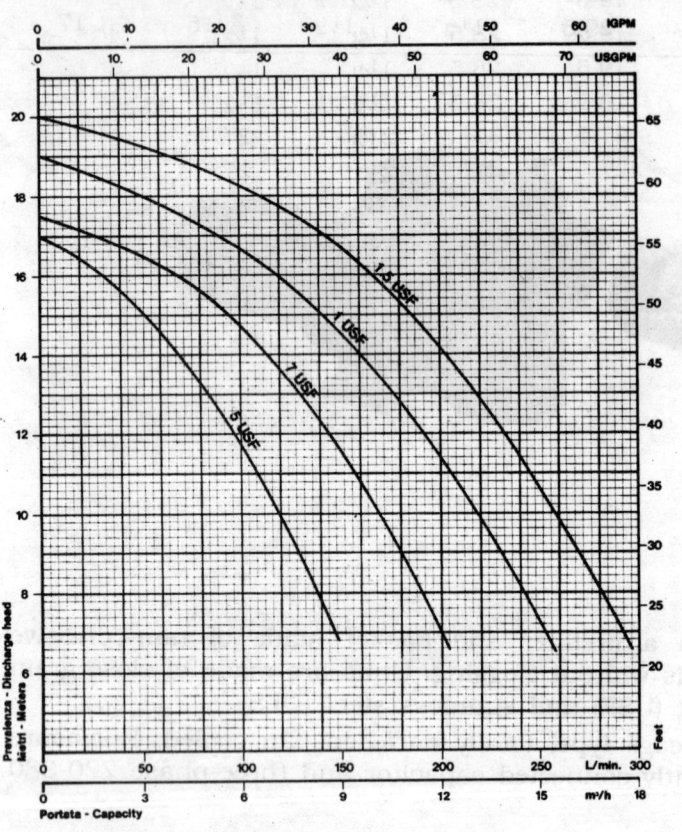

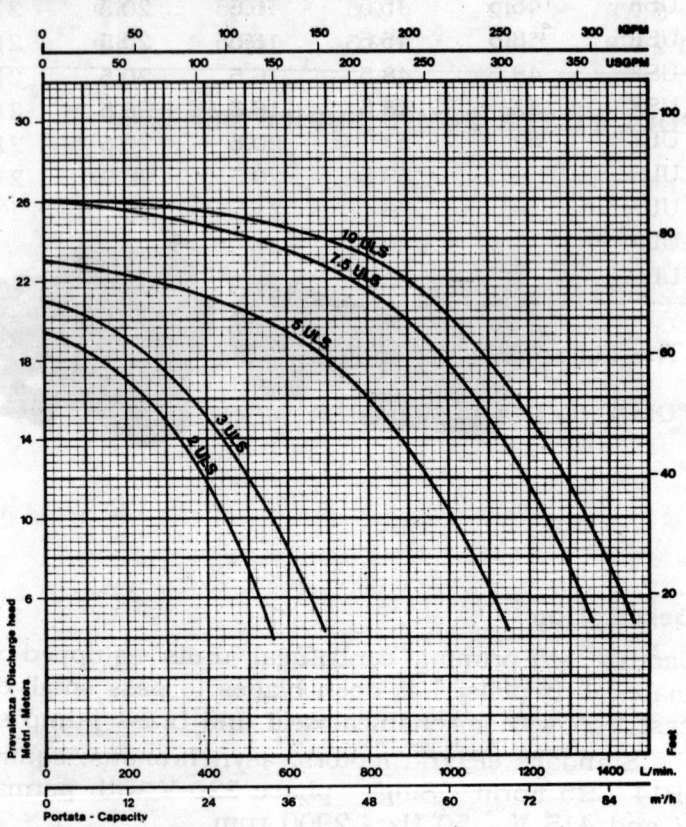

DIMENSIONI D'INGOMBRO in cm

OVER-ALL DIMENSIONS in cm

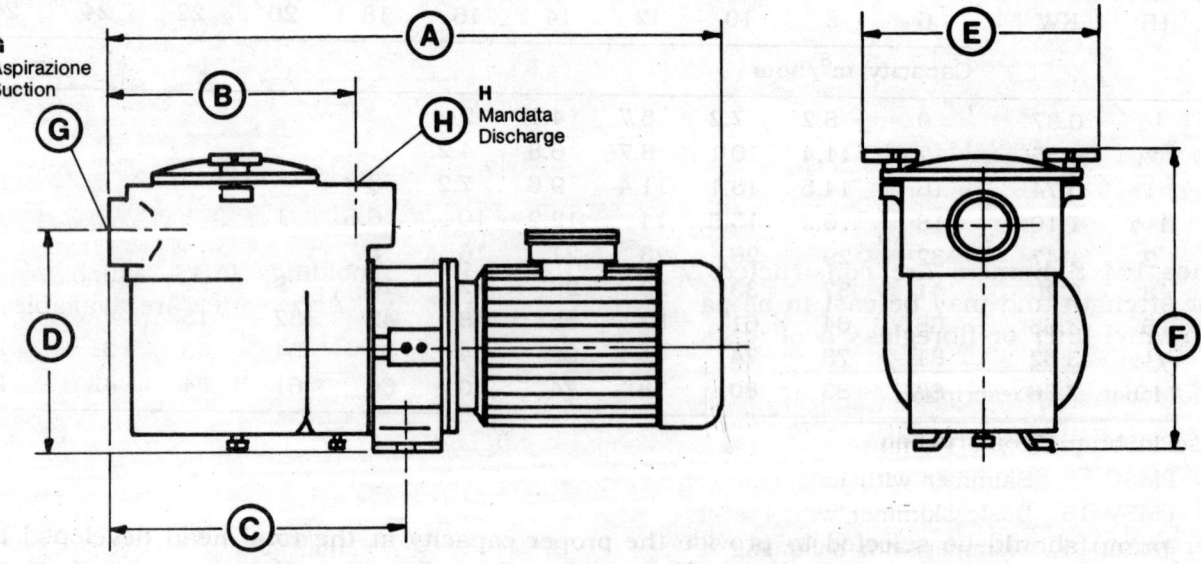

PERFORMANCE CURVES
Overall dimensions in cm

Model	A		B	C	D	E	F	G	H	Ø Prefilter
	1 Phase	3 Phase								
5 USF	45.5	45.5	16.5	20.5	21	19.5	28.5	1½″	1½″	4″
7 USF	48.5	45.5	16.5	20.5	21	19.5	28.5	1½″	1½″	4″
1 USF	48.5	48.5	16.5	20.5	21	19.5	28.5	1½″	1½″	4″
1.5 USF	52	48.5	16.5	20.5	21	19.5	28.5	1½″	1½″	4″
2 ULS	59	57	20.5	26.5	21	21.5	31	2½″	2″	6″
3 ULS	–	59.5	20.5	26.5	21	21.5	31	2½″	2″	6″
5 ULS	–	76	31	39	29.5	30	41	3″	3″	8″
7.5 ULS	–	80	31	39	29.5	30	41	3″	3″	8″
10 ULS	–	80	31	39	29.5	30	41	3″	3″	8″

The above data are indicative only.

AUTOMATIC POOL SKIMMER SERIES PM

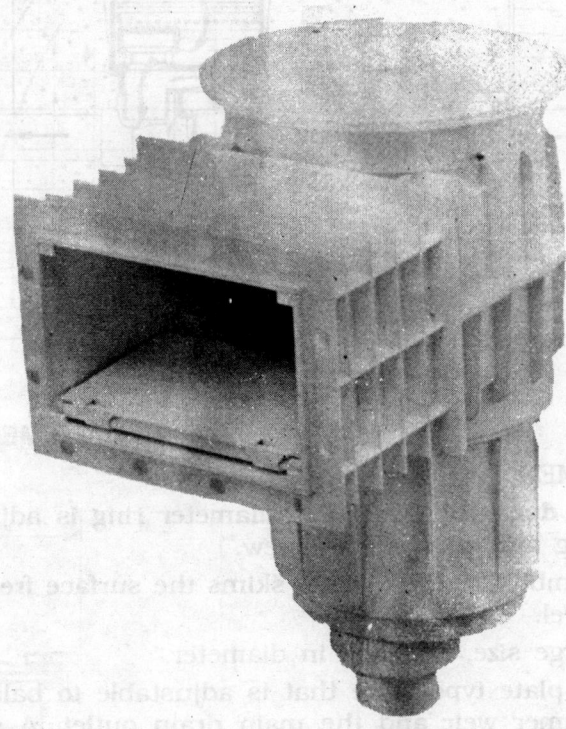

Description

The Jacuzzi series PM Skimmers are constructed of white thermoplastic moldings (ABS) which are heavily ribbed for strength and may be cast in as part of a concrete pool wall. Accessories are available to allow use with vinyl liner or fibreglass pool walls.

Model	Description
PMSV	Basic skimmer
PMSC	Skimmer with balancing valve
PMSV 15	Basic skimmer with over-flow outlet 1½″ dia.
PMSC 15	Skimmer with balancing valve and over-flow outlet ½″ dia.

ACCESSORIES FOR SKIMMER

Throat extension : model PMPS, used with liner or fibreglass pools. Allows skimmer to be located 18 cm away from pool wall.

Face flange : model FTK-L (plastic) used with liner or fibreglass pools. Allows watertight seal between skimmer flange and pool wall by use of bolted flange.

Skimmer to hose adapter : model PHMS, allows pool vacuum hose to be directly connected to pump outlet opening of skimmer.

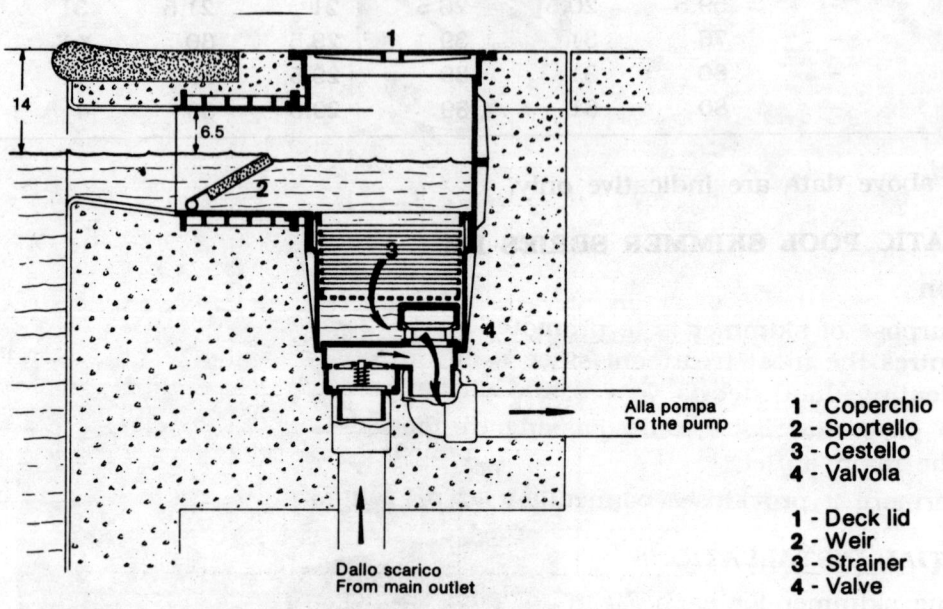

Alla pompa
To the pump

Dallo scarico
From main outlet

1 - Coperchio
2 - Sportello
3 - Cestello
4 - Valvola

1 - Deck lid
2 - Weir
3 - Strainer
4 - Valve

A COMPLETE PM SKIMMER INCLUDES

Mounting ring with deck lid : 27.2 cm diameter ring is adjustable to match exact deck height. Lid is twist-locked to ring and secured by screw.

Weir : flap type assembly, 20.5 cm wide, skims the surface freely and continuously through a 11.4 cm variation in water level.

Strainer basket : large size, 18.3 cm in diameter.

Balancing valve : a plate type valve that is adjustable to balance the flow of water taken by the pump suction from skimmer weir and the main drain outlet. A special float valve will automatically shut-off skimmer if unit become dry to eliminate loss of pump prime. This feature is optional.

Equalizing outlet : opening 1½″ for connection to main drain of pool. Flow of water is controlled by balancing valve (installation with one skimmer only).

Pump outlet : opening 1½″ for connection to recirculation pump suction.

Over-flow outlet : opening 1½″ located at normal water level for drainage of excess pool water. This feature is optional.

Trimmer valve : small adjustable plate to restrict pump outlet and balance flow between multiple skimmers installed in one pool.

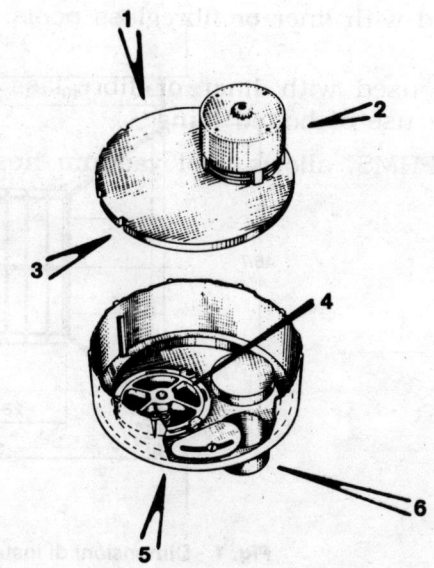

Application

The purpose of skimmer is to promote surface flow across the pool. In addition to drawing in water which requires the most treatment since it is continuously exposed to sun, rain and wind, the skimmer removes floating dust, debris, leaves and oil films before they can accumulate on floors and walls.

It also proportions the water passing through the filter between the flows across the surface and through the main outlet.

Furthermore it provides a connection for a vacuum cleaner hose.

RESIDENTIAL INSTALLATIONS

Use one skimmer for each 70 m² of pool surface or any part remaining thereof.

Flow rate through skimmer is 1.2 m³h minimum and 12 m³h maximum.

Piping for skimmers must be sized to handle at least 50% of filter flow rate.

COMMERCIAL INSTALLATIONS

Use one skimmer for each 45 m² of pool surface or any part remaining thereof.

Flow rate through skimmer is 1.2 m³h minimum and 12 m³h maximum.

Piping for skimmers must be sized to handle at least 80% of filter flow rate.

Use of equalizing outlet in pool wall is recommended.

BELOW DECK EQUIPMENT

MAIN DRAINS

Jacuzzi series MO and PM Main Drains are constructed of white thermoplastic moldings which are ribbed for strength and may be cast in as part of a concrete pool bottom.

Design includes modern anti-vortex cover which snaps on to main body.

Models are available to allow use with vinyl liner or fibreglass pools. The liner or pool bottom is clamped between two gaskets and secured to drain outlet body by eight stainless steel screws.

Optional hydro-static relief valves are available. Installation allows high ground water levels to enter if pool has been drained preventing the pool shell from 'floating' up out of the ground.

MOD. MO 20 Concrete pools - Dimensions in mm Prefabricated pools - Dimensions in mm

MOD. PM 550 Concrete pools - Dimensions in mm Prefabricated pools - Dimensions in mm

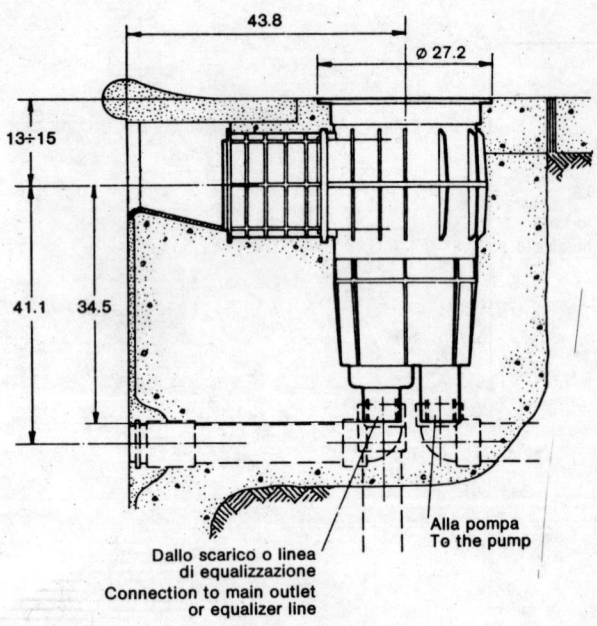

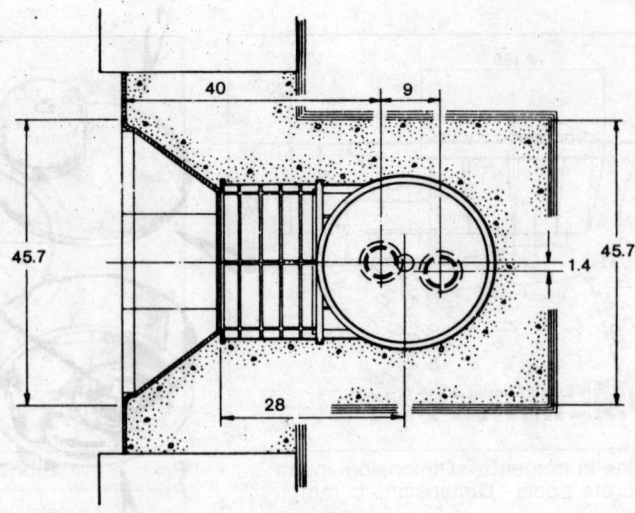

Dallo scarico o linea
di equalizzazione
Connection to main outlet
or equalizer line

Alla pompa
To the pump

Fig. 1 - Dimensioni di installazione
per piscine costruite in
cemento
Dimensions for installation
in concrete pool

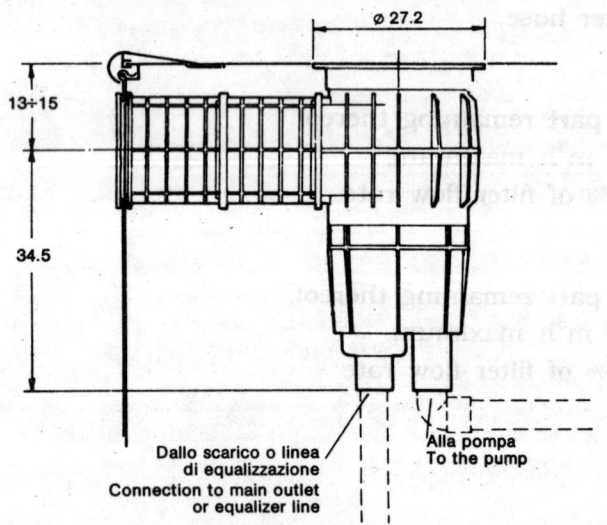

Dallo scarico o linea
di equalizzazione
Connection to main outlet
or equalizer line

Alla pompa
To the pump

Parete | Pool wall

Cornice
Face flange

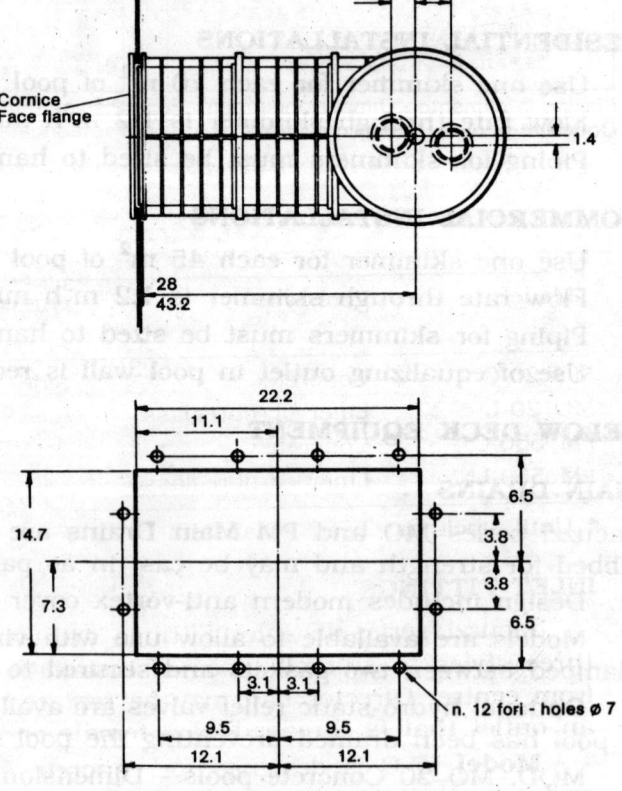

n. 12 fori - holes Ø 7

Fig. 2 - Dimensioni di installazione
per piscine prefabbricate

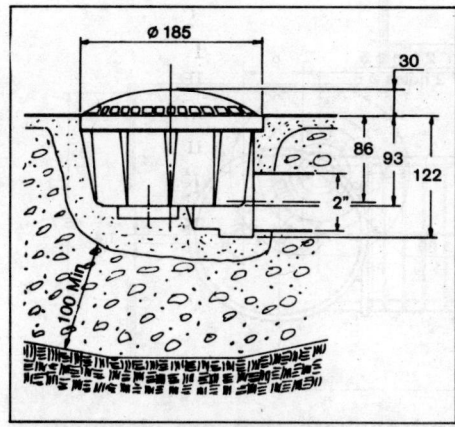

Piscine in cemento - Dimensioni in mm
Concrete pools - Dimensions in mm

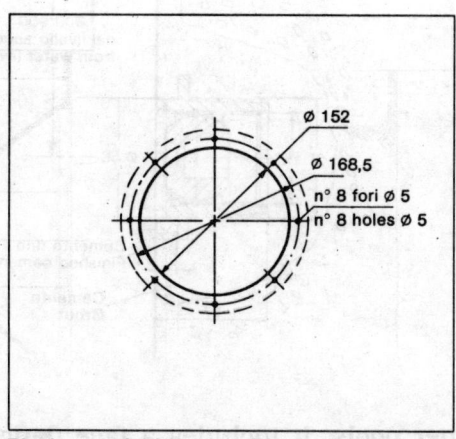

Piscine prefabbricate - Dimensioni in mm
Prefabricated pools - Dimensions in mm

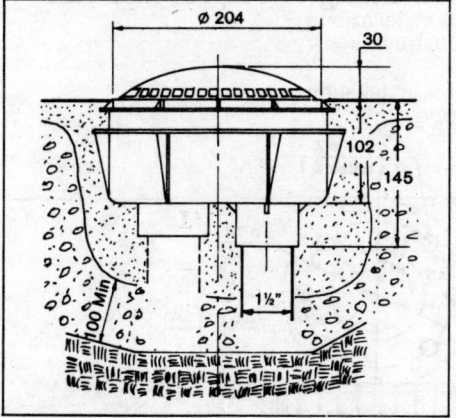

Piscine in cemento - Dimensioni in mm
Concrete pools - Dimensions in mm

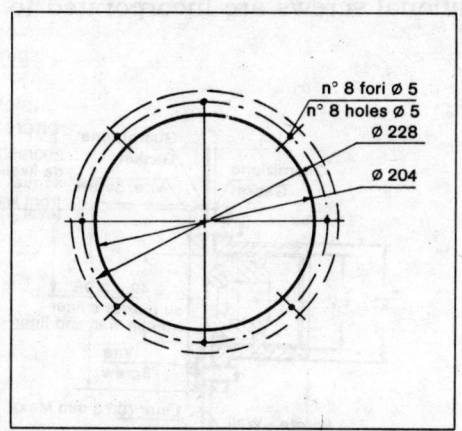

Piscine prefabbricate - Dimensioni in mm
Prefabricated pools - Dimensions in mm

Model	Pool type	Outlet location	Max. flow rate m^3h	Size outlet	Hydro-relief valve
MO 20-C	Concrete	Side	15	2"	Mod. MOHRV
MO 20-L	Liner or fibreglass	Side	15	2"	Mod. MOHRV
PM 550-C*	Concrete	Bottom	12	1½"	Mod. RVP 150
PM 550-L*	Liner or fibreglass	Bottom	12	1½"	Mod. RVP 150

* Until stock has been exhausted.

INLET FITTINGS

Jacuzzi series IF Inlet Fittings are constructed of white thermoplastic moldings and available in three styles to suit all types of installations and flow rates. Flow direction is adjustable up to 30°C from centre. Director ball may be replaced with 1½" plug for pressure testing of piping. Fittings have an outlet that is threaded 1½" female or 2" male for optional pipe sizes.

Model IF-C is designed for concrete pools. It includes a no-leak flange with two mounting holes and a protection cap over the face.

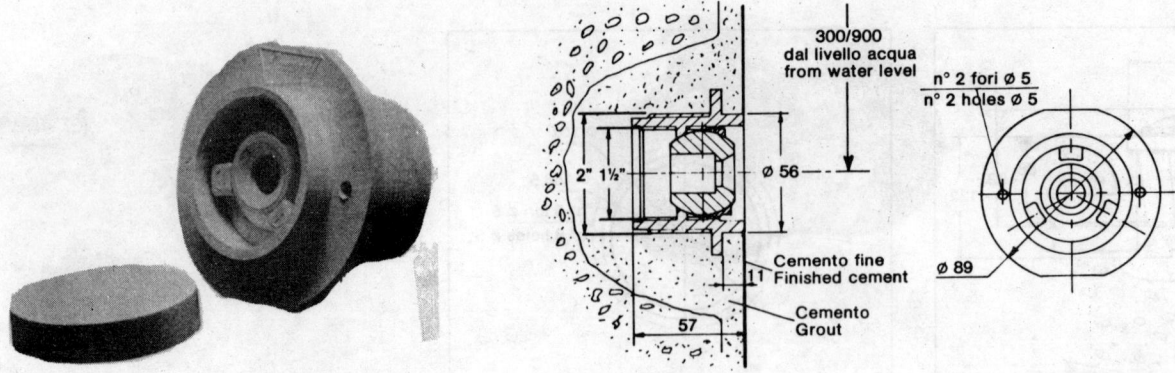

Model IF-L is designed for vinyl liner pools. It includes a face flange and gaskets that are clamped on both sides of the liner by four stainless steel screws. This design allows the liner to be removed without disturbing the fitting. Two additional screws are incorporated to allow fitting to be pre-mounted on pool walls before installation.

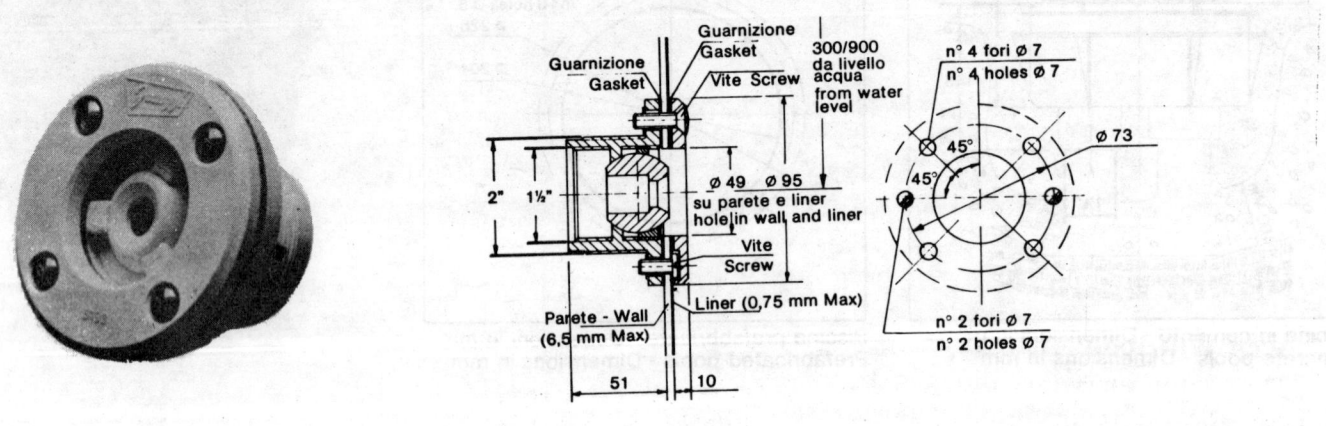

Model IF-A is designed for fibreglass pools. It includes a molded face flange and gaskets that are clamped on both sides of the wall by a draw nut. Installation quick, utilizing only one drilled hole.

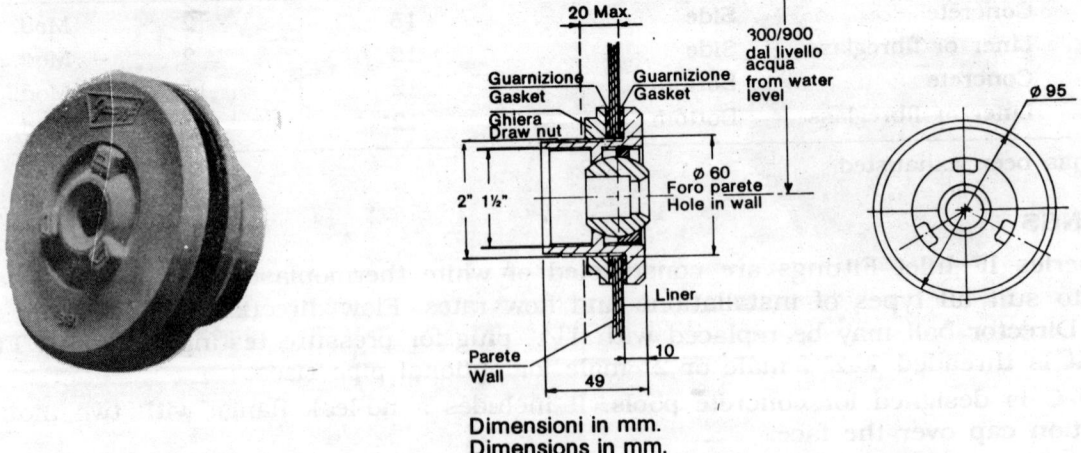

Dimensioni in mm.
Dimensions in mm.

LRF 40 Chromed bronze wall fitting for concrete pools; with grate. Capacity 40 m^3/h.

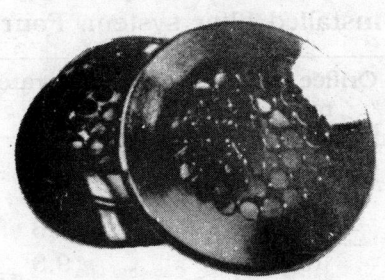

AFR - F20 Chromed bronze adjustable floor inlets; 12 mm flange on bottom; 136 mm diameter. Capacity 20 m^3/h

HTA Whirlpool fittings Jacuzzi's HTA fittings are made of ABS thermoplastic material.

The water/air jet is directed by means of an adjustable nozzle, which has a 15° movement in all directions with respect to its horizontal axis. The intensity is controlled by turning the knurled ring anti-clockwise to increase the jet, clockwise to reduce it.

Capacity 150 litres/minute at a pressure of 8 mt.

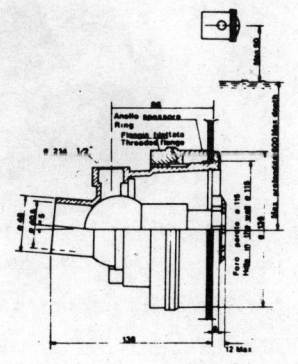
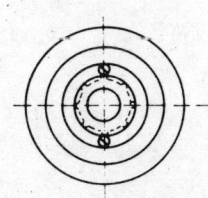

Flow rate selection chart

The orifice size of the inlet fittings should be selected to pass the amount of water equal to the desired filter-rate and pump capacity of the installed filter system. Four sizes are available.

Model	Type of pool	Orifice size mm	Max. flow rate m³h	Recommended pipe size
IF-C50	Concrete	13	3.6	1½″ M
IF-C75	Concrete	19	9.3	1½″ M
IF-L50	Liner	13	3.6	1½″ M
IF-L75	Liner	19	9.3	1½″ M
IF-A50	Fibreglass	13	3.6	1½″ M
IF-A75	Fibreglass	19	9.3	1½″ M
HTA	Concrete	30	9	1½″
LRF 40	Concrete	–	40	3″ M
AFR - F 20	Concrete	–	20	M 2″ - F ½″

VACUUM FITTINGS

Jacuzzi series VF Vacuum Fittings are constructed of white thermoplastic moldings and include a 1½″ plug that may be removed for insertion of HW 15 hose to wall adapter.

Installation features are same as series IF fittings.

Model	Type of pool	Pipe size	
		Female	Male
VF - C	Concrete	1½″	2″

ROPE ANCHORS

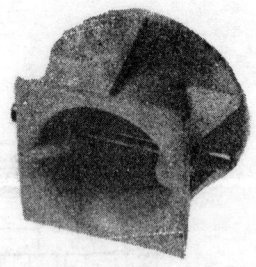

Model PMCA rope anchor is constructed of white thermoplastic molding with stainless steel pin. It is suitable for concrete pools.

LRF - P

Louver type inlet fitting suitable for concrete pools has female connections 1½" - 2" with plugs. Length 25 cm.

BELOW DECK EQUIPMENT

UNDERWATER LIGHTS

Jacuzzi series UL 300 Underwater Lights are constructed of white thermoplastic moldings and may be cast in as part of a concrete pool wall. Models are available to allow use with vinyl liner of Fibreglass pools. The liner or pool wall is clamped between two gaskets and secured to the light nitch by stainless steel screws.

A light installation includes the following components :

    a) **Light assembly** complete with nitch, 300 Watt - 12 Volt, lamp and conduit.

    b) **Deck box** for connection between conduit and electrical service.

    c) **Transformer** for underwater lights (for 1,2 or 3 lights).

220 V - 50 Hz - 12 V.

115 V - 60 Hz - 12 V.

    d) **Trim ring** of stainless steel is optional.

Model	Description	Voltage	Outlet size
UL 300 - C	Light assembly for concrete	12 V	¾″
UL 300 - L	Light assembly for liner	12 V	¾″
SD	Deck box	–	¾″ × ¾″
LV 220/1/2/3	Step-down transformer	220/12 V	–
LV 115/1/2/3	Step-down transformer	115/12 V	–
LR 300	Trim ring stainless steel	–	–

OVERALL DIMENSIONS in cm

PISCINE IN CEMENTO   **CONCRETE POOLS**

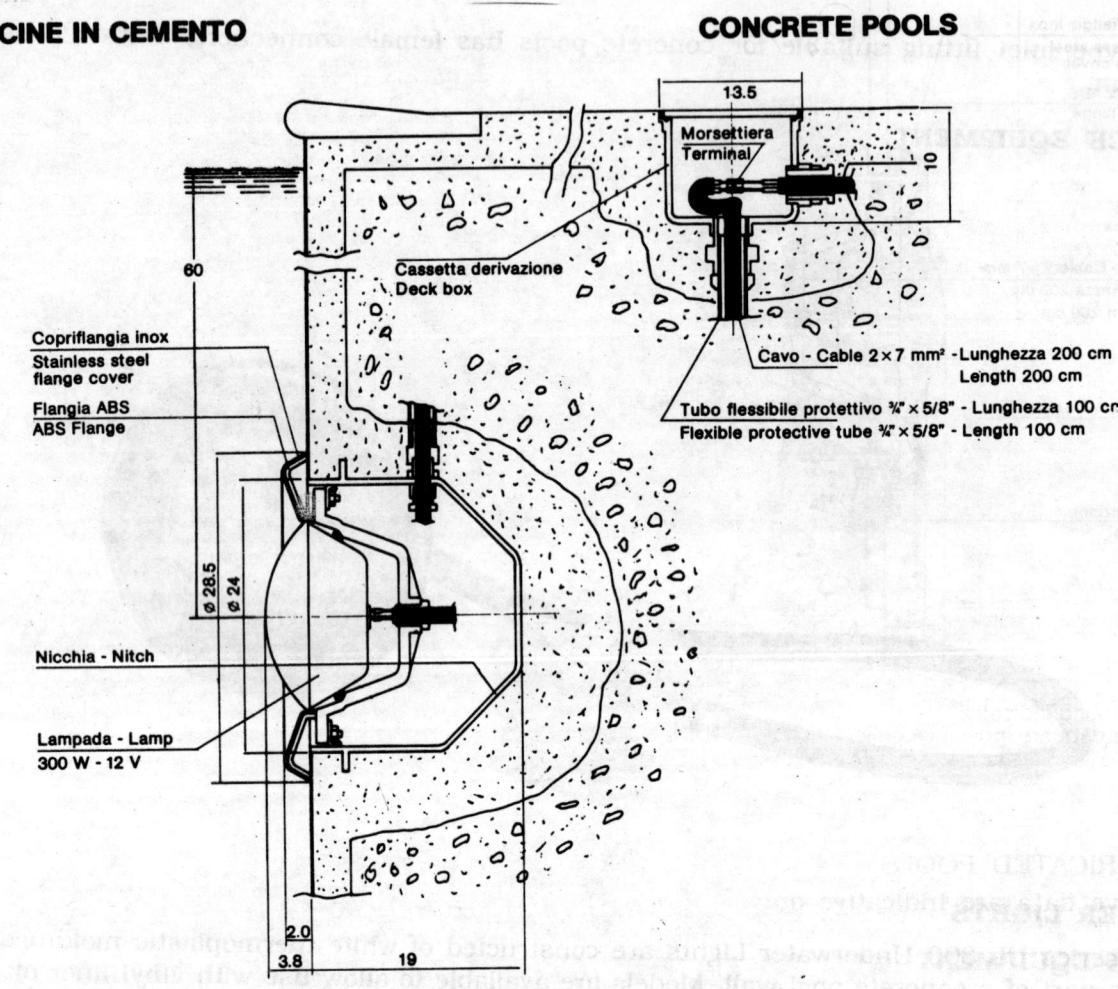

Morsettiera
Terminal

Cassetta derivazione
Deck box

Coprilfangia inox
Stainless steel
flange cover

Flangia ABS
ABS Flange

Cavo - Cable 2 × 7 mm² - Lunghezza 200 cm
Length 200 cm

Tubo flessibile protettivo ¾″ × 5/8″ - Lunghezza 100 cm
Flexible protective tube ¾″ × 5/8″ - Length 100 cm

Nicchia - Nitch

Lampada - Lamp
300 W - 12 V

CONCRETE POOLS
NOTE :

Leave sufficient length of cable on inside of nitch to substitute lamp removing it from water (about 110 cm).

PISCINE PREFABBRICATE

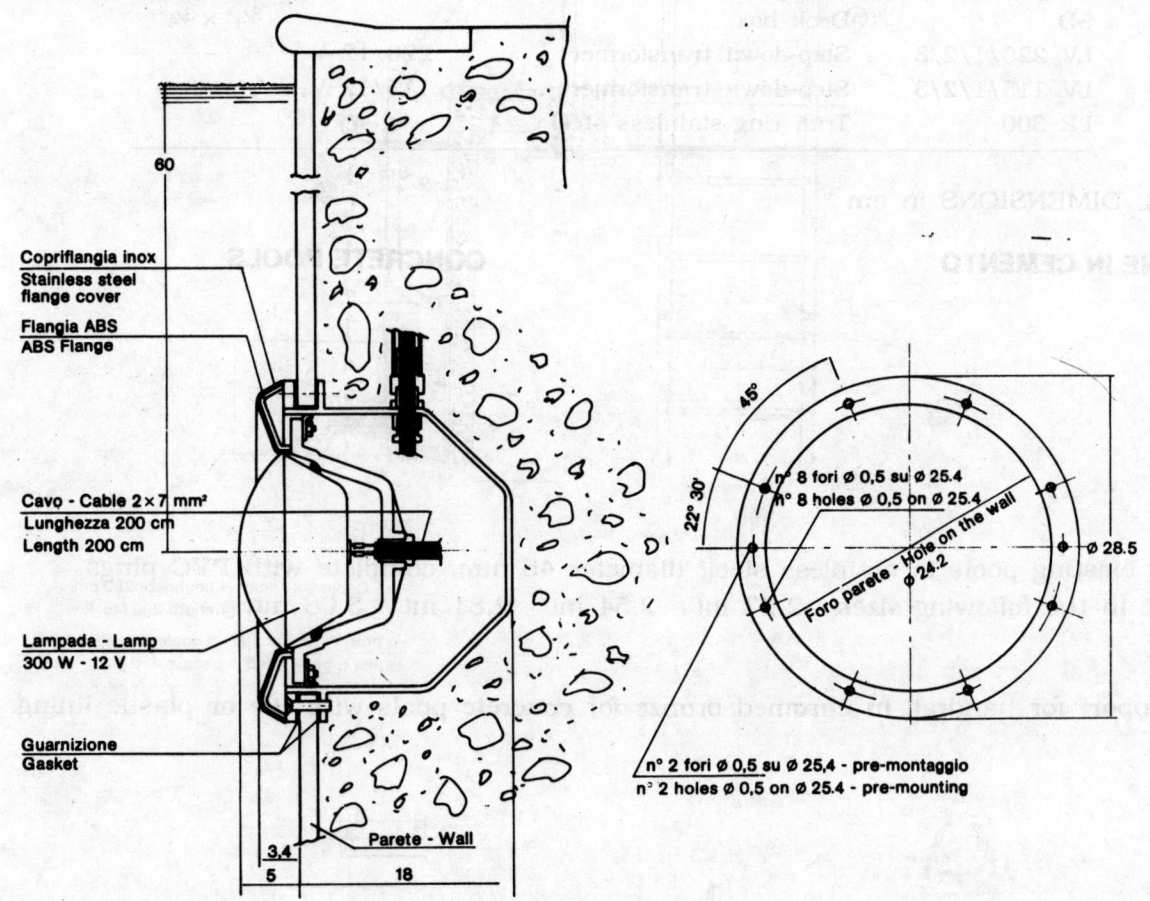

Copriflangia inox
Stainless steel
flange cover

Flangia ABS
ABS Flange

Cavo - Cable 2 × 7 mm²
Lunghezza 200 cm
Length 200 cm

Lampada - Lamp
300 W - 12 V

Guarnizione
Gasket

Parete - Wall

n° 8 fori Ø 0,5 su Ø 25.4
n° 8 holes Ø 0,5 on Ø 25.4

Foro parete - Hole on the wall
Ø 24.2

Ø 28.5

n° 2 fori Ø 0,5 su Ø 25,4 - pre-montaggio
n° 2 holes Ø 0,5 on Ø 25.4 - pre-mounting

Tutti i dati sono indicativi.
The above data are indicative only.

PREFABRICATED POOLS
The above data are indicative only.

POOL DECK EQUIPMENT

LADDERS

Jacuzzi series L ladders are constructed of high grade, polished stainless steel. The rails are 43 mm diameter seamless one piece tubing. The treads are of skid resistant design with stainless fastenings.

Ladder assembly includes wedge anchors and covers.

Models 2L two steps, 3L three steps and 4L four steps are available.

Overall dimensions in cm

Model	A	B	C
2L	134	70	25
3L	159	95	25
4L	184	120	25

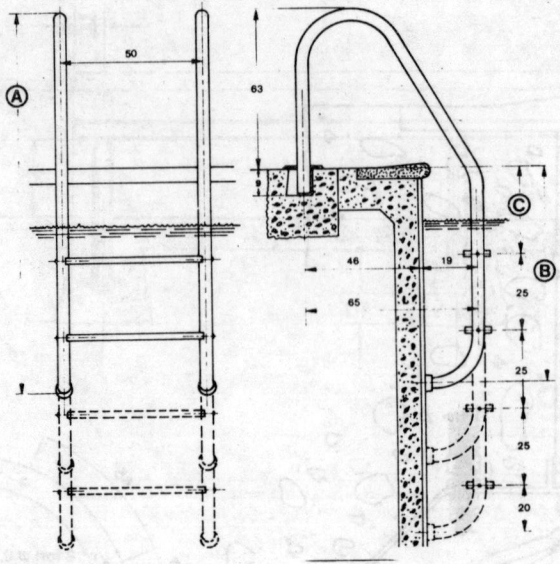

WH

Handrail for existing pools in stainless steel, diameter 45 mm, complete with **PVC** plugs.

Available in the following sizes : 2.27 mt - 2.54 mt - 2.81 mt - 3.08 mt.

WHS

Eye type support for handrail in chromed bronze for concrete pools with tile or plastic lining.

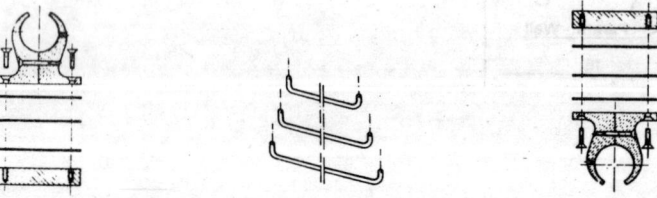

The above data are indicative only.

DIVING BOARDS

Jacuzzi series TR diving boards are constructed of Fibreglass with wooden support inserts. The top is finished in anti-skid, sand coating.

The Lux models utilize top grade kim dried Duglass fit for longer life under heavy use.

Models TR include board, stainless steel tube type support standards, mounting pads and fittings.

Model	Description	Length	
		metres	feet
TR240	Standard construction	2.40	8

SB

Made from AISI 316 stainless steel, complete with built-in handrail and non-slip fibreglass board.

Fixed type.

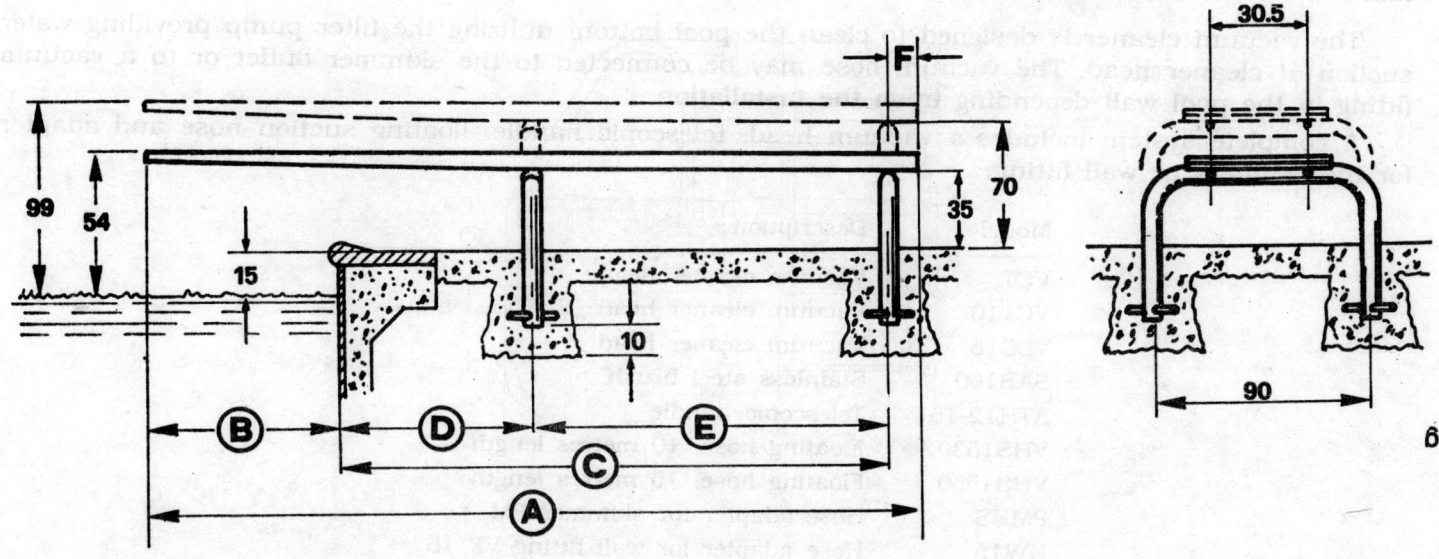

Overall dimensions in cm

Model	A	B	C	D	E	F
TR240	240	75	165	50	115	10.2

The above data are indicative only.

SAFETY AND MAINTENANCE EQUIPMENT

VACUUM CLEANING EQUIPMENT

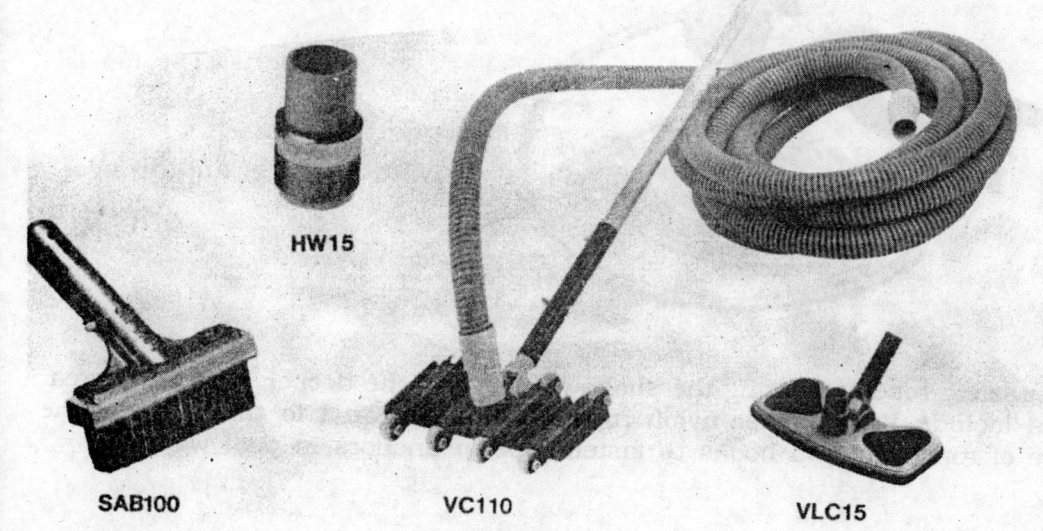

HW15

SAB100   VC110   VLC15   VCT

The vacuum cleaner is designed to clean the pool bottom utilizing the filter pump providing water suction at cleaner head. The vacuum hose may be connected to the skimmer outlet or to a vacuum fitting in the pool wall depending upon the installation.

A complete system includes a vacuum head, telescopic handle, floating suction hose and adapter for the skimmer or wall fitting.

Model	Description
VCT	Vacuum cleaner head
VC110	Vacuum cleaner head
VLC15	Vacuum cleaner head
SAB100	Stainless steel brush
ATH12-15	Telescopic handle
VHS1530	Floating hose, 10 metres length
VHS1550	Floating hose, 15 metres length
PMHS	Hose adapter for skimmer PM
HW15	Hose adapter for wall fitting VF 15

MAINTENANCE EQUIPMENT

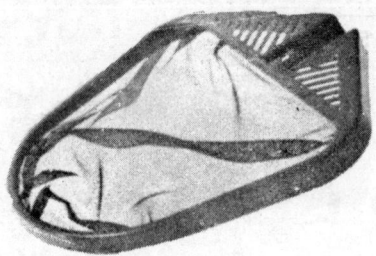 

Model CWBN 100 wall brush is 30 cm long nylon brush to be attached to vacuum cleaner handle. Use to assist in wall cleaning where vacuum cleaner use is inconvenient.

Model LS200 leaf skimmer is a molded plastic net to be attached to vacuum cleaner handle. Use to remove floating leaves and debris on pool surface.

SAFETY EQUIPMENT

Normally a pool will have a safety rope separating the shallow area from the deeper swimming area. A safety rope assembly should include ¾″ diameter nylon rope in a length equal to the width of the pool, rope floats (25 per metre of rope) and two hooks to fasten rope in anchors in pool wall.

Model	Description	Size
PR 75	Nylon rope	¾"
G20 - A	Rope floats	¾"
RH 119	Rope hook	¾"

Rope anchors must be installed during pool construction. See 'Below deck equipment' for available models.

WATER TEST KIT

Good pool maintenance includes two separate operations, pool filtration and proper balance of pool water chemistry. The pool must be kept in a neutral zone between acidity and alkalinity as measured on the pH scale by the direct addition of selected chemicals. The water must also be kept free of algae by the addition of chlorine chemical. Use a water test kit daily to identify the requirements for chemicals in each pool.

Model WTK104 includes test solutions, vials and comparator color tubes. Kit indicates chlorine standards on a scale of 0.1 through 1.2 and acidity (pH) standard on a scale of 6.8 through 8.2.

CHEMICAL FEED PUMPS

Chemical solutions required to maintain proper pool water chemistry may be conveniently added to the recirculation system by the use of chemical feed pumps. These pumps are intended primarily to serve as automatic hypochlorinators for handling calcium or sodium hypochlorite solutions. They perform equally well with other chemical solutions such as soda ash, alum and inorganic acids.

The units are of the positive-displacement, diaphragm type. Pump head and fittings are molded or machined from chemical- resistant PVC thermoplastic. Capacity of discharge is adjustable in terms of percentage (Mod. C 1005). Maximum suction lift is one metre. Motor is heavy-duty, life-time lubricated, 220 Volt - 50/60 Hz, with over-load fuse.

Pump includes pump-motor assembly, suction filter, inlet fitting with non-return valve and 2 metres of plastic tubing. To complete system, select solution tank and pump mounting.

Model	Description	Max. capacity l/h	Max. head Ate	Motor power Watt	Outlet size
C 101.5	Chemical pump	1.5	10	35	3/8"
C 1005	Chemical pump	5	10	35	3/8"
C 100	Solution tank	100 litres	–	–	–

SHOWERS

D1 Single arm stainless shower with cold water tap and footwash tap.

D2 Twin arm stainless shower with 2 cold water taps and 2 footwash taps.

PD Available fibreglass shower trap.

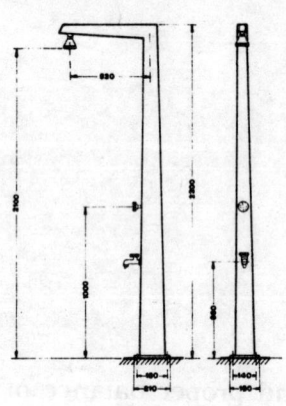 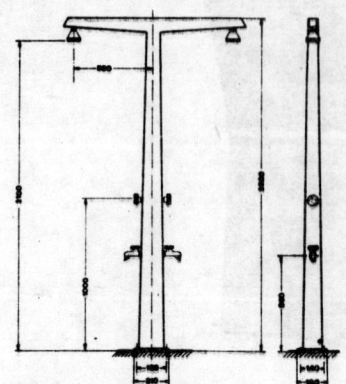 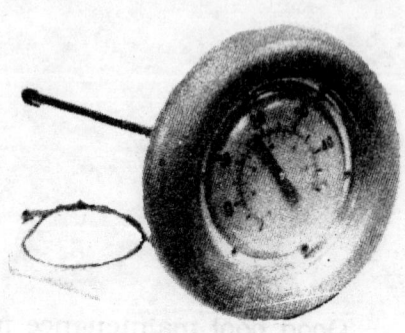

THERMOMETERS TG

Floating thermometer with blue rubber rings, diameter 18 cm; 30 cm shaft; complete with plastic cord and snap ring for fitting to edge or ladder.

–10°C to +50°C; 40 to 120°F.

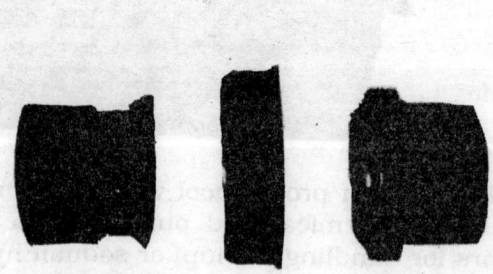

Pipe union

Articulated joint, easy to install, of multiple use.

Available either threaded 1½" gas or smooth ∅ 50 glued.

COMMERCIAL POOL FITTINGS

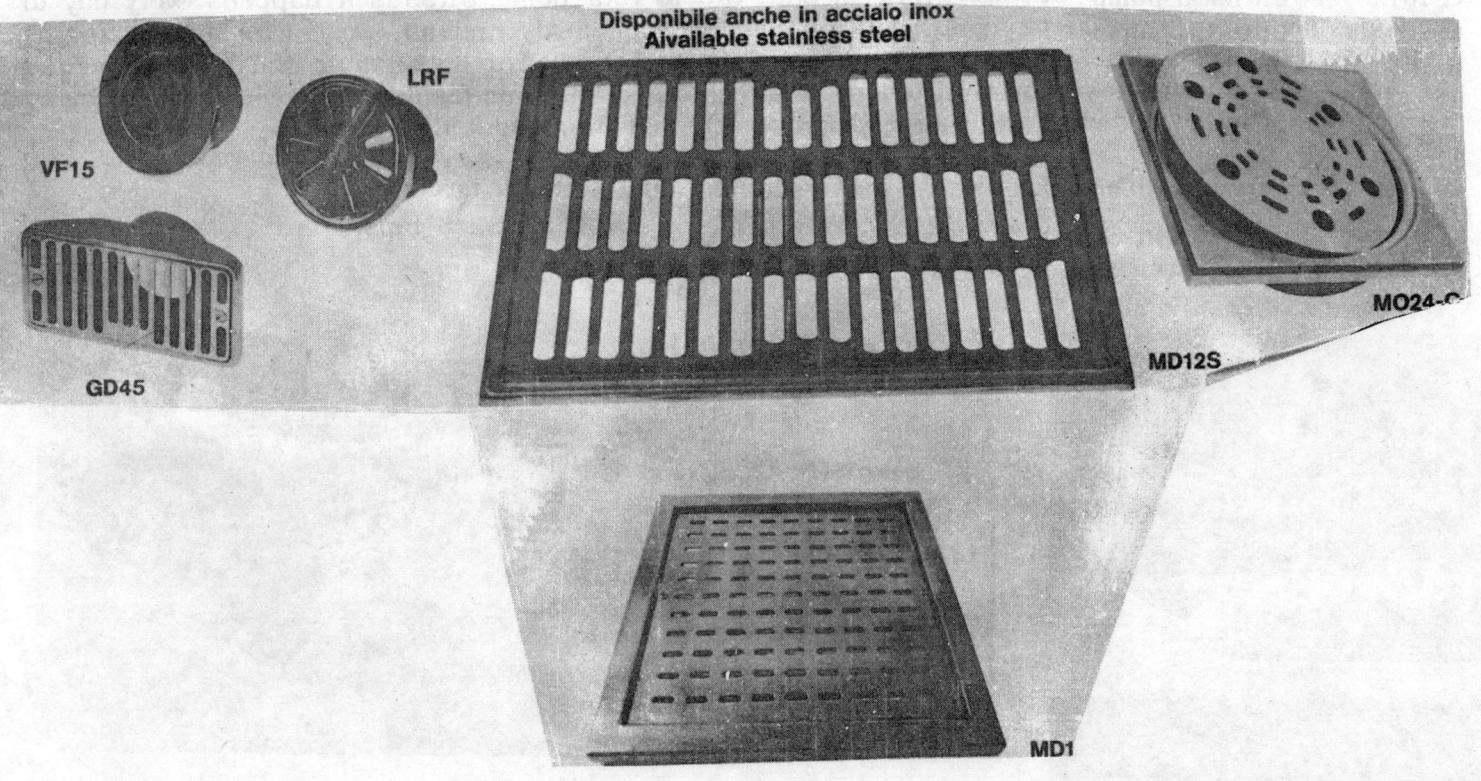

Available stainless steel

These special Jacuzzi fittings are constructed of cast bronze, chrome plated and polished to meet engineering specifications for use in concrete commercial sized pools with scum gutters.

Model	Description	Max. flow rate m^3h	Outlet size
MD12S	Main drain 12″ square (dm² 4.64)	74	Max. 4″ *
MD14S	Square grate stainless steel 35 × 35 cm	25	2″
MO24-C	Square grate (ABS material) 24.5 × 24.5 cm	11	2″
GD45	Gutter drain (dm² 0.40)	20	2″
LRF15	Inlet, lowered wall mount	12	1½″
LRF20	Inlet, lowered wall mount	20	2″
VF15	Vacuum fitting	–	1½″

* Frame and grate only, sump outlet provided by others.

MC pump card

Complete whit ground fault circuit breaker and 33 metres of rubbered electric cable.

JACUZZI THE SUPERIOR ONE

You, too, can now have a Jacuzzi whirlpool bath tub in your home. Just as it happens every day to famous people in the entertainment, cultural and sporting fields and all those who appreciate the benefits of hydromassage. A range of bathtubs, especially studied to incorporate the famous Jacuzzi hydromassage principal; with elegantly shaped and contoured lines and fashionable colours which satisfy every need. All this added to our wide experience achieved throughout the world.

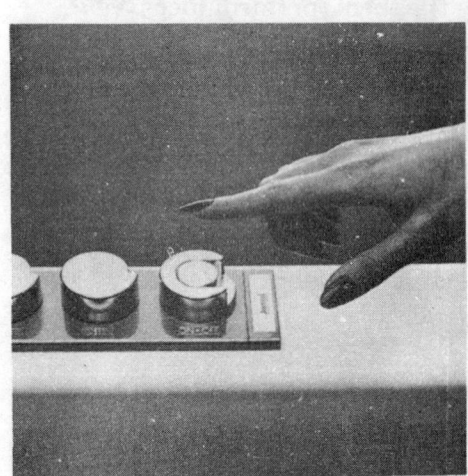

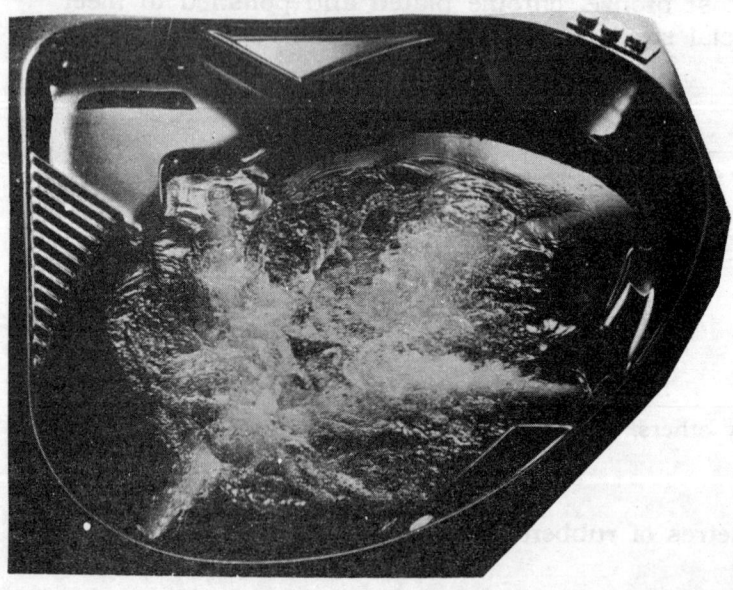

The caressed, stimulating and relaxing action of the hydromassage makes a daily routine one of the most pleasant moments of the day.

Jacuzzi Whirlpool Baths are a world of air and water to keep you in shape.

The Jacuzzi method was designed and patented over forty years ago, and gives a perfect mixture of air and water. Jacuzzi's whirlpool jets, with their adjustable flow and directional setting, really relax you or tone you up.

Jacuzzi Whirlpool Baths can be installed easily in a new bathroom or as a replacement for an existing bath, and are available in a wide range of models, sizes and colours.

To choose Jacuzzi is to choose the best.

The best of built in

Great whirlpool performance has to be designed in. It can't be added on! So we build every bath part-by-part - from the ground up. The result - a bath that gives you the best performances year after year. A bath we're proud to put our name on. For forty years we've refined the art of whirlpool engineering. We don't cut any corners. If we did, you'd feel the difference.

Fitted with larger and more powerful jets than any other system on the market. Our patented jets make our massage unique.

IMMEDIATE EFFECT

General feeling of well being

Rest and relaxation

Relief from stress

Re-vitalizing.

EFFECT AFTER REGULAR USE

Invigorates the muscular structure

Rebalances the metabolism of tissues

Increases vasodilatory action

Improves blood circulation.

JACUZZI THE BENEFITS

The beneficial effects of the whirlpool therapy bath have gained widespread recognition in the physiotherapy field and this treatment is now widely used in hospitals across U.S., Canada ad Europe.

The soothing treatment of Jacuzzi hydromassage eases the symptoms of hypertension and stress. Due to hydrotherapy's beneficial effects in increasing circulation and reducing muscle spasm and pain it has been used in the treatment of arthritis and rheumatism. Psychologically, the patient's sense of well-being invariably increases with Jacuzzi whirlpool hydrotherapy.

For a better knowledge of hydromassage and its beneficial effect, we present below a medical report made by a man who, day after day, devotes himself to the rehabilitation and recovery of physical fitness. Doctor Campacci is the head physician of the Functional Recovery and Re-education Service of Rome's Hospital and Clinical Centre, Verona (Italy).

The total immersion hydromassage, or, better yet - the hydro-pressure massage bathtub consisting of an intelligent and perfectly calibrated mixing of air and water, represents one of the most, pure practices, totally free from complications, for a correct, stimulating and invigorating treatment all over the body.

While more or less useful, so called Alternative Medicine methods are developing (acupuncture, various type of massage, physical exercises, etc.), the hydro-pneumatic massage provides, in all its simplicity, results definitely better than those boasted by other techniques — perhaps more sophisticated — and requiring qualified assistance.

The idea of replacing the household bathtub with a tub incorporating delivery outlets, adjustable for both water direction and flow with an accurate mixing, is interesting and captivating. Such an installation is a simple and intelligent addition in the health-care area.

We deal here with a particular type of micromassage based on the microaction that thousands of small air bubbles carry out on the body. A suitable device is therefore necessary to direct the air to the center of the water jet to deliver it, without any loss, to the desired area. A beneficial action is caused by a heat generated, although small, caused in turn by the transformation of the kinetic energy of the gaseous particles and by a vibrating action exercising a mild increase of pressure on the specific tissue area.

This action, therefore, results in the well known peripheral vasodilatory effect, including a positive balance of tissue metabolism, if not a real increase of it.

These modifications cause a deep invigorating effect to the muscular system and an increase of the muscular-trophy.

These stimuli combine with the effect of the water temperature (35°-36°C) to create a total relaxing action.

The hydromassage is, therefore, recommended as an effective hygienic practice to be used daily to help recover lost energy.

It should also be noted that this method can represent a particularly useful tool in case of light traumas such as contusions or sprains to the locomotory limbs.

THE TECHNOLOGY OF JACUZZI

Three key factors work together to give you a superior massage. First, the unique whirlpool action of our patented jets creates a broad circular pattern of bubbles as the air/water mixture flows into your bath, providing a luxurious, deeply penetrating massage. Second, these jets are directionally adjustable, allowing you to direct the massage to where you want it. Additionally, you can adjust the air/water mixture until it feels best to you for a truly custom massage. Finally, our baths house the jets comfortably in their own recesses. With our whirlpool action jets, air is injected from all sides, allowing a complete mixture for a better massage.

Other jets bring air in only from the top. Whether you're interested in the largest or smallest Jacuzzi whirlpool Bath, you can be certain that all genuine Jacuzzi baths are fitted with the largest and most powerful jets available so as to provide the most relaxing foot and back massage.

The jets, for instance, are fully recessed for maximum comfort. Whilst the almost silent Jacuzzi whirlpool pump, developed specifically for each individual bath size and mounted on a unique, anti-reverberation base, operates with unprecedented safety and smooth power.

You know it's genuine by the feel and the seal.

Unlike our competitors, each Jacuzzi Whirlpool Bath is a completely engineered, non-conductive, noncorrosive unit, with every part designed and built specifically for its function within that unit.

The Jacuzzi Whirlpool Bath is the result of over 40 years research and experience in design, engineering and manufacturing of quality whirlpool products for the home.

All of the following features and advantages are incorporated as standard characteristics of every Jacuzzi bath and they combine to deliver hydromassage qualities which distinguish a Jacuzzi product from those of other manufacturers.

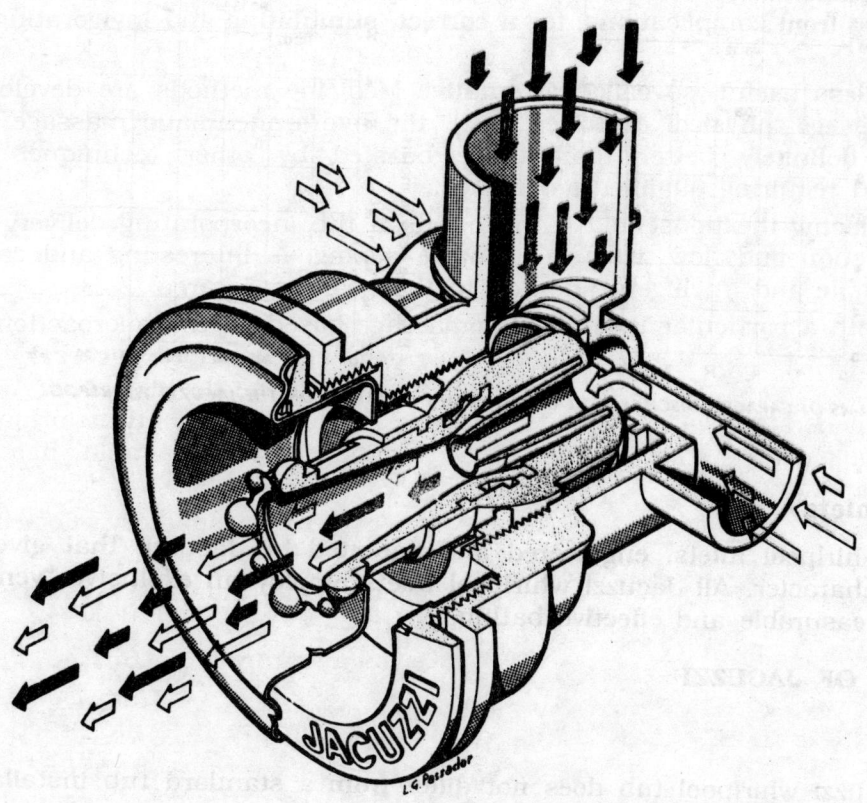

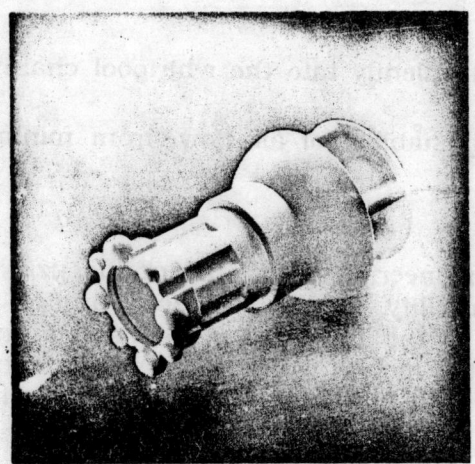

The Venturi is an integral part of the whirlpool jet.

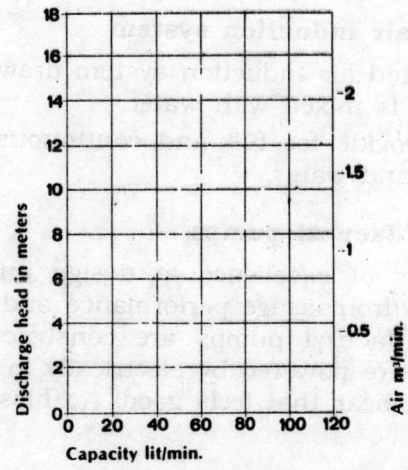

Jacuzzi guarantees a hydromassage between a minimum jet flow of 0.5 ATE and a maximum 1 ATE.

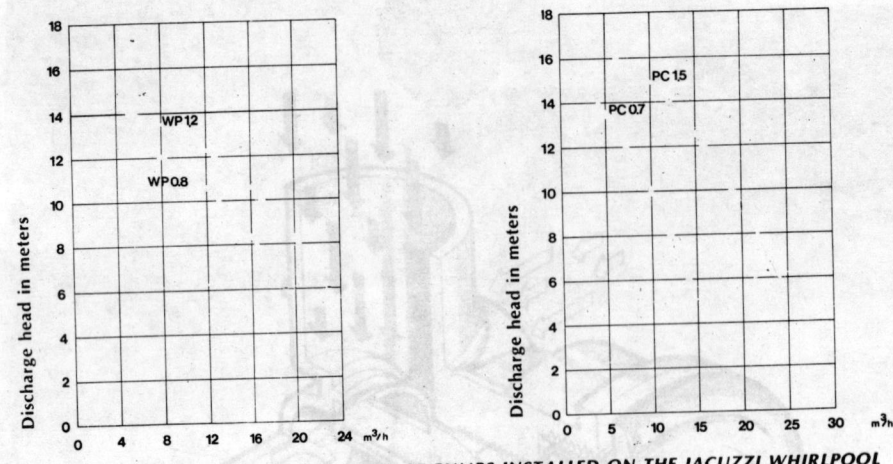

DIAGRAMS OF CAPACITY-DISCHARGE HEADS OF PUMPS INSTALLED ON THE JACUZZI WHIRLPOOL BATH.

Jacuzzi whirlpool inlets

It is the recessed whirlpool inlets, engineered and patented by Jacuzzi, that give our technique its unique feeling and character. All Jacuzzi whirlpool inlets employ an exclusive "venturi" design to mix air with water for pleasurable and effective bathing.

THE TECHNOLOGY OF JACUZZI

Simple installation

Installation of a Jacuzzi whirlpool tub does not differ from a standard tub installation, except for its connection to the electrical mains.

Really, a Jacuzzi Whirlpool tub is like any other tub and, if you wish, it can be used in the same way.

You can use the hydromassage whenever you wish. Just press the start button when the water level is over the water jets.

The water connections are standard; there is no need for a further support, preparation or operation.

Jacuzzi quiet air induction system

Jacuzzi's patented air induction system draws air silently into the whirlpool channeling it to the inlet heads where it is mixed with water.

Controls provide for full and continuous adjustability of air flow, from minimal air to an equal mixture of air and water.

World famous Jacuzzi pumps

Over fifty years of excellence in design and engineering stand behind Jacuzzi pumps. Specifically designed for hydromassage performance and durability, the high capacity pumps recirculate the bath water rapidly. Jacuzzi pumps are constructed of the highest quality non-conductive, non-corrosive materials, and are powered by electrically insulated motors with special safety features. Sometimes it's what you don't hear that feels good! No hissing, no whine. You can speak in a whisper even with the whirlpool on.

The Venturi principle

Within the inlets, the air is "drawn" into the water from both sides, evenly and naturally. As this air/water mixture flows into your bath, it makes a broad, circular pattern of air bubbles that are the right dimension for a uniquely luxurious and beneficial experience.

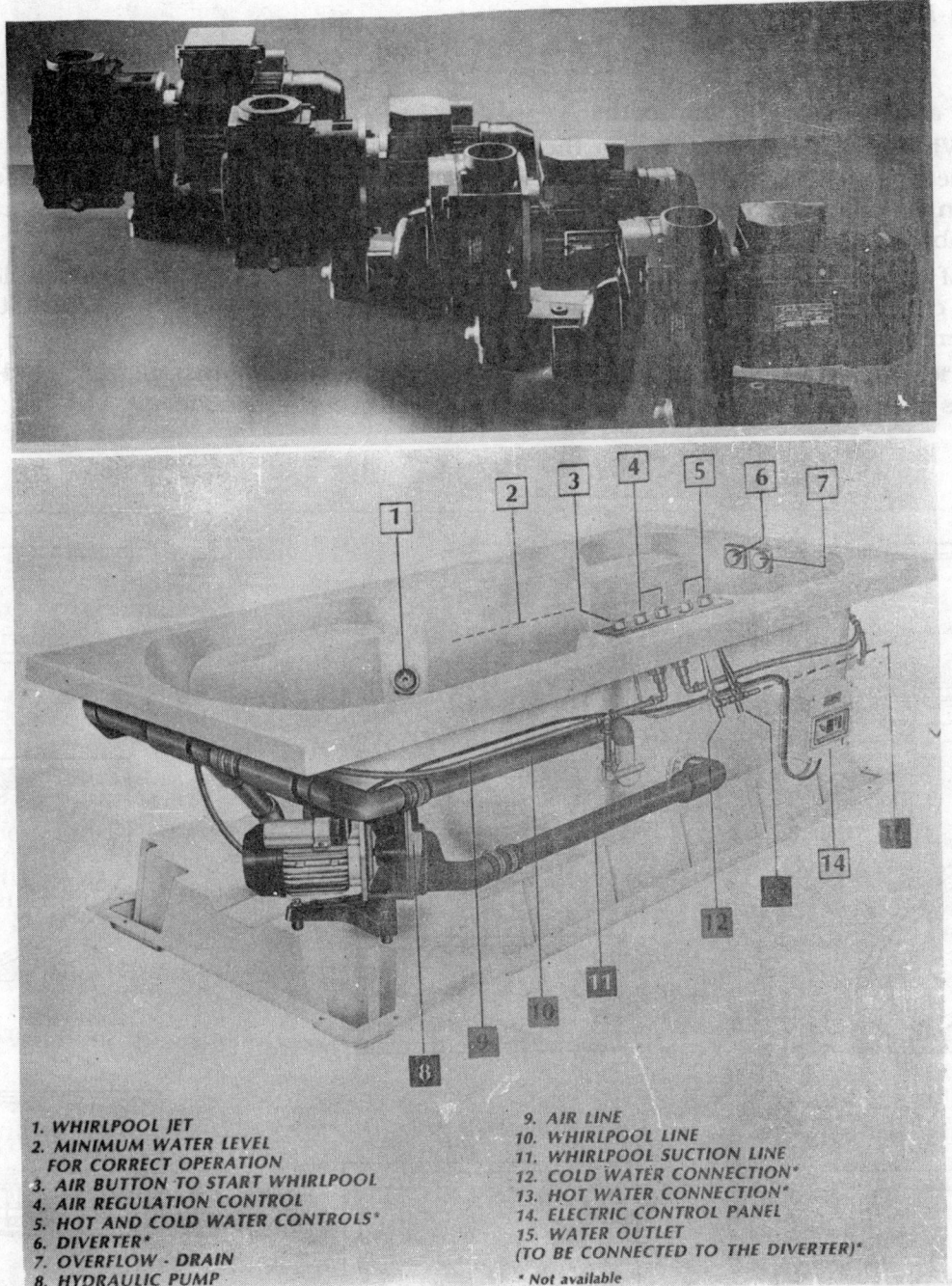

1. WHIRLPOOL JET
2. MINIMUM WATER LEVEL
   FOR CORRECT OPERATION
3. AIR BUTTON TO START WHIRLPOOL
4. AIR REGULATION CONTROL
5. HOT AND COLD WATER CONTROLS*
6. DIVERTER*
7. OVERFLOW - DRAIN
8. HYDRAULIC PUMP

9. AIR LINE
10. WHIRLPOOL LINE
11. WHIRLPOOL SUCTION LINE
12. COLD WATER CONNECTION*
13. HOT WATER CONNECTION*
14. ELECTRIC CONTROL PANEL
15. WATER OUTLET
(TO BE CONNECTED TO THE DIVERTER)*

* Not available

Our baths are equipped with fully adjustable HJE inlets. HJE inlets rotate in a full circle, within a 30 degree angle, enabling you to easily direct the impact to different areas of your body. They also allow independent, full adjustment of the amount of flow from each inlet.

Simple operation

Jacuzzi's Whirlpool Baths are the result of technology and experience that are unique in this field.

The hydromassage system is started by pressing the pneumatic button located on the rim of the bath. The quantity of air is controlled by the knobs positioned next to the pneumatic button.

The jet should be adjusted to the desired flow and intensity. Different settings may be selected simultaneously for lumbar and foot regions.

Factory tested

Jacuzzi Whirlpool Baths come to you ready to install. The bathtubs are completely pre-plumbed at the factory. All baths are factory tested before delivery to meet quality standards for optimum performance.

MATERIAL AND DESIGN

The shell

We have chosen acrylic sheet for our baths.

All Jacuzzi Whirlpool Baths consist of a thermoformed acrylic sheet shell reinforced with fibreglass, which has excellent insulating properties and maintains the temperature of the water for a long time. Then mounted on its own GRP base. The bath is light but strong, and is extremely easy to clean (using any liquid non-abrasive cleaner).

The range of dimensions satisfies any requirement and the design is fashionable and elegant. Jacuzzi's colour chart takes into consideration the major standard colour range. Additionally, special colour requirements can be produced to order.

The accessories are available in several finishes : chrome-plated, brass-plated, gold-plated or in a variety of colours also front and end panels are available for certain models.

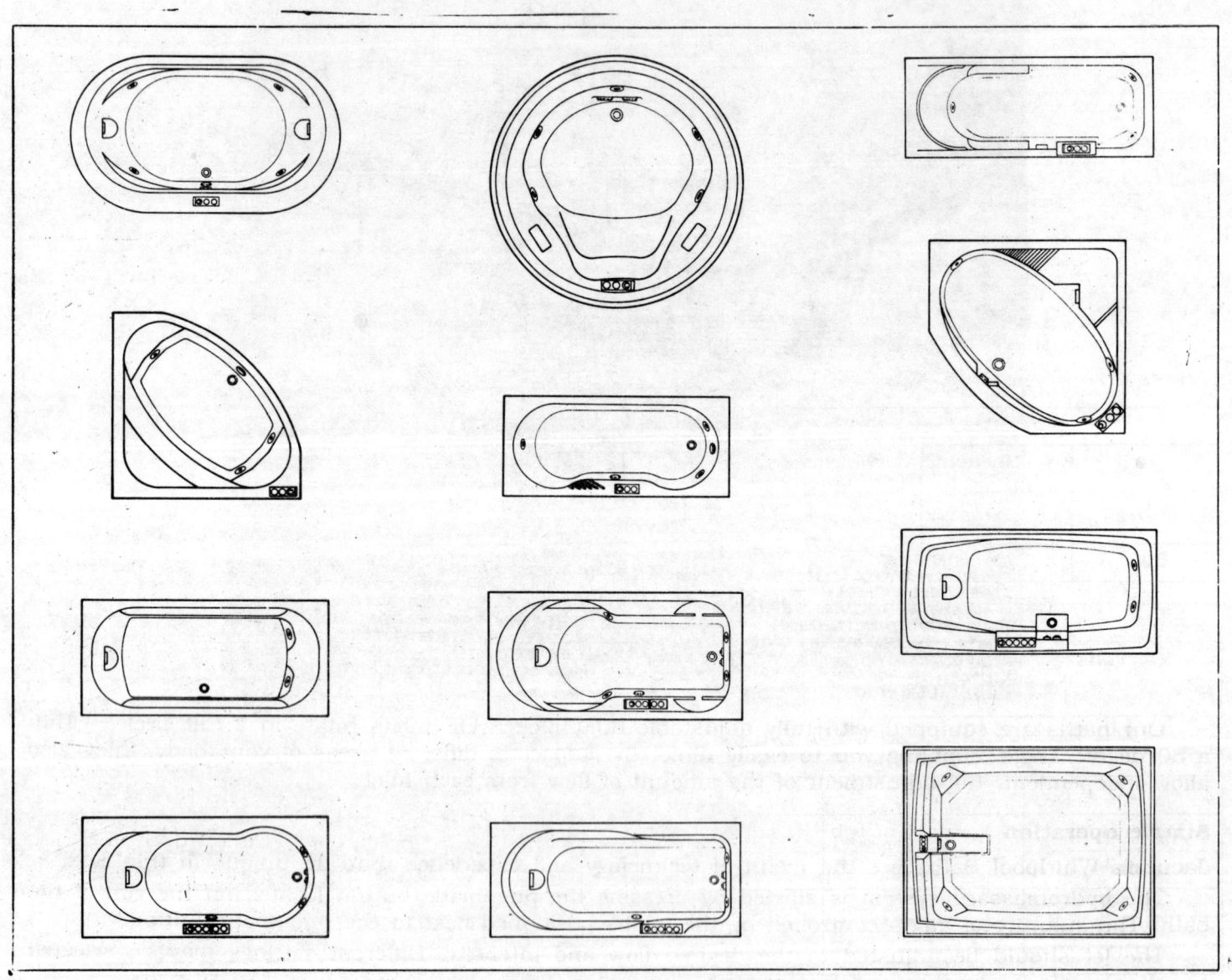

Wide choice of sizes and designs

Jacuzzi has designed and contoured each bath for maximum body comfort and offers you superior hydromassage action in a wide variety of designs from a single person bath to one large enough for a family of four.

Although several deluxe models, are extra-large multi-person units, other tubs are designed to fit standard bathroom dimensions, replacing the traditional tubs.

OPALIA

Simply elegant, wide and deep, this oval shaped, designer bath will enhance the most refined decor. One or two persons cannot sail to enjoy the comfort of its smooth contours. Three jets have been placed at both ends of the bath to ensure maximum personal massage for each user.

The bath has six inlet fittings, one suction fitting and is powered by a 1.2 HP (0.88 kW) pump. For the most effective hydromassage, approximately 230 litres of water should be used.

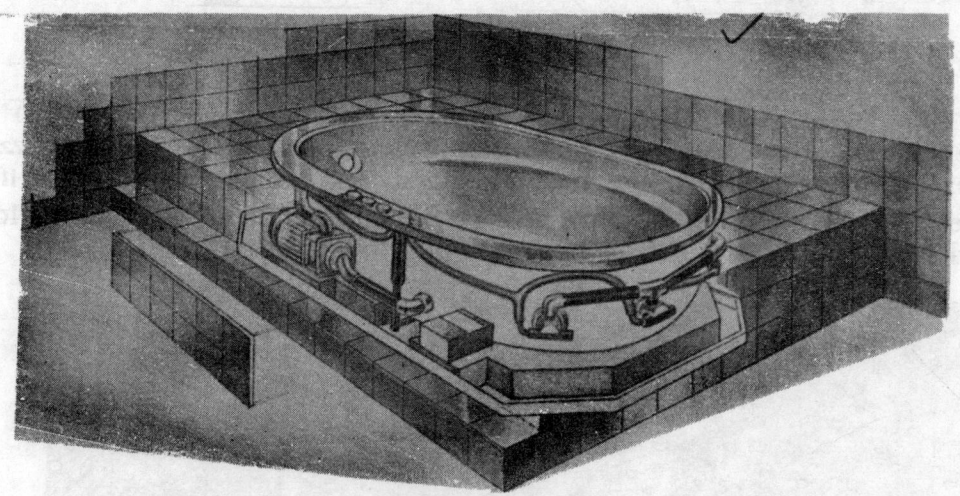

HYDRAULIC DETAILS							
Capacity litres (*)	N. Suction fittings	N. whirlpool fittings	Water capacity lit/min	Air capacity lit/min	Tap units		Drain element
					hot/cold connection	fill-spout/ diverter	
290	1	6	400	120	–	–	1½″

ELECTRICAL DETAILS				
Volts	Amperes	Electric pump		
		CV	kW	Hz
240 V	6.0 (18.0)	1.2	0.88	50

DIMENSIONS					
Dimensions	Net weight of tub	Load on floor	Shipping weight	Shipping volume	
cm	kg	kg/m^2	kg	cm	m^3
L 190 W 110 H 60	68	207	100	L 205 W 124 H 75	1.91

(*) Water capacity when filled to over-flow level.

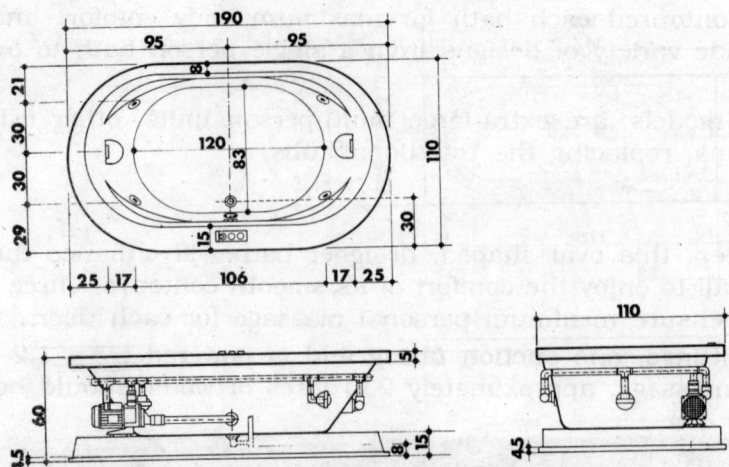

ATLANTA

The compact dimensions and clever design of the Atlanta whirlpool tub enables Jacuzzi hydromassage to be introduced into the smallest of bathrooms. Front and end panels are available if required.

There are three inlet fittings, one suction fitting and 0.8 HP (0.59 kW) pump. Ideally, the bath should be filled with 140 litres of water.

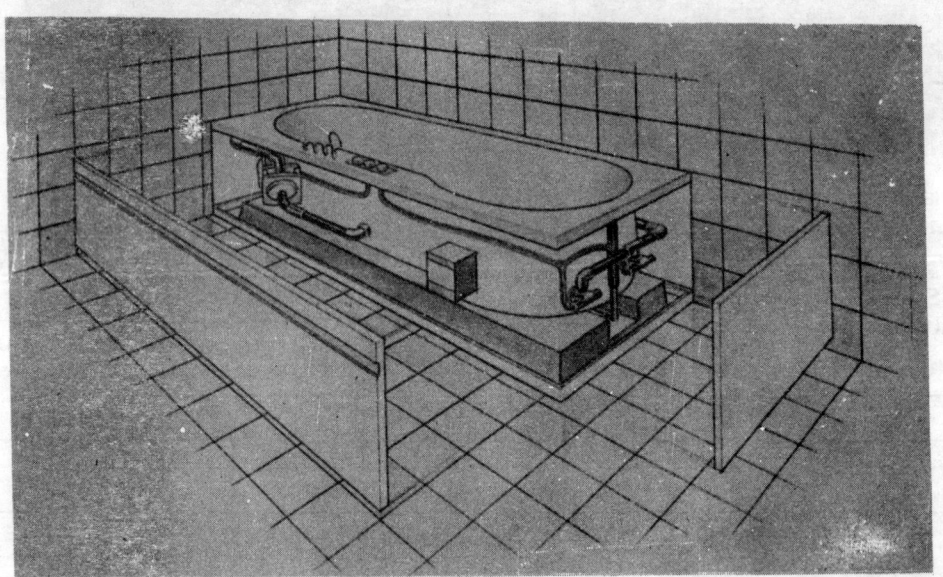

					HYDRAULIC DETAILS			
Capacity litres (*)	N. Suction fittings	N. whirlpool fittings	Water capacity lit/min	Air capacity lit/min	Tap units		Drain element
					hot/cold connection	fill-spout/ diverter	
200	1	3	240	72	–	–	1½"

		ELECTRICAL DETAILS		
Volts	Amperes	Electric pump		
		CV	kW	Hz
240 V	4.5 (13.5)	0.8	0.59	50

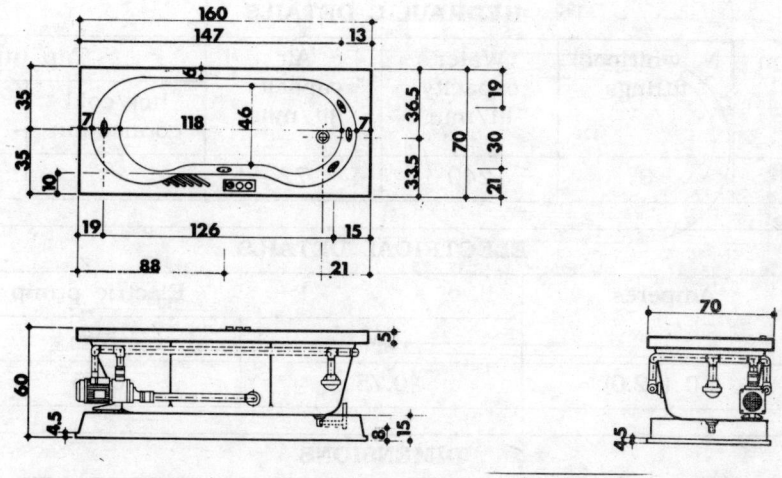

DIMENSIONS					
Dimensions	Net weight of tub	Load on floor	Shipping weight	Shipping volume	
cm	kg	kg/m^2	kg	cm	m^3
L 160 W 70 H 60	53	226	70	L 170 W 80 H 74	1.00

(*) Water capacity when filled to over-flow level.

SOLUZIONE

The Soluzione is a standard sized bathtub with all the convenience and features of the larger models. A comfortable headrest is incorporated in the design allowing you to lay back and enjoy the relaxation of the Jacuzzi hydromassage sensation.

It consists of three whirlpool fittings, one suction fitting and is actioned by a 0.75 HP (0.55 kW) pump.

The bath is available with front and side panels. For the most effective hydromassage, the bath should be filled with 110 litres of water.

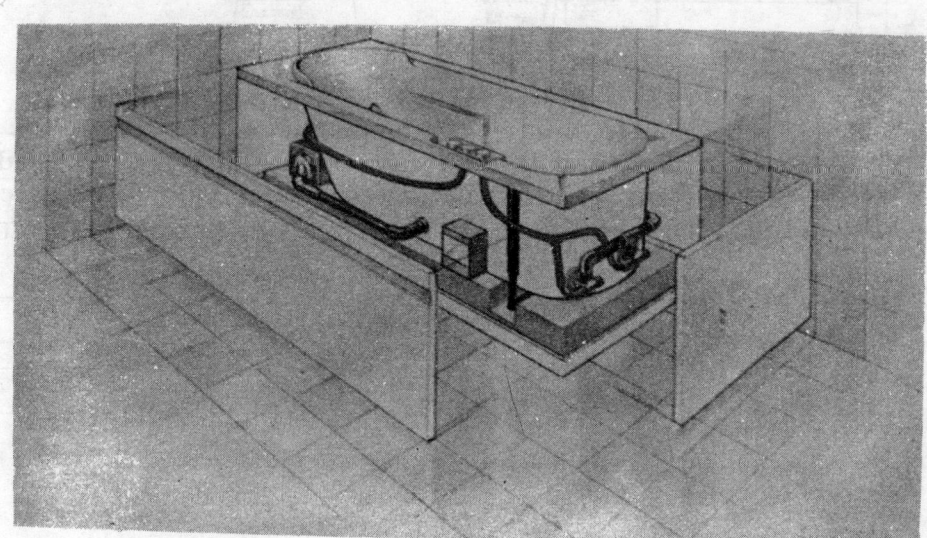

HYDRAULIC DETAILS

Capacity litres (*)	N. Suction fittings	N. whirlpool fittings	Water capacity lit/min	Air capacity lit/min	Tap units		Drain element
					hot/cold connection	fill-spout/ diverter	
170	1	3	240	72	–	–	1½"

ELECTRICAL DETAILS

Volts	Amperes	Electric pump		
		CV	kW	Hz
240 V	4.0 (12.0)	0.75	0.55	50

DIMENSIONS

Dimensions	Net weight of tub	Load on floor	Shipping weight	Shipping volume	
cm	kg	kg/m^2	kg	cm	m^3
L 170 W 70 H 60	48	183	70	L 180 W 80 H 74	1.06

(*) Water capacity when filled to over-flow level.

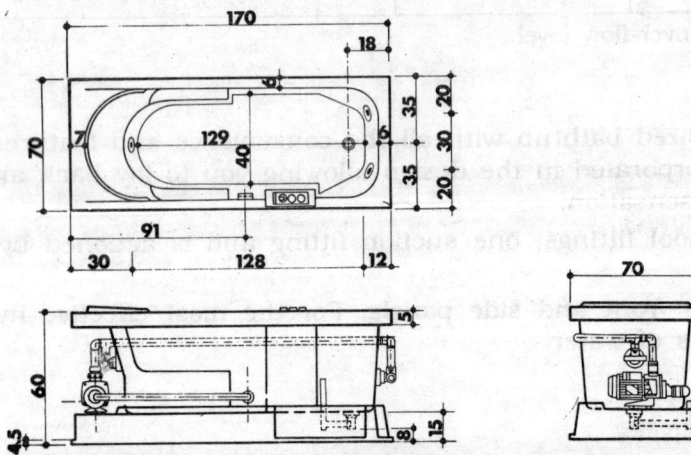

RIVA

Smooth and sculptured lines with double armrests reminiscent of the art deco era, unusual double-ended design and extra width, make the Riva the most versatile bath ever made. Because of the centrally positioned waste, one or two people can be accommodated with equal ease.

The bath is fitted with three whirlpool jets and a 0.75 HP (0.55 kW) pump. 120 litres of water is recommended for an average adult.

HYDRAULIC DETAILS							
Capacity litres (*)	N. Suction fittings	N. whirlpool fittings	Water capacity lit/min	Air capacity lit/min	Tap units		Drain element
					hot/cold connection	fill-spout/ diverter	
175	1	3	240	72	–	–	1½″

ELECTRICAL DETAILS				
Volts	Amperes	Electric pump		
		CV	kW	Hz
240 V	4.5 (13.5)	0.8	0.59	50

DIMENSIONS					
Dimensions	Net weight of tub	Load on floor	Shipping weight	Shipping volume	
cm	kg	kg/m^2	kg	cm	m^3
L 170 W 80 H 60	47	163	70	L 180 W 90 H 74	1.20

(*) Water capacity when filled to over-flow level.

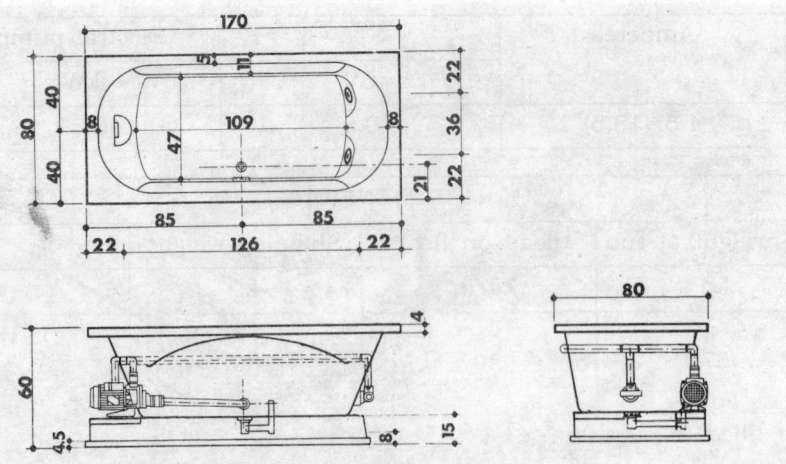

LIRIA

Its trim and pleasing lines, extra-wide shower area and the pleasure of Jacuzzi hydromassage are the qualities which guarantee that the Liria bath will be a splendid addition to any bathroom.

Liria has three inlet fittings, one suction fitting and a 0.8 HP (0.59 kW) pump.

Additionally, front and end panels are available upon request.

Ideally, the bath should be filled with 160 litres of water for an average adult.

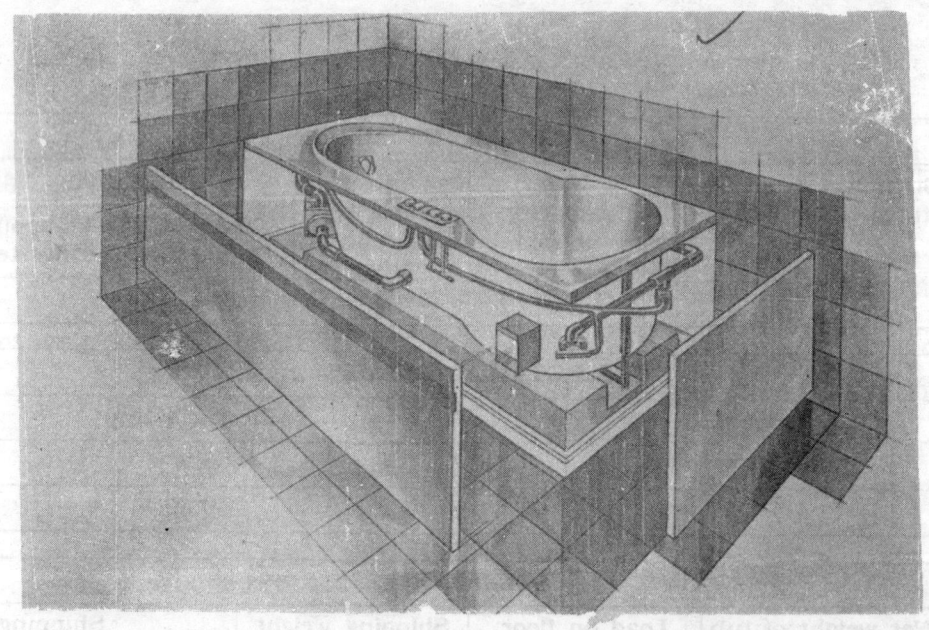

					Tap units		
HYDRAULIC DETAILS							
Capacity litres (*)	N. Suction fittings	N. whirlpool fittings	Water capacity lit/min	Air capacity lit/min	hot/cold connection	fill-spout/ diverter	Drain element
225	1	3	240	72	Ø 14	¾″	1½″

ELECTRICAL DETAILS				
Volts	Amperes	Electric pump		
		CV	kW	Hz
240 V	4.5 (13.5)	0.8	0.59	50

DIMENSIONS					
Dimensions	Net weight of tub	Load on floor	Shipping weight	Shipping volume	
cm	kg	kg/m^2	kg	cm	m^3
L 170 W 80 H 60	57	207	80	L 180 W 90 H 74	1.20

(*) Water capacity when filled to over-flow level.

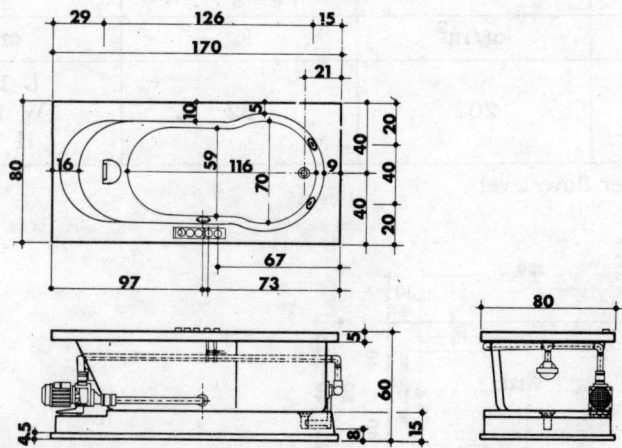

ASTREA

For those seeking the Jacuzzi experience in a conventionally shaped generously sized bath, incorporating within its stylish lines a backrest for maximum comfort. This bath has three inlet fittings, one suction fitting and a 0.8 HP (0.59 kW) pump.

Front and end panels are available, if required. 170 litres of water will give the most effective massage for an average adult.

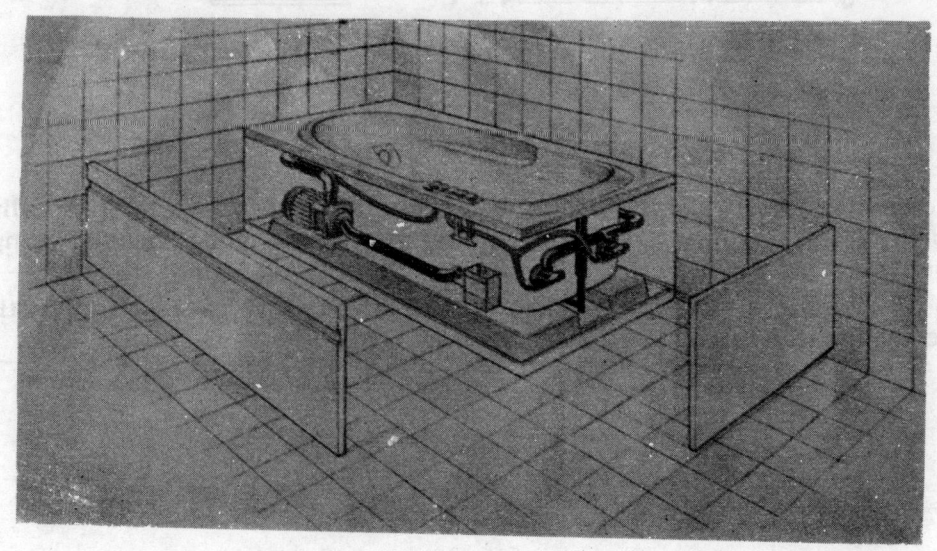

HYDRAULIC DETAILS

Capacity litres (*)	N. Suction fittings	N. whirlpool fittings	Water capacity lit/min	Air capacity lit/min	Tap units		Drain element
					hot/cold connection	fill-spout/ diverter	
235	1	3	240	72	Ø 14	¾"	1½"

ELECTRICAL DETAILS

Volts	Amperes	Electric pump		
		CV	kW	Hz
240 V	4.5 (13.5)	0.8	0.59	50

DIMENSIONS

Dimensions	Net weight of tub	Load on floor	Shipping weight	Shipping volume	
cm	kg	kg/m²	kg	cm	m³
L 180 W 80 H 60	58	203	82	L 192 W 101 H 74	1.44

(*) Water capacity when filled to over-flow level.

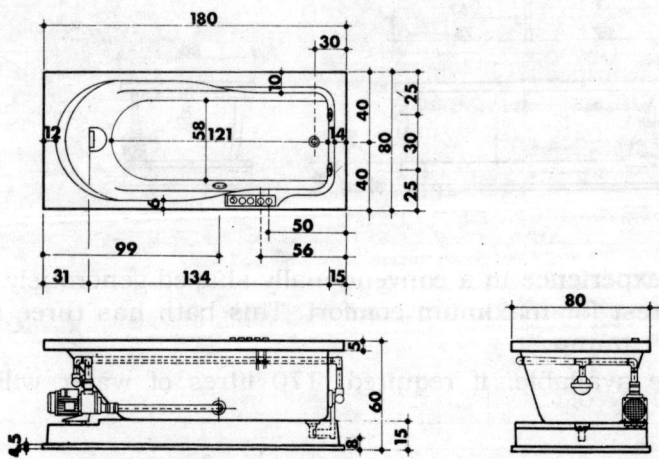

THAIS

Beautifully designed, modern bath giving maximum comfort and relaxation for the individual user. The five jets are positioned scientifically to soothe aches and pains just where needed. Ample capacity and a particular spacious shower area offer the ultimate in bathtime pleasure.

A 1.2 HP (0.88 kW) pump is provided together with five inlet fittings and one suction fitting. Front and end panels are available. Water capacity for an average adult is 150 litres.

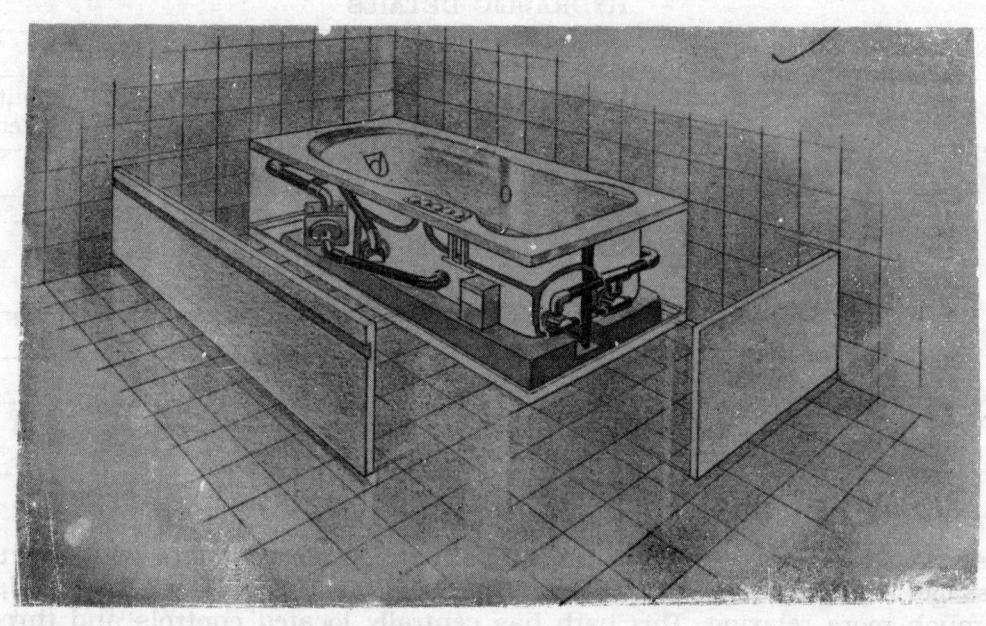

HYDRAULIC DETAILS

Capacity litres (*)	N. Suction fittings	N. whirlpool fittings	Water capacity lit/min	Air capacity lit/min	Tap units		Drain element
					hot/cold connection	fill-spout/ diverter	
205	1	5	400	120	Ø 14	¾"	1½"

ELECTRICAL DETAILS

Volts	Amperes	Electric pump		
		CV	kW	Hz
240 V	6.0 (18.0)	1.2	0.88	50

DIMENSIONS

Dimensions	Net weight of tub	Load on floor	Shipping weight	Shipping volume	
cm	kg	kg/m^2	kg	cm	m^3
L 180 W 90 H 60	62	165	90	L 192 W 101 H 74	1.44

(*) Water capacity when filled to over-flow level.

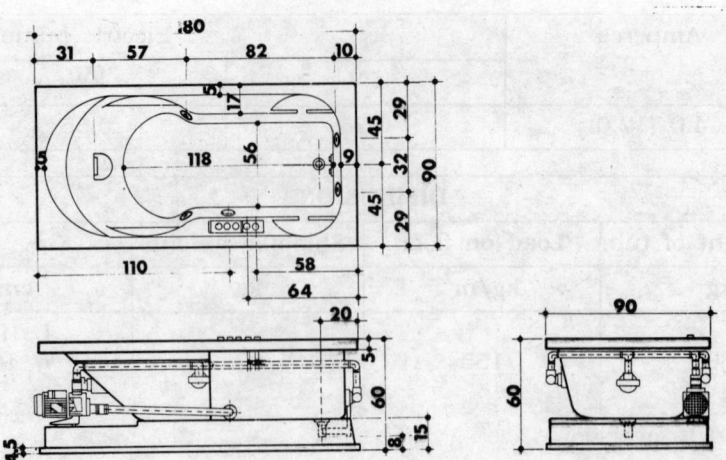

PRIMA

Perhaps the most classic of the Jacuzzi Whirlpool Bath range, Prima offers you everything you could ask of hydromassage. And, of course, with more than generous width, it makes the Jacuzzi whirlpool experience that much more relaxing. This bath has centrally located controls and three powerful jets, one suction fitting and a 0.75 HP (0.55 kW) pump.

Matching front and side panels constitute optional and attractive extras. For the most effective hydromassage, fill the bath with 140 litres of water.

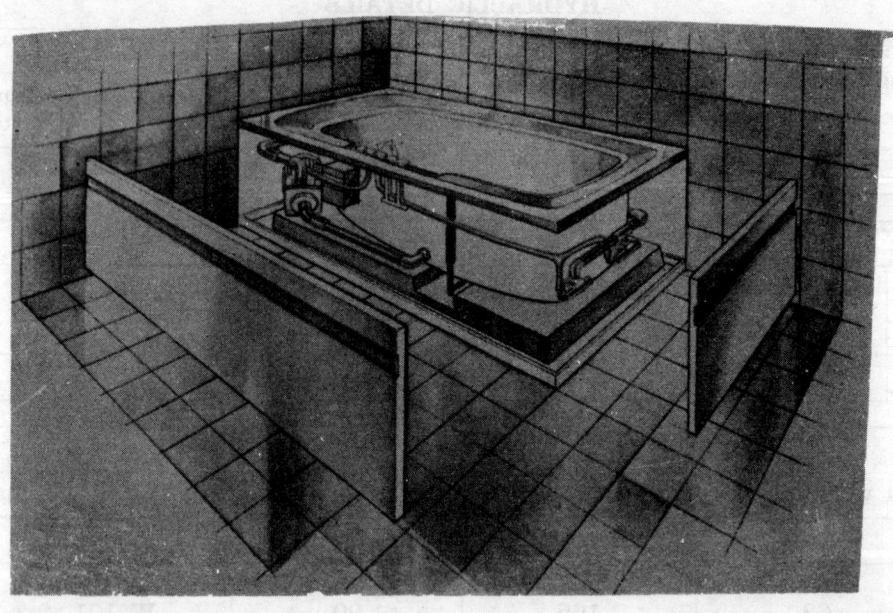

HYDRAULIC DETAILS							
Capacity litres (*)	N. Suction fittings	N. whirlpool fittings	Water capacity lit/min	Air capacity lit/min	Tap units		Drain element
					hot/cold connection	fill-spout/ diverter	
195	1	3	240	72	Ø 14	¾"	1½"

ELECTRICAL DETAILS

Volts	Amperes	Electric pump		
		CV	kW	Hz
240 V	4.0 (12.0)	0.75	0.55	50

DIMENSIONS

Dimensions	Net weight of tub	Load on floor	Shipping weight	Shipping volume	
cm	kg	kg/m^2	kg	cm	m^3
L 180 W 90 H 60	60	155	83	L 192 W 101 H 74	1.44

(*) Water capacity when filled to over-flow level.

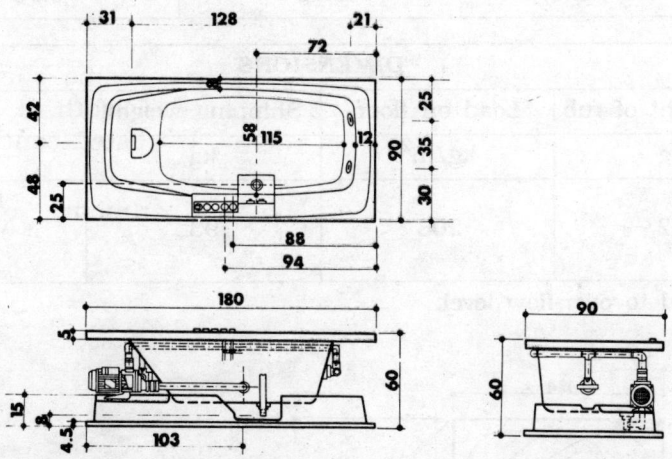

CLIO

The Clio is a corner bath tub of compact dimensions. Measuring a mere 1350 mm × 1350 mm, it is both space saving yet spacious and is very suitable for the smaller bathroom. However, the superb

utilisation of the internal space and its attractive shape, assures an effective hydromassage.

Clio has three inlet fittings, one suction fitting and a 0.75 HP (0.55 kW) pump. 150 litres of water will give an effective hydromassage. A front skirt is available for this model.

HYDRAULIC DETAILS

Capacity litres (*)	N. Suction fittings	N. whirlpool fittings	Water capacity lit/min	Air capacity lit/min	Tap units		Drain element
					hot/cold connection	fill-spout/ diverter	
215	1	3	240	72	–	–	1½″

ELECTRICAL DETAILS

Volts	Amperes	Electric pump		
		CV	kW	Hz
240 V	4.0 (12.0)	0.75	0.55	50

DIMENSIONS

Dimensions	Net weight of tub	Load on floor	Shipping weight	Shipping volume	
cm	kg	kg/m^2	kg	cm	m^3
L 134 W 134 H 60	52	206	93	L 155 W 155 H 76	1.83

(*) Water capacity when filled to over-flow level.

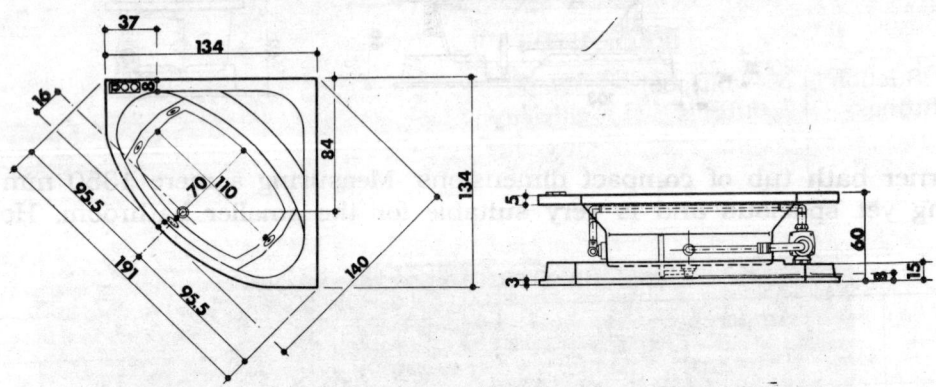

STELLARIA

An extremely comfortable and attractive corner bath, Stellaria lends itself to the most original furnishing ideas. The ample internal dimensions make the best use of available space and offers a delightfully stimulating massage.

Suitable for one or two persons, this model has five inlet fittings, one suction fitting and a 1.2 HP (0.88 kW) pump. Approximately 190 litres of water should be used for an efficient hydromassage. A front skirt is available as an optional extra.

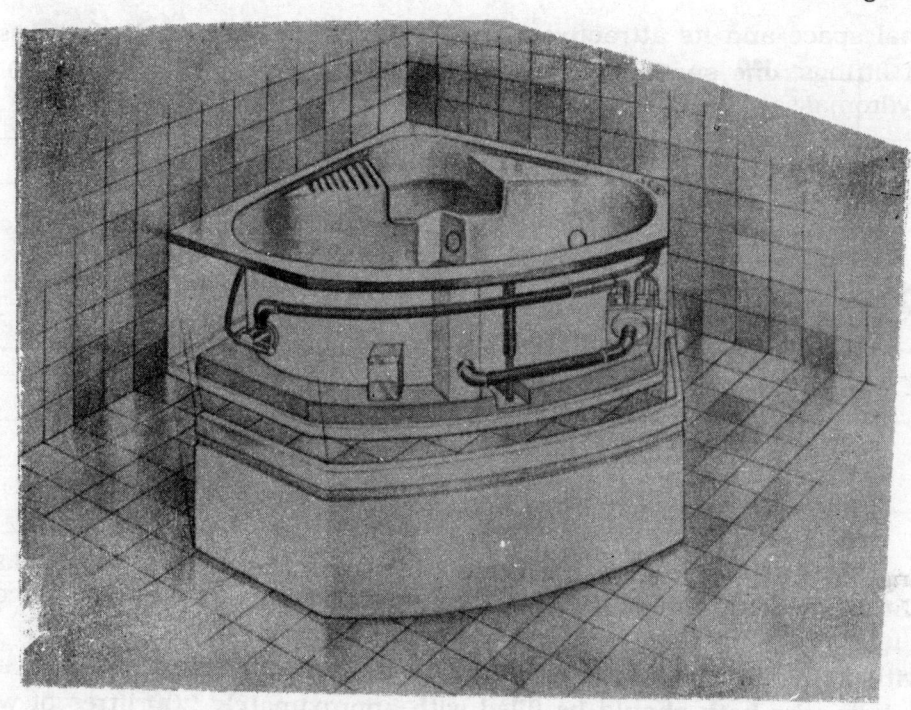

HYDRAULIC DETAILS

Capacity litres (*)	N. Suction fittings	N. whirlpool fittings	Water capacity lit/min	Air capacity lit/min	Tap units		Drain element
					hot/cold connection	fill-spout/ diverter	
252	1	5	400	120	–	–	1½″

ELECTRICAL DETAILS

Volts	Amperes	Electric pump		
		CV	kW	Hz
240 V	6.0 (18.0)	1.2	0.88	50

DIMENSIONS

Dimensions	Net weight of tub	Load on floor	Shipping weight	Shipping volume	
cm	kg	kg/m²	kg	cm	m³
L 140 W 140 H 60	62	195	103	L 155 W 155 H 76	1.83

(*) Water capacity when filled to over-flow level.

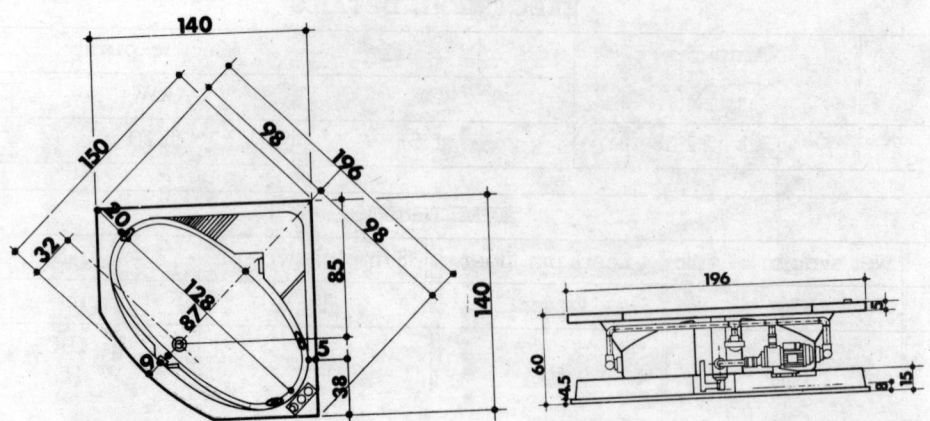

ATHENA

Dramatic and imposing designed for use in the grandest surroundings. This is the largest of all the Jacuzzi whirlpool bathtubs and an exhilarating all-over massage can be enjoyed in complete comfort by two three or even four people.

Athena is fitted with four powerful whirlpool jets for optimum back and foot massage.

For four average adults, the bath should be filled with approximately 200 litres of water. The pump is 1.5 HP (1.10 kW) and there are four inlet fittings plus one suction fitting.

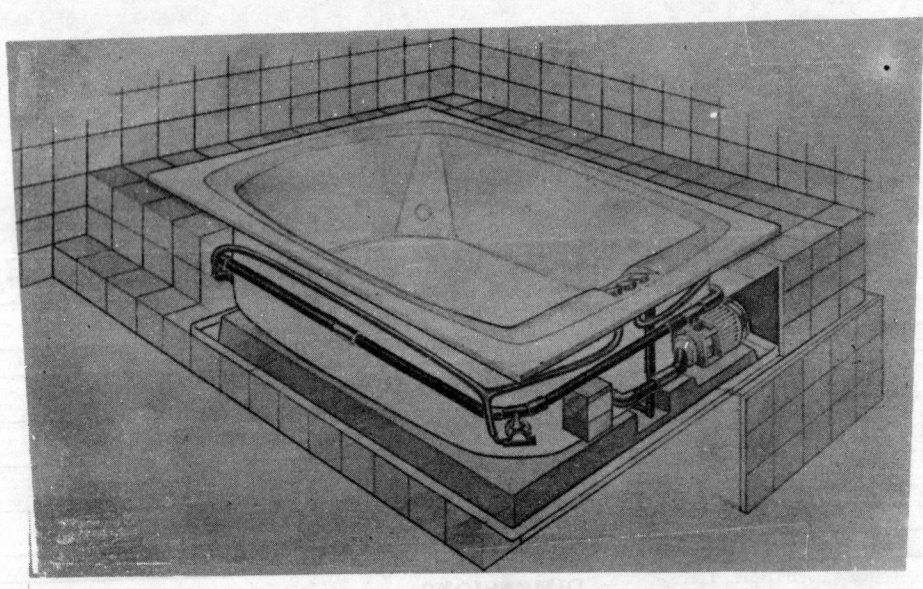

HYDRAULIC DETAILS							
Capacity litres (*)	N. Suction fittings	N. whirlpool fittings	Water capacity lit/min	Air capacity lit/min	Tap units		Drain element
					hot/cold connection	fill-spout/ diverter	
440	1	4	320	96	–	–	1 ½"

ELECTRICAL DETAILS

Volts	Amperes	Electric pump		
		CV	kW	Hz
240 V	7.1 (21.3)	1.5	1.10	50

DIMENSIONS

Dimensions	Net weight of tub	Load on floor	Shipping weight	Shipping volume	
cm	kg	kg/m^2	kg	cm	m^3
L 180 W 140 H 60	90	297	117	L 190 W 151 H 74	2.12

(*) Water capacity when filled to over-flow level.

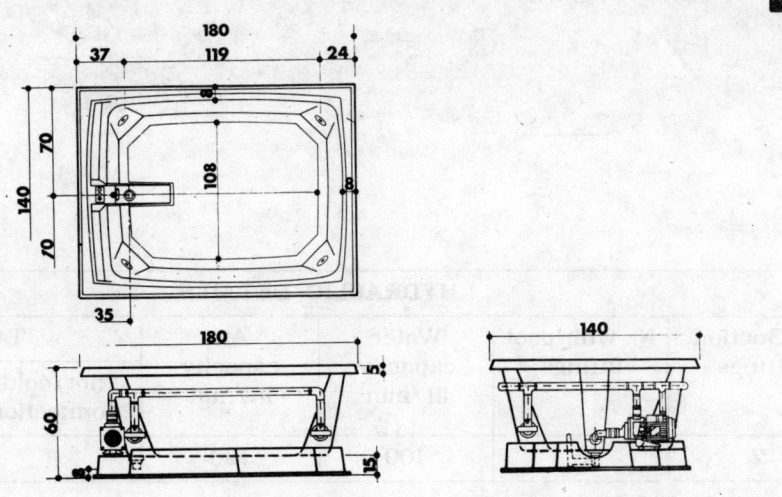

GEMINI

Undoubtedly, one of the most glamorous baths in the Jacuzzi range. This spectacular circular bath has soft contours with profiled seats and can accommodate up to a family of four for a relaxing and luxurious hydromassage. Gemini has five inlet fittings, two suction fittings and a 1.5 HP (1.10 kW) pump. 250 litres is a suitable amount of water for two persons.

HYDRAULIC DETAILS

Capacity litres (*)	N. Suction fittings	N. whirlpool fittings	Water capacity lit/min	Air capacity lit/min	Tap units		Drain element
					hot/cold connection	fill-spout/ diverter	
365	2	5	400	120	–	–	1½"

ELECTRICAL DETAILS

Volts	Amperes	Electric pump		
		CV	kW	Hz
240 V	7.1 (21.3)	1.5	1.10	50

DIMENSIONS

Dimensions	Net weight of tub	Load on floor	Shipping weight	Shipping volume	
cm	kg	kg/m^2	kg	cm	m^3
L W Ø 181 H 60	83	174	120	L 190 W 190 H 74	2.67

(*) Water capacity when filled to over-flow level.

Extract Ref. — JACUZZI ITALIA

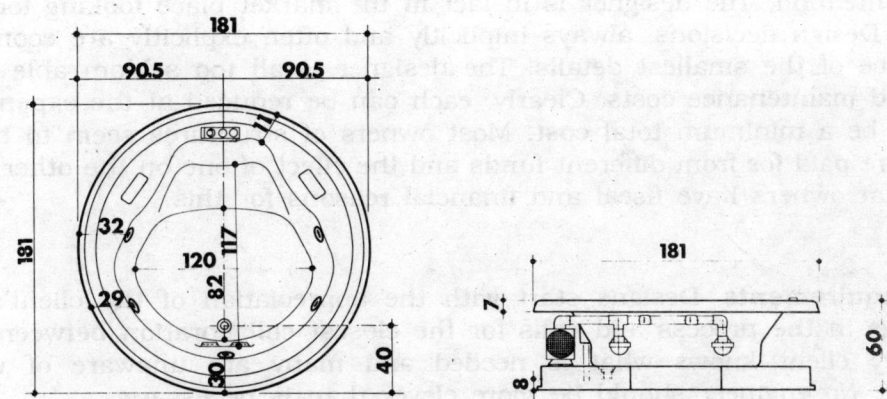

ENGINEER IN LAW

(GENERAL INFORMATION CONTINUED)

THE DESIGNER AND THE PROFESSION BUILDING ENGINEERING

Structural Design

Design dominates structural engineering; there is nothing in a structure not defined in the design. In the exercise of structural design many talents are needed; its execution engages many persons, plants and large sums of money; heavy responsibilities are undertaken. It is governed by the general desire to have what is needed at minimum cost; it is an art concerned with the adequate. Its history shows a perpetual extension and refinement of the knowledge of what is needed of a structure and how the needs be satisfied at least expense of human effort and wealth.

Factors in design

Function The structure must fulfill its intended purpose and function. This may be complex, concerned not only with the support of load but with water tightness, thermal and sound insulation, fire resistance, chemical resistance. Load itself may be complex and show not a single design load but a wide spectrum of loads and their combinations occurring with varying frequency. Moreover, as soon as structure is built, the influences of weather, ageing, fatigue, use, ill-use, accident, and other causes of deterioration commence to act upon it. The designer seeks to ensure the continued serviceability of his structure during its intended life.

Safety The structure must be safe. Most structures involve the public; there are few to whose strength someone does not at some time entrust his life. The consequences of collapse are grave and the possibility of collapse must be remote. The designer's estimate of adequate safety is partly analytical and partly a matter of judgement the accident prone structure may be difficult to define but can be recognised by the experienced.

Economy The structure must be of least cost. The client having decided, doubtless in consultation with the designer on the standard of strength, durability and other qualities that he requires, the design will proceed with this intention. The designer is in fact in the market place looking for the best value for the client's money. Design decisions, always implicitly and often explicitly are economic decisions, right down to the choice of the smallest details. The designer is all too seldom able to consider the balance of first cost and maintenance costs. Clearly, each can be reduced at the expense of the other; somewhere there must be a minimum total cost. Most owners of structures seem to be so organized, however, that the two are paid for from different funds and the effect of one on the other is not revealed. It must be assumed that owners have fiscal and financial reasons for this.

Process of design

Appreciation of requirements Designs start with the appreciation of the client's requirements. This is the critical stage in the process and calls for the closest collaboration between the client and the designer. Not every client knows what is needed and many are unaware of what structural engineering can provide. No engineer should be more clever than is necessary.

Formulation of schemes As many facts and factors as possible are collected and assimilated. The mind concentrates upon them, immerses itself in them. By some process of the intellect or imagination, often in moments of relaxation, concepts of the general structure suggest themselves, sometimes vague, sometimes surprisingly complete.

Appraisal of schemes These concepts are then developed approximately and appraised by the designer. The amount of mathematical analysis at this stage will depend on the designer's experience and often no more is done than is enough to determine general dimensions and the designer is likely to concentrate more on functional suitability, ease of erection and economy. Good designers are recognised as much by their discrimination at this stage of design as by their inventiveness in the previous one. To think up clever schemes is not an unusual ability; to choose between them, to recognise difficulties and advantages is perhaps less common but not less important. Indeed, the more experienced the designer, the more ruthless that is in checking its schemes against the problems that the designer foresees will arise on site and in use. These problems themselves will prompt further developments of the designer's scheme. At the end of this appraisal, the designer returns to the original requirements, perhaps to the client. The process may have thrown up unexpected possibilities that modify the requirements, alternatively what was though feasible may be found to be no so. The economic aspect may need rethinking.

The spiral process of collection of facts, conception, appraisal and return may be repeated several times with occasionally a complete new leap; it may often be only after much laboured thought and sketching that quite suddenly the simple, the obvious answer occurs. The designer must give thought to the method of erection right at the beginning; if the designer fails to do so, construction may be costly or even impossible. There is a dilemma here; the designer should not normally remove from the constructor the choice of method although the designer is obliged to assume one in order to be able to develop the scheme. In practice, if the designer can think of one method, the designer can think of five and the contractor will frequently invent a sixth; if he can think of none, none there probably is. But having adopted one method, all details must be consistent with it; a coherent erection scheme will usually permit of a variant, which may then be readily assessed. Moreover, the economical job is usually the one with a straight forward, clear cut method of erection with a minimum of separate processes following one another without mutual interference. In considering cost, it is more useful to concentrate on operations than on minutiae of quantities and rates whose apparent precision may mislead. With the choice of structural scheme, implying a material and a method of erection, the decision of major effect on the cost and utility of the final structure has been made; keen analysis will have only a small effect.

Analysis of final scheme It is now that a full structural analysis is made. This is frequently referred to as 'design'; in fact, the significant design decisions have already been taken and the operation of analysis is much more a process of checking the scheme for soundness in service and during erection. Clearly, what the designer does not know about, cannot check; no analysis will tell the designer of anything the designer is not looking for. Here again knowledge of difficulties and dangers to be guarded against is a mark of competence.

Details and specifications Detailing must be seen as an integral and important part of design; it is a matter of which the designer cannot afford to lose control because bad detailing can lead to disaster just as surely as a defective overall scheme — in fact, it is much the more frequent cause of trouble; bad detailing can greatly increase cost; it is easy without close supervision for details to creep in which are not consistent with the rest of the structure and which can render impossible an otherwise carefully considered erection scheme. Specifications are a part of the design, complementary to the detail drawings and not merely a means of regulating contractual relations. These directions to the constructor are necessarily complex. One of the reasons why innovation in structure is evolutionary rather than revolutionary is that too many unfamiliar features in one job can so overload communication that cost rises and the risk of failure increases.

Supervision of construction Supervision should be under the designer's control; the designer's task is not completed when the designer has handed over the bundle of documents. Construction is rarely without incident requiring decision, the consequences of which the designer alone may be able to appreciate.

Philosophy of safety

General Absolute freedom from danger cannot be attained, no matter how much money be spent on the structure. Since, however, probability has been found to be a common measure of the diverse circumstances affecting safety, the probabilities that a structure will be understrength or overloaded can be estimated. If the gravity of the consequences of inadequacy be expressed as an acceptable probability, the combined risks of understrength and overload can be compared against it. The practical designer has always varied factors of safety according to the designer's estimate, often subconscious, of the danger. Current knowledge makes this estimate explicit; it can be analysed and its separate parts assessed. The margin of ignorance is thereby narrowed and structures may be built of consistent safety; wasteful overprovision is avoided and more can be built for the same money.

Strength of structure The designer's estimate of the strength of the structure is based on two successive assessments of probability.

The first is of the strength of the elements of structure as built. No small part of the classic factor of safety was intended (and rightly so) to cover the risks, of material being less strong or dimensions smaller than specified. Both these risks, however are open to assessment by processes of sampling, testing and checking; certain forms of prefabrication permit of so dealing with complete structural members whilst a method of construction such as prestressing incorporates an automatic acceptance test of its materials. The strength of a structural element may thus be expressed statistically and the cost of a programme of sampling, testing and supervision may be balanced against the economics obtainable by a higher probability of adequate strength in the structural members.

The second assessment is of the strength of the structure as a whole. Based on the previous estimate of the strength of its members, the behaviour of the structure is analysed, typically, at two limiting states : the state at which it ceases to be serviceable and the state at which it collapses. The techniques of analysis available to the designer, whether mathematical or experimental, have been highly developed. They are powerful. Judgement is still needed in their use. There is no automatic means of calculating structural strength and even the most refined methods have their own inexactitudes.

Loading Much attention is given to the understanding of the loads occurring in engineering structures; it is regrettable that though many buildings are, in fact, major engineering structures, the loads to which they are subject have not received such attention. Understanding is important : if the loading is underestimated or if some effect of loading has escaped attention, the structure is in hazard; if the loading assumed is greatly in excess of that occurring in reality, the structure is wasteful. It is necessary to know the association between value of load and the frequency with which it occurs. With structures exposed to wind or wave, for instance, it is often difficult to determine the extreme value and it is necessary to estimate a maximum value the likelihood of exceeding which is sufficiently small, albeit always possible. With other structures, it is impossible much to exceed the load to which the structure is permanently subjected. With others again, combinations of loads may be critical; if for instance, there are two independent forms of loading each with a wide scatter, a reasonable estimate of necessary strength may be made only by calculations of the probability of occurrence of dangerous combinations. The life of a structure is significant; the longer it is, the higher the probability of an extreme load occurring. There remains an accident that are likely, such as vehicle striking the central pier of a bridge over a motorway; there are those that are credible but unlikely, such as wind velocity substantially higher than any hitherto recorded; there are those so unlikely as to justify ignoring them, such as a meteorite striking a dwelling house. A design based on probability incorporates all these risks into a rational analysis of safety.

Consequences of structural inadequacy Inadequacy may be of two types : the structure may cease to be serviceable or it may collapse. The acceptability of the risk will depend on its consequences. Normally, the loss resulting from unserviceability will be much less than that from collapses, though in many industrial structures interruption of production places a heavy penalty upon unserviceability. The response of a structure to the failure of a member may be one of the following :

a. the structure will collapse immediately as with a pin jointed truss : if any member of such a truss fails, the truss collapses. This is the statically determinate or isostatic situation and is that covered by classic factor of safety :

b. the structure will not only collapse but entrain the collapse of adjacent structures 'progressive collapses'. An example is a masonry multiple arch viaduct the failure of one arch removes the support for the thrust from its neighbours which collapse in turn;

c. the structure may continue to stand, though with reduced strength, due to redistribution of load. This is the situation of a structure specifically designed to be 'fail safe' and sometimes, but not always, of a statically indeterminate or 'hyperstatic' structure.

The practical consequences of any of these three types of failure will vary widely and must be assessed in a given case and set against the risk of failure assessed as above. Clearly, if the failure of a structural member leads to a widespread collapse, the consequences will be graver than if collapse were limited to the structure of which it forms part, but, provided that the designer ensures that the risk is adequately remote, such a structure may properly be adopted — multiple arch masonry via-ducts do not have a conspicuous history of failure. In practical design, the three broad aspects of strength, loading, consequences of inadequacy, are expressed by factors relating to each aspect separately. The overall factor against collapse would thus be a function of the factor expressing the probability of obtaining the foreseen strength, the factor expressing the probability that a certain load will not be exceeded and an additional factor expressing the consequences of collapse.

The designer and the profession

General In practice, the designer is not in a void alone with the client, the constructor and his conscience. Structural engineers form a community working within society as a whole and the individual designer may expect to receive aid from other colleagues through professional institutions and learned societies. The aid relevant to this report that the designer receives from the colleagues acting corporately is as follows The expression of a philosophy of design, the accumulation of facts, recommendations concerning sound and tried procedures, evaluation of innovation, assistance in standardising widely used materials and components.

Not all of these functions are beyond the wit of individual engineer and a person of acute and powerful intellect could perhaps dispense with help on most. Nor is there any sanctity in numbers — an assembly of ill-informed persons can reach only an ill informed conclusion. Nevertheless, it is clearly to the advantage of all that the profession should place special skill, experience and knowledge of some at the disposal of many.

Philosophy of design A philosophy is needed to formulate rational criteria of adequacy; lacking such, structures are either wildly expensive or stand up, if at all, only by accident. The greatest service an institution or a learned society can offer is to enable a philosophy of design to be elaborated, discussed, confirmed and developed. Papers are the classic means of transmitting experiencing in design and construction, presenting new findings and proposing new departures, all of which form the raw material of a philosophy. The method of presentation of a paper emphasizes discussions, both verbal and written and makes of it a forum.

Recommended design procedures Many structures executed at any time have much in common; if the accepted rules of a mature technique whose value has been confirmed by experience may be set out by the profession, it will be a help to many. Codes of practice are the chosen instrument for this purpose. They consist of a compendium of good practice; drawn up by engineers of authority, they are submitted for approval to the profession at large before publication. Their drafting, publishing and amending place a heavy burden on the profession, they are intended as a guide for chartered engineers experienced in the field in question and combine the functions of presenting fact and recommendation for the design of discrete structural elements, the overall structural concept lying outside their scope. They are not intended to be mandatory. By and large they have succeeded in the purpose described. They have also been successful in establishing a vernacular of sound detailing. The junior levels of the profession rapidly become skilled in detailing according to the code; skill so gained is not wasted when the junior changes employment; seniors know by and large what they may expect from juniors. Moreover, the ability to find a summary guide within small compass is a great saver of drawing office time. Codes contain both fact and recommendation and it is important to distinguish between them. Fact is verifiable, recommendation must usually be accepted on authority. Too much insistence on conformity with the recommendations of other persons can screen the designer from reality; loss of contact with reality in any branch of engineering is finally disastrous. The use of codes is dangerous if their function is misunderstood and they are treated not as aids but as complete instructions. Under the impulse of economic competition, the code is sometimes used as a check list; satisfaction of its items is deemed automatically to ensure an adequate structure. This is an illusion. No rules can replace the skills and conscience of the designer; too many can destroy them.

Presentation of facts The accumulation of fact is continual; its presentation in accessible form is an uninspiring but necessary task. Many tables of mathematical functions of structural constants, of properties of materials, have become indispensable tools of design; new facts must be similarly collected.

Evaluation of innovation Everything in engineering changes; not only do knowledge and methods improve, but needs, available labour and materials, relative costs, all change continuously; innovation in design is a necessary response to these changes. When new materials, new methods of design, new structural forms arise, they are usually discussed and appraised in papers to the institution. It is then the institution's practice to produce reports upon them to facilitate more general adoption. These reports are partly factual description and partly a guide to good practice in so far as it has evolved — they constitute, in fact, preliminary codes of practice awaiting wider experience and approval. A newcomer to a building world is the agreement bond which produces temporary certificates for new building materials, devices and components.

The designer and society

Building regulation The works of structural engineer are of major consequences to society, which can and should define the functional standards of building. These standards are expressed in building regulations which have a large effect in determining the general cost of building. Since not all structures are designed and built with professional assistance, the regulations contain a mass of detailed requirements, whose necessity and value in certain branches of building, as at present carried out, cannot be questioned. It has, however, been found convenient to incorporate clauses which say that structures built in accordance with certain codes of practice are deemed to satisfy the regulations. The current practice of the profession is thereby recognised as acceptable; there are, however, objections :

The codes become in effect mandatory. There is a gulf between recommending an action and commanding it. Codes are explicitly recommendations; were the profession to draw up documents intended to be mandatory, their contents would be different. Codes are guides to chartered engineers experienced in the field in question; they are unsuitable for embodiment in regulations which permit the design of structures by the unqualified and inexperienced. No regulations, however complex or all embracing, can guarantee safe structures and rather than by ever tighter rules, public safety will be better served by a clear cut definition of responsibility. Consideration should be given to developing a way in which such responsibility can be carried by the person or firm preparing a structural design.

Forms of contract

Over a period of time contract procedure has become standardized so that throughout the country three forms of contract are in current use :

1. The Royal Institute of British Architects (RIBA) form of contract;
2. The CCC/Wks/i form of contract used by government departments;
3. The Institution of Civil Engineers (ICE) form of contract.

The RIBA form of contract is used specifically for general building construction. The contract is one between a client and a builder who for a sum of money, paid to him/constructor by the client, erects a building to the requirement of the client as expressed in drawings prepared by the client's architect. The work is carried out to the specifications of the architect and the architect has a separate contract for services to the client.

The CCC/Wks/i form of contract is usually made with the ministry of public Building and works; occasionally with other government departments. This ministry is responsible for preparing plans and the work is carried out to the specifications of the Superintending officer.

Under the ICE form of contract, work of civil engineering construction is carried out, to the plans prepared by an engineer on behalf of the client, by a contractor for a sum of money paid to the contractor by the client. The work is carried out to the specification of the client's engineer.

Local authorities use for their contracts either the RIBA standard form or the ICE standard form as is appropriate or, occasionally, their own. In the CCC/Wks/i form of contract and the ICE form of contract powers are given to the engineer (Supervising Officer in the CCC/Wks/i form) to issue instructions to the contractor to carry out the work in a certain manner. The engineer is also required to issue certificates covering completion of the work and payments due to the contractor. In the RIBA form of contract a similar function is performed by the architect. Notwithstanding the common use of RIBA form of contract for **structural engineering work**, the fact remains that **no powers or duties under the contract between the client and the contractor** are given to the structural engineer, who has **no powers of supervision**, nor is the work required to **structural engineer's satisfaction**.

Aims of structural design

A good design has certain typical features — simplicity, unity and necessity. Simplicity is common to excellence in all the arts that impression of ease which is usually the result of intense effort. The structure in which evident difficulties have been painfully overcome lacks excellence. simple does not mean elementary; a structure suited to its purpose and easily constructed may be of great analytical complexity.

Unity means a single thing, not just a heap. The unity of a small span bridge will be quite other than of a power station; but both are formed of separate parts which are made to serve the whole and if whole and parts are well designed, the unity will be seen. There is a superficial unity which is to be avoided. How easy to avoid a chaos of unrelated parts behind a screen of a bold eye-catching pattern. And how easily — and with what disastrous results — the parts can show through. The last feature, necessity, is perhaps the key. How magnificent the result if there is nothing in a structure but that which is necessary. This is not a paradox; the mind is at rest in the acceptance of necessity, but is uneasy in the presence of the wasteful, the superfluous, the factitious. Not that characteristics are to be aimed for directly. They are that by which good design is recognised; they are attained by the single-minded pursuit of the specific purpose of a structure — its function, safety and economy. A design that satisfies functional requirements practically and economically is the least that is expected of the designer. The very magnitude of designer's responsibilities, however, imposes on him the ceaseless pursuit of excellence.

(*Source :* Aims of structural design, A report of The Institution of Structural Engineers, August 1969. UK. II, Upper Belgrave Street, London SW i X 8 BH) For information exchange to readers of India.

For the general information to the reader

Professional bodies with statutory status constitutionally/in legal frame work involved in the building industry in India and United Kingdom

India	1.	The Institution of Engineers (India)	Indian Companies Act of 1913

India	1.	The Institution of Engineers (India) 8, Gokhale Road, Calcutta - 700 020, India	Established 1920 9th September **Incorporated by Royal Charter**, 9th Sep. 1935
		Status of the corporate members **Chartered Engineer (India)** FIE/MIE/AMIE	
	2.	The Council of Architecture 8 B, Shanker Market Connaught Place, New Delhi - 110001 Grants certificate of registration as **Architect**	Established under 1972 Architects Act
U.K.	1.	The Institution of Civil Engineers 1-7, Great George Street London, SW I P 3 A A	Established 1818 **Incorporated by Royal Charter 1828**
		Status of the corporate members **Chartered Engineer** FICE/MICE	
	2.	The Chartered Institute of Buildings Englemore, Kings Ride Ascot, Berks SL 5 8 BJ	Established 1834 **Incorporated by Royal Charter 1980**
		Status of the corporate members **Chartered Builder** FCIOB/MCIOB	
	3.	The Institution of Structural Engineers I I, Upper Belgrave Street, London SW I X 8 BH	Established 1908 **Incorporated by Royal Charter 1934**
		Status of the corporate members **Chartered Engineer** FIStrtE/MIstrtE	
	4.	The Royal Institute of British Architects 66, Portland Place, London W IN 4 AD Status of the corporate members **Architect** FRIBA/MRIBA or simply RIBA	
	5.	The Royal Institute of Chartered Surveyors 12, Great George Street London SW IP 3 AD Status of the corporate members ARICS	
	6.	The Royal Institute of Building Services 222, Balham High Road London SW 12, 9 BS	

Membership of the RIBA

Qualification for **corporate** membership

A person may qualify for election as a corporate member if the candidate :

a. Passes or received exemption from parts 1,2, and 3 of the RIBA Examination in Architecture

b. Have otherwise obtained a qualification in architecture outside the United Kingdom which in the opinion of the Special Entry Committee indicates the possession of the requisite knowledge and experience of architectural practice to make the candidate a fit person for admission to membership.

Institution of Structural Engineers

Qualification for the **corporate** class MIstrtE

An engineering bachelor degree (Civil/structural) for the exemption from the Institution's Part 1 and 2 examinations

A further structural Engineering experience of at least 5 years for the candidate's eligibility to take Institution's Part 3 Examination. Having qualified the part 3 examination, the candidate shall be eligible for the admission as a Corporate member, **Chartered Structural Engineer.**

Institution of Civil Engineers

Qualification for the **corporate** class MICE

An approved academic Bachelor degree in civil engineering

Achieved the Institution's core training objectives

Undertaken a programme on continuing education

Had responsible practical experience in one of the many branches of civil engg.

The candidate shall submit to a **Chartered Professional Review** (CPR) and having got through that shall be admissible to the institution as a **Chartered Civil Engineer** MICE

(Details in further pages)

The Chartered Institute of Buildings

Qualification for the **corporate** class MCIOB

Not less than twenty three years of age

Pass or be exempt from the Member Examination parts I and II

Practical experience of a satisfactory nature and duration of minimum 3 years in a professional level as exhibited in the details of the institution.

Must pass the Professional Interview based on the candidates experience.

The Institution of Engineers (India)

Qualification for the **corporate** class AMIE

Not less than 27 yrs on application day

Passed Section A and B of Institution Examination or equivalent degree of engg. bachelor level recognised by the **National Council of the Institution**.

Minimum of 5 years of professional engineering experience in a position of responsibility in the execution and design of engineering works or their operations or in recognised teaching profession in engg.

At the discretion of the Institution council, the candidate may have to appear in a written or oral or both tests called Section 'C' and if so shall be admitted to the corporate class only after passing that.

Qualification for MIE

Not less than 32 yrs on application day

Academical qualification same as for AMIE class

Experience Minimum 10 years in a responsible position in engg.

Qualification for FIE

Not less than 40 Yrs on application day

Academic qualification same as for AM and M

Experience Minimum 15 Yrs in position of high responsibility or 18 yrs aggregate

(See the details in the enclosed bye laws)

The Institution of Engineers (India) holds the following branches of engineering :

1. Civil Engineering Division
2. Mechanical Engineering
3. Electrical Engineering
4. Electronics and Telecommunication Engg
5. Mining Engineering
6. Metallurgical and Material Science Engg

7. Chemical Engg
8. Environmental Engg
9. Agricultural Engg
10. Textile Engg
11. Aerospace Engg
12. Architectural Engg
13. Marine Engg
14. Production Engg
15. Computer Science Engg

Council of Architecture — a body Corporate under Architects Act 1972

Any recognised architecture bachelor degree or equivalent duly published in Gazette.

Authorised Style and Title — **Architect**

Enclosures for informations

(1) Bye laws — The Institution of Engineers (India)
(2) The Architects Act 1972
(3) Routes to Chartered Engineer (Institution of Civil Engineers)
(4) Institution of Buildings.

Quotation from page 1 of the Volume 41 Number 10 January 1992 of Journal of The Institution of Engineers (India) in Presidential Address by President G.P. Lal, Fellow

THE INSTITUTION

As mentioned in the industrial commission Report of 1916-1918, the need to establish an Institution was felt from 1916. However, the Institution was founded on September 13, 1920, under the Indian Companies Act of 1913. Later it was granted the Royal Charter in 1935. This was the first Royal Charter given to a body with its origin and functions in India.

The object of the Institution as defined in the Royal Charter is to promote the general advancement of engineering and engineering science and their applications in India and to facilitate dissemination and exchange of information and ideas on those subjects among the members attached to the Institution.

THE INSTITUTION OF ENGINEERS (INDIA)
8, Gokhale Road Calcutta 700 020

BYE-LAWS *
(*Effective 30 May, 1994*)

* As amended by the Corporate Members at the Extraordinary General Meeting held at Ahmedabad on May 30, 1994.

INTERPRETATIONS

1. In these Bye-Laws, unless the context otherwise requires, expressions shall have the same meaning as in the Charter, words importing males shall include females and words implying the singular number shall include the plural number and vice versa, and words importing individual person or persons shall include body or bodies corporate. Furthermore

   (a) 'The Institution' means The Institution of Engineers (India) incorporated by Royal Charter dated 9th September, 1935.

   (b) 'The Charter' means the Royal Charter of the Institution dated 9th September, 1935.

   (c) 'The Bye-Law' means the Bye-Laws of the Institution for the time being in force.

   (d) 'The Council' means the Council of the Institution constituted under these bye-laws.

(e) 'The President' means the President of the Institution.

(f) 'The Secretary and Director General' means the Secretary and Director General of the Institution as may be appointed by the Council from time to time.

(g) 'Regulations' means the Regulations of the Council for the time being in force, made pursuant to Bye-Law 26.

(h) 'Financial Year' means the twelve months commencing on the 1st day of April of a year and ending with the 31st day of March of the following year.

(i) 'Session' means the period of time between the date of one Annual General Meeting and the date of the succeeding Annual General Meeting.

(j) 'The Roll' means the register of members of the Institution.

(k) 'Recorded Address' means the address of a member as given in the proposal for his election or transfer, or subsequently varied by notice in writing to the Secretary and Director General and as recorded in the Roll.

(l) 'The Statues' means the rules of the Board of Governors of the Engineering Staff College of India made pursuant to Bye-Law 98 and Regulation 75.

(m) 'Academically Admissible Discipline' means the disciplines being taught and pursued at under-graduate level in several engineering institutions.

HEADQUARTERS

2. The Headquarters of the Institution shall be situated in Calcutta. The Headquarters may be shifted to any other place as may be decided by the two-thirds majority of Corporate Members on the Roll by postal ballot.

MANAGEMENT

3. The government and control of the Institution and its affairs shall vest with its Council and the Council shall exercise all the powers of the Institution and do all duties of the Institution with intent to give effect to the provisions of the Charter and the Bye-Laws for the time being of the Institution except so far as the same are by the Charter or by the Bye-Laws for the time being expressly required to be done in General Meetings of Corporate Members or by the President or the Secretary and Director General of the Institution.

THE COUNCIL

4. (a) The Council shall consist of :
   (i) The President;
   (ii) The two immediate Past Presidents;
   (iii) One Corporate Member from each of the State Centres, whose Corporate Membership is not less than one and half percent of the total strength of Corporate Members on the Roll as on 31st day of March of the year immediately preceding the year in which the election falls due, elected by the Corporate Members attached to the respective State Centres from amongst themselves;
   (iv) The representatives of the Divisions elected by the Corporate Members attached to the respective Divisions from amongst themselves as per following norms :

Strength of Corporate Members attached to Divisions as a percentage of the total Corporate Membership strength of the Institution as on 31st March preceding the year in which the election is due	No. of Representatives
upto 2%	1
above 2% but upto 5%	2
above 5% but upto 10%	3

above 10% but upto 20% — 4

above 20% — 5

(b) Following shall be *ex-officio* members of the Council;
  (i) The Chairman of the State Centres.
  (ii) The Chairman of the Local Centres whose Corporate Membership is not less than one percent of the total strength of Corporate Members on roll as on 31st March immediately preceding the Annual General Meeting of the Corporate Members of the Institution.

Note : Total Corporate Membership strength shall only be rounded off to the nearest hundred as per IS : 2-1960 and thereafter the required percentage be arrived at and compared with the respective strength of Corporate Membership either of State or Local Centre or Division, as the case might be, to determine the eligibility.

(c) The Council may, if considered necessary, also co-opt not more than four outstanding engineering personalities upon the recommendation of the Committee comprising five Past Presidents of the Institution as may be appointed by the Council for the purpose.

(d) No act of the Council or any body set-up under the Bye-Laws or Regulations of the Council shall be invalid merely by reasons of :
  (i) any vacancy in or defect in the constitution thereof;
  (ii) any defect in the election, nomination or appointment of a person acting as a member thereof.

5. The Council shall elect the President from amongst the Fellows in the Council but excluding the Past Presidents.

6. The term of the elected members of the Council, other than the President and two Immediate Past Presidents shall be for four sessions.

7. The term of the President shall be for one session and he shall retire at the Annual General Meeting of the Corporate Members at the termination of his session of office, but shall be eligible for re-election for further one session as President and after completion of his period of office (of one or two successive sessions, as the case may be) he shall not be eligible for re-election at any time.

8. The term of the members co-opted in accordance with the provisions of Bye-Laws 4(c) and 12(a) and 12(b) shall be one session only and they shall retire at the Annual General Meeting of the Corporate Members on termination of the session. They shall, however, be eligible for co-option to the Council for further session.

9. In the event of the President being unable to perform the duties of his office due to death, resignation or any other cause, or in the case of his prolonged absence from India for any reason, the senior-most member of the Council shall assume all the duties of the President till the Council elects a President from amongst themselves in accordance with the provision of Bye-Law 5 for the residual part of the session provided however, that the residual part of the session is not less than two months. The seniority of members of the Council shall be based upon the continuous number of years that the members concerned have been on the Roll as Corporate Members.

10. The President shall be ex-officio member of all Boards and Committees of the Institution.

11. The Secretary and Director General shall, soon after the 31st March of the year in which the elections under Bye-Law 4(a) is due, notify the Honorary Secretaries of the State Centres of the members of Council who will retire at the Annual General Meeting and shall call upon the State Centres to hold election under Regulation 14 to fill the vacancies in the Council.

12. (a) The Council may fill by co-option any casual vacancy arising from death or resignation or any other cause of an elected member of the Council which may occur between one Annual General Meeting and the succeeding Annual General Meeting. Resignation from the Council, including resignation from Chairmanship of a State Centre and Local centre, shall become effective from the date of resignation and shall not be withdrawn except by special permission

of the Council.

(b) The Council may also fill by co-option the vacancies caused by election of one member of the Council as President under Bye- Law 5 and by reason of one or two of the elected members of the Council being member or members of the Council as immediate Past President under Bye-Law 4(a).

13. If a Chairman of a State or Local Centre is already a member of the Council, then the Council shall co-opt one Corporate Member attached to the State or Local Centre as member of the Council during the period of office of the Chairman upon the recommendation of the State or Local Centre as the case may be.

POWERS, PROCEDURES AND DUTIES OF THE COUNCIL

14. Pending the election of a Council according to the provisions of these Bye-Laws, the existing Council shall be the Council of the Institution and shall exercise all powers of the Institution subject to the provisions of the charter.

15. The Council shall direct and manage the affairs and property of the Institution, subject to the provisions of the Charter, the Bye-Laws and the Resolutions of General Meetings of Corporate Members which have been duly summoned and held in accordance with the Charter and the Bye-Laws and duly recorded in the minutes of the meeting. Subject as aforesaid, the Council shall further exercise all powers of the Institution not otherwise specifically provided for, provided such powers are not by the Charter or the Bye-Laws required to be exercised by the Corporate Members at a General Meeting.

16. The decisions of the Council on all matters dealt with by it in accordance with the provisions of the Charter, the Bye-Laws and the Regulations, and such Resolutions of General Meetings of Corporate Members as aforesaid, shall be final and binding on all classes of members.

17. The Council may appoint committee or committees and may delegate any of its powers to any such committees except as stated in Bye-Law 18. The Council may, in addition, delegate any of its powers to the President, the Chairman of state or Local Centres, the Chairman of Board of Governors or Director of the Engineering Staff College of India, the Chairmen of the Divisions, the Secretary and Director General or such other person or persons or bodies as it may appoint or constitute from time to time for the efficient working of the Institution. All such committees or persons shall in the exercise of powers delegated to them by the Council and in the transaction of business conform to any directions that may be given to them by the Council and subject thereto may regulate their proceeding as they think fit.

18. None of the Powers or functions of the Council under Bye-Laws 74 and 75 shall, however, be delegated to a committee save only the hearing of any explanation or defence given orally.

19. The Council shall meet as often as the business of the Institution may require, but not less than four times a year. At every meeting of the Council twenty members shall constitute a quorum and at a meeting of the Council held after adjournment — partly or wholly — members present shall form the quorum. A meeting of the Council at which a quorum is present shall be competent to exercise all or any of the powers or discretions vested in or exercisable by the Council, notwithstanding any vacancy in the body. If at any meeting there is no quorum, the Council shall stand adjourned for twenty-four hours and shall meet at the same place on expiry of twenty-four hours and transact the business of the meeting with the members present as the quorum.

20. To summon a meeting of the Council, the Secretary and Director General shall issue the notice of meeting to each member of the Council together with the agenda not less than fourteen days before the date of the meeting. This period of fourteen days shall be exclusive of the day on which the notice is issued, but inclusive of the day for which the notice is given. This period shall not apply to supplementary agenda which may be issued as the exigencies of the case may require.

21. The Secretary and Director General shall summon a special meeting of the Council on receipt of a written requisition which shall contain the specific matters desired to be discussed at the

special meeting and shall be signed by at least ten members of the Council and within thirty days of the receipt of the requisition. The notice of such special meeting shall state the purpose for which the meeting is called, and its period shall be at the same as in Bye-Law 20. At such meeting no business other than specified in the requisition shall be transacted.

22. At any meeting of the Council, each member of the Council present in person shall have one vote. Except as provided in Bye-Laws 74 and 75 all questions shall be decided in the Council by a majority of those present and having a right to vote. In the case of equality, the President or other person presiding shall have a second or casting vote. A postal vote of the Council shall, however, be taken whenever :

    (a) Any six present in person at the meeting shall demand it; or,

    (b) Any twelve, whether present at the meeting or not, shall by notice in writing delivered to the Secretary and Director General before the hour fixed for the meeting, demand it; or,

    (c) The meeting of the Council shall so direct.

23. The Council shall cause a statement of the funds of the Institution, and of the income and expenditure during the past year, terminating 31st March, to be made and verified and signed by the Auditors. The Council shall cause such accounts to be presented at the Annual General Meeting.

24. The Council shall draw up an Annual Report of the Council on the state of the Institution, and shall present it at the Annual General Meeting.

25. The Council may make, amend or rescind Bye-Laws provided that the same be not repugnant to the Charter and no such addition, amendment or rescission shall come into operation until the same has been approved by an Extraordinary General Meeting of Corporate Members.

26. The Council may make, amend or rescind Regulations provided that the same be not repugnant to the Charter and the Bye-Laws, and no such addition, amendment or rescission shall come into operation until the same has been approved by a Special General Meeting of Corporate Members.

27. The Council may, upon receipt of a request to that effect from any society with objects kindred to those of the Institution, arrange for the amalgamation of such society with the Institution, and may also extent such concessions as it thinks fit to members of such society at the time of amalgamation to facilitate their admission into the Institution. But no such amalgamation shall be effective until it is approved by a Special General Meeting of Corporate Members.

SECRETARY AND DIRECTOR GENERAL

28. The Secretary and Director General of the Institution shall be appointed by the Council at such remuneration and on such other terms as it may think fit.

29. (i) The Secretary and Director General shall carry out all his duties under the direction of the Council. He shall, unless exempted from the provisions hereof by the Council, devote his whole time to the business of the Institution and shall not engage any other business or profession. He shall be in administrative charge of all the employees at the headquarters, State Centres, Local Centres and such other Centre or Centres as may be established in future.

    (ii) He shall be responsible for the conduct of all correspondence and office work of the Headquarters and the keeping of accounts of the Institution; he shall maintain Roll of all classes of members and of their addresses; he shall attend all meetings of the Council and shall have the business thereat correctly and fully recorded and reported; he shall attend all meetings of the Board of Governors of the Engineering Staff College of India as *ex-officio* member; he shall superintend the publications and the examinations of the Institution; he shall have charge of the libraries of the Institution; he shall have charge of all the properties of the Institution save only the funds and moneys in the custody of the Bankers; he shall engage, subject to the approval of the Council, all persons employed under him and arrange their work and duties, and he shall generally conduct the ordinary business of the Institution in accordance with the Charter, the Bye-Laws, the Regulations and the directions of the Council, and shall refer to the President in any matters of importance or difficulty requiring urgent decision.

(iii) The powers of the Secretary and Director General shall be generally described in the Regulations, and may be varied by the Council from time to time.

GENERAL MEETINGS

30.   The General Meetings of the Corporate Members shall be of the following classes :

(a) The Annual General Meeting of Corporate Members only for the business prescribed in the Charter and the Bye-Laws, as detailed in Bye-Law 31(a).

(b) Special General Meeting of Corporate Members for the purpose of :

(i) considering addition, amendment, recession of the Regulations made by the Council under the provisions of Bye-Law 26;

(ii) considering any resolution duly passed by the Council other than those in respect of addition, amendment or rescission of Bye-Laws, and requiring the confirmation of the Corporate Members; and

(iii) considering the specific matters to be moved at such Special General Meeting pursuant to any requisition signed by not less than five hundred Corporate Members and submitted to the Secretary and Director General.

(c) Extraordinary General Meeting Corporate Members for the purpose of making, amending or rescinding any Bye-Law of the Institution or approving the resolution of the Council under Bye-Law 25 in respect of addition, amendment or rescission of any Bye-Law of the Institution.

31   (a)   (i) The Annual General Meeting shall be held at such place as the Council shall determine, but in the month of November or December in every year. But if not held due to unavoidable reasons, the Councils shall have the power to hold the Annual General Meeting in any other month, and the reasons for the postponement shall be announced at the Annual General Meeting. Only Corporate Members shall be entitled to attend such Annual General Meeting. The business of the Annual General Meeting shall be to receive the Audited Accounts of the Institution and the Annual Report of the Council for the preceding financial year and the report on the elections to the Council and to appoint Auditors and fix their remuneration.

(ii) In case the Council fails to lay the Audited Accounts of the Institution at a duly convened Annual General Meeting, the Annual General Meeting shall be adjourned and shall take place at a later date for the purpose of laying the said Audited Accounts and the place, date and time for such adjourned Annual General Meeting shall be decided by the Chairman of the Annual General Meeting and declared along with the announcement for adjournment and no business of the Corporate Members performed at the Annual General Meeting shall be invalid merely by reason of adjournment of the Annual General Meeting for not laying Audited Accounts.

(b)   (i) A Special General Meeting of the Corporate Members may be called at any time by the Council at such place as the Council may decide for the purpose.

(ii) The Council shall also be bound to call a Special General Meeting whenever a requisition signed by at least five hundred Corporate Members is made and delivered to the Secretary and Director General, specifying the matter to be moved at such Special General Meeting and shall cause the notice of the Special General Meeting to be issued within ninety days of the receipt of the requisition by the Secretary and Director General. At such Special General Meeting only the special matters of which notice has been given or such questions as necessarily arise thereof shall be considered.

(c)   (i) An Extraordinary General Meeting of Corporate Members may be called at any time by the Council at such place as the Council may decide for the purpose.

(ii) The Council shall also be bound to call an Extraordinary General Meeting whenever a requisition signed by at least five hundred Corporate Members is made and delivered to the Secretary and Director General specifying the matter to be moved at such an Extraordinary General Meeting and shall cause the notice of the Extraordinary General Meeting to be issued, within ninety days of the receipt of requisition by the Secretary and Director General. These provisions shall not, however, apply to Bye-Law 2 which shall be operated only according to its own provisions.

32. Notice of the making, amending or rescinding of any Bye-Law or Regulation shall be published in the appropriate publication of the Institution as soon as it is reasonably practicable after the same shall have been passed in accordance with the provisions of the Charter and the Bye-Laws.

33. At least twenty-five days notice shall be given of all General Meetings of Corporate Members. This period of twenty-five days shall be exclusive of the day on which the notice is deemed to be given, but inclusive of the day for which the notice is given. In the case of a Special General Meeting and an Extraordinary General Meeting convened the instrument for appointing a proxy as described in Bye-Law 40 shall accompany the notice.

34. Service of a notice shall be deemed to be effected by properly addressing, pre-paying and posting the letter containing the notice. The accidental omission to give notice of a General Meeting of Corporate Members to or the non-receipt of notice of such a meeting by any Corporate Member entitled to receive notice shall not invalidate the proceedings of that meeting.

PROCEEDINGS OF GENERAL MEETINGS

35. The President, when present, shall be the Chairman at all General Meetings, and in his absence the chair shall be taken by a member of the Council in order of their seniority. But if no member of the Council is present and willing to act, the meeting shall elect a Chairman from the Corporate Members present at the meeting.

36. No business shall be transacted at any General Meeting of Corporate Members unless a quorum is present when the meeting proceeds to business. The Corporate Members present shall be the quorum for the Annual General Meeting, fifty Corporate Members for a Special General Meeting, and one hundred Corporate Members for an Extraordinary General Meeting.

37. If within thirty minutes after the time appointed for a Special General Meeting or an Extraordinary General Meeting, or at any point of time during a Special General Meeting or an Extraordinary General Meeting, the requisite quorum is not present, the meeting shall be dissolved and the Chairman with the consent of the meeting and if so directed by the meeting, shall adjourn the meeting by twenty-four hours. The adjourned meeting shall be held at the same place on expiry of twenty-four hours and the Corporate Members present at the adjourned meeting shall form the quorum. No business shall be transacted at any adjourned meeting other than the business left unfinished at the meeting from which adjournment took place. It shall not be necessary to give any notice of an adjournment or of the business to be transacted at an adjourned meeting.

38. (a) At all General Meetings of Corporate Members, questions shall be decided according to the majority of votes properly given thereat by show of hands :

   (i) Unless a poll is, before or on the declaration of the results by the show of hands, demanded by at least ten Corporate Members present, provided no poll shall be demanded on the election of a Chairman or on a question of adjournment, and

   (ii) Unless instruments of proxy have been deposited with the secretary and Director General in accordance with the provision of Bye-Law 41 in which case due account shall be taken of them.

   (b) In the case of an equality of votes the Chairman of the meeting shall both on a show of hands and at a poll have a casting vote. The acceptance or rejection of votes by the Chairman shall be conclusive for the purpose of the decision of the matter in respect of which the votes are tendered.

39. Votes may be given at a General Meeting of Corporate Members either personally or by proxy. Each Corporate Member present in person or by proxy shall have one vote. No person shall be

appointed a proxy to vote at any meeting who is not a Corporate Member entitled in his own right to vote at such meeting.

40. The instrument appointing a proxy shall be in writing under the hand of the appointer and shall, as nearly as circumstances will admit, be in the form or to the effect following :

The Institution of Engineers (India)

I..of.....................being a Corporate Member of the above Institution hereby appoint..............................of..or failing him.. of ....as my proxy at (Special General Meeting or Extraordinary General Meeting, as the case may be) of the Institution to be held on the.........day of..................19.........and at any poll held in connection therewith. As witness my hand...........day of...........19.........

Signature....................................

Class of membership in I.E. (I)..............No. ....

41. The instrument appointing a proxy shall be deposited with the Secretary and Director General not less than three days before the hour fixed for the meeting at which the person named in the proxy proposes to vote, but no instrument appointing a proxy shall be valid after expiration of three calendar months from its date.

42. A poll shall be taken by means of postal voting papers and shall be a poll of all Corporate Members who at the time of despatch of such shall be entitled to receive notice of a meeting :

   (a) When Corporate Members demand it under the provisions of Bye- Law 38(a) (i), and

   (b) When in its opinion, the Council considers it in the interest of the Institution that an appeal should be made on any question to the whole body of Corporate Members.

   The form and contents of the voting paper shall specify the subject matter to be voted upon and the date of return thereof, and shall be issued at least thirty days, exclusive of the day on which the papers are deemed to be issued, but inclusive of the day of return, prior to the date of return.

MEMBERSHIP

43. The Institution shall consist of members in the following orders :Honorary, Corporate and Non-Corporate. The Honorary order shall comprise the classes of Honorary Fellows and Honorary Life Fellows. Corporate Members shall comprise the classes of Fellows, Members, Associate Members and Non-Corporate Members shall comprise the classes of Associates. Affiliate Member, Senior Technician Members, Technician Members, Students, Institutional Members and Donor Members. The names and addresses of all members shall be entered on the Roll of the Institution.

QUALIFICATIONS FOR MEMBERSHIP

Honorary Fellows

44. Honorary Fellows shall be the President of India, the Vice- President of India, the Prime Minister of India and Governors of States or equivalent Heads of States and Union Territories as may be invited by the Council to be Honorary Fellows for the period of their respective tenures of office.

Honorary Life Fellows

45. Honorary Life Fellows shall be elected by the Council from persons in India and abroad for their high eminence in engineering or the sciences and every such election shall be announced at the next Annual General Meeting, provided that there shall not be more than twenty-five Honorary, Life Fellows at any one time. The Council shall frame rules for the election of such Honorary Life Fellows and in regard to the manner of bestowing the honour on them.

Fellows

46. Every candidates for election as a Fellow or for transfer to the class of Fellowship shall satisfy the Council that he possesses the following qualifications :

   (i) *Age :* He shall have attained the age of forty on the date of his application for election or transfer.

   (ii) *Occupation :* He shall be engaged in the profession of engineering in a position of high responsibility, or shall have retired from the profession after having held, before his application for election or transfer, a position of high responsibility.

   (iii) *Attainments :* He shall have one of the following qualifications lettered respectively (a) or (b) or (c) :

   (a) He shall either be a Member or shall have fulfilled the conditions necessary for Membership; and he shall have had at least fifteen years employment in the aggregate in a position of responsibility in the design or execution of important engineering works.

   For the purpose of this Bye-Law, employment as a senior member of the engineering staff in an engineering college or institute which has regular courses of study leading to an educational qualification recognised by the council as exempting from Sections A and B of the Institution Examinations or employment in a responsible position on important engineering research may be accepted by the Council in place of employment in the design or execution of important engineering works.

   (b) He shall have had at least eighteen years' employment in the aggregate in a position of responsibility as described in Bye-Law 46 (iii) (a).

   (c) He shall have high educational qualifications and shall occupy a prominent position in the profession of engineering to the satisfaction of the Council.

Members

47. Every candidate for election as Member or for transfer to the class of Member shall satisfy the Council that he possesses the following qualifications :

   (i) *Age :* He shall have attained the age of thirty two years on the date of his application for election or transfer.

   (ii) *Occupation :* He shall have been engaged in a position of responsibility in the design and execution or operation of engineering works. For the purpose of this Bye-Law, employment as a teacher of engineering or in a likewise capacity in an engineering college or institute which has regular courses of study leading to an educational qualification recognized by the Council or employment in engineering research may be accepted by the Council in place of employment in the design and execution or operation of engineering works.

   (iii) *Examination :* He shall have passed Sections A and B of the Institution Examinations prescribed by the Council or possesses an educational qualification recognized by the Council as exempting therefrom.

   (iv) *Training :* He shall have received engineering training in a regular course of study in an engineering college or institute leading to an educational qualification recognized by the Council as exempting from Sections A and B of the Institution Examinations or as a pupil or apprentice or assistant in an engineering office or works as would provide him with engineering training to the satisfaction of the Council.

   (v) *Experience :* He shall have had further at least ten years professional engineering experience in a position of responsibility. In the case of a candidate who has passed Sections A and B of the Institutions Examinations, the Council may, at its discretion, take into account periods of responsible employment prior to his passing Section B.

Associate Members

48. Every candidate for election as an Associate Member or for transfer to the class of Associate Membership shall satisfy the Council that he possesses the following qualifications :

(i) *Age :* He shall have attained the age of twenty seven years on the date of his application for election or transfer.

(ii) *Occupation :* He shall have been engaged in position of responsibility in the design and execution or operation of engineering works. For the purpose of this Bye Law, employment as a teacher of engineering or in a likewise capacity in an engineering college or institute which has regular courses of study leading to an educational qualification recognized by the Council or employment in an engineering research may be accepted by the Council in place of employment in the design and execution or operation of engineering works.

(iii) *Examination :* He shall have passed Sections A and B of the Institution Examinations prescribed by the Council or possesses an educational qualification recognized by the Council as exempting therefrom.

(iv) *Training :* He shall have received engineering training in a regular course of study in an engineering college or institute leading to an educational qualification recognized by the Council as exempting from Sections A and B of the Institution Examinations, or as a pupil or apprentice or assistant in an engineering office or works as would provide him with engineering training to the satisfaction of the Council.

(v) *Experience :* He shall have had further at least five years professional engineering experience in a position of responsibility. In the case of a candidate who has passed Sections A and B of the Institution Examinations, the Council may, at its discretion, take into account periods of responsible employment prior to his passing Section B. The Council may, at its discretion, at its discretion, arrange for candidates for election as Associate Members to be examined by a written or an oral test or both called Section C of the Institution Examinations in order that they may be satisfied that such candidates have acquired during their practical training and professional engineering experience adequate engineering knowledge.

Associate

49. Every candidate for election as an Associate or for transfer from Student or Technician Member or Senior Technician Member to Associate shall satisfy the Council that he possesses the following qualifications :

(i) *Age :* He shall have attained the age of twenty two years on the date of his application for election or transfer.

(ii) *Occupation :* He shall have engaged in the engineering profession and shall in the opinion of the Council be making satisfactory progress in the acquisition of qualifications for Associate Membership provided that temporary unemployment is no bar to his election.

(iii) *Examination :* He shall have passed Sections A and B of the Institution Examinations prescribed by the Council or possesses an educational qualification recognised by the Council as exempting therefrom.

(iv) *Training :* He shall have received engineering training in a regular course of study in an engineering college or institute leading to an educational qualification recognized by the Council as exempting from Sections A and B of Institution Examinations, or as pupil or apprentice or assistant in an engineering office or works as would provide him with engineering training to the satisfaction of the Council.

Affiliate Members

50. Every candidate for election as an Affiliate Member or for transfer from any other grade to Affiliate Member shall satisfy the Council that he possesses the following qualifications :

(i) *Age :* He shall have attained the age of thirty years on the date of his application for election or transfer.

(ii) *Occupation :* He shall have engaged as a teacher in a statutory engineering college or technical institution or engaged in engineering profession in a position of responsibility.

(iii) *Examination :* He shall have passed a degree in Science and/or Social Sciences awarded by any statutory university but does not hold any engineering qualification recognized as exempting from Sections A and B of the Institution Examinations.

Senior Technician Members

51. Every candidate for election as a Senior Technician Member or for transfer from Technician Member or Student to Senior Technician Member shall satisfy the Council that he possess the following qualifications :

    (i) *Age* : He shall have attained the age of eighteen years on the date of his application for election or transfer.

    (ii) *Occupation* : He shall be engaged in the engineering profession as a pupil or apprentice or assistant under a Corporate Member or person possessing the qualification required to be elected as a Corporate Member and shall in the opinion of the Council be making satisfactory progress in the acquisition of qualifications for Corporate Membership provided that temporary unemployment is no bar to his election.

    (iii) *Examination* : He shall have passed (a) either an accredited diploma examination in engineering or technology or its equivalent as recognised by the Council, or (b) such other examination recognized by the Council as exempting from passing Section A of the Institution Examination in non-diploma stream.

Technician Members

52. Every candidate for election as a Technician Member or for transfer from Student to Technician Member shall satisfy the Council that he possesses the following qualifications :

    (i) *Age* : He shall have attained the age of eighteen years on the date of his application for election or transfer.

    (ii) *Examination* : He shall have passed the Studentship Examination as conducted by the Institution or shall possess an educational qualification recognised by the Council as sufficient for exemption from passing the said Studentship Examination.

    (iii) He shall in the opinion of the Council be making satisfactory progress in the acquisition of qualification for Corporate Membership.

Student

53. Every candidate for election as Student shall satisfy the Council that he possesses the following qualifications :

    (i) *Age* : He shall have attained the age of sixteen years on the date of his application for election.

    (ii) *Academic training* : (a) He shall be receiving training in regular, course of study in an engineering or technical institute leading to an educational qualification recognised by the Council, or (b) shall be qualifying himself for passing the Studentship Examination as conducted by the Institution.

Institutional Members and Donor Members

54. The Council may, at its discretion, attach to the Institution as a Institutional or Donor Member any Public or Local Body, Registered Company or Firm, or an individual who may desire to be so attached.

Professional Engineer

55. The Council may upon an application from a candidate who is a Fellow or a Member of the Institution, approve that he may be entitled to describe himself as a Professional Engineer so long as he is a Fellow or a Member of the Institution provided he satisfies the following requirements (i), (ii) and (iii).

    (i) He shall have attained 35 years of age.

    (ii) He shall have compounded his subscription for life as a Fellow or as a Member of the

Institution.

(iii) He shall have experience of not less than ten years in the profession of engineering in a position of responsibility covering one or more areas in research, design, development, manufacturing, maintenance, engineering services, execution or operation of engineering projects and/or consultancy for engineering research/design projects, teaching together with involvement in practical design/consultancy works.

Notwithstanding anything contained in this Bye Law a person enrolled as a Professional Engineer shall cease to have any right to designate himself as a Professional Engineer in case he ceases to be a Fellow or a Member of the Institution for any reason whatsoever.

ELECTION AND TRANSFER

56. Any person desirous of election as a Corporate Member or transfer to any class of Corporate Membership shall be proposed and recommended from personal knowledge in the prescribed form by : (a) three Fellows in case of application for election as Fellow or for transfer to the class of Fellow, (b) at least one Fellow and two Members in case of application for election as Member or for transfer to the class of Member, and (c) by two Corporate Members of any class in case of application for Associate Member or transfer to the class of Associate Member of whom at least one shall be a Fellow.

Any person seeking entitlement as a 'Professional Engineer' shall be proposed and recommended from personal knowledge in the prescribed form by three Fellows. The proposal form shall be accompanied by such fees as are prescribed in the Regulation and shall be delivered at the Headquarters of the Institution.

57. Any person desirous of election as Affiliate Member or Associate shall be proposed and recommended from personal knowledge in the prescribed form by at least two Corporate Members, of whom at least one shall be a Fellow or Member. The proposal form shall be accompanied by such fees as prescribed in the Regulations and shall be delivered to the Secretary and Director General.

58. Any person desirous of election as Senior Technician Member, Technician Member or Student shall be proposed and recommended from personal knowledge in the prescribed form by two Corporate Members. The proposal form shall be accompanied by such fees as are prescribed in the Regulations, and shall be delivered to the Secretary and Director General.

59. Any applicant desirous of attachment as Institutional Member and/or Donor Member shall deliver his application direct to the Secretary and Director General with such fees as are prescribed in the Regulations.

60. (a) On each proposal form received from an applicant normally resident in India for election to the class of Fellow, the Secretary and Director General shall record his opinion as to whether or not the qualifications of the candidate appear to be in accordance with the provisions of the appropriate Bye-Law.

    (b) The Secretary and Director General shall then forward the proposal with his note aforesaid to the Honorary Secretary to the State Centre within the boundaries of which the candidate resides or practices or carries on business.

    (c) The Honorary Secretary of the State Centre shall obtain the opinion of the Committee of the State Centre, record it on the aforesaid note of the Secretary and Director General in the space set apart for this purpose and return it to the Secretary and Director General within the period of time provided in Regulation 13.

    (d) The Secretary and Director General shall place the proposals to the Membership Committee consisting of the President, Chairman of three Division Boards to be nominated by the Council and the Secretary and Director General for decision, and then shall declare the proposal as approved or not approved by the Membership Committee on behalf of the Council, as the case may be.

61. On each proposal form received from an applicant normally resident abroad for election, to the

class of Fellow, the Secretary and Director General shall record his opinion as to whether or not the qualifications of the candidate appear to be in accordance with the provisions of the appropriate Bye-Law and then place the proposal to the Membership Committee for final decision.

62. (a) Each proposal form for member, Associate member, Associate, Senior Technician Member, Technician Member, or Student shall be scrutinised by the Secretary and Director General as to whether or nor the qualifications of the candidate are in accordance with the appropriate Bye-Laws.

(b) The Secretary and Director General shall then on behalf of the Council declare the proposal as approved by the Council or as not approved by the Council as the case may be.

(c) If the candidate has the requisite educational qualification, but the nature of his training is such as can be best verified by the State Centre within the boundaries of which he resides, the Secretary and Director General shall follow the procedure set out in Bye-Law 60 (a), (b), (c) and 62 (b).

63. (a) Each proposal form for Affiliate Member or Institutional Member or Donor Members shall be scrutinized by the Secretary and Director General as to whether or not it is desirable to attach the applicant to the grade applied for.

(b) If the applicant, in the opinion of the Secretary and Director General is suitable, the Secretary and Director General shall on behalf of the Council declare the proposal as duly approved by the Council, or if not suitable, as not approved by the Council.

64. Every candidate for the class of Fellow, Member, Associate Member, Associate, Affiliate Member, Senior Technician Member, Technician Member, or Student shall on his election or transfer, be forthwith notified of the fact by the Secretary and Director General. But no person, although duly elected in accordance with the Bye-Laws, shall be entitled to any of the rights and privileges of the Institution nor shall his name be entered on the Roll, until he shall have signed and delivered to the Secretary and Director General an undertaking in the prescribed form that he will be governed by the provisions of the Bye-Laws and the Regulations in force from time to time and that he will accept as final and binding the decisions of the Council.

65. Every applicant for the class of Institutional Member or Donor Member shall on his attachment, be forthwith notified of the fact by the Secretary and Direct General.

66. The Secretary and Director General shall give notice of every such election or attachment to the Honorary Secretary of the State Centre within the boundaries of which the candidate or applicant resides or practices or carries on business.

RIGHTS AND PRIVILEGES

67. Subject to the provisions of the Charter and to the restrictions contained in the Bye-laws :

(a) Fellows shall enjoy all the rights and privileges of the Institution, but in accordance with the provisions in the Charter, Bye-Laws and the Regulations.

(b) Members and Associate Members shall enjoy the rights and privileges of the Institution, but in accordance with the provisions in the Charter, Bye-Laws and the Regulations and save that they shall not be eligible to hold office as President. Chairman of a State Centre or Chairman of a Local Centre in accordance with the provision in Regulation 29(f).

(c) Associates, Affiliate Members, Senior Technician Members, Technician members and students shall enjoy the rights and privileges of the Institution, but in accordance with the provisions in the Bye-Laws and the Regulations and save that they shall not be eligible to hold office and shall not have any right of voting except in cases as may be provided in the rules framed by the Council for the time being.

68. The rights and privileges of every Fellow, Member, Associate member, Associate, Affiliate Member, Senior Technician Member, Technician Member or Student shall be personal to himself as such and shall not be transmissible by his own act or by operation of law.

69. (i) Every Fellow, Member and Associate Member is, and is entitled to describe himself as a Chartered Engineer, and in using that description after his name as C. Eng. (I) shall place

it before that designation of the class in the Institution to which he belongs, stated in accordance with the following abbreviated forms, namely, F.I.E., M.I.E., A.M.I.E.

(ii) A Fellow or Member or Associate Member practising in partnership with any person who is not a Fellow or Member or Associate Member under the title of a firm shall not use or permit to be used after the title of such firm the designation Chartered Engineer or Chartered Engineers, or describe or permit the description of such firm in any way as Chartered Engineers.

(iii) A Fellow or Member or Associate Member practising or acting in a professional capacity under the title of, or as a director, officer or employee of a company, whether such company shall be authorized or not to carry on the profession or business of an engineer in all or any of its branches shall not use or permit to be used after the title of such company the designation of Chartered Engineer or Chartered Engineers, or describe or permit the description of such company in any way as Chartered engineers.

(iv) Every Associate, Affiliate Member, Senior Technician Member and Technician Member shall be entitled to place, after his name the designation of the class in the Institution to which he belongs, stated in accordance with the following abbreviated forms, namely A.I.E., Aff. I.E., Sr. Tech. I.E. and Tech. I.E. respectively.

(v) Every Fellow or Member entitled to described himself as a Professional Engineer can use that description after his name as P. Eng.

RESIGNATION, REMOVAL AND REINSTATEMENT

70. Every person elected to any class of membership shall cease to be member in case of death or under the conditions as provided in Bye-Laws 71 and 72.

    Provided further that every person elected to the class of Senior Technician Member, Technician Member or Student shall cease to be a member of that class on qualifying himself for election or transfer to the next higher class of membership.

71. (a) Every Fellow, Member, Associate Member on the Roll of the Institution as on the 31st day of March 1989 shall be liable to pay his appropriate annual subscription as indicated in Table 1 of the Regulation 48.

    (b) Every person elected directly to the class of Fellows, Members, Associate Members, Associates, Senior Technician Members, Technician Members and Students enrolled for appearing in Studentship examination shall be liable to pay to the Institution the appropriate sum of money as Composite subscription in the manner as prescribed in the Regulation 50.

72. (a) If any member of any class liable to pay annual subscription as per Bye-Law 71 (a) shall leave his subscription in arrears for two years, and shall fail to pay such arrears within three months after a notice has been sent to him by the Secretary and Director General, his name shall be removed from the Roll, and he shall thereupon cease to have any rights or privileges, but he shall nevertheless continue to be liable to pay the arrears of subscription due at the time of his name being removed from the Roll. The Diploma or Certificate of every such person shall be returned to the Secretary and Director General.

    (b) If any member of any class liable to pay Composite subscription shall fail to pay full amount of composite subscription within the period stipulated in the Regulations his name shall be removed from the Roll and he shall thereupon cease to have any rights or privileges. The Diploma or Certificate of every such person shall be returned to the Secretary and Director General.

    (c) Every Corporate Member elected before 31st March 1989 shall be liable to pay his appropriate annual subscription until he shall have signified in writing to the Secretary and Director General his desire to resign, having previously paid all arrears including the current year's subscription, or until he has forfeited his right to remain in the Institution, provided that a letter of resignation may, with the consent of the Council, be cancelled at the request of the sender within six months of its receipt by the Secretary and Director General but on payment of all sums which he would have been liable to pay had his connection with the Institution not been interrupted. He shall thereupon recover all his former rights and privileges without re-election.

(d) If a member of any class elected on or after 1st April 1989 shall have signified in writing to the Secretary and Director General his desire to resign or he has forfeited his right to remain as a member of the Institution, his name shall be removed from the Roll of the Institution and the amount of Composite subscription paid by him to the Institution in part or in full shall be forfeited by the Institution. Provided, however, that a letter of resignation may, with the consent of the Council, be cancelled at the request of the sender within six months of its receipt by the Secretary and Director General but on payment of all sums which he would have been liable to pay had his connection with the Institution not been interrupted. He shall thereupon recover all his former rights and privileges without re-election.

73. The Council may, if they find good reason to do so, reinstate on such conditions as they may deem fit, any person who has been a member of any class and whose name has been removed form the Roll on account of resignation or non-payment of subscriptions. These cases shall be considered and reported upon to the Council by a Committee appointed by the Council for the purpose.

EXPULSION AND DISCIPLINARY ACTION

74. (i) The Council shall have the right to expel from the Institution any member who shall have, in the opinion of the Council, willfully acted in contravention of the Charter or the Bye-Laws, or the Regulations from time to time in force under the Bye-Laws, or who, in the opinion of the Council, shall have been guilty of violation of the Code of Ethics as framed by the Council under the provision of Regulation 47 or such conduct as shall render him unfit to remain a member of the Institution, provided that :

(a) The meeting of the Council at which the resolution shall be passed shall be specially convened for the purpose, and the person concerned shall have been given clear thirty days' notice that his conduct is to be enquired into, and is given an opportunity of stating his case to the Council.

(b) The resolution shall be passed by a majority of two-thirds of those present.

(ii) Upon the resolution of expulsion being passed, the name of the person concerned shall be removed from the Roll and he shall cease to have any connection with the Institution. Neither the Council collectively nor any member of the Council individually shall be made liable for anything done under this Bye-Law. Every candidate applying for election to the Institution shall be deemed in so applying to agree to accept as final any decision of the Council under this Bye-Law.

(iii) Every person who has ceased to belong to the Institution shall be called upon by the Secretary and Director General to return immediately his Diploma or Certificate to the Secretary and Director General, and he shall not be entitled any longer to make use of any designation implying past connection with the Institution.

75. The Council shall also have the right to take any disciplinary action against any member who shall have, in the opinion of the Council, acted in such a manner as to warrant disciplinary action, but not expulsion. The procedure laid down in Bye Law 74 (i) (a) and (b) shall be followed in such cases also, and the decision of the Council shall be final and binding upon the member concerned.

STATE CENTRES AND LOCAL CENTRES

76. The Council shall establish one Centre of the Institution in each State and each Union Territory of the Indian Union and it shall be known as State Centre.

77. Each State Centre shall be constituted and its affairs shall be managed and carried on in accordance with the provisions laid down in the Bye-Laws, Regulations and other rules duly framed and approved from time to time by the Council.

78. Notwithstanding anything contained in Bye-Laws 76 and 77, the Council may, if thought fit, permit establishment and functioning of one State Centre for two or more geographically adjoining States, Union Territories or both of the Indian Union provided such arrangement is mutually agreed to by and between the Corporate Members of the concerned States, Union Territories or both of the Indian Union as the case may be.

    The provisions of Bye-Laws 75 and 76 shall not apply to the State Centre which are already in existence for two or more States, Union Territories or both of the Indian Union.

79. The Council may also establish one or more Local Centres of the Institution within the geographical boundaries of a State Centre. The affairs of such Local Centres shall be carried on in accordance with the provisions of the Bye-Laws, Regulations and other rules duly framed and approved from time to time by the Council.

TECHNICIANS CHAPTERS

80. (a) (i) The Council may establish on the recommendation of a State Centre one or more Technicians Chapters within the geographical boundaries of a State Centre with geographical boundaries of such Chapters defined and notified.

    (ii) All Senior Technician Members and Technician Members of the Institution having their recorded addresses within the geographical boundaries of a Technicians Chapter shall be attached to the Technicians Chapter.

    (b) The Council may also establish Technicians Chapters at engineering and technical establishments recognised by the Council provided total number of Senior Technician Members and Technician Members engaged or employed in the said establishment is not less than twenty.

    (c) The affairs of the Technicians Chapter shall be managed in accordance with the rules framed and approved by the Council from time to time.

81. The Council shall cause formation of an All India Technicians Committee, with a Chairman who shall be a member of the Council nominated by the Council from amongst themselves and members not exceeding ten in number nominated or elected in accordance with the rules framed and approved by the Council from time to time.

STUDENTS CHAPTERS

82. (a) (i) The Council may establish on the recommendation of a State Centre one or more Students Chapters within the geographical boundaries of a State Centre with geographical boundaries of such Chapters defined and notified.

    (ii) All students of the Institution having their recorded addresses within the geographical boundaries of a Students Chapter shall be attached to the Students Chapter.

    (b) (i) The Council may also establish Engineering College Students Chapters or Polytechnic Students Chapters at engineering colleges or polytechnics conducting courses recognised by the Institution with the concurrence of the managing authorities of the respective engineering colleges or polytechnics and notify the establishment to the State Centre within the geographical boundaries of which such Chapter will be located.

    (ii) All Students who are students of such an engineering college or a polytechnic shall be eligible for attachment to such Chapter and the affairs of such Chapter shall be managed in accordance with the rules framed and approved by the Council from time to time under the guidance of the Adviser who shall be a member of the faculty nominated by the managing authorities of the respective engineering colleges or polytechnics.

83. The Council shall cause formation of an All India Students Committee with a Chairman who shall be a member of the Council nominated by the Council from amongst themselves and members not exceeding fifteen in number nominated or elected in accordance with the rules framed and approved by the Council from time to time.

OVERSEAS CHAPTERS

84. The Council may establish Overseas Chapters in overseas countries with which India has

diplomatic relationship for the benefit of the Corporate Members residing and practising the profession in those countries. The affairs of such Overseas Chapters shall be carried on in accordance with the rules framed and approved by the Council from time to time.

DIVISIONS

85. The Council may, if deemed fit, establish Divisions in academically admissible disciplines of engineering for the advancement of the respective branches of engineering provided the strength of Corporate Members likely to be attached to the proposed division is not less than one percent of the Corporate Members on the Roll of the Institution.

86. The Council may also establish Groups of Engineering within any Division established in accordance with the provisions of Bye Law 85 for the advancement of the respective disciplines of engineering.

87. The Council may also abolish any Division or Group if in its opinion the number of Corporate Members attached to the Division or Group fall below one percent of the Corporate Members on Roll of the Institution as on the 31st day of March of the year.

88. Corporate Members shall be attached to one Division only based on their educational qualifications approved by the Council and the field of engineering in which they practice at the time of their admission to the Institution, but if so qualified for attachment to more than one Division, a Corporate Member shall be allowed to select and change to the Division to which he would like to be attached. The Council may, however, at its discretion, allow the transfer of a Corporate Member from one Division to another provided he establishes to its satisfaction that he deserves such transfer on account of change in the field of his practice of engineering subsequent to his election as a Corporate Member.

89. No member of any class other than those who are Corporate Members in terms of Bye Law 43 shall be attached to any Division.

90. The affairs of each Division shall be carried on by a Division Board constituted as per provision of Bye-Law 91 in accordance with the rules laid down from time to time by the Council. The Council shall have the power to vary the same as they may deem fit, subject always to the provisions of the Charter and the Bye-Laws.

91. Members of the Council attached to a particular Division shall constitute the Division Board of the Division. If the number of members in any Division Board is less than three, then that Division Board may co-opt additional members to make up that number. Such co-opted members shall not, however, be members of the Council.

92. (a) The Council shall nominate a Chairman for each Division from among the members attached to the Divisions for a period as the Council may deem fit but not exceeding two sessions.

    (b) The Council may, if thought necessary, nominate a Chairman for a Division Board whose number of members is less than three from amongst the Fellows attached to that Division and in such event the Fellow so elected shall be a member of the Division Board.

93. Any casual vacancy of a Chairman of a Division arising from death or resignation or any other cause, which may occur between one Annual General Meeting and the succeeding Annual General Meeting, shall be filled by the Council.

ADVANCEMENT OF TECHNOLOGY AND ENGINEERING

94. Whilst the affairs of each Division pertaining to that Division shall be carried on by the Division Board, their effort in common and inter-disciplinary aspects shall be supplemented by continuous and special attention to the advancement of Science, Technology and engineering.

95. For carrying out the above activities the Council shall constitute a Committee for the Advancement of Technology and Engineering (CATE), and shall frame rules for the constitution, define the powers and functions of the Committee and shall have power from time to time to add to vary, or rescind any such rules. The strength of this committee shall not exceed five excluding the Chairman of the Division Boards, Chairman of All India Technicians Committee

and Chairman of All India Students Committee, who will be ex-officio members. The Committee shall have the power to co-opt experts for specific periods for specific purposes.

96. The Committee for the Advancement of Technology and Engineering shall give particular attention to : Promotion of Research; Promotion of Development; Promotion of Appropriate Technology; Building up Design Talent; Development and Promotion of Engineering Information Services; Formulation and Implementation of Norms and Standards for Technical Activities including publications of the Institution; Continuous vigilance on behalf of the Institution on Science and Technology Policies of the Nation; Coordination of Technical Education with Research and Industrial Development; and Co-ordination of Inter- disciplinary work.

ENGINEERING STAFF COLLEGE OF INDIA

97. The Council shall establish the Engineering Staff College of India at suitable location for imparting continuing education to the practising engineers and the engineering community.

98. The affairs of the Engineering Staff College of India shall be managed and carried on by a Board of Governors duly appointed by the Council and in accordance with the provisions of Bye-Laws Regulations and the Statutes as may be framed and approved from time to time by the Council.

EDUCATION, EXAMINATIONS AND ACCREDITATION

99. (i) The Council shall cause the Institution Examinations in Sections A and B to be held for election as Members or Associate Members and the Studentship Examination to be held for transfer from the class of Student to the class of Technician Member in the cases in which such Examination is or may be required.

(ii) Further, the Council shall also cause to be held Section C of the Institution Examinations for the purpose stated in Bye-Law 48 (v).

100. The Council shall cause formation of an Education, Examinations and Accreditation Committee with the President as its Chairman and six members nominated by the Council from amongst the members of the Council for a period not less than two sessions and shall define the powers and functions of the Committee. The Council shall have power from time to time to add, vary, or rescind any such rules.

101. The Council shall on the recommendation of the Education, Examination and Accreditation Committee frame rules in respect of such examinations, the syllabi, list or recommended text and other books for study of the candidates, define the places and programmes for holding such examinations, the fees to be paid by the candidates, select the paper-setters and the examiners of different subject, define the conditions under which the candidates may be admitted to such examinations.

102. The Council shall on the recommendation of the Education, Examinations and Accreditation Committee have the power to recognise such university degrees and collegiate or other diplomas or certificates as after scrutiny they may deem to prove a sufficient standard of attainment in the subjects referred to; and may exempt graduates or holders of such diplomas or certificates from passing in whole or in part the aforesaid examinations appointed and directed by the Council.

FINANCES

103. The revenue of the situation shall form the general funds of the Institution and all its property, income and effects, of whatsoever kind, are vested in the Corporate Members as such Corporate Members for the furtherance of the objects of the Institution as defined in the Charter.

104. Under no pretence whatever shall any such property, income, revenue or effects of the Institution, derived howsoever, be paid or transferred directly or indirectly, by way of dividend or bonus or otherwise by way of profit to the members of the Institution except in the case of and as a salaried officer or employee of the Institution. Provided always that nothing herein contained shall prevent—

(a) the payment by the Institution in good faith of reasonable and proper remuneration to any member of the Institution for any services rendered to the Institution;

(b) the payment by the Institution of interest on money lent to the Institution by any such member or reasonable or proper rent for premises demised or let by any such member to the Institution; or

(c) the giving by Institution to any such member of prizes whether in cash or otherwise and scholarships for the furtherance of the objects of the Institution.

105. The Council shall constitute a Finance Committee from amongst its members, and the strength of the Finance Committee shall be five. Every vacancy that arises shall be filled by the Council by election from amongst it members.

106. (i) All compounding fees, composite subscription and composite journal fees shall be regularly invested as soon as possible after the receipt thereof in short-term deposit or in fixed deposit with one or more Scheduled Banks as defined in Bye- Law 110 or in one or more forms or modes of investing or depositing the moneys as per provisions of the Income Tax Act, 1961 as amended from time to time and only the interest accruing therefrom shall be available for the general expenditure of the Institution. Such moneys and investments shall form the Permanent Reserve Fund (Corpus Fund).

(ii) All entrance fees and transfer fees shall be treated as Capital receipts and shall only be spent for capital expenditure.

(iii) The Council shall be at liberty :

(a) To vary the investment at their discretion either in fixed deposit or in short term deposit with the State Bank of India or with one or more Scheduled Banks as defined in Bye-Law 110 or in approved securities;

(b) To deposit all or any of the securities representing the Permanent Reserve Fund (Corpus Fund) wit the bankers of the Institution to secure an overdraft of Current Account not exceeding Rs. 50,000 at any time, and

(c) To withdraw a portion not exceeding twenty percent of the Permanent Reserve Fund (Corpus Fund) under extraordinary circumstances with prior approval of the Corporate Members upon the recommendation of the Council.

(d) Notwithstanding anything contained in Bye-Law 105 (iii) (c) the approval of the Corporate Members on Roll shall be obtained in the manner as provided in Bye-Law 105(iv). The recommendation of the Council shall be valid provided three-fourth of the members of the Council present at its meeting where the proposal shall be placed for consideration shall vote in favour of the proposal.

Further the recommendation on the proposal by the Council and its approval by the Corporate Members made under Bye-Law 105 (iii) (c), shall stipulate the period by which the sum withdrawn shall be deposited to the Permanent Reserve Fund (Corpus Fund).

(iv) Save as aforesaid, no portion of the Permanent Reserve Fund shall be alienated without the consent of the Corporate Members to be obtained by a postal vote on which vote the majority shall be not less than three-fourth of the effective votes received.

(v) The Council shall have the power to make from time to time such additions as they consider desirable to the Permanent Reserve Fund from the general funds of the Institution and the provisions of the Bye-Laws shall apply to these additions exactly as if they had formed part of the receipt required by the Bye- Laws to be invested for forming that fund.

107. All the securities and moneys forming the assets of the Institution shall be lodged with the Institution's Bankers in the name of the Institution. Any surplus funds not required for the current expenditure and not carried by the Council to the Permanent Reserve Fund shall be invested from time to time in short-term deposits or in fixed deposits with one or more Scheduled Banks as defined in Bye-Law 110 or in one or more securities as per provisions of the Income Tax Act, 1961 as amended from time to time and no securities forming part of this fund shall be sold or otherwise disposed of except by order of the Council.

108. (i) The Institution shall keep proper books of accounts with respect to :

(a) all sums of money received and expended by the Institution and the matters in respect of which the receipts and expenditure take place;

(b) the assets and liabilities of the Institution.

(ii) For the purpose of Bye-Law 107(i) proper books of accounts shall not be deemed to be kept with respect to the matters specified therein if there are not kept such books as are necessary to give a true and fair view of the state of affairs of the Institution and to explain its transactions.

(iii) The accounts of receipts and payments shall be considered at regular intervals by the Finance Committee.

(iv) The accounts shall be audited once in a year by the Auditor/Auditors of the Institution to be appointed as per provision of Bye-Law 111.

(v) The audited accounts of the preceding Financial Year shall be presented by the Council for the consideration of the Corporate Members at the Annual General Meeting and they shall comprise :

(a) Balance Sheet as on 31st March,

(b) Income and Expenditure account for the year ended 31st March, and

(c) Separate accounts of Endowment and Trust Funds created for specific purposes.

109. (a) The short-term deposit accounts and the fixed deposit account shall be operated jointly by any two of the following : (i) a member of the Finance Committee nominated by the Council, (ii) the Secretary and Director General and (iii) the Head of the Finance Department of the Institution.

(b) No payment from the moneys or funds of the Institution shall be made except under the expressed or implied sanction of the Council, and all cheques shall be signed jointly by any two of the following : (i) a member of the Finance Committee nominated by the Council, (ii) the Secretary and Director General and (iii) the Head of the Finance Department of the Institution.

Provided, however, that in case of bank account opened by the Institution for a special activity of he Institution, the same shall be operated by any two or three persons nominated by the Council of whom shall be a member of the Finance Committee.

(c) No payments from the moneys or funds of the State Centres and Local Centers shall be made except under the expressed or implied sanction of their respective Committees and all bank accounts shall be operated jointly by the Honorary Secretary and any one of either the Chairman or Immediate Past Chairman, of the State Centres and Local Centres.

110. The names of all persons who have made any voluntary contributions to the funds of the Institution shall be published in the Annual Report of the Council circulated to the Corporate Members for consideration at the Annual General Meeting.

BANKERS

111. The Bankers of the Institution and its State Centres, Local Centres, Engineering Staff College of India and similar Centres of activities of the Institution shall be one or more Scheduled banks (as defined by the Reserve Bank of India Act, 1934) in India as may be determined by the Council, Committee of the State Centres, Local Centres, Governing Body of Engineering Staff College of India and Committees of other centres of activities of the Institution as the case may be.

AUDITORS

112. (i) At each Annual General Meeting shall be appointed one or more properly qualified Auditor or Auditors to hold office until the next Annual General Meeting.

(ii) Every such Auditor or Auditors shall be a member of the Institute of Chartered Accountants of India, and shall hold a Certificate of Practice granted by that Institute.

(iii) The Council may fill any casual vacancy in the office of Auditor to hold office until the next

Annual General Meeting, but while any such vacancy continues the surviving or continuing Auditor or Auditors (if any) may act.

(iv) The remuneration of the Auditors shall be fixed at the Annual General Meeting at which they are appointed except that the remuneration of any Auditor appointed to fill any casual vacancy may be fixed by the Council.

NOTICES

113. A notice may be served by the Council or the Secretary and Director General upon any member of any class either personally or by sending it whether as a separate communication or included in or with one of the publications of the Institution, prepaid through the post addressed to such person at his recorded address.

114. Any notice served personally or sent by post shall be deemed to have served or delivered at the expiration of ninety-six hours after it was posted.

115. A member whose recorded address is not in India shall not be entitled to any notice, and all proceedings may be had and taken without notice to such a member in the same manner as if he had received one such notice.

COPYRIGHT

116. Each paper presented to the Institution and accepted for reading or publication in full or in abstract be the property of the Institution. The Council, in such cases as they may think fit, shall have power to release or surrender the rights of the Institution in respect of any such paper or the copying thereof.

INDEMNITY

117. Each member of the Council and the Secretary and Director General shall be indemnified out of the funds of the Institution and to such extent as the Council shall approve from and against such costs, charges, damages and expenses as he may sustain by reason of his acting in execution of the duties or powers imposed upon or given to him by the Charter or the Bye-Laws.

118. Each member of the Council shall not be accountable and shall not incur any personal liability in respect of any loss or damage incurred through any act, matter or thing done, authorised or suffered, being done in good faith for the benefit of the Institution, although in excess of his legal power or incurred through any omission, error of judgement, or oversight on his part.

COMMON SEAL

119. The Council shall provide a Common Seal of the Institution. The Seal shall be in the custody of the Secretary and Director General and be fixed to such documents as in law are required to be sealed, but only in the presence of the President or a member of the Council. The Secretary and Director General and the President or a member of the Council shall sign every instrument to which the seal of the Institution is so affixed.

INSTITUTION BADGE

120. The Council shall approve the design of a badge to be worn by all Corporate and non-Corporate Members.

THE ARCHITECTS ACT, 1972.

(20 of 1972)

[31st May, 1972]

An Act to provide for the registration of architects and for matters concerned therewith

Be it enacted by Parliament in the Twenty-third Year of the Republic of India as follows :

CHAPTER I
Preliminary

1. Short title, extent and commencement.—(1) This Act may be called the Architects Act, 1972.

(2) It extends to the whole of India.

(3) It shall come into force on such date as the Central government may, by notification in the official Gazette, appoint.

Comment

"Shall".—When the Legislature employs the expression "shall" it must normally be construed to mean "shall" and not "may".

2. Definitions.—In this Act, unless the context otherwise requires,—

(*a*) "architect" means a person whose name is for the time being entered in the register;

(*b*) "Council" means the council of the Architecture constituted under Sec. 3;

(*c*) "Indian Institute of Architects" means the Indian Institute of Architects registered under the Societies Registration Act, 1860;

(*d*) "recognized qualification" means any qualification in architecture for the time being included in the schedule or notified under Sec. 15;

(*e*) "register" means the register of architects maintained under the Sec. 23;

(*f*) "regulation" means a regulation made under this Act by the council;

(*g*) "rule" means a rule made under this Act by the Central Government.

CHAPTER II
Council of Architecture

3. Constitution of Council of Architecture.—(1) The Central Government shall, by notification in the official Gazette, constitute, with effect from such date as may be specified in the notification, a council to be known as the Council of Architecture, which shall be a body corporate, having perpetual succession and a common seal, with power to acquire, hold and dispose of property, both moveable and immoveable, and to contract, and may by that name sue or be sued.

(2) The Head Office of the Council shall be at Delhi or at such other place as the Central Government may, by notification in the official Gazette, specify.

(3) The Council shall consist of the following members, namely :

(*a*) five architects possessing recognized qualifications elected by the Indian Institute of Architects from among its members;

(*b*) two persons nominated by the All-India Council for Technical Education established by the Resolution of the Government of India in the late Ministry of Education No. F. 16-10/44-E-III, dated 30th November, 1945;

(*c*) five persons elected from among themselves by heads of architectural institutions in India imparting full-time instruction for recognized qualifications;

(*d*) the Chief Architects in the Ministries of the Central Government to which the Government business relating to defence and railways has been allotted and the head of the Architectural Organization in the Central Public Works Department, *ex- officio*;

(*e*) one person nominated by the Central Government;

(*f*) an architect from each State nominated by the Government of the state;

(*g*) two persons nominated by the Institution of Engineers (India) from among its members; and

(*h*) one person nominated by the Institution of Surveyors of India from among its members.

Explanation.—For the purposes of this sub-section,—

(*a*) "Institution of Engineers (India)" means the Institution of Engineers (India) first registered in 1920 under the Indian Companies Act, 1913 and subsequently incorporated by the Royal Charter in 1935;

(*b*) "Institution of surveyors of India" means the Institution of Surveyors registered under the Societies Registration Act, 1860.

(4) Notwithstanding anything contained in Cl. (*a*) of sub-section (3), the Central Government may, pending the preparation of the register, nominate to the first Council, in consultation with the Indian Institute of Architects persons referred to in the said Cl. (*a*) who are qualified for registration under Sec. 25, and the persons so nominated shall hold office for such period as the Central Government may, by notification in the official Gazette, specify.

(5) Notwithstanding anything contained in Cl. (*f*) of sub-section (3), the Central Government may, pending the preparation of the register nominate to the first Council, in consultation with the State Governments concerned, persons referred to in said Cl. (*f*), who are qualified for registration under Sec. 25, and the persons so nominated shall hold office for such period as the Central Government may, by notification in the official Gazette, specify.

Comment

Explanation.—It is now well settled that an Explanation added to a statutory provision is not a substantive provision in any sense of the term but as the plain meaning of the word itself shows it is merely meant to explain or clarify certain ambiguities which may have crept in the statutory provision.

4. President and Vice-President of Council.—(1) The President and the Vice-President of the Council shall be elected by the members of the Council from among themselves;

Provided that on the first constitution of the Council and until the President is elected, a member of the Council nominated by the Central Government in this behalf shall discharge the functions of the President.

(2) An elected President or Vice-President of the Council shall hold office for a term of three years or till he ceases to be a member of the Council, whichever is earlier, but subject to his being a member of the Council, he shall be eligible for re- election :

Provided that—

(*a*) The President or the Vice-President may, by writing under his hand addressed to the Vice-President or the President, as the case may be, resign his office;

(*b*) The President or the Vice-President shall notwithstanding the expiry of his term of three years, continue to hold office until his successor enters upon office.

(3) The President and the Vice-President of the Council shall exercise such powers and discharge such duties as may be prescribed by regulations.

5. Mode of elections.—(1) Elections under this chapter shall be conducted in such manner as may be prescribed by rules.

(2) Where any dispute arises regarding any such election, the matter shall be referred by the Council to a Tribunal appointed by the Central Government by notification in the official Gazette in this behalf, and the decision of the Tribunal shall be final :

Provided that no such reference shall be made except on an application made to the Council by an aggrieved party within thirty days from the date of the declaration of the result of the election.

(3) The expenses of the Tribunal shall be borne by the Council.

Comment

The word "shall"—Meaning of.—It has been laid down consistently by the Supreme Court that the mere use of the word "shall" by itself in the Statute does not make the provision mandatory, but it is the duty of the courts of justice to try to get at the real intention of the Legislature by carefully attending to the whole scope of the statute to be construed. In each case, one has to look to the subject-matter, consider the importance of the provisions and the relations of that provisions with the general object intended to be occurred by the Act and upon the review of the case in that aspect decide whether the enactment is mandatory or only directory.

6. Terms of office and casual vacancies.—(1) Subject to the provisions of this section, an elected or nominated member shall hold office for a term of three years from the date of his election or nomination or until his successor has been duly elected or nominated, whichever is later.

(2) An elected or nominated member may, at any time resign his membership by writing under his hand addressed to the President, or in his absence, to the Vice-President, and the seat of such member shall thereupon become vacant.

(3) A member shall be deemed to have vacated his seat—

(*i*) if he is absent without excuse, sufficient in the opinion of the Council, from three consecutive ordinary meetings of the Council; or

(*ii*) if he ceases to be a member of the body referred to in Cl. (*a*), Cl. (*g*) or Cl. (*h*) of sub section (3) of Sec. 3 by which he was elected or nominated, as the case may be; or

(*iii*) in the case where he has been elected under Cl. (*c*) of sub-section (3) of Sec. 3, if he ceases to hold his appointment as the head of an institution referred to in the said clause.

(4) A casual vacancy in the Council shall be filled by fresh election or nomination, as the case may be, and the person so elected or nominated to fill the vacancy shall hold office only for the remainder of the term for which the member whose place he takes was elected or nominated.

(5) Members of the Council shall be eligible for re-election or renomination, but not exceeding three consecutive terms.

7. Validity of act or proceeding of Council, Executive Committee or other committees not to be invalidated by reason of vacancy, etc.—No act or proceeding of the Council or the Executive Committee or any other committee shall be invalid merely by reason of—

(*a*) any vacancy in, or defect in the constitution of, the Council, the Executive Committee or any other committee, or

(*b*) any defect in the election or nomination of a person acting as a member thereof, or

(*c*) any irregularity in procedure not affecting the merits of the case.

8. Disabilities.—A person shall not be eligible for election or nomination as a member of the Council, if he—

(*a*) is an undischarged insolvent; or

(*b*) has been convicted by a court in India for any offence and sentenced to imprisonment for not less than two years, and shall continue to be ineligible for a further period of five years since his release.

Comment

Some disabilities have been provided under this section for election or nomination of the members of the Council. *Firstly*, an undischarged insolvent cannot seek election or nomination as a member of the Council. *Secondly*, the person convicted for two years or more shall not be also eligible for such election and the disability shall continue for further five years since his release.

9. Meetings of Council.—(1) The Council shall meet at least once in every six months at such time and place and shall observe such rules of procedure in regard to the transaction of business at its meetings as may be prescribed by regulations.

(2) Unless otherwise prescribed by regulations, nine members of the Council shall form a quorum, and all the acts of the Council shall be decided by a majority of the members present and voting.

(3) In the case of an equal division of votes, the President, or in his absence, the Vice-President or, in the absence of both, the member presiding over the meeting, shall have and exercise a second or casting vote.

10. Executive committee and other committees.—(1) The Council shall constitute from among its members an Executive Committee, and may also constitute other committees for such general or special purposes as the Council deems necessary to carry out its functions under this Act.

(2) The Executive Committee shall consist of the President and the Vice-President of the Council who shall be members *ex- officio* and five other members who shall be elected by the Council from among its members.

(3) The President and the Vice-President of the Council shall be the Chairman and Vice-Chairman respectively of the Executive Committee.

(4) A member of the Executive Committee shall hold office as such until the expiry of his term as a member of the Council but subject to his being a member of the Council, he shall be eligible for re-election.

(5) In addition to the powers and duties conferred and imposed on it by this Act, the Executive Committee shall exercise such powers and discharge such duties as may be prescribed by regulations.

11. Fees and allowances to President, Vice-President and members.—The President, the Vice-President and other members of the Council shall be entitled to such fees and allowances as the Council may, with the previous sanction of the central Government, fix in this behalf.

12. Officers and other employees.—(1) The Council shall—

(a) appoint a Registrar who shall act as its secretary and who may also act, if so decided by the Council, as the treasurer;

(b) appoint such other officers and employees as the Council deems necessary to enable it to carry out its functions under this Act;

(c) with the previous sanction of the Central Government, fix the pay and allowances and other conditions of service of officers and other employees of the Council.

(2) Notwithstanding anything contained in Cl. (a) of sub-section (1), for the first three years from the first constitution of the Council, the Registrar of the Council shall be a person appointed by the Central Government, who shall hold office during the pleasure of the Central Government.

(3) All the persons appointed under this section shall be employees of the Council.

13. Finances of Council.—(1) There shall be established a fund under the management and control of the Council into which shall be paid all moneys received by the Council and out of which shall be met all expenses and liabilities properly incurred by the Council.

(2) The Council may invest any money for the time being standing to the credit of the fund in any Government security or in any other security approved by the Central Government.

(3) The Council shall keep proper accounts of the fund distinguishing capital from revenue.

(4) The annual accounts of the Council shall be subject to audit by an auditor to be appointed annually by the Council.

(5) As soon as may be practicable at the end of each year, but not later than the thirtieth day of September of the year next following, the Council shall cause to be published in the official Gazette a copy of the audited accounts and the report of the Council for that year and copies of the said accounts and report shall be forwarded to the Central Government. (6) The fund shall consist of—

(a) all moneys received from the Central Government by way of grant, gift or deposit;

(b) any sums received under this Act whether by way of fee or otherwise.

(7) All moneys standing at the credit of the Council which cannot immediately be applied shall be deposited in the State Bank of India or in any other bank specified in Col. 2 of the First Schedule to the Banking Companies (Acquisition and Transfer of Undertaking) Act, 1970.

14. Recognition of qualifications granted by authorities in India.—(1) The qualifications included in the Schedule or notified under Sec. 15 shall be recognized qualification for the purposes of this Act.

(2) Any authority in India which grants an architectural qualification not included in the Schedule may apply to the Central Government to have such qualification recognized, and the Central Government, after consultation with the Council, may, by notification in the official Gazette, amend the Schedule so as to include such qualification therein and any such notification may also direct an entry shall be made in the Schedule against such architectural qualification declaring that it shall be a recognized qualification only when granted after a specified date;

Provided that until the first Council is constituted, the Central Government shall, before issuing a notification as aforesaid, consult an expert committee consisting of three members to be appointed by the Central Government by notification in the official Gazette.

15. Recognition of architectural qualifications granted by authorities in foreign countries.—(1) The Central Government may, after consultation with the Council, direct by notification in the official Gazette, that an architectural qualification, granted by any university or other institution in any country outside India in respect of which a scheme of reciprocity for the recognition of architectural qualification is not in force, shall be recognized qualification for the purpose of this Act or, shall be so only when granted after a specified date or before a specified date :

Provided that until the first Council is constituted the Central Government shall, before issuing any notification as aforesaid, consult the expert committee set up under the proviso to sub-section (2) of Sec. 14.

(2) The Council may enter into negotiations with the authority in any State or country outside India, which by the law of such State or country is entrusted with the maintenance of a register of architects, for settling of a scheme of reciprocity for the recognition of architectural qualifications, and in pursuance of any such scheme, the Central Government may, by notification in the Official Gazette, direct that such architectural qualification as the Council has decided should be recognized, shall be deemed to be a recognized qualification for the purpose of this Act, and any such notification may also direct that such architectural qualification shall be also recognized only when granted after a specified date or before a specified date.

Comment

The word "may" should not be read as "shall". The word "shall" does not make the section permissible.

16. Power of Central Government to amend Schedule.— Notwithstanding anything contained in sub-section (2) of Sec. 14, the Central Government, after consultation with the Council, may, by notification in the official Gazette amend the Schedule by directing that an entry be made therein respect of any architectural qualification.

17. Effect of recognition.—Notwithstanding anything contained in any other law, but subject to the provisions of this Act, any recognized qualification shall be sufficient qualification for enrolment in the register.

18. Power to require information as to courses of study and examinations.—Every authority in India which grants a recognized qualification shall furnish such information as the Council may, from time to time, require as to the courses of study and examinations to be undergone in order to obtain such qualification, as to the ages at which such course of study and examinations are required to be undergone and such qualification is conferred and generally as to the requisites for obtaining such qualification.

19. Inspection of examinations.—(1) The Executive Committee shall subject to regulations, if any, made by the Council, appoint such number of inspectors as it may deem requisite to inspect any college or institution where architectural education is given or to attend any examination held by any college or institution for the purpose of recommending to the Central Government recognition of architectural qualifications granted by that college or institution.

(2) The inspectors shall not interfere with the conduct of any training or examination, but shall report to the Executive Committee on the adequacy of the standards of architectural education including staff, equipment, accommodation, training and such other facilities as may be prescribed by regulations

for giving such education or on the sufficiency of every examination which they attend.

(3) The Executive Committee shall forward a copy of such report to the college or institution and shall also forward copies with remarks, and, of the college or institution thereon, to the Central Government.

Comment

The Executive Committee shall appoint number of Inspectors for inspection of any college or institution where architectural education is provided or to attend any examination held by any institution. The inspectors shall do it for the purpose of recommending to the Central Government recognition of architectural qualifications granted by that college. The report of the Inspectors shall be forwarded to the Executive Committee on the point of standard, staff, accommodation, other equipments and training and other facilities prescribed by regulations for providing such architectural education. The Inspector shall not interfere with the conduct of any training or examination. On receiving such report the Executive Committee shall forward a copy of such report to the institution concerned and shall also forward copies with the remarks, if any, to the Central Government.

20. Withdrawal of recognition.—(1) When upon report by the Executive Committee it appears to the Council—

(a) that the courses of study and examination to be undergone in, or the proficiency required from the candidates at any examination held by, any college or institution, or

(b) that the staff, equipment, accommodation, training and other facilities for staff and training provided in such college or institution,

do not conform to the standards prescribed by regulations, the Council shall make a representation to that effect to the appropriate Government.

(2) After considering such representation the appropriate Government shall forward it along with such remarks as it may choose to make to the college or institution concerned, with an intimation of the period within which the college or institution, as the case may be, may submit its explanation to the appropriate Government.

(3) On receipt of the explanation or where no explanation is submitted within the period fixed, then on the expiry of that period, the State Government, in respect of the college or institution referred to in Cl. (b) of sub-section (5), shall make its recommendations to the Central Government.

(4) The Central Government—

(a) after making such further enquiry, if any, as it may think fit, in respect of the college or institution referred to in sub-section (3), or

(b) on receipt of the explanation from a college or institution referred to in Cl. (a) of sub-section (5), or where no explanation is submitted within the period fixed, then on the expiry of that period,

may, by notification in the official Gazette, direct that an entry, shall be made in the Schedule against the architectural qualification awarded by such college or institution, as the case may be, declaring that it shall be a recognized qualification only when granted before a specified date and the Schedule shall be deemed to be amended accordingly.

(5) For the purposes of this section, "appropriate Government" means—

(a) in relation to any college or institution established by an Act of Parliament or managed, controlled or financed by the Central Government, the Central Government, and

(b) in any other case, the State Government.

21. Minimum standard of architectural education.—The Council may prescribe the minimum standards of architectural education required for granting recognized qualifications by colleges or institutions in India.

22. Professional conduct.—(1) The Council may by regulations prescribe standards of professional conduct and etiquette and a code of ethics for architects.

(2) Regulations made by the Council under sub-section (1) may specify which violations thereof shall constitute infamous conduct in any professional respect, that is to say, professional misconduct,

582 Civil Engineering Building Practice

and such provision shall have effect notwithstanding anything contained in any law for the time being in force.

CHAPTER III
Registration of Architects

23. Preparation and maintenance of register.—(1) The Central Government shall, as soon as may be, cause to be prepared in the manner hereinafter provided a register of architects for India.

(2) The Council shall upon its constitution assume the duty of maintaining the register in accordance with the provisions of this Act.

(3) The register shall include the following particulars, namely;

(*a*) the full name with date of birth, nationality and residential address of the architect;

(*b*) his qualification for registration, and the date on which he obtained that qualification and the authority which conferred it;

(*c*) the date of his first admission to the register;

(*d*) his professional address; and

(*e*) such further particulars as may be prescribed by rules.

24. First preparation of register.—(1) For the purpose of preparing the register of architects for the first time, the Central Government shall, by notification in the official Gazette, constitute a Registration Tribunal consisting of three persons who have, in the opinion of the Central Government, the knowledge of, or experience in, architecture; and the Registrar appointed under Sec. 12. shall act as Secretary of the Tribunal.

(2) The Central Government shall, by the same or a like notification, appoint a date on or before which application for registration, which shall be accompanied by such fee as may be prescribed by rules, shall be made to the Registration Tribunal.

(3) The Registration Tribunal shall examine every application received on or before the appointed day and if it is satisfied that the applicant is qualified for registration under Sec. 25, shall direct the entry of the name of the applicant in the register.

(4) The first register so prepared shall thereafter be published in such manner as the Central Government may direct and any person aggrieved by a decision of the Registration Tribunal expressed or implied in the register so published may, within thirty days from the date of such publication, appeal against such decisions to an authority appointed by the Central Government in this behalf by notification in the official Gazette.

(5) The authority appointed under sub-section (4) shall, after giving the person effected an opportunity of being heard and after calling for relevant records, make such order as it may deem fit.

(6) The Registrar shall amend, where necessary, the register in accordance with the decisions of the authority appointed under sub-section (4).

(7) Every person whose name is entered in the register shall be issued a certificate of registration in such form as may be prescribed by rules.

(8) Upon the constitution of the Council, the register shall be given into its custody, and the Central Government may direct that the whole or any specified part of the application fees for registration in the first register shall be paid to the credit of the Council.

25. Qualification for entry in register.—A person shall be entitled on payment of such fee as may be prescribed by rules to have his name entered in the register, if he resides or carries on the profession of architect in India and—

(*a*) hold a recognized qualification, or

(*b*) does not hold such qualification but, being a citizen of India, has been engaged in practice as an architect for a period of not less than five years prior to the date appointed under sub-section (2) of Sec. 24, or

(*c*) possesses such other qualifications as may be prescribed by rules :

Provided that no person other than a citizen of India shall be entitled to registration by virtue of a qualification—

(a) recognized under sub-section (1) of Sec. 15 unless by the law and practice of a country outside India to which such person belongs, citizens of India holding architectural qualification registrable in that country, are permitted to enter and practise the profession of architect in such country, or

(b) unless the Central Government has, in pursuance of a scheme of reciprocity or otherwise, declared that qualification to be a recognized qualification under sub-section (2) of Sec. 15.

26. Procedure for subsequent registration.—(1) After the date appointed for the receipt of applications for registration in the first register of architects, all applications for registration shall be addressed to the Registrar of the Council and shall be accompanied by such fee as may be prescribed by rules.

(2) If upon such application the Registrar is of opinion that the applicant is entitled to have his name entered in the register he shall enter thereupon the name of the applicant;

Provided that no person, whose name has under the provisions of this Act been removed from the register, shall be entitled to have his name re-entered in the register except with the approval of the Council.

(3) Any person whose application for registration is rejected by the Registrar, may, within three months of the date of such rejection, appeal to the Council.

(4) Upon entry in the register of a name under this section, Registrar shall issue a certificate of registration in such form as may be prescribed by rules.

27. Renewal fees.—(1) The Central Government may, by notification in the official Gazette, direct that for the retention of a name in the register after the 31st day of December, of the year, following the year in which the name is first entered in the register, there shall be paid annually to the Council such renewal fee as may be prescribed by rules and where such direction has been made, such renewal fee shall be due to be paid before the first day of April of the year to which it relates.

(2) Where the renewal fee is not paid before the due date, the Registrar shall remove the name of the defaulter from the register :

Provided that a name so removed may be restored to the register on such conditions as may be prescribed by rules.

(3) On payment of the renewal fee, the Registrar shall, in such manner as may be prescribed by rules, endorse the certificate of registration accordingly.

28. Entry of additional qualifications.—An architect shall, on payment of such fee as may be prescribed by rules, be entitled to have entered in the register any further recognized qualification which he may obtain.

29. Removal form register.—(1) The Council may, by order, remove from the register the name of any architect—

(a) from whom a request has been received to that effect, or

(b) who has died since the last publication of the register.

(2) Subject to the provisions of this section, the Council may order that the name of any architect shall be removed from the register where it is satisfied, after giving him a reasonable opportunity of being heard and after such further inquiry, if any, as it may think fit to make,—

(a) that his name has been entered in the register by error or on account of misrepresentation or suppression of a material fact; or

(b) that he has been convinced of any offence which, in the opinion of the Council, involves moral turpitude; or

(c) that he is an undischarged insolvent; or

(d) that he has been adjudged by a competent court to be of unsound mind.

584 Civil Engineering Building Practice

(3) An order under sub-section (2) may direct that any architect whose name is ordered to be removed from a register shall be ineligible for registration under this Act for such period as may be specified.

(4) An order under sub-section (2) shall not take effect until the expiry of three months from the date thereof.

30. Procedure in inquiries relating to misconduct.—(1) When on receipt of a complaint made to it, the Council is of opinion that any architect has been guilty of professional misconduct which, if proved, will render him unfit to practice as an architect, the Council may hold an inquiry in such manner as may be prescribed by rules.

(2) After holding the inquiry under sub-section (i) and after hearing the architect the Council may, by order reprimand the said architect or suspend him from practice as an architect or remove his name from the register or pass such other order as it thinks fit.

31. Surrender of certificates.—A person whose name has been removed from the register sub-section (2) of Sec. 27, sub-section (1) or sub-section (2) of Sec. 29 or sub-section (2) of Sec. 30, or where such person is dead, his legal representative, as defined in Cl. (11) of Sec. 2. of the Code of Civil procedure, 1908, shall forthwith surrender his certificate of registration to the Registrar, and the name so removed shall be published in the official Gazette.

32. Restoration to register.—The Council may, at any time, for reasons appearing to it to be sufficient and subject to the approval of the Central Government, order that upon payment of such fee as may be prescribed by rules, the name of the person removed from the register shall be restored thereto.

33. Issue of duplicate certificates.—Where it is shown to the satisfaction of the Registrar that certificate of registration has been lost or destroyed, the Registrar may on payment of such fee as may be prescribed by rules, issue a duplicate in the form prescribed by rules.

34. Printing of register.—As soon as may be after the 1st day of April in each year, the Registrar shall cause to be printed copies of the register as it stood on the said date and such copies shall be made available to persons applying therefor on payment of such fee as may be prescribed by rules and shall be evidence that on the said date the persons whose name are entered therein were architects.

35. Effect of registration.—(1) Any reference in any law for the time being in force to an architect shall be deemed to be a reference to an architect registered under this Act.

(2) After expiry of two years from the date appointed under sub- section (2) of Sec. 24, a person who is registered in the register shall get preference for appointment as an architect under the Central or State Government or in any other local body or institution which is supported or aided from the public or local funds or in any institution recognized by the Central or State Government.

CHAPTER IV

Miscellaneous

36. Penalty for falsely claiming to be registered.—If any person whose name is not for the time being entered in the register falsely represents that it is so entered or uses in connection with his name or title any words or letters reasonably calculated to suggest that his name is so entered, he shall be punishable with fine which may extend to one thousand rupees.

37. Prohibition against use of title.—(1) After the expiry of one year from the date appointed under sub-section (2) of Sec. 24, no person other than a registered architect, or a firm of architects shall use the title and style of architect :

Provided that the provisions of this section shall not apply to—

(a) practice of the profession of an architect by a person designated as a "landscape architect" or "naval architect";

(b) a person who, carrying on the profession of an architect in any country outside India, undertakes the function as a consultant or designer in India for a specific project with the prior permission of the Central Government.

Explanation.—For the purposes of Cl. (*a*),—

(*i*) "landscape architect" means a person who deals with the design of open spaces relating to plants, trees and landscape.

(*ii*) "naval architect" means a person who deals with the design and construction of ships.

(2) If any person contravenes the provisions of sub-section (1), he shall be punishable on first conviction with fine which may extend to five hundred rupees and on any subsequent conviction with imprisonment which may extend to six months, or with fine not exceeding one thousand rupees or with both.

Comments

Proviso.—A proviso is intended to limit the enacted provision so as to except something which would have otherwise been within it or in some measure to modify the enacting clause. Sometimes a proviso may be embedded in the main provision and becomes an integral part of it so as to amount to a substantive provision itself.

Provision.—Exclusion of the jurisdiction is not to be readily inferred and there is a presumption against exclusion of jurisdiction of civil court by a statute. The exclusion of the jurisdiction of civil courts to entertain civil cases will not be assumed unless the relevant statutes contain an express provision to that effect or leads to a necessary and inevitable implication of that nature. The mere fact that a special statute provides for certain remedies may not by itself necessarily exclude the jurisdiction of the civil courts to deal with a case brought before it in respect of the matters covered by the said statute.

38. Failure to surrender certificate of registration.—If any person whose name has been removed from the register fails without sufficient cause forthwith to surrender his certificate of registration, he shall be punishable with fine which may extend to one hundred rupees, and, in the case of a continuing failure, with an additional fine which may extend to ten rupees for each day after the first during which he has persisted in the failure.

Comment

"Person".—The word "person" has been used to make it clear that in order to exercise the powers of a controller under the Act, the statutory functionary has to be duly appointed by the Government and that he is a persona designata or designated person.

39. Cognizance of offences.—(1) No court shall take cognizance of any offence punishable under this Act, except upon complaint made by order of the Council or a person authorized in this behalf by the Council.

(2) No Magistrate other than a Presidency Magistrate or a Magistrate of the first class shall try any offence punishable under this Act.

40. Information to be furnished by Council and publication thereof.—(1) The Council shall furnish such reports, copies of its minutes, and other information to the Central Government as that Government may require.

(2) The Central Government may publish, in such manner as it may think fit, any report, copy or other information furnished to it under this section.

41. Protection of action taken in good faith.—No suit, prosecution or other legal proceeding shall lie against the Central Government, the Council or any member of the Council, the Executive Committee or any other committee or officers and other employees of the Council for anything which is in good faith done or intended to be done under this Act or any rule or regulation made thereunder.

Comment

The section provides protection to the Central Government, the Council or any member of the Council, the Executive Committee, Officers or other employees from facing the prosecution or legal proceedings in connection with the work done in good faith or intended to be done under this Act or any rule or regulation made thereunder. The section prohibits the prosecution or legal proceedings against the officers of Central Government.

42. Members of Council and officers and employees to be public servants.—The members of the Council and officers and other employees of the Council shall be deemed to be public servants within the meaning of Sec. 21 of the Indian Penal Code.

43. Power to remove difficulties.—(1) If any difficulty arises in giving effect to the provisions of this Act, the Central Government may by order published in the official Gazette, make such provisions, not inconsistent with the provisions of this Act, as appear to it to be necessary or expedient for removing the difficulty :

Provided that no such order shall be made under this section after the expiry of two years from the date of commencement of this Act.

(2) Every order made under this section shall, as soon as may be after it is made, be laid before each House of Parliament and the provisions of sub section (3) of Sec. 44 shall apply in respect of such order as it applies in respect of a rule made under this Act.

44. Power of Central Government to make rules.—(1) The Central Government may, by notification in the Official Gazette, make rules to carry out the purposes of this Act.

(2) In particular and without prejudice to the generality of the foregoing power, such rules may provide for all or any of the following matters, namely :

(a) the manner in which elections under Chapter II shall be conducted, the terms and conditions of service of the member of the Tribunal appointed under sub-section (2) of Sec. 5 and the procedure to be followed by the Tribunal;

(b) the procedure to be followed by the expert committee constituted under the proviso to sub-section (2) of Sec. 14 in the transaction of the business and the powers and duties of the expert committee and the travelling and daily allowances payable to the members thereof;

(c) the particulars to be included in the register of architects under sub-section (3) of Sec. 23.

(d) the form in which certificate of registration is to be issued under sub-section (5) of Sec. 24, sub-section (4) of Sec. 26 and Sec. 32;

(e) the fee to be paid under Secs. 24, 25, 26, 27, 28, 32 and 33;

(f) the conditions in which name may be restored to the register under the proviso to sub-section (2) of Sec. 27;

(g) the manner of endorsement under sub-section (3) of Sec. 27;

(h) the manner in which the Council shall hold an enquiry under Sec. 30;

(i) the fee for supplying printed copies of the register under Sec. 34;

(j) any other matter which is to be or may be provided by rules under this Act.

(3) Every rule made under this section shall be laid, as soon as may be after it is made, before each House of Parliament, while it is in session, for a total period of thirty days which may be comprised in one session or in two or more successive sessions, and if, before the expiry, of session, immediately following the session or the successive sessions aforesaid, both Houses agree in making any modification to the rule or both Houses agree that the rule should not be made, the rule shall thereafter have effect only in such modified form or be of no effect, as the case may be; so, however, that any such modification or annulment shall be without prejudice to the validity of anything previously done under that rule.

Comment

Retrospectivity of an amendment.—If the suit was pending in the date when the amendments in the principal Act were brought into force the amended provisions of the Act will govern the disposal of the suit.

45. Power of Council to make regulations.—(1) The Council may, with the approval of the Central Government by notification in the official Gazette make regulations not inconsistent with the provisions of this Act, or the rules made thereunder to carry out the purposes of this Act.

(2) In particular and without prejudice to the generality of the foregoing power, such regulations may provide for—

(*a*) the management of the property of the Council;

(*b*) the powers and duties of the President and the Vice-President of the Council;

(*c*) the summoning and holding of meetings of the Council and the Executive Committee or any other committee constituted under Sec. 10, the times and places at which such meetings shall be held, the conduct of business thereat and the number of persons necessary to constitute a quorum;

(*d*) the functions of the Executive Committee or of any other committee constituted under Sec. 10;

(*e*) the courses and periods of the study and of practical training, if any, to be undertaken, the subjects of examinations and standards of proficiency therein to be obtained in any college or institution for grant or recognized qualifications;

(*f*) the appointment, powers and duties of Inspector;

(*g*) the standard of staff, equipment, accommodation, training and other facilities for architectural education;

(*h*) the conduct of professional examinations, qualifications of examination and the conditions of admission to such examination;

(*i*) the standards of professional conduct and etiquette and code of ethics to be observed by architects;

(*j*) any other matter which is to be or may be provided by regulations under this Act and in respect of which no rules have been made.

(3) Every regulation made under this section shall be laid, as soon as may be after it is made, before each House of Parliament while it is in session, for a total period of thirty days which may be comprised in one session or in two or more successive sessions, and if, before the expiry of the session immediately following the session or the successive sessions aforesaid, both Houses agree in making any modification in the regulation or both Houses agree that the regulation should not be made, the regulation shall thereafter have effect only in such modified form or be of no effect, as the case may be; so, however, that any such modification or annulment shall be without prejudice to the validity of anything previously done under that regulation.

Comment

"May" and "shall".—Where the Legislative uses two words "may" and "shall" in two different parts of the same provision *prima facie* it would appear that the Legislature manifested its intention to make one part directory and another mandatory. But that by itself is not decisive. The power of the Court still to ascertain the real intention of the Legislature by carefully examining the scope of the statute to find out whether the provision is directory or mandatory remains unimpaired even where both the words are used in the same provision.

- - - - -

THE SCHEDULE

(*See* Sec. 14)

Qualifications

1. Bachelor Degree in Architecture awarded by Indian Universities established by an Act of the Central or State Legislature.

2. National Diploma (formerly All-India Diploma) in Architecture awarded by All-India Council for Technical Education.

3. Degree of Bachelor of Architecture (B. Arch) awarded by the Indian Institute of Technology Kharagpur.

4. Five-year full-time diploma in architecture of the Sir J.J. School of Art, Bombay, awarded after 1941.

5. Diploma in Architecture awarded by the State Board of Technical Education and Training of the Government of Andhra Pradesh with effect from 1960 (for the students trained at the Government College of Arts and Architecture, Hyderabad).

6. Diploma of the Architecture awarded by the Government College of Arts and Architecture, Hyderabad, till 1959, subject to the condition that candidates concerned have subsequently passed a special final examination in Architecture held by the State Board of Technical Education, Andhra Pradesh and obtained a special certificate.

7. Diploma in Architecture awarded by the University of Nagpur with effect from 1965, to the students at the Government Polytechnic Nagpur.

8. Government Diploma in Architecture awarded by the Government of Maharashtra (or the former Government of Bombay).

9. Diploma in Architecture of Kalabhavan Technical Institute, Baroda.

10. Diploma in Architecture awarded by the School of Architecture Ahmedabad.

11. Membership of the Indian Institute of Architects.

The Institution of Civil Engineers, London (UK) is the senior professional engineering institution in the United Kingdom. It was established as a Society in 1818 and granted a Royal Charter in 1828 with Thomas Telford as its first President.

An extract from the Royal Charter reads as follows :

'... a Society ... for promoting the acquisition of that species of knowledge which constitutes the profession of a Civil Engineer ... as applied in the construction of roads, bridges, aqueducts, canals, river navigation and docks ... in the construction of ports, harbours, moles, breakwaters and lighthouses ... and in the drainage of cities and towns ... many important and public and private works and services in the United Kingdom and overseas which contributes to the well-being of mankind are dependent on Civil Engineers and call for a high degree of professional knowledge and judgement in making the best use of scarce resources in care for the environment and in the interest of public health and safety it is accordingly of importance that there should be a ready means heretofore of ascertaining persons who by proper training and experience are qualified to carry out such works.'

The institution has always responded to the responsibilities placed on it by virtue of its unique history since its foundation in a coffee house in London 1818, and although the needs of society today are much more complex and subtle than in the nineteenth century, the philosophy, and objectives underlying the Royal Charter still apply.

Thus apart from its very extensive learned society activities the Institution also has responsibility for regulating and testing the education, training and practical experience of candidates seeking professional qualifications.

The armorial bearings of the Institution are headed by a representation of Eddystone Lighthouse, the work of John Smeaton, the first to describe himself as a civil engineer. The beaver is remarkable for its ingenuity in constructing its home from tree trunks, stones and mud, and in damming streams in order to preserve its water supply. The crane is a heraldic symbol of vigilance — a quality of special significance for the Institution in its duties of regulating the profession and promoting the advancement of engineering science. On the shield the natural forces of heat, electricity and water are symbolized respectively by the sun, thunderbolt, and fountain, while the annulets represent mathematics — a science by the aid of which civil engineers exercise their art with skill.

THE INSTITUTION OF CIVIL ENGINEERS

Great George Street, Westminster, London SW 1P 3AA, United Kingdom

CHAPTER 1

Membership of the Institution of Civil Engineers

1.1 Candidates for corporate membership of the Institution and the Chartered Civil Engineer status which it confers must :

    (i) hold an approved academic qualification;

    (ii) have achieved the Institution's core training objectives;

    (iii) have undertaken a programme of continuing education;

    (iv) have had responsible practical experience in one of the many branches of civil engineering.

1.2 The recommended way of achieving the core training objectives is through Approved Training under Agreement. As an alternative to Approved Training candidates may apply for membership on the basis of having achieved the training objectives through a period of practical experience (see Chapter 3).

1.3 To verify that candidates have satisfactorily completed all the requirements for membership they must submit themselves to a Chartered Professional Review (CPR).

1.4 Before proceeding to the CPR candidates must demonstrate that they have achieved the required training objectives. In the case of trainees under Agreement this will be by means of a Training Review and Completion Certificate (ICE 123). Candidates who have not trained under Agreement are required to apply for an Experience Appraisal (see Chapter 4).

1.5 The CPR takes place after candidates have gained practical experience in a position of responsibility and takes the form of a rigorous review of their careers to date to verify that their education, training and experience will enable them to discharge, in full the responsibilities of a Chartered Professional Review, candidates are admitted to corporate membership (MICE).

1.6 Nearly all candidates for corporate membership will pass through the two intermediate grades of membership (Student and Graduate), but certain candidates may proceed directly to a CPR; these are :

    (i) mature candidates (see ICE 104);

(ii) Chartered Engineers of other British professional engineering Institutions;

(iii) candidates with approved academic qualifications and a minimum of eight years' practical experience in civil engineering.

CHAPTER 2

Approved academic qualifications

2.1 The approved academic qualification required of candidates for corporate membership is satisfied by holders of :

(a) an accredited degree on the list maintained by the Institution; or

(b) a United Kingdom degree in a subject related to engineering together with passes in papers from the Engineering Council (EC) Examination Part 2 as stipulated by the Institution; or

(c) an acceptable overseas degree together with, if necessary, passes in papers from the Engineering Council (EC) Examination Part 2 as stipulated by the Institution; or

(d) a pass in the EC Examination Parts 1 and 2, provided that the subjects are taken from the Institution's approved list.

2.2 On payment of a fee (see ICE 202) a UK or overseas degree not on the list of accredited degrees may be assessed individually (ICE 111) and candidates will be advised which, if any, supplementary papers from the EC Examination Part 2 they must pass in order to meet the approved academic requirement. Alternatively, they will have the choice of taking single subjects in final papers of accredited degree courses where such facilities exist.

CHAPTER 3

Training, practical experience and continuing education

3.1 The Institution strongly recommends that candidates for corporate membership should achieve their training objectives by entering into a formal Training Agreement whereby their training is structured to meet core and specific objectives, and carried out under the guidance of, a Supervising Civil Engineer (SCE) through an employer on the index maintained by the Institution. The core objectives have been devised by the Institution and are issued with the Training Record (ICE 107). The specific objectives are drawn up by SCEs on behalf of the candidates' employing organisations and are intended to cover the requirements and practices of the organisations and the job-related needs of the trainees. (See Appendix A for details concerning the design of training schemes).

3.2 Candidates undergoing Approved Training may expect to take between two and four years to meet the core and specific objectives. Completion is verified by means of the Training Review (TR).

3.3 Candidates who have not trained under Agreement but have gained at least four years' relevant practical experience may demonstrate that they have achieved the Institution's core training objectives through an Experience Appraisal (EA).

3.4 Training Agreements between trainees and their employers are normally entered into for a three-year period. However, the Training Review may be taken at any time if the training objectives are achieved prior to completion of three years. Training Agreements may be extended in order to complete the training objectives.

3.5 Experience gained before or during the obtaining of an approved academic qualification may count towards the period of Approved Training and/or practical experience in lieu. This also applies to experience gained in engineering research, in the teaching of engineering, and on relevant full-time postgraduate courses. Such experience will be assessed in relation to the core training objectives and so will determine the appropriate length of a Training Agreement. It will also be taken into account at an Experience Appraisal.

3.6 If, through a previous period of training or practical experience, a trainee has already achieved some of the Institution's core objectives, the balance of the training may be undertaken through a Training Agreement of less than three years, subject only to a Training Agreement being of not less than twelve months' duration.

3.7 If a Supervising Civil Engineer is able to offer training leading to only a partial completion of the training objectives, then a Training Agreement of less than three years may be undertaken, subject only to a minimum of twelve months. Candidates must achieve the balance of the training objectives through : an extension of their current Agreement; a further Training Agreement; or through practical experience in lieu followed by an Experience Appraisal.

3.8 The Training Record (ICE 107) is available from the Institution and is a document which all candidates trained under Agreement must complete and includes as part of their submission for the Training Review. Candidates whose practical experience is not covered by a Training Agreement are strongly advised to maintain such a Training Record because it will greatly strengthen their presentation at the Experience Appraisal.

3.9 No matter which route candidates have taken prior to the Chartered Professional Review, they are required to show evidence of continuing education which should be recorded in the individual's Record of Continuing Professional Development (CPD record — ICE 108). Candidates who have completed a period of Approved Training or experience in lieu are required to show that they have undertaken continuing education for at least 30days during their preparation for the CPR, of which a minimum of 15 days must have been gained prior to the Training Review or Experience Appraisal (See Appendix B).

CHAPTER 4

Training Review and Experience Appraisal

Training Review

4.1 The purpose of the Training Review is to verify that candidates who have trained under Agreement have achieved the core and specific objectives as defined in their approved training schemes, and met the continuing education requirement. Responsibility for certifying on form ICE 123 that this stage has been successfully completed rests with the Supervising Civil Engineer (SCE) in collaboration with the Regional Training Officer (RTO).

Experience Appraisal

4.2 Candidates who have not trained under Agreement must make a submission for an Experience Appraisal to be carried out by an independent Assessor appointed by the Institution. This submission provides candidates with the opportunity to demonstrate that the practical experience they have gained in lieu of Approved Training has enabled them to achieve the core objectives and that they have met the continuing education requirement (See 3.9).

4.3 The submission for an Experience Appraisal shall comprise the following documents :

(i) a 2000-word typewritten réport on the candidate's experience to date, which also sets out how the core training objectives have been achieved. (The guidance notes for the preparation of the 2000-word report on training and practical experience for the CPR should be followed — see Appendix C1).

(ii) the candidate's record of continuing education to date (a photocopy of the CPD Record ICE 108).

(iii) a list of the Institution's core objectives with the candidate's personal assessment of their level of competence and the date it was achieved.

(iv) documents which demonstrate that the candidate has met the Institution's core objective to identify, define and solve an engineering problem of some complexity. All the documents must relate to work which the candidate has produced during the normal course of employment.

(v) a completed Training Record. (This requirement is desirable rather than mandatory).

4.4 Prior to making a decision whether to certify on form ICE 124 that the submission meets the Institution's requirements, the Assessor may decide to invite the candidate to an interview at an appointed centre. No candidate will be rejected without an interview.

CHAPTER 5

The Chartered Professional Review

5.1 The Chartered Professional Review (CPR) is a review of candidates' engineering careers to ascertain the quality of their professional experience in civil engineering.

5.2 The time taken to achieve the standards required for corporate membership will vary from candidate to candidate but the normal period to cover the training and practical experience to qualify for corporate membership is four years.

5.3 As indicated in section 3.9 one of the requirements for candidates applying for the CPR is that they must have undertaken at least 30 days' continuing education. (See Appendix B).

5.4 The review procedures are designed to enable candidates to demonstrate that during their employment they have :

(i) developed and proved their technical competence, including the exercising of independent technical judgement requiring both practical experience and the application of engineering principles; and

(ii) acquired an understanding of financial, commercial, statutory, safety and environmental considerations.

5.5 The evidence used to reach a decision whether or not candidates have reached the required standard comes from the following sources :

(i) a 2000-word report on training and practical experience (see Appendix C); and

(ii) a 4000-word project report (see Appendix C); and

(iii) the candidate's record of continuing education to date (a photocopy of the CPD Record ICE 108);

(iv) an interview at an appointed centre by two senior Corporate Members of the Institution acting as Reviewers;

(v) two essays written under examination conditions (see sections 5.7 to 5.10).

5.6 Candidates who have completed a period of Approved Training are also required to submit their Training Record (ICE 107).

5.7 At the interview, candidates will be given 15 minutes in which to make an informal presentation of their project report. Following this presentation they will be invited to respond to such questioning by the Reviewers as is required to determine that the criteria indicated in section 5.4 have been satisfied. The interview will normally last from 45 to 60 minutes.

5.8 Later the same day candidates will be required to write two essays, both of which are intended to test their ability to communicate in good written English, and to marshal their thoughts and express them on paper in a clear and concise manner.

5.9 The first essay will be on one of two technical subjects set by the Reviewers in the context of the 2000-word report, and the interview. The purpose is to allow candidates to expand upon particular aspects of their experience and technical knowledge.

5.10 The second essay will be on one of two topics selected by the Reviewers from a list published by the Institution. The purpose of this essay is to enable candidates to demonstrate that they have thought sufficiently about the role of the civil engineer in the community to be able to form a broad view of the social value of their work, and to demonstrate that their knowledge of modern management concepts is sufficient to provide a foundation for their developing leadership roles.

5.11 Candidates are allowed 1½ hours for each essay.

5.12 On successful completion of the CPR, candidates' applications for admission as corporate members of the Institution will be placed before the appropriate committee for approval.

APPENDIX A

Training schemes

A1 Training schemes should recognise that there is a need for a core of knowledge and achievement common to all trainees, regardless of the type of work undertaken or the specific direction in which their interests take them. Therefore the Institution has drawn up a list of core objectives which must form an integral part of all approved Training Schemes.

A2 The core objectives have been prepared so that :

(a) they encompass and relate to all types of civil engineering work; and

(b) they do not depend on time-serving as a measure of achievement; and

(c) the stated achievement criteria are, as far as possible, capable of objective assessment.

The definitions of and the achievement criteria for these core objectives are published in full in the Training Record (ICE 107).

A3 The core objectives in themselves do not provide the total framework within which training should take place. They must be supplemented by objectives written by the Supervising Civil engineer for the employing organisation in order to cover the particular activities, and the administrative and management practices, of that organisation. These specific objectives must follow the pattern established for the core objectives, in that they should be stated in a similar style and have the same achievement criteria.

A4 The schedule of training objectives formed by combining the core and specific objectives is intended to provide the basis for ensuring that trainees achieve the level of competence required by the Institution before applying for a Training Review. The schedule should be devised and executed so that specialists are catered for as well as those undertaking a more conventional career in civil engineering.

A5 As illustrated by Figure A the schedule has three components :

(a) professional and general

(b) an engineering solution

(c) the implementation process.

The engineering solution and implementation process components of the schedule are intended to develop the trainees' ability to apply their academic knowledge and subsequent training to the solution of a practical civil engineering problem of some complexity.

A6 For further details see Appendix D.

Schedule of training objectives

Professional and general	Engineering solution	Implementation process
Core objectives e.g.	Core objectives e.g.	Core objectives e.g.
ICE activities	Defining the problem	The contract and its operation
Civil engineering procedures	Designing a solution	Drawings and instructions
Current affairs	Specification and measurement	Methods and plant
Communication and reports	Project costing	Construction materials
Regulations and legislation	Safety in design	Safety at work
The professional team		
	Specific objectives	Specific objectives
Specific objectives	Tailored to the particular	Tailored to the particular
Tailored to the particular	employment experience to be	employment experience to be
employment experience to be	undertaken	undertaken
undertaken		

Achievement verified at Training Review or Experience Appraisal

APPENDIX B

Continuing education

B1 As noted in section 3.9, all candidates for the Chartered Professional Review who have completed a period of Approved Training or experience in lieu are required to show that they have undertaken continuing education for the equivalent of at least 30 days during their preparation for the CPR, of which a minimum of 15 days must have been gained prior to the Training Review/Experience Appraisal.

B2 Continuing education may include courses, technical conferences, seminars, symposia, organised site visits, and meetings of professional bodies. The programme of continuing education prior to the Training Review/Experience Appraisal should place the emphasis on technical content, whilst in the later stages the emphasis should move on to managerial/professional aspects.

B3 In order to achieve a well balanced programme, candidates are advised that the 30 days should be made up of 10 days spent on technical subjects, 10 days on managerial/professional subjects and a further 10 days on either of these forms of continuing education. The overall programme must also include one day on safety issues. Achievement of the mandatory core objective requiring trainees to develop and maintain an interest in Institution affairs will contribute three days. No single course may count for more than five days of the programme.

B4 A day must have not less than six hours of lectures/supervised tuition, etc.; it may also comprise two half-days, each being of not less than three hours, or two evening sessions.

B5 The continuing education programmes for trainees under Agreement should be arranged by the Supervising Civil Engineer after discussion with the trainee, and details should be included in the Record of Continuing Professional Development (ICE 108). the responsibility for certifying that this first part of the continuing education requirement has been satisfied rests with the Supervising Civil Engineer.

B6 Although Training Agreements do not extend beyond the Training Review it is expected that employers will continue to provide guidance and assistance in order to facilitate candidates' acquisition of the balance of their continuing education requirements before the CPR.

B7 Candidates who do not receive Approved Training under Agreement must make their own arrangements for satisfying the continuing education requirement, and attendance must be certified by their employers.

B8 To assist in identifying suitable courses the Institution publishes the 'Guide to continuing education courses for civil engineers' (ICE 105). In addition, the Local Associations of the Institution may assist in advice on courses which are available in their areas, including in-house courses run by employers for the benefit of their own staff but which are open to non-employees.

B9 Appropriate distance learning or correspondence courses may be used especially by overseas candidates and those in isolated locations who have difficulty in meeting the requirements. The advice of the Institution should be sought.

B10 The candidates' record of all continuing education activities as maintained in the CPD Record (ICE 108) forms part of the submission documents for the CPR.

APPENDIX C

Guidance notes for preparation of reports

Training and practical experience report

C1 The 2000-word typewritten report (which must not be a mere inventory of work prepared and executed) must describe the tasks on which the candidates have been employed, whether in investigation, planning, design, construction, or research. The account should set out the candidates' training and practical experience in the development of their careers to date, and should explain clearly the precise positions they have occupied, together with the degree of responsibility assigned to them. Reports should indicate how the core training objectives were achieved. Candidates should enlarge on any special problems they have met and on any areas where they have obtained extensive experience. Where appropriate, some indication of the size and costs of the works should be given. Candidates should emphasise their personal experience, bringing out the principal lessons learnt, and avoid extensive job specifications and descriptions.

C2 Reports should bear the names and signatures of the candidates.

Project Report

C3 The purpose of the 4000-word typewritten report is to demonstrate candidates' technical and professional competence. The report should describe particular projects or parts of projects in which the candidate has played a major role during the periods of training and practical experience. The candidate should indicate the role played in the development of these projects, and give the background to any important decisions for which they were responsible. The report should incorporate numerical analyses, drawing and/or other illustrations as appropriate, and should include cost data to show that the candidate has an adequate understanding of the financial implications of the decisions taken.

C4 Reports should be certified as the work of the candidate by the person to whom the candidate was responsible when the work was done.

APPENDIX D

Administrative arrangements

Training Agreements : registration

D1 Within six months of the start of the period of approved training the SCE must send the formal Agreement (ICE 114) to the Institution together with the appropriate fee. Trainees should check that this has been done. It trainees are not already Graduate or Student members of the Institution their membership application forms must accompany the Agreement. The Institution will then register the agreement and return the form to the SCE.

D2 As indicated in section 3.7 candidates may enter into separate Agreements with SCEs for periods of not less than twelve months each. Such Agreements need not be consecutive, nor with the same SCE, but the overall training and practical experience must match up to that which would be gained in a normal three-year Agreement. Each Agreement must be registered with the Institution.

D3 Any Training Agreement may be extended with the consent of the trainee, the SCE and the Institution.

Training Agreements : transfer

D4 All parties to a Training Agreement should recognise the professional obligation entered into and that valid reasons for a transfer can arise. Trainees under Agreement contemplating a change of employment should notify their potential employer of the existence of the Training Agreement. All parties should agree to the transfer. The potential employer must be on the Index of Employers Approved for Training (ICE 106) in order to transfer an agreement. When a trainee moves employment the new SCE and the first employer should complete and sign a partial Completion Certificate (ICE 123) recording the training completed.

Training Record : monitoring and completion

D5 Trainees under Agreement must keep and make regular entries in a Training Record (ICE 107). This should contain :
  (i) the approved training scheme with achievements certified by the SCE;
  (ii) the trainees' quarterly reports on the practical experience that has been gained.

D6 Details of continuing education activities must be recorded in the trainee's CPD Record (ICE 108).

D7 Training Records form an essential part of the monitoring process which is carried out by the

RTOs who visit employers on the Index at least once a year and meet registered trainees as necessary. During these visits the RTOs will expect the Training and CPD Records to be available and up to date. This will enable them to discuss trainees' progress and asses the general operation of Training Agreements.

D8 When all the training objectives as outlines in the Agreement have been met, the SCE will complete a report on the trainees' performance using form ICE 123 which must be verified by the RTO.

Application for Experience Appraisal in the UK

D9 The following documents, with appropriate fees (see ICE 202), should be forwarded to the Institution :

(i) two copies of 'Application for Experience Appraisal' ICE 124);

(ii) evidence of the candidate's academic qualifications (unless already a Graduate member);

(iii) two copies of a 2000-word report on practical experience;

(iv) a photocopy of the relevant pages of the CPD Record containing details of continuing education to date;

(v) the list of core objectives, as indicated in 4.3(iii);

(vi) documents relating to the solution of an engineering problem (see 4.3 (iv);

(viii) three self-addressed but unstamped envelopes, size DL (220 mm × 110 mm).

The above package of documents should not exceed 1 kg in weight, and may be sent by post or delivered by hand. The package should be marked in the top left-hand corner 'Experience Appraisal Documents.' Dates for submission are given in Table 1.

Application for the CPR in the UK

D10 The following documents, with appropriate fees (see ICE 202), should be forwarded to the Institution :

(i) two copies of 'Application for CPR' (ICE 130);

(ii) three copies of 'Application for admission as a Member' (ICE 132);

(iii) a photocopy of the relevant pages of the CPD Record containing details of continuing education to date;

(iv) three self-addressed but unstamped envelopes, size DL (220 mm × 110 mm).

(v) form ICE 123 if the candidate trained under Agreement;

(vi) evidence of the candidate's academic qualifications unless he or she is a Graduate member or has passed the Experience Appraisal;

(vii) one copy of the report on training and practical experience (see 5.5 (i);

(viii) one copy of the project report (see 5.5(ii).

D11 Applications must be forwarded in sufficient time to reach the Institution by the appropriate closing date (see Table 1), and may be sent by post or delivered by hand. The parcel should be marked at the top left-hand corner 'Chartered Professional Review (CPR) documents'. Candidates resident outside the United Kingdom at the time of their applications for CPR and who wish to be interviewed in London are requested to provide a UK address on form ICE 130 and their envelopes, and to give a UK telephone number of a contact.

D12 In addition the following documents should be forwarded to each Reviewer :

(i) one copy of the report on training and practical experience;

(ii) one copy of the project report;

(iii) a photocopy of the relevant pages of the CPD Record containing details of continuing education to date.

D13 Those candidates who have completed a period of Approved Training should send a copy of their Training Record (ICE 107) to each reviewer. The plastic folder should not be sent but the pages of the Record should be adequately secured together.

D14 Candidates will be advised of the addresses of their two Reviewers approximately one month before the date of their interviews. Copies of the documents may be sent by recorded or registered post; they must reach the Reviewers at least ten days before the interview, failing which the interview may be cancelled. The package must not weigh more than 1 kg and should be marked at the top left-hand corner 'Chartered Professional Review (CPR) documents'.

Dates, venues and results for the CPR

D15 The dates, venues and results for the CPR and Experience Appraisal are set out in Table 1.

D16 After approval by the Council of the Institution the result of the Chartered Professional Review will be sent to each candidate. In no circumstances will results be given over the telephone or by telex.

D17 Unsuccessful candidates will be advised why they have not satisfied the Reviewers and given guidance about preparation for retaking the CPR. There is no appeal against the results, but for this reason there is no limit to the number of times candidates may apply for the CPR.

Overseas candidates

D18 An overseas candidate training under Agreement will be put in touch with a local Corporate Member who will act in place of the RTO and collaborate with the SCE in the Training Review as detailed in 4.1.

D19 The requirements and procedures applying to candidates wishing to applying for an Experience Appraisal or the CPR at overseas centres are virtually identical to those to be complied with by UK candidates; the only differences concern the arrangements for submission of documents and application dates for the CPR.

Experience Appraisal (see 4.2 to 4.4)

D20 Overseas candidates for an Experience Appraisal should forward to the Institution, with appropriate fees, the documents listed in section D9. An acknowledgment of the arrival of the documents will be sent within twenty eight days. The Institution will, if necessary, send all documents to a local Corporate Member who is an Assessor in order that an interview can be arranged. (See 4.4).

CPR (see Chapter 5)

D21 Overseas interviews for the CPR will be held once a year (see Table 2), in October, at those centres where there is sufficient demand. Applications to take the CPR must be despatched in time to arrive at the Institution as soon as possible after the opening date and before the closing date. In order to retain equity with the same interview dates in the UK, practical experience (as shown on ICE 132 and in the report on training and practical experience) may be aggregated to 15 August.

D22 Candidates will receive an acknowledgment of their applications usually within twenty-eight days of the arrival of their application at the Institution. By the end of September they will hear from the local Corporate Member who will be making the arrangements for their CPR and with whom they should communicate. The Corporate Member will give the candidate instructions as to where the remaining documents (two each of the documents listed in D12 and D13 if relevant) are to be sent.

D23 Arrangements will be made for candidates to take the CPR at the centre nearest to their place of employment or residence at the date of application. If this is not convenient, the only alternative will be to travel to London to attend the review in either April or October. Candidates resident overseas will not be allowed to attend the CPR at any of the centres in the UK provinces. Candidates who move after the overseas closing date may transfer their review only to the London centre, and then provided notice is given which arrives at the Institution no later than 1 September. Candidates will not be allowed to transfer to another overseas centre other than by cancelling their current application and reapplying the following year. Candidates should expect arrangements for their interview to be made for the convenience of their Reviewers rather than themselves. Arrangements will not be reorganised to suit an individual candidate.

D24 The submission documents should be grouped together as shown below and forwarded to the Institution and the local Corporate Member as indicated.

D25 Send to the Institution with the appropriate fees (see ICE 202);

    (i) two copies of 'Application for CPR' (ICE 130);

    (ii) three copies of 'Application for admission as a Member' (ICE 132);

    (iii) a photocopy of the relevant pages of the CPD Record containing details of continuing education to date;

    (iv) three self-addressed but unstamped envelopes, size DL (220 mm × 110 mm);

    (v) form ICE 123 if the candidate trained under Agreement;

    (vi) evidence of the candidate's academic qualifications unless he or she is a Graduate member or has passed the Experience Appraisal.

  Send to the local Corporate Member :

    (vii) two copies of the report on training and practical experience (see 5.5 (ii);

    (viii) two copies of the project report (see 5.5 (ii).

Cancellations or rejections

D26 In the event of a cancellation or rejection an administrative charge of two thirds of the fee will be deducted from the amount returned to the candidate.

APPENDIX E

Booklets, forms and information sheets

Booklets

ICE 100	Grades of Membership
ICE 101	Routes to Membership — Chartered Engineer
ICE 102	Routes to Membership — Incorporated Engineer
ICE 103	Routes to Membership — Engineering Technician
ICE 104	Routes to Membership — Mature Candidates Review
ICE 105	Guide to Continuing Education Courses
ICE 106	Index of Employers Approved for Training
ICE 107	Training Record
ICE 108	Record of Continuing Professional Development

Forms

ICE 110	Application for admission as a Student member
ICE 111	Academic assessment
ICE 112	Application to be placed on the Index of Approved Employers
ICE 113	Application for admission as a Graduate member
ICE 114	Training Agreement
ICE 115	Transfer of Training Agreement
ICE 117	Application to attend Technician Professional Review
ICE 118	Application for admission as a Technician member
ICE 120	Application to attend Incorporated Professional Review
ICE 121	Application for courses to be included in the Guide
ICE 123	Training Review and Completion Certificate
ICE 124	Application for Experience Appraisal
ICE 125	Application for admission as an Associate Member

ICE 126	Resit Certificate
ICE 130	Application to attend Chartered Professional Review
ICE 131	Application to attend Mature Candidate Review
ICE 132	Application for admission as a Member
ICE 133	Questionnaire for CPR sponsor
ICE 134	Application for membership (MCR)
ICE 135	Questionnaire for mature candidate sponsors
ICE 140	Application for transfer of a Member to the class of Fellow
ICE 141	Application for direct admission as a Fellow
ICE 142	Application for admission as a Fellow (Eminent Route)
ICE 143	Application for admission as a Companion
ICE 144	Application for admission as an Affiliate

Information sheets

ICE 200	Failures and Resits
ICE 201	Reciprocal arrangements with Institution of Professional Engineers, New Zealand
ICE 202	Education, Training and Membership Fees
ICE 203	CPR essay topics : A Civil Engineer's Responsibilities

Interview centre	Application dates Opening	Closing	Month of interview	Results published	Admission dates for those passing the CPR leading to MICE
London	1 January	15 February	April	30 June	1 June
	1 July	15 August	October	31 December	1 December
Belfast Edinburgh Manchester	1 January	15 February	May	30 June	1 June
Cardiff Durham Glasgow	1 July	15 August	November	31 December	1 December
All overseas centres	1 March	15 May	October	28 February	1 December

Table 1 : Dates, venues and results for the CPR and Experience Appraisal. The closing date for an overseas candidate applying for an Experience Appraisal is 15 February to allow the candidate to go forward to the CPR in the autumn if successful.

Country	Interview centre	Country	Interview centre	Country	Interview centre
Australia	Adelaide Brisbane Melbourne Perth Sydney	Jordan Kenya Malawi Malaysia	Amman Nairobi Blantyre Kuala Lumpur	South Africa	Bloemfontein Cape Town Durban Johannesburg Port Elizabeth Pretoria
Brazil	Sao Paulo	New Zealand	Auckland Christchurch	Sri Lanka	Colombo
Brunei	Brunei			Sudan	Khartoum
Canada	Calgary Halifax Toronto Vancouver	Nigeria	Ibadan Kaduna Lagos	Tanzania	Dar es Salaam
		Pakistan	Lahore	UAE	Dubai

Cyprus	Nicosia	Papua New Guinea	Lae	Venezuela	Valencia
Ghana	Accra	Saudi Arabia	Riyadh	West Indies	Barbados
Hong Kong	Hong Kong	Sierra Leone	Freetown		Jamaica
India	New Delhi	Singapore	Singapore	Zambia	Lusaka
Indonesia	Jakarta			Zimbabwe	Harare

Table 2 : Overseas interview centres

APPLICATION FOR CPR IN THE UK

INTERIM AMENDMENT NO. 1

Please note the following amendments to ICE 101 — Autumn 1990 :—

D10 The following documents, with appropriate fees (see ICE 202), should be forwarded to the Institution :

(i) one copy of "Application for Professional Review" (ICE 160);

(ii) three copies of "Application for admission as a Member" (ICE 132);

(iii) a photocopy of the relevant pages of the CPD Record containing details of continuing education to date;

(iv) three self-addressed, but unstamped, envelopes, size DL (220 mm × 110 mm);

(v) form ICE 123 if the candidate trained under Agreement;

(vi) evidence of the candidate's academic qualifications unless he or she is a Graduate member or has passed the Experience Appraisal;

(vii) a one-page summary of your Project Report, outlining in simple layman's terms a description of the work and its purpose.

D11 Applications must be forwarded in sufficient time to reach the ICE by the appropriate date at the very latest (see Table 1) and may be sent by post or delivered by hand. The parcel should be marked at the top left-hand corner "Chartered Professional Review (CPR) documents". Candidates resident outside the United Kingdom at the time of their applications for CPR and who wish to be interviewed in London are requested to provide a UK address on form ICE 160 and their envelopes, and to give a UK telephone number of a contact.

D12 In addition the following documents should be forwarded to each Reviewer, when advised :

(i) one copy of the report on training and practical experience [see 5.5 (i)];

(ii) one copy of the project report [see 5.5 (ii)];

(iii) a photocopy of the relevant pages of the CPD Record (ICE 108) containing details of continuing education to date.

APPENDIX E

ICE 133 Please note that ICE 45/Q will be accepted in place of ICE 133. These should be distributed to each of your sponsors and returned by them directly to the Professional Reviews Department clearly marked "Private and Confidential".

JULY 1992

THE INSTITUTION OF CIVIL ENGINEERS
GUIDANCE NOTES
ON
THE ASSESSMENT AND GRADING OF ESSAYS

For clarity, the candidate is referred to in the singular. This consequently necessitates the use of the masculine gender in the singular. It should be remembered, however, that many civil engineers are women and that they are not excluded by references to "he", "his" and "him".

1 The essay performs three functions. Firstly it is a test of the candidate's knowledge of, and ability to communicate in, acceptable English. Secondly, it tests his ability to marshal his thoughts and express them on paper in a clear and concise manner. Finally, being set in the context of the candidate's 2,000-word report, it must display his knowledge of the subject so it will be one with which he is familiar. Where the candidate is required to express an opinion, he will not be penalized should his opinion not agree with that of his Reviewers, provided that the argument supporting it is logical. The finished essay is not required to be a polished article; the standard should be consistent with that of a tidy first draft, but in acceptable English and with no ambiguities.

2 The essay should follow an ordered structure, displaying an awareness of the importance of division into suitable paragraphs. The candidate is encouraged to include references to his reading within the text. Rough notes should be written on the page designated for that purpose, not on a separate sheet. A dictionary may be used, but not other books or notes are allowed at the candidate's desk. One and a half hours are allowed for the work, from the commencement of the reading of the topic to the collection of the script. It is stressed that failure in this part of the Review may lead to failure of the CPR as a whole. Reviewers will judge and mark the essays against the criteria given in paragraphs 3 to 7 below. They will not, however, "tot up" points in each of the three sections; the questions in each are to clarify their thoughts in deciding the three grades required by paragraph 8.

3 Knowledge of the subject and
Relevance of the candidate's answer to the topic set
   a) Does the candidate exhibit a reasonable (depth and breadth of) knowledge of the topic, bearing in mind his experience as demonstrated by his submitted documents and his interview?
   b) Has he augmented his experience by reading widely on the subject?
   c) Has he understood the subject?
   d) Does the essay cover the whole subject set or only a part of it?
   e) Does the essay keep to the point or does it tend to wander from it?
   f) Has the candidate refrained from padding the essay with irrelevant or repetitive material?

4 Grammar and Syntax
   a) Does every sentence contain a finite verb, and are the tenses correct?
   b) Is there agreement between nouns and verbs with regard to singular and plural cases?
   c) Are the spelling and punctuation reasonably correct?
   d) Has the candidate avoided jargon and the use of catch phrases, or apparently meaningless abbreviations? (Abbreviations when first used should be accompanied by a full spelling.)
   e) Are most sentences easy to understand?
   f) Are the sentences constructed correctly?

5 Clarity of argument and Presentation
   a) Has the candidate a good vocabulary and has he avoided the use of malapropisms?
   b) Is his phraseology sufficiently mature for a graduate of his age?
   c) Are his ideas expressed in a logical manner?

d) Is it easy to grasp the candidate's argument?

e) Does the essay have a discernible framework or pattern?

f) Are paragraph divisions sensibly chosen?.

g) Is the presentation reasonably tidy and is the handwriting legible?

h) Is the essay generally a pleasure to read?

6 The finished essay is not required to be a polished article; the standard should be consistent with that of a tidy first draft, but in acceptable English and with no ambiguities.

7 Where a candidate is required to express an opinion, he will not be penalized should his opinion not find favour with his Reviewers, provided that the argument supporting it is logical.

8 Each essay will be assessed as follows :—

KNOWLEDGE & RELEVANCE	GRAMMAR & SYNTAX	CLARITY & PRESENTATION
A Very good	A Very good	A Very good
B Satisfactory	B Satisfactory	B Satisfactory
C Unsatisfactory	C Unsatisfactory	C Unsatisfactory
D Very bad	D Very bad	D Very bad

A candidate will fail if, of the three marks possible for either essay, he obtains either two or more grades C or one grade D.

9 Where a candidate obtains three grades A, he will automatically receive a commendation, and notice of this will be given in the published pass list.

10 If a candidate receives a commendation, he becomes eligible for consideration for the Renee Redfern Hunt Memorial Prize.This is awarded by the Institution to the candidate selected as writing the best essays. The Prize is presented by the President personally and the recipient's name is recorded in the ICE Yearbook.

* The Renee Redfern Hunt Memorial Prize was instituted in memory of the long and devoted service of Miss RR Hunt MBE, Professional Examinations Officer at the ICE from 1945 until within a few months of her death in 1981.

The Chartered Institute of Building (UK)

INTRODUCTION

The Chartered Institute of Building was formed in 1834, incorporated under the Companies Acts in 1884 and granted a Royal Charter in 1980. It is registered as an educational charity and is a professional institution of over 33,000 members engaged in building. The members hold administrative, commercial, educational, managerial, scientific or technical positions, or are undergoing training in a firm, consultant practice, department or other organisation engaged in the construction, alteration, maintenance, repair, provision, design, inspection or management of buildings, in other construction work, in building education or research.

The Institute's objects are :

the promotion, for the public benefit, of the science and practice of building. The advancement of education in the science and practice of building including all necessary research and the publication of the results of all such research. The establishment and maintenance of appropriate standards of competence and conduct of those engaged, or about to engage, in the science and practice of building.

CONTRIBUTION TO THE PUBLIC AND THE INDUSTRY

The Institute's examinations and practical experience requirements ensure that an individual member has a sound education and training in building appropriate to the needs of the industry. Particular emphasis is placed on management subjects and the development of management skills. The Rules of

Professional Conduct require members to discharge their duties to an employer or client with fidelity and probity. They also require members to keep themselves informed of new thought and development, and the Institute provides a wide range of services to assist members to raise their standards of knowledge and skill. Influence on the quality of building is exercised by raising standards of building education through the CIOB's own examinations and by a system of exemptions for courses and qualifications of appropriate level and scope.

The Institute's membership provides an unequalled body of knowledge and experience of building practice, education, training and research. A great deal of the resources available are devoted to the dissemination of this knowledge in the form of codes of practice,technical meetings, conferences and seminars, technical papers and other publications for the benefit of members and others in the industry. The Institute also represents the building industry on a very wide range of bodies concerned with building education and practice. It offers advice to government departments and other agencies on many aspects of building so as to improve the climate in which building is carried on and to secure the adoption of good building practice. The Institute grants scholarships to fund individual investigation and research. It also makes annual awards for outstanding performance in the management of building projects.

SINGLE EUROPEAN ACT

Freedom of movement for people to work throughout the European Community is one of the essential rights being incorporated in the Single European Market. The Chartered Institute of Building has been engaged with the UK Government in consultations leading to the final wording of a directive on a General System for the Recognition of Higher-Education Diplomas (89/48/EEC). Within the context of this directive, member of the Chartered Institute of Building are recognised as pursuing a professional activity. This is by virtue of the self-regulation imposed by possession of a Royal Charter which promotes and maintains a high standard in the professional field of building and which, to achieve that purpose, is recognised in a special form by the United Kingdom Government and operated through the offices of the Privy Council.

The CIOB has been designated as a 'Competent Authority' within the meaning of the directive. It has established a mechanism whereby professionals from the other EC Countries will be able to take an aptitude test or undergo an adaptation period necessary to achieve mutual recognition of their qualifications in the United Kingdom. These professionals, on payment of the due subscription, will be able to obtain membership of the CIOB with the right to use the designations and titles according to the class for which they qualify.

In a similar manner the CIOB is engaged in the process of establishing contact with its Continental counterparts in order to achieve mutual recognition of the qualifications of the Institute's members.

The CIOB is unusual among chartered institutions in the United Kingdom in that it has retained its technicians and higher technicians in membership whereas most institutions have only corporate, professional level, members. The CIOB thus provides a natural home for technicians and higher technicians in building ad will be able to do so for their counterparts operating at that level on the Continent. Accordingly, the Institute has been engaged with the UK Government in the process of developing the Second General Directive which deals with the freedom of movement of technicians and higher technicians. This second Directive is intended to be complementary to the General Directive and the CIOB expects to operate in a similar manner under the Second Directive as it does under the First.

Those professionals, higher technicians and technicians operating in the management of the whole building process on the Continent who wish to establish their right to operate in the UK are recommended to contact the Chartered Institute of Building.

MEMBERSHIP OF THE INSTITUTE

Admission to, and progression in, membership is based upon both academic attainment and practical experience in building of an approved nature, duration and level. There are six classes of membership in three groups as follows :

Fellows
Members } Professional level

| Associates Licentiates } | Higher technician and technician level |

Associates
Licentiates } Higher technician and technician level

Graduates
Students } Those undergoing training and obtaining practical experience

Fellows and Members are entitled to use the designation FCIOB and MCIOB respectively and the description 'Chartered Builder.' Associates and Licentiates are entitled to use the description in full...of the Chartered Institute of Building. Associates who have been admitted to a Cadet Scheme are allowed to use the designation ACIOB.

A 'ladder of opportunity' is provided to enable a member, following additional study, examination and experience, to progress from one class of membership to the next 'step.' A separate route is available for mature candidates.

A diagram showing the principal routes to the Institute's membership is at Figure 1.

THE BENEFITS OF MEMBERSHIP

The tangible benefits of membership include :

The right as a corporate member to use an Institute designation signifying a high standard of competence and integrity and an academic level which equates to that of a first degree.

The availability of technical information, publications and other services. The Technical Information Service which includes papers on a wide range of subjects such as estimating, maintenance, site management and surveying, Building Research Establishment Digests, Building Management Abstracts, photo-copying and library facilities.

Access to professional activities, including conferences, seminars, lectures, discussion meetings and visits and overseas tours at national, regional and local levels.

The facility to participate at national, regional and local levels in developments aimed at improving the science and practice of building.

Members who make their own contribution to the corporate efforts of raising the standard of building which is the Institute's prime aim obtain the most value and satisfaction from their membership.

ORGANISATION

In the United Kingdom and the Republic of Ireland the Institute is organised at national, regional and local levels. The governing body of the Institute is the National Council. England and Wales are divided into 11 Regions and there is a Branch for Ireland and for Scotland. Every Region and Branch has a number of local centres, of which there are more than 50, each administered by a committee. Local members elect each centre committee which has a number of seats on its appropriate Regional or Branch Council. These in turn have a number of seats on the National Council which elects the National Officers. Overseas a member is appointed as overseas representative to look after the Institute's interests in countries where there is a significant number of Institute members. Centre committees are established if the membership is large enough. There are currently two overseas Branches — Hong Kong and Malaysia. In addition there is one centre— in Jamaica and two group committees, one in Guernsey and one in South Africa. It is the Institute's policy to encourage the development of independent institutions overseas and there are autonomous institutes in the following countries :

Australia
Kenya
New Zealand
Nigeria
Singapore
South Africa
United States of America

SOCIAL ACTIVITIES

A wide range of social activities is available for members and their guests at all levels. The Institute's National Annual Dinner is held, by courtesy of the Corporation of London, at the Guildhall in the City and is attended by many distinguished persons, representative of the national life of the country, and leaders of organisations concerned with building. Events are organised locally and dinners, dances and open days for members and their guests are held from time to time at 'Englemere' the institute's headquarters. The estate with its swimming pool and tennis courts is available for members and their families.

CAREER OPPORTUNITIES

A career in building offers first class rewards for young men and women with opportunities to work both at home and overseas. Responsibility can come early with leadership potential and promotion prospects are excellent. People will always need buildings in which to live, work and play. We cannot exist without them and we cannot exist without the industry that builds them. There will always, therefore, be a future for the professional who is at the centre of that industry — the Chartered Builder.

FURTHER INFORMATION

Qualifications for Admission

The qualifications for admission to the various classes of membership are based upon the requirement to demonstrate the achievement of both an academic standard and practical experience in the discipline of building. Full details are contained in Membership Regulations which can be obtained free on request from the Institute.

Fees and Subscriptions

The current scales of admission, transfer and examination fees and membership subscriptions are available from the Institute on request.

The science and practice of building is nearly as old as man himself. As primitive man emerged from his caves he needed other shelter from the elements and protection against predatory beasts and his fellow men.

As camps, settlements, towns and cities developed, those engaged in the science and practice of building progressively increased in importance. The quality of the built environment has never been more significant in its social and economic impact on life of the country than it is today.

Modern society needs more — and more complex — buildings of every kind : homes, schools, shops, hospitals, offices, factories and an increasing variety of new building types. It is the task of the building industry to satisfy these needs with speed and economy.

The total activity of the construction industries in the UK is over £ 22,000 million a year. Some three-quarters of this output is building, ranging from small alterations, maintenance and repair of existing buildings to the erection of complex multi-storey structures.

To meet the demands being placed upon it, the building industry must have more well-trained and qualified managers, technologists and technicians committed to the pursuit of the highest standards of competence.

Those already in established positions need to keep abreast of new thought and developments. The industry also needs a steady flow of young men and women for training. Flexibility of outlook is essential to meet the continually changing face of building.

These are the prime concerns of The Chartered Institute of Building.

The Chartered Institute of Building

The Chartered institute of Building was formed in 1834, incorporated under the Companies Acts in 1884 and by Royal Charter in 1980 and is registered as an educational charity.

It is the professional institution for those concerned with the construction, alteration, maintenance and repair of buildings, those engaged in teaching building and those in building research. it has some 31,000 members.

Its objects are :

The promotion for the public benefit of the science and practice of building.

The advancement of education in the science and practice of building, including all necessary research and the publication of the results of all such research.

The establishment and maintenance of appropriate standards of competence and conduct of those engaged or about to engage in the science and practice of building.

The governing body of the Institute is its National Council. England and Wales are divided into eleven regions each with a regional council. There are branch councils for Ireland and Scotland and overseas in Hong Kong and Malaysia. Each region or branch has a number of local centres, a total of 55 in the UK and the Republic of Ireland.

There are overseas units of the Institute in Canada, Jamaica and South Africa. Separate autonomous Institutes of Building or Construction, with which the Institute has close relations, exist in the following countries :

America Australia Kenya New Zealand Nigeria Singapore South Africa.

The Benefits of Membership...

The Institute provides a wide range of services to assist embers to raise their standards of knowledge and skill.

The following is a summary of the benefits of membership.

- Recognised qualifications in building.
- Monthly journal Chartered Builder.
- Technical Information Service, for which a nominal charge is made, comprising a series of information papers covering primarily estimating, maintenance, site management and surveying, papers dealing with management and education are issued from time to time.
  Building Management Abstracts
  Building Research Establishment Digests
  Photocopying service
  Library and technical information enquiry service.
- A voice in the industry's and the Institute's affairs.
- The opportunity to belong to a body of importance and repute. The opportunity to participate in developments aimed at improving building education and practice.
- The opportunity to create friendships and to exchange knowledge and experience with other members.
- Personal advice and guidance on career development
- Activities (conferences, seminars, lecture and discussion meetings, visits, overseas tours and social functions) — a total of over 300 per year.
- A wide variety of Institute publications at 20% discount — see Books for Building.
- Benefits arising from the Institute's efforts to improve building education and practice, e.g.,
  Site management education and training scheme.
  Safety training scheme
  Building legislation
  Codes of good practice
  Materials control and waste in building
  Careers Service
  Advice to government, other institutions and associations.
- Sporting, recreational and social activities at Englemere and elsewhere.
- Right to participate in Private Medical Insurance scheme, insurance scheme, mortgage scheme, etc.

and your route to...

membership

Admission to and progression in membership is based upon academic attainment and practical experience. of an approved nature, duration and level.

There are six classes of membership in three groups as follows :

Fellows
Members } Professional level

Associates
Licentiates } Higher technician and technician level

Graduates
Students } Those undergoing training and obtaining practical experience

Fellows and Members are entitled to use the designation FCIOB and MCIOB respectively and the description 'Chartered Builder.'

Associate members who have been approved for the Cadet Scheme may use the designation "ACIOB."

Associates, Licentiates, Graduates and Students are entitled to use the description in full '......of the Chartered Institute of Building.'

A 'staircase of opportunity' is provided to enable a member following additional study, examination and experience to progress from one class of membership to the next with 'landings' and appropriate services available for those who decide not to seek higher progression.

Detailed membership requirements are contained in *Membership Regulations* obtainable free on request.

Examinations

The Institute is the qualifying association for the building industry. Its examinations are taken in June by some 3,500 candidates at centres in the UK and overseas. Re-sit facilities are available in certain circumstances in November.

The Associate examination, at higher technician level, consists of the following subjects :

Building Construction

Building Materials and Environmental Science

Building Services

Quantity Surveying and Estimating

Industrial Studies

Building Law

*Computing

*Site Surveying

* Note. The Institute does not offer examinations in Computing or Site Surveying. These subjects are assessed by approved colleges, who issue a "Certificate of Competence" under a scheme monitored by the Institute.

The Member Examination, at degree level, consists of the following subjects :

Part I   Building Technology I

        Building Management I

        Quantity Surveying and Estimating

        Economics

        Building Law

Part II   Building Technology II

        Building Management II

Contract Administration

Building Economics and Finance

*Project Evaluation and Development

* Note. The Institute does not offer an examination in Project Evaluation and Development. This subject will be taught and assessed by teaching institutions under a scheme monitored by the Institute.

Details of the examinations are given in Examination Admission Regulations and Syllabuses and the awards giving exemption from the whole or part of the examinations are given in Recognised Exempting Awards both obtainable free on request.

The current structure of building education and of the Institute's own examinations are designed to produce the general practitioner in building. For whose who subsequently specialise, the Institute offers specialist qualifications in maintenance management and site management.

Modified forms of examination for admission or transfer to the classes of Member and Associate are provided for those over 35 years of age who have had appropriate practical experience in building. Applicants aged 38 years and over may offer a thesis as an alternative to sitting the examination. Details are given in the Direct Membership Regulations and Regulations for the Direct Associateship Examination obtainable free on request.

Practical Experience

Experience in one or more of the following functions in a firm, consultant practice, department or other organisations engaged in the construction, alteration, maintenance, repair, provision, design, inspection or management of buildings, other construction work, or in building education is acceptable :

Building Control	General management	Quantity surveying
Building asset management	Inspection	Research
Building surveying	Maintenance management	Site engineering
Construction management	Planning	Site management
Cost and production control	* Project management	Teaching
Design	Purchasing	Training
Estimating	Quality Control	

Or any other function acceptable to the Institute which requires the exercise of professional judgement relating to some aspect of building construction.

For admission/transfer to the Member class a minimum of three years practical experience at professional level in one or more of the above-mentioned functions is required. Candidates are assessed at the Professional Interview which consists of a report on training and experience and attendance at an interview.

For admission/transfer to the classes of Licentiate and Associate a minimum of two years practical experience at technician or higher technician level in one or more of the above-mentioned functions is required.

(Note : Experience in an operative or first line supervisory position does not count towards the required minimum period).

Coverage of these regulations

The regulations prescribe the rules for admission or transfer into Institute membership in the classes of Student, Graduate, Licentiate, Associate and Member and for transfer from Member to Fellow. They include the regulations for the Professional interview.

The regulations and syllabuses for the Associate Examination and the Member Examination Parts I and II are contained in Examination Admission Regulations and Syllabuses obtainable from the Institute free on request.

Section 1

Student Class

Purpose

1. The Student class is provided for those undergoing instruction. It enables students to participate in Institute activities, to benefit from its publications and services and to keep in touch with new developments while undertaking further education courses and gaining the necessary practical experience for admission to the classes of Licentiate, Associate or Member.

2. Except for a candidate already in membership in a higher class, Student membership is compulsory for admission to the Institute's Associate Examination and Member Examination Parts I and II.

Qualifications for admission

3. For admission to Student membership a candidate must be at least 16 years old and have been accepted for, be engaged in, or have completed a course of general building studies recognised by the Institute.

How to apply

4. Application for admission to Student membership must be made on Form M1 obtainable from the Institute, and be accompanied by the admission fee and one year's subscription.

Fees and subscriptions

5. The current admission fee and annual subscription for Student membership are detailed in the loose-leaf insert to these regulations.

   Note : The whole of the annual subscription is allowable as a deductible expense for the purposes of Income Tax Schedule E assessment under Inland Revenue Technical Division letter reference CI/SUB/248/NC dated 8 October 1980.

6. An applicant who fails to secure admission is entitled to a refund of the fee and subscription.

Section 2

Graduate Class

Purpose

7. The Graduate class is provided for those who are academically qualified for the Member class while they are acquiring the relevant practical experience for admission to the Professional Interview leading to the Member class.

Qualifications for admission or transfer

8. For admission or transfer to the Graduate class a candidate must pass or be exempt from the Member Examination Parts I and II.

Exempting awards

9. Details of examination qualifications recognised for exemption from the Member Examination Parts I and II are given in Recognised Exempting Awards obtainable free on request.

How to apply

10. Application for admission or transfer to the Graduate class must be made on Form MA obtainable from the Institute, and be accompanied by the admission or transfer fee and subscription.

11. Applicants claiming exemption from the Member Examination Parts I and II on the basis of an exempting award must submit to the Institute a photographic copy of the examination certificate (or the original thereof) upon which the application is based.

    Note : Original certificates are submitted at the applicant's own risk and should be given proper protection in the post.

Fees and subscriptions

12. The current admission and transfer fees and annual subscriptions are detailed in the loose-leaf insert to these regulations.

whole of the annual subscription is allowable as a deductible expense for the purposes of Income Tax Schedule E assessment under Inland Revenue Technical Division letter, reference CI/SUB/248/NC dated 8. October 1980.

13. An applicant for admission to the Graduate class must submit with the application the admission fee and one year's subscription.

14. An applicant for transfer from Student membership must submit the transfer fee with the application. No additional subscription is payable in the year of transfer.

15. An applicant who fail to secure admission or transfer is entitled to a refund of the admission or transfer fee and if applicable, subscription.

Section 3

Licentiate Class

Qualifications for admission or transfer

16. For admission or transfer to Licentiate membership a candidate must :
    a. be not less than twenty-one years of age
    b. hold an approved examination qualification and
    c. have had practical experience of a satisfactory nature and duration (see regulation 19).

Approved examination qualifications

17. The Institute does not offer an examination for Licentiate membership. Among the qualifications approved for this class are :
    a. The National Certificate or Diploma in Building Studies of the Business and Technician Education Council.
    b. The Certificate or Diploma in Building of the Scottish Vocational Education Council.

18. Details of other approved qualifications are given in *Recognised Exempting Awards* obtainable free on request.

Experience of a satisfactory nature and duration

19. An applicant must have had a minimum of two years experience at technician level in one or more of the following functions in a firm, consultant practice, department or other organisation engaged in the construction, alteration, maintenance, repair, provision, design, inspection or management of buildings, other construction work or in building education.

Building Control	General management	Quantity surveying
Building asset management	Inspection	Research
Building surveying	Maintenance management	Site engineering
Construction management	Planning	Site management
Cost and production control	* Project management	
Design	Purchasing	Teaching
Estimating	Quality Control	Training

Any other function acceptable to the Institute which requires the exercise of professional judgement relating to some aspect of building or construction.

Note : Experience in an operative, or a first line supervisory position cannot be counted towards the requirement.

* **Project management** is defined by the Institute as :

'**The overall planning and co-ordination of a project from inception to completion aimed at meeting a client's requirements and ensuring completion on time within cost and to required quality standards.'**

How to apply

20. Application for admission or transfer to Licentiate membership must be made on Form MA, obtainable from the Institute and be accompanied by the relevant fee and if applicable, subscription.

21. Applicants must submit to the Institute a photographic copy of the examination certificate (or the original thereof) upon which the application is based.

Notes :

    a. Original certificates are submitted at the applicant's own risk and should be given proper protection in the post.

    b. The provision of full and precise details of the nature, duration and level of duties and responsibilities is of great importance if the application is to succeed.

    c. If a candidate is in doubt whether he/she complies with either the Institute's education or experience requirements, advice should be sought from the Institute before completing the application form.

Fees and subscriptions

22. The current admission and transfer fees and annual subscriptions are detailed in the loose-leaf insert to these regulations.

    Note : The whole of the annual subscription is allowable as a deductible expense for the purposes of Income Tax Schedule E assessment under Inland Revenue Technical Division letter, reference CI/SUB/248/NC dated 8 October 1980.

23. An applicant for *admission* to the Licentiate class must submit with the application the admission fee and one year's subscription.

24. An applicant for *transfer* from Student to Licentiate membership must submit the transfer fee with the application. No additional subscription is payable in the year of transfer.

25. An applicant who fails to secure admission or transfer is entitled to a refund of the admission or transfer fee and if applicable, subscription.

Section 4

Associate Class

Qualifications for admission or transfer

26. For admission or transfer to Associate membership a candidate must :

    a. be not less than twenty-one years of age

    b. pass or be exempt from the Associate Examination and

    c. have had practical experience of a satisfactory nature and duration (see regulation 29).

The Associate Examination

27. Details of the Associate Examination are contained in *Examination Admission Regulations and Syllabus* obtainable free on request.

Exempting awards

28. Details of external awards recognised for exemption are given in *Recognised Exempting Awards* obtainable free on request.

Experience of a satisfactory nature and duration

29. An applicant must have had a minimum of two years experience at higher technician level in one or more of the following functions in a firm, consultant practice, department or other organisation engaged in the construction, alteration, maintenance, repair, provision, design, inspection or management of buildings, other construction work or in building education :

Building Control	General management	Quantity surveying
Building asset management	Inspection	Research

Building surveying	Maintenance management	Site engineering
Construction management	Planning	Site management
Cost and production control	* Project management	
Design	Purchasing	Teaching
Estimating	Quality Control	Training

Any other function acceptable to the Institute which requires the exercise of professional judgement relating to some aspect of building or construction.

Note : Experience in an operative, or a first line supervisory position cannot be counted towards the requirement.

*Project management is defined by the Institute as :

'The overall planning and co-ordination of a project from inception to completion aimed at meeting a client's requirements and ensuring completion on time within cost and to required quality standards.'

Experience gained on a sandwich course

30. An applicant who has undergone practical training or planning experience while engaged on a sandwich course may be permitted to count up to a maximum of one year of such training towards the two years required, provided the experience gained is of a satisfactory nature and level.

How to apply

31. Application for admission or transfer to Associate membership must be made on Form MA obtainable from the Institute, and be accompanied by the relevant fee and if applicable, subscription.

32. Applicants claiming exemption from the Associate Examination on the basis of an exempting award must submit to the Institute a photographic copy of the examination certificate (or the original thereof) upon which the application is based.

Notes :

    a. Original certificates are submitted at the applicant's own risk, and should be given proper protection in the post.

    b. The provision of full and precise details of the nature, duration and level of duties and responsibilities is of great importance if the application is to succeed.

    c. If a candidate is in doubt whether he/she complies with either the Institute's education or experience requirements, advices should be sought from the Institute before completing the application form.

Fees and subscriptions

33. The current admission and transfer fees and annual subscriptions are detailed in the loose-leaf insert to these regulations.

Note : The whole of the annual subscription is allowable as a deductible expense for purposes of Income Tax Schedule E assessment under Inland Revenue Technical Division letter, reference CI/SUB/248/NC dated 8 October 1980.

34. An applicant for *admission* to Associate membership must submit with the application the admission fee and one year's subscription.

35.   a. An applicant for *transfer* from Student or Licentiate membership must submit the transfer fee with the application. No additional subscription is payable in the year of transfer.

    b. An applicant who fails to secure admission or transfer is entitled to a refund of the admission or transfer fee and if applicable, subscription.

Associate Cadet Scheme

36. Associates who undertake to progress to corporate membership within a seven year period, if accepted on to the Cadet Scheme, will be entitled to use the abbreviated designation ACIOB. Further details are available on request.

Section 5

Member Class

Qualifications for admission or transfer

37. For admission or transfer to the Member class a candidate must :
    a. be not less than twenty-three years of age
    b. pass or be exempt from the Member Examination Parts I and II
    c. have had practical experience of a satisfactory nature and duration (see regulation 40).
    d. pass the Professional Interview.

The Member Examination Parts I and II

38. Details of the Member Examination Parts I and II are contained in *Examination Admission Regulations and Syllabuses* obtainable free on request.

    #### Exempting Awards

39. Details of external awards recognised for exemption from the Member Examination Parts I and II are given in *Recognised Exempting Awards* obtainable free on request.

Experience of a satisfactory nature and duration

40. An applicant must have had a minimum of three years experience at professional level in one or more of the following functions in a firm, consultant practice, department or other organisation engaged in the construction, alteration, maintenance, repair, provision, design, inspection or management of buildings, other construction works or in building education :

Building Control	General management	Quantity surveying
Building asset management	Inspection	Research
Building surveying	Maintenance management	Site engineering
Construction management	Planning	Site management
Cost and production control	* Project management	Teaching
Design	Purchasing	Training
Estimating	Quality Control	

Any other function acceptable to the Institute which requires the exercise of professional judgement relating to some aspect of building or construction.

Note : Experience in an operative, or a first line supervisory position cannot be counted towards the requirement.

*Project management is defined by the Institute as 'The overall planning, control and co-ordination of a project from inception to completion aimed at meeting a client's requirements and ensuring completion on time within cost and to required quality standards.'

Experience gained on a sandwich course

41. An applicant who has undergone practical training or planned experience while engaged on a sandwich course may be permitted to count up to a maximum of one year of such training towards the three years required, provided that the experience gained is of a satisfactory nature and level.

Post graduate experience

42. A graduate candidate must have had at least one year's experience of the three years required, subsequent to the award of the degree.

The Professional Interview

43. Details of the Professional Interview including entry procedure are given in Section 6 of these regulations on page 11.

How to apply for membership

44. Upon successful completion of the Professional Interview application for admission or transfer to the Member class must be made by submitting Form M5 accompanied by the relevant fee and if

applicable, subscription.

Fees and subscriptions

45. The current admission and transfer fees and annual subscriptions are detailed in the loose-leaf insert to these regulations.

   Note : The whole of the annual subscription is allowable as a deductible expense for the purposes of Income Tax Schedule E assessment under Inland Revenue Technical Division letter, reference CI/SUB/248/NC dated 8 October 1980.

46. An applicant for *admission* to the Member class must submit with the application the admission fee and one year's subscription.

47. An applicant for *transfer* from another class to the Member class must submit the transfer fee with the application. No additional subscription is payable in the year of transfer.

48. An applicant who fails to secure admission or transfer is entitled to a refund of the admission or transfer fee and if applicable, subscription.

Section 6

The Professional Interview

The examination

49. The Professional Interview comprises :
   a. a report upon the candidate's training and experience (see Regulations 56-62)
   b. an interview (see Regulations 63-69)

Entry procedure

50. A candidate may be admitted to the examination at any time after the requirements detailed in regulation 37, a,b, and c have been complied with.

51. The Institute may, at its discretion, modify the practical experience requirements for entry in respect of a member of the Institute who is 35 years of age or over.

52. A candidate's submission must include :
   a. a completed application form X4/A and three further copies of the completed form.
   b. four copies of his/her report upon training and experience.
   c. the examination fee (see the loose-leaf insert to these regulations).

Notes :
   a. A candidate who is not in membership at the date of making application must submit a photographic copy of the award upon which exemption from the Member Examination Parts I and II is claimed.
   b. If a candidate is in doubt whether he/she complies with either the Institute's education or experience requirements, advice should be sought from the Institute before commencing the report.

Expenses

53. All expenditure, including travelling and subsistence, incurred by the candidate shall be borne by the candidate.

Notification of result

54. A candidate will be informed of the examination result by post normally within fourteen days of attending the professional interview and, if successful, he/she will be invited to apply for admission or transfer to the Member class.

55. A candidate who is deferred will be given advice and guidance concerning a subsequent re-examination.

Report on training and experience

56. The report is the basis of the Professional Interview and the candidate should therefore give careful thought to its structure and content, before commencing its preparation. The candidate is also

recommended to note the purpose of the Professional Interview (Regulations 63-69) because it should be possible, in a carefully prepared report, to go some way towards satisfying the members of the interviewing panel that the experience gained meets the Institute's requirements.

57. The report should be objective and should :

    a. be concise and not more than 2000 words in length.

    b. provide clear evidence of the nature of the candidate's experience, the level of personal duties and responsibilities, and personal development and career progression.

    c. contain details, including the names, professional qualifications and appointments of immediate superiors under whom training and experience has been gained.

    d. emphasise the experience gained in the required period of three years.

    e. include, where applicable, examples indicative of the range, type and cost of building projects in which the candidate has been personally involved.

    f. highlight problems encountered in particular projects and solutions adopted.

    g. include a detailed account of at least one aspect of his/her experience in which he/she was centrally involved.

58. The report should not be confined to factual details, or be a catalogue of tasks performed by the candidate as these details appear elsewhere in the completed application form.

59.   a. The first page of the Report must be set out as a title pages as follows :

'The Chartered Institute of Building

Professional Interview

Report on Training and Experience'.

    b. The title page must also include the candidate's full name and the date.

60. The paper must be A4 size and each copy must be bound or stapled on the long side. Typing must be double-spaced and on one side of the paper only. Pages should be numbered centrally at the bottom. Margins should be 4 cms on the left and at least 1 cm on the right.

61. Candidates are to complete, sign and date the front page of the X4/A.

62. The candidate will be informed in writing whether his/her report on training and experience is acceptable and, if acceptable, he/she will be advised of the venue, date and time of the interview.

The Interview

63. The purpose of the interview is to enable the interviewing panel to satisfy themselves that the candidate fulfils the requirements for admission to corporate membership of the Institute.

64. The interview is not intended to be a test of the theoretical knowledge, which will have been previously tested, but the candidate will be expected to :

    a. Show that the duties and responsibilities being or having been undertaken comply with the Institute's requirements concerning practical experience.

    b. Demonstrate a knowledge and understanding of the building process generally.

    c. Indicate an ability to apply theoretical knowledge to problems in the candidate's own field of building experience.

    d. Display the capacity to accept professional responsibility and the leadership/managerial qualities expected of a corporate member of the Institute.

    e. Have the ability to communicate.

65. The interview may be held at the headquarters of the Institute or at other centres in the United Kingdom or overseas at the Institute's discretion.

66. The candidate will normally receive not less than 21 days notice of the date, time and place of the interview and must inform the Institute of his/her intention to attend not less than 10 days before the date of the interview. If the candidate is unable to attend an alternative date will be offered.

67. The interviewing panel will normally consist of three corporate members.

68. Each interview will last approximately 35 minutes and the candidate is required to be available not less than 10 minutes before the start of the interview.

69. The submission of examples of the candidate's work is not a requirement of the interview. A candidate may, however, bring examples of work, or other relevant items, to the interview if he/she believes that they will assist the interviewing panel to determine the quality of his/her experience and level of duties and responsibilities. It is essential that such items are examples of the candidate's own work, that they are relevant and that they are capable of inspection within the limited time available.

Section 7

Admission of Mature Candidates

General

70. There are two methods of entry to membership for mature candidates : the Direct Membership and the Direct Associateship Examinations.

Direct Membership and Associateship Examinations

71. These examinations are provided as the means whereby entry to the Institute may be achieved by those who do not possess the necessary academic awards but who have acquired an acceptable breadth and level of knowledge as a result of long experience in building practice in technical or managerial posts.

Applicants must be 33 years of age or over on the date of making application and have had at least 7 years experience of a satisfactory nature in an approved firm or organisation for the Direct Associateship Examination, or 35 years of age and have had ten years professional level experience for the Direct Membership Examination.

The Direct Associateship Examination consists of a report upon the candidate's training and experience and on oral examination.

The Direct Membership Examination consists of Part A, a report upon the candidate's training and experience and a Professional Interview. The successful completion of Part A enables the candidate to proceed to Part B, a written examination based upon the syllabuses for the Member Examination Parts I and II. Candidates over 38 years of age can offer a thesis as an alternative to sitting the examination for Part B.

72. Full details are contained in Regulations for the Direct Associate Examination or The Direct Membership Examination as appropriate which can be obtained free on request.

Section 8

Fellow Class

Qualifications for transfer

73. For transfer from the Member class to Fellowship a candidate must :

    a. be not less than thirty-five years of age

and

    b. have achieved and maintained a very responsible position in his chosen field of building for not less than five years

and

    c. i. have demonstrated high qualities of leadership or of managerial or technical ability in his chosen field of building

or

       ii. have made a significant contribution to building outside his firm or organisation (e.g., in the Institute or a trade association or in the field of industrial relations, education, training or research) or

      iii. have made a significant contribution to building knowledge (by papers, books, lectures or contributions to the work of committees, study groups, etc.).

Note : A candidate will be expected to show that in the period since election as a Member he has undertaken continuing professional development as recommended by the Institute.

How to apply

74. Application for transfer to Fellowship must be made on Form M6 obtainable from the Institute.

Fees and subscriptions

75. The current admission and transfer fees and annual subscriptions are detailed in the loose-leaf insert to these regulations.

Note : The whole of the annual subscription is allowable as a deductible expense for purposes of Income Tax Schedule E assessment under Inland Revenue Technical Division letter, reference CI/SUB/248/NC dated 8 October 1980.

76. An applicant for transfer from the Member class to Fellowship must submit the transfer fee with the application. No additional subscription is payable in the year of transfer.

77. An applicant who fails to secure transfer is entitled to a refund of the fee.

Building Practice Operations

In this book I have included a few small and medium size to considerable size home profiles/designs from general viewpoint maintaining the significant role of the structural formation mechanism. The designs are based on the provisions of building bye-laws. I desire that the student or the associate in building engineering should do the work according to the professional sanctions using the intuitive skills. So when an owner or a client inquires you about the scheme/plan for his/her use (naxha) and invites you to get a design for his/her requirements for possessed plot or for buying the same, you shall interrogate the client about the plot of land records namely site layout approved or unapproved, sales deed, General power of attorney, transfer of ownership, zoning, validity of the allotment for the completion, permissible coverage, permissible height, vistas, set backs, floor area ratio, provision of cellar/basement, etc., you shall then ask the needs of the client and fix the contract agreement.

Very Important Discussion

For general information of the reader I shall state that according to the international standards of building statutes/bye-laws both civil engineer and the architect are authorised to plan the schemes. Architects are authorised to do work on the general building schemes while the civil engineers are authorised to work on all civil engineering projects involving all kinds of structures.

In National Capital Territory of Delhi in India at the manuscript writing time, the competencies and the qualifications of building engineering personnel are ridiculous as given hereinafter (Ref. 6.6 Appendix 'P' unified building bye laws)

ARCHITECT

Qualifications — The qualification for licensing of Architect will be the Associate Membership of the Indian Institute of Architects or such degree or Diploma which makes him eligible for such membership or such qualification listed in schedule XIV of Architects Act 1972 and shall be registered under the council of Architecture as per Architects Act 1972.

Competence — The licensed architect is competent to carry out work related to Building permit as given below and will be entitled to submit —

      i. All plans and related information connected with building permit

      ii. Structural details and calculations for buildings on plots upto 500 sqm and upto 4 storeys and

      iii. Certificate of supervision of all buildings

      iv. All layout plans.

ENGINEER

Qualifications — The qualification for licensing of Engineer will be the corporate membership (Civil) of the Institution of Engineers (India) or such degree or diploma in Civil or Municipal or Structural Engineering which makes him eligible for such membership, or which is recognised by union public service commission for the post of an Assistant Engineer.

Competence — The licensed engineer is competent to carry out the work related to Building permit as given below and will be entitled to submit —

i. All plans and related information connected with building permit on plots up to 500 sqm and upto four storeys.

ii. Structural details and calculations for all buildings and

iii. Certificate of supervision of all buildings

iv. All layout plans

v. Sanitary/water supply works for all type of buildings.

SUPERVISOR

Qualifications — The qualifications for licensing of Supervisor will be —

i. Three years architectural assistantship or intermediate in Architecture with two years experience

ii. Diploma in Civil Engineering with two years experience or

iii. Draftsman in Civil Engineering from ITI with five years experience under architect/engineer

from a recognised institution which would enable him for the post of Supervisor recognised by Union Public Service Commission.

Competence — The Supervisor will be entitled ~~will be entitled~~ to submit —

All plans and related information connected with Building permit on plots upto 200 sqm and upto two storeys.

GROUP OR AGENCY

When an agency or group comprising of qualified architect and engineer is practising then the qualification and competence of work will be combination of the individual qualifications and competence, given under the relevant bye laws and the agency shall be licensed by the Authority.

LICENSING

Technical Personnel to be licensed — The qualified technical personnel or group shall be licensed with the Authority and the license shall be valid for one calendar year ending 31 December after which it shall be renewed annually.

Comments Regarding Qualifications

ARCHITECT

Section 2 (a) Chapter 1 of Architect Act defines 'architect' means a person whose name is for the time being entered in the register of Architects maintained under section 23 of the Architects Act 1972 (20 of 1972)

Section 35 of Architects Act 1972 (20 of 1972)

Effect of registration (I) Any reference in any law for the time being in force to an architect shall be deemed to be a reference to an architect registered under Architects Act 1972.

Section 36 of Architects Act 1972.

Penalty for falsely claiming to be registered — If any person whose name is not for the time being entered in the register falsely represents that is so registered or uses in connection with his name or title any words or letters reasonably calculated to suggest that his name is so entered, he shall be punishable with fine which may extend to one thousand rupees.

ENGINEER

Corporate membership (Civil) of the Institution of Engineers (India) is acceptable as this is statutory and constitutional authorisation to use the style and title of Chartered Engineer (India) [Ref. to bye law 68 (i) Rights and Privileges]

Institution of Engineers Bye-laws and the Regulation of the Council

According to the Institution of Engineers (India) statute there is no degree or diploma granted by any of engineering educational institutions in India which shall make the holder of such degree or diploma eligible for the corporate membership of the institution of Engineers (India) but those educational qualifications shall help the holders to exemption from section A and B only. Involvement of the union public service commission for the designation of Indian Chartered Engineer is irrelevant and outside the UPSC scope/Objective and further more ridiculously Municipal corporation of Delhi has equated the designation of statutory engineer to the designation of an assistant engineer (bad provision).

SUPERVISOR

This is the designation of the senior technician of the IEI

GROUP/AGENCY

Acceptable

Competencies

ARCHITECT

'All plans' should be replaced by all architectural plans and details. Structural designs upto 4 storeys on 500 sqm plot should be deleted as a fact of non inclusion of the compulsory structural frames theory and design practice along with the general structural design adequacy and stability criteria with the Bachelor programme in architecture. In modern and post modern era no building can be designed free of complex structurally indeterminate frames of high degree of redundancy from parapet to the foundation formation which only an experienced Civil/Structural engineer can do.

Complete supervision authorised to architect also which is in contravention to the existing international conventions and practice wherein all stresses bearing complex and indeterminate structures are supervised only by the competent civil/structural engineers as the proper execution of the building structures and the structural stability. The plans of the multistoreyed and tall buildings submission authorised to the architects which is not practically feasible due to the involvement of complex structural preliminary stage design conception and the decision and visualisation before the commencement of the planning involving the thorough understanding the estimates of prediction of forces on the three dimensional structures shall meet or bear during the construction and the service life of the building namely dead load, imposed load, wind load, snow load, explosion, impact, sonic boom, seismic or earthquake forces, lightning, vibrations, fire effects, fire attack, structural fire protection and resistance, structural deflections, relaxation of overall structures and elements and elastic recovery along with the material hysteresis considerations, etc. The anticipated redistribution of stresses due to continuous and prolonged strains leading to alteration of forces in various elements of structures due to the generation of creep, the shear walls behaviours and the cores of the buildings, etc., which shall continue to impart their unfavourable effects even after a span of 50 years of construction and service, can be understood within certain tolerances by a competent civil engineer or structural engineer and his knowledge is required to be included and applied in the preliminary design considerations which shall almost decide the positioning and sizes of various frame elements and ultimately lead to the severity of effecting the alterations of the unrealistic hypothetical planning by the architect alone and if this structural concept is not given due weight, the centre of gravity/the centre of rotation of the building may be out and reduce the three dimensional structural life of the structure and due to additional torsional stresses, the overall structural behaviour shall be badly effected , therefore, it is very important from public safety viewpoint. The building permit authority must demand a structural design adequacy certificate while accepting multistoreyed/tall buildings and other schemes/Plans for sanction, duly attested or signed by Chartered Engineer India Civil otherwise the agency or group should submit the scheme.

Structural supervision should be withdrawn from the authority of the architect.

ENGINEER

The space planning competency of Chartered Engineer Civil is confined to 500 sqm plot and upto 4 storeys only which is absolutely baseless and arbitrary and not in conformity to the international standards set by the Institution of Civil Engineers London and the Institution of the Structural Engineers London which grant full competencies to the Chartered civil or structural engineers to plan any civil engineering project.

The instant sanction of 500 sqm plot in the approved residential zones should also be applicable to the Chartered Engineer (India) Civil due to the holding superior title to architect and by virtue of being superior, this imparts a professional right with the Chartered Engineer (India) Civil automatically.

Prior to including the contents of the proposals I filed with the Municipal Corporation of Delhi as objections and suggestions to

the draft bye law 1992 in compliance to the Delhi high court order by (Justice B.N. Kirpal) to replace the unified building bye laws 1983 by new bye laws for the National Capital Territory of Delhi to make these laws comprehensive and implementable, I like to quote the definition of bye law/statute/delegated legislature.

DELEGATED LEGISLATION

Parliament is out supreme law making body but it is not the only body capable of making laws. However, laws can be made by a body other than parliament if the parliament itself has given it that power. Parliament delegates its law making power to these inferior bodies and the laws which they make are known as delegated legislation which are three types as follows —

(i) Orders in the council

(ii) Ministerial Regulation

(iii) Bye-laws.

(i) Orders in the council — These are items of legislation which are passed by the body of senior government ministers which have delegated powers by the parliament and need no Royal assent (In India presidential approval)

(ii) Ministerial Regulation — Government ministers are given powers in parliament acts to make laws which supplement the provision of the Act (Parent Acts)

(iii) Bye laws — A bye law is a law made by a local authority or the institution or some other public body or individual. Parliament gives such bodies the power to make laws to regulate those matters for which the bodies are responsible.

This power is given through the Royal Charter. The delegated legislation will only be valid if parliament has given authority for the type of law made. The laws that the body makes if exceeds its delegated authority or are repugnant to the provisions in the charter, that bye law can be challenged in the court and the court will declare that void.

So the question of licensing the Chartered Engineers (India) or the registered architects India does not arise whether to be done or not to be done by any authorities in India other than the relevant authorities constituted by law of India i.e. The Institution of Engineers (India) and the council of architecture for the Chartered Engineers (India) and the Architects India respectively as no other body has been given any authorisation for this purpose by the parent acts of those bodies and as such those statutory bodies have been abusing this power unconstitutional in nature and arbitrary and therefore, ultra vires the authorities given to the local bodies to make laws for the purpose of urban developments and space setting designs in the form of building bye laws and hence the provisions made for the registrations of the building engineering professionals should be treated as void, due to their being no provision in the parent acts of these bodies. Similarly, fixing the competencies of the building engineering professionals/personnel is the job of relevant institution at the international level and not that of these small statutory bodies meant to operate the building legislature in co-operation with the building engineering professionals/personnel only i.e. Chartered Engineers (India) FIE/MIE/AMIE and the Architects India and Group/Agency.

The Institution of Engineers (INDIA) bye law 68 (ii) under Rights and Privileges of members provides that the corporate members i.e. 'A Fellow or Member or Associate Member practising in partnership with any person who is not a Fellow or Member or Associate Member under the title of a firm shall not use or permit to be used after the title of such firm the designation Chartered Engineer or Chartered Engineers, or describe or permit the description of such firm in any way as Chartered Engineers. and bye law 68 (iii) A Fellow or Member or Associate Member practising or acting in a professional capacity under the title of, or as a director, officer or employee of a company, whether such company shall be authorised or not to carry on the profession or business of an engineer in all or any of its branches shall not use or permit to be used after the title of such company the designation of Chartered Engineer or Chartered Engineers, or describe or permit the description of such company in any way as Chartered Engineers.'

Indian Penal Code 1860 (I.P.C. 45 of 1860) Section 19, the definition of Judge includes other than the judicial officers and the heads of statutory bodies. Any one of the body of persons authorised by law to legislate or to decide on civil or on criminal matters and this part of the definition of a Judge applies to the statutory designation or the class of corporate members of The Institution of Engineers (India) FIE/MIE/AMIE and the section 21 sixth and third of I.P.C. incorporates them as public servants.

So the bodies like Municipal Corporation of Delhi or (HUDA/Haryana Urban Development Authority, etc., never applied mind on this issue/subject and issued the licences by illegal registrations of Indian state authorities or officials. These bodies are not even allowed to use the designation of Chartered Engineer (India) after the title of the body while surprisingly these bodies are issuing licences as engineers. Why do not these local bodies issue licences to the Chartered Accountant, advocates, medical practitioners, etc., if competent to do for Engineer. I mean to explain here is only to communicate to my colleagues in the profession not to submit to these illegal licences but to apply the institutional authorities and the Indian Constitutional designation of Chartered Engineer (India) and find status in the society as a professional judge in engineering. For the junior colleagues I extend the meaning of the designation CHARTERED according to law 'Raaj Adhikaar Patra Dooara Diya Gaya' In case of the denial by the official of the statutory body to honour the designation of the profession, the relevant concerned official of the body may be prosecuted under the provision of the IPC' refusing to accept the statutory document' and denying the acceptance of any provisions of the statute/law. According to IPC 1860 provisions the word 'Election; whenever used means 'election to a legislature' and the corporate members of the Institutions of Engineers (India) are statutorily elected under by law 55 of IEI statute 'Election and Transfer' and a Diploma is granted to the elected corporate member under regulation 45 Bye law 26 which remain the property of the IEI to be returned on demand under regulation 46 bye law 26. The Institution of Engineers (India) Annual Report 1992-93 regarding membership — corporate member is automatically designated as chartered engineer as per provisions of the Royal charter.

Ref. I.E.I. DIRECTORY 1992 page 3

ROYAL CHARTER

AT THE COURT AT BUCKINGHAM PALACE

The 13th day of August, 1935

PRESENT

THE KING'S MOST EXCELLENT MAJESTY IN COUNCIL

(Royal Charter composed of the 22 Presents incorporating the Institution of Engineers (India) was granted by King George V. The essence of the Royal Charter is given hereunder).

The objects and purposes for which the Institution of Engineers (India) (hereinafter called "the Institution") is hereby constituted are to promote the general advancement of engineering and engineering science and their application in India and to facilitate the exchange of information and ideas on those subjects amongst the Members of and persons attached to the Institution and otherwise, and for that purpose.

(a) To promote and advance the science, practice and business of Engineering in all its branches (hereinafter referred to as "Engineering ") in India.

(b) To establish, subsidise, promote, form and maintain local Associations of members belonging to the Institution and others engaged or interested in Engineering so as to assure to each individual member as far as may be possible equal opportunity to enjoy the rights and privileges of the Institution.

(c) To diffuse among its members information on all matters affecting Engineering and to encourage, assist and extend knowledge and information connected therewith by establishment and promotion of lectures, discussions or correspondence; by the holding of conferences; by the publication of papers, periodicals or journals, books, circulars and maps or other literary undertaking; by encouraging research work; or by the formation of a library or libraries and collection of models, designs drawings, and other articles of interest in connection with Engineering or otherwise howsoever.

(d) To promote the study of Engineering with a view to disseminate the information obtained for facilitating the scientific and economic development of Engineering in India.

(e) To establish, acquire, carry on, control or advise with regard to colleges, schools or other educational establishments where students and apprentices may obtain a sound education and training in Engineering on such terms as may be settled by the Institution.

(f) To encourage, regulate and elevate the technical and general knowledge of person engaged in or about to engage in Engineering or any employment manual or otherwise in connection therewith and with a view thereto provide for the holding of classes and to test by examination or otherwise the competence of such persons and to institute and establish professorships, studentships, scholarships, rewards and other benefactions and to grant certificates of competency whether under any Act of the Government of India or Local Governments regulating the conduct and qualifications of Engineers or otherwise howsoever.

(g) To give the Government of India, the Local Governments and Municipalities and other public bodies and others, facilities for conferring with and ascertaining the views of Engineers as regards matters directly or indirectly affecting Engineering and to confer with the said Governments, Municipalities and other public bodies and others in regard to all matters affecting Engineering.

(h) To encourage inventions and investigate and make known their nature and merits.

(i) To arrange and promote the adoption of equitable forms of contracts and other documents used in Engineering and to encourage the settlement of disputes by arbitration and to act as or nominate arbitrators and umpires on such terms and in such cases as may seem expedient.

(j) To promote efficiency and just and honourable dealing and to suppress malpractice in Engineering.

(k) To do all such other acts and things as are incidental or conducive to the attainment of the above objects or any of them.

The government and control of the Institution and its affairs shall be vested in the Council subject to the provisions of these Presents and to the Bye-laws for the time being of the Institution. The business of the Council shall be conducted in such manner as the Council may from time to time prescribe.

All the powers of the Institution shall be vested in and exercisable by the Council except so far as the same are by these Presents or by the Bye-laws for the time being of the Institution expressly required to be exercised by the Institution in general meeting.

A Member of the Institution shall be entitled to the exclusive use after his name of the abbreviated designation "M.I.E. (Ind.)"; an Associate Member of the abbreviated designation "A.M.I.E. (Ind.)"; an Honorary Life Member of the abbreviated designation "Hon. Life M.I.E. (Ind.)"; and an Honorary Member of the abbreviated designation "Hon. M.I.E. (Ind.)".

Every person being at any time a Corporate Member of the Institution may so long as he shall be a Corporate Member take or use the name or title of "Chartered Engineer (India)"

Note — F.I.E. was created in 1971.

IEI Bye-laws 66 RIGHTS AND PRIVILEGES

Subject to the provisions of the charter and to the restrictions contained in byelaws.

(a) Fellows shall enjoy all the rights and privileges of the Institution, but in accordance with the provisions in the charter, Bye-laws and the Regulations.

(b) Member and Associate Members shall enjoy the rights and privileges of the Institution, but in accordance with the provisions in the charter, Bye-laws and the Regulations and save that they shall not be eligible to hold office as President, Chairman of a State Centre or Chairman of a Local Centre in accordance with the provision in Regulation 29 (f).

Note : The essence of the Royal Charter of incorporation was traced by me on 1.1.1994 at the Eight Indian Engineering Sales counter in Vigyan Bhawan so that could not be included in preceding discussions due to being a supplement manuscript part.

Comments :

Article (f) has enough statutory force to register the professional engineers in India and no other authority exists for this job.

Article (g) has enough statutory force to make all the engineering agencies in India to employ the chartered Engineers (India) in all responsible and very responsible offices.

Article (i) has enough statutory force to be enjoyed by Chartered Engineers (India) to draw documents of contracts under the purview of Indian Contract Act 1872 and I.P.C. document definition for the purpose of engineering and therefore leaves no other authority in India to decide on the competencies of professional engineers and all the municipalities/other bodies whoever it may be that register the engineers or decide the competencies is unconstitutional. The article authorises the corporate members to act as arbitrators according Arbitration act 1940 and covers the corporate member under the sections 19 and 21 (3rd and 6th) of the Indian penal codes i.e. Judges and Public servants.

Article (j) This authorises the corporate member to use the authority to remove corruption on his job or otherwise I mean to say that a corporate member could use the power to arrest if the evidences are available according the provisions of I.P.C. and Indian evidence act 1872.

Article (k) Confers unlimited authority within the provisions of law on the Institution and therefore on the corporate members to attain the engineering objectives in India for which this body has been constituted by law.

Note : Relevance to constitution of India came into force on 26.1.1950 subject entry 65 a in the union list which provides for the professional Institution partly or wholly financed by the government of India and established by an act of parliament or statute before the constitution came in force.

The section 57 chapter III ON PROOF, facts which need not be proved and the facts of which court must take judicial notice 1, 2 and 6 of Indian Evidence Act, 1872 as on 1991 apply to this Royal Charter being an act of Parliament/UK and Indian law. All laws in India, all public acts passed by UK parliament and all seals which any person is authorised to use by the constitution or UK parliament act or an Indian law must be taken judicial note of. And under section 74 of evidence act the roll of the Institution of Engineers (India) and the diploma of corporate membership, etc., are public documents.

Now I inform to you some of the suggestions regarding professional personnel filed with the Municipal Corporation of Delhi to incorporate with the unified building bye laws of the national capital territory of Delhi in India.

ARCHITECTURAL ENGINEER

Qualifications — Registered with the council of architecture of India, Chartered Engineer (India) Architectural Engineering division

Competencies — Shall carry out the following gross operations

1. To submit all the plans of architectural disciplines and the architectural details
2. Structural details and calculations for all non structural framed buildings/simple structures of general buildings and all load bearing walls
3. Plans of all the cultural structure with the supporting signatures of a Chartered Engineer (India) Civil Engineering division in case the plan involves complicated structural analysis.
4. Certificate of supervision for all buildings other than the buildings involving the complicated structural frames of reinforced concrete, structural steel or the pre-stressed concrete or other high technology structures.
5. All layout plans
6. Sanitary works/Water supply works for all buildings.

CIVIL ENGINEER

Qualifications — Chartered Engineer (India) Civil Engineering division
Competencies :

1. To submit all plans of structural framed buildings, design calculations of all buildings and all related engineering details other than the electrical air conditioning, mechanical, telecommunication, etc., disciplines
2. Certificate of supervision for all types of buildings and infrastructural net works.
3. All layout plans
4. Drainage, sanitary, water supply works for all types of buildings and infrastructure
5. Structural stability responsibilities for all types of buildings/structures

SERVICES ENGINEERS

Qualifications — Chartered Engineer (India) Electrical/Mechanical/Electronics division
Competencies — To carry out all the works of the relevant disciplines in co-operations/cosigning with the architectural engineer and civil engineer in respect to the schemes.

GROUP/AGENCY/ASSOCIATION/SYNDICATE/COLLABORATION/PROJECT ENGINEER/ENGINEER in the international terms.

Qualifications — Partners on schemes — Architectural Engineer, Civil Engineer, Services Engineers and other associates depending on the requirements but legally these partners shall be liable.

Competencies — All operations on the schemes for the complete project/job. Architectural Engineer, Civil Engineer and the Service Engineers shall be authorised by the relevant statutory professional body/bodies or council/s and the number of enrollment in respect of the professional engineers of the professional bodies gazette/directory or the Charter shall be mentioned with the signature of the professional on the schematic documents and in case of the gross misconduct or misrepresentation or the non professional performance by the professional engineer, the matter shall be referred to the relevant statutory professional body or the council or the institution as the case be with the technocrat/s in question/subject but before that an ordinary explanation/s may/shall be called by the plan approving body to avoid any overlooking/condoning/sighting, etc., and allow a period of at least two months for snags/lapse remedials if possible.

ENGINEERS' ASSOCIATES/SUPERVISORS

Qualifications — Associates of the Institution of Engineers (India) A.I.E. Civil/Architectural engineering division/Electrical/Mechanical/Electronics divisions.

Competencies — All plans on plots of maximum size 500 sqm under the direction/guidance of either of Architectural engineer/Civil Engineer/Services Engineer

JUNIOR SUPERVISOR

Qualifications — Senior Technician of the Institution of Engineers (India) Sr. Tech. IE of the relevant discipline

Competences — Upto the area of 200 sqm of plots all buildings under the directions of Engineer's Associates/Supervisor or Architectural and Civil Engineers.

The object to propose the foregoing competencies for the Architect and Engineers is that the existing bye laws are contradictory in this relation. According to existing competencies engineer can make all layouts but barred to make plans for plots more than 500 sqm and upto 4 storeys which is contradictory. While architect can make all plans means architect can make structural plans also. In another competency architect can design only upto 500 sqm and 4 storeys, these both competencies contradict. Architect can make all services plans and civil Engineer can work only on plumbing and sanitary works. Architect and engineer both have full competencies on supervision.

Now the question arises that the engineer shall only independently do the execution of plumbing and sanitary works and cannot win a client for a project due to being barred from submission of schemes of bigger sizes than 500 sqm plots and upto 4 storeys. It is very much clear that the person/legislator who decided these competencies firstly exceeded the powers given to him/her/judge by the parliament act and secondly did not at all apply mind and kept the public off obtain the real professional services. I also opine that the relevant professional statutory bodies too did not care about this matter which should have been done long back, the consequence of that is that the structures built are not safe generally and mostly the engineers are not aware about the competencies given to them by the law of India and the status they should enjoy. It is really surprising that an architect who does not excellently understand structure/s can supervise any building structure according to the existing law. I really condemn this as it is against the public interest and public/user is made at risk of life and money as happened in various failures during occurrence of earth quakes. I expect the engineer colleagues to condemn this provision against the general safety of the public. In practice in the advanced countries and the middle east, no architect is allowed to supervise the structural frames but only coordinations during constructions. I shall be happy should the Group/Agency practice be adopted but the building permit granting local bodies do not accept that, however, the provision exists in the law. I also suggested to abolish the need of C and D certificates/Forms meant for inspections of underground drainage sanitary and other pipes' lincs before covering the lines and for the final inspections of sanitary and water supply respectively as there shall be a separate provision of the final inspection and testing of all the service at the time of the practical completion in the presence of the representatives of the local bodies, contractor's engineers and the consultants/Engineer.

I also suggested a form to replace the present certificate of supervision and building fitness signed by the supervising professional/for the occupation and use to be submitted to the local bodies by the owner/user at the time of applying to obtain the occupation certificate or building commissioning/authorisation which reads as follows : (Existing) "I hereby certify that the erection/re-erection or material alteration in/at building no. ... on/in plot no. ... Block no. ... situated at scheme ... has been supervised by me and has been completed on ... according to the plans sanctioned, vide office communication no. ... dt. ... The work has been completed to my best satisfaction, the workmanship and all the materials (type and grade) have been used strictly in accordance with general and detailed specifications. No provisions of the building bye laws, no requisition made, conditions prescribed or orders issued thereunder have been transgressed in the course of the work. The building is fit for use for which it has been erected/re-erected or altered, constructed and enlarged.

Signature of Architect/Engineer/Group/Supervisor

Name ... ...

License No. ... ...

Address ... ...

Suggested Substitution

1. Scheme-Erection/Re-erection/Alteration/Addition
2. Situation Address
3. Plot No.
4. Block No.
5. Permit No.
6. Supervising technocrat
7. Description of the professional technological codes involved
8. Project scheme record/s reference
9. Certificate as follows

Certified that the above scheme has been successfully supervised during all the way of construction executions from the setting out of the layouts to the last end of the day of the practical completion/s of all the work elements as covered per sanctioned plans and the details to the above reference permit.

The works have been completed to the least of the minimum technological professional standards which must be expected of the sanctioned plans and the details of the scheme designs and complete technological directions applied on the construction execution, saving all the infringements of the bye laws and no aberrations allowed and incorporated with the execution practically without the knowledge and approval by the building permit sanctioning body.

The complete professional technological controls and checks of all the structural elements have been observed, satisfactorily tested and recorded from the soil investing/investigation and soil stability, reaffirmation of the designs of all structures, from laying of the foundations to the top of the parapet or any other structural element, etc., and no room has been left for the probability of controllable structural design's inadequacy based on the informations of various parameters available and all the structural informations accurately and efficiently communicated, standard checks carried out during the supervisions of the execution controls of all the structural elements and all the required tests on site.

The constructed building/s store/s all the resistances in it for sustaining all the worst conditions working loads for which the use of the building/s was approved. In legal term, the constructed building/s is commissioned fit for its meant use.

Signatures

Chartered Engineer (India) / Group partners

I also suggested to include about the adoption of management contract and the execution contracts possible formats with the proposed bye laws for the benefit of the general public and the engineering professionals as the Indian Contract Act 1872 does not provide any standard form for that purpose but the provisions of Indian Contract Act 1872 should be followed while drawing the contract provisions to keep the contract valid and legally enforceable. In case any of the provisions in the agreement goes repugnant to the provisions of the Indian Contract Act 1872 or any other law for the time being in force, that provision in the contract shall render the part of the contract void/or in certain cases may, cause/considerable damage to either of the contracting parties. The details of these suggested forms have been incorporated with the guide to site management. Furthermore, the absolute responsibility of engineering services in the management contract goes to the Engineer appointed and constituted by the contract under the Indian Contract Act and that shall not be shared by the building permit sanctioning local body or the municipality as the local body always stands indemnified by the owner of the building of all the expenses or consequences to appear in the court of law or before any other authority which it might incur due to the results of the building plans/scheme sanctions and the consequences but the Engineer shall be liable in all respects and likely to be prosecuted in case of failure of the structure/s or any other adverse situation or inclement design conditions and all the properties of the Engineer will be attached to compensate the damages caused to the user by the bad design of the project.

Copyright — For the purpose of information to my colleagues engineers and architects, I explain the provisions in relation to architectural engineering designs and documents incorporated with the Copyright Act 1957 with rules 1958 Section 2 (b) 'Architectural work of art' means any building or structure having an artistic character or design or any model for such building or structure

(c) "Artistic work" means

(i) A painting , a sculpture, a drawing (including a diagram, map, chart, or plan) an engraving or a photograph, whether or not any such work possesses artistic quality.

Section 3 — Meaning of publication — For the purpose of the copyright act "publication" means

(a) In the case of literary, dramatic, musical or artistic work, the issue of copies of the work either in whole or in part to the public in a manner sufficient to satisfy the requirement of the public having regard to the nature of the work but does not include in the case of a work of sculpture or an architectural work of art, the issue of photographs and engraving of such work.

Section 13 (5) of works in which Copyright subsists —

In the case of architectural work of art, copyright shall subsist only in the artistic character and design and shall not extend to the processes or methods of construction.

Section 15 — Special provision regarding copyright in designs registered or capable of being registered under the Designs Act 1911.

(1) Copyright shall not not subsist under the copyright act in any design which is registered under the designs act 1911 (2 of 1911)

(2) Copyright in any design, which is capable of being registered under the Design Act 1911 (2 of 1911) but which has not been so registered, shall cease as soon as any article to which the design has been applied has been reproduced more than fifty times by an industrial process by the owner of the copyright, or with his license by any other person.

For the purpose of the ownership of the copyright and the rights of the owner section 17 chapter IV dd — In case of a work made or first published by or under the direction or control of any public undertaking, such public undertaking shall in the absence of any agreement to the contrary, be the first owner of the copyright therein and for the purpose of this clause and section 28 A which reads as Term of copyright in works of public undertakings — In the case of a work, where a public undertaking is first owner of the copyright therein, copyright shall subsist until sixty years from the beginning of the calendar year next following the year in which the work is first published. Public undertaking means

(i) An undertaking owned or controlled by government or

(ii) A government company as defined in section 617 of Companies act 1956 (I of 1956)

(iii) or A body corporate established by or under any central, provincial or state Act.

Section 59 — Restriction on remedies in case of works of architecture —

(1) Notwithstanding anything contained in the Specific Relief Act 1963 (47 of 1963), where the construction of a building or other structure which infringes or which if completed would infringe the copyright in some other work has been commenced, the owner of the copyright

shall not be entitled to obtain an injunction to restrain the construction of such building or structure or to order its demolition

(2) Nothing in section 58 shall apply in respect of the construction of a building or other structure which infringes or which if completed would infringe the copyright in some other work.

Section 58 — Rights of owner against persons possessing or dealing with infringing copies —

All infringing copies of any work in which copyright subsists and all plates used or intended to be used for the production of such infringing copies, shall be deemed to be the property of the owner of the copyright, who accordingly may take proceedings for the recovery of possessions thereof or in respect of the conversion thereof;

Provided that the owner of the copyright shall not be entitled to any remedy in respect of the conversion of any infringing copies if the opponent proves —

a. That he was not aware and had no reasonable ground to believe that copyright subsisted in the work of which such copies are alleged to be infringing copies, or

b. That he had reasonable grounds for believing that such copies or plates do not involve infringement of the copyright in any work.

Comments — Though the architectural design on building or structures are protected in the copyright act 1957/58 but practically there is no remedy in law in case another building of the same design as of original design by the copyright owner is constructed without obtaining a license by the constructor/owner of the copied or reproduced structure or building from the copyright owner i.e., engineer or architect. The copyright owner could sue the engineer or the architect of the copied or reproduced building, had the law made the engagement of an engineering professional mandatory. I suggest to the engineering professionals of buildings to organise to save the copyrights on infringement as there is no remedy for the professional on that in the copyright act for because according to Indian Law there is no mandatory provision to employ the engineer on every building or structure. If the law is made mandatory to engage designer and supervisor for every building or structure, the owner shall not be at liberty and the copyright owner, the designer (engineer or architect) shall claim the royalty from the owner's copying or reproducing engineer or the architect or the copyright owner shall be requested by an application of the copying professional or the owner or both for obtaining a license under the provisions of copyright act section 30 chapter VI for one reproduction or multiple reproductions as the case be and the royalty paid accordingly as agreed between the copyright owner and the copying person.

For making the professional's engagement mandatory I have written to the Indian constitutional head of state and the Executive head of the state under copy to the Indian body of official Engineers The Institution of Engineers (India) the extract of that I reproduce below for the information of the reader.

SKY ASSOCIATES

Building Engineers Agency/Group under the provisions of Building legislature

Legal authorities — Suraj Singh Chartered Engineer (India) M.I.E.

K.K. Jaitly Architect

Office : 3373 Delhi Gate, New Delhi 110002

Date 1.10.1993

MEMORANDUM

Your excellency,

The president of India and the honorary Fellow The Institution of Engineers (India)

The 30th September 1993 seismic effects shocked the entire world and elevated thousands of structures to ruins with cost of thousands of lives. Every one terms it a natural disaster but notwithstanding the occurrence of more than six Richter scale seismic force/Earth quake the collapse of the structures cannot be exclusively incorporated with the natural disaster or calamity. In India people do not bother about the value and the power of building engineering sciences. The owners rely on the masons (Raj mistries) or the unqualified professionals or the architects who in fact are not competent to produce worst conditions forces resistant structures.

There are laws in the building legislatures for engaging civil engineers to incept, design, supervise and certify the structural stability around the world and also in India but someone hardly bothers in India to concern about this fact and the owners commit this invisible crime. I hold the state also responsible for such happenings.

According the Indian Constitution pre independence provisions The Institution of Engineers (India) was incorporated by the Royal Charter on 13th August 1935 with the objective responsibility of promoting the general advancement of engineering and engineering sciences and their applications in India along with the dissemination of free exchange of engineering knowledge.

The corporate members of this Institution of the statutory class of grade FIE/MIE/AMIE are by Indian law designated/Entitled as Chartered Engineer (India) of the relevant discipline.

I shall suggest to your excellency that every building in the country must be designed and supervised by and under the legal authority of Chartered Engineer/S (India) Civil come what may and howsoever the small size of the building may be, for which the designated Chartered Engineer India shall issue the certificate and declare the structure stable and fit for occupation and also in case of the failure of the structure, the engineer shall be liable and punishable by law. No building should be licensed to be erected otherwise. All buildings being constructed or constructed without the statutory supervision and authenticity should be tested by the Chartered Engineers of India as required by laws now and if discovered not to the standards, should be either restored or demolished from the public safety view point. the owners not complying with the above should be prosecuted. The personnel impersonating as qualified engineers without being entitled must be barred. An ordinance may be promulgated to this effect.

Suraj Singh

Chartered Engineer (India)

Seal

Scope of the Consumer Protection Act 1986 with ordinance of 1993. According to this amendment ordinance the architectural building engineering services too have been included with the act's objectives to provide remedies to the building/services consumers/owners who may sue the building engineering professional architect or the civil or the structural engineer whoever may be contractually responsible for the design or the supervision of the building in case of the defective professional services rendered by the professional to the owner for the production of making of the building but within a period of limitation of filing the complaint with the consumer protection forum, of one year of the day of the completion of the building. Though there was no bar to any owner of a building who got professional services of engineering to sue the designer or the supervisor before this amendment ordinance came into force, in any civil court of justice by paying court fee and within a period specified in the Limitation act 1963 provisions. In this amendment there is only one favouring point for the building engineering professional that there is limitation of one year if the owner/client/consumer intends to sue, which in fact has been condoned by the law makers either due to lack of knowledge about the building engineering profession or that the government controlled state departments have been producing buildings for the sale to the general public who got/the status of a consumer under the provisions of the consumer protection act and the government shall be liable for all the damages claimed and ordered by the forum under the act. I add that the very important responsibility on any structure or building is the structural stability and it is the only civil or the structural engineer/Chartered Engineer (India) civil engg. divn. who shall be absolutely the owner for that due to being legal designer and the supervisor of the structure if appointed contractually under the Indian Contract Act 1872 for the proposed building. It is the engineer who declares the building fit for use and based on this declaratory certificate the local bodies/municipalities issue the completion/Occupancy/commissioning/Authorisation certificate along with other fulfillment of the statutory formalities by the owner/consumer/user. Local bodies have no statutory responsibility or the liability for the accuracy of the structural controls exercised by the statutory contractual engineer/ Chartered Engineer (Civil or Architectural). By virtue of this authority vested with Chartered Engineer (India), the engineer can cross examine the architects' plannings and amend them to the structural needs and help the owner in safeguarding the public interest. There shall be arguments and personality clashes but there is no alternative as structural decisions shall dominate the show.

For the general information of the colleagues The Contract Labour (Regulation and abolition) Act 1970 and Rules 1971 shall frequently apply on the big projects which shall run into years more than two.

Some of the communications made in professional statutory connections are reproduced hereinafter for the readers' reference.

SKY ASSOCIATES

Building Engineers

Office : 3373, Delhi Gate, New Delhi - 110002.

Your Ref. No. Nil

Our Ref. No. SA/G/HO/92   Date 1.6.92

The Commissioner,

M.C.D., Town Hall, Chandni Chowk, Delhi - 110006

Sub : Building engineering profession and the bye-laws
Clarifications for the qualifications to license the technical personnel to prepare
building schemes to obtain building permit and supervise.

Our previous references - SA/G/HO/92 dt. 1.5.92; 22.4.92; 27.11.91 (enclosed)

Dear Sir,

We regret to advise you that no response has come from your authority regarding the subject matter even after a course of couple of months. Does it resemble the sloppy departmental functioning of the authority or particularly, this matter is levelled on the low profiles ?

As far as we value our concern, it amounts to a significant degree since the matter involves the professional rights and working authorities of the engineering technocrats and directly relates to the interests and rights of the common people to obtain utilisation of the authentic engineering services and related technology for which they remain ignorant despite making payments for the required specialist services, to the architects and the authority M.C.D. per conditions 4 and 7 of Appendix 'E' form 1 bye law 6.7.1 remains indemnified against any claim on account of infringement of bye-laws if occurred and to keep itself harmless from all proceedings in courts and before any other authority of all the liabilities as a consequences of the sanctions of the building plans accorded by the authority M.C.D.

We are sure that the authority owns enough calibre to decide about the technical policy matter in relation to the legal sanctity constitutionally. In case we don't get any response now immediately, we shall construe that the authority has nothing to say in this matter and our queries stand accepted and tacitly clarified.

Head (Suraj Singh Chartered Engineer)

SKY ASSOCIATES
Building Engineers
Office : 3373, Delhi Gate, New Delhi - 110002.

Your Ref. No. Nil

Our Ref. No. SA/G/HO/92

Date 1.5.92

The Commissioner,
Municipal Corporation of Delhi,
Town Hall, Chandni Chowk,
Delhi - 110006.

Sub : Building engineering profession and the bye-laws
Clarifications for the qualifications to license the technical personnel to prepare
building schemes to obtain building permits and supervise.

Our previous ref. - SA/G/HO/92 dt. 22/4/92 (Copy enclosed)

Dear Sir,

Please find enclosed herewith a copy of the authority that is an issue by the Institution of Engineers (I), 8 Gokhale Road, Calcutta 700 020, recently received by us/me from the august professional institution of India, to support the contents for the clarification moved/requested by from the executive body/authority/MCD.

Hope that this document shall provide you with sufficient logic to help you clarify the queries contained in the above reference letter.

Thanks.

Suraj Singh, Chartered Engineer (India), MIE

Enclosure : Copy of certificate issued by IEI authorising to use the style and title of Chartered Engineer (India)

SKY ASSOCIATES
Building Engineers
Office : 3373, Delhi Gate, New Delhi - 110002

Your Ref. No. Nil

Our Ref. No. SA/G/HO/92

Date 22.4.92

The Commissioner,
MCD/Town Hall, Delhi - 6.

Sub : Building engineering profession and the bye-laws
Clarifications for the qualifications to license the technical personnels prepare
building schemes to obtain building permits and supervise.

Dear Sir,

We forwarded to you a copy of the memorandum (enclosed) communicated to the Delhi court justice B.N. Kirpal and Arun Kumar pointing out certain facts about the proof but we did not receive any response from your authority to that.

Would you clarify the following points —

Bye-law P-2.2 (Competency of architect)

All plans and related informations connected with the building permit which should be amended as

All architectural arrangements connected with the building permit

Certificate of supervision for all buildings

Which should be amended as

Certificate of supervision for all buildings excluding the structure

Structural stability certificate —

Are you contemplating to reform the stability recording forms/process related to structural designs and execution supervision.

Registration —

1. Architect — You are allowing to practice on the registration with the council of architecture according to the architect act 1972.

2. Engineer — Do you allow engineer (civil) on election as a corporate member duly elected by the council of engineers of the Institution of Engineers (I) established on 13.9.1920 under Companies Act of 1913 and granted with Royal Charter in 1935 according to the bye-laws 68 (i) of the Institution per act 1935, every corporate member is, and is entitled to describe himself as a Chartered Engineer and stated in accordance with the classes as named FIE; MIE; AMIE, as it is ridiculous and step motherly treatment with the chartered engineer to obtain additional registration from the authority while that does not hold good for the architect and do you contemplate on enhancing the competencies of the chartered engineers experienced on buildings only, as also according to the practice provision in the forms of the Institution of Civil Engineers (London) whereon the engineer is fully authorised to plan and supervise all the civil engineering projects in addition to form of contracts of the Royal Institute of British Architects (RIBA) and CCC Wki

The RIBA form of contract is used specifically for the general building construction. The contract is one between a client and a builder who for a sum of money paid to him by the client, erects a building to the clients requirements as expressed in drawing prepared by the client's architect. The work is carried out to the specifications of architect and the architect has a separate contract for the services to the client.

The CCC/Wks/i form of contract is usually made with the ministry of public buildings and works, occasionally with government departments. The ministry is responsible for preparing plans and the work is carried out to the satisfaction of the superintending officer.

Under the ICE form of contract work of civil engineering construction is carried out to the plans prepared by the engineer on behalf of the client, by a contractor for a sum of money paid to him by the client. The work is carried out to the specifications of the client's engineer who has a separate contract for the services to the client.

Local authorities use for their contracts either RIBA standards form or ICE standards form as an appropriate or occasionally their own CCC/Wks/i.

In the CCC/Wks/i form of contract and the ICE form of contract powers are given to the engineer (superintending/supervising officer) to issue instructions to the contractor to carry out the work in a certain manner. The engineer is also required to issue certificates covering completion of works and payments due to the contractors.

In the RIBA form of contract a similar function is performed by architect.

(Above is an abstract form a report of the Institution of structural engineers London U.K. August 1969)

3. Group/Agency — Do you contemplate to accept Group/agency operations as permissible per unified bye-law P-5

4. Services engineer — Do you contemplate to introduce and include the competencies and the qualification for the allied building services engineer/general building services for the inclusion of their services in the preparation of the building schemes.

Thanks.

Sincerely, yours,

Suraj Singh C.E. MIE (I)

Copy of the memo enclosed

SKY ASSOCIATES
Building Engineers
Office : 3373, Delhi Gate, New Delhi - 110002.

Our Ref. No. SA/HO/CI/91 Date 27 Nov. 91

His Excellency the Prime Minister of India
His Excellency the Home Minister of India/Urban Development Minister of India
His Excellency the Lt. Governor of Union Territory of Delhi
The Commissioner of the authority the Municipal Corporation of Delhi
The Secretary and the Director General of the Institution of Engineer (I)
The Editor, Hindustan Times/TOI/IE

Sub : Building engineering profession and the bye-laws.

To the excellencies is hereby copied the memorandum communicated to the honourable Delhi high court Justice B.N. Kirpal and Justice Arun Kumar in connection with the subject of the responsibilities/ obligations of the engineering professional on the building projects in India and abroad in general and the suggestions to be considered for being incorporated with the bye-laws for the betterment of the building engineering profession and to be in the interest of the general public.

Thanks. With regards.

Head Experienced abroad on U.K. System

Specially copied to the opposition leader L.K. Advani and Ex-Prime Minister V.P. Singh with a request to pursue the matter with the government in the public interest.

Memo to Justice

To Justice B.N. Kirpal and Justice Arun Kumar of Delhi High court

Memorandum in connection to the report published in the news media/HT on the recommendations of the high power committee appointed by the high court for the check on the illegal building constructions dt. 17 and 18 November 1991.

Your honour,

In the general interest of the subject we would like to bring certain facts of the building engineering profession to the court's notice for the inclusion in the considerations by the court if found to some of the uses of the court.

In the international sphere for the building projects and building schemes, the user/promoter/ builder/the government appoints an engineer (qualified architecture engineer, civil engineer, structural engineer) and this appointment in fact resembles to the agency/group consultants/building engineers/ building engineering professional consultants comprising of all the engineers of the relevant engineering disciplines incorporated/involved with the building designs and the constructions and could be classified as architecture engineer; civil engineer; structural engineer; plumbing and drainage engineer; mechanical engineer (including air-conditioning); electrical engineer and other specialist engineers if any depending on the building's nature. For the building schemes in general the duties and the obligations of the appointed engineer's performances on the scheme comprises of the following operations –

1. Schematic layouts of different residential, commercial and all other use building and the components of the scheme including the infrastructure as the case be
2. Schematic architectural layout plans for all the building and structures
3. Architectural and structural developments and the projections of the overall formation mechanisms of the relevant individual buildings
4. Preparation of the contract forms and the legal contract documents for the project
5. Preparation of the management guide intended to be used on the project scheme for the direct and indirect management and the administration control on the site
6. Site land survey supervision and interaction with the soil investigation operation

7. Review the layout plans and adjustments per settings and modellings for approval
8. Schematic structural layout plans
9. Architectural development of plans, elevations, sections and general details
10. Developments of structural plans and the general details
11. Architectural and structural working details duly co-ordinated to each other
12. Complete allied services to buildings schematic details for the preparation of shop and working drawings to be produced before the real execution depending on the availability of the materials and the system designs relevant to the various disciplines to be approved by the engineer/group as a part of execution
13. Foundation designs and the details
14. Submission of the project with the related authority to obtain a building permit for the client i.e. the user according the prevailing bye-laws
15. Quantity engineering details for the purpose of tenders and detailed project run
16. Project site works execution supervision and administrative control till the practical completion of all the operations involved per project details/contract documents as agreed and extras if any
17. Post practical completion period supervision of the maintenance works till the grant of the final completion and the completed contract.
18. Grant of the final completion to the project contractor/contract completion
19. To assist obtain an occupation certificate from the concerned authority by supporting the user's application with the clearances from the fire fighting, lift control, electrical, telephones, public health departments and enclosed with the structural stability certificate, etc.
20. Issue of the record drawings/as built drawings to the user and the related authorities for the post completion use for maintenances references
21. Clearance of any extra completion period guaranty/warranty given by the suppliers specialists manufactures and contractors of any special items or components included in the scheme's systems, etc.

The concept in India being adopted in general is still an old and obsolete which was fit to the use 2-3 decades in the past when the high building structural technologies were not brought in common use in the structural development works in this country. The concept in general the common men understand/construe that the architect is the person who makes the buildings and they seldom know the significance and importance of the contribution by the other relevant engineers who are equally or more responsible in their role to the building designs and the construction execution. In those days the buildings used to be designed on the restricted elevations/height and corresponded to the monumental and the cultural type and not very much commercial in nature wherein the thoughts of the decorations and the cultural ekistics used to play a dominant role in the design concept and the planning of the structures, of course, forming major part of architecture engineer as artist of fine arts to let the dream come true. But in the post modern architecture era on buildings and the complexes designs based completely on the integrated planning of the layouts forming a part of the town planning process taking elevations to the considerable heights involving the applications of the detailed and elaborate knowledge of the expertise in the structural engineering of the buildings' frames and the thorough knowledge of soil engineering.

Should you refer to the bye-laws passed by the government of India vide no. ministry of home affairs F 27 (6I) Judl II dt. 2/8/61 and Delhi administration ref. no. 1983/2/181/75-LSG/3876. dt 23.6.83 bye-law 6.6 appendix 'P' titling the qualifications of licensing the technical personnel for the preparations of the building schemes for obtaining building permit and the supervision wherein the competencies of the personnel are described as below —

P-2.2 Architect

1. All plans and related information connected with the building permit
2. Structural details and calculations for buildings on plot upto 500 sqm and upto 4 storeys

3. Certificate of supervision for all buildings

4. All layout plans

P-3.2 Engineer

1. All plans and related informations connected with the building permit on plot upto 500 sqm and upto 4 storeys

2. Structural details and calculations for all buildings

3. Certificate of supervision for all buildings

4. All layout plans

5. Sanitary/water supply works for all types of buildings

P-5 Group/Agency

Joint operations carried out by one qualified engineer and one qualified architect to work on the combined competencies but as inquired has been obsolete now from registration.

The architecture engineer/architect has been empowered to design all the buildings of any height which is not virtually feasible as no architect can conceive or conceptualise the high rise and the multistoreyed structural formation mechanisms on major plots independently due to the involvement of the high technologies of the reinforced concrete; pre-stressed concrete and the structural steels as the case be. The architect cannot finalise optimally the details of any high rise or multistoreyed structure of considerable magnitudes and massive construction without involving the conception of the civil/structural engineer for the structural formation mechanism/structural frame/skeleton. The best solution is the joint planning by the architect and the structural engineer of the building engineering group/agency so the visualisation shall include all the expected forces to which the structure shall be exposed to during the life span of the building structure including all the structural loadings such as dead load, live load, wind load, seismic resistance, fire resistance, structural flexibility, foundation behaviours pattern, soil structure and mechanism and many other factors to be involved and shall also anticipate the probable clashes met in the practical co-ordinations of the various disciplines during executions. The project schematic details should be prepared by the joint efforts of the agency/group building engineers incorporating all the aforesaid details and factors to make the other relevant engineering professional feel involved and responsible.

The application to obtain the occupation certificate from the authority is supported with one of the various required documents that is the structural stability certificate. It is obvious that the structural stability corresponds to the stable structural design and the complete safety of the constructed structure at the time of the applications of all the forces prescribed and for the worst situation structural safety the structure should be designed keeping in view all the related loadings. Should the structure yield the whole building shall collapse or imminent to a collapse or the lost strength case shall cost excessively for the restorations of the required strength. The term structural stability therefore should be collectively defined as the process of the structural design analysis, detailing, communication of the informations and the degree of accuracy at the time of the construction executions according to the specified working standards acceptable per codes and the practice prevalent. Therefore, the overall supervision of the construction execution should go into the legal obligations and professional responsibilities of the civil/structural/building engineer's structural consultant.

A little light on the registration of the professional engineers —

1. Architect engineer

   Presently allowed to practice on the registration with the council of architecture per architect act 1972

2. Engineer

   In India there is only one professional apex lawful body for the registration of the practising engineers/professional engineers of all disciplines named as Institution of Engineers (I) incorporated by the Royal Charter/Society in 1935. The Institution comprises the divisions of all the engineering disciplines and all the professional engineers to practice on the building engineering profession should be allowed on the registration with the relevant division of the Institution of Engineers which shall enable the engineer to the entitlement of using the suffix as CHARTERED ENGINEER. Therefore, there should be no need of any additional registration elsewhere with the authority as is prevailing at present.

3. For the major projects there should be a third registration council for the registration of the agencies/group of the building engineers comprising with the qualified engineer and the qualified architect engineer which at present does not exist through the bye-law makes the provision for that registration per P-5

4. The term building engineer should refer to the collective title of joint responsibilities of all the relevant engineering professional disciplines met in the building project engineering profession from the conception to the completion.

5. The competence of very experienced chartered civil/structural engineer having experiences of 10 years or more in senior levels in the international sphere or in India exclusively on the building projects should be enhanced/enriched to full to assist in the augmentation of the better control and the practice of formation of the building engineering groups/agencies officially be brought in force for the generalisation of the good controlled projects for the qualitative production of the buildings in the Indian building industry.

6. The agency/group/individual architect/engineer should not issue the completion and the structural stability certificate in case the building has been constructed sub-standard and violating the building permit drawings and the aberrations referred to the authority concerned by the 'engineer'

7. The management contract between the promoter/builder and the agency/group/individual practitioners should exist involving the prescribed professional responsibilities and the obligations on the part of the appointed 'engineer' group/individual practitioner to keep on binding both the contract parties.

8. No professional engineer should be allowed to practice without being registered with either of the councils/Institutions mentioned earlier to rationalize the standards of the building engineering professional practice in India in parallel to those in existence in all the advanced countries the world over. In case of any default by the group/individual or gross misconduct professionally, the matter be referred to the relevant registration council for the initiation for the proper action by the engineering forum of the discipline against the erring professional and based on that the claims by the user be framed against the 'engineer' in legal terms for any damages or losses caused to the user and that shall also keep the 'engineer' protected against the harassment by the user.

9. The time has arrived that India must rationalize in the engineering professional practice altogether to reach the world standards and the best qualitative control and professional performances marked to the standards if not super.

Head Experienced abroad on U.K. Systems.

Addendum to the suggestions communicated to the high court justice B.N. Kirpal and Justice Arun Kumar dt. 27 Nov. 1991 (New Delhi high court)

Bye law 6.2.4 — Building plans shall

    a. include floor plans of all the floors together with covered area clear by indicating the size and spacings of all framing members and sizes of rooms and positions and width of staircase, ramps and other exit ways, liftwells, lift machine rooms and lift pit details.

    d. include sectional drawings showing clearly the sizes of the footings, thickness of basement walls, wall construction, size and spacing of framing members, floor slabs and roof slabs, with their material. The section shall indicate the height of buildings and rooms and also the height of parapet and the drainage and the slope of the roof. At least one section should be taken through the staircase, kitchen and toilet, bath and WC.

6.2.4.I

    i. Details of building services : air conditioning system with position of dampers, mechanical ventilation system, electrical services, boiler, gas pipes, etc.;

    k. Location of the generator, transformer and switch gear room;

l. Smoke exhauster system, if any;

m. Details of fire alarm system network;

n. Location of centralised control, connecting all fire alarm system built in fire protection arrangement and public address system;

o. Location and dimensions of static water storage tank and pump room;

p. Location and details of fixed fire protection installations such as sprinklers wet risers, hose reels, trenches, CO_2 installation; and

q. Location and details of first aid fire fighting equipments/installations.

6.2.5

Service plans — Plans, elevations and sections of private water supply and sewage disposal system and details of building services, where required by the authority.

The responsibilities as included in the foregoing portion of 6.2.4 and in the contents of 6.2.4.1 as aforesaid are in relation to the building permit standard details for all building and fall inside the scope of the licensed/registered architect per competencies P-2.

Construe the competencies

Architect allowed to work on the structural details and supervision of structural elements per p-2.2 (I and 5)

Virtually Engineer debased from his competencies as the client lies outside his power and he has to work at the mercy of the architect who is in fact not competent for he engineer's operations as without the safe and durable structure architecture cannot exist. Engineer cannot even supervise the jobs as that part has also been covered in the competencies of the architect. Layout plans are out of question without the client. And the engineer has been left with the public health/water supply and the plumbing works only and that part also parts as in the related details for the building permit and architect can plan therefore, the engineer of India is practically a plumber with engineer license.

So the client appoints you the building engineer and you have to render the professional engineering services. First enter an agreement called the management contract and let the client apprise of the execution contract and all about the management guide. Go to the site and make a survey in details about the plot or the land sizes, the nature of soil, history of the area, soil investigation if that is required according to your judgement of the soil/visual inspection, and about all the relevant local sources of materials and manpowers, etc., from the point of view of making the construction execution economical, easier and better depending upon the pragmatic approach of the professional toward the engineering. Layout and all the other details regarding services infrastructure, local connections, state arrangements, civic amenities, etc., that you consider necessary or helping in the planning part or operation of the scheme on desk in the design office. You can inquire about the data regarding soil, footings existing, depth of foundations, water level fluctuations, winds, storms, solar temperature during the year, information or statistics of seismology of the area from the relevant institution or from the codes. Then concentrate on the imagination and immerse yourself in the hypothetism and try to make out of your intuition and the capacity of mind jointly certain possibilities and communicate them to the paper in the form of the line plans and their three dimensional sketches in the isometric form or so called bird's eyes view so that the client shall be able to know what you want to get on the plot. Let the client select a few of your imaginary creations and hold meetings with the client and your responsible colleagues and discuss the merits and demerits of all the proposals and respecting the budget and the time factor reach an agreement of the proposal by taking an approval from the client formally. In case the statutory provisions require the local body's approval make the proposal for making the submission and get the client approval of the layout of the scheme. This operation is very important factor when integrated planning is done for the promotion of the new or the virgin townships for urbanisation of the area wherein you shall deal with a number of authorities and the boards and the ministerial officials and shall attend various meetings to make them apprise of what the layout shows, has been done according to the provision of the master plans or the national or the state policy of the nation or the same is according to the needs of the people which is really a hard task and needs thorough knowledge of the building engineering and town planning, landscaping, etc. You shall be asked to produce many models and bring on the desk your modelled imagination. To accomplish this activity you must be practically an excellent engineer with structural and architectural conceptions and the understanding of town planning. My statement does not qualify that you should be a titled town planner or an architect but must have the relevant pragmatism and knowledge to utilise them as media for your application/performance of the engineering skills. Sometimes, this may not be required due to the site located outside the limits of the applicability of the municipal bye laws. After obtaining the approval of the layout proposal from the client and the authorities proceed for the production of the tender/Contract documents.

STATUTORY

Quotations/References

The Indian Penal Code (Act No. 45 of 1860)

[As on 1st December 1992]

Section 12 "Public" — The word public includes any class of public or any community

Section 19 "Judge" — The word "Judge" denotes not only every person who is officially designated as a Judge but also every person,

Who is empowered by law to give, in any legal proceeding, civil or criminal, definitive judgement or a judgement which, if not appealed against, would be definitive, or a judgement which is confirmed by some other authority, would be definitive, or who is one of a body of person, which body of person is empowered by law to give such a judgement.

Section 20 "Court of Justice" — The words "Court of Justice" denote a judge who is empowered by law to act judicially alone, or a body of judges who is empowered by law to act judicially as a body, when such judge or body of judges is acting judicially.

Section 21 "Public Servants" — The words "Public Servant" denote a person under any of the descriptions hereinafter following namely

1. Omitted by A.O. 1950

2. Every Commissioned officer in the military, [Naval or Air] forces [of India]

3. Every judge including any person empowered by law to discharge, whether by himself or as a member of any body of persons, any adjudicatory functions.

4. Every officer of a court of Justice [including a liquidator, receiver or a commissioner] whose duty it is as such officer, to investigate or report on any matter of law or fact, or to make, authenticate, or keep any document, or to take charge or dispose of any property or to execute any judicial process, or to administer any oath, or to interpret or to preserve order in the court, and every person specially authorised by a court of justice to perform any of such duties.

5. Every jury man, assessor, or member of a panchayat assisting a court of Justice or a Public Servant.

6. Every arbitrator or other person to whom any cause or a matter has been referred for decision or a report by any court of Justice, or by any other competent public authority.

7. Every person who holds any office of virtue of which he is empowered to place or keep any person in confinement.

8. Every officer of the Government whose duty it is, as such officer, to prevent offences, to give information of offences, to bring offenders to justice, or to protect the public health, safety or conveyance.

9. Every officer whose duty it is, as such officer, to take, receive, keep or expend any property on behalf of the Government, or to make any survey, assessment or contract on behalf of the Government, or to execute any revenue process, or to investigate or to report, on any matter affecting the pecuniary interest of the Government or to make, authenticate or keep any document relating to the pecuniary interest of the Government or to prevent the infraction of any law of the protection of the pecuniary interests of the Government.

10. Every officer whose duty it is, as such officer, receive, keep or expand any property, to make any survey or assessment or to levy any rate or tax for any secular common purpose of any village, town or district, or to make, authenticate or keep any document for the ascertaining of the rights of the people of any village, town or district.

11. Every person who holds any office in virtue of which he is empowered to prepare, publish, maintain or revise an electoral roll or to conduct an election or part of an election.

12. Every person —

    (a) in the service or pay of the Government or remunerated by fees or commission for the performance of any public duty by the Government

    (b) in the service or pay of local authority, a corporation established by or under a central, provincial or State Act or a Government company as defined in section 617 of the Companies Act 1956 (1 of 1956)

Persons falling under any of the above descriptions are public servants whether appointed by government or not. Whatever the words 'public servant' occur, they shall be understood of every person who is in actual possession of the situation, of public servant, whether legal defect there may be in his right to hold that situation. The word 'election' denotes an election for the purpose of selection members of any legislative, municipal or other public authority, of whatever character, the method of selection to which is by or under, any law prescribed as by election.

Section 29 — 'Document' — The word 'document' denotes any matter expressed or described upon any substance by means of letters, figures, or marks, or by more than one of those means, intended to be used or which may be used, as evidence of that matter. It is immaterial by what means or upon what substance the letters, figures or marks are formed, or whether the evidence in intended, for or may be used in, court of justice, or not. A writing expressing the terms of a contract, which may be used as evidence of the contract, is a document. A power of attorney is a document.

A cheque upon a banker is a document.

A map or plan which is intended to be used or which may be used as evidence, is a document.

A writing containing directions or instructions is a document.

Section 41 — Special law — A special law is a law applicable only to a particular subject

Section 42 — Local law — A local law is a law applicable only to a particular part of India.

Section 51 — Oath — The word oath includes a solemn affirmation substituted by law for an oath, and any declaration required or authorised by law to be made before a public servant or to be used for the purpose of proof, whether in a court of justice or not.

Section 166 — Public servant disobeying law, with intent to cause injury to any person - whoever, being a public servant knowingly disobeys any direction of the law as to the way in which he is to conduct himself as such public servant intending to cause, to knowing it to be likely that he will, by such disobedience, cause injury to any person, shall be punished with simple imprisonment for a term which may extend to one year, or with fine or with both.

Section 167 — Public servant framing an incorrect document with intent to cause injury - whoever, being a public servant, and being, as such public servant, charged with the preparation or translation of any document, frames or translates that document in a manner which he knows or believes to be incorrect, intending thereby to cause or knowing it to be likely that he may thereby cause injury to any person, shall be punished with imprisonment of either description for a term which may extend to three years or with fine or with both.

Section 170 — Personating a public servant — Whoever pretends to hold any particular office as a public servant knowing that he does not hold such office or falsely personates any other person holding such office, and in such assumed character does or attempts to do any act under colour of such office, shall be punished with imprisonment of either description for a term which may extend to two years, or with fine, or with both.

Section 174 — Non attendance in obedience to an order from public servant — whoever, being legally bound to attend in person or by an agent at a certain place and time in obedience to a summons, notice, order or proclamation proceeding from any public servant legally competent, as such public servant to issue the same, intentionally omits to attend at that place or time, or departs from the place where he is bound to attend before the time at which it is lawful for him to depart, shall be punished with simple imprisonment for a term which may extend to one month or with fine which may extend to five hundred rupees, or with both.

Section 175 — Omission to produce document to public servant by person legally bound to produce it — whoever being legally bound to produce or deliver up any document to any public servant, as such, intentionally omits so to produce or deliver up the same, shall be punished with simple imprisonment for a term which may extend to one month, or with fine which may extend to five hundred rupees, or with both.

Section 176 — Omission to give notice or information to public servant by person legally bound to give it — whoever, being legally bound to give any notice or to furnish information on any subject to any public servant as such intentionally omits to give such notice or to furnish such information in the manner and at the time required by law shall be punished with simple imprisonment for a term which may extend to one month or with fine which may extend to five hundred rupees or with both, or if the notice or information required to be given respects the commission of an offence, or is required for the purpose of preventing the commission of an offence, or in order to the apprehension of an offender, with simple imprisonment for a term which may extend to six months, or with fine which may extend to one thousand rupees or with both.

Section 177 — Furnishing false information whoever, being legally bound to furnish information on any subject to any public servant, as such furnishes, as true, information on the subject which he knows or has reason to believe to be false, shall be punished with simple imprisonment for a term which may extend to six months or with fine which may extend to one thousand rupees, or with both or if the information which he is legally bound to give respects the commission of an offence, or is required for the purpose of preventing the commission of an offence, or in order to the apprehension of an offender, with imprisonment of either description for a term which may extend to two years, or with fine or with both.

Section 178 — Refusing oath or affirmation when duly required by public servant to make it — whoever refuses to bind himself by an oath or affirmation to state the truth, when required so to bind himself by a public servant legally competent to require that he shall so bind himself, shall be punished with simple imprisonment for a term which may extend to six months, or with fine which may extend to one thousand rupees or with both.

Section 179 — Refusing to answer public servant authorised to question — whoever, being legally bound to state the truth or any subject to any public servant, refuses to answer any question demanded of him touching that subject by such public servant, in the exercise of the legal powers of such public servant shall be punished with simple imprisonment for a term which may extend to six months, or with fine which may extend to one thousand rupees, or with both.

Section 180 — Refusing to sign statement — whoever refuses to sign any statement made by him, when required to sign that statement by a public servant legally competent to require that he shall sign that statement, shall be punished with simple imprisonment for a term which may extend to three months or with fine which may extend to five hundred rupees, or with both.

Section 181 — False statement on oath or affirmation to public servant or person authorised to administer an oath or affirmation — whoever, being legally bound by an oath or affirmation to state the truth on any subject to any public servant or other person authorised by law to administer such oath or affirmation, makes to such public servant or other person as aforesaid, touching that subject, any statement which is false, and which he either knows or believes to be false or does not believe to be true, shall be punished with imprisonment of either description for a term which may extend to three years and shall also be liable to fine.

Section 186 — Obstructing public servant in discharge of public function — whoever voluntarily obstructs any public servant in the discharge of his public functions, shall be punished with imprisonment of either description for a term which may extend to three months or with fine which may extend to five hundred rupees or with both.

Section 189 — Threat of injury to public servant — whoever hold out any threat of injury to any public servant or to any person in whom he believes that public servant to be interested, for the purpose of inducing that public servant to do any act, or to for bear or delay to do any act connected with the exercise of the public function of such public servant, shall be punished with imprisonment of either description for a term which may extend to two years, or with fine, or with both.

Section 191 — Giving false evidence — whoever, being legally bound by an oath or by an express provision of law to state the truth, or being bound by law to make a declaration upon any subject, makes any statement which is false, and which he either knows or believes to be false or does not believe to be true, is said to give false evidence. A statement is within the meaning of this section, whether it is made verbally or otherwise.

Section 192 — Fabricating false evidence — whoever, causes any circumstances to exit or makes any false entry in any book or record or makes any document containing a false statement, intending that such circumstance, false entry or false statement may appear in evidence in a judicial proceeding, or in a proceeding taken by law before a public servant as such, or before an arbitrator, and that such circumstance, false entry or false statement so appearing in evidence, may cause any person who in such proceeding is to form an opinion by the evidence, to entertain an erroneous opinion touching any point material to the result of such proceeding is said to fabricate false evidence.

Section 193 — Punishment for false evidence — whoever intentionally gives false evidence in any stage of a judicial proceeding, or fabricates false evidence for the purpose of being used in any stage of a judicial proceeding shall be punished with imprisonment of either description for a term which may extend to seven years and shall also be liable to fine and whoever intentionally gives or fabricates false evidence in any other case, shall be punished with imprisonment of either description for a term which may extend to three years, and shall also be liable to fine.

Section 196 — Using evidence known to be false — whoever corruptly uses or attempts to use as true or genuine evidence any evidence which he knows to be false or fabricated, shall be punished in the same manner as if he gave a fabricated false evidence.

Section 197 — Issuing or signing false certificate — whoever issues or signs any certificate required by law to be given or signed, or relating to any fact of which such certificate is by law admissible in evidence, knowing or believing that such certificate is false in any material point, shall be punished in the same manner as if he gave false evidence.

Section 198 — Using as true a certificate known to be false — whoever, corruptly uses or attempts to use any such certificate as a true certificate, knowing the same to be false in any material point, shall be punished in the same manner as if he gave false evidence.

Section 199 — False statement made in declaration which is by law receivable as evidence — whoever, in any declaration made or subscribed by him, which declaration any court of justice, or any public servant or other person, is bound or authorised by law to receive as evidence of any fact, makes any statement which is false, and which he either knows or believes to be false or does not believe to be true, touching any point material to the object for which the declaration is made or used, shall be punished in the same manner as if he gave false evidence.

Section 200 — Using as true such declaration knowing it to be false — whoever corruptly uses or attempts to use as true any such declaration, knowing the same to be false in any material point, shall be punished in the same manner as if he gave false evidence.

The Indian Evidence Act 1872 (Act 1 of 1872)

[As on 1st August 1990]

Section 3 — Evidence - means and includes

(1) all statements which the court permits or requires to be made before it by witnesses, in relation to matters of the fact under inquiry, such statements are called oral evidence.

(2) all documents produced for the inspection of the court, such documents are called documentary evidence,

Section 57 Part II Chapter III

Facts which need not be proved

Facts of which court must take a judicial notice —

The court shall take judicial notice of the following facts —

(1) All laws in force in Indian territory

(2) All Public Acts passed or hereafter to be passed by parliament of United Kingdom and all local and personal Acts directed by parliament of the United Kingdom to be judicially noticed

(3) Articles of war for Indian Army, Navy or Air Force

(4) The course of proceeding of parliament of the United Kingdom, of the Constituent Assembly of India, of Parliament and of the legislatures established under any law for the time being in force in a province or in the state.

(5) The accession and the sign manual of the sovereign for the time being of the United Kingdom of Great Britain and Ireland.

(6) All seals of which English courts take judicial notice; the seal of all the courts in India and of all courts out of India established by the Authority of the Central Government or the crown Representative; the seals of courts of Admiralty and Maritime Jurisdiction and of Notaries Public, and all seals which any person is authorised to use by the constitution or an Act of Parliament of the United Kingdom or an Act or Regulation having the force of law in India.

(7) The accession to office, names, titles, functions and signatures of the persons filling for the time being any public office in any state, if the fact of their appointment to such office is notified in any official Gazette.

(8) The existence, title and national flag of every state or sovereign recognised by the Government of India.

(9) The divisions of time, the geographical divisions of the world, and public festivals, fasts, and holidays notified in the official Gazette.

(10) The territories under the dominion of the Government of India.

(11) The commencement, continuance and termination of hostilities between the Government of India and any other state or body of persons.

(12) The names of the members and officers of the court and of their deputies and subordinate officers and assistants, and also of all officers acting in execution of its process, and of or all advocates, attorneys, proctors, vakils, pleaders and other persons authorised by law to appear or act before it;

(13) The rule of the road on land or at sea.

(14) In all these cases and also all matters of public history, literature, science or art, the court may resort for its aid to appropriate books or documents of reference.

Public Documents

Section 74 — The following documents are public documents —

(1) documents forming the acts, or records of the acts

(i) of the sovereign authority

(ii) of the official bodies and tribunals and

(iii) of public officers, legislative, judicial and executive, of any part of India or of the commonwealth or of a foreign country.

(2) Public records kept in any state of private documents

Section 75 — Private documents — All other documents are private.

Section 78 — Proof of other official documents — The following documents may be proved as follows —

(1) Acts, orders or notifications of the central government in any of its departments, or the crown representative or any state government or any department of any state government by the records of departments, certified by the heads of those departments respectively, or by the

document purporting to be printed by order of any such government or as the case may be, of the crown representative.

(2) The proceeding of the legislatures — by the journals of those bodies respectively, or by published Acts or abstracts or by copies purporting to be printed by order of the government concerned.

(3) Proclamation, orders or regulations issued by her Majesty or by the privy council, or by any department of her Majesty's government — by copies or extracts contained in the London Gazette, a purporting to be printed by the Queen's printer.

(4) The acts of the Executive or the proceeding of the legislature of a foreign country — by journals published by their authority or commonly received in that country as such or by a copy certified under the seal of the country or sovereign or by a recognition thereof in some central Act.

(5) The proceeding of a municipal body in a state — by a copy of such proceedings, certified by a legal keeper there of, or by a printed book purporting to be published by the authority of such body.

(6) Public documents of any other class in a foreign country — by the original, or by a copy certified by the legal keeper thereof, with a certificate under the seal of a Notary Public, or of an Indian consul or diplomatic agent that the copy is duly certified by the officer having the legal custody of the original and upon proof of the character of the document according to the law of the foreign country.

Presumption as to Documents

Section 79 — Presumption as to genuineness of certified copies — The court shall presume to be genuine every document purporting to be certified copy or other document which is by law declared to be admissible as evidence of any particular fact and which purports to be duly certified by any officer of the central government or of a state government or by any officer in the state of Jammu & Kashmir who is duly authorised thereto by the central government. Provided that such document is substantially in the form and purports to be executed in the manner directed by law in that behalf. The court shall also presume that any officer by whom any such document purports to be signed or certified held when he signed it, the official character which he claims in such paper.

Section 80 — Presumption as to documents produced as records of evidence — whenever any document is produced before any court, purporting to be a record or memorandum of the evidence or of any part of the evidence, given by a witness in a judicial proceeding or before any officer authorized by law to take such evidence or to be a statement or confession by any prisoner or accused person taken in according with law, and purporting to be signed by any judge or magistrate or by any such officer as aforesaid, the court shall presume that the document is genuine; that any statement as to the circumstances under which it was taken, purporting to be made by the person signing it, are true, and that such evidence, statement or confession was duly taken.

Section 81 — Presumption as to Gazettes, newspapers, private Acts of Parliaments and other documents —

The court shall presume the genuineness of every document purporting to be the London Gazette or any official Gazette, or the Government Gazette of any colony, dependency or possession of the British Crown, or to be a newspaper or journal, or to be a copy of a private act of parliament of the United Kingdom printed by the Queen's printer and of every document purporting to be a document directed by law to be kept by any person, if such document is kept substantially in the form required by law and is produced from proper custody.

Section 82 — Presumption as to document admissible in England without proof of seal or signature — when any document is produced before any court, purporting to be document which, by the law in force for the time being in England or Ireland, would be admissible in proof of any particular in any court of justice in England or Ireland, without proof of the seal or stamp or signature authenticating it, or of the judicial or official character claimed by the person by whom it purports to be signed, the court shall presume that such seal, stamp or signature is genuine and that the person signing it held, at the time when he signed it, the judicial or official characters which he claims, and the document shall be admissible for the same purpose for which it would be admissible in England or Ireland.

Section 83 — Presumption as to maps or plans by authority of Government — The court shall presume that maps or plans purporting to be made by the authority of the central Government or any state government were so made, and are accurate, but maps or plans made for the purposes of any cause must be proved to be accurate.

Section 85 — Presumption as to power of attorney — The court shall presume that every document purporting to be a power of attorney and to have been executed before, and authenticated by, a Notary Public, or any court of Justice, Judge, Magistrate, Indian Consul or Vice-Consul, or representative of the central government, was so executed and authenticated.

Section 90 — Presumption as to documents thirty years old-where any document, purporting or proved to be thirty years old, is produced from any custody when the court in the particular case considers proper, the court may presume that the signature and every other part of such document, which purports to be in the hand writing of any particular person, is in that person's handwriting, and in the case of a document executed or attested, that it was duly executed and attested, by the persons by whom it purports to be executed and attested.

The foregoing quotations have been given to let the engineering professionals apprise themselves with various provisions and to make use of those provisions whenever required.

CONSTITUTION OF INDIA

FUNDAMENTAL RIGHTS

Article 12. Definition—In this part, unless the context otherwise requires, 'the State' includes the Government and Parliament of India and the Government and the Legislature of each of the States and all Local and other Authorities within the territory of India or under the control of the Government of India.

Article 13 (3). (a) 'Law' includes any Ordinance, Order, Bye-Law, Rule, Regulation, Notification, Custom or usage having in the territory of India the force of law.

(b) "Law in force" includes law passed or made by a Legislature or other Competent Authority in the territory of India before the commencement of this Constitution and not previously repealed, notwithstanding that any such law or any part thereof may not be then in operation either at all or in particular areas.

Article 18. Abolition of titles—(I). No title, not being a military or academic distinction, shall be conferred by the State.

Article 19. Right to Freedom—Protection of certain rights regarding freedom of speech, etc., (g) to practice any profession, or to carry on any occupation, trade or business. (6) Nothing in sub clause (g) shall affect the operation of any existing law in so far as it imposes, or prevents the State from making any Law imposing, in the interests of the general Public, reasonable restrictions on the exercise of the right conferred by the said sub clause and in particular, nothing in the said clause shall affect the operation of any existing law in so far as it relates to, or prevent the state from making any law relating to—

(i) the Professional or Technical qualifications necessary for practising any Profession or carrying on any occupation, trade or business or

(ii) the carrying on by the State, or by a Corporation owned or controlled by the State, of any trade, business, industry or service, whether to the exclusion, complete or partial of citizens or otherwise.

Article 51 A. Fundamental Duties—It shall be the duty of every citizen of India—

a. to abide by the Constitution and respect its ideals and Institutions, the National Flag and the National Anthem.

b. to cherish and follow the noble ideals which inspired our National struggle for freedom.

c. to uphold and protect the sovereignty, unity and integrity of India.

d. to defend the country and render national service when called upon to do so

e. to promote harmony and the spirit of common brotherhood amongst all people of India transcending religious, linguistic and regional diversities, to renounce practices derogatory to the dignity of women

f. to value and preserve the rich heritage of our Composite Culture

g. to protect and improve the natural environment including forests, lakes, rivers and wildlife, and to have compassion for living creatures

h. to develop the Scientific temper, humanism and the Spirit of inquiry and reform

i. to safeguard public property and to abjure violence

j. to strive towards excellence in all spheres of individual and collective activity so that the Nation constantly rises to higher levels of endeavour and achievement.

Article 53. Executive power of the Union—1. The executive power of the Union shall be vested in the President and shall be exercised by him either directly or through officers subordinate to him in accordance with this Constitution.

2. Without prejudice to the generality of the foregoing provision, the supreme command of the Defence Forces of the Union shall be vested in the President and the exercise thereof shall be regulated by law.

3. Nothing in this article shall—

a. be deemed to transfer to the President any functions conferred by any existing law on the Government of any State or other authority; or

b. prevent Parliament from conferring by law functions on authorities other than the president.

Article 73. The Union—The Executive—Extent of executive power of the Union

1. Subject to the provision of this constitution, the executive power of the Union shall extend—

a. to the matters with respect to which the Parliament has power to make laws; and

b. to the exercise of such rights, authority and jurisdiction as are exercisable by the Government of India by virtue of any treaty or agreement :

provided that the executive power referred in sub clause 'a' shall not, save as expressly provided in this Constitution or in any law made by Parliament, extend in any State to matters with respect to which the legislature of the State has also power to make laws.

2. Until otherwise provided by Parliament, a State or any officer or authority of a State may, notwithstanding anything in this article, continue to exercise in matters with respect to which Parliament has power to make laws for that State such executive power or functions as the State or officer or authority thereof could exercise immediately before the commencement of this Constitution.

Article 131. Union Judiciary—Original jurisdiction of the Supreme Court—Subject to the provisions of this Constitution, the Supreme Court shall, to the exclusion of any other Court, have original jurisdiction in any dispute—

a. between the government of India and one or more States or

b. between the Government of India and any State or state on one side and one or more States on the other or

c. between two or more States

if in so far as the disputes involve any question whether of law or fact on which the existence or extent of a legal right depends, provided that the said jurisdiction shall not extend to a dispute arising out of any treaty, agreement, covenant, engagement, sanad or other similar instrument which, having been entered into or executed before the commencement of this Constitution continuous in operation after such commencement or which provides that the said jurisdiction shall not extend to such a dispute.

Article 138. Enlargement of the jurisdiction of the Supreme Court—1. The Supreme Court shall have such further jurisdiction and powers with respect to any of the matters in the Union List as Parliament may by law confer.

2. The Supreme Court shall have such further jurisdiction and powers with respect to any matter as the Government of India and the Government of any State may by special agreement confer if

Parliament by law provides for the exercise of such jurisdiction and powers by the Supreme Court.

Article 140. Ancillary powers of Supreme Court—Parliament may by law make provisions for conferring upon the Supreme Court such supplemental powers not inconsistent with any of the provisions of this Constitution as may appear to be necessary or desirable for the purpose of enabling the court more effectively to exercise the jurisdiction conferred upon it by or under this Constitution.

Article 147. Interpretation—In this chapter and in the chapter V of part VI, references to any substantial question of law as to the interpretation of this Constitution shall be construed as including references to any substantial question of law as to the interpretation of the Government of India Act 1935 including any enactment amending or supplementing that Act, or of any order in Council or order made thereunder or of the Indian Independence Act 1947 or of any Order made thereunder.

Article 254. Relation between the Union and the States—Inconsistency between laws made by Parliament and laws made by the Legislature of States.

1. If any provision of a law made by the Legislature of a State is repugnant to any provision of a law made by Parliament which Parliament is competent to enact, or to any provision of an existing law with respect to one of the matters enumerated in the Concurrent List, then, subject to the provisions of clause 2 the law made by the Legislature of such State, or, as the case may be, the existing law, shall prevail and the law made by the Legislature of the State shall, to the extent of repugnancy, be void.

2. Where a law made by a State Legislature with respect to one of the matters enumerated in the Concurrent List contains any provision repugnant to the provisions of an earlier law made by Parliament or an existing law with respect to that matter, then the law so made by the State Legislature, if it has been reserved for the consideration of the President or has received his assent, prevail in that State provided that nothing in this clause shall prevent Parliament from enacting at any time any law with respect to the same matter including a law adding to amending or repealing the law so made by the State Legislature.

Article 261. Administrative Relations—Public acts, records and judicial proceedings—1. Full faith and credit shall be given throughout Indian territory to public acts, records and judicial proceedings of the Union and of every State.

2. The manner in which and the conditions under which the acts, records and proceedings referred to shall be proved and effect determined shall be as provided by law made by Parliament.

Article 365. Miscellaneous—Effect or failure to comply with, or to give effect to, directions given by the Union — where any State has failed to comply with, or to give effect to, any directions given in the exercise of the executive power of the Union under any of the provisions of this constitution, it shall be lawful for the President to hold that a situation has arisen in which the Government of the State cannot be carried out in accordance with the provisions of this Constitution.

Article 366. Definitions—10. 'Existing Law' means any law, Ordinance, Order, Bye-Laws, Rules, or Regulation passed or made before the commencement of this Constitution by any Legislature, Authority or Person having power to make such a law, ordinance, order, bye-law, rule or regulation.

Article 367. Interpretation—1. Unless the context otherwise requires the General Clause Act 1897, shall subject to any adaptations and modifications that may be made therein under article 372, apply for the interpretation of this Constitution as it applies for the interpretation of an Act of the Legislature of the Dominion of India.

2. Any reference in this Constitution to Acts or Laws of, or made by, Parliament, or to Acts or Laws of, or made by, the State Legislature, shall be construed as including a reference to an ordinance made by the President or, to an Ordinance by a Governor as the case may be.

Article 372. Continuance in force of existing laws and their adaptations—1. Notwithstanding the repeal by this Constitution of the enactments referred to in article 395 but subject to the other provisions of this Constitution, all the laws in force in the territory of India immediately before the commencement of this Constitution shall continue in force therein until altered or repealed or amended by a competent Legislature, or other competent authority.

2. For the purpose of bringing this provisions of any law in force on the territory of India, into accord with the provisions of this Constitution, the President may by order make such adaptations and modifications of such law, whether by way of repeal or amendment, as may be necessary or expedient, and provide, that the law shall, as from such date as may be specified in the order, have effect subject to the adaptation and modifications so made, and any such adaptation or modification shall not be questioned in any Court of Law.

3. Nothing in clause 2 shall be deemed—

a. to empower the President to make any adaptation or modification of any law after the expiration of three years from the commencement of this Constitution or

b. to prevent any Competent Legislature or other competent authority from repealing or amending any law adapted or modified by the President under the said clause.

Explanation I—The expression "law in force" in this article shall include a law passed or made by a Legislature or other competent authority in the territory of India before the commencement of this Constitution and not previously repealed, notwithstanding that it or parts of it may not be then in operation at all or in particular areas.

Explanation II—Any law passed or made by a Legislature or other competent authority in the territory of India which immediately before the commencement of this Constitution had extra-territorial effect as well as effect in the territory of India shall, subject to any such adaptations and modifications as aforesaid, continue to have such extra-territorial effect.

Explanation III—Nothing in this article shall be construed as continuing any temporary law in force beyond the date fixed for its expiration or the date on which it would have expired if this Constitution had not come in force.

Article 372 A. Power of the President to adapt laws—1. For the purposes of bringing the provisions of any law in force in India or in any part thereof, immediately before the commencement of this Constitution 7th amendment Act 1956, into accord with the provisions of this Constitution as amended by that Act, the President may by order made before the 1st day of November, 1957, make such adaptations and modifications of the law, whether by way of repeal or amendment, as may be necessary or expedient, and provide that the law shall, as from such date as may be specified in the order, have effect subject to the adaptations and modifications so made, and such adaptation or modification shall not be questioned in any court of law.

2. Nothing in clause 1 shall be deemed to prevent a competent Legislature or other competent authority from repealing or amending any law adapted by the President under the said clause — Ref. Adaptation of Laws Order, 1950, dated the 26th January, 1950, Gazette of India, Extraordinary, p. 449, etc.

Article 375. Temporary, Transitional and Special Provisions—Courts, authorities and officers to continue to function subject to the provisions of the Constitution — All Courts of Civil, Criminal and Revenue Jurisdiction, all authorities and all officers, judicial executive and ministerial, throughout the territory of India, shall continue to exercise their respective functions subject to the provisions of this Constitution.

Article 393. Short title—This Constitution may be called the Constitution of India.

Article 394. Commencement—This article and articles 5, 6, 7, 8, 9, 60, 324, 366, 367, 379, 380, 388, 391, 392 & 393 shall come into force at once and the remaining provisions of this Constitution shall come into force on the 26th January 1950, which day is referred to in this Constitution as the commencement of this Constitution.

Article 395. Repeals—The Indian Independence Act 1947 and the Government of India Act 1935, together with all enactments amending or supplementing the later Act, but not including the Abolition of Privy Council Jurisdiction Act 1949, are hereby repealed.

Union List — Entry 65—Union agencies and institutions for—
a. professional, vocational or technical training, including the training of the police officers or
b. the promotion of special studies or research or
c. scientific or technical assistance in the investigating or detection of crime.

Concurrent List — Entry 26—Legal, medical and other professions.

Comments

Article 12—Local or other authorities include statutory, constitutional and Chartered bodies and IEI belongs to this category.

Article 13 (3) a & b—Royal Charter IEI comes under this provision.

Article 18—The title Chartered Engineer India and Professional Engineer and other statutory designations come under this provision.

Article 19—This provision confers on IEI the legislative powers to control, superintend and direct the engineering profession in India as in accordance with the objectives of the Royal Charter IEI 1935 constitution of which makes the IEI legally obliged to dispose its Constitutional and Statutory duties in Engineering as defined in the Charter for serving the Nation and also a Fundamental duty and responsibility collectively as a body. Ref. Article 51 A.

Article 51 A—This is self-explanatory and very important for every member of the engineering and the other professions.

Article 53—The engineering executive functions of the Union in accordance with the Royal Charter provisions have been vested in the IEI by the people of the Republic of the Indian Union, as deemed w.e.f. 26.11.49.

Article 131—Provides teeth with the IEI to exercise the powers of the engineering professional apex court of the State India subject to the provisions of article 138 of enlargement of Supreme Court Jurisdiction and article 140 together.

Article 147—This applies to IEI also.

Article 254—This article automatically repeals all the (void) State Legislature/Municipal bye-laws for the regulation of engineers and the engineering professional competencies and all other matters due to the provisions already incorporated with the IEI Royal Charter 1935 now implied as IEI Constitutional Charter of India.

Article 261—All the IEI functions come under this provision.

Article 366—IEI Royal Charter comes under this provision.

Article 366 (10)—IEI Royal Charter comes under this provision.

Article 367—IEI Royal Charter comes under this provision as the designations and titles are the Constitutional references under the Presidential protection.

Article 372/372 A—Since IEI Royal Charter continues in force without being repealed or amended, the provision applies to it.

Article 365—If IEI being a State remains unfunctional in parts, may be a subject of this provision of the Constitutional article.

Article 375—IEI being an Authority, shall be covered under this provision too. Constitutional Amendment Act article 243 W, 12th schedule enlisting the functions of the constituted municipalities and municipal corporations does not include the provisions or power of regulating the Engineers or Engineering Profession and the relevant competencies, saving regulating land use/town planning and construction of buildings, implying constitutional annulment/enquash of the building bye-laws made by municipal corporations regulating the engineers and engineering competencies of professional engineers, as effective from 20th April 1993 and as in the year 1994.